University of Tehran Science and Humanities Series

Series Editor

Central Acquisitions Office, University of Tehran, Tehran, Iran

The *University of Tehran Science and Humanities Series* seeks to publish a broad portfolio of scientific books, basically aiming at scientists, researchers, students and professionals. The series includes peer-reviewed monographs, edited volumes, textbooks, and conference proceedings. It covers a wide range of scientific disciplines including, but not limited to Humanities, Social Sciences and Natural Sciences.

Morteza Akbari · Leila Azimi ·
Matin Hemmati
Editors

Sustainable Entrepreneurship

Springer

Editors
Morteza Akbari
Faculty of Entrepreneurship
University of Tehran
Tehran, Iran

Leila Azimi
Faculty of Entrepreneurship
University of Tehran
Tehran, Iran

Matin Hemmati
Graduate School of Management
and Economics
Sharif University
Tehran, Iran

ISSN 2367-1092 ISSN 2367-1106 (electronic)
University of Tehran Science and Humanities Series
ISBN 978-3-032-04908-7 ISBN 978-3-032-04909-4 (eBook)
https://doi.org/10.1007/978-3-032-04909-4

This Springer imprint is published by the registered company Springer Nature Switzerland AG
The registered company address is: Gewerbestrasse 11, 6330 Cham, Switzerland

Martyr Sayyid Morteza Avini:

"…We are the heirs of the prophets,
And the divine signs of human creation
Find their meaning in the mirror of our being.

We do not fear death —
For our death is martyrdom,
And martyrdom is life.
In the presence of the Almighty.

How shall veiled minds,
And the tarnished mirrors of the Western soul,
Ever grasp the secret of life besides God?

That life —
Life with the Divine —
Is the final summit of humanity's ascension,
A height that none may reach
But through the passage of martyrdom.

It is we
Who have shouldered the weight of history,
So that we might carry the world
Toward its destined truth.

Our crimson blood
Is the blush of dawn —
A herald of the sun of justice

*That soon shall rise upon the heavens
of fate.*

*We have stood
In the ranks of Imam Hussein, peace be
upon him.
We have repaid
The treachery of the faithless Kufans..."*

Introduction

The dawn of the twenty-first century has brought with it a confluence of global challenges—climate change, resource scarcity, social inequities, and economic volatility—that demand a redefinition of entrepreneurial practice. As the editors of *Sustainable Entrepreneurship*, we are driven by a profound conviction that entrepreneurship can transcend its traditional boundaries of profit-driven innovation to become a transformative force for the greater good of society and the environment. This book represents a collective effort to articulate a new paradigm—sustainable entrepreneurship—that integrates economic viability with social responsibility and environmental stewardship. Drawing on the expertise of a diverse group of scholars, as reflected in the chapters of this volume, this work aims to provide a comprehensive framework for understanding and implementing sustainable entrepreneurial practices in a rapidly changing world.

Sustainable entrepreneurship is not merely an academic concept but a practical imperative for addressing the interconnected crises facing humanity. The United Nations' Sustainable Development Goals (SDGs) underscore the urgency of fostering inclusive, equitable, and environmentally conscious progress, and entrepreneurs are uniquely positioned to drive this agenda forward. Unlike conventional entrepreneurship, which often prioritizes short-term financial gains, sustainable entrepreneurship seeks to create long-term value by balancing the triple bottom line: people, planet, and profit. This book is both a scholarly exploration and a practical guide, offering insights for entrepreneurs, policymakers, academics, and students who aspire to build ventures that contribute to a regenerative and just future.

Context and Motivation

The motivation for this book stems from a recognition of the crucial role entrepreneurship can play in addressing global challenges. The environmental degradation caused by unchecked industrialization, coupled with persistent social inequalities, necessitates a shift toward business models that prioritize sustainability. Our journey as scholars and observers of entrepreneurial ecosystems has revealed the transformative potential of ventures that align economic goals with social and environmental objectives. This book synthesizes cutting-edge research and real-world applications to provide a holistic understanding of sustainable entrepreneurship, drawing on the diverse perspectives of its contributing authors.

The chapters of this book are organized into eight parts, each addressing a critical facet of sustainable entrepreneurship. From foundational theories to innovative interdisciplinary approaches, the volume offers a comprehensive exploration of how entrepreneurs can navigate the complexities of sustainability while fostering innovation and resilience. As editors our aim is to bridge the gap between theory and practice, providing readers with both a conceptual framework and actionable strategies for building sustainable enterprises.

Structure and Scope of the Book

The book is structured to guide readers through a logical progression of ideas, from foundational principles to future research directions.

- **Foundations of Sustainable Entrepreneurship** lays the groundwork by exploring the core concepts of sustainability (Chapter "Foundations of Sustainability"), entrepreneurship (Chapter "Foundations of Entrepreneurship"), and their integration into sustainable entrepreneurship (Chapter "Sustainable Entrepreneurship"). These chapters establish the theoretical underpinnings that inform the rest of the book, emphasizing the importance of aligning entrepreneurial activities with environmental and social goals.
- **Sustainable Business Design and Development** focuses on practical strategies for creating sustainable ventures. Chapter "Sustainable Business Model" examines sustainable business models, while Chapter "Sustainable Product and Service Development" explores the development of eco-friendly products and services. Chapter "Key Performance Indicators (KPIs) for Sustainable Entrepreneurship" introduces key performance indicators (KPIs) for measuring sustainability, providing entrepreneurs with tools to assess their impact.
- **Technology, Innovation, and Digitalization** highlights the role of technological advancements in driving sustainable outcomes. Chapter "Role of Technology and Innovation in Sustainable Entrepreneurship" discusses how innovation can support sustainable entrepreneurship, and Chapter "Digital Sustainable Entrepreneurship" explores the emerging field of digital sustainable entrepreneurship, reflecting the growing influence of technology in shaping business practices.
- **Policy, Marketing, and Human Resources** addresses the broader ecosystem in which sustainable entrepreneurship operates. Chapter "Entrepreneurship Policies and Strategies for Sustainable Development" examines supportive policies and strategies, while Chapter "Enhancing Sustainable Entrepreneurship with Sustainable Supply Chain Strategies" focuses on sustainable supply chain management. Chapter "Branding Challenges for Sustainable Entrepreneurship" addresses branding challenges, and Chapter "Sustainable Job Crafting: Introducing a Concept in Sustainable Human Resource Management" introduces the concept

of sustainable job crafting, highlighting the role of human resource management in sustainability.

- **Sustainable Entrepreneurship in a National and International Context** provides a contextual analysis, with Chapter "Sustainable Entrepreneurship Development in Iran with an Emphasis on Environmental Issues" focusing on sustainable entrepreneurship in Iran and Chapter "International Agencies/Institutions to Promote Sustainable Entrepreneurship" exploring the role of international agencies in promoting sustainable ventures globally.
- **Special Issues** delves into specialized topics, including sustainable entrepreneurship in small and medium enterprises (Chapter "Sustainable Entrepreneurship in Small and Medium Enterprises (SMEs)"), financing models (Chapter "Financing Sustainable Entrepreneurship: Models, Challenges, and Strategic Innovations"), women's contributions to the field (Chapter "Women and Sustainable Entrepreneurship"), social entrepreneurship (Chapter "Sustainable Social Entrepreneurship: A Multifaceted Approach to Addressing Complex Global Challenges"), ethical considerations (Chapter "Ethics and Sustainable Entrepreneurship"), leadership (Chapter "The Role of Leadership in Sustainable Entrepreneurship"), and crisis management (Chapter "Crisis Management and Sustainable Entrepreneurship"). These chapters highlight the diverse applications of sustainable entrepreneurship across different contexts and challenges.
- **Innovative and Interdisciplinary Approaches** pushes the boundaries of the field by exploring cognitive neuroscience (Chapter "Cognitive Neuroscience of Sustainable Entrepreneurship: Cognition and Emotion in Action"), entrepreneurial ecosystems (Chapter "Sustainable Entrepreneurial Ecosystems: Building a Green Future for Business Innovation"), and behavioral dimensions (Chapter "Behavioral Dimensions of Sustainable Entrepreneurship"). These chapters underscore the interdisciplinary nature of sustainable entrepreneurship, drawing on insights from psychology, neuroscience, and systems thinking.
- Finally, **Future Research Directions** provides a comprehensive review of the field in Chapter "A Holistic Review of the Sustainable Entrepreneurship", identifying gaps in current knowledge and proposing avenues for future research.

Purpose and Contribution

The purpose of *Sustainable Entrepreneurship* is to advance both the academic discourse and the practical application of sustainable entrepreneurship. By bringing together contributions from leading scholars, the book provides a multidimensional perspective on how entrepreneurs can address global challenges through innovative and sustainable practices. It serves as a resource for academics seeking to deepen their understanding of the field, practitioners aiming to build sustainable ventures, and policymakers working to create enabling environments for such enterprises.

As the editors, our vision is to inspire a new generation of entrepreneurs to view their work as a means of creating shared value for society and the environment. The diverse chapters of this book reflect the collaborative spirit of this endeavor, drawing on the expertise of contributors who share a commitment to sustainability. We hope that this volume will serve as both a catalyst for change and a practical guide for those who seek to build businesses that endure, uplift, and regenerate.

In conclusion, *Sustainable Entrepreneurship* is a call to action for entrepreneurs to reimagine their role in a world of finite resources and interconnected systems. The chapters that follow provide a roadmap for navigating the complexities of sustainable entrepreneurship, offering insights, tools, and inspiration for creating ventures that thrive economically while contributing to a sustainable and equitable future.

Morteza Akbari
Leila Azimi
Matin Hemmati

Contents

Foundations of Sustainable Entrepreneurship

Foundations of Sustainability

Mohammad Sharif Sharifzadeh and Maryam Tajeri Moghadam

Abstract

In recent decades, sustainability has gained prominence as a key concern, attracting increasinging interest from researchers, business leaders, and policymakeworldwide This chapter provides a comprehensive study of the foundations of sustainability, a concept that, while long a cornerstone of developers' discourse and a vital criterion for the practical thinking and intervention of development actors, has always been dynamically interpreted and evolving. The chapter begins by reviewing relevant literature to explore the concept of sustainability and traces its roots through the history of sustainability. A significant portion is dedicated to integrating sustainability and development, introducing sustainable development and its three interconnected dimensions: environmental sustainability, economic sustainability, and social sustainability. Furthermore, it clarifies the critical roles of institutions, technology, and innovation in fostering sustainability, as well as the importance of multi-level governance in achieving sustainable development goals. The discussion then moves to methods for sustainability assessment and addresses global sustainability challenges. Finally, the chapter addresses sustainability in Iran explicitly, outlining key initiatives and primary concerns in this context. This chapter provides an essential framework for understanding the complexities and requirements of sustainability in the contemporary world.

Keywords

Sustainability · Sustainable development · Environmental sustainability · Social sustainability · Sustainability assessment

1 Learning Objectives

Upon completion of this chapter, readers should be able to:

1. Understand the foundational concept of sustainability, including its historical evolution and interconnected environmental, economic, and social dimensions.
2. Analyze the integration of sustainability and development, with a particular focus on the concept of sustainable development.
3. Explain the pivotal roles of institutions, technology, innovation, and multi-level governance in achieving sustainability goals.
4. Familiarize yourself with sustainability assessment methods and identify and analyze the significant global sustainability challenges.

M. S. Sharifzadeh (✉) · M. T. Moghadam
Department of Agricultural Extension and Education, Faculty of Agricultural Management, Gorgan University of Agricultural Sciences and Natural Resources, Gorgan, Golestan, Iran
e-mail: sharifzadeh@gau.ac.ir

M. Akbari et al. (eds.), *Sustainable Entrepreneurship*, University of Tehran Science and Humanities Series, https://doi.org/10.1007/978-3-032-04909-4_1

5. Examine and evaluate the state of sustainability in Iran, including key initiatives and primary concerns within the country.

2 Introduction

In the contemporary era, marked by unprecedented challenges such as global climate change, the depletion of natural resources, and the rise of "failing states" that can no longer meet the basic needs of their people, the critical tasks of "saving civilization" and "saving the planet" have been entrusted to the concept of sustainability (Purvis et al., 2019; Thiele, 2013). Learning how to live and act sustainably is undoubtedly one of the most significant challenges of our time. But what exactly is sustainability? Is it merely a moral ideal to strive for, or a scientific effort to efficiently manage and preserve natural resources? Do we have the right to burden our children with economic debt, social disorder, resource scarcity, environmental degradation, and climate instability? What are our responsibilities to future generations? These are fundamental questions that make attention to the issue of sustainability essential.

If current trends persist, future generations will face a less hospitable planet, with a warmer and more unstable climate, a more crowded and polluted world, rising and acidifying oceans, drastically reduced biodiversity, scarce natural resources, and ultimately, a more unsustainable existence (IPCC, 2021).

Advocates of sustainability emphasize the well-being of future generations, asserting that we shouldn't deprive them of the chance to experience a healthy and diverse environment. However, prioritizing future generations doesn't mean dictating their way of life; instead, it means "preserving options" for them. Our present-day lives and businesses impact the prospects of our children, and practicing sustainability means living and working in a way that doesn't strip them of meaningful and abundant choices. As Al Gore, former U.S. Vice President and Nobel Peace Prize laureate, stated, "The future whispers while the present screams." Giving an equal voice to the future isn't easy because humans are often quite short-sighted; the comforts and pleasures of the present loom large, while distant consequences remain hidden. Nevertheless, the capacity for self-sacrifice on behalf of our children is a fundamental characteristic that can steer us toward a more sustainable future (Boston & Stuart, 2015; Thiele, 2013).

Since the release of the Brundtland Report in 1987, sustainability has been understood as fulfilling current needs in a way that does not jeopardize the ability of future generations to fulfill theirs. In recent decades, the volume of publications on "sustainability" has grown significantly, leading to the recognition of "sustainability science" as an independent academic discipline (Kajikawa et al., 2014; Komiyama & Takeuchi, 2006; Kuhlman & Farrington, 2010; Nagabhatla, 2024; Purvis et al., 2019; Schoolman et al., 2012). Although sustainability has gained increasing prominence, it remains a flexible concept with diverse interpretations shaped by specific contexts (Jagani & Saboori-Deilami, 2025; Purvis et al., 2019). It is commonly framed around three core pillars—environmental, social, and economic—typically illustrated as overlapping circles, with true sustainability emerging at their intersection. These three pillars form the foundation of the sustainable development concept and are globally accepted (Ebaid, 2024; Kosmopoulos, 2024; Purvis et al., 2019; Sarabdeen, 2024).

The United Nations (UN) explicitly incorporated these three pillars into the formulation of the Sustainable Development Goals (SDGs) (Purvis et al., 2019; UN, 2012). The Sustainable Development Goals (SDGs), comprising 17 goals, 169 targets, and 232 unique indicators, are promoted as a comprehensive framework for assessing the level of sustainable development and addressing diverse social, economic, and environmental challenges (Chen, 2021; Gao et al., 2021; Ogunmakinde et al., 2022; Scrucca et al., 2023; UN, 2022). Achieving these goals requires action across multiple dimensions, including the exploration and optimization of technological innovations' potential (Cancino et al., 2018). Technological innovation plays a vital role in addressing the challenges of sustainable development by promoting economic growth, safeguarding the environment, and advancing social well-being. It offers new pathways for the efficient and effective utilization of limited resources (Institute of Sustainability Studies, 2025; Klewitz & Hansen, 2014; Omri, 2018).

According to the United Nations Industrial Development Organization (UNIDO), technology and innovation have the potential to simultaneously address all three pillars of sustainable development (Omri, 2018). Numerous studies (Boons et al., 2013; Bekhet & Latif, 2018; Omri, 2018) have demonstrated the crucial role of institutional quality and the importance of technological innovation in achieving sustainability and balancing environmental protection with economic growth, topics that are extensively discussed in this chapter.

Today, sustainability grapples with challenges such as environmental balance, social equity, and economic stability (Cancino et al., 2018; Klewitz & Hansen, 2014; Vaikma, 2025). Social issues encompass inadequate living conditions and inequality; environmental concerns include climate change and resource depletion; and economic instability primarily refers to financial crises and poor resource management. The dimensions of sustainability are profoundly interconnected, meaning strategies aimed at improving one dimension can have both positive and negative impacts on the others (Vaikma, 2025).

This chapter embarks on a comprehensive journey, delving into the deep and multifaceted foundations of sustainability. Our goal is to provide a comprehensive understanding of the dimensions, requirements, and processes that underpin a transition toward a more sustainable world. In this chapter, we'll not only explore the concept and history of sustainability but also clarify the vital role of institutions, technology, and innovation in achieving it, alongside the significance of multi-level governance for sustainability in reaching sustainable development goals. Additionally, this chapter will discuss sustainability assessment methods and global sustainability challenges. Ultimately, it will focus specifically on sustainability in Iran, addressing key initiatives and primary concerns in this area. Offering the theoretical and practical underpinnings of sustainability, this chapter aims to serve as an invaluable guide for students, researchers, and professionals committed to comprehending and advancing a sustainable future. We hope this comprehensive exploration proves engaging and informative, equipping you with the knowledge and tools necessary for active participation in this vital field.

3 The Concept of Sustainability

The word "sustainability" originates from the Latin "sustinere," meaning "to hold up" or "to maintain" (Adekeye & Niyi, 2013). Today, this concept has evolved beyond a mere scientific term to become the central theme of global discourse regarding the future of our planet and humanity. In its broadest sense, sustainability refers to a system's (such as a society or an ecosystem) ability to continue functioning in the future without being weakened by resource depletion or excessive burden. In other words, a system's sustainability is entirely dependent on its capacity for adaptation, transformation, and responsiveness to its external environment. As the environment itself is constantly changing, this process of adaptation must be dynamic and sensitive (Zahedi, 2007).

However, understanding and defining this concept have evolved significantly over time. One of the primary challenges in sustainability research is its inherent ambiguity and polysemy; multiple, diverse, and sometimes contradictory definitions make it difficult to select an appropriate conceptual framework (Adekeye & Niyi, 2013; Purvis et al., 2019). This breadth of interpretations has transformed sustainability into an open and context-dependent concept, with each field having its specific understanding (Purvis et al., 2019).

Broadly, the term sustainability can be understood through four main applications, each revealing distinct facets of the concept across theoretical, practical, and analytical contexts (Adekeye & Niyi, 2013):

1. **Sustainability as a Set of Socio-Environmental Criteria**: This perspective emphasizes principles that guide human actions to maintain socio-environmental balance.
2. **Sustainability as a Vision for Humanity**: In this approach, sustainability refers to an ideal state for humanity, achieved through the convergence of social and environmental goals within a specific system.
3. **Sustainability as a Systemic Phenomenon**: This interpretation views sustainability as a characteristic or property within specific socio-environmental systems, indicating their durability and continued existence.
4. **Sustainability as an Analytical Approach**: This application entails incorporating social and environmental variables into the study or implementation of a human activity, process, or product.

The concept of sustainability, as a policy framework, originates from the Brundtland Report of 1987. This seminal document addressed the inherent tension between humanity's aspirations for a better life and the limitations imposed by nature (Kuhlman & Farrington, 2010). It classically defined sustainability as "meeting the needs of the present without compromising the ability of future generations to meet their own needs" (Brundtland, 1987). This definition is considered a pivotal moment in sustainability literature (Kosmopoulos, 2024; Purvis et al., 2019; Sarabdeen, 2024).

Although sustainability is widely recognized as a multifaceted concept, its various dimensions (environmental, social, and economic) have historically prompted different discourses and often been examined in isolation. Today, the concept of sustainability extends beyond mere economics and environment to embrace the social dimension. This tripartite approach focuses on balancing social, economic, and environmental objectives. Sustainability aims to meet a wide range of human needs, including health, literacy, and equity. Sustainability refers to a system's (such as an ecosystem or society) ability to endure and persist in the long term. In this state, human life and social welfare are preserved, supported by a sustainable economy and a healthy ecosystem (Sharifzadeh et al., 2017). In other words, sustainability isn't just about managing natural resources; it also concerns fostering intergenerational social justice and promoting individual freedom to meet needs while considering the needs of others (Dresner, 2012; Schaltegger & Burritt, 2005). These three dimensions of sustainability are interconnected and overlapping. For instance, sustainable economic growth must be achieved through methods that don't degrade the environment or exacerbate social inequalities (López & Toman, 2006; Vaikma, 2025). Therefore, sustainability means balancing the environmental, social, and economic pillars, aiming to enhance human quality of life and preserve ecosystems simultaneously (Elkington, 1998; Mathur & Vayas, 2013). Achieving sustainability demands comprehensive co-operation among nations, communities, and individuals. Broadly, promoting sustainability addresses various global issues while simultaneously strengthening global peace (Sarabdeen, 2024).

Despite a significant increase in publications on sustainability and the emergence of "sustainability science" as an independent discipline, sustainability remains a complex, multidisciplinary, and relative concept. Due to the multiplicity of factors, the interaction of forces, and continuous changes within systems, providing a standard and universal definition of sustainability is a challenging task. Instead, it's better to view sustainability as a dynamic set of values in the realm of development, emphasizing characteristics like ambiguity, complexity, flexibility, diversity, and relativity. These features themselves embody the ideal of sustainability, guiding us toward continuous learning, adaptation, and innovation to manage crises and achieve shared goals (Sharifzadeh et al., 2017).

4 History of Sustainability

The motivation for sustainable living has deep roots in human history, stretching back to ancient times. There have always been two opposing views on humanity's relationship with nature: one emphasizing harmony and adaptation with the natural world, and the other seeing nature as something to be conquered and exploited (Kuhlman & Farrington, 2010). While the latter view has been somewhat dominant, especially in recent centuries in Western civilization, the former has never entirely disappeared.

The concept of sustainability, though not always explicitly termed as such, has always been a natural subject for economists, primarily because resource scarcity is a core concern of the discipline (Kuhlman & Farrington, 2010). Initially, this concept gained prominence in forestry, where it meant that one should never harvest more timber than the forest could regenerate through new growth (Wiersum, 1990).

4.1 Rise of Environmental Awareness and "The Limits to Growth"

Widespread and systematic efforts to promote sustainability only began a few decades ago. Their emergence can be attributed to a growing awareness of global environmental and development issues in the 1950s and 1960s. This awareness stemmed from the detrimental environmental and socioeconomic impacts of unbalanced industrial and conventional development. Early influential actions promoting environmental preservation and sustainability included the UNESCO Biosphere Conferences (Paris, 1968), the 1960s environmental movements, and the publication of impactful books like Rachel Carson's "Silent Spring" and Paul Ehrlich's "The Population Bomb." Furthermore, the metaphor of "Spaceship Earth" by Kenneth Boulding and Edward Mishan, and Herman Daly's concept of a "Steady-State Economy," significantly bolstered this growing awareness (Dresner, 2002).

A pivotal moment during this period was the publication of "The Limits to Growth" report by the Club of Rome in 1972 (Meadows et al., 1972). This report predicted that many natural resources vital for human survival would be depleted within one or two generations. This stark warning, although initially perceived as unsuitable for public policy, served as a serious alarm for the world and laid the groundwork for new thinking on the topic of sustainability. Considering the origin of such initiatives, it can be concluded that in the 1970s, the environmental movement was primarily a Western approach, receiving less attention in developing countries (Dresner, 2002).

4.2 The Brundtland Report and the Definition of Sustainable Development

The evolution of the concept of sustainability can be divided into three distinct periods: before the Stockholm Conference, from the Stockholm Conference to the publication of the Brundtland Report, and the period thereafter (Mebratu, 1998).

In the first period, religious and traditional beliefs played a special role in environmental protection, often endorsing humanity's harmony with nature.

The second period in the evolution of sustainability began with the United Nations Conference on the Human Environment in Stockholm (1972). This was the first major international gathering to examine the detrimental effects of human activities on the environment and the threats they posed to human life. As a direct result of this pivotal conference, the United Nations Environment Programme (UNEP) was established (Mebratu, 1998).

However, the pinnacle in the history of sustainability was the publication of "Our Common Future" in 1987 by the World Commission on Environment and Development (WCED), widely known as the Brundtland Report (WCED, 1987). This report, following four years of extensive international research and dialogue, addressed a fundamental question: "How can the aspirations of the world's nations for a better life be reconciled with limited natural resources and the risks of environmental degradation?" (WCED, 1987).

Their answer was the concept of sustainable development, which the Commission famously defined as: "Development that meets the needs of the present without compromising the ability of future generations to meet their own needs" (WCED, 1987).

4.3 Evolution Post-Brundtland and New Challenges

The third period began with the United Nations Conference on Environment and Development (the Earth Summit) in Rio de Janeiro (1992). This summit, the largest international conference of its time, led to the publication of the "Rio Declaration" and "Agenda 21." It placed greater emphasis on balancing economic development with environmental protection (United Nations Conference on Environment and Development, 1992). This event brought global attention to sustainability to its peak, positioning it at the forefront of development agendas in many countries.

In 2002, the World Summit on Sustainable Development in Johannesburg adopted the slogan "People, Planet, Prosperity" (Rao, 2000). This emphasized that sustainable development isn't just about balancing economic growth and environmental protection; it also includes social development as its third, vital pillar. Since then, socioeconomic and socio-cultural linkages have become increasingly prominent, and social concerns have entered the policymakers' lexicon (George & Kirkpatrick, 2006; White & Lee, 2009). This highlights not only the importance of intergenerational ethics but also the value of public participation as a prerequisite for achieving sustainable development.

More recently, with the adoption of the Sustainable Development Goals (SDGs) by the United Nations in 2015, the concept of sustainability has found a prominent position in global policymaking and research. These 17 global goals, which aim to reduce poverty, protect the planet, and promote prosperity for all, reflect the continuous evolution of our understanding of sustainability and its vital importance in the contemporary world. Furthermore, the Paris Agreement (COP21) in 2015 marked another significant step in this journey, aiming to reduce carbon emissions and combat climate change, thereby striving to end the global economy's reliance on fossil fuels (United Nations, 2015).

In essence, the history of sustainability illustrates its evolution from a localized concern in forestry to a multifaceted global framework. It consistently aims to strike a sustainable balance between human needs and planetary limits, a process that's constantly evolving.

5 Merging Sustainability and Development: Sustainable Development

The concepts of development and underdevelopment can be viewed as two interconnected facets of a single entity. This relationship, although embedded in the structures of the modern global order and the frameworks governing international development since the end of World War II, is deeply rooted in a longer historical trajectory. Indeed, human civilization has consistently encountered—and will continue to face—various forms of change.

This duality of progress and stagnation is deeply embedded in the current global structure, although its roots extend far into the past, well before the post-World War II international frameworks.

The quest for improvement and progress is as ancient as humanity itself. As societies have gradually and organically advanced, the momentum of growth and development has increased, reshaping not only economic structures but also values, perceptions, and expectations. A common thread in these efforts has been the persistent endeavor to elevate human living standards by building upon past achievements and optimizing available resources.

Although widely used in both everyday conversations and academic discourse, the concept of development remains complex and evolving. Its multifaceted nature and context-dependent interpretations have hindered the emergence of a universally accepted definition. Still, in practice and policy, there is a need for a shared understanding based on essential, mutually accepted traits.

To define development, multiple aspects must be examined. First, development must be seen as a value-based construct; second, as a multi-layered and complex process; and third, as a concept intimately tied to comprehensive and continuous progress.

Development is often perceived through two lenses. One approach emphasizes economic growth, defined by sustained increases in output, productivity, and income per capita. In contrast, the second—promoted by the Human Development Report and numerous scholars—views development as a process that expands individual freedoms and broadens meaningful life choices. This latter perspective integrates economic and social advancement with cultural values. Here, poverty is not just material deprivation but also a lack of freedom to choose a more productive, fulfilling, and meaningful life.

This perspective encourages societies to pursue development strategies grounded in values that may diverge from those of the wealthiest nations, where per capita income is often used as the primary measure of success.

With rising awareness of the adverse environmental and social impacts of past development models, the urgency to redefine development through a sustainable lens has grown substantially.

Building on this broader understanding—combining economic expansion and human freedom—it becomes essential to ensure the long-term viability of such development. While improved living standards are desirable, they have at times come at a significant cost to natural ecosystems and future generations. This realization has led to the rise of sustainable development, which seeks a balanced integration of economic growth, environmental preservation, and social equity, ensuring that current needs are met without jeopardizing future generations' ability to meet their own needs.

At its core, sustainable development represents a global commitment to maintaining economic progress, fostering social fairness, and safeguarding environmental resources for both present and future generations. It embodies a holistic vision for humanity's shared future and promotes policies and actions that do not trade off tomorrow's well-being for today's gains.

In recent decades, sustainable development has emerged as a key focus in academia, policy-making, and global governance. It requires not only a conceptual grasp but also mastery of analytical tools and models for practical implementation.

Sustainable development remains a dynamic and contested concept. Some define it as the intersection of environmental and developmental goals, while others frame it as a process where economy, environment, and society function in a complementary, not competitive, relationship. A recurring element in most definitions is environmental preservation as the basis for all development, made possible through balanced interaction among its three core pillars: economic, social, and environmental. If this equilibrium is disrupted, true sustainability cannot be achieved.

A critical milestone was the 1972 Stockholm Conference, organized by the United Nations—the first international gathering addressing environmental concerns. In 1980, the IUCN released its "World Conservation Strategy", advocating for development that balances economic, social, and environmental goals (Bochańczyk-Kupka & Pęciak, 2015).

The most influential and widely cited definition of sustainable development was presented in the 1987 Brundtland Report by the World Commission on Environment and Development (WCED):

"Development that meets the needs of the present without compromising the ability of future generations to meet their own needs." While similar terminology appeared earlier (e.g., IUCN, UNEP, and WWF in 1980), the Brundtland Report significantly popularized the concept and introduced two critical principles:

1. **Needs**: prioritizing the essential needs of the world's poor. This aspect focuses on the continuous provision of needs and the maximization of individual well-being and satisfaction, alongside an improved quality of human life. Brundtland emphasized that to achieve sustainable development, we must promote values that

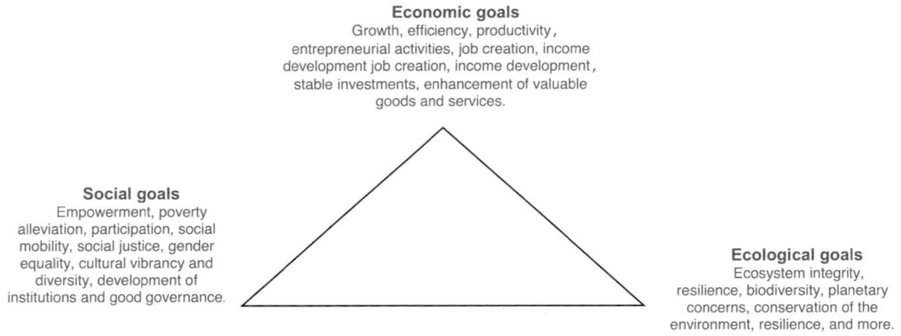

Fig. 1 Main themes of the sustainable development goals

shape consumption patterns to remain within ecological limits, making them reasonably attainable for everyone.

2. **Limitations**: acknowledging the technological and social constraints on the environment's capacity to meet present and future needs (WCED, 1987).

The emergence of sustainability as a key concept in the 1970s was the result of growing global awareness of environmental degradation and uneven development. This awakening, intensified by the post-industrial environmental crises of the 1950s and 60 s, led to a series of pivotal conferences and reports throughout the following decades.

Through the Brundtland Report, sustainability was reframed not just as an ecological matter but as a cross-cutting issue encompassing economic and social dimensions. It outlined a roadmap for change in resource use, technological direction, institutional structure, and investment orientation to support long-term goals (White & Lee, 2009).

Despite some criticisms of its broad and compromise-based approach (Tulloch, 2013), sustainable development gained importance rapidly throughout the 1990s, becoming a mainstay in global political agendas and academic research (Gatto, 1995). Today, it is widely seen as the essential pathway to a viable global future.

The goals of sustainable development are commonly structured around three interrelated dimensions:

- **Economic**: productivity, innovation, stable investment, job creation, and growth of goods and services.
- **Social**: equity, poverty reduction, participation, gender equality, governance, cultural vitality, and social justice.
- **Environmental**: resilience, biodiversity, ecosystem health, climate stability, and resource conservation.

Achieving sustainability requires an integrated strategy that respects the interdependence of these dimensions (Fig. 1).

Despite various interpretations, the "three pillars" model—economic, social, environmental—is one of the most commonly accepted frameworks (Giddings et al., 2002; Purvis et al., 2019). These pillars are not independent; instead, they interact continuously, where progress in one can positively or negatively affect the others (Elkington, 1998; Schaltegger & Burritt, 2005). Development is only sustainable when all three dimensions are jointly addressed and harmonized (Fig. 2).

In some frameworks, additional pillars such as institutional (Spangenberg et al., 2002), cultural (Soini & Birkeland, 2014), and technological (Hill & Bowen, 1997) are also considered. This reflects the complex, interdisciplinary nature of sustainability, involving environmental, social, economic, legal, cultural, political, and psychological domains.

Critics argue that while the theory of sustainable development has advanced, practical action has lagged. They call for a critical reassessment

Fig. 2 Interaction of sustainable development dimensions

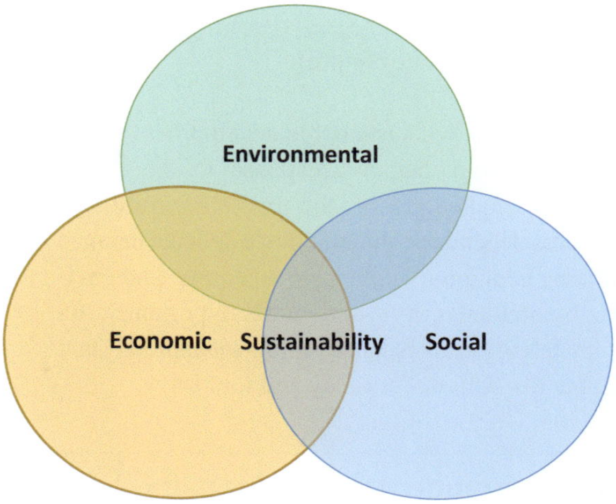

and activation of sustainability thinking—a transition from theory to practice. This has led to the idea of sustainable development praxis, which integrates (Fig. 3):

- **Theoretical insight**: critical awareness of sustainability concepts and scientific progress.
- **Practical engagement**: real-world collective actions to promote sustainable initiatives (spreading and diffusing ideas, innovations, and initiatives of sustainable development within the society).
- **Conscious practitioner efforts**: deliberate and reflective development of actionable

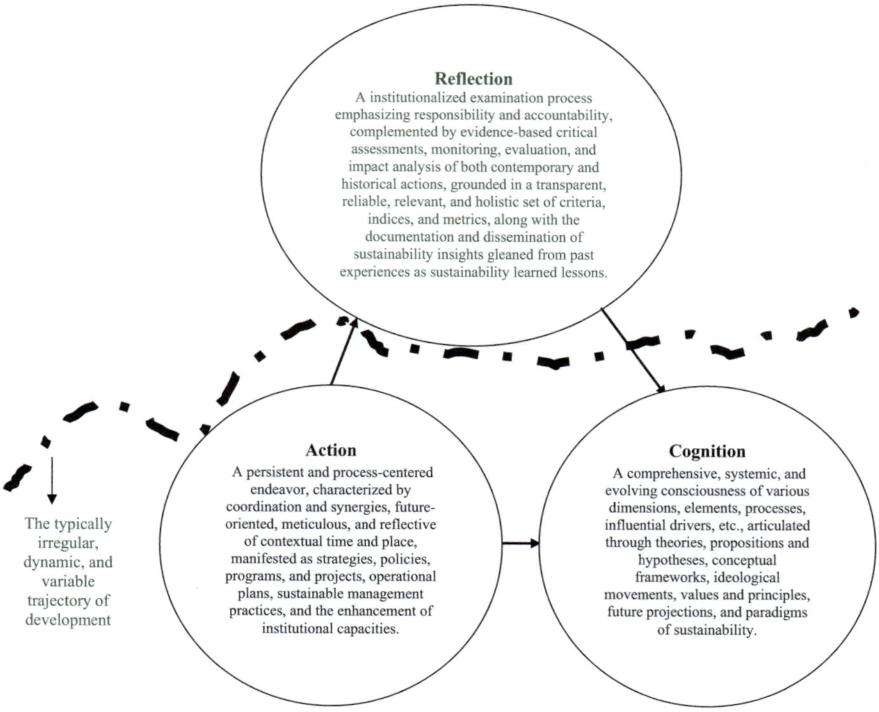

Fig. 3 Triple helix of sustainable development praxis

sustainability strategies (learning by developing for a better future).

In conclusion, sustainable development represents a comprehensive response to the multifaceted challenges of the modern world. Its success depends on maintaining a delicate balance among economic vitality, social equity, and environmental integrity—and more importantly, on translating this balance into widespread and impactful actions for today and tomorrow.

6 Environmental Sustainability (Environment and Sustainable Development)

Environmental sustainability focuses on preserving and enhancing the quality of the environment, as this quality is essential for both economic activities and human well-being (Klarin, 2018). This comprehensive dimension includes ecosystem integrity, carrying capacity (the environment's ability to absorb pollution and produce resources), and biodiversity (Morelli, 2011; Vos, 2007). The primary goal is to protect natural capital as a source of economic inputs and a capacity for waste absorption (Khan, 1995; Schramski et al., 2015). This means that renewable resources should not be consumed faster than their regeneration capacity, and waste emissions should not exceed the environment's absorptive capacity (Khan, 1995).

Environmental sustainability is a conservation concept, meaning it integrates the services and resources needed by current and future generations without jeopardizing the health of the ecosystems that provide these services. In essence, this dimension aims to enhance human well-being by protecting natural capital, such as the atmosphere and soil, and by reducing carbon footprints, packaging waste, and water consumption.

To achieve environmental sustainability, adherence to the following principles is essential (Morelli, 2011):

- **Biodiversity and Energy Resources Conservation**: This involves preserving biodiversity and utilizing sustainable energy sources to promote environmental sustainability.
- **Regenerative Capacity**: Renewable resources should be utilized at a rate that allows them to replenish naturally, and non-renewable resources must be consumed in a manner that ensures the development of renewable alternatives occurs faster than their depletion.
- **Waste Management**: Practices such as reuse and recycling should be encouraged to reduce waste and greenhouse gas emissions. Furthermore, the release of waste into the environment should not exceed the ecosystem's ability to absorb it.
- **Consumption Limits**: The human economic system's scale must remain within its carrying capacity, encompassing both per capita consumption and technological advancements.
- **Raw Material Quality and Product Life Cycle**: The environmental characteristics of raw materials must be considered, and decisions regarding product manufacturing and management should fully account for their environmental impacts throughout their life cycle.

Therefore, by focusing on improving human well-being through the conservation of natural capital, such as the atmosphere and soil, and by reducing carbon footprints, packaging waste, and water consumption, environmental sustainability plays a pivotal role in ensuring a healthy future for all.

7 Economic Sustainability

Economic sustainability, as one of the main pillars of sustainable development, encompasses a set of strategies and policies aimed at ensuring economic growth and prosperity without harming the social, environmental, and cultural aspects of a society (Klarin, 2018). This approach emphasizes the continuous improvement of economic performance, effective process management, and the creation of balance at both macro and microeconomic levels. In this pursuit, factors such as value creation, productive employment, optimal resource allocation, and intelligent investment play a key role (Markulev & Long, 2013). This significance is evident in the United Nations Sustainable Development Goal 8 (SDG 8), which aims to foster inclusive and sustainable economic growth, ensure productive employment, and promote decent work opportunities for everyone (United Nations, 2015).

This goal comprises twelve sub-targets, encompassing economic diversification, innovation, productivity, youth employment, protection of workers' rights, sustainable tourism development, and improved access to financial services.

In the past, economic growth was primarily measured by Gross Domestic Product (GDP); however, a singular focus on GDP has led to inappropriate policies due to its neglect of consequences like environmental degradation and increasing inequality (Leamer, 2009). For this reason, modern approaches emphasize the development of alternative indicators such as Gross Ecosystem Product (GEP), an index that attempts to monetarily estimate the value of nature's services to the economy and human well-being (e.g., clean water, food security, and health) (Kosmopoulos, 2024; Ouyang et al., 2020).

The concept of economic sustainability, rooted in Hicks's ideas (1946), emphasizes that economic growth should not come at the expense of environmental or social capital. In other words, the development of economic capital should not lead to a reduction in natural or social capital. In this context, intergenerational equity and long-term resource allocation are fundamental components of this approach (Anand & Sen, 2000; Markulev & Long, 2013). The goal is to build a production system that ensures a high level of well-being for both current and future generations.

In practice, achieving this sustainability hinges on the preservation and strengthening of natural, social, and human capital; these are the fundamental pillars of a society's income and living standards. Efficient resource utilization, sustained profitability of economic enterprises, and risk reduction through optimal asset management are the operational cornerstones of this sustainability (Jeronen, 2023; Klarin, 2018).

However, the differing stages of development among countries lead to varied approaches to economic sustainability. In industrialized nations, where resource consumption, particularly energy, is high, sustainability means a sustained reduction in consumption through increased efficiency and lifestyle changes. These countries, leveraging their financial and technological capabilities, can lead the way in developing clean technologies and play a supportive role for developing nations.

In contrast, in poorer countries, the priority shifts to alleviating absolute poverty and improving living standards. Providing basic necessities, ensuring equal access to resources, and empowering various segments of the population to participate in the economic growth process are prerequisites for achieving economic sustainability in these nations. This is because, without meeting minimum living standards, there will be little attention paid to environmental protection or the future of the planet (Anand & Sen, 2000; Sachs, 2015).

Finally, environmental economists define economic sustainability in terms of the non-depletion of assets, particularly natural capital. Accordingly, two perspectives have emerged regarding the substitutability of assets:

- In weak sustainability, it is assumed that different forms of capital (e.g., natural, economic, human) can substitute for one another.
- However, in strong sustainability, these assets are considered irreplaceable, emphasizing the need for their direct protection (Neumayer, 2010).

Despite advancements in science and technology, some resources, such as essential environmental services, have no substitutes and require cautious and responsible use to avoid compromising the well-being of future generations (Jeronen, 2023).

8 Social Sustainability

Social sustainability is one of the three key dimensions of sustainable development that is often overlooked in comparison to its economic and environmental counterparts (Dempsey et al., 2011). This poses a serious challenge to sustainable development, as changing behavior and fostering sustainable attitudes among people requires understanding, participation, and a sense of social belonging. Attempts have been made to define social sustainability based on quality of life, but these very efforts have led to increased complexity and ambiguity in understanding the concept (Vallance et al., 2011). Researchers believe that development leading to the strengthening of collective life, social justice, and sustainable participation is the accurate measure of sustainable development (Ayodele & Ogunlola, 2016; Chiu, 2003).

Social sustainability is a dynamic concept, contingent on time and place. Consequently, various definitions have emerged, all emphasizing key elements such as access to public services, preservation of cultural heritage, adherence to ethical principles, creation of healthy work environments, and local political participation (Chiu, 2003; Morelli, 2011). Furthermore, a sense of belonging, human health, social infrastructure, and transparency in local decision-making are other essential criteria for evaluating a socially sustainable structure (Ayodele & Ogunlola, 2016).

Despite the increasing importance of this dimension, social sustainability has received less attention in policy-making. A singular focus on environmental or economic indicators has often led to the failure of actual sustainable development implementation (Alonso-Morales et al., 2025; Hueting & Reijnders, 2004). Disregarding social justice, rising poverty, the growth of informal settlements, and gender and class inequality all threaten social cohesion and the sustainable future of cities (Sachs, 2015; Wu et al., 2018).

Social sustainability emphasizes intangible aspects of life such as security, a sense of belonging, human dignity, and social participation, striving for an equitable distribution of well-being across time and space (Summers & Smith, 2014). The United Nations' Sustainable Development Goals officially integrate the three dimensions of economic, social, and environmental sustainability (United Nations, 2015). However, researchers classify social goals differently; some place Goals 1–5 and 11 in this dimension (Leal Filho et al., 2018), while others, like Bali Swain and Yang-Wallentin (2020), also consider Goals 1, 2, 3, 12, 13, and 14 as social. Another group aligns Goals 4, 5, 10, 11, 16, and 17 with this domain, drawing parallels to Maslow's hierarchy of needs (Wu et al., 2018).

Social sustainability is crucial not only for improving the quality of life and ensuring equitable access to resources but also for preserving a society's social capital. This capital includes public trust, social networks, public ethics, and institutional participation (Kahn, 1995; Klarin, 2018). Creating equal opportunities, fostering intergenerational equity, respecting cultural diversity, upholding social freedoms, and ensuring the fair distribution of resources are all fundamental to social sustainability (Dresner, 2012; Morelli, 2011). Sustainable development will only be truly achieved when this dimension is fully and equally considered alongside the other pillars.

9 Institutions and Sustainability

Institutions, as the fourth pillar of sustainability, play a fundamental role in achieving sustainable development alongside the environmental, economic, and social dimensions (Valentin & Spangenberg, 2000). Numerous studies have demonstrated that the quality of institutions has a significant influence on economic performance, poverty alleviation, and the establishment of sustainable social structures (Acemoglu et al., 2001; North, 1990). These institutions—both formal and informal—constitute the frameworks of rules, procedures, and norms that guide human behavior and interaction in social, economic, and environmental spheres (Dovers, 2001).

Institutional sustainability refers to the ability of institutions to perform effectively over the long term, adapt to change, learn from experience, and foster inclusive social participation.

To achieve this, desirable principles of institutional design—such as revisability, robustness, value sensitivity, transparency, and experimental diversity—are essential for improving governance and enhancing institutional resilience (Dovers, 2001). By defining roles, regulating relationships, managing resources, and enabling participation, institutions strengthen the foundation of sustainable development (Matutinovic, 2007).

Furthermore, civil institutions such as Non-Governmental Organizations (NGOs) play a critical role in facilitating the implementation of development programs. By promoting voluntary engagement, enhancing social accountability, and fighting corruption, these institutions help bridge the gap between policy and practice, especially in developing countries. Their presence empowers local communities and enhances the legitimacy and inclusiveness of development efforts (Elobeid, 2012a).

Institutional relationships, policies, rules, regulations, and governance frameworks have a significant impact on the efficacy of development initiatives. In other words, the effectiveness, durability, and sustainability of development programs are largely dependent on the institutional capacities of both implementing actors and local communities, as well as how their relationships are institutionalized—whether through coordinated, interactive, or synergistic mechanisms. This implies that institutional development is not only a means to an end but also a goal and responsibility in itself for many development stakeholders.

Institutions are thus understood as collective patterns that establish public order by setting and enforcing rules, norms, laws, and regulations. These institutions govern the behaviors and actions of a wide array of actors, including government bodies, private enterprises, NGOs, and civil organizations, across various levels and domains. They determine the rules of participation, interaction, and behavior in public life, thereby shaping how societies function and evolve.

The configuration of institutions varies across societies and is influenced by the extent to which these structures are differentiated and functionally specialized. In more developed contexts, institutions tend to be more clearly demarcated and autonomous, leading to scenarios in which they do not merely coexist or clash, but interact cooperatively or competitively in ways that reinforce one another. This differentiation fosters harmonious institutional interplay, contributing to a more robust and adaptive governance ecosystem.

Ultimately, institutions play a pivotal role in creating synergy among the pillars of sustainability. By shaping human interaction, enhancing governance, enabling innovation, and promoting social justice, they form a critical prerequisite for realizing a sustainable and equitable future, for both present and future generations (Hesen et al., 2022; Quiroz-Niño & Murga-Menoyo, 2017).

10 Technology, Innovation, and Sustainability

Technology and innovation are recognized as key driving forces in achieving sustainable development, playing a pivotal role in addressing the growing environmental, economic, and social challenges. Technological solutions, such as advancements in renewable energy, intelligent resource management, and the implementation of circular economy principles, represent some of the most effective mechanisms for reducing environmental risks, enhancing efficiency, and improving socioeconomic conditions (Dornfeld, 2014; Huang, 2021). Notably, advancements in energy-efficient appliances, precision agriculture, and sustainable infrastructure have laid the foundation for more targeted resource utilization, facilitating a transition toward more sustainable patterns of production and consumption (Siriram, 2022).

At a macro level, technological innovation is regarded not only as a strategic imperative in corporate and industrial policy-making but also as an effective instrument for redefining traditional economic structures, transforming business models, and generating long-term social and environmental value. The integration of emerging technologies—such as digitalization, biotechnology, the Internet of Things (IoT), and artificial intelligence (AI)—into production and distribution processes has had a significant impact

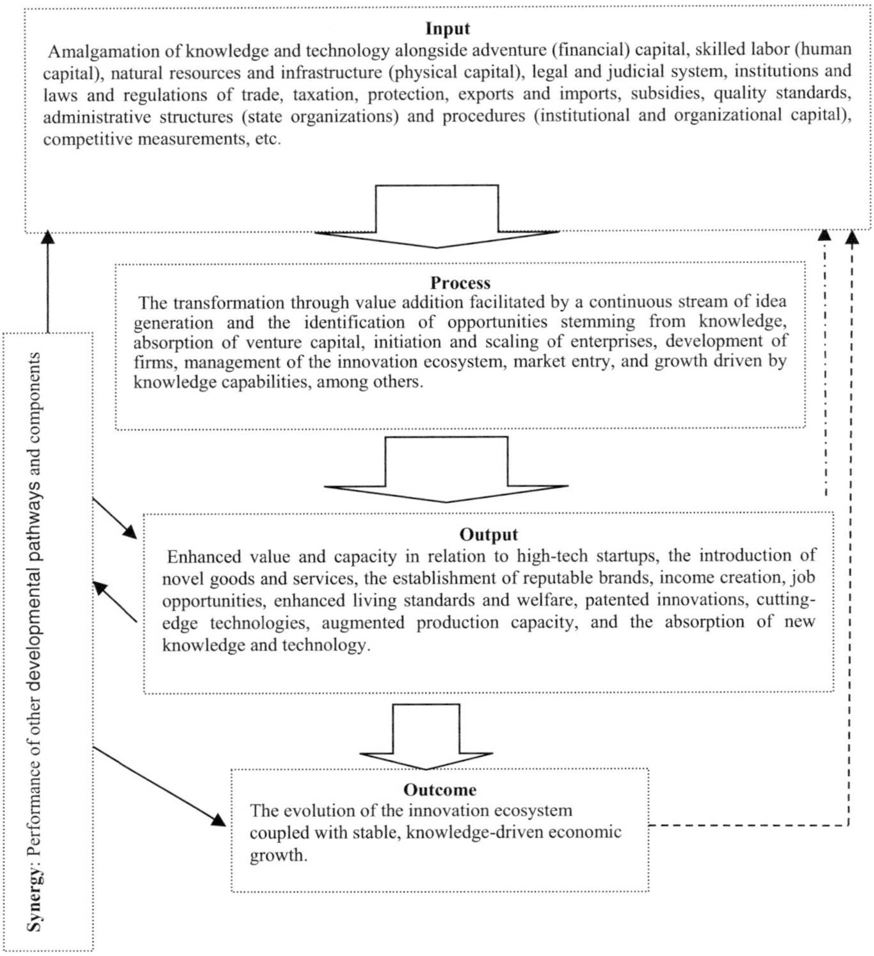

Fig. 4 Logical framework of knowledge-based economic development

on energy efficiency, waste reduction, health and safety, and sustainable infrastructure design. This technological convergence has prompted a reexamination of the relationship between innovation and sustainability (Kiron et al., 2012; Nidumolu et al., 2009).

The logical framework of knowledge-based economic development (Fig. 4) offers a structured approach for understanding the pathways through which technology and innovation contribute to sustainability outcomes.

It begins with Input, where the amalgamation of knowledge, financial capital, human capital, infrastructure, institutions, and governance frameworks forms the foundation. These inputs support the Process phase, characterized by continuous idea generation, venture capital absorption, and the evolution of innovation ecosystems. Through this transformation, new knowledge and capabilities are translated into high-impact initiatives.

The Output stage reflects tangible gains such as enhanced technological capacity, the emergence of startups, patented innovations, and improved standards of living. These, in turn, contribute to the Outcome, which is defined as the evolution of a resilient innovation ecosystem and sustained, knowledge-driven economic growth. This framework underscores the centrality of institutional quality and systemic coordination in ensuring that technological innovations lead to sustainable, long-term impact.

However, the effective utilization of technological capacity in support of sustainable development requires the integration of sustainability principles across all stages of technology design, development, and implementation. Empirical evidence from emerging technologies such as biofuels and nanomaterials illustrates that, in the absence of comprehensive environmental and social impact assessments, even seemingly green technologies can produce adverse consequences for human health and ecosystems (Huang, 2021; Omri, 2020).

Comparative studies indicate that the effectiveness of technological innovation in achieving sustainability goals is highly dependent on the quality of governance institutions, the level of economic development, and the availability of technological infrastructure. Specifically, in advanced economies, innovations tend to simultaneously enhance all three dimensions of sustainable development, whereas in middle- and low-income countries, the impacts are primarily confined to economic and environmental domains (Liu & Liang, 2024). This highlights the crucial need for investment in technological governance, education, and institutional capacity-building, especially in developing countries (Bakht & Abdul Latif, 2018).

Emerging technologies—such as smart agriculture, smart grids, 3D printing, data mining, and advanced resource management systems—enable the optimal utilization of natural resources, reduction of waste, and increased efficiency in key sectors, including agriculture, energy, and transportation. These technologies not only hold significant potential for mitigating environmental impacts but also contribute to sustainable development by improving supply chain effectiveness and promoting equity in resource access (Everett & Kemp, 2022).

Additionally, as visualized in the framework, technological advancements generate synergy with other developmental pathways and governance components. By enhancing institutional performance and supporting interactive innovation networks, technology amplifies the systemic capacity for sustainable transformation.

Ultimately, the future of sustainable development depends on societies' ability to design, develop, and scale technologies that holistically address long-term environmental, social, and economic objectives. Governments and organizations that place sustainable innovation at the core of their strategic agendas will not only play an active role in mitigating global crises but also strengthen their social legitimacy and competitive advantage in the knowledge-based economy (Kiron et al., 2012).

11 Sustainability Multi-level Governance

In the face of interlinked global crises—such as poverty, inequality, environmental degradation, and threats to peace and security—rethinking conventional governance structures has become imperative. In this context, Multi-level Governance for Sustainability (MLG) has emerged as a pivotal approach, emphasizing horizontal and vertical collaboration across government levels (local, national, and international) and among non-state actors, including civil society, businesses, and local communities (Hoogeveen et al., 2017; Poyraz & Szalmáné Csete, 2022).

Unlike traditional top-down models, Multi-level Governance for Sustainability (MLG) underscores inclusive decision-making, policy coherence, public participation, and institutional accountability. It recognizes that complex issues such as climate change and resource management require integrated responses that transcend administrative boundaries and involve coordinated action across sectors and scales (Bulkeley & Betsill, 2013; Pangalos, 2023). From the World Bank's perspective, good governance encompasses the government's capacity to effectively design and implement policies, ensuring transparency, participation, and accountability—factors that are increasingly viewed as equally important as sound macroeconomic management in advancing sustainable development initiatives (Elobeid, 2012b). In this sense, Multi-level Governance for Sustainability (MLG) is not only a governance structure but also a mechanism for building institutional capacities to implement policy innovation and sustainability (Weibust & Meadowcroft, 2014).

Key components of multi-level governance for sustainability include:

1. **Inter-level collaboration**: Facilitating the exchange of information, resources, and expertise among different governance levels to foster coordinated sustainability efforts (Hoogeveen et al., 2017).
2. **Policy alignment**: Ensuring coherence between local and national policies to support shared sustainability goals (Marks, 1993).
3. **Stakeholder participation**: Engaging diverse actors—including civil society, the private sector, and local communities—in policy formulation to enhance policy relevance and legitimacy (Selin & VanDeveer, 2012).
4. **Accountability and transparency**: Establishing mechanisms for monitoring and evaluating policy outcomes to promote public trust and improve policy effectiveness (Zahdi & Ebrahimipour, 2012).
5. **Capacity-building and innovation**: Strengthening institutional and community capacity to design and implement innovative sustainability policies (Geels, 2010).

The Multi-Level Perspective (MLP), originating from innovation studies and evolutionary economics, provides a systemic framework for understanding transitions toward sustainability. It conceptualizes the dynamic interactions between niche innovations, socio-technical regimes, and broader landscapes, offering insights into how systemic change can be steered over time (Geels, 2010).

The concept of multi-level governance, originally developed within the context of European integration to describe the interplay between local, national, and supranational decision-making (Hooghe & Marks, 2001), has since been adopted in environmental governance and climate policy literature to conceptualize complex actor and institutional linkages (Weibust & Meadowcroft, 2014).

Ultimately, multi-level governance for sustainability offers not only an operational tool but also an analytical lens for examining the interconnectedness of actors and systems across scales. A successful transition to sustainability necessitates shifting from a growth-centered paradigm to a sustainability-centered one, guided by principles of inclusiveness, accountability, and ecological balance (Poyraz & Szalmáné Csete, 2022; Zahdi & Ebrahimipour, 2012).

12 Sustainability Measurement/ Assessment

With the paradigm shift from a traditional, economy-centered development approach to an alternative, sustainability-focused model, the need for revising planning, management, and performance evaluation methods has become more evident than ever. In this new approach, the assessment and measurement of development components are recognized as essential tools for interpreting and explaining sustainability (Dresner, 2008).

Sustainability assessment is a multidimensional process that utilizes indicators, frameworks, and tools to evaluate the environmental, social, and economic impacts of activities. Commonly used instruments include the Sustainability Development Goals (SDGs), Life Cycle Assessment (LCA), Environmental Impact Assessment (EIA), and Social Justice Measurement (Devuyst et al., 2001; Ness et al., 2006).

The primary goal of these assessments is to provide a scientific basis for policy formulation, identify strengths and weaknesses, and enable continuous monitoring of progress toward sustainable development. Additionally, models such as the Sustainability Assessment Model (SAM) provide mechanisms for stakeholder engagement and support organizational decision-making (Bebbington & Frame, 2003).

Institutions such as the World Bank define sustainable development as the management of a portfolio of assets across human, natural, and physical capital. For measuring sustainability, indicators such as Adjusted Net Savings have been developed, based on the premise that no single form of capital can substitute another and each must be preserved independently (World Bank, 2006).

At the implementation level, sustainability assessment frameworks—such as the Triple Bottom Line (TBL), circular economy models, LCA, ecological footprint, and indicator-based models—evaluate development performance across its three core dimensions. These tools assist decision-makers by clarifying trends, benchmarking performance, and designing effective interventions (Poveda, 2017; Shortall et al., 2015).

Modern sustainability evaluation frameworks, such as the strategic model proposed by Hacking and Guthrie (2008), emphasize three key features: comprehensiveness (covering economic, social, and environmental aspects), integration (bridging diverse dimensions), and strategic orientation (focusing on long-term goals). Unlike one-dimensional and growth-oriented models, these frameworks adopt integrated analyses across different decision-making levels.

Despite the advantages of current tools, challenges remain, including a lack of integration, difficulty in measuring non-market goods, and limitations in combining quantitative and qualitative indicators. These issues have been noted in the literature (Brandon & Lombardi, 2011). In response, initiatives such as the European Union's "Sustainability A-Test" project have emerged to develop coherent frameworks that integrate tools like multi-criteria analysis, scenario modeling, and composite indicators.

Ultimately, the success of sustainability assessment hinges on three core components: stakeholder participation, transparency of evaluation principles, and the use of reliable data. For instance, engaging the public and experts in designing indicators can facilitate the implementation of development actions (Gibson et al., 2010).

In conclusion, sustainable development without systematic and scientific assessment frameworks risks devolving into a trial-and-error approach. Therefore, enhancing the quality of evaluation models and integrating environmental, social, human, and physical capital should remain at the center of future development policy-making.

13 Global Sustainability Challenges

Global sustainability is an interdisciplinary concept encompassing the environmental, economic, social, and institutional dimensions of development. Confronted with escalating pressure on natural resources, pervasive social inequalities, and evolving institutional challenges, the global community faces a complex array of crises that threaten the future of sustainable development (Edwards, 2019; Hariram et al., 2023; Jeronen, 2020; UN, 2015).

The first and perhaps the most significant challenge is climate change and environmental degradation. The release of greenhouse gases has contributed to global warming, rising sea levels, and an increase in the frequency of extreme weather events, including floods, droughts, and storms. These phenomena not only threaten biodiversity but also jeopardize human livelihoods (Hariram et al., 2023; IPCC, 2023). If current trends persist, irreversible damage will be inflicted upon natural ecosystems.

From another perspective, the overconsumption of natural resources—including water, forests, fossil fuels, and minerals—has also created a crisis. Unsustainable consumption patterns in both developed and developing countries have led not only to the depletion of natural reserves but also to the disruption of ecological balances (World Bank, 2022). This overexploitation poses a grave threat to the future of global food and water security.

From a social perspective, economic inequality and poverty are fundamental obstacles to sustainable development. Disparities in income, education, and health impede social progress, leading to instability, migration, and social unrest (United Nations, 2020; UNDP, 2020). Sustainability is impossible without social justice, as unequal societies cannot fully leverage their human potential, hindering collective well-being and societal advancement.

Furthermore, weak institutions and ineffective governance also represent a fundamental challenge. In many countries, governmental bodies grapple with issues such as corruption, a lack of transparency, and insufficient accountability. These factors erode public trust and undermine the successful implementation of sustainable policies (World Bank, 2021b). Good governance is a foundational element for achieving sustainability, as it strengthens institutional integrity and fosters public participation (OECD, 2017; UN, 2015).

Moreover, the decline in biodiversity resulting from human activities such as habitat destruction, overhunting, and environmental pollution poses a significant threat to the health of ecosystems. Biodiversity is considered a fundamental pillar of natural resilience against climate change and environmental shocks (Cardinale et al., 2012; Oliver et al., 2015). This loss diminishes the capacity of ecosystems to adapt and provide essential services, thereby increasing global vulnerability to environmental shifts (Barnosky et al., 2012; Pörtner et al., 2021).

Finally, unplanned urbanization and unregulated urban development have increasingly led to significant challenges, including the proliferation of informal settlements, escalating pollution, severe traffic congestion, and the inefficiency of public services (Gencer, 2018). If urban development is not guided by sustainability principles, social and environmental crises in urban areas will intensify, undermining the well-being and livability of cities (OECD, 2017).

In summary, global sustainability is attainable only through the simultaneous understanding of and coordinated action across its four main dimensions: environmental, economic, social, and institutional. Disregarding any of these dimensions will pose a significant obstacle on the path toward sustainable development.

14 Sustainability in Iran: Main Measures and Key Concerns

Sustainability in Iran has gained increasing importance as a crucial topic in the context of social, economic, and environmental development. Given its abundant natural resources and rich cultural diversity, Iran faces unique challenges and opportunities on its path toward achieving sustainability.

One of Iran's pivotal initiatives in the pursuit of sustainability is its focus on renewable energy. Given the country's substantial solar and wind potential, the Iranian government has approved ambitious programs aimed at reducing reliance on fossil fuels and increasing the share of renewables in the national energy mix (Zarif & Hossain, 2021). These measures are not only instrumental in mitigating greenhouse gas emissions (IPCC, 2023) but also contribute to creating new employment opportunities and enhancing national energy security (Fattahizadeh & Andarkhord, 2025). This strategic shift aligns with global objectives for combating climate change and transitioning towards a green economy, a model that emphasizes sustainable growth and environmental protection (UNEP, 2011).

In addition, Iran is actively working to improve its economic and social conditions in line with sustainability principles through international collaborations and the adoption of innovative technologies. These collaborations encompass knowledge and technology exchange with other nations and international organizations, which can significantly contribute to elevating environmental and economic standards (Bashiriyeh, 2020; UN, 2021). Furthermore, efforts to foster a knowledge-based economy and strengthen high-productivity sectors are aimed at economic diversification and reducing reliance on oil revenues (Miraskari Jildani et al., 2024; Naghavi & Asnaashari, 2022). This strategic shift is crucial for building a more resilient and sustainable economic future.

However, despite these efforts, Iran faces numerous concerns and significant challenges on its path toward achieving sustainability, which demand particular attention:

1. **The Imperative for Shifting Mindsets and Cultural Paradigms**

A primary challenge lies in the need to transform existing mindsets and cultural patterns, as well as elevate societal awareness regarding sustainability. This requires a precise understanding of the concept of sustainability and a profound appreciation for the environmental and social implications of economic activities (Asgari, 2024). Enhancing public literacy and awareness in environmental issues and sustainable development can be achieved through effective educational programs that integrate both knowledge and practical skills (Bazm, 2021; UNESCO, 2025). This broader societal shift is crucial for fostering a collective commitment to sustainable practices and overcoming cultural barriers that may impede environmental initiatives.

2. **Management of Limited Resources**

Iran, due to its arid and semi-arid climate, faces increased pressure on its water and energy resources, particularly in the agricultural and urban sectors (Rahimi et al., 2020; WHO, 2022). Critical issues such as the depletion of groundwater resources, changing energy consumption patterns, and biodiversity loss pose severe threats to the country's sustainability. Over-extraction of groundwater, in particular, and the pollution of water sources further intensify these challenges. Therefore, the optimal management of natural resources through the adoption of novel technologies in energy efficiency, water conservation, promotion of sustainable agriculture, and preservation of biodiversity is crucial for achieving sustainability goals. These interventions are vital for ensuring long-term resource security and ecological balance in the face of climate change and population growth (IEA, 2024).

3. **Environmental Pollution**

Environmental pollution, notably air pollution from industrial and household activities in major Iranian cities, poses a significant threat to public health and ecosystems. High levels of pollutants, often exacerbated by the use of low-quality fuels and unfavorable meteorological conditions, lead to an increased incidence of respiratory and cardiovascular diseases. Beyond human health, this pollution damages soil, water, and biodiversity, necessitating urgent and effective solutions (WHO, 2022).

4. **Lack of Institutional Coordination and Governance**

A significant impediment to sustainability in Iran is the insufficient coordination among government policies, public institutions, and the private sector (Asgari, 2024). Effective collaboration and synergy between the government, industry, and civil society are paramount for implementing sustainable programs and strengthening good governance (OECD, 2017; World Bank, 2021b). The absence of a unified approach often leads to fragmented efforts, hindering the systemic changes necessary for sustainable development (UN, 2015).

5. **Financial Challenges and Resource Provision**

Sustainable initiatives demand substantial initial investments. Despite their clear long-term benefits, securing adequate financing for sustainable projects in Iran faces considerable hurdles and limitations (Asgari, 2024; World Bank, 2021a). Key strategies to address this include meticulous financial planning, providing government incentives and support, encouraging sustainable investments, and employing creative financial management approaches (Gholami et al., 2022; UNEP, 2021). These measures are crucial for mobilizing the necessary capital to drive the transition towards a sustainable economy.

In summary, Iran's sustainable future hinges on the sustainable performance of its government, institutions, and companies, coupled with effective policy-making and a shared culture of sustainability within society. To achieve these goals, a concerted focus is needed on research and development of novel technologies, raising public education and awareness, enacting robust environmental laws and regulations, encouraging sustainable investments, and strengthening inter-institutional collaboration. These actions are paramount for fostering a truly sustainable trajectory for the nation.

15 Conclusions: Sustainability as a Common Human Developmental Trajectory

One of the first and arguably most pressing challenges facing contemporary societies is responding to crises whose dimensions transcend both geographical and generational boundaries. In this context, sustainability is no longer merely an abstract ideal; rather, it has become a vital and multidimensional imperative for the survival of humanity. This necessity has gained a central position in the discourse on human development, particularly in response to unprecedented challenges such as climate change, the depletion of natural resources, and growing inequalities.

Sustainability, rooted in the cautious management of resources, entered a new phase with the landmark Brundtland Report in 1987, which described it as a broad concept focused on fulfilling current needs while ensuring that future generations can also meet theirs. Today, this concept is understood as an interdisciplinary and threefold approach encompassing environmental, economic, and social dimensions—pillars that have been thoroughly addressed in the United Nations' Sustainable Development Goals (SDGs) (Brundtland, 1987; Purvis et al., 2019).

Sustainability is described as a broad concept focused on fulfilling current needs while ensuring that future generations can also meet theirs.

However, an examination of both global and national contexts reveals that sustainability often remains the missing link in the development pathway. While many institutions and companies acknowledge its importance, in practice, only a small percentage have systematically integrated it into their policies. This gap between understanding and implementation stems from barriers such as a short-term focus on profit, competitive pressures, and resource limitations. Despite this, the long-term benefits of sustainable strategies—from improved efficiency and cost reduction to increased stakeholder satisfaction and intelligent risk management—are clear evidence of this path's value.

Achieving sustainable development necessitates the synergy of governments, the private sector, civil society institutions, and society as a whole. Disregarding any of the dimensions of this triad, especially in today's world, which faces environmental threats, institutional challenges, and social inequalities, can severely disrupt the development process. Technology and innovation are essential tools, but they will only lead to inclusive outcomes within a framework of responsible, transparent, and accountable governance.

Ultimately, sustainability is not merely a destination, but a shared path for societies to ensure the well-being of current generations and protect the future. This journey requires cultural shifts, increased public awareness, and targeted investment in innovative technologies, education, and environmental policy-making. Iran's sustainable future also hinges on the commitment and cooperation of its institutions, companies, and all members of society; a future where human development and planetary health are achieved in a balanced and sustainable connection.

16 Questions

To assess your understanding of the chapter on sustainability, consider these questions:

1. How has the historical evolution of sustainability, from its roots to the Brundtland Report, shaped our understanding of the interconnectedness of its environmental, economic, and social dimensions?
2. What does the concept of "preserving options" for future generations mean in sustainability? How can this approach influence our current decisions?
3. How can technology and innovations both serve as a solution to sustainability challenges and potentially exacerbate them?
4. What is meant by multi-level governance of sustainability? How is the role of different institutions (government, private sector, civil

society) defined in this structure, and what are the consequences of their lack of coordination?
5. What are the most significant challenges to sustainability in the world and Iran, particularly in the implementation of sustainable projects?
6. What requirements does this perspective create for flexibility and adaptability in facing future challenges, given that sustainability is not a "destination" but a "journey"?
7. How can changes in thought patterns and cultural values contribute to promoting sustainability within society?

References

Acemoglu, D., Johnson, S., & Robinson, J. A. (2001). The colonial origins of comparative development: An empirical investigation. *American Economic Review, 91*(5), 1369–1401.

Adekeye, I., & Niyi, W. (2013). Grassroots participation in decision-making process and development programmes as correlate of sustainability of community development programmes in Nigeria. *Journal of Sustainable Development, 6*(3), 47–57.

Alonso-Morales, N., Sáez-Martín, A., & Haro-de-Rosario, A. (2025). The role of local public spending in the achievement of the social dimension of sustainable development: An empirical analysis. *Sustainable Futures, 9*, 100475.

Anand, S., & Sen, A. (2000). Human development and economic sustainability. *World Development, 28*(12), 2029–2049.

Asgari, A. (2024). Sustainability; The missing link in development in Iran and the world. Available at: https://B2n.ir/hk5721

Ayodele, T., & Ogunlola, A. (2016). Implementation of sustainable development: The impact of social sustainability. *International Proceedings of Chemical, Biological and Environmental Engineering, 93*, 145–151.

Bakht, R., & Abdul Latif, R. (2018). Institutional quality and technological innovation: Evidence from Malaysia. *Journal of Sustainable Development, 11*(3), 72–81.

Bali Swain, R., & Yang-Wallentin, F. (2020). Achieving sustainable development goals: Predicaments and strategies. *International Journal of Sustainable Development & World Ecology, 27*(2), 96–106.

Barnosky, A. D., Hadly, E. A., Bascompte, J., Berlow, E. L., Brown, J. H., Fortelius, M., Smith, A. B., et al. (2012). Approaching a state shift in Earth's biosphere. *Nature, 486*(7401), 52–58.

Bashiriyeh, H. (2020). *The transformation of Iranian foreign policy: From political Islam to pragmatism.* Palgrave Macmillan.

Bazm, S. (2021). Education for sustainable development (ESD) in Iran. *Journal of Environmental Health and Sustainable Development, 6*(3), 1141–1148.

Bebbington, J., & Frame, R. (2003). Moving from SD reporting to evaluation: The sustainability assessment model. *Chartered Accounting Journal of New Zealand, 82*(7), 11–13.

Bekhet, H. A., & Latif, N. W. A. (2018). The impact of technological innovation and governance institution quality on Malaysia's sustainable growth: Evidence from a dynamic relationship. *Technology in Society, 54*, 27–40.

Bochańczyk-Kupka, D., & Pęciak, R. (2015). Institutions in the context of sustainable development. *The Macrotheme Review, 4*(5), 29–41.

Boons, F., Montalvo, C., Quist, J., & Wagner, M. (2013). Sustainable innovation, business models and economic performance: An overview. *Journal of Cleaner Production, 45*, 1–8.

Boston, J., & Stuart, T. (2015). Protecting the Rights of Future Generationsare constitutional mechanisms an answer? *Policy Quarterly, 11*(2), 60–71.

Brandon, P. S., & Lombardi, P. (2011). *Evaluating sustainable development in the built environment* (2nd ed.). Wiley-Blackwell.

Brundtland, G. H. (1987). Our common future—Call for action. *Environmental Conservation, 14*(4), 291–294.

Bulkeley, H., & Betsill, M. (2013). Revisiting the urban politics of climate change. *Environmental Politics, 22*(1), 136–154.

Cancino, C. A., La Paz, A. I., Ramaprasad, A., & Syn, T. (2018). Technological innovation for sustainable growth: An ontological perspective. *Journal of Cleaner Production, 179*, 31–41.

Cardinale, B. J., Duffy, J. E., Gonzalez, A., Hooper, D. U., Perrings, C., Venail, P., Loreau, M., et al. (2012). Biodiversity loss and its impact on humanity. *Nature, 486*(7401), 59–67.

Chen, Y. (2021). Progress toward the sustainable development goals: A global perspective. *Sustainable Development, 29*(5), 857–869.

Chiu, R. L. H. (2003). Social sustainability, sustainable development and housing development. *Housing and the Built Environment, 18*(2), 223–237.

Dempsey, N., Bramley, G., Power, S., & Brown, C. (2011). The social dimension of sustainable development: Defining urban social sustainability. *Sustainable Development, 19*(5), 295–300.

Devuyst, D., Hens, L., & Lannoy, W. (2001). *How green is the city?* Columbia University Press.

Dornfeld, D. A. (2014). Moving towards green and sustainable manufacturing. *International Journal of Precision Engineering and Manufacturing-Green Technology, 1*(1), 63–66.

Dovers, S. (2001). Institutions for sustainability. *The Ecological Economics of Sustainability*, 417–442.

Dresner, S. (2002). In M. D. Khaki., S. Dehghanian, & F. Dinghezeli (Eds.), *Foundation of sustainability*. Jahad Daneshgahi Mashhad.
Dresner, S. (2008). *The principles of sustainability* (second ed.). Earthscan. ISBN: 9781844074969.
Dresner, S. (2012). *The principles of sustainability*. Routledge.
Ebaid, M. (2024). Interconnected dimensions and guiding principles of sustainability: A comprehensive review. *Journal of Al-Azhar University Engineering Sector, 19*(71), 623–641.
Edwards, G. I. (2019). Multidisciplinary approach to environmental problems and sustainability. In W. Leal Filho (Ed.), *Encyclopedia of sustainability in higher education*. Springer.
Elkington, J. (1998). Partnerships from cannibals with forks: The triple bottom line of 21st century business. *Environmental Quality Management, 6*, 37–51.
Elobeid, A. (2012a). Institutional factors and sustainable development in developing countries. *International Journal of Development Issues, 11*(1), 58–77.
Elobeid, E. (2012b). *Good governance and its impact on sustainable development*. World Bank Publications.
Everett, T., & Kemp, R. (2022). Technological innovation and environmental sustainability: Pathways and policy perspectives. *Sustainability Science, 17*(3), 657–671.
Fattahizadeh, A., & Andarkhord, S. (2025). *Renewable energies and Iran's energy security strategies*. Iranian Energy Economics.
Gao, Y., Huang, H., & Zhou, B. (2021). China's progress towards the sustainable development goals: a comprehensive assessment. *Sustainable Development, 29*(5), 903–918.
Gatto, M. (1995). Sustainability: Is it a well defined concept? *Ecological Applications*, 1181–1183.
Geels, F. W. (2010). Ontologies, socio-technical transitions (to sustainability), and the multi-level perspective. *Research Policy, 39*(4), 495–510.
Gencer, M. (2018). Informal settlements and urban sprawl: A critical review of challenges for sustainable urban development. *Journal of Architecture and Urbanism, 42*(2), 140–151.
George, C., & Kirkpatrick, C. (2006). Environmental assessment and sustainable development: An evaluation of international experience. *Environmental Impact Assessment Review, 26*(2), 159–173.
Gholami, M., Ahmadi, F., & Fereidooni, F. (2022). Investigating challenges and opportunities for financing sustainable development in Iran. *Journal of Sustainable Development Management, 10*(2), 1–18.
Gibson, R. B., Hassan, S., Holtz, S., Tansey, J., & Whitelaw, G. (2010). *Sustainability assessment: Criteria and processes*. Earthscan.
Giddings, B., Hopwood, B., & O'brien, G. (2002). Environment, economy and society: Fitting them together into sustainable development. *Sustainable Development, 10*(4), 187–196.
Hacking, T., & Guthrie, P. (2008). A framework for clarifying the meaning of triple bottom-line, integrated and sustainability assessment. *Environmental Impact Assessment Review, 28*(2), 73–89.
Hariram, N. P., Mekha, K. B., Suganthan, V., & Sudhakar, K. (2023). Sustainalism: An integrated socio-economic-environmental model to address sustainable development and sustainability. *Sustainability, 15*(13), 10682.
Hesen, M., Debray, B., & Söderholm, P. (2022). Integrated sustainability: Social, environmental and economic. *Journal of Sustainability Studies, 5*(1), 45–63.
Hicks, J. R. (1946). *Value and capital* (2nd ed.). Clarendon Press.
Hill, R. C., & Bowen, P. A. (1997). Sustainable construction: Principles and a framework for attainment. *Construction Management and Economics, 15*(3), 223–239.
Hoogeveen, L., Runhaar, H., van der Heijden, J., van Zanten, B., & Termeer, C. (2017). Multi-level governance for sustainability: Advancing the agenda in the EU. *Environmental Science & Policy, 67*, 103–113.
Hooghe, L., & Marks, G. (2001). *Multi-level governance and European integration*. Rowman & Littlefield.
Huang, Y. (2021). Technology, innovation and the sustainability challenge: Strategic implications. *Journal of Cleaner Production, 279*, Article 123728.
Hueting, R., & Reijnders, L. (2004). Broad sustainability contra sustainability: The proper construction of sustainability indicators. *Ecological economics, 50*(3–4), 249–260.
IEA (International Energy Agency). (2024). *Renewables—Iran*. Retrieved from https://www.iea.org/countries/iran/renewables
Institute of Sustainability Studies. (2025). *How technology drives resource efficiency in business*. Retrieved from https://instituteofsustainabilitystudies.com/insights/lexicon/how-technology-drives-resource-efficiency-in-business/
Intergovernmental Panel on Climate Change (IPCC). (2021). Climate change 2021: The physical science basis. In *Contribution of Working Group I to the Sixth Assessment Report of the Intergovernmental Panel on Climate Change*. Cambridge University Press.
Intergovernmental Panel on Climate Change (IPCC). (2023). Climate change 2023: Synthesis report. In Core Writing Team, H. Lee, & J. Romero (Eds.), *Contribution of Working Groups I, II and III to the Sixth Assessment Report of the Intergovernmental Panel on Climate Change*. IPCC. https://www.ipcc.ch/report/ar6/syr/
Jagani, S., & Saboori-Deilami, V. (2025). Sustainability orientation, sustainability implementation, and brand image in service firms. *Business Strategy and the Environment*, 1–12.
Jeronen, E. (2020). Sustainable development. In: S. Idowu, R. Schmidpeter, N. Capaldi, L. Zu, M. Del Baldo, & R. Abreu, (Eds.), *Encyclopedia of sustainable management*. Springer.
Jeronen, E. (2023). Economic sustainability. In: S. O. Idowu, R. Schmidpeter, N. Capaldi, L. Zu, M. Del Baldo, & R. Abreu (Eds.), *Encyclopedia of sustainable*

management. Springer. https://doi.org/10.1007/978-3-031-25984-5_197

Kahn, M. (1995). Concepts, definitions, and key issues insustainable development: The outlook for the future. In *Proceedings of the 1995 International SustainableDevelopment Research Conference* (pp. 2–13), Manchester, England.

Kajikawa, Y., Tachikawa, M., & Takeda, Y. (2014). A long-term vision of sustainability science. *Sustainability Science, 9*(3), 253–261.

Khan, A. (1995). *Sustainable development: A policy framework*. Routledge.

Kiron, D., Kruschwitz, N., Reeves, M., Haanaes, K., & Goh, E. (2012). The benefits of sustainability-driven innovation. In *Own the future: 50 ways to win from the Boston Consulting Group* (pp. 119–123).

Klarin, T. (2018). The concept of sustainable development: From its beginning to the contemporary issues. *Zagreb International Review of Economics & Business, 21*(1), 67–94.

Klewitz, J., & Hansen, E. G. (2014). Sustainability-oriented innovation of SMEs: A systematic review. *Journal of Cleaner Production, 65*, 57–75.

Komiyama, H., & Takeuchi, K. (2006). New paradigm for sustainability science. *Environmental Science & Technology, 40*(13), 3862–3864.

Kosmopoulos, P. (2024).The way forward for sustainable development. In *Planning and management of solar power from space* (pp. 173–189) (Chapter 10).

Kuhlman, T., & Farrington, J. (2010). What is sustainability? *Sustainability, 2*(11), 3436–3448.

Leal Filho, W., Salvia, A. L., Brandli, L. L., Azeiteiro, U. M., Berg, A. M., & Pace, P. (2018). Implementing the United Nations' sustainable development goals: A challenge for research and higher education. *Sustainability, 10*(11), 3843.

Leamer, E. E. (2010). Tantalus on the road to asymptopia. *Journal of Economic Perspectives 24*(2), 31–46.

Liu, G., & Liang, K. (2024). The role of technological innovation in enhancing resource sustainability to achieve green recovery. *Resources Policy, 89*, Article 104659.

López, R., & Toman, M. A. (2006). *Economic development and environmental sustainability*. Oxford University Press.

Marks, G. (1993). Structural policy and multi-level governance in the EC. In A. Cafruny & G. Rosenthal (Eds.), *The state of the European community*. Longman.

Markulev, A., & Long, A. (2013). *Sustainable economic development: Initiatives, programs and strategies*. Australian Government Productivity Commission.

Mathur, A., & Vyas, D. (2013). Socio-ecological issues-eco-economic and sustainability status. *Journal of Ecology and Environmental Sciences, 4*(1), 87–90.

Matutinovic, I. (2007). Worldviews, institutions and sustainability: An introduction to the special issue. *Ecological Economics, 60*(1), 1–5.

Meadows, D. H., Meadows, D. L., Randers, J., & Behrens, W. W., III. (1972). *The limits to growth*. Potomac Associates, New American Library.

Mebratu, D. (1998). Sustainability and sustainable development: Historical and conceptual review. *Environmental Impact Assessment Review, 18*(6), 493–520.

Miraskari Jildani, S. R., Masoumi, S. S., & Asgharpour, M. H. (2024). Investigating strategies of knowledge-based economy development in Iran. *Majlis and Economy, 2*(4), 375–401.

Morelli, J. (2011). Environmental sustainability: A definition for environmental professionals. *Journal of Environmental Sustainability, 1*(1), 1–9.

Nagabhatla, N. (2024). Revisiting the sustainability science research agenda. *Sustainability Science, 19*(7), 1–13.

Naghavi, S., & Asnaashari, H. (2022). Investigating the role of knowledge-based economy in sustainable development. In *National Conference on Knowledge-Based Agricultural Economy and Entrepreneurship, Sari. Civilica*. https://civilica.com/doc/1615205

Ness, B., Urbel-Piirsalu, E., Anderberg, S., & Olsson, L. (2006). Categorising tools for sustainability assessment. *Ecological Economics, 60*(3), 498–508.

Neumayer, E. (2010). *Weak versus strong sustainability: Exploring the limits of two opposing paradigms*. Edward Elgar Publishing.

Nidumolu, R., Prahalad, C. K., & Rangaswami, M. R. (2009). Why sustainability is now the key driver of innovation. *Harvard Business Review, 87*(9), 56–64.

North, D. C. (1990). *Institutions, institutional change and economic performance*. Cambridge University Press.

OECD. (2017). *Good Governance for sustainable development*. OECD Publishing.

Ogunmakinde, O. E., Olanrewaju, A., & Al-Awadi, A. (2022). Sustainable development goals and the built environment: A systematic review of literature. *Sustainable Cities and Society, 83*, Article 103986.

Oliver, T. H., Roy, D. B., Hill, J. K., Brereton, T., Thomas, C. D., & Purvis, A. (2015). Biodiversity and the resilience of ecosystems. *Science, 348*(6233), 1258611.

Omri, A. (2018). Entrepreneurship, sectoral outputs and environmental improvement: International evidence. *Technological Forecasting and Social Change, 128*, 46–55.

Omri, A. (2020). Technological innovation and sustainable development: Does the stage of development matter? *Environmental Impact Assessment Review, 83*, Article 106398.

Ouyang, Z., Zheng, H., Zhang, L., Xiao, Y., Polasky, S., Liu, J., Daily, G. C., et al. (2020). Gross ecosystem product: Concept, accounting framework and case study. *Proceedings of the National Academy of Sciences, 117*(51), 32283–32292.

Pangalos, C. (2023). *Tackling climate change through multi-level governance, systems thinking and policy coherence*. United Nations System Staff College (UNSSC). Retrieved from https://www.unssc.org/news-and-insights/blog/tackling-climate-change-through-multi-level-governance-systems-thinking-and

Pörtner, H. O., Roberts, D. C., Tignor, M., Poloczanska, E. S., Mintenbeck, K., Alegría, A., & Poloczanska, E. S. (Eds.). (2021). Climate change 2021: The physical science basis. In *Contribution of Working Group I to the Sixth Assessment Report of the Intergovernmental Panel on Climate Change*. Cambridge University Press.

Poveda, C. A. (2017). *Sustainability assessment: A rating system framework for best practices. CA Poveda* (p. 38) Emerald Publishing Limited.

Poyraz, A. Y., & Szalmáné Csete, M. (2022). Multi-level governance of sustainability transition in the European Union. In W. Leal Filho, M. A. P. Dinis, S. Moggi, E. Price, & A. Hope (Eds.), *SDGs in the European region: Implementing the UN sustainable development goals—Regional perspectives*. Springer. https://doi.org/10.1007/978-3-030-91261-1_88-1

Purvis, B., Mao, Y., & Robinson, D. (2019). Three pillars of sustainability: In search of a conceptual origin. *Sustainability Science, 14*(3), 681–695.

Quiroz-Niño, C., & Murga-Menoyo, M. Á. (2017). Social sustainability: A competing concept to sustainable development? *Sustainability, 9*(12), 2132.

Rahimi, S., Abbaspour, A., & Vaghefi, M. (2020). Water resources management in Iran: Challenges and opportunities for sustainable development. *Environmental Science & Policy, 106*, 11–20.

Rao, P. K. (2000). *Sustainable development: Economics and policy*. Blackwell Publishers.

Sachs, J. D. (2015). *The age of sustainable development*. Columbia University Press.

Sarabdeen, J. (2024). The role of government in driving sustainability: A public policy perceptive. *Emerging Science Journal, 8*(3), 1184–1200.

Schaltegger, S., & Burritt, R. (2005). Corporate sustainability. In *The international yearbook of environmental and resource economics 2005/2006*. Edward Elgar.

Schoolman, E. D., Sperling, J., & Borie, M. (2012). The emergence of sustainability science. *Environmental Science & Policy, 23*, 1–9.

Schramski, J. R., Dell'Angelo, J., & D'Odorico, P. (2015). The anthropocene and the great acceleration: Ecological impacts of human activities. *Global Environmental Change, 31*, 1–7.

Scrucca, F., Cetrulo, S., & Baffo, G. (2023). Progress towards the sustainable development goals in European countries: A comprehensive assessment. *Sustainable Development, 31*(1), 123–138.

Selin, H., & VanDeveer, S. D. (2012). Federalism, multilevel governance and climate change politics across the Atlantic. In P. V. Steinberg & S. D. VanDeveer (Eds.), *Comparative environmental politics: Theory, practice and prospects* (pp. 341–368). MIT Press.

Sharifzadeh, M. S., Abdollahzadeh, G. H., Akbari, M., & Mobini Dehkordi, A. (2017). *Entrepreneurship and sustainability: Towards sustainable entrepreneurship*. University of Tehran Branch.

Shortall, R., Davidsdottir, B., & Axelsson, G. (2015). Geothermal energy for sustainable development: A review of sustainability impacts and assessment frameworks. *Renewable and Sustainable Energy Reviews, 44*, 391–406.

Siriram, R. (2022). Technological and non-technological innovation effects on firm performance. *The Journal of High Technology Management Research, 33*(2), Article 100429.

Soini, K., & Birkeland, I. (2014). Exploring the scientific discourse on cultural sustainability. *Geoforum, 51*, 213–223.

Spangenberg, J. H., Pfahl, S., & Deller, K. (2002). Towards indicators for institutional sustainability: Lessons from an analysis of Agenda 21. *Ecological Indicators, 2*(1–2), 61–77.

Summers, J. K., & Smith, L. M. (2014). The role of social and intergenerational equity in making changes in human well-being sustainable. *Ambio, 43*(6), 718–728.

Thiele, L. P. (2013). *Sustainability*. Polity Press.

Tulloch, L. (2013). On science, ecology and environmentalism. *Policy Futures in Education, 11*(1), 100–114.

UNEP (United Nations Environment Programme). (2021). *UNEP finance initiative*. Retrieved from https://www.unepfi.org/

UNESCO. (2025, May 20). *UNESCO programs in Iran to include environmental education*. Tehran Times. https://www.tehrantimes.com/news/513299/UNESCO-programs-in-Iran-to-include-environmental-education

United Nations (UN). (2012). *The future we want. Resolution adopted by the general assembly on 27 July 2012 (A/RES/66/288)*. United Nations, New York.

United Nations. (2015). *Transforming our world: The 2030 agenda for sustainable development*. United Nations.

United Nations. (2020). *World social report: Inequality in a rapidly changing world*. United Nations Department of Economic and Social Affairs.

United Nations. (2021). *United Nations sustainable development co-operation framework for the Islamic Republic of Iran (2023–2027)*. https://unsdg.un.org/sites/default/files/2022-11/UNSDCF_Iran%20Republic%20of-2023-2027.pdf

United Nations (UN). (2022). *The sustainable development goals report 2022*. United Nations.

United Nations Conference on Environment and Development. (1992). *Agenda 21*.

United Nations Environment Programme (UNEP). (2011). *Towards a green economy: Pathways to sustainable development and poverty eradication*. UNEP. https://www.unep.org/resources/report/towards-green-economy-pathways-sustainable-development-and-poverty-eradication

United Nations Development Programme. (2020). *Poverty and inequality*. https://www.undp.org/eurasia/our-focus/Poverty-and-inequality

Vaikma, H. (2025). Unveiling the mindset: Measuring consumer perception towards the dimensions of sustainability. *Sustainable Futures, 9*(100616), 1–12.

Valentin, A., & Spangenberg, J. H. (2000). A guide to community sustainability indicators. *Environmental Impact Assessment Review, 20*(3), 381–392.

Vallance, S., Perkins, H. C., & Dixon, J. E. (2011). What is social sustainability? *A Clarification of Concepts. Geoforum, 42*(3), 342–348.

Vos, R. O. (2007). Defining sustainability: A conceptual orientation. *Journal of Chemical Technology and Biotechnology, 82*(4), 334–339.

Weibust, I., & Meadowcroft, J. (2014). *Multi-level environmental governance: Managing water and climate change in Europe and North America.* Edward Elgar.

White, L., & Lee, G. J. (2009). Operational research and sustainable development: Tackling the social dimension. *European Journal of Operational Research, 193*(3), 683–692.

Wiersum, K. F. (1990). Social and cultural aspects of forest protection and forest management in developing countries. *Forest Ecology and Management, 31*(1–4), 1–12.

World Bank. (2006). *Where is the wealth of nations? Measuring capital for the 21st century.* World Bank. Retrieved from https://openknowledge.worldbank.org/handle/10986/7505

World Bank. (2021a). *Climate change and development report: Islamic Republic of Iran.* World Bank Group. https://www.worldbank.org

World Bank. (2021b). *Worldwide governance indicators.* Retrieved from https://info.worldbank.org/governance/wgi/

World Bank. (2022). *Global material flows and resource productivity.* World Bank.

World Commission on Environment and Development (WCED). (1987). *Our common future.* Oxford University Press.

World Health Organization. (2022). *Health and climate change country profile 2022: Islamic Republic of Iran.* World Health Organization. https://iris.who.int/bitstream/handle/10665/352630/WHO-HEP-ECH-CCH-22.01.01-eng.pdf?sequence=1

Wu, M., Lu, Y., & Liu, Y. (2018). The social dimension of sustainability: A review of literature and research agenda. *Sustainability, 10*(3), 725.

Zahedi, S. (2007). *Sustainable development.* SAMT.

Zahedi, S., & Ebrahimpour, H. (2012). *Sustainability-based governance with an emphasis on environmental protection.* Samt Publications.

Zarif, M., & Hossain, K. (2021). Renewable energy in Iran: Opportunities and challenges. *Energy Policy Journal, 149*, Article 112021.

Foundations of Entrepreneurship

Vajihe Baghersad, Elham Garmroudesfandiari, and Seyed Mostafa Razavi

Abstract

This chapter provides an extensive overview of the foundational principles and theories of entrepreneurship, exploring key concepts such as opportunity recognition, risk management, and innovation. It delves into the entrepreneurial mindset, emphasizing the importance of traits like creativity, resilience, and adaptability. The chapter discusses the stages of the entrepreneurial process, from idea generation and business planning to resource mobilization and execution. Additionally, it highlights the role of entrepreneurial ecosystems and support structures, such as accelerators, incubators, and government policies, in fostering a conducive environment for startups. Real-world case studies illustrate practical applications and the influence of entrepreneurship on economic growth and societal progress. Furthermore, the chapter addresses the challenges and ethical dilemmas entrepreneurs might face, providing strategies to overcome these obstacles effectively. By integrating theoretical insights with practical strategies, this chapter equips aspiring entrepreneurs by providing essential expertise and practical resources to effectively manage the challenges of launching and expanding a business, ultimately driving innovation and transformation in various industries.

1 Learning Objectives

By the end of this chapter, readers will be able to:

1. Understand the foundational principles and theories of entrepreneurship, including key concepts such as opportunity recognition, risk management, and innovation.
2. Identify and apply the stages of the entrepreneurial process, from idea generation and business planning to resource mobilization and execution.
3. Recognize the importance of traits like creativity, resilience, and adaptability in developing an entrepreneurial mindset.
4. Understand the role of entrepreneurial ecosystems and support structures, such as incubators and accelerators, in fostering a conducive startup environment.
5. Gain insights into entrepreneurship's practical applications and impact on economic growth

V. Baghersad (✉)
Department of Management, Technical and Vocational University (TVU), Tehran, Iran
e-mail: vbaghersad@tvu.ac.ir

E. Garmroudesfandiari
Department of Corporate Entrepreneurship, Faculty of Entrepreneurship, University of Tehran, Tehran, Iran
e-mail: eliesfandiary1998@gmail.com

S. M. Razavi
Faculty of Management, University of Tehran, Tehran, Iran
e-mail: mrazavi@ut.ac.ir

M. Akbari et al. (eds.), *Sustainable Entrepreneurship*, University of Tehran Science and Humanities Series, https://doi.org/10.1007/978-3-032-04909-4_2

and societal progress through real-world case studies.

2 Introduction

In a rapidly evolving world, the essence of entrepreneurship lies in recognizing opportunities and driving innovation, serving as a catalyst for economic growth and societal transformation (Zelin et al., 2021). In this chapter, we'll explore the rich and multifaceted realm of entrepreneurship, unraveling its core principles, theories, processes, and essential skills that underpin success. At its heart, entrepreneurship is about identifying and creating business opportunities to meet market needs while taking on financial risks in the quest for profit. It plays a crucial role in spurring innovation, managing risks, and making strategic decisions (Chevrollier et al., 2025; Günar, 2025). Entrepreneurs possess a strategic ability to propel economic advancement through the innovation of novel products, services, and business frameworks (Shahid et al., 2023; Velinov, 2025).

In recent decades, the domain of entrepreneurship has undergone substantial advancements, marked by notable theoretical and practical evolution. Recent studies highlight the importance of entrepreneurs in efficient resource allocation and market dynamics (Osunmuyiwa & Ahlborg, 2022; Ranabahu & Tanima, 2022). Modern theories build upon the ideas of Schumpeter and Kirzner, emphasizing how entrepreneurship drives market transformations through innovation and opportunity recognition (Calderini et al., 2023; Venâncio & Jorge, 2022). This chapter aims to comprehensively explore entrepreneurship and its critical role in fostering economic and social progress. Key focuses include the significance of innovation, opportunity recognition, and risk management within the entrepreneurial journey.

Have you ever wondered how some trailblazers reshape entire industries and leave a profound impact with their innovative ideas and fearless risk-taking? Entrepreneurship isn't only about creating wealth and jobs; it also addresses social issues by introducing pioneering business models and solutions (Kharche et al., 2025; Turker & Ozmen, 2021). In the following sections, we'll cover essential concepts, theories, stages of the entrepreneurial process, and the vital skills every entrepreneur needs. We'll also look into how entrepreneurial ecosystems and support structures like incubators and accelerators play a pivotal role in nurturing startups.

Entrepreneurship is a dynamic process encompassing opportunity recognition, business creation, and innovation to generate value (Alekseevna, 2023; Sihombing et al., 2025). We'll explore various theories, including Schumpeter's paradigm of disruptive innovation and Kirzner's framework of entrepreneurial opportunity identification, focusing on innovation-driven market transformation and opportunity recognition, respectively (Attarit, 2024). The entrepreneurial journey unfolds through distinct stages: recognizing opportunities, planning, mobilizing resources, executing plans, and achieving growth (Cicchiello et al., 2021). Tools like the Business Model Canvas and Lean Startup methodologies help bring these ideas to life (Attarit, 2024; Pustovrh et al., 2020).

Successful entrepreneurs stand out due to their unique characteristics and skills, including creativity, resilience, risk-taking, and quick decision-making (Mahajan & Bandyopadhyay, 2021; Thompson & Bolton, 2024). Strong management, financial, and marketing skills are also indispensable for entrepreneurial success (Velinov, 2025). Entrepreneurial ecosystems, consisting of interconnected elements working together, are essential for supporting entrepreneurial activities (Adegbile, 2018; Kansheba & Wald, 2020). These ecosystems provide critical resources, expert advice, and networking opportunities to help entrepreneurs navigate challenges and succeed (Hernández & Virgüez, 2024).

Entrepreneurship research is vital as it sheds light on the factors contributing to successful ventures and uncovers the challenges and obstacles entrepreneurs face. This knowledge can guide policy-making, support structures, and educational initiatives aimed at fostering entrepreneurship. Furthermore, the research emphasizes the role of entrepreneurship in addressing societal issues, promoting inclusive growth, and driving sustainable development (Sihombing et al., 2025; Wasnik & Jain, 2023). This chapter offers a holistic view of entrepreneurship by combining theoretical insights with practical applications and examining both individual entrepreneurial traits and broader supportive ecosystems. Real-world case studies and examples enrich the practical relevance of the content, making it an invaluable resource for students and practitioners alike (Bellotti et al., 2014).

Entrepreneurship is more than an economic activity; it is a force for social progress and innovation. This chapter will provide you with an in-depth exploration of foundational entrepreneurial concepts and methodologies, equipping you with the knowledge and tools needed to thrive in this dynamic field. I hope you find this exploration engaging and insightful.

3 Introduction to Entrepreneurship

Entrepreneurship is universally acknowledged as a pivotal catalyst for economic expansion, societal evolution, and technological innovation (Audretsch, 2006; Schumpeter, 1934). It represents the process of identifying opportunities, mobilizing resources, and implementing strategies to create value through innovative solutions (Drucker, 1986; Shane & Venkataraman, 2000). This value can take multiple forms, such as economic profit, societal impact, or cultural advancement, underscoring its multidimensional nature (Van Praag & Versloot, 2007). The entrepreneurial journey, defined by resilience, creativity, and vision, has shaped societies across time (Schumpeter, 1934; Shane, 2003). As a dynamic field, entrepreneurship has evolved significantly, adapting to the rapid technological, social, and economic shifts of the modern era (Audretsch, 2006; Hisrich et al., 2017).

At its core, entrepreneurship involves creating new opportunities and resolving challenges through innovation (Drucker, 1986; Schumpeter, 1934). Entrepreneurs, often described as agents of change, are distinguished by their ability to envision possibilities beyond the status quo (Sarasvathy, 2001; Shane & Venkataraman, 2000). This vision enables them to spot gaps in markets, create new products or services, and address existing inefficiencies. Historically, entrepreneurship has played a pivotal role in economic development, from the early trade systems and industrial innovations of the eighteenth century to the digital and technological transformations of the twenty-first century (Baumol, 1996; Drucker, 1986). During the Industrial Revolution, entrepreneurs such as James Watt and Andrew Carnegie laid the foundation for modern industries, while in the twentieth century, visionaries like Steve Jobs and Bill Gates revolutionized computing and communications. This historical trajectory reflects not only the adaptability of entrepreneurial practices but also their integral role in shaping contemporary economies and societies (Audretsch, 2006; Hisrich et al., 2017). The significance of entrepreneurship is most evident in its contributions to modern economic and social systems. One of its most notable impacts is fostering economic growth and generating employment. By establishing businesses, entrepreneurs create direct job opportunities and stimulate the broader economy.

For instance, the entrepreneurial ventures of the late twentieth century and the early years of the twenty-first century have generated unprecedented wealth and employment opportunities. Technology-based startups, particularly in Silicon Valley, have revolutionized industries such as transportation, healthcare, and entertainment, contributing to the global economy's dynamism (Audretsch, 2006; Audretsch & Thurik, 1999). Entrepreneurship is not limited to creating economic value; it also drives innovation (Drucker, 1986; Van Praag & Versloot, 2007). Entrepreneurs address inefficiencies and introduce groundbreaking solutions to global challenges (Schumpeter, 1934; Shane, 2003). For example, the development of renewable energy technologies is a direct response to climate change and resource scarcity (Muhamed Yunus & Jolis, 1999; Van Praag & Versloot, 2007). Similarly, advancements in digital healthcare and education have made these services more accessible and efficient, particularly in underserved areas. The emergence of disruptive business models, such as ride-sharing platforms and online marketplaces, further exemplifies the innovative capacity of entrepreneurship (Drucker, 1986; Van Praag & Versloot, 2007). In addition to its economic and innovative contributions, entrepreneurship has profound societal and cultural implications (Austin, 2006). Social entrepreneurship, a subset of the broader entrepreneurial paradigm, emphasizes creating sustainable solutions to pressing global issues (Muhamed Yunus & Jolis, 1999). Unlike traditional entrepreneurs, social entrepreneurs prioritize social value over profit.

Initiatives such as microfinance programs, grassroots education projects, and renewable energy solutions for marginalized communities demonstrate how entrepreneurship transcends profit motives to deliver social impact (Martin & Osberg, 2007). Nobel laureate Muhammad Yunus, for example, pioneered the concept of microcredit to empower low-income individuals, particularly women, in developing countries (Bornstein, 2007; Muhamed Yunus & Jolis, 1999). Such endeavors highlight the capacity of entrepreneurship to serve as a catalyst for systemic change, fostering inclusion, equity, and empowerment (Martin & Osberg, 2007; Muhammad Yunus, 2007). Cultural entrepreneurship is another dimension that showcases the transformative potential of entrepreneurial activities. Cultural entrepreneurs promote and preserve cultural heritage while innovating ways to engage new audiences. From supporting local artisans to launching global movements that celebrate cultural diversity, these initiatives enhance the richness of the human experience. By blending economic activities with cultural preservation, entrepreneurship contributes to a more vibrant and inclusive global culture (Austin, 2006; Leadbeater, 1997).

Despite its transformative potential, entrepreneurship is accompanied by considerable challenges. Among the most significant obstacles are financial constraints, particularly for early-stage ventures with limited access to capital. A significant number of entrepreneurs encounter challenges in obtaining the necessary financial resources to translate innovative concepts into sustainable and competitive business ventures (Mickiewicz et al., 2017; Müller-Wieland et al., 2019; O'Gorman, 2009). Traditional funding mechanisms such as bank loans may not always be accessible, prompting entrepreneurs to turn to alternative sources, including venture capital, crowdfunding, and government grants (Israel Kirzner, 1973). Beyond financial barriers, entrepreneurs must navigate competitive markets that demand constant innovation and adaptability. Market dynamics often favor established players, requiring newcomers to differentiate themselves through unique value propositions. Regulatory complexities, ranging from licensing requirements to compliance standards, add further layers of difficulty. Entrepreneurs must also contend with workforce management challenges, building teams that align with their vision while balancing the demands of leadership and personal life (Sarasvathy, 2001; Stevenson & Gumpert, 1985).

Despite these hurdles, successful entrepreneurs share distinctive traits that enable them to navigate challenges and achieve their goals. Visionary thinking is a hallmark of entrepreneurial success, allowing individuals to anticipate trends and seize opportunities. This vision is often paired with resilience, as setbacks and failures are inevitable in any entrepreneurial journey (Addy et al., 2024; Alekseevna, 2023; Celestin & Vanitha, 2018). Risk tolerance is another key characteristic, as entrepreneurs must weigh the potential for failure against the promise of success. Creativity, meanwhile, empowers entrepreneurs to devise innovative solutions that address complex problems and disrupt conventional industries. Prominent examples include Elon Musk's ventures into electric vehicles and space exploration and Muhammad Yunus's transformative work in social finance (Shane, 2003). Entrepreneurship remains an essential force in advancing economic, social, and cultural progress. Entrepreneurship shapes a dynamic and interconnected global landscape by fostering innovation, driving growth, and addressing systemic challenges. Its ability to adapt to changing circumstances ensures its relevance in addressing both present and future challenges.

From the Industrial Revolution to the digital age, entrepreneurship has continuously redefined the limits of what is possible. As this chapter delves deeper into the theoretical and practical dimensions of entrepreneurship, it aims to provide a comprehensive understanding of its significance in contemporary contexts (Audretsch, 2006; Hisrich et al., 2017).

3.1 Definitions and Key Concepts

Entrepreneurship is conceptualized as a systematic approach to recognizing, initiating, and overseeing business ventures, strategically aligning them with market demands while navigating financial uncertainties to ensure profitability. It involves innovation, risk management, and strategic decision-making to capitalize on opportunities in competitive markets (Mazzarol et al., 1999; Thompson & Bolton, 2024). Entrepreneurs often play a key role in driving economic progress through the introduction of innovative products, services, and business models. Entrepreneurship is often compared with startups and small businesses, though they have distinct differences. A startup is typically a newly founded business focused on rapid growth and scalability, usually driven by technological advancements and disruptive innovation (Fritsch & Storey, 2014; Schiavone et al., 2020). On the other hand, small businesses are independently owned enterprises that serve local markets with limited growth ambitions (Wennekers & Thurik, 1999). While both startups and small businesses may be entrepreneurial, entrepreneurship itself is characterized by the pursuit of opportunity, innovation, and risk management, regardless of the business size or age.

The historical evolution of entrepreneurship has significantly shaped modern economic theories. The term originates from the French word *entreprendre*, meaning "to undertake," and was first associated with risk-taking merchants and traders during the eighteenth century. Early economic thinkers such as Richard Cantillon (1730) (Brown & Thornton, 2013) and Smith (1776) emphasized the entrepreneur's role in market efficiency and resource allocation. In the twentieth century, Joseph Schumpeter introduced the concept of *creative destruction* (Schumpeter, 1942), highlighting how entrepreneurs disrupt existing markets with innovative products and services (Fritsch & Storey, 2014). This theory has become central in modern entrepreneurship studies, particularly with the rise of the technology-driven economy. Entrepreneurship is a critical driver of economic growth, job creation, and technological advancement. It contributes to market competition, efficiency, and the introduction of groundbreaking innovations. Entrepreneurs not only generate wealth but also help address global challenges through social entrepreneurship, promoting sustainability and positive societal impact (Turker & Ozmen, 2021; Wennekers & Thurik, 1999). Entrepreneurial dynamics play a pivotal role in the economic development of emerging markets, wherein small and medium-sized enterprises (SMEs) constitute a considerable proportion of both the labor force and overall economic activity, fostering innovation, competitiveness, and sustainable growth. (Blackburn & Kovalainen, 2009). Governments worldwide recognize this impact and have prioritized policies that support entrepreneurship through education, financial incentives, and regulatory frameworks to stimulate economic development (Cicchiello et al., 2021; Henrekson & Sanandaji, 2014).

3.2 Theoretical Foundations

Entrepreneurship is an important driver of economic innovation, market evolution, and societal progress. Theoretical explorations have delved into its multifaceted nature, emphasizing different dimensions, from innovation to market equilibrium. This section examines key theories, compares their perspectives, and highlights the key role of entrepreneurship in sustainable development and economic growth. The foundational theories of entrepreneurship, developed by prominent scholars, provide diverse insights into entrepreneurial processes and their economic implications.

1. **Joseph Schumpeter** (1934)—**Theory of Creative Destruction**: Schumpeter identified entrepreneurs as innovators who disrupt existing markets through creative destruction. This process entails introducing groundbreaking products, services, or technologies that replace outdated systems, driving structural transformation and economic progress. Schumpeter emphasized the revolutionary role of entrepreneurship in fostering technological advancement and reshaping industries (Schumpeter, 1934).
2. **Israel Kirzner** (1973)—**Theory of Entrepreneurial Discovery**: Kirzner viewed entrepreneurs as agents of market efficiency, focusing on opportunity recognition rather than innovation. His theory highlights the entrepreneur's alertness to profit opportunities arising from market imperfections. Kirzner argued that entrepreneurs bridge gaps, restore equilibrium, and enhance market performance, aligning entrepreneurship with efficiency and incremental gains (Kirzner, 1973).
3. **Classical Theories: Cantillon and Smith**: Cantillon (1755) conceptualized the entrepreneur as a risk-taker navigating uncertainty, laying the groundwork for understanding entrepreneurship as a distinct economic function. Smith (1776) expanded on this idea by emphasizing the entrepreneur's role in capital accumulation, resource allocation, and productivity. These classical theories underline entrepreneurship's role in competitive markets and economic organization (Table 1).

Schumpeter's perspective aligns with radical innovation and large-scale transformation, often associated with high-growth ventures. Conversely, Kirzner emphasizes the entrepreneur's role in everyday market activities, highlighting incremental improvements and smaller-scale initiatives. Together, these perspectives provide a comprehensive understanding of entrepreneurship.

Entrepreneurship contributes significantly to the achievement of sustainable development goals (SDGs) and economic growth:

1. **Innovation and Technological Progress**: Entrepreneurs introduce innovative products, services, and processes that drive productivity gains and technological advancement (Schumpeter, 1934).
2. **Employment and Income Generation**: Entrepreneurial ventures, particularly SMEs, serve as critical sources of employment and income generation, especially in developing economies (Wennekers & Thurik, 1999).
3. **Market Efficiency and Optimization**: By addressing market inefficiencies, entrepreneurs enhance resource allocation and market functioning (Kirzner, 1973).
4. **Social and Environmental Impact**: Social entrepreneurship applies entrepreneurial principles to address global challenges, for example, poverty, education, climate change, advancing SDGs, and promoting inclusive growth.
5. **Policy and Economic Resilience**: Policymakers support entrepreneurship through education, funding, and regulatory frameworks, recognizing its vital role in fostering economic resilience and sustainable innovation (Henrekson & Sanandaji, 2014).

Table 1 Schumpeter and Kirzner represent contrasting paradigms in entrepreneurial thought

Aspect	Schumpeter (1934)	Kirzner (1973)
Focus	Innovation and market disruption	Opportunity recognition and discovery
Market impact	Disrupts equilibrium drives change	Restores equilibrium, corrects inefficiencies
Scale of impact	Transformative, industry-wide	Incremental, smaller-scale

4 Characteristics and Skills of Entrepreneurs

4.1 Core Characteristics of Successful Entrepreneurs

Entrepreneurs play a critical role in driving innovation, economic growth, and societal development. Their success is often attributed to a combination of unique personality traits, essential skills, and strategic actions (Thompson & Bolton, 2024). Understanding these core characteristics provides valuable insights into what differentiates successful entrepreneurs from others. Entrepreneurs are known for their willingness to take calculated risks, enabling them to navigate uncertain environments and exploit opportunities (Nave et al., 2024). Risk propensity, as highlighted by Rauch and Frese (2007), is a key factor in entrepreneurial success. For example, Elon Musk's bold decision to invest in cutting-edge technologies like electric vehicles and space exploration reflects his ability to balance high stakes with potential rewards. Coupled with risk-taking is their innovativeness, a trait emphasized by Schumpeter (1934), who viewed entrepreneurs as agents of change capable of disrupting markets with transformative ideas. Musk's innovation at Tesla and SpaceX exemplifies this disruptive approach.

Resilience and adaptability are also integral traits of successful entrepreneurs. The entrepreneurial journey is often fraught with setbacks and challenges, requiring individuals to persist despite failures and adjust their strategies in response to changing circumstances (Nave et al., 2024). This resilience is often linked to what Duckworth et al. (2007) describe as "grit," a combination of passion and perseverance. Oprah Winfrey's rise from a challenging upbringing to building a multimedia empire highlights how adaptability and resilience can propel entrepreneurial success. Similarly, decisiveness is critical for navigating competitive markets. Entrepreneurs must make timely and informed decisions under pressure, a vital skill for sustaining their ventures (Baron, 2008; Thompson & Bolton, 2024). In addition to personality traits, successful entrepreneurs demonstrate proficiency in essential skills such as management, marketing, and finance. Effective leadership enables them to build cohesive teams, allocate resources efficiently, and set a clear vision for their enterprises (Nave et al., 2024). For instance, Jeff Bezos demonstrated exceptional leadership by transforming Amazon into a global e-commerce powerhouse through strategic planning and resource management. Marketing expertise is equally critical, as it allows entrepreneurs to understand market dynamics, anticipate customer needs, and position their offerings effectively. Kotler et al. (2016) underscore the importance of market orientation in fostering entrepreneurial growth. Financial literacy, on the other hand, ensures sound budgeting, cash flow management, and long-term profitability (Cicchiello et al., 2021). Entrepreneurs like Warren Buffett exemplify how mastery of financial principles can drive sustainable success. Case studies further illustrate these characteristics in practice. Jack Ma, the founder of Alibaba, demonstrated vision and perseverance, overcoming multiple failures to establish a global e-commerce giant. His strategic thinking and ability to inspire teams were pivotal to Alibaba's success. Similarly, Oprah Winfrey's resilience and innovative approach to brand-building highlight the importance of adaptability and vision in entrepreneurship. Elon Musk's ventures in Tesla and SpaceX reflect a unique blend of risk-taking, innovation, and bold decision-making that reshaped entire industries.

In summary, the success of entrepreneurs lies in the interplay of distinctive personality traits and essential skills. Leadership, marketing, and financial acumen complement their ability to take risks, innovate, and adapt. These qualities, as exemplified by figures like Elon Musk, Oprah Winfrey, and Jack Ma, showcase the profound the influence of entrepreneurial activities in fostering innovation and driving economic expansion.

4.2 Entrepreneurial Mindset and Motivation

Entrepreneurs possess a distinct mindset that differentiates them from non-entrepreneurs, shaping their approach to risk, problem-solving, and innovation (Thompson & Bolton, 2024). This entrepreneurial mindset encompasses a proactive and opportunity-focused outlook, enabling individuals to identify and exploit market gaps while navigating uncertainties. Coupled with this mindset are powerful motivational factors that drive entrepreneurs to persist in their ventures and create impactful outcomes. The entrepreneurial mindset is characterized by traits such as a tolerance for ambiguity, creativity, and resilience. Entrepreneurs tend to view challenges as opportunities and are more comfortable operating in uncertain environments than their non-entrepreneurial counterparts. According to Krueger and Brazeal (1994), entrepreneurs exhibit a proactive disposition, allowing them to anticipate trends and act decisively. Unlike non-entrepreneurs, who often prioritize stability and risk avoidance, entrepreneurs embrace change and uncertainty as integral to the innovation process. Dweck's (2006) theory of a growth mindset further aligns with entrepreneurial behavior, emphasizing learning from failures and striving for continuous improvement (Thompson & Bolton, 2024). This mindset fosters innovation and adaptability, distinguishing entrepreneurs from individuals with fixed or risk-averse approaches.

Motivation plays a pivotal role in driving entrepreneurial behavior, with several key factors influencing individuals to pursue and sustain entrepreneurial ventures (Turker & Ozmen, 2021). The need for achievement, as identified by McClelland (1961), is a fundamental motivator for entrepreneurs. This need reflects a desire to accomplish challenging goals and measure success through tangible outcomes. Entrepreneurs are inherently motivated by a profound aspiration for autonomy and self-determination, seeking autonomy in decision-making and control over their professional lives (Van Gelderen et al., 2006). This desire often stems from dissatisfaction with hierarchical structures in traditional employment, motivating entrepreneurs to establish ventures where they can operate freely. Financial incentives also contribute to entrepreneurial motivation, as the potential for wealth creation and economic rewards encourages individuals to take calculated risks. However, financial gain is not always the primary driver (Cicchiello et al., 2021). Many entrepreneurs are motivated by intrinsic factors such as a passion for a particular field or the desire to address societal challenges (Turker & Ozmen, 2021). Social entrepreneurship, for instance, is often fueled by the ambition to create positive social and environmental impacts, as highlighted by Dees (1998). The opportunity to make a meaningful difference further sustains entrepreneurial efforts. An essential determinant of entrepreneurial motivation is self-efficacy, defined as the individual's conviction in their capacity to achieve success and effectively overcome challenges. Bandura (1986) identified self-efficacy as a core determinant of entrepreneurial persistence, enabling individuals to navigate setbacks and uncertainties with confidence. Entrepreneurs often exhibit high levels of self-efficacy, which fuels their determination to overcome obstacles and achieve their goals (Thompson & Bolton, 2024). The combination of an entrepreneurial mindset and intrinsic and extrinsic motivational factors underpins entrepreneurial success. This mindset allows entrepreneurs to navigate the complexities of innovation and risk, while motivational drivers such as achievement, independence, and purpose sustain their efforts. Together, these elements offer a comprehensive understanding of what propels entrepreneurs to create and grow ventures, contributing significantly to economic and societal advancement.

5 Entrepreneurial Process

5.1 Stages of the Entrepreneurial Process

The entrepreneurial process involves a series of interconnected stages through which an idea transforms into a thriving business (Cicchiello et al., 2021). These stages—opportunity recognition, business planning, resource mobilization, execution, and growth—are essential for guiding entrepreneurs from conceptualization to sustained success. At each stage, specific tools and techniques are pivotal in ensuring effective implementation. The process begins with opportunity recognition, where entrepreneurs identify market gaps, unmet needs, or innovative possibilities. This stage is driven by creativity and critical thinking, supported by tools such as market research, SWOT analysis, and customer feedback mechanisms. Sarasvathy (2001) introduced the concept of effectuation, emphasizing how entrepreneurs leverage available resources to discover and shape opportunities rather than waiting for perfect conditions. Techniques like brainstorming and customer interviews further refine potential ideas. Once an opportunity is identified, entrepreneurs transition to business planning, a stage that provides a structured roadmap for the venture. A robust business plan outlines goals, market strategies, operational structures, and financial projections (Attarit, 2024). Tools such as the Business Model Canvas Osterwalder (2010) and Lean Startup methodologies (Ries, 2011) are widely employed to visualize value propositions, revenue streams, and customer segments. Financial modeling and competitor analysis are critical in this phase, helping entrepreneurs assess the feasibility of their ideas and refine their strategies.

Following the planning stage is resource mobilization, which involves securing the financial, human, and physical resources needed to bring the plan to life (Ferreira, 2024). Entrepreneurs often seek funding through venture capital, angel investors, or crowdfunding platforms (Koziol et al., 2025). Tools such as pitch decks and financial forecasting software facilitate investor engagement. At the same time, effective team building relies on recruitment platforms and network mapping, ensuring the right talent is in place to support the venture (Stahl et al., 2012). The execution stage marks the transformation of plans into action. Entrepreneurs launch their products or services and focus on building efficient operations. Tools like project management software (e.g., Trello, Asana) and customer relationship management (CRM) systems streamline workflows and enhance customer interactions. Lean and agile methodologies are often applied to ensure responsiveness to market feedback and adaptability in evolving conditions. Finally, the process culminates in the growth and scaling stage, where entrepreneurs expand their market presence, diversify offerings, and enhance operational efficiency. Advanced analytics platforms, enterprise resource planning (ERP) systems, and branding strategies support this phase. Entrepreneurs also focus on building strong customer loyalty and establishing strategic partnerships to sustain long-term success. The entrepreneurial process, with its distinct stages and tailored tools, provides a comprehensive framework for turning ideas into impactful ventures. From identifying opportunities to scaling businesses, each stage is a critical step toward achieving sustainable success.

5.2 Opportunity Recognition and Evaluation

Recognizing and evaluating opportunities is a critical phase in the entrepreneurial process, enabling entrepreneurs to identify viable ideas and assess their feasibility. This process blends creative insight with analytical rigor, ensuring that opportunities are both innovative and practical (Muhammad et al., 2025).

Opportunity Recognition begins with observing market trends, technological shifts, and changing consumer needs. Entrepreneurs employ environmental scanning and trend analysis tools to identify gaps or emerging demands (Audretsch et al., 2022). Networking and stakeholder interactions provide valuable insights into hidden opportunities, while creative methods like brainstorming and design thinking help refine ideas into potential business ventures. As Drucker (1986) suggests, innovation often emerges from analyzing unexpected occurrences or inefficiencies within industries. Ardichvili et al. (2003) emphasize the role of prior knowledge and entrepreneurial alertness in identifying promising opportunities, suggesting that entrepreneurs with industry expertise are better positioned to spot gaps.

Once opportunities are identified, evaluation becomes essential to determine their feasibility. Market analysis examines the size, growth potential, and dynamics of the target market. Techniques such as customer segmentation help entrepreneurs identify high-potential market segments, while competitor analysis evaluates existing players and their market strategies (Sihombing et al., 2025). Tools like Porter's Five Forces framework assess industry attractiveness by analyzing competitive intensity, buyer and supplier power, and potential threats from new entrants or substitutes (Porter, 2008). Shane and Venkataraman (2000) highlight the importance of evaluating opportunities based on their alignment with the entrepreneur's resources, skills, and goals. To ensure systematic evaluation, entrepreneurs employ analytical frameworks such as SWOT analysis, which serves as a strategic tool for assessing an organization's internal strengths and weaknesses alongside external opportunities and threats, thereby facilitating informed decision-making and strategic planning. Similarly, the PESTEL analysis offers a comprehensive approach to examining macro-environmental factors—including political, economic, social, technological, environmental, and legal influences—that shape business operations and impact long-term sustainability. Sarasvathy (2001) advocates for effectuation as a method for evaluating opportunities by leveraging available resources and partnerships rather than relying solely on prediction. These tools enable entrepreneurs to identify risks, understand market conditions, and align opportunities with strategic goals. Combining creative exploration with structured analysis, opportunity recognition, and evaluation provides entrepreneurs with a solid foundation for decision-making. This process minimizes risks and increases the likelihood of success, ensuring that identified opportunities align with market demands and competitive realities.

5.3 Resource Mobilization

Resource mobilization is a critical element of entrepreneurial success, requiring the effective acquisition and utilization of human and financial resources, as well as leveraging external networks. Entrepreneurs must strategically navigate these domains to ensure their ventures thrive in competitive environments. Financial resources are fundamental to covering operational costs, innovation, and scaling efforts. Entrepreneurs employ various strategies to secure funding, including self-financing through personal savings, debt financing via loans, and equity-based funding from venture capital investments, angel investments, and crowdfunding. For instance, informal entrepreneurs often rely on social and human capital to access financing in resource-constrained contexts (Zehra, 2018). Additionally, utilizing network diversity enhances access to complementary resources, ultimately improving business performance (Adegbile, 2018).

Human resources play an equally crucial role in driving innovation and efficiency. Entrepreneurs must attract skilled employees who are aligned with their vision and strategy. Social networks are vital for identifying and recruiting talent, as highlighted by research showing the role of human capital in resource mobilization and opportunity recognition in industries like India's handloom sector (Bhagavatula et al., 2010). Networking is indispensable for mobilizing resources beyond internal capacities. Building and maintaining diverse social and professional networks allows entrepreneurs to access critical resources, mentorship, and market opportunities. Entrepreneurs often use their networks to mobilize social, human, and financial capital, turning challenges into opportunities (Clough et al., 2019).

Furthermore, innovative approaches like bricolage enable entrepreneurs to creatively utilize available resources in constrained environments, particularly in social entrepreneurship (Desa, 2012). External resources, such as business incubators, academic institutions, and government programs, also provide advisory support, infrastructure, and funding opportunities. For example, incubators not only help entrepreneurs secure financial backing but also offer management expertise to enhance resource mobilization efforts (Suk & Mooweon, 2006). By combining strategic financial planning, human resource acquisition, and networking, entrepreneurs can effectively mobilize resources to ensure their ventures' growth and sustainability.

6 Entrepreneurial Ecosystems and Support Structures

6.1 Role of Entrepreneurial Ecosystems

Entrepreneurial ecosystems are dynamic environments that consist of interconnected stakeholders, resources, and institutions working collectively to support entrepreneurial activity (Lu et al., 2025). These ecosystems play a vital role in fostering innovation, enabling business creation, and contributing to regional and global economic growth. By offering critical resources and promoting collaboration, entrepreneurial ecosystems provide entrepreneurs with the tools they need to overcome challenges and achieve long-term success (Stam & Spigel, 2016). An entrepreneurial ecosystem is defined as a network of interrelated elements, including entrepreneurs, investors, educational institutions, support organizations, and government bodies, which function cohesively to promote entrepreneurship. These systems facilitate resource sharing, knowledge exchange, and collaboration, enabling entrepreneurs to thrive. The significance of entrepreneurial ecosystems lies in their ability to promote economic development by driving innovation, creating jobs, and enhancing the competitiveness of regions in global markets (Malecki, 2018). Entrepreneurial ecosystems rely on several interrelated components that collectively support entrepreneurial activities. Securing adequate financial capital is fundamental for funding business development and growth. These ecosystems provide diverse funding options such as venture capital, angel investments, crowdfunding, and government grants, ensuring that entrepreneurs have the capital needed for different stages of their ventures (Spigel & Harrison, 2018). Experienced mentors and advisors play a key role in entrepreneurial ecosystems by guiding entrepreneurs through challenges and providing strategic advice. Their expertise helps refine business strategies and fosters valuable connections within the ecosystem (Harima et al., 2020). Educational institutions, such as universities and research organizations, contribute significantly to ecosystems by equipping individuals with the skills and knowledge necessary for entrepreneurship and serving as hubs for innovation and collaboration (Clark et al., 2021). Government initiatives, such as tax incentives, startup-friendly regulations, and infrastructure investments, are also critical in creating a supportive environment for entrepreneurship. Policies designed to foster innovation and reduce barriers to business formation enhance the effectiveness of ecosystems and attract talent and investment (Stam & Spigel, 2016). Networking and collaboration within entrepreneurial ecosystems connect entrepreneurs with key stakeholders such as suppliers, investors, and customers, enabling resource sharing, partnerships, and mutual growth (Kansheba & Wald, 2020). Cultural elements also play an essential role in the success of these ecosystems. Societies that value entrepreneurship, risk-taking, and innovation create an enabling ecosystem for entrepreneurs to pursue innovative ventures and contribute to economic progress (Spigel & Harrison, 2018). Entrepreneurial ecosystems provide a collaborative framework that supports entrepreneurs by integrating critical components and fostering innovation. These systems benefit individual businesses and contribute to broader economic development, technological advancement, and societal growth.

6.2 Support Structures and Institutions

Government policies and initiatives play a major role in fostering entrepreneurship by creating a conducive environment for business creation and growth. Governments enhance the entrepreneurial ecosystem's efficiency through supportive measures such as tax incentives, startup grants, regulatory frameworks, and infrastructure development. For example, policies aimed at reducing bureaucratic hurdles and providing access to seed funding have been instrumental in fostering entrepreneurial activity in countries like Singapore and Israel (Wasnik & Jain, 2023).

Support structures like incubators and accelerators further enhance entrepreneurial ecosystems. Incubators provide startups with access to office spaces, mentorship, and essential services during their early stages, ensuring they establish a solid foundation for growth. **Accelerators**, on the other hand, focus on scaling businesses rapidly by offering intensive mentorship programs, funding access, and networking opportunities over a shorter time frame (Pustovrh et al., 2020). These support structures act as intermediaries, bridging the gap between entrepreneurs and critical resources, thereby reducing the risks associated with starting and growing a business (Goswami et al., 2018). Successful entrepreneurial ecosystems around the world demonstrate the value of these support structures. Silicon Valley, for instance, thrives on a robust network of venture capitalists, accelerators like Y Combinator, and collaboration between academia and industry. Similarly, Bangalore in India has emerged as a hub for technology-driven startups due to targeted government initiatives, access to skilled labor, and accelerators' contributions to scaling innovation-driven enterprises (Brown et al., 2019). These ecosystems highlight the interplay of policies, institutions, and culture in fostering entrepreneurial success.

By integrating effective government policies and leveraging support structures like incubators and accelerators, entrepreneurial ecosystems create environments where businesses can innovate, scale, and contribute to economic growth.

7 Practical Implications

Entrepreneurs rely on various practical tools and strategies to navigate challenges and achieve success in competitive markets. These strategies combine technological, financial, and managerial methods to enhance decision-making and execution in real-world scenarios. One essential tool for entrepreneurs is business modeling frameworks, such as the Business Model Canvas, which helps define and structure key aspects of a venture, including value propositions, revenue streams, and customer segments. Using this visual framework, entrepreneurs can better align their strategies with market needs (Pustovrh et al., 2020). Lean Startup methodologies focus on continuous product improvement and user insights, enabling entrepreneurs to enhance their offerings effectively (Neck & Greene, 2011). Digital tools and platforms play a vital role in modern entrepreneurship. For instance, Customer Relationship Management (CRM) software like Salesforce enables entrepreneurs to manage interactions with clients and improve sales processes. Similarly, project management tools such as Trello or Asana assist in organizing tasks and enhancing team collaboration (Foster & Yaoyuneyong, 2016). These tools bridge the gap between planning and execution, making it easier to implement entrepreneurial concepts in practice. Networking and collaboration are integral strategies for entrepreneurs. Incubators and accelerators provide structured environments where entrepreneurs can access mentorship, funding, and networking opportunities to scale their ventures rapidly (Hernández & Virgüez, 2024). These programs often include workshops and real-world case studies to prepare entrepreneurs for market challenges (Bellotti et al., 2014). Entrepreneurs also use data-driven approaches to optimize their operations and make informed decisions. Analytics tools such as Google Analytics or Tableau offer valuable data on customer interactions, market fluctuations, and business success, enabling entrepreneurs to refine their strategies in a flexible and informed manner. Real-world applications of entrepreneurial strategies emphasize adaptability, resource optimization, and innovation. Entrepreneurs who integrate these tools and techniques effectively are better positioned to seize opportunities and overcome challenges in dynamic business environments.

8 Conclusion

Understanding the foundational principles of entrepreneurship is essential for aspiring entrepreneurs and business leaders. This chapter has explored the multifaceted nature of entrepreneurship, including its theoretical underpinnings, practical applications, and the critical role of ecosystems and support structures.

Integrating these elements has laid a strong foundation for understanding how entrepreneurship drives innovation, economic growth, and societal transformation.

The discussion began with an overview of entrepreneurship, differentiating it from related concepts such as startups and small businesses. Key theories, such as Schumpeter's concept of creative destruction (Schumpeter, 1934) and Kirzner's theory of entrepreneurial discovery (Kirzner, 1973), provided valuable insights into the entrepreneurial process. These theories highlight the importance of both innovation and market efficiency, offering complementary perspectives on how entrepreneurs contribute to economic development. Entrepreneurial ecosystems emerged as vital frameworks that support entrepreneurs through access to capital, mentorship, educational resources, and collaborative culture (Stam & Spigel, 2016). Examples such as Silicon Valley and Bangalore demonstrate how well-developed ecosystems can foster innovation and accelerate entrepreneurial success (Mason & Brown, 2014). Government policies and support structures, including incubators and accelerators, further enhance these ecosystems, enabling entrepreneurs to overcome barriers and scale their ventures effectively (Pustovrh et al., 2020). The practical tools and strategies discussed in this chapter emphasize the actionable aspects of entrepreneurship. Techniques such as business modeling, Lean Startup methodologies (Neck & Greene, 2011), and digital tools empower entrepreneurs to translate concepts into reality. Furthermore, leveraging data-driven approaches and collaborative networks equips entrepreneurs to adapt to dynamic market conditions and seize emerging opportunities (Goswami et al., 2018). Entrepreneurship is not merely a business activity but a transformative force that reshapes industries and addresses societal challenges. By fostering creativity, resilience, and innovation, entrepreneurs contribute to sustainable economic growth and social progress (Bellotti et al., 2014). Aspiring entrepreneurs and business leaders can draw on the principles and insights outlined in this chapter to navigate the complexities of entrepreneurship and make a meaningful impact.

In conclusion, the knowledge of entrepreneurship principles equips individuals to pursue opportunities with confidence and purpose. By combining theoretical understanding with practical application, entrepreneurs are well-positioned to thrive in an ever-evolving global economy. This chapter serves as a guide for understanding the essential components of entrepreneurship, empowering readers to embark on their entrepreneurial journeys with clarity and determination.

9 Case Study

Entrepreneur X, Alex Morgan, is the founder and CEO of InnovateAI, a leading tech startup specializing in artificial intelligence solutions for businesses. Founded in 2016 in San Francisco, InnovateAI aimed to bridge the gap between cutting-edge AI technologies and their practical applications in industries such as healthcare, logistics, and customer service. Alex, a computer science graduate from Stanford University with experience working in Silicon Valley, combined his technical expertise and entrepreneurial vision to create a company that now serves over 200 global clients. The journey began in 2015 when Alex recognized a growing demand for AI-driven solutions to improve operational efficiency across industries. During his tenure as a software engineer, he observed companies struggling to integrate AI tools into their workflows due to complexity and high costs. Identifying this market gap, Alex founded InnovateAI with the mission of simplifying AI adoption for SMEs. Starting with limited resources, Alex bootstrapped the company, leveraging his savings and contributions from close friends. His first breakthrough came when he pitched InnovateAI at a startup accelerator program, securing $500,000 in seed funding. With this initial support, Alex assembled a small team of developers and launched the company's first product, AIFlow, a user-friendly AI automation tool designed for SMEs.

Through continuous iterations based on client feedback, InnovateAI expanded its product line to include healthcare diagnostics tools and predictive analytics platforms. By 2020, InnovateAI had raised $20 million in Series A funding, enabling the company to scale operations globally. Several strategic decisions contributed to Alex's success. First, his focus on solving a specific pain point—making AI accessible and affordable—allowed him to differentiate InnovateAI in a competitive market. By targeting underserved SMEs, Alex carved out a niche where InnovateAI could thrive without directly competing with larger, established firms.

Second, Alex prioritized building a strong organizational culture centered on innovation and customer focus. His leadership style emphasized collaboration, creativity, and responsiveness to customer needs, fostering a motivated and agile team. Third, Alex's decision to participate in a startup accelerator program proved pivotal. The program provided not only funding but also mentorship and access to a robust entrepreneurial network. This network played a significant role in securing early clients and additional funding rounds. Finally, Alex embraced a data-driven approach to product development and marketing. InnovateAI's emphasis on gathering user feedback and analytics helped the company refine its offerings and build customer loyalty. Strategic partnerships with larger corporations also enabled InnovateAI to scale faster by leveraging existing distribution channels and client bases. Alex Morgan's entrepreneurial journey exemplifies the power of recognizing market opportunities, leveraging networks, and maintaining a customer-focused approach. By combining technical expertise, strategic vision, and resilience, Alex transformed InnovateAI into a globally recognized brand. His success highlights critical lessons for aspiring entrepreneurs, including the importance of solving real-world problems, fostering innovation, and staying adaptable in a dynamic business environment.

10 Questions

To assess your understanding of the chapter on entrepreneurship, consider these questions:

1. What are the key traits that distinguish successful entrepreneurs from others?
2. Explain the concept of "creative destruction" as introduced by Joseph Schumpeter.
3. Analyze the different stages of the entrepreneurial process.
4. How would you use the Business Model Canvas to develop a business plan for a new startup?
5. Apply Sarasvathy's concept of effectuation to identify and shape entrepreneurial opportunities in a resource-constrained environment.
6. How can networking and collaboration within an entrepreneurial ecosystem enhance a startup's chances of success? Provide examples.
7. What are the potential challenges entrepreneurs might face when securing funding through alternative sources like crowdfunding and venture capital?
8. Examine how social entrepreneurship contributes to solving global issues like poverty and climate change. How does it differ from traditional entrepreneurship?

References

Addy, W. A., Ajayi-Nifise, A. O., Bello, B. G., Tula, S. T., Odeyemi, O., & Falaiye, T. (2024). Entrepreneurial leadership in high-tech industries: A review of key traits and success strategies. *GSC Advanced Research and Reviews, 18*(2), 286–296.

Adegbile, A. S. (2018). *Linking resource mobilization approaches and performance in entrepreneurial ventures: A social network perspective.* Europa-Universität Viadrina Frankfurt.

Alekseevna, S. S. (2023). *The role of entrepreneurial imaginativeness in sustaining business success through crisis: The case of Russian business owners.*

Ardichvili, A., Cardozo, R., & Ray, S. (2003). A theory of entrepreneurial opportunity identification and development. *Journal of Business Venturing, 18*(1), 105–123.

Attarit, T. (2024). An analysis of structural causal relationship of corporate governance on sustainable growth through firm value in the covid-19 pandemic of sustainable investment group in the stock exchange of Thailand. *Journal of Law and Sustainable Development, 12*(12), e4186–e4186.

Audretsch, D., Belitski, M., Caiazza, R., Günther, C., & Menter, M. (2022). From latent to emergent entrepreneurship: The importance of context. *Technological Forecasting and Social Change, 175*, Article 121356.

Audretsch, D. B. (2006). *Entrepreneurship and economic growth*. Oxford University Press.

Audretsch, D. B., & Thurik, R. (1999). Linking entrepreneurship to economic growth. *Advances in the Study of Entrepreneurship, Innovation, and Economic Growth, 11*, 1–28.

Austin, J. (2006). Social and commercial entrepreneurship: Same, different, or both. *Entrepreneurship Theory and Practice*.

Bandura, A. (1986). *Social foundations of thought and action* (Vol. 2, pp. 23–28). Prentice Hall.

Baron, R. A. (2008). The role of affect in the entrepreneurial process. *Academy of Management Review, 33*(2), 328–340.

Baumol, W. J. (1996). Entrepreneurship: Productive, unproductive, and destructive. *Journal of Business Venturing, 11*(1), 3–22. https://doi.org/10.1016/0883-9026(94)00014-X

Bellotti, F., Berta, R., De Gloria, A., Lavagnino, E., Antonaci, A., Dagnino, F., Ott, M., Romero, M., Usart, M., & Mayer, I. S. (2014). Serious games and the development of an entrepreneurial mindset in higher education engineering students. *Entertainment Computing, 5*(4), 357–366.

Bhagavatula, S., Elfring, T., Van Tilburg, A., & Van De Bunt, G. G. (2010). How social and human capital influence opportunity recognition and resource mobilization in India's handloom industry. *Journal of Business Venturing, 25*(3), 245–260.

Blackburn, R., & Kovalainen, A. (2009). Researching small firms and entrepreneurship: Past, present and future. *International Journal of Management Reviews, 11*(2), 127–148.

Bornstein, D. (2007). *How to change the world: Social entrepreneurs and the power of new ideas*. Oxford University Press.

Brown, C., & Thornton, M. (2013). How entrepreneurship theory created economics. *Quarterly Journal of Austrian Economics, 16*(4).

Brown, R., Mawson, S., Lee, N., & Peterson, L. (2019). Startup factories, transnational entrepreneurs and entrepreneurial ecosystems: Unpacking the lure of startup accelerator programs. *European Planning Studies, 27*(5), 885–904.

Calderini, M., Chiodo, V., Gerli, F., & Pasi, G. (2023). The centrality of social-tech entrepreneurship in an inclusive growth agenda. In *Social economy science* (pp. 284–308). Oxford University Press.

Cantillon, R. (1755). Essai sur la nature du commerce en général. *Fetcher Gyler*.

Celestin, M., & Vanitha, N. (2018). Ten essential habits of successful entrepreneurs: A guide for the next generation. *International Journal of Applied and Advanced Scientific Research (IJAASR), 3*(2), 56–64.

Chevrollier, N., Argyrou, A., & Jeurissen, R. (2025). Introduction: Radical business perspectives for sustainability transitions. In *Radical business perspectives for sustainability transitions* (pp. 1–10). Edward Elgar Publishing.

Cicchiello, A. F., Pietronudo, M. C., Leone, D., & Caporuscio, A. (2021). Entrepreneurial dynamics and investor-oriented approaches for regulating equity-based crowdfunding. *Journal of Entrepreneurship and Public Policy, 10*(2), 235–260.

Clark, D. N., Reboud, S., Toutain, O., Ballereau, V., & Mazzarol, T. (2021). Entrepreneurial education: An entrepreneurial ecosystem approach. *Journal of Management & Organization, 27*(4), 694–714.

Clough, D. R., Fang, T. P., Vissa, B., & Wu, A. (2019). Turning lead into gold: How do entrepreneurs mobilize resources to exploit opportunities? *Academy of Management Annals, 13*(1), 240–271.

Dees, J. G. (1998). *The meaning of social entrepreneurship*. Draft Report for the Kauffman Center for …, Stanford University.

Desa, G. (2012). Resource mobilization in international social entrepreneurship: Bricolage as a mechanism of institutional transformation. *Entrepreneurship Theory and Practice, 36*(4), 727–751.

Drucker, P. F. (1986). *Innovation and entrepreneurship: Practice and principles*. Harper & Row.

Duckworth, A. L., Peterson, C., Matthews, M. D., & Kelly, D. R. (2007). Grit: Perseverance and passion for long-term goals. *Journal of Personality and Social Psychology, 92*(6), 1087.

Dweck, C. S. (2006). *Mindset: The new psychology of success*. Random House.

Ferreira, M. N. (2024). Conservation priorities mapping—A first step toward building area-based strategies. In *Frontiers in science* (Vol. 2, p. 1440501). Frontiers Media SA.

Foster, J., & Yaoyuneyong, G. (2016). Teaching innovation: Equipping students to overcome real-world challenges. *Higher Education Pedagogies, 1*(1), 42–56.

Fritsch, M., & Storey, D. J. (2014). Entrepreneurship in a regional context: Historical roots, recent developments, and future challenges. *Regional Studies, 48*(6), 939–954.

Goswami, K., Mitchell, J. R., & Bhagavatula, S. (2018). Accelerator expertise: Understanding the intermediary role of accelerators in the development of the Bangalore entrepreneurial ecosystem. *Strategic Entrepreneurship Journal, 12*(1), 117–150.

Günar, D. Ö. (2025). AI innovations and silicon valley: How Wall Street is changing. In *Economic and political consequences of AI: Managing creative destruction* (pp. 205–232). IGI Global Scientific Publishing.

Harima, J., Freiling, J., & Harima, A. (2020). Public accelerators-developing resources in nascent entrepreneurial ecosystems. *Academy of Management Proceedings, 2020*(1), 21463.

Henrekson, M., & Sanandaji, T. (2014). Small business activity does not measure entrepreneurship. *Proceedings of the National Academy of Sciences, 111*(5), 1760–1765.

Hernández, M. A., & Virgüez, J. E. B. (2024). Innovative technological tools for teaching situated entrepreneurship: A case study in higher education. *International Journal of Professional Business Review*, *9*(3), 1.

Hisrich, R. D., Peters, M. P., & Shepherd, D. A. (2017). *Entrepreneurship*. McGraw-Hill Education.

Kansheba, J. M. P., & Wald, A. E. (2020). Entrepreneurial ecosystems: A systematic literature review and research agenda. *Journal of Small Business and Enterprise Development, 27*(6), 943–964.

Kharche, N., Mir, S. M., & Dias, R. (2025). The future of indigenous inclusivity: Unlocking finance for economic development. In *The future of indigenous inclusivity: Unlocking finance for economic development* (pp. 185–200). IGI Global Scientific Publishing.

Kirzner, I. (1973). *Competition and entrepreneurship*. University of Chicago Press.

Kotler, P., Keller, K. L., Brady, M., Goodman, M., & Hansen, T. (2016). *Marketing management 3rd edn PDF eBook*. Pearson Higher Ed.

Koziol, K., Schmitz, M., & Bort, S. (2025). Gender differences in entrepreneurial equity financing—A systematic literature review. *Small Business Economics*, 1–56.

Krueger, N. F., Jr., & Brazeal, D. V. (1994). Entrepreneurial potential and potential entrepreneurs. *Entrepreneurship Theory and Practice, 18*(3), 91–104.

Leadbeater, C. (1997). The rise of the social entrepreneur. *Demos*.

Lu, Y., Muhamad, N. S., & Hanafiah, M. H. (2025). Opportunity recognition in digital entrepreneurship through digital capabilities: A conceptual framework for sustainable entrepreneurship to achieve SDG. *Journal of Lifestyle and SDGs Review, 5*(1), e04265–e04265.

Mahajan, R., & Bandyopadhyay, K. R. (2021). Women entrepreneurship and sustainable development: Select case studies from the sustainable energy sector. *Journal of Enterprising Communities: People and Places in the Global Economy, 15*(1), 42–75.

Malecki, E. J. (2018). Entrepreneurship and entrepreneurial ecosystems. *Geography Compass, 12*(3), Article e12359.

Martin, R. L., & Osberg, S. (2007). *Social entrepreneurship: The case for definition*. Stanford Social Innovation Review Stanford.

Mason, C., & Brown, R. (2014). Entrepreneurial ecosystems and growth-oriented entrepreneurship. *Final Report to OECD, Paris, 30*(1), 77–102.

Mazzarol, T., Volery, T., Doss, N., & Thein, V. (1999). Factors influencing small business start-ups: A comparison with previous research. *International Journal of Entrepreneurial Behavior & Research, 5*(2), 48–63.

McClelland, D. C. (1961). *The achieving society* (p. 196l). D. Van Norstrand Company. Inc.

Mickiewicz, T., Nyakudya, F. W., Theodorakopoulos, N., & Hart, M. (2017). Resource endowment and opportunity cost effects along the stages of entrepreneurship. *Small Business Economics, 48*, 953–976.

Muhammad, M., Suleiman, W., Al Mamun, M. A., Jewel, R. M., & Mozumder, M. A. S. (2025). Mediating effect of appraisal fairness between the relationship of appraisal satisfaction and employee performance appraisal: A case of Bauchi State Teachers Service Commission, Nigeria. *Cogent Business & Management, 12*(1), 2450297.

Müller-Wieland, R., Muschner, A., & Schraudner, M. (2019). Academic entrepreneurship: Phase-specific constraints and needs. *Journal of Enterprising Communities: People and Places in the Global Economy, 13*(3), 353–371.

Nave, E., Ferreira, J., & Marques, L. M. (2024). Business environment reforms effect on entrepreneurial activities of high-income economies: Panel data evidence. *Cross Cultural & Strategic Management, 31*(1), 26–49.

Neck, H. M., & Greene, P. G. (2011). Entrepreneurship education: Known worlds and new frontiers. *Journal of Small Business Management, 49*(1), 55–70.

O'Gorman, C. (2009). The financial requirements of early-stage entrepreneurs. In *Entrepreneurship and growth in local, regional and national economies*. Edward Elgar Publishing.

Osterwalder, A. (2010). *Business model generation: A handbook for visionaries, game changers, and challengers*. Wiley.

Osunmuyiwa, O. O., & Ahlborg, H. (2022). Stimulating competition, diversification, or re-enforcing entrepreneurial barriers? Exploring small-scale electricity systems and gender-inclusive entrepreneurship. *Energy Research & Social Science, 89*, Article 102566.

Porter, M. E. (2008). The five competitive forces that shape strategy. *Harvard Business Review, 86*(1), 78.

Pustovrh, A., Rangus, K., & Drnovšek, M. (2020). The role of open innovation in developing an entrepreneurial support ecosystem. *Technological Forecasting and Social Change, 152*, Article 119892.

Ranabahu, N., & Tanima, F. A. (2022). Empowering vulnerable microfinance women through entrepreneurship: Opportunities, challenges and the way forward. *International Journal of Gender and Entrepreneurship, 14*(2), 145–166.

Rauch, A., & Frese, M. (2007). Let's put the person back into entrepreneurship research: A meta-analysis on the relationship between business owners' personality traits, business creation, and success. *European Journal of Work and Organizational Psychology, 16*(4), 353–385.

Ries, E. (2011). How today's entrepreneurs use continuous innovation to create radically successful businesses. *The Lean Startup*.

Sarasvathy, S. D. (2001). Causation and effectuation: Toward a theoretical shift from economic inevitability to entrepreneurial contingency. *Academy of Management Review, 26*(2), 243–263.

Schiavone, F., Tutore, I., & Cucari, N. (2020). How digital user innovators become entrepreneurs: A sociomaterial analysis. *Technology Analysis & Strategic Management, 32*(6), 683–696.

Schumpeter, J. (1934). *The theory of economic development*. Harvard University Press.

Schumpeter, J. (1942). *Capitalism, socialism, and democracy*. Routledge.

Shahid, M. S., Hossain, M., Shahid, S., & Anwar, T. (2023). Frugal innovation as a source of sustainable entrepreneurship to tackle social and environmental challenges. *Journal of Cleaner Production, 406*, Article 137050.

Shane, S. A. (2003). *A general theory of entrepreneurship: The individual-opportunity nexus*. Edward Elgar.

Shane, S., & Venkataraman, S. (2000). The promise of entrepreneurship as a field of research. *Academy of Management Review, 25*(1), 217–226.

Sihombing, S. H., Sitanggang, L., & Astuti, D. (2025). Analisis Studi Kelayakan Bisnis Kopi kalean Pada Aspek Aspek Pemasaran. *Jurnal Ekonomi Dan Bisnis Digital, 2*(3), 1613–1617.

Smith, A. (1776). *The wealth of nations* (Vol. 11937).

Spigel, B., & Harrison, R. (2018). Toward a process theory of entrepreneurial ecosystems. *Strategic Entrepreneurship Journal, 12*(1), 151–168.

Stahl, G., Björkman, I., Farndale, E., Morris, S. S., Paauwe, J., Stiles, P., Trevor, J., & Wright, P. (2012). Six principles of effective global talent management. *Sloan Management Review, 53*(2), 25–42.

Stam, E., & Spigel, B. (2016). *Entrepreneurial ecosystems*. USE Discussion paper series.

Stevenson, H. H., & Gumpert, D. E. (1985). *HBR*.

Suk, J. Y., & Mooweon, R. (2006). Resource mobilization and business incubation: The case of Korean incubators. *Development and Society, 35*(1), 29–46.

Thompson, J., & Bolton, B. (2024). *Entrepreneurs: Talent, temperament, opportunity, and mindset*. Taylor & Francis.

Turker, D., & Ozmen, Y. S. (2021). How do social entrepreneurs develop technological innovation? *Social Enterprise Journal, 17*(1), 63–93.

Van Gelderen, M., Thurik, R., & Bosma, N. (2006). Success and risk factors in the pre-startup phase. *Small Business Economics, 26*, 319–335.

Van Praag, C. M., & Versloot, P. H. (2007). What is the value of entrepreneurship? A review of recent research. *Small Business Economics, 29*(4), 351–382.

Velinov, E. (2025). Automotive industry and global diversity management: An Eastern European perspective. In *Research handbook on global diversity management* (pp. 319–330). Edward Elgar Publishing.

Venâncio, A., & Jorge, J. (2022). The role of accelerator programs on the capital structure of startups. *Small Business Economics, 59*(3), 1143–1167.

Wasnik, A. P., & Jain, A. (2023). *Government support for startups: A comprehensive analysis of funding initiatives and the role of the Indian government in nurturing the startup ecosystem.*

Wennekers, S., & Thurik, R. (1999). Linking entrepreneurship and economic growth. *Small Business Economics, 13*, 27–56.

Yunus, M., & Jolis, A. (1999). *Banker to the poor: Micro-lending and the battle against world poverty public affairs*. Amazon Publishers.

Yunus, M. (2007). Creating a world without poverty: Social business and the future of capitalism. *Public Affairs*.

Zehra, K. (2018). *Resource mobilization among informal entrepreneurs*.

Zelin, Z., Caihong, C., XianZhe, C., & Xiang, M. (2021). The influence of entrepreneurial policy on entrepreneurial willingness of students: The mediating effect of entrepreneurship education and the regulating effect of entrepreneurship capital. *Frontiers in Psychology, 12*, 1–10. https://doi.org/10.3389/fpsyg.2021.592545

Sustainable Entrepreneurship

Leila Azimi, Matin Hemmati, and Morteza Akbari

Abstract

This chapter provides a comprehensive review of sustainable entrepreneurship, a rapidly evolving field at the intersection of business innovation and sustainable development. Against the backdrop of escalating global challenges—environmental degradation, social inequality, and resource scarcity—the chapter explores how sustainable entrepreneurship integrates economic, social, and environmental objectives to create value for both present and future generations. Tracing the intellectual and practical evolution of sustainability and entrepreneurship, the chapter distinguishes sustainable entrepreneurship from related concepts, such as social and environmental entrepreneurship, by highlighting its unique focus on the Triple Bottom Line and intergenerational well-being. The chapter synthesizes key theoretical frameworks, including the influence of traditional entrepreneurship theories and the emergence of the Triple Bottom Line paradigm, to elucidate how sustainable entrepreneurship has matured into a distinct academic and practical discipline. It examines the essential role of innovation in sustainable entrepreneurship, particularly in developing products, processes, and services that address systemic societal and ecological challenges. The review also analyzes the unique institutional, financial, and informational barriers sustainable entrepreneurs face, emphasizing the heightened complexity of balancing profit, purpose, and accountability. By evaluating current research trends, conceptual ambiguities, and practical applications, the chapter identifies critical gaps and future research directions, such as the need for standardized definitions, robust impact assessment methodologies, and sector-specific studies. Ultimately, it argues that sustainable entrepreneurship is essential for catalyzing systemic change, fostering resilience, and advancing a more equitable and sustainable global economy. This chapter serves as a foundational resource for scholars, practitioners, and policymakers seeking to understand and promote the transformative potential of sustainable entrepreneurship.

Keywords

Sustainable entrepreneurship · Triple bottom line · Social entrepreneurship · Eco-entrepreneurship · Sustainability innovation · Hybrid entrepreneurship · Sustainable Development Goals (SDGs)

L. Azimi · M. Akbari (✉)
Faculty of Entrepreneurship, University of Tehran, Tehran, Iran
e-mail: mortezaakbari@ut.ac.ir

L. Azimi
e-mail: L.Azimi@ut.ac.ir

M. Hemmati
Graduate School of Management and Economics, Sharif University of Technology, Tehran, Iran
e-mail: Matin.Hemmati@gsme.sharif.edu

M. Akbari et al. (eds.), *Sustainable Entrepreneurship*, University of Tehran Science and Humanities Series, https://doi.org/10.1007/978-3-032-04909-4_3

1 Learning Objectives

By the end of this chapter, readers will be able to:

1. Understand the core definition of sustainable entrepreneurship
2. Distinguish sustainable entrepreneurship from related concepts
3. Trace the historical development of sustainable entrepreneurship
4. Assess the importance of sustainability innovation in sustainable entrepreneurship
5. Identify the unique institutional, financial, and informational barriers faced by sustainable entrepreneurs.

2 Introduction

Sustainable entrepreneurship refers to entrepreneurial activities that aim to protect the natural environment, life support systems, and local communities while simultaneously creating value. This approach involves balancing economic, social, and environmental objectives, often referred to as the Triple Bottom Line. It is considered the only method capable of integrating economic, social, and ecological value creation with a focus on the welfare of future generations. Recognizing entrepreneurship as a solution rather than a contributor to environmental harm and social inequality has led to the emergence of a new category of entrepreneurial activity: sustainable entrepreneurship. Essentially, sustainable entrepreneurship is a specific type of social or environmental entrepreneurship. Although still in its early stages, the core principle of sustainable entrepreneurship is that entrepreneurial activities must not damage the ecological and social environments in which they operate; ideally, they should restore or support these environments to maintain a balance between nature, society, and economic activity (Parrish, 2010; Schaefer et al., 2015). Sustainable entrepreneurs aim to merge the advantages of both worlds by initiating activities and processes that foster profitable opportunities while promoting sustainable development (Lans et al., 2014). This form of entrepreneurship addresses social and environmental challenges through successful for-profit ventures. It focuses on preserving nature, life support, and communities while pursuing opportunities to create future products, processes, and services that generate broad gains—both economic and non-economic—for individuals, the economy, and society (Shepherd & Patzelt, 2011, p. 137). The link between traditional entrepreneurship, society, and the environment has given rise to this new entrepreneurial category, termed sustainable entrepreneurship (Cohen & Winn, 2007; Gibbs, 2006) or sustain-entrepreneurship (Aghelie et al., 2016). Related concepts such as Social Entrepreneurship and Environmental Entrepreneurship are connected to sustainable entrepreneurship; however, sustainable entrepreneurship is viewed as a distinct perspective that integrates environmental, social, and economic value creation with a focus on safeguarding the well-being of future generations (Anderson, 1998). Entrepreneurs who act sustainably are understood to meet the needs of the present without compromising the ability of future generations to meet theirs by incorporating economic, social, and environmental pillars. In this regard, initiatives like those from the United Nations encourage entrepreneurs to generate social and ecological benefits through responsible practices (Larsson et al., 2016).

Most research on social entrepreneurship focuses on addressing social challenges (Corner & Ho, 2010; Perrini et al., 2010), with some studies also examining environmental issues (Robinson, 2006). Therefore, sustainable entrepreneurship can be considered a subset of social entrepreneurship (Kury, 2012). While recognizing the overlap between social and sustainable entrepreneurship, distinctions exist regarding (1) the diversity of goals, (2) the concept of equity, and (3) the types of organizations studied. Sustainable entrepreneurs are increasingly recognized for tackling pressing social and environmental issues (Hockerts & Wüstenhagen, 2010; York & Venkataraman, 2010; Zahra et al., 2009). These socially and environmentally aware individuals play a crucial role by providing solutions to complex societal problems that are often neglected or inadequately addressed by governments, established businesses, or civil society groups (Defourny & Nyssens, 2006; Elkington & Hartigan, 2008; Kerlin, 2009; Nicholls, 2006). Currently, challenges such as environmental degradation, wealth inequality, and unequal access to opportunities and resources are escalating. These issues have amplified the demand for sustainable entrepreneurship, which involves adopting sustainable business practices. Entrepreneurship is central to the transition toward a more sustainable future, making the alignment of social, economic, and ecological goals essential. Ecological entrepreneurs, in particular, play a pivotal role in this transformation.

Considering the significant influence of businesses on society and the effects of their operations, entrepreneurship is vital for advancing a sustainable future. It encompasses both economic and non-economic activities that stimulate job creation and increase the demand for improved products and services worldwide (Shepherd & Patzelt, 2011). Sustainable entrepreneurship represents more than just a specialized branch of social or environmental entrepreneurship; rather, it constitutes a holistic framework that merges economic, social, and ecological considerations. As a distinct sub-discipline within entrepreneurship, sustainable entrepreneurship has evolved to a point where it is increasingly recognized as a mature and legitimate field of study. However, despite this progress, a concern remains that the field may face premature conceptual stagnation—a phenomenon previously noted by scholars such as (Martí and Parkinson 2006; Parkinson & Howorth, 2008). Contemporary research characterizes sustainable entrepreneurs as transformative actors who strive to harmonize economic profitability, societal well-being, and ecological preservation (Belz & Binder, 2017; Muñoz & Dimov, 2015). Positioned at the convergence of sustainable management and entrepreneurial studies, sustainable entrepreneurship is an emerging area of academic inquiry that emphasizes systemic transformation over gradual, incremental progress.

The connection between entrepreneurship and ethical conduct is well-documented. Traits commonly associated with entrepreneurial success—such as creativity, innovation, and perceptiveness—are also integral to ethical decision-making processes (Buchholz & Rosenthal, 2005). This intersection is particularly evident in the case of ecopreneurs, a distinct category of entrepreneurs identified by Rodgers (2010). These individuals and organizations function as advocates for environmentally conscious change, disseminating sustainable innovations through both market-driven and non-market strategies.

Although definitions of sustainable entrepreneurship vary, a recurring theme is its alignment with the triple bottom-line framework, which integrates economic, social, and environmental objectives (Belz & Binder, 2017; McMullen & Warnick, 2016). This conceptualization is rooted in the foundational principles of sustainable development articulated in the Brundtland Report (Jolink & Niesten, 2015), which emphasizes meeting present needs without jeopardizing the capacity of future generations to fulfill their own (WCED, 1987). Building on this, Shepherd and Patzelt (Shepherd & Patzelt, 2011) describe sustainable entrepreneurship as a form of entrepreneurship that places a strong emphasis on protecting natural ecosystems, maintaining essential life-support systems, and enhancing community well-being, all while seeking opportunities to develop innovative products, services, and processes. Notably, the concept of "gain" in this context encompasses not only financial profit but also broader social and environmental benefits.

In contrast to the conventional perspective that views entrepreneurship as primarily driven by the pursuit of profit (Parrish, 2010), sustainable entrepreneurs intentionally tackle urgent social and environmental issues through sustainable business models. For these entrepreneurs, sustainability is a core aspect of their business strategy, with their long-term financial success closely linked to their positive social and ecological contributions. As a result, sustainable entrepreneurship marks a fundamental change in entrepreneurial practice, redefining success to encompass the creation of multidimensional value rather than focusing solely on economic returns.

3 Introduction to Sustainability

Due to the multitude of theoretical approaches that have examined the concept of sustainability from various angles, offering a complete overview of the literature on this topic within a single chapter would be unfeasible. Indeed, as Kidd in his work (Kidd, 1992) has pointed out, "the body of work on sustainability is so extensive that a thorough analysis would be impractical—and even if it were possible, the effort might not be justified." Rather than striving for an all-encompassing review, this chapter highlights several pivotal studies that have explored sustainability from different viewpoints to identify the central discourses that have shaped discussions on the subject. Sustainability has emerged as one of the most pervasive buzzwords of the last twenty years. It appears that no concept or practice is beyond being labeled as 'sustainable': virtually anything can be either combined with or modified by the term. Examples encompass sustainable cities, economies, resource management, businesses, and livelihoods, with sustainable development standing out as particularly significant. Sustainability has evolved into what Gieryn (1999) describes as a boundary term—a concept situated at the crossroads of science and politics, where both fields interact and mutually shape each other.

The *boundary work* surrounding sustainability—the effort to foster shared understandings and collaborative commitments among diverse groups, particularly in connecting environmental and economic development priorities—has grown into a global central preoccupation. Over the last twenty years, numerous stakeholders have built networks, formed partnerships, established institutions and organizations, developed projects, and directed growing financial investments, all under the umbrella of sustainability. It is within this complex intersection of science and politics that boundary work takes place, where terminology, frequently ambiguous and contested, serves a vital political function in influencing policy-making and development initiatives.

3.1 A Brief Intellectual History

However, like many popular terms, the concept of sustainability carries a historical background. Its meaning has evolved and is not always as weighty as it is today. The word itself originated centuries ago, introduced by the German forestry expert Hans Carl von Carlowitz in his 1712 work *Sylvicultura Oeconomica*, where he outlined principles for the long-term management of forests. Despite this early usage, the term did not gain widespread traction until the 1980s.

The emergence of the modern environmental movement during the late 1960s and 1970s, combined with growing awareness of the Earth's limited resources, led environmentalists to link ecological issues with larger development goals. This conversation reached a significant milestone in the mid-1980s through the efforts of the Brundtland Commission, chaired by Gro Brundtland, the former Prime Minister of Norway. The commission's landmark 1987 publication, Our Common Future, introduced what is now widely accepted as the definition of sustainable development: "Sustainable development is the development that meets the needs of the present without compromising the ability of future generations to meet their own needs" (WCED, 1987, p. 43).

The concepts of "sustainability" and "sustainable development" have deep roots in interdisciplinary dialogue. From the 1980s onward, these ideas fueled a wave of scholarly and policy debates worldwide, particularly in the period leading up to the 1992 United Nations Conference on Environment and Development (UNCED), also called the Rio Earth Summit (Johnson, 2001).

Ecologists have long studied how ecosystems respond to and recover from disturbances and stressors. The discipline of mathematical ecology expanded notably in the 1970s and 1980s, with foundational contributions from researchers like Buzz Holling and Bob May, who explored the stability and resilience of both abstract and real ecological systems (Holling, 1973; May, 1977). In this context, sustainability is understood as a system's ability to absorb shocks and pressures while remaining stable or returning to equilibrium.

In economics, neo-classical theorists used the idea of substitutable capital to define a "weak" form of sustainability. This approach generated significant debate among economists, some of whom argued for a "strong" sustainability model that stressed the non-replaceable nature of essential natural capital (Pearce & Atkinson, 1993). Conversely, ecological economics aimed to forge stronger links with environmental values, resulting in tools such as life cycle analysis, ecological footprint measurement, and alternative approaches to economic accounting (Common & Stagl, 2005).

These themes also influenced the business world, where the "triple bottom line" concept became prominent, making sustainability a central aim alongside traditional corporate objectives. This evolution led to new accounting and auditing systems that incorporated sustainability into business strategies and financial disclosures (Elkington, 1997). The creation of the World Business Council for Sustainable Development during the Rio Summit highlighted this shift, drawing major companies and marking a growing institutional dedication to sustainability (Schmidheiny & Timberlake, 1992).

Beyond economics and business, political theorists such as Dobson (1999) engaged with sustainability by incorporating environmental concerns into broader political discourse, framing sustainability as a central component of social justice and political transformation. Other scholars sought to synthesize these diverse perspectives, bridging economic, ecological, and socio-political dimensions into an interdisciplinary approach that Kates and colleagues termed "sustainability science" (Kates et al., 2001). This integrative framework addresses sustainability as a complex, multifaceted challenge that requires coordinated analysis across multiple disciplines.

By the 1990s, sustainability had evolved into a concept with numerous interpretations, ranging from broad to narrow, strong to weak, and "big S" to "small s" sustainability, among other distinctions (Scoones, 2007). These varying technical definitions emerged alongside competing visions of how sustainable development should be conceptualized and implemented (Connelly, 2007). The resulting discourse was dynamic, if at times bewildering, as different schools of thought vied for dominance. Yet a critical question remained: How would these theoretical debates manifest in real-world policy? The year 1992 proved pivotal in shaping this transition from theory to practice.

Despite its widespread use and practical significance, the precise meaning of sustainability remains a subject of contention. Some critics (Caradonna, 2022) have dismissed it as "one of the least meaningful and most overused words in the English language," given its frequent and often vague application. However, the solution to such criticism is not to abandon the term but rather to refine its definition while ensuring its implementation is both measurable and consequential. One effective way to clarify sustainability is to frame it in negative terms, that is, by identifying what it is *not* (Gorissen et al., 2024). A practice, relationship, or institution is unsustainable if it degrades the very social, economic, or environmental foundations upon which its existence depends. For example, withdrawing water from rivers, lakes, and aquifers at a rate exceeding natural replenishment is unsustainable, as it leads to water scarcity and ecological collapse (Musa et al., 2024). Similarly, agricultural practices that deplete topsoil faster than they can regenerate result in barren land, food insecurity, and, ultimately, systemic failure. Likewise, a business that consistently operates at a loss, with expenditures outpacing revenues, is unsustainable by definition—its inevitable outcome being insolvency (Linna, 2020).

At its core, sustainability is about avoiding systemic breakdown. Many practitioners define it as "meeting present needs without compromising future well-being," a definition that, while serviceable, remains incomplete. It fails to address a fundamental question: whose needs are being prioritized, and whose future is being safeguarded (Henderson & Loreau, 2023)? As explored in this discussion, sustainability encompasses more than just the longevity of a particular practice, institution, or social arrangement. It demands a broader ethical and practical perspective—one that extends consideration beyond immediate participants to include all affected parties, both human and non-human, whether they exist in the present or in distant times and places. True sustainability, therefore, requires an inclusive approach that takes into account the far-reaching consequences of our actions on ecosystems, communities, and future generations (Dasht Peyma, 2025).

3.2 Evolution of Sustainable Entrepreneurship

3.2.1 Historical Context and Evolution of Sustainable Entrepreneurship

Traditionally, the success of an enterprise was evaluated primarily through the lens of economic performance. Entrepreneurship research focused predominantly on generating financial returns or, in specific contexts, on job creation. These factors were long considered the primary measures of entrepreneurship's contribution to regional and national development (Amit et al., 2001; Davidsson & Wiklund, 2001). Consequently, value creation was narrowly defined in economic and financial terms, assessed through metrics such as revenue, profitability, and return on investment, with an underlying assumption that the ultimate goal was individual profit maximization (Schlange, 2006). In this paradigm, entrepreneurship was viewed primarily as a driver of economic growth and wealth accumulation (Kirzner, 2015; Schumpeter & Backhaus, 2003), while environmental and social considerations were largely overlooked.

In the 2010s, sustainable entrepreneurs often faced hostile entrepreneurial ecosystems, requiring them to act as institutional entrepreneurs-pioneers who challenged existing norms and created new opportunities for others. As these efforts matured, the entrepreneurial ecosystem evolved, becoming more supportive of sustainability-oriented startups. However, even in more favorable environments, sustainable entrepreneurs continue to play a crucial role in shaping institutions and driving systemic change (Carle & Rayna, 2024).

Since 2015, there has been a notable increase in academic and practical interest in sustainable entrepreneurship, driven by global initiatives such as the United Nations' Sustainable Development Goals (SDGs). The field has expanded rapidly, with a growing body of research analyzing trends, influential authors, and the impact of sustainable entrepreneurship across sectors. Technological innovation, changing consumer preferences, and policy shifts have further accelerated the integration of sustainability into mainstream entrepreneurship (Alkathiri et al., 2024).

However, in recent decades, a growing body of research has emerged examining the relationship between entrepreneurship and sustainable development (Miah et al., 2024). Sustainable entrepreneurship has evolved into a prominent field of study, building upon earlier concepts such as sustainable management and eco-conscious business initiatives. Over time, it has attracted scholarly attention from diverse academic disciplines, each contributing unique perspectives to the discourse (Binder & Belz, 2015; Thompson et al., 2011) (Table 1).

Table 1 Key milestones in the evolution

1960s–1980s	The rise of environmentalism and social responsibility in business led to the emergence of early social enterprises (Cochran, 2007)
1987	The Brundtland Report introduced the concept of "sustainable development," which has influenced business and policy (Hajian & Kashani, 2021)
1990s–2000s	Sustainable business practices become more widespread; the triple bottom line framework gains traction (Larson & Johnson, 2008)
2010	Sustainable entrepreneurship faces ecosystem challenges but is beginning to influence institutional change (Carle & Rayna, 2024)
2015–Present	The surge in research, policy support, and practical adoption of sustainable entrepreneurship aligned with the SDGs (Alkathiri et al., 2024)

3.2.2 The Expanding Scope of Corporate Responsibility

Discussions surrounding the social and environmental responsibilities of businesses are not new; they have been debated since at least the late twentieth century. Some scholars have long warned that the planet's capacity to sustain current economic activities is finite (Assessment, 2001). Nevertheless, in the past decade, interest in understanding the broader societal impact of businesses has intensified significantly. Certain theorists even argue that a fundamental shift in economic paradigms is underway (Pigou, 1951, 2017). The conventional interpretation of value creation—once confined to financial profit—has expanded to encompass non-economic benefits, including ecological preservation and social well-being (Dorfman, 1993).

Initially, sustainable entrepreneurship was primarily concerned with environmental sustainability; however, it has since evolved into a multidimensional discipline that incorporates social and institutional dimensions (Shepherd & Patzelt, 2011). The field has gained prominence due to increasing awareness of the interdependence between economic, social, and ecological objectives in business. Over the past thirty years, SE has evolved from a niche concept to a mainstream area of research, driven by pressing global challenges such as climate change, resource scarcity, and widening social disparities (Rosário et al., 2022).

3.2.3 Theoretical Foundations and Growth of Sustainable Entrepreneurship

The evolution of sustainable entrepreneurship demonstrates a steady integration of sustainability concepts into conventional business practices. A major factor driving this change has been the Triple Bottom Line framework, developed by John Elkington in the late 1990s, which promotes assessing business achievements through social, environmental, and economic outcomes (Dixon & Clifford, 2007).

Academic interest in sustainable entrepreneurship has grown substantially, evidenced by the publication of over 280 peer-reviewed articles and multiple dedicated journal issues since 2000. Despite this expansion, ecopreneurship—a closely related concept—remains an emerging area with a relatively nascent body of literature (Linnanen, 2002; Pastakia, 1998; Schaltegger & Burritt, 2018).

Researchers (Shepherd & Patzelt, 2011) define sustainable entrepreneurship as the pursuit of opportunities that preserve nature, life support systems, and communities while introducing future products, processes, and services. They emphasize that the notion of "gain" extends beyond financial profit to include broader societal and environmental benefits. Although concepts like social entrepreneurship and environmental entrepreneurship are related, sustainable entrepreneurship stands out for its comprehensive approach to generating environmental, social, and economic value, prioritizing the well-being of future generations (Anderson, 1998).

To facilitate the growth of sustainable entrepreneurship, governments and regulatory bodies must actively diminish structural obstacles, introduce financial and regulatory incentives for sustainable initiatives, and enhance public education and awareness. Entrepreneurs should recognize sustainable development not merely as an ethical obligation but as a strategic advantage, one that enables the shift from conventional economic systems to sustainable alternatives (Crals & Vereeck, 2005). This transition addresses pressing social and environmental challenges while fostering equitable economic progress (Schaltegger & Wagner, 2011). Sustainable entrepreneurship is fundamentally characterized by a dedication to ethical conduct, economic advancement, and the well-being of employees, communities, and future generations (Crals & Vereeck, 2005).

Bibliometric analyses reveal exponential growth in sustainable entrepreneurship research since 2020, with annual publication rates increasing by 62% between 2020 and 2025 (Ebabu et al., 2025). Key developments include:

Interdisciplinary Expansion: Integration of technological perspectives (42% of recent studies) with traditional business models (Olurina et al., 2023).

Methodological Diversification: Increased use of mixed-methods approaches (33% of post-2023 studies) combining quantitative impact assessments with qualitative case studies (Abbas & Bulut, 2024).

Geographic Shifts: Growing research output from developing economies (28% of 2024–25 publications), particularly in renewable energy and waste management sectors (Ebabu et al., 2025).

4 Theoretical Foundations of Sustainable Entrepreneurship

4.1 Influence of Traditional Entrepreneurship

The document establishes connections between sustainable entrepreneurship and classical theories of entrepreneurship, such as Schumpeter's notion of "creative destruction," highlighting how both frameworks emphasize innovation and systemic transformation. Historically, entrepreneurship has been examined, conceptualized, and practiced as a means of creating self-employment opportunities that yield financial returns (Segal et al., 2005) or as a mechanism for job creation (Sarango-Lalangui et al., 2018). In contrast to conventional entrepreneurs, sustainable entrepreneurs often prioritize enduring societal and environmental value over immediate profit maximization. Consequently, initial scholarly investigations into sustainable entrepreneurship have concentrated on identifying the defining characteristics of these entrepreneurs and determining how they differ from their traditional counterparts. This approach mirrors the early development of entrepreneurship research as a discipline, which initially emphasized personality-based analyses (Gartner, 1988) before shifting toward cognitive frameworks rooted in psychological theory (Mitchell et al., 2002).

Over time, the conventional understanding of entrepreneurship, which primarily emphasizes economic value creation, has expanded to incorporate non-financial benefits, including social and environmental impacts (Shepherd & Patzelt, 2011; Urbaniec, 2018). This broader perspective has prompted researchers (Cohen & Winn, 2007; O'Neill Gerald et al., 2006) to explore the relationship between entrepreneurship and sustainable development, ultimately giving rise to the concept of sustainable entrepreneurship (Muñoz & Cohen, 2018a), sometimes referred to as "sustainopreneurship" (Aghelie et al., 2016). Unlike traditional entrepreneurial models, sustainable entrepreneurship deliberately embeds sustainability principles into the business's foundational strategy. Whereas conventional entrepreneurship primarily drives economic growth, sustainable entrepreneurship seeks to harmonize social, economic, and environmental priorities (Belz & Binder, 2017), positioning itself as a holistic approach to value creation.

Schaltegger and Wagner (2011) researchers characterize sustainable entrepreneurship as an organizational strategy wherein firms adopt sustainable practices to enhance efficiency and market competitiveness while harmonizing ecological, economic, and social outcomes. This approach entails identifying and capitalizing on opportunities that yield both societal and environmental value, thereby advancing broader sustainability objectives. Sustainable entrepreneurs distinguish themselves by rejecting outdated, resource-intensive practices in favor of innovative solutions that deliver superior social and environmental benefits.

4.2 The Transition Toward a Sustainable Economy

The shift toward a sustainable economy is broadly recognized as essential for addressing global challenges, especially climate change. In this context, sustainable entrepreneurs are seen as key contributors as they develop innovative market-based solutions that combine ecological, social, and economic value (Diepolder et al., 2021). Although scholarly interest in sustainable entrepreneurship has grown over the last decade, the field is still emerging, and there is not yet a universally agreed-upon definition (Vallaster et al., 2018).

Scholars from various fields have introduced multiple terms to describe the intersection of entrepreneurship and sustainable development, such as ecological and social entrepreneurship (Schaltegger & Wagner, 2011). Nevertheless, sustainable entrepreneurship is set apart from these and other entrepreneurial forms by its core motivation and main business aim, which center on sustainability. While traditional entrepreneurship focuses mainly on generating economic value (Kirzner, 2015; Schumpeter, 1942), systematic literature reviews by Greco and de Jong (2017) and Binder and Belz (2017) indicate that SE is best understood as the process of discovering, developing, and leveraging business opportunities to create and market innovative products and services that deliver ecological, social, and economic value (Cohen & Winn, 2007; Pacheco et al., 2010). A central element in this field is the Triple Bottom Line framework introduced by Elkington (1997), which calls for evaluating business performance across three dimensions: people, profit, and planet. This comprehensive perspective stresses the need to balance financial achievement, social responsibility, and environmental care. Unlike conventional entrepreneurship, which is primarily concerned with profit, sustainable entrepreneurship embeds sustainability at the heart of its activities.

The World Commission on Environment and Development (WCED, 1987) defines sustainable development as meeting present needs without compromising the ability of future generations to meet theirs. Within this sustainability context, related ideas such as green entrepreneurship, environmental entrepreneurship, eco-entrepreneurship, social entrepreneurship, and corporate social responsibility (CSR) have emerged. These concepts represent ongoing efforts to encourage both startups and established companies to adopt more responsible business practices (Kraus et al., 2017). Although these concepts overlap, they are not synonymous. Environmental entrepreneurship, ecological entrepreneurship, and eco-entrepreneurship are often grouped under green entrepreneurship, as their main focus is on ecological issues (Antolin-Lopez et al., 2019). Building on this, sustainable entrepreneurship can be described as the process by which individuals identify, develop, and exploit opportunities to create future products and services that provide economic, social, and environmental value (Cohen & Winn, 2007). The core of entrepreneurship, including its sustainable form, lies in recognizing, developing, and capitalizing on such opportunities.

The benefits in economic, social, and ecological terms correspond to the triple bottom-line framework, which requires balancing these three aspects, even when there are conflicts, uncertainties, or trade-offs (Dyllick & Hockerts, 2002; Hahn et al., 2010). This balancing act highlights the inherent complexity of sustainable entrepreneurship, as it involves managing competing demands to achieve long-term sustainability in all three areas.

The expanding literature demonstrates a growing appreciation for the role of sustainable entrepreneurship in driving systemic change. However, the absence of a clear, standardized definition reflects the ongoing development of this field. Continued research is necessary to clarify its conceptual boundaries and to enhance understanding of how sustainable entrepreneurs support long-term ecological and social resilience while sustaining economic performance.

5 Intersection of Sustainable Entrepreneurship with Social Entrepreneurship and Eco Entrepreneurship

5.1 Social Entrepreneurship

Social entrepreneurship is fundamentally oriented toward advancing social equity and enhancing community welfare while also emphasizing ecological sustainability. Although the concept has gained traction in both academic discourse and practical business applications, scholarly debates persist regarding its precise definition. This lack of consensus is not unique to social entrepreneurship but mirrors broader definitional ambiguities within the general entrepreneurship literature (Peredo & McLean, 2006). As Venkatarman

(1997) observed, there are fundamentally divergent interpretations of the entrepreneur and the entrepreneurial function, making it unlikely that a universally accepted definition of the field will ever emerge.

To clarify the concept, Austin et al., (2006) proposed a distinction between two forms of entrepreneurship. Commercial entrepreneurship, in their framework, involves the identification, assessment, and pursuit of opportunities that generate financial profit. In contrast, social entrepreneurship focuses on identifying and leveraging opportunities that create societal benefits rather than private wealth. The ability to perceive and act upon such opportunities hinges on an entrepreneur's awareness of unmet needs in either supply or demand for value-generating products or services (Kirzner, 1973). Social entrepreneurs exhibit a heightened sensitivity to societal gaps and respond by devising innovative organizational solutions to address these gaps. This emphasis on social value creation is a common thread among various definitions of social entrepreneurship (Peredo & McLean, 2006; Shaw & Carter, 2007). Apart from this key distinction—prioritizing social impact over financial gain the definitions of commercial and social entrepreneurship share many similarities. These parallels support the (Dees, 1998) assertion that "social entrepreneurs represent one variant within the broader category of entrepreneurs."

Entrepreneurs operating in different environments encounter distinct sustainability challenges, shaped by their varying levels of engagement with external stakeholders to secure both tangible and intangible resources necessary for competitive survival. Sustainable entrepreneurship is characterized by its commitment to a triple bottom line, which integrates economic, social, and ecological objectives. Social entrepreneurship, by comparison, typically pursues a dual bottom line, balancing social and financial goals. The additional focus on environmental sustainability sets sustainable entrepreneurship apart from its social counterpart (Thompson et al., 2011). The pursuit of multiple, sometimes competing, objectives introduces greater complexity into the entrepreneurial process, influencing how opportunities are identified, developed, and exploited within the framework of sustainable development.

Furthermore, the term "sustainable entrepreneurship" originates from the broader concept of sustainable development, which the World Commission on Environment and Development (WCED, 1987, p. 43) defined as "development that fulfills present needs without compromising the ability of future generations to meet their own."

Austin offers a more formal definition of social entrepreneurship as "innovative, social value-creating activity that can occur within or across the nonprofit, business, or government sectors" (Austin et al., 2006). Two critical aspects of this definition warrant attention. First, it explicitly underscores the importance of innovation. Social entrepreneurship inherently involves applying novel technologies or methodologies to generate social value, aligning with the Schumpeterian perspective that positions innovation as central to entrepreneurial activity. In this view, social entrepreneurs function as agents of social innovation (Casson, 2005). Reinforces this notion by describing social entrepreneurs as (Dees, 1998) "change agents in the social sector who continually innovate, adapt, and learn." Second, the definition acknowledges the diverse institutional settings in which social entrepreneurship may manifest, whether through individual actors, newly established or pre-existing organizations (nonprofit or for-profit), or governmental initiatives.

5.2 Eco Entrepreneurship

Entrepreneurship is widely recognized as a multifaceted process encompassing the identification, development, and utilization of opportunities (Bygrave, 2004; Venkataraman, 1997). This concept is elaborated on by Venkataraman, who describes it as encompassing "all the functions, activities, and actions related to recognizing opportunities and establishing organizations to pursue them." This perspective highlights that entrepreneurial behavior is fundamentally driven by the recognition of opportunities, which subsequently motivates individuals to take action. Thus, entrepreneurship is inherently action-oriented, involving a dynamic interplay of creative, strategic, and organizational efforts (Moroz & Hindle, 2012). Entrepreneurs are characterized by their proactive, innovative, and risk-taking behaviors (Runyan et al., 2008). Importantly, such proactive and creative actions are not exclusive to new ventures but can also occur within established small and medium-sized enterprises (SMEs).

Building on this understanding, we define ecologically sustainable entrepreneurship as the systematic process of identifying, evaluating, and leveraging entrepreneurial opportunities that minimize a venture's ecological footprint, thereby generating societal and local community benefits. Eco-entrepreneurship is particularly noteworthy because it merges seemingly contradictory logics—market-driven objectives and social activism, offering entrepreneurial activists a means to promote ecological causes without relying on large bureaucratic institutions or government support (Marens, 2008). Consequently, eco-entrepreneurship represents a significant yet under-researched dimension of the environmental movement, one that emphasizes market-based strategies and neoliberal policies rather than confrontational activism against governmental and corporate structures.

This shift reflects the broader normalization of environmental consciousness over recent decades, which has facilitated the adoption of once-controversial practices such as recycling (Lounsbury et al., 2003) and the integration of corporate social and environmental responsibility initiatives (Hoffman, 1997). Our objective is to enrich the discourse on eco-entrepreneurship by offering nuanced perspectives that may inspire deeper academic explorations related to sustainable entrepreneurship. This approach broadens the conceptualization of entrepreneurship and invites critical reflection on how market mechanisms and activist principles can coexist within the evolving landscape of sustainable business practices (Mallika, 2024). By doing so, we hope to contribute to a growing body of research that examines the transformative potential of eco-entrepreneurship in fostering both economic and environmental value.

5.3 Hybrid Entrepreneurship

Sustainable entrepreneurship involves the concurrent pursuit of social, environmental, and economic objectives, frequently facilitated by innovative business structures such as Benefit Corporations (B Corps). Scholars have established that, much like social entrepreneurs, sustainable entrepreneurs are motivated by multiple goals beyond mere profit maximization—a departure from traditional economic theories of entrepreneurship that emphasize rent-seeking behavior (De Clercq & Voronov, 2011). However, while there are overlaps between social and sustainable entrepreneurship, key distinctions remain. This form of entrepreneurship fosters the creation of economically viable enterprises that simultaneously promote environmental sustainability and social equity (Muñoz & Dimov, 2015). According to researchers (Doherty et al., 2014), this blending of objectives is a defining characteristic of emerging entrepreneurial models. They describe hybrid organizations as entities that incorporate values and operational elements from multiple, often competing, domains (Niesten & Jolink, 2015). This dual focus, balancing ecological concerns with market demands, is further emphasized in the business models adopted by sustainable entrepreneurs, who navigate the tension between planetary well-being and profitability.

Sustainability-driven entrepreneurs (Parrish, 2010) leverage innovation not only to develop market-competitive products and services but also to embed social and environmental considerations at the heart of their business strategies (Schaltegger & Wagner, 2011). Benefit Corporations serve as a prime illustration of enterprises that harmonize financial success with measurable societal and ecological benefits. By introducing novel organizational frameworks, sustainable entrepreneurs significantly mitigate adverse environmental and social externalities (Lee &

Fu, 2024) while reconciling the often conflicting demands of ecological responsibility, social ethics, and economic viability (Parrish, 2010). These hybrid ventures (Belz & Binder, 2017; McMullen & Warnick, 2016) merge market-driven strategies with socially conscious imperatives, embedding sustainability into their core value-creation processes.

This synthesis of market logic and social purpose enables sustainable entrepreneurs to redefine traditional business paradigms, ensuring that environmental stewardship and social justice are not peripheral concerns but foundational to organizational success.

6 Sustainability Innovation

Role of Innovation in Sustainable Entrepreneurship

Sustainability innovation refers to the development and implementation of new products, processes, business models, or organizational approaches that create value while addressing environmental and social challenges. This type of innovation seeks to strike a balance between economic growth and the preservation of natural resources, as well as the well-being of society, aligning closely with the United Nations' Sustainable Development Goals (SDGs) (Surana et al., 2020). Sustainable entrepreneurship is fundamentally connected to advancing sustainability innovation, which entails creating products, processes, and services designed to tackle pressing environmental and societal issues (Krara et al., 2025). Innovation for sustainability frequently depends on collaboration among diverse stakeholders, including businesses, governments, academia, and civil society. Such partnerships foster knowledge sharing, co-creation, and the scaling of sustainable solutions. However, research indicates that academic-corporate collaboration is still limited and often sector-specific, suggesting room for growth in cross-sector partnerships (Khan & Majava, 2021). To successfully promote sustainable innovation, businesses must overcome the innovation-adoption gap—a disconnect between the rapid evolution of eco-friendly technologies and an organization's ability to effectively adopt and utilize them (Slaper & Hall, 2011). By closing this gap, companies can not only meet their sustainability goals but also strengthen their market position in an economy where environmental consciousness is becoming a decisive competitive factor (Slaper & Hall, 2011). As the pace of technological progress intensifies, organizations must place greater emphasis on seamlessly incorporating these innovations into their operations to maintain relevance and market leadership.

At its core, sustainable entrepreneurship involves the development and large-scale implementation of sustainability innovations that deliver widespread societal benefits. Sustainable entrepreneurs often introduce radical innovations that address the unmet needs of a broader range of stakeholders (Ayandibu & Ayandibu, 2024). Stakeholders, defined as individuals or groups who influence or are impacted by a firm's activities (Freeman, 1984), extend beyond shareholders to include diverse societal and environmental interests. These stakeholder demands, which surpass purely financial considerations, serve as key drivers of entrepreneurial opportunities in sustainability innovation (Figge et al., 2002). The identification and pursuit of these opportunities lie at the heart of sustainable entrepreneurship (Dean & McMullen, 2007). This perspective aligns with contemporary research suggesting that market failures, such as inefficiencies in addressing social and environmental challenges, are primary motivators for entrepreneurial initiatives focused on societal betterment and ecological progress (Srivastava et al., 2025).

The intersection of sustainable entrepreneurship and sustainability innovation is essential to examine because their underlying principles differ significantly. In the context of innovation, the key considerations revolve around private and social benefits, which can be linked to a structured progression of product and process innovations, the availability of complementary assets, and strategies to safeguard innovations from unintended knowledge leakage (Teece, 1986; Utterback, 1994; Utterback & Abernathy, 1975). In contrast, entrepreneurship is primarily guided by the logic of opportunity recognition and exploitation (Shane, 2000, 2003), with innovativeness essentially being an individual-level trait (Kuckertz & Wagner, 2010). Thus, while sustainability innovation emphasizes systemic technological and procedural advancements, sustainable entrepreneurship centers on the entrepreneurial vision to identify and capitalize on opportunities that align with broader societal and environmental imperatives.

7 Barriers and Challenges in Sustainable Entrepreneurship

Sustainable entrepreneurs aim to balance environmental, social, and economic impacts within their businesses, as demonstrated by their commitment to achieving triple-bottom-line objectives (Schaltegger & Wagner, 2011). They possess the innovative capacity to both recognize and realize opportunities that promote sustainability, thereby fostering sustainable socioecological systems (Muñoz & Cohen, 2018b). In this study, sustainable entrepreneurship refers to entrepreneurial endeavors driven primarily by the goal of significantly advancing sustainable development. Entrepreneurs focused on sustainability create ventures mainly aimed at enhancing both environmental quality and social well-being, ensuring that these objectives support and reinforce each other (Parrish & Foxon, 2006). Although sustainable entrepreneurship holds promise for addressing social, ecological, and economic issues, studies indicate that its implementation is challenging and faces various obstacles. To promote sustainable entrepreneurship, scholars investigate the challenges that sustainable entrepreneurs face and the strategies they use to overcome them. Research into the sustainable entrepreneurship process has highlighted several barriers (Hoogendoorn et al., 2019). While these obstacles are often discussed in the context of launching sustainable businesses, they can also persist throughout the later stages of a company's development. Barriers are typically grouped into market entry barriers (Pinkse & Groot, 2015), institutional barriers (Pacheco et al., 2010), and stakeholder management barriers (Nguyen et al., 2020). Market entry barriers relate to entering markets dominated by established players with new sustainable products or services (Schaltegger & Wagner, 2011), targeting customers who may be unwilling to pay higher prices for sustainable options, or encouraging shifts towards sustainable practices (Hoogendoorn et al., 2019). Institutional barriers necessitate the development of updated regulations, policies, and industry standards to facilitate sustainable solutions for users (Pacheco et al., 2010).

7.1 Institutional and Market Barriers

Sustainable entrepreneurs—individuals who establish businesses to pursue both personal and societal objectives by tackling unmet social and environmental needs—encounter distinct challenges compared to conventional entrepreneurs. These difficulties stem from the inherent tension between generating private value (financial returns) and social value (community and environmental benefits). Research indicates that sustainable entrepreneurs face three primary obstacles: (1) constrained financial resources and limited access to initial funding, (2) intricate regulatory and bureaucratic systems, and (3) inadequate informational guidance for sustainability-focused initiatives (Dean & McMullen, 2007; Mair & Marti, 2006; Pacheco et al., 2010).

Achieving meaningful sustainability demands coordinated efforts across all societal actors. The rise of service-based business models has accelerated the adoption of sustainable entrepreneurship, yet entrenched institutional frameworks continue to favor traditional, less sustainable industries (Stabler et al., 2024). Consequently, sustainable entrepreneurs (or "ecopreneurs") must often act as catalysts for systemic change, challenging and reshaping existing economic paradigms.

A defining feature of sustainable entrepreneurship is the need to reconcile conflicting priorities, including financial viability, ecological preservation, and social justice. This dual focus complicates decision-making, as sustainable entrepreneurs must align profit-driven strategies with broader ethical and environmental commitments (Alabi, 2024). The additional layers of accountability—to both shareholders and society—intensify the complexity of managing such enterprises. Thus, while traditional entrepreneurs primarily navigate market risks, sustainable entrepreneurs must also mitigate institutional resistance, stakeholder skepticism, and the challenges of measuring social impact alongside financial performance.

7.1.1 High Costs and Limited Access to Finance

Sustainable entrepreneurs often face significant financial barriers, including higher upfront costs for sustainable technologies and limited access to funding specifically tailored for green ventures. Many investors remain hesitant to support sustainability-focused startups due to perceived risks and longer return horizons. Small and medium-sized enterprises (SMEs), in particular, often lack the financial capacity to invest in research and development for sustainability initiatives (Burch et al., 2022).

7.1.2 Short-Termism in Business Planning

A focus on short-term profits and survival, particularly among small and medium-sized enterprises (SMEs), can hinder the adoption of long-term sustainability strategies. This short-term outlook is often reinforced by hierarchical management cultures and a lack of time or skills dedicated to preparing for environmental and social challenges (Burch et al., 2022).

7.1.3 Lack of Influence Over Policy and Regulatory Frameworks

Sustainable entrepreneurs frequently lack the power to influence or change existing regulations and market rules that favor traditional, less sustainable business models. As a result, they must collaborate with NGOs, policymakers, and other stakeholders to advocate for policy changes, which can be a slow and uncertain process (Veleva, 2021).

7.2 Risk Perception and Unique Challenges

Sustainable entrepreneurs tend to perceive greater personal risks, including reputational damage and uncertainties related to stakeholders. Research indicates that while sustainable and conventional entrepreneurs share similar risk perceptions regarding financial uncertainties, sustainable entrepreneurs exhibit greater apprehension toward personal failure (Hoogendoorn et al., 2019). This distinction arises from their broader accountability to diverse stakeholders, including environmental and social interests. Such findings hold critical implications for policymakers and investors, underscoring the need for institutional reforms that encourage sustainable ventures.

Unlike traditional entrepreneurs, they must often confront and reshape existing institutional frameworks, including policies, societal norms, and legal structures (Dean & McMullen, 2007; Hockerts & Wüstenhagen, 2010). Consequently, they are likely to hold more pessimistic views regarding the availability of financial, administrative, and informational support. While all entrepreneurs assume risks, the nature of these risks varies depending on the type of venture they are involved in. Prior studies suggest that social and sustainable entrepreneurs are particularly concerned with non-financial risks, such as diminished local credibility or the erosion of personal networks (Block et al., 2015; Shaw & Carter, 2007). However, empirical research comparing the risk perceptions of sustainable and conventional entrepreneurs remains limited. However, empirical research comparing the risk perceptions of sustainable and conventional entrepreneurs remains limited. Addressing this gap is essential for advancing our understanding of how risk perception shapes entrepreneurial decision-making and outcomes in sustainability-oriented ventures. Future research should systematically investigate the cognitive and affective processes underlying risk assessment among different types of entrepreneurs, considering not only financial risks but also the broader spectrum of social, reputational, and personal risks that are particularly salient in sustainable entrepreneurship (Bregman, 2020; L'hocine Walid, 2022).

Moreover, as sustainable entrepreneurs often operate in complex institutional environments and face unique challenges, such as balancing the triple bottom line and navigating stakeholder expectations, scholars should explore how these contextual factors influence risk tolerance and opportunity recognition (Greco & De Jong, 2017).

Finally, there is a pressing need to develop standardized frameworks and measurement tools for assessing risk perception in the context of sustainable entrepreneurship. Such advancements would not only enhance the comparability of research findings but also inform policy interventions and support mechanisms tailored to the distinctive needs of sustainable entrepreneurs (Aghelie et al., 2016). By addressing these research gaps, future scholarship can contribute to the development of more resilient, impactful, and sustainable entrepreneurial ecosystems.

7.2.1 Heightened Perception of Financial and Market Risks

Sustainable entrepreneurs often perceive greater financial risks compared to conventional ventures. This is due to the higher initial costs of adopting sustainable technologies and practices, combined with uncertainty about market demand for green products or services. Many customers are unwilling to pay premium prices for sustainable alternatives, and established competitors dominate the market, making entry into the market more difficult for new, sustainability-focused businesses. These factors increase the perceived risk of business failure and can deter potential entrepreneurs from pursuing sustainable ventures (Karani & Mshenga, 2021).

7.2.2 Uncertainty Due to Evolving Regulations and Policy Environments

Sustainable entrepreneurs face unique challenges stemming from the rapidly changing landscape of environmental regulations and industry standards. The lack of clear, consistent, and supportive policies creates uncertainty, making it difficult for entrepreneurs to plan long-term investments in sustainability. This regulatory unpredictability not only raises perceived risk but also hampers the ability to secure funding and stakeholder support for sustainable initiatives. As a result, entrepreneurs may be reluctant to innovate or scale their sustainable business models (Odeyemi et al., 2024).

7.2.3 Slow Adoption and Implementation of Sustainability Practices

Compared to larger corporations, smaller enterprises tend to be slower in developing and implementing ambitious sustainability interventions. This lag is due to constraints in resources, expertise, and the absence of supportive ecosystems that encourage experimentation and capacity-building for sustainability.

These challenges underscore the necessity for supportive policy frameworks, targeted financial instruments, and collaborative ecosystems to foster sustainable entrepreneurship in diverse economic contexts (Burch et al., 2022).

8 Conclusion

The advancement of an emerging research field, such as sustainable entrepreneurship, cannot rely exclusively on building upon pre-existing knowledge. Instead, it is crucial to actively seek opportunities to expand and diversify its theoretical foundations. Sustainable entrepreneurship presents a distinctive platform for bridging entrepreneurship and sustainability theory—two disciplines that themselves draw upon a wide array of concepts and perspectives from other academic domains (Markman et al., 2016). To foster the growth of sustainable entrepreneurship, scholars from diverse backgrounds should intensify their efforts to contribute novel insights from their respective fields. Furthermore, the development of an emerging discipline requires not only the construction of knowledge but also social engagement, ensuring that not just ideas converge but that researchers themselves interact and collaborate. To facilitate this, several strategies can be implemented: (1) expanding the number of academic and professional gatherings (such as conferences, specialized tracks, and workshops) centered on sustainable entrepreneurship; (2) promoting the formation of interdisciplinary research teams when exploring SE-related topics; (3) incorporating theories and methodologies from other disciplines while simultaneously disseminating SE knowledge to broader academic and practitioner communities; and (4) sustaining ongoing scholarly discourse on the evolution of sustainable entrepreneurship.

As Birkeland (Birkeland, 2022) asserts, a key tenet of sustainability thinking is that true sustainability extends beyond mere eco-efficiency—reducing energy consumption, resource use, and waste—and instead necessitates profound transformations at the individual, societal, and institutional levels. Ecopreneurial ventures and their founders may serve as catalysts for such systemic change (Ridderstråle & Engström, 2000). Further, it is suggested that, even within a business framework, individuals must ultimately take ownership of their autonomy, and with that freedom comes accountability, including responsibility for the ethical dimensions of their professional and personal lives. This insight is particularly relevant to ecopreneurs, who invest not just capital but their core values and identities into their enterprises. These businesses are not merely profit-driven entities but manifestations of personal conviction, reflecting deeply held ethical principles. Consequently, they possess a resilience and influence that extends beyond conventional commercial enterprises. In an era marked by volatility, Friend (2009) argues that sustainability offers a stabilizing force—a means of navigating uncertainty. If this holds true, ecopreneurial businesses are uniquely positioned to thrive. Yet, as (Woolf, 1928) poignantly observed, the relentless pursuit of wealth for its own sake is ultimately hollow when basic well-being can be secured through far simpler means. Ecopreneurs, by contrast, appear to recognize that genuine fulfillment arises from aligning economic activity with broader societal and environmental imperatives.

Moving forward, research in sustainable entrepreneurship must place greater emphasis on harmonizing economic, social, and environmental objectives. As the paradigm shifts from narrow profit maximization toward a more comprehensive triple-bottom-line framework, scholars must explore how businesses can effectively reconcile these often-competing priorities (Kareem et al., 2025). This transition may catalyze the emergence of innovative business models that not only advance sustainability but also foster economic resilience and social equity.

References

Abbas, M. H., & Bulut, M. (2024). Navigating the landscape of sustainable entrepreneurship research: A systematic literature review. *Discover Sustainability*. https://api.semanticscholar.org/CorpusID:271581875

Aghelie, A., Sorooshian, S., & Azizan, N. A. (2016). Research gap in sustainopreneurship. *Indian Journal of Science and Technology, 9*(12), 1–6.

Alabi, M. (2024). *Social entrepreneurship: Balancing social impact and financial sustainability*.

Alkathiri, N. A., Said, F. Ben, Meyer, N., & Soliman, M. (2024). Knowledge management and sustainable entrepreneurship: a bibliometric overview and research agenda. *Journal of Innovation and Entrepreneurship, 13*, 1–24. https://api.semanticscholar.org/CorpusID:270555218

Amit, R., MacCrimmon, K. R., Zietsma, C., & Oesch, J. M. (2001). Does money matter?: Wealth attainment as the motive for initiating growth-oriented technology ventures. *Journal of Business Venturing, 16*(2), 119–143.

Anderson, A. R. (1998). Cultivating the Garden of Eden: Environmental entrepreneuring. *Journal of Organizational Change Management, 11*(2), 135–144.

Antolin-Lopez, R., Martinez-del-Rio, J., & Cespedes-Lorente, J. J. (2019). Environmental entrepreneurship as a multi-component and dynamic construct: Duality of goals, environmental agency, and environmental value creation. *Business Ethics: A European Review, 28*(4), 407–422.

Assessment, M. E. (2001). *Millennium ecosystem assessment* (Vol. 2). Millennium Ecosystem Assessment.

Austin, J., Stevenson, H., & Wei-Skillern, J. (2006). Social and commercial entrepreneurship: Same, different, or both? *Entrepreneurship Theory and Practice, 30*(1), 1–22.

Ayandibu, A. O., & Ayandibu, E. O. (2024). The role of innovation in sustainable entrepreneurship. In

Sustainable Development in Industry and Society 5.0: Governance, Management, and Financial Implications (pp. 205–221). IGI Global.

Belz, F. M., & Binder, J. K. (2017). Sustainable entrepreneurship: A convergent process model. *Business Strategy and the Environment, 26*(1), 1–17. https://doi.org/10.1002/bse.1887

Binder, J. K., & Belz, F.-M. (2015). Sustainable entrepreneurship: what it is. In *Handbook of entrepreneurship and sustainable development research* (pp. 30–72). Edward Elgar Publishing.

Birkeland, J. (2022). Nature positive: Interrogating sustainable design frameworks for their potential to deliver eco-positive outcomes. *Urban Science, 6*(2), 35.

Block, J. H., Kohn, K., Miller, D., & Ullrich, K. (2015). Necessity entrepreneurship and competitive strategy. *Small Business Economics, 44*, 37–54.

Bregman, M. (2020). *Perceptual variables as drivers of sustainable entrepreneurship: A comparative Study.*

Buchholz, R. A., & Rosenthal, S. B. (2005). Toward a contemporary conceptual framework for stakeholder theory. *Journal of Business Ethics, 58*, 137–148.

Burch, S., DiBella, J., Wiek, A., Schaltegger, S., Stubbs, W., Farrelly, M., Ness, B., & McCormick, K. (2022). Building urban resilience through sustainability-oriented small- and medium-sized enterprises. *Urban Transformations, 4*(1), 12. https://doi.org/10.1186/s42854-022-00041-9

Bygrave, W. D. (2004). The entrepreneurial process. *The Portable MBA in Entrepreneurship, 2*, 1–26.

Caradonna, J. L. (2022). *Sustainability: A history.* Oxford University Press.

Carle, A., & Rayna, T. (2024). Exploring the coevolution of entrepreneurial ecosystems and institutional entrepreneurship: The case of sustainable entrepreneurship. *Creativity and Innovation Management*. https://api.semanticscholar.org/CorpusID:272788765

Casson, M. (2005). Entrepreneurship and the theory of the firm. *Journal of Economic Behavior & Organization, 58*(2), 327–348.

Cochran, P. L. (2007). The evolution of corporate social responsibility. *Business Horizons, 50*(6), 449–454.

Cohen, B., & Winn, M. I. (2007). Market imperfections, opportunity and sustainable entrepreneurship. *Journal of Business Venturing, 22*(1), 29–49.

Common, M., & Stagl, S. (2005). *Ecological economics: An introduction.* Cambridge University Press.

Connelly, S. (2007). Mapping sustainable development as a contested concept. *Local Environment, 12*(3), 259–278.

Corner, P. D., & Ho, M. (2010). How opportunities develop in social entrepreneurship. *Entrepreneurship Theory and Practice, 34*(4), 635–659.

Crals, E., & Vereeck, L. (2005). The affordability of sustainable entrepreneurship certification for SMEs. *The International Journal of Sustainable Development & World Ecology, 12*(2), 173–183.

Dasht Peyma, N. (2025). Themes in Eco-Literature. In *Eco-Literature and Environmental Advocacy: Bridging Theory and Practice* (pp. 45–94). Springer.

Davidsson, P., & Wiklund, J. (2001). Levels of analysis in entrepreneurship research: Current research practice and suggestions for the future. *Entrepreneurship Theory and Practice, 25*(4), 81–100.

De Clercq, D., & Voronov, M. (2011). Entrepreneurial legitimacy as a set of discursive practices. In *Second Conference on Rhetoric and Narratives in Management Research: Management and Persuasion. Barcelona: ESADE* (pp. 125–146).

Dean, T. J., & McMullen, J. S. (2007). Toward a theory of sustainable entrepreneurship: Reducing environmental degradation through entrepreneurial action. *Journal of Business Venturing, 22*(1), 50–76.

Dees, J. G. (1998). Enterprising nonprofits: What do you do when traditional sources of funding fall short. *Harvard Business Review, 76*(1), 55–67.

Defourny, J., & Nyssens, M. (2006). Defining social enterprise. *Social Enterprise: At the Crossroads of Market, Public Policies and Civil Society, 7*, 3–27.

Diepolder, C. S., Weitzel, H., & Huwer, J. (2021). Competence frameworks of sustainable entrepreneurship: A systematic review. *Sustainability (Switzerland), 13*(24). https://doi.org/10.3390/su132413734

Dixon, S. E. A., & Clifford, A. (2007). Ecopreneurship—A new approach to managing the triple bottom line. *Journal of Organizational Change Management, 20*(3), 326–345.

Dobson, A. (1999). *Fairness and futurity: Essays on environmental sustainability and social justice.* OUP Oxford.

Doherty, B., Haugh, H., & Lyon, F. (2014). Social enterprises as hybrid organizations: A review and research agenda. *International Journal of Management Reviews, 16*(4), 417–436.

Dorfman, R. (1993). *Some concepts from welfare economics.*

Dyllick, T., & Hockerts, K. (2002). Beyond the business case for corporate sustainability. *Business Strategy and the Environment, 11*(2), 130–141.

Ebabu, E. A., Yu, H., Zou, W., & Gong, H. (2025). Economic and social impacts of sustainable entrepreneurship: Systematic review and bibliometric study of global evidence. *Sustainable Development.* https://api.semanticscholar.org/CorpusID:277294410

Elkington, J. (1997). The triple bottom line. *Environmental Management: Readings and Cases, 2*, 49–66.

Elkington, J., & Hartigan, P. (2008). *The power of unreasonable people: How social entrepreneurs create markets that change the world.* Harvard Business Press.

Figge, F., Hahn, T., Schaltegger, S., & Wagner, M. (2002). The sustainability balanced scorecard–linking sustainability management to business strategy. *Business Strategy and the Environment, 11*(5), 269–284.

Freeman, R. E. (1984). Strategic management: A stakeholder theory. *Journal of Management Studies, 39*(1), 1–21.

Friend, G. (2009). *The truth about green business.* Que Publishing.

Gartner, W. B. (1988). "Who is an entrepreneur?" Is the wrong question. *American Journal of Small Business, 12*(4), 11–32.

Gibbs, D. (2006). Sustainability entrepreneurs, ecopreneurs and the development of a sustainable economy. *Greener Management International, 55*, 63–78.

Gieryn, T. F. (1999). *Cultural boundaries of science: Credibility on the line*. University of Chicago Press.

Gorissen, K., Weijters, B., & Deltomme, B. (2024). Green versus grey framing: Exploring the mechanism behind the negative footprint illusion in environmental sustainability assessments. *Sustainability, 16*(4), 1411.

Greco, A., & De Jong, G. (2017). *Sustainable entrepreneurship: Definitions, themes and research gaps*.

Hahn, T., Figge, F., Pinkse, J., & Preuss, L. (2010). Trade-offs in corporate sustainability: You can't have your cake and eat it. *Business Strategy and the Environment, 19*(4), 217–229.

Hajian, M., & Kashani, S. J. (2021). Evolution of the concept of sustainability. From Brundtland Report to sustainable development goals. In *Sustainable resource management* (pp. 1–24). Elsevier.

Henderson, K., & Loreau, M. (2023). A model of sustainable development goals: challenges and opportunities in promoting human well-being and environmental sustainability. *Ecological Modelling, 475*, Article 110164.

Hockerts, K., & Wüstenhagen, R. (2010). Greening Goliaths versus emerging Davids—Theorizing about the role of incumbents and new entrants in sustainable entrepreneurship. *Journal of Business Venturing, 25*(5), 481–492.

Hoffman, J. (1997). Can we define sovereignty? *Politics, 17*(1), 53–58. https://doi.org/10.1111/1467-9256.00034

Holling, C. S. (1973). *Resilience and stability of ecological systems*. International Institute for Applied Systems Analysis Laxenburg.

Hoogendoorn, B., Van der Zwan, P., & Thurik, R. (2019). Sustainable entrepreneurship: The role of perceived barriers and risk. *Journal of Business Ethics, 157*, 1133–1154.

Johnson, S. P. (2001). *The Earth Summit: The United Nations Conference on Environment and Development (UNCED)*.

Jolink, A., & Niesten, E. (2015). Sustainable development and business models of entrepreneurs in the organic food industry. *Business Strategy and the Environment, 24*(6), 386–401.

Karani, C., & Mshenga, P. (2021). Steering the sustainability of entrepreneurial start-ups. *Journal of Global Entrepreneurship Research,* 11(1), 223–239. https://doi.org/10.1007/s40497-021-00279-w

Kareem, S., Fehrer, J. A., Shalpegin, T., & Stringer, C. (2025). Navigating tensions of sustainable supply chains in times of multiple crises: A systematic literature review. *Business Strategy and the Environment, 34*(1), 316–337.

Kates, R. W., Clark, W. C., Corell, R., Hall, J. M., Jaeger, C. C., Lowe, I., McCarthy, J. J., Schellnhuber, H. J., Bolin, B., & Dickson, N. M. (2001). Sustainability science. *Science, 292*(5517), 641–642.

Kerlin, J. A. (2009). *Social enterprise: A global comparison*. UPNE.

Khan, I. S., & Majava, J. (2021). Stakeholders interdependencies and their role in sustainable business model innovation. ArXiv Preprint arXiv: 2103.16281.

Kidd, C. V. (1992). The evolution of sustainability. *Journal of Agricultural and Environmental Ethics, 5*, 1–26.

Kirzner, I. M. (1973). The entrepreneur. *Competition and Entrepreneurship*, 30–87.

Kirzner, I. M. (2015). *Competition and entrepreneurship*. University of Chicago Press.

Krara, W., Alzubi, A., Khadem, A., & Iyiola, K. (2025). The nexus of sustainability innovation, knowledge application, and entrepreneurial success: Exploring the role of environmental awareness. *Sustainability, 17*(2), 716.

Kraus, M. W., Park, J. W., & Tan, J. J. X. (2017). Signs of social class: The experience of economic inequality in everyday life. *Perspectives on Psychological Science, 12*(3), 422–435.

Kuckertz, A., & Wagner, M. (2010). The influence of sustainability orientation on entrepreneurial intentions—Investigating the role of business experience. *Journal of Business Venturing, 25*(5), 524–539.

Kury, K. W. (2012). Sustainability meets social entrepreneurship: A path to social change through institutional entrepreneurship. *International Journal of Business Insights & Transformation, 4*.

Lans, T., Blok, V., & Wesselink, R. (2014). Learning apart and together: Towards an integrated competence framework for sustainable entrepreneurship in higher education. *Journal of Cleaner Production, 62*, 37–47.

Larson, A., & Johnson, R. R. (2008). *An overview of the historical context for sustainable business in the United States, 1960–2000*.

Larsson, G., Maire, M., & Shakhnarovich, G. (2016). Learning representations for automatic colorization. In *Computer Vision–ECCV 2016: 14th European Conference, Amsterdam, The Netherlands, October 11–14, 2016, Proceedings, Part IV 14* (pp. 577–593).

Lee, C.-W., & Fu, M.-W. (2024). Conceptualizing sustainable business models aligning with corporate responsibility. *Sustainability, 16*(12), 5015.

L'hocine Walid, H. P. (2022). Entrepreneurial risk perception and sustainable entrepreneurship intention among SMEs in Algeria: A multidimensional approach. *Development, 1*(2), 7–15.

Linna, T. (2020). Business sustainability and insolvency proceedings—the EU perspective. *Journal of Sustainability Research, 2*(2), Article e200019.

Linnanen, L. (2002). An insider's experiences with environmental entrepreneurship. *Greener Management International*, 71–80. https://api.semanticscholar.org/CorpusID:154840397

Lounsbury, M., Ventresca, M., & Hirsch, P. M. (2003). Social movements, field frames and industry emer-

gence: A cultural–political perspective on US recycling. *Socio-Economic Review, 1*(1), 71–104.
Mair, J., & Marti, I. (2006). Social entrepreneurship research: A source of explanation, prediction, and delight. *Journal of World Business, 41*(1), 36–44.
Mallika, D. (2024). The role of social entrepreneurship in promoting environmental sustainability: A green marketing perspective. *Educational Administration: Theory and Practice, 30*(6), 1658–1665.
Marens, R. (2008). Recovering the past: Reviving the legacy of the early scholars of corporate social responsibility. *Journal of Management History, 14*(1), 55–72.
Markman, G. D., Russo, M., Lumpkin, G. T., Jennings, P. D., & Mair, J. (2016). Entrepreneurship as a platform for pursuing multiple goals: A special issue on sustainability, ethics, and entrepreneurship. *Journal of Management Studies, 53*(5), 673–694.
May, R. M. (1977). Togetherness among schistosomes: Its effects on the dynamics of the infection. *Mathematical Biosciences, 35*(3–4), 301–343.
McMullen, J. S., & Warnick, B. J. (2016). Should we require every new venture to be a hybrid organization? *Journal of Management Studies, 53*(4), 630–662.
Miah, M. T., Lakner, Z., & Fekete-Farkas, M. (2024). Addressing poverty through social entrepreneurship for sustainable development: A comprehensive bibliometric analysis. *Administrative Sciences, 14*(1), 16.
Mitchell, R. K., Busenitz, L., Lant, T., McDougall, P. P., Morse, E. A., & Smith, J. B. (2002). Toward a theory of entrepreneurial cognition: Rethinking the people side of entrepreneurship research. *Entrepreneurship Theory and Practice, 27*(2), 93–104.
Moroz, P. W., & Hindle, K. (2012). Entrepreneurship as a process: Toward harmonizing multiple perspectives. *Entrepreneurship Theory and Practice, 36*(4), 781–818.
Muñoz, P., & Cohen, B. (2018a). Sustainable entrepreneurship research: Taking stock and looking ahead. *Business Strategy and the Environment, 27*(3), 300–322. https://doi.org/10.1002/bse.2000
Muñoz, P., & Cohen, B. (2018b). Sustainable entrepreneurship research: Taking stock and looking ahead. *Business Strategy and the Environment, 27*(3), 300–322. https://doi.org/10.1002/bse.2000
Muñoz, P., & Dimov, D. (2015). The call of the whole in understanding the development of sustainable ventures. *Journal of Business Venturing, 30*(4), 632–654.
Musa, I. O., Samuel, J. O., Adams, M., Abdulsalam, M., Nathaniel, V., Maude, A. M., Adedayo, O. A., & Tiamiyu, A. T. (2024). Soil erosion, mineral depletion and regeneration. In *Prospects for soil regeneration and its impact on environmental protection* (pp. 159–172). Springer.
Nguyen, Q., Nguyen, P. T., Huynh, V. D. B., & Nguyen, L. T. (2020). Application Chang's extent analysis method for ranking barriers in the e-learning model based on multi-stakeholder decision making. *Universal Journal of Educational Research, 8*(5), 1759–1766.
Nicholls, A. (2006). *Social entrepreneurship: New models of sustainable social change*. Oxford University Press.
Niesten, E., & Jolink, A. (2015). The impact of alliance management capabilities on alliance attributes and performance: A literature review. *International Journal of Management Reviews, 17*(1), 69–100.
Odeyemi, O., Usman, F. O., Mhlongo, N. Z., Elufioye, O. A., & Ike, C. U. (2024). Sustainable entrepreneurship: A review of green business practices and environmental impact. *World Journal of Advanced Research and Reviews, 21*(2), 346–358.
Olurina, J. O., Gidiagba, J. O., Ehiaguina, V. E., Ndiwe, T. C., Ayodeji, S. A., Banso, A. A., Tula, O. A., & Ojo, G. G. (2023). Engineering innovations and sustainable entrepreneurship: A comprehensive literature review. *Materials & Corrosion Engineering Management*. https://api.semanticscholar.org/CorpusID:267181929
O'Neill Gerald D, J., Hershauer, J. C., & Golden, J. S. (2006). The cultural context of sustainability entrepreneurship. *Greener Management International, 55*, 33–46.
Pacheco, D. F., Dean, T. J., & Payne, D. S. (2010). Escaping the green prison: Entrepreneurship and the creation of opportunities for sustainable development. *Journal of Business Venturing, 25*(5), 464–480.
Parkinson, C., & Howorth, C. (2008). The language of social entrepreneurs. *Entrepreneurship and Regional Development, 20*(3), 285–309.
Parrish, B. D. (2010). Sustainability-driven entrepreneurship: Principles of organization design. *Journal of Business Venturing, 25*(5), 510–523. https://doi.org/10.1016/j.jbusvent.2009.05.005
Parrish, B. D., & Foxon, T. J. (2006). Sustainability entrepreneurship and equitable transitions to a low-carbon economy. *Greener Management International, 55*, 47–62.
Pastakia, A. (1998). Grassroots ecopreneurs: Change agents for a sustainable society. *Journal of Organizational Change Management, 11*(2), 157–173.
Pearce, D. W., & Atkinson, G. D. (1993). Capital theory and the measurement of sustainable development: An indicator of "weak" sustainability. *Ecological Economics, 8*(2), 103–108. https://doi.org/10.1016/0921-8009(93)90039-9
Peredo, A. M., & McLean, M. (2006). Social entrepreneurship: A critical review of the concept. *Journal of World Business, 41*(1), 56–65.
Perrini, F., Vurro, C., & Costanzo, L. A. (2010). A process-based view of social entrepreneurship: From opportunity identification to scaling-up social change in the case of San Patrignano. *Entrepreneurship & Regional Development, 22*(6), 515–534.
Pigou, A. C. (1951). Some aspects of welfare economics. *The American Economic Review, 41*(3), 287–302.
Pigou, A. C. (2017). *The economics of welfare*. Routledge.
Pinkse, J., & Groot, K. (2015). Sustainable entrepreneurship and corporate political activity: Overcoming market barriers in the clean energy sector. *Entrepreneurship Theory and Practice, 39*(3), 633–654.
Ridderstråle, J., & Engström, P. (2000). *Toward a knowledge-based theory of organizational design*. Citeseer.

Robinson, J. A. (2006). Economic development and democracy. *Annual Review of Political Science, 9*(1), 503–527.

Rodgers, C. (2010). Sustainable entrepreneurship in SMEs: A case study analysis. *Corporate Social Responsibility and Environmental Management, 17*(3), 125–132.

Rosário, A. T., Raimundo, R. J., & Cruz, S. P. (2022). Sustainable entrepreneurship: A literature review. *Sustainability, 14*(9), 5556.

Runyan, R., Droge, C., & Swinney, J. (2008). Entrepreneurial orientation versus small business orientation: What are their relationships to firm performance? *Journal of Small Business Management, 46*(4), 567–588.

Sarango-Lalangui, P., Santos, J. L. S., & Hormiga, E. (2018). The development of sustainable entrepreneurship research field. *Sustainability, 10*(6), 2005.

Schaefer, L. M., Burke, N. L., Thompson, J. K., Dedrick, R. F., Heinberg, L. J., Calogero, R. M., Bardone-Cone, A. M., Higgins, M. K., Frederick, D. A., Kelly, M., Anderson, D. A., Schaumberg, K., Nerini, A., Stefanile, C., Dittmar, H., Clark, E., Adams, Z., Macwana, S., Klump, K. L., Swami, V., et al. (2015). Development and validation of the sociocultural attitudes towards appearance questionnaire-4 (SATAQ-4). *Psychological Assessment, 27*(1), 54–67. https://doi.org/10.1037/a0037917

Schaltegger, S., & Burritt, R. (2018). Business cases and corporate engagement with sustainability: Differentiating ethical motivations. *Journal of Business Ethics, 147*, 241–259.

Schaltegger, S., & Wagner, M. (2011). Sustainable entrepreneurship and sustainability innovation: Categories and interactions. *Business Strategy and the Environment, 20*(4), 222–237.

Schlange, L. E. (2006). What drives sustainable entrepreneurs? In *3rd Applied Business and Entrepreneurship Association International (ABEAI) Conference*. Kona, Hawaii.

Schmidheiny, S., & Timberlake, L. (1992). *Changing course: A global business perspective on development and the environment* (Vol. 1). MIT Press.

Schumpeter, J. A. (1942). Capital. In *Socialism, and democracy*. Harper.

Schumpeter, J. A., & Backhaus, U. (2003). Joseph Alois Schumpeter. *The European Heritage in Economics and the Social Sciences, the European Heritage in Economics and the Social Sciences, 1*, 61–116.

Scoones, I. (2007). Sustainability. *Development in Practice, 17*(4–5), 589–596.

Segal, G., Borgia, D., & Schoenfeld, J. (2005). The motivation to become an entrepreneur. *International Journal of Entrepreneurial Behavior & Research, 11*(1), 42–57.

Shane, S. (2000). Prior knowledge and the discovery of entrepreneurial opportunities. *Organization Science, 11*(4), 448–469.

Shane, S. (2003). *A general theory of entrepreneurship: The individual-opportunity nexus*. Edward Elgar Publishing.

Shaw, E., & Carter, S. (2007). Social entrepreneurship: Theoretical antecedents and empirical analysis of entrepreneurial processes and outcomes. *Journal of Small Business and Enterprise Development, 14*(3), 418–434.

Shepherd, D. A., & Patzelt, H. (2011). The new field of sustainable entrepreneurship: Studying entrepreneurial action linking "what is to be sustained" with "what is to be developed." *Entrepreneurship Theory and Practice, 35*(1), 137–163.

Slaper, T. F., & Hall, T. J. (2011). The triple bottom line: What is it and how does it work. *Indiana Business Review, 86*(1), 4–8.

Srivastava, S. P., Srivastava, S., Varshney, H., & Paigude, S. (2025). Sustainable entrepreneurship in emerging economies (BRICS): unveiling opportunities amid challenges. In *The synergy of sustainable entrepreneurship* (pp. 255–284). IGI Global Scientific Publishing.

Stabler, D., Hakala, H., Huikkola, T., & Mention, A.-L. (2024). Aligning servitization and circularity: The role of institutional confluence in sustainable business models. *Journal of Cleaner Production, 462*, Article 142666.

Surana, K., Singh, A., & Sagar, A. D. (2020). Strengthening science, technology, and innovation-based incubators to help achieve sustainable development goals: Lessons from India. *Technological Forecasting and Social Change, 157*, Article 120057.

Teece, D. J. (1986). Profiting from technological innovation: Implications for integration, collaboration, licensing and public policy. *Research Policy, 15*(6), 285–305.

Thompson, N., Kiefer, K., & York, J. G. (2011). Distinctions not dichotomies: Exploring social, sustainable, and environmental entrepreneurship. In *Social and sustainable entrepreneurship* (pp. 201–229). Emerald Group Publishing Limited.

Urbaniec, M. (2018). Sustainable entrepreneurship: Innovation-related activities in European enterprises. *Polish Journal of Environmental Studies, 27*(4), 1773–1779.

Utterback, J. M. (1994). One point of view: Radical innovation and corporate regeneration. *Research-Technology Management, 37*(4), 10.

Utterback, J. M., & Abernathy, W. J. (1975). A dynamic model of product and process innovation. OMEGA. *The International Journal of Management Science, 3*(6), 639–656.

Vallaster, C., Kraus, S., Kailer, N., & Baldwin, B. (2018). Responsible entrepreneurship: Outlining the contingencies. *International Journal of Entrepreneurial Behavior & Research, 25*(3), 538–553.

Veleva, V. (2021). The role of entrepreneurs in advancing sustainable lifestyles: Challenges, impacts, and future opportunities. *Journal of Cleaner Production, 283*, Article 124658. https://doi.org/10.1016/j.jclepro.2020.124658

Venkataraman, S. (1997a). Value at risk for a mixture of normal distributions: The use of quasi-Bayesian estimation techniques. *Economic Perspectives-Federal Reserve Bank of Chicago, 21*, 2–13.

Venkatarman, S. (1997b). The distinctive domain of entrepreneurship research: An editor's perspective. *Advances in Entrepreneurship, Firm Emergence, and Growth, 3*, 119–138.

WCED, S. W. S. (1987). World commission on environment and development. *Our Common Future, 17*(1), 1–91.

Woolf, L. (1928). *Imperialism and civilization.* Harcourt, Brace.

York, J. G., & Venkataraman, S. (2010). The entrepreneur–environment nexus: Uncertainty, innovation, and allocation. *Journal of Business Venturing, 25*(5), 449–463.

Zahra, S. A., Filatotchev, I., & Wright, M. (2009). How do threshold firms sustain corporate entrepreneurship? The role of boards and absorptive capacity. *Journal of Business Venturing, 24*(3), 248–260.

Sustainable Business Design and Development

Sustainable Business Model

Mohammad Azizi and Mohsen Akhondi

Abstract

Sustainable entrepreneurship naturally requires a sustainable business model. Considering the recent and extensive sustainable business models, as well as the patterns that lead to sustainability, studying and developing sustainable business models is important in both theory and practice. This research was conducted as a review of the sustainable business models of 50 companies globally that provide sustainability reports following international GRI standards. In this research, with the help of 8 sustainable business models that are considered as references in sustainability literature and two models of "taking regulatory and guiding role" and "re-targeting for society and the environment" are the two best models of business model sustainability that have executive ability in the business model of a wide range of companies and globally the most important models. For sustainability, the business model. The relationship between variables was investigated using the regression method using SPSS software. The results of this study confirmed the relationship between sustainable social performance and financial performance, and the relationship between financial performance and sustainable environmental performance was rejected.

M. Azizi (✉) · M. Akhondi
Faculty of Entrepreneurship, University of Tehran, Tehran, Iran
e-mail: m_azizi@ut.ac.ir

Keywords

Sustainable entrepreneurship · Social sustainability · Environmental sustainability · Economic performance · Innovation

1 Key Learning Objectives

1. Knowing the key pillars of a sustainable business model.
2. Knowing the status of the selected top companies in terms of their attention to sustainable business models.
3. Learn the challenges of implementing a sustainable business model.
4. Explore the relationship between social, environmental, and financial sustainability in top companies.

2 Introduction

The adoption of diverse business models allows financial markets to generate intrinsic value for their stakeholders while simultaneously fostering economic development and fulfilling their societal responsibilities. Nevertheless, the evolving influence of technological innovation, global capital flows, and regulatory shifts continually reshape the financial landscape, introducing heightened uncer-

M. Akbari et al. (eds.), *Sustainable Entrepreneurship*, University of Tehran Science and Humanities Series, https://doi.org/10.1007/978-3-032-04909-4_4

tainty. In light of these challenges, conventional business model frameworks prove inadequate, as they fail to address both present and anticipated market demands. Consequently, a fundamental restructuring of business model design is imperative. Amid such volatility, maintaining a "business as usual" approach is unsustainable; instead, financial markets must prioritize resilient and adaptive business models that ensure long-term strategic stability and sustained value creation.

The interplay between business models and corporate performance, particularly in the context of sustainability, has been well-documented in the scholarly literature. Recent research in sustainable entrepreneurship further explores how such models contribute to systemic transformations in both markets and broader societal structures (Schaltegger et al., 2016). According to researchers (Eriksson & Penker, 2000), the business model serves as a conceptual framework for analyzing organizational logistics, optimizing operational processes, and articulating the structural foundations of an enterprise. Furthermore, it functions as a strategic competitive dimension, enabling firms to establish and sustain a differential advantage within their industry. (Hamel & Trudel, 2001). Innovation in business models is more important than product innovation (Zott et al., 2010). While product innovations are inherently susceptible to replication, business model innovation offers a more sustainable mechanism for preserving or enhancing an organization's competitiveness. The incorporation of sustainability principles into corporate business models necessitates an inclusive stakeholder framework, wherein ecological and social considerations are accorded parity with economic objectives in value distribution (Bocken et al., 2014). A company's business model needs to be changed so that it can be sustainable or newly developed to create full value that benefits all stakeholders of the company (Drimmelen, 2013).

In an era of increasing environmental degradation, resource scarcity, and social inequality, the need for sustainable business practices has become more critical than ever. While many top companies have recognized the importance of integrating sustainability into their operations, a significant gap remains between intention and implementation. Despite the growing adoption of corporate social responsibility (CSR) initiatives and environmental, social, and governance (ESG) frameworks, many organizations struggle to develop and maintain truly sustainable business models that balance profitability with long-term ecological and social impact.

This raises the question: How can top companies design, implement, and scale sustainable business models that not only drive financial success but also contribute positively to environmental preservation and social well-being? The challenge lies in identifying innovative strategies, operational frameworks, and measurable metrics that enable these companies to transition from traditional profit-driven models to ones that prioritize sustainability without compromising competitiveness in the global market. Addressing this issue is essential for ensuring the long-term viability of businesses and the planet.

3 Literature

3.1 Business Model Definition

The existence of multiple business model definitions in academic literature has created conceptual ambiguity regarding their fundamental purpose. To address this ontological challenge, scholars have sought to demonstrate its practical utility through empirical applications. Contemporary research positions the business model as an analytical framework for examining value exchange dynamics between an organization and its stakeholder ecosystem (Gordijn & Akkermans, 2001), which can help improve decision-making (Hayes & Finnegan, 2005). A business model can be used as a unit of analysis to examine a company's business logic. Table 1 presents various definitions of a business model.

The conceptual heterogeneity surrounding business model definitions reveals a persistent lack of consensus regarding its precise conceptualization, notwithstanding its widely acknowledged theoretical and practical significance. Scholarly attempts to define the construct

Table 1 Some definitions of business model

Author (year)	Definition
(Timmers, 1998)	The business model architecture encompasses the structural design of product, service, and information flows within an organizational ecosystem. This framework systematically identifies and characterizes: (1) the participating business entities and their respective functions, (2) the value proposition and anticipated benefits for each stakeholder, and (3) the revenue generation mechanisms and monetization strategies
(Hamel & Trudel, 2001)	A business model represents the operationalization of a business concept, which comprises four fundamental elements: (1) the core strategic orientation, (2) strategic resource allocation, (3) customer engagement mechanisms, and (4) value network architecture. This framework transforms theoretical business propositions into executable organizational processes
(Johnson et al., 2008)	"A business model consists of four interconnected elements (customer value proposition, profit formula, key resources, and key processes) that together create and deliver value."

have bifurcated into two predominant approaches: (1) component-based frameworks that delineate specific structural elements, and (2) holistic conceptualizations that articulate generalized phenomenological understandings. For example,some researchers such as, Rayport et al. (2005), Hedman and Kalling (2003) and Johnson et al. (2008), highlighted the components of a business model in their definitions. However, it seems to be a gap in the definitions of business model (Hamel, 2001). Other authors define business models as architectures, ideas, representations, tools, or frameworks. We can conclude that a business model is a multidimensional concept that describes how a business creates, delivers, and captures value.

3.2 Sustainable Business Model

A company must continuously create value for its shareholders, customers, and other stakeholders to succeed (Mahadevan, 2002).

In completing the Theory of Sustainable Entrepreneurship (Dean et al., 2007) and the framework developed by field researchers (Schaltegger & Wagner, 2011), Business models can be systematically categorized along a sustainability integration spectrum based on their strategic prioritization of environmental and social objectives:

1. *Compliance-Oriented Models*: Characterized by the explicit exclusion of sustainability from strategic objectives, where environmental and social considerations are treated as:
   - Peripheral regulatory obligations
   - Custodial responsibilities rather than value drivers
   - Managed through reactive compliance mechanisms.

   Typical of organizations that institutionally marginalize sustainability, adopting purely formalistic approaches to meet minimum legal requirements.
2. *Complementary Integration Models*: Feature partial sustainability embedding, where environmental and social goals:
   - Coexist with (but remain subordinate to) core economic objectives
   - Are pursued primarily when they create ancillary business benefits
   - Demonstrate opportunistic rather than systemic integration.
3. *Transformative Models*: Exhibit deep structural integration of sustainability through:
   - Strategic alignment of all operations with sustainability principles
   - Internalization of environmental/social goals as primary performance metrics
   - Institutionalization of sustainable practices as core competitive advantages.

This represents a paradigmatic shift where sustainability becomes the central organizing logic.

To date, no conclusive evidence has been produced regarding the successful implementation of the sustainable business model (Evans et al., 2017), which examines the relationship between sustainable social initiatives and economic performance (Schaltegger & Synnestvedt, 2002). The knowledge gap between social sustainability and economic success has typically led companies to adopt general solutions or, at best, consider them as an add-on to the business principle rather than confronting social sustainability issues.

Other studies have shown that improved social sustainability can enhance financial performance by reducing costs or increasing revenue. In fact, on the one hand, higher job satisfaction and a sense of belonging lead to lower levels of absenteeism, voluntary resignation, and increased productivity (Pinzone et al., 2019). On the other hand, the development of human resources is conducive to higher quality and innovation in processes and products, as well as increased capacity to understand the context in which a firm operates (Baumgartner, 2014). Among the different approaches used to connect business models to sustainability, the "Business Case for Sustainability" addresses the reciprocal relationships between voluntary environmental and social activities and corporate economic success, which "can bring management, advanced, or innovation to improve economic success through voluntary social and environmental activities"(Schaltegger & Wagner, 2011).

3.3 Core Pillars of Sustainable Business Models

Businesses do not operate in a vacuum; instead, they exist within an environment. It can be said that companies rely on the external environment for at least two key reasons: First, they obtain raw materials and essential resources from the environment, and second, they sell their products back to it. Therefore, environmental sustainability is crucial for businesses not only from a social and ethical responsibility perspective but also from an economic standpoint. A sustainable business model (SBM) is a framework that integrates environmental, social, and economic considerations into a company's operations, ensuring long-term value creation while minimizing negative impacts on the planet and society. Unlike traditional models focused solely on profit, it balances three pillars of sustainability.

1. **Triple Bottom Line Approach**:

   In the traditional approach, businesses operate primarily to generate profit, with all their activities geared towards that goal. Meanwhile, SBMs focus on creating value across three dimensions: economic, environmental, and social (Mazur et al., 2024). This approach ensures that businesses do not prioritize profit at the expense of environmental and social responsibilities.
2. **Stakeholder Integration**:

   Businesses have different stakeholders. In the governance of a business, it is necessary to consider all of these stakeholders. SBMs consider the interests of a wide range of stakeholders, including customers, employees, suppliers, and the broader community (Dembek & York, 2019). Although the stakeholders of a business are numerous and diverse, the degree to which they influence and are influenced by businesses varies. Effective stakeholder management and engagement are crucial for addressing diverse concerns and achieving sustainability goals (Rana et al., 2017). Businesses that prioritize stakeholders must be able to balance the conflicting demands of their stakeholders.
3. **Value Creation and Delivery**:

   SBMs emphasize creating and delivering value in a sustainable manner, often through innovative practices such as eco-innovations, product-service systems, and circular economy principles (Goni et al., 2021). The Sustainable Value Exchange Matrix is one tool that helps visualize and discuss value creation from a multi-stakeholder perspective (Sousa-Zomer & Cauchick-Miguel, 2019).

4 Methodology

The best source for identifying a sustainable business model and ultimately developing one is to review the research conducted by experts and study the relevant literature. In this case, with the help of research conducted by Bocken et al. (2014), which introduces eight basic models of sustainable business model, to classify companies leading in the field of sustainability reporting according to the GRI standard, a detailed picture of the sustainable business model and its indicator elements in practice was found to develop a sustainable business model for use. Homegrown businesses utilize it in the field of entrepreneurship, and with the help of reviewing the relevance of patterns in companies across a wide range of activities, they have achieved this important milestone.

To answer the questions in the literature about insufficient clarity and the need to investigate further the impact of social and environmental goals on overall performance to attach sustainable business model patterns to businesses, the effect of pursuing social and environmental goals along with economic goals in the overall performance of companies should be examined separately to determine the effectiveness of each.

The research is considered a combination of both quantitative and qualitative approaches for conducting the study. Statistical analysis of the research was performed using SPSS software. The research was conducted using secondary data, and the type of research applied was identified. Studies aim to discover the relationships and interactions between economic, social, and environmental variables in the business models of companies and their impact on selecting, developing, and implementing a sustainable business model.

The research was conducted using a systematic review approach to apply business model innovation categories in promoting sustainability and to study such developments further, aiming to develop a sustainable business model while examining its impact on the company's performance. This research offers eight sustainable business model pillars based on previous studies (Bocken et al., 2014).

Patterns are categorized in groups that describe the main type of business model innovation: technological, social, and organizational innovations. The grouping of technological innovations includes patterns with a dominant innovative component (e.g., product manufacturing and redesign process), and social grouping contains patterns with a dominant social innovative component (such as innovation in consumer presentation or changing consumer behavior). In contrast, patterns in organizational grouping have a dominant element of organizational innovation (e.g., business financially trusted change). The following patterns are categorized based on two key areas: (1) **Technology** (Maximizing material and energy efficiency, Creating value from waste, and Replacing with renewable and natural processes); (2) **Social** (Providing performance instead of ownership to customers, taking over the supervisory role, and providing enough encouragement) and (3) **Enterprise** (Re-targeting for society and the environment, Development of high-scale solutions).

The statistical population of the study comprises a selection of sustainability reports from 50 international companies, all of which have been prepared and developed in accordance with the GRI Sustainability Reporting Standards (Global Reporting Initiative). In line with the longitudinal quality of the research, all reports have been reviewed over the 2 years (2019–2020). In selecting companies, efforts have been made to use as many well-known companies as possible with tangible work areas to better generalize sustainable concepts in businesses that have similar domestic ones.

Companies are investigated based on eight sustainable business models in research (Bocken et al., 2014), which serves as the basis for this research. With the help of a scoring system, companies' actions were ranked to comply with these eight patterns. If the prerequisites for each pattern are fulfilled, 1 point will be awarded to the company; otherwise, 0 points will be awarded. The sum of these points determines the depth of sustainability of each company's business model, based on the sustainability patterns. With the help of these privileges, companies are classified based on the privileges they have obtained to study the trends of the sustainability business model according to their field of activity. In the study of each pattern, exemplary companies that have exemplary or far-fetched activity will be listed.

Then, the multiplicity of pattern repetitions across a total of 50 companies will be analyzed to determine the most frequent patterns. In the following, based on the ease of integration of the model with the business model and its advantages for the business, the best patterns for developing the business model in sustainable entrepreneurship will be introduced. In this way, the primary research question will be addressed. To find the answers to the research sub-questions and to investigate the impact of social performance and economic performance on the sustainability of the business model, all social activities related to sustainability for 50 companies in 2019 and 2020 will be presented in a table alongside their economic performance at the same time.

5 Results

This part of the research, considering that it seeks to predict the dependent variable's behavior in terms of financial performance as a result of the impact of two independent variables—environmental performance and social performance—a multiple regression test has been used. In this regard, first, two main prerequisites for implementing regression analysis, i.e., nonlinearity of independent variables and nonlinear errors, are investigated. Then, the hypothesis test is examined through regression analysis.

The 50 international companies operating in the country of origin or secondary countries that publish their sustainability reports under the international GRI standard have been investigated. These companies are analyzed based on the eight sustainable business models mentioned above, and by studying their reports from 2019 and 2020, they receive 1 point for each pattern adhered to. In the event of non-compliance with any pattern, 0 points will be awarded. The total score of each company indicates the degree of integration of the eight sustainable business models in the business model it uses. After calculating the points, companies will be categorized and separated, and more critical and influential patterns will be identified, as well as patterns that have been difficult to observe in the business model. Then, the ideal model for incorporating into the business model to achieve a sustainable and optimal business model will be proposed. It should be noted that observing even a model means having a sustainable business model, and more numbers merely indicate a deeper level of business sustainability activities.

According to the relatively similar formatting of reports, comparison with eight reference patterns is based on GRI standards and descriptions of patterns, and there can be significant differences in the reports of previous or future years. Companies from various fields of activity and from around the world have been selected to provide a comprehensive view of sustainable business models and related elements (Table 2).

Reviewing Sustainable Business Model Archetypes in Companies:

The average score obtained by companies based on sustainable business model patterns is 5.3 out of 8 patterns (5.3.8 points). This means that the business models of the companies studied have, on average, at least five sustainability patterns. It should be noted that to develop a sustainable business model in entrepreneurship, adding only one of the eight patterns to the business model will lead to sustainability, albeit on a small scale. In this regard, the simplest and most optimal model is identified and integrated into a straightforward yet efficient business model to form a new, sustainable business model (Table 3).

Overall, "Adopt a stewardship role" and "Repurpose the business for society/environment" were recognized as the most frequent sustainable business model archetypes that the companies under study adhered to. These models are almost necessary for companies to have a sustainable business model in the international arena. While there is no inherent superiority in these two patterns, the ease of implementing, reporting, and monitoring them has led to their frequent occurrence. In line with the research objective of "developing a business model in sustainable

Table 2 Sustainable business model archetype implementation by companies

	Company name	Technological			Social			Organizational		Score
		Maximise material and energy efficiency	Create value from 'waste'	Substitute with renewables and natural processes	Deliver functionality rather than ownership	Adopt a stewardship role	Encourage sufficiency	Re-purpose the business for society/ environment	Develop scale-up solutions	
1	Alcoa Corporation	1	1	1	0	1	0	1	0	5
2	BCE	1	1	0	1	1	1	1	0	6
3	BHP	0	1	0	0	1	0	1	1	4
4	BMW	1	1	1	1	1	1	1	1	7
5	Carl Zeiss	1	0	0	0	1	1	1	0	4
6	CEZ Group	1	1	1	0	1	0	1	0	5
7	CIMB	0	1	1	0	1	0	1	1	5
8	China Airlines	1	0	1	1	1	0	1	0	5
9	Cisco	1	1	1	1	1	1	1	1	8
10	CCI	1	1	1	0	1	0	1	1	6
11	Dagonte Cement	0	0	1	0	1	0	1	1	4
12	Dufry	1	0	0	0	1	0	1	0	3
13	Epson	1	1	1	1	1	0	1	1	7
14	Ericsson	0	1	1	0	1	1	1	0	5
15	Ferrari	1	0	1	0	1	1	1	1	6
16	Fisher & Paykel	0	1	1	0	1	0	0	0	3
17	GAZPROM	1	0	0	1	1	0	1	0	4
18	General Motors	1	1	1	0	1	0	1	1	6
19	H&M Group	1	1	1	0	1	0	1	0	5
20	HTC	1	1	1	0	1	0	1	0	5
21	Ingram Micro	1	1	0	1	1	0	1	0	6
22	KBC Group	0	0	1	1	1	1	1	1	6
23	Kingston	1	1	0	0	1	0	1	0	4

(continued)

Table 2 (continued)

	Company name	Technological			Social			Organizational		Score
		Maximise material and energy efficiency	Create value from 'waste'	Substitute with renewables and natural processes	Deliver functionality rather than ownership	Adopt a stewardship role	Encourage sufficiency	Re-purpose the business for society/environment	Develop scale-up solutions	
24	Lockheed Martin	1	0	1	1	1	1	1	1	7
25	LSR Group	1	0	1	0	1	1	1	0	5
26	Mazda Motors	1	1	1	0	1	1	1	1	7
27	Micron	1	1	1	0	1	0	1	1	6
28	Migros Ticaret	0	1	1	0	1	0	1	0	4
29	MTR Corporation	1	0	0	0	1	1	1	0	4
30	Nammo Group	0	0	0	1	0	0	1	0	2
31	Neinver	0	0	1	1	0	0	1	1	4
32	Nissan	1	1	1	0	1	1	1	1	7
33	Norma Group	1	0	0	0	1	0	1	0	3
34	Nova Chemicals	1	1	1	0	1	1	1	1	7
35	Novatek	1	1	1	1	1	0	1	0	6
36	Orbia	1	1	1	1	1	0	1	1	7
37	OTP Bank	0	1	0	1	1	1	1	1	6
38	Pruksa	0	1	1	0	1	1	1	0	5
39	Qualcomm	0	1	0	0	1	1	1	1	5
40	Royal Dutch Shell	1	1	0	1	1	0	1	0	5
41	Scania	0	0	1	0	1	1	1	1	5
42	Singapore Airlines	1	1	1	1	1	1	1	1	8

(continued)

Table 2 (continued)

	Company name	Technological			Social			Organizational		Score
		Maximise material and energy efficiency	Create value from 'waste'	Substitute with renewables and natural processes	Deliver functionality rather than ownership	Adopt a stewardship role	Encourage sufficiency	Re-purpose the business for society/ environment	Develop scale-up solutions	
43	ST Microelectronics	1	1	1	0	1	0	1	1	6
44	Telekom Austria	0	1	1	1	1	0	0	1	5
45	SAIB	0	1	0	1	1	0	1	0	4
46	TIM Group	0	1	1	1	1	0	1	1	6
47	Wilmar International	0	1	1	0	1	0	1	1	5
48	XPO Logistics	0	1	1	1	1	0	1	0	5
49	Yokogawa	1	1	1	0	1	0	1	0	5
50	Yamaha	1	1	1	0	1	1	1	1	7
		Total implementation of each sustainable business model pattern by companies								
		32	36	36	20	48	19	48	26	Average score 5.3

Table 3 Repetition of sustainable business model archetypes

8 Points (2)	7 Points (8)	6 Points (11)	5 Points (16)	4 Points (9)	3 Points (3)	2 Points (1)
Cisco	BMW	BCE	Alcoa Corporation	BHP	Dufry	Nammo Group
Singapore Airlines	Epson	CCI	CEZ Group	Carl Zeiss	Fisher & Paykel	
	Lockheed Martin	Ferrari	CIMB	Dagonte Cement	Norma Group	
	Mazda Motors	General Motors	China Airlines	GAZPROM		
	Nissan	Ingram Micro	Ericsson	Kingston		
	Nova Chemicals	KBC Group	H&M Group	Migros Ticaret		
	Orbia	Micron	HTC	MTR Corporation		
	Yamaha	Novatek	LSR Group	Neinver		
		OTP Bank	Qualcomm	SAIB		
		ST Microelectronics	Pruksa			
		TIM Group	Royal Dutch Shell			
			Scania			
			Telekom Austria			
			Wilmar International			
			XPO Logistics			
			Yokogawa			

entrepreneurship", adding these two patterns can be a "good start" towards sustainability. It is worth noting that other patterns can also be employed to establish a sustainable business model, depending on the nature of the business and its environmental needs. On the next page, the dispersion of company points based on observing sustainable business model patterns is displayed.

Investigation of Economic Performance in Comparison with Sustainable Social Performance:

In published reports from companies based on international GRI standards published annually and by paying the relevant costs to the organization, in addition to reporting on the environmental aspect of the sustainability of reports, there are also sections related to social performance issues and internal mechanisms such as how to operate, value chain and organizational chart. The economic information in the reports is brief, providing the reader with a general overview. For full information on economic performance, we should refer to the companies' "annual reports". However, the social issues that were less discussed in the previous section of the research and the eight models of the sustainable business model have been presented in the GRI reports.

The main headings of social activities reported in the company's internal environment include equality in the workplace, human rights, fair working hours, social education, and transparency in legal and financial issues. At the community level, this includes educational programs, charity, contributions to regional and national infrastructure, stakeholder empowerment, cooperation with non-profit organizations, public health promotion, volunteer work, and cultural activities. Can be counted.

Because reports using the GRI framework are not published by all companies every year, the research has only reviewed companies that published their reports in 2019 and 2020. It is worth noting that, due to the global coronavirus pandemic in 2020, financial and social information has undergone significant changes. This has been particularly the case with the economic performance of companies and the decline in financial turnaround, often influenced by the coronavirus pandemic and the slowdown in global economic growth. Nevertheless, there has been no direct mention in the reports about the impact of pandemics on economic performance, accompanied by numbers and figures.

To facilitate the classification of companies and also to perform statistical analysis concerning the extent of social sustainability activities, by group integration of activities into four main areas of social sustainability, including diversity and equality, charity, education, health, and public health, in 50 companies studied. Using these groups, social sustainability activities were rated on a scale of 0 to 4 to provide a more accurate statistical analysis of the relationship between economic performance and corporate social performance.

This part of the research, considering that it seeks to predict the dependent variable's behaviour in terms of financial performance as a result of the impact of two independent variables—environmental performance and social performance —a multiple regression test has been used. In this regard, first, two main prerequisites for implementing regression analysis, i.e., nonlinearity of independent variables and nonlinear errors, are investigated. Then, the hypothesis test is examined through regression analysis.

In the form of three indices of torsion or inflation variance (VIF), tolerance, and condition index, the nonlinearity of independent variables in the regression model is investigated. Table 4 examines the nonlinearity of independent variables.

The variance inflation factor should be less than 2 for regression models; the closer it gets to 2, the more linear the model becomes with respect to other variables. In this regard, according to the obtained results (value 1.15), it is clear that the two independent variables have little linearity with each other. Tolerance = 1/VIF is the VIF reverse tolerance index. The closer this value approaches zero, the more linear it appears to be. In this regard, according to the obtained results (value 0.869), it is clear that the two independent variables have little linearity with each other.

Table 4 VIF factor

Independent variables	Variance inflation factor (VIF)	Tolerance	Condition index
Environmental sustainability performance	1.15	0.869	8.666
Social sustainability performance	1.15	0.869	10.165

If the condition index is greater than 15 (CI > 15), a line is present, and if it is greater than 30, the line is critical, and the regression test is invalid. In this regard, according to the obtained results, the conditional index shows a lack of linearity, because the obtained values are smaller than 15.

Then, before presenting the results of the hypothesis test, the variance is analysed. Analysis of variance refers to the analysis of changes performed using the Fisher index (F). Fisher Index (F) is a coefficient that shows the quality of the model and partially replaces the fit index in structural equations. The significance of this coefficient ($p < 0.05$) indicates that the regression model has a good quality in predicting the behaviour of the dependent variable. Table 5 analyses the variance.

As shown in the table above, the value obtained for Sig is less than 0.05, indicating that the regres-

Table 5 Sig and Fisher index

	Degree of freedom	F index	Sig
Regression	2	3.512	0.038
Residual	47		
Total	49		

Table 6 Hypothesis test

Hypothesis	β	*P* value/ Sig	t-value	Result
H1	0.083	0.573	0.568	Rejected
H2	0.322	0.032	2.208	Confirmed

sion model is of good quality in predicting the behavior of the dependent variable. Finally, in Table 6, the research assumptions are tested.

In the first hypothesis of the research on the effect of business model stability on financial performance, considering the *p*-value, which is greater than 0.05, and the t-value in the range of − 1.96 to 1.96, it can be said that the sustainability of the business model has no significant effect on financial performance.

In the second hypothesis of the research on the effect of social performance on financial performance, considering the *p*-value, which is less than 0.05, and the t-value outside the range (− 1.96 and 1.96), it can be stated that social performance has a significant effect on financial performance. Now, we will examine the quality (intensity and direction) of this influence. Since the β value is 0.322, it can be stated that the social performance variable has a 32% effect on financial performance, and the direction of this effect is positive, as indicated by the β sign. Therefore, social performance has a positive and significant effect on financial performance.

The relationship between the integration of social sustainability and financial performance is neither automated nor direct. However, the nature of the relationship itself seems different when considered from a different horizon. Additionally, the inclusion of new key variables enhances the understanding of the relationships that underlie the evolution of a sustainable business model. Integrating social sustainability with the business model affects the characteristics and effectiveness of the innovation process. Companies that integrate social goals as a core part of their business model are more likely to develop effective activism strategies.

Nevertheless, the intense attention focused on social sustainability in business model design should be supported by smart activities based on the development and renewal of policies and tools because they can combine social sustainability continuously and effectively at different stages of the business. Ultimately, social sustainability enhances business resilience.

6 Conclusion

This research was conducted using eight main models of the sustainable business model (Bocken et al., 2014). This study aimed to examine the sustainability reports of a group of international companies and identify key elements in the sustainable business model. Based on the success of these models in enhancing the performance of the companies, it sought to identify superior patterns for developing the business model in sustainable entrepreneurship and propose them for implementation. The top two models that are simpler to implement, do not cost much, and have a tremendous impact on business sustainability are "taking a regulatory and guiding role" and "re-targeting for society and the environment". Other models can be integrated into the business model in line with sustainability, given the business conditions, although they can be complex and challenging to implement.

The outbreak of the coronavirus pandemic in recent years has led to dramatic changes in economic conditions and the mechanisms employed by businesses for decades. A large number of countries are still suffering from pandemic-induced recessions. For this reason, research conducted in a similar field will be subject to the existing special conditions and will not provide a 100% accurate picture. It is recommended that a similar study be conducted in the future with a larger statistical sample and a more comprehensive longitudinal design to obtain more accurate results. Under normal conditions, the relationships between environmental sustainability and financial performance, as well as between social sustainability and social performance, may be confirmed. Finally, the sustainable business models of the selected index are identified as models that have been used in the statistical population and have demonstrated favourable performance, but are not necessarily superior to other executive patterns.

In the subject of social sustainability, this research has implications for both academic and practical discussions. Theoretically, the findings help raise awareness about sustainable business models, sustainable innovation, and the relationship between financial performance and sustainable social performance, contributing to the existing literature on the subject. Sustainable business models must integrate social sustainability considerations from their initial design phase, supported by comprehensive management systems that facilitate the ongoing monitoring and achievement of social objectives throughout the organizational lifecycle. Through proactive design principles and voluntary governance mechanisms, such models can maintain dynamic adaptability, enabling systematic responses to evolving socio-environmental conditions.

Sustainable business models are crucial for addressing modern-day environmental, social, and economic challenges. While significant progress has been made, ongoing research and innovation are needed to overcome existing barriers and fully realize the potential of SBMs across various sectors.

Current research has yet to establish comprehensive empirical evidence regarding either (1) the effective implementation of sustainable business models (Evans et al., 2017). The extant literature has not yet yielded definitive empirical evidence establishing a causal relationship between sustainability-oriented initiatives and measurable economic performance outcomes. Current research remains inconclusive regarding whether and to what extent such initiatives systematically impact financial indicators at the organizational or operational level. (Schaltegger & Synnestvedt, 2002; Sch€onborn, 2010; Wagner, 2007) to show. Statistical findings show the lack of relationship between environmental sustainability and financial performance. However, it is possible to associate business financial performance with social performance in the field of sustainability with an acceptable approximation. In the future, using a longitudinal approach to assess how a sustainable business model affects corporate financial success over long-term periods can provide significant information, aligning with the analysis of how social activities impact financial performance.

This research helps enrich the current knowledge of sustainable business models by filling the gap in understanding the methods and tools that companies may use in the innovation process for their sustainable business model. In this way, companies can prioritize sustainable business model patterns in selecting and implementing them in a more measured way. While productivity and quality improvement may have easily meant more profits in the past, it's not always so clear how delivering social and environmental value might lead to profits and competitive advantage for businesses. Knowing that in the results of the research, the relationship between social performance and financial performance was confirmed and the relationship between financial performance and environmental performance was rejected, however, the growing attention to the business model in literature and practice shows that this is a useful framework for corporate innovation of any size and scale, and hence it is possible to advance sustainability innovation and sustainable models of the case. to be used.

6.1 Challenges in Implementing SBMs

Given the complexity and multidimensionality of a sustainable business model, its implementation will always face challenges and difficulties. So, implementing a sustainable business model can present several challenges, the most important of which are:

1. **Complexity and Integration**:

   Integrating sustainability into business functions is a complex process that requires a systemic and holistic approach. Firms must balance conflicting objectives across the three dimensions of sustainability, which can be a challenging task.

2. **Cultural and Structural Changes**:

   Adopting SBMs often requires significant internal changes, including the development of new structural and cultural capabilities. Organizations must transform their business system to prioritize sustainability over traditional profit-maximizing models.
3. **Measurement and Assessment**:

   Assessing the sustainability impact of business models requires comprehensive evaluation methods, such as the Business Model Canvas for Sustainability (BMCS). Tools like value mapping can help identify and measure the sustainability impacts of business activities.

6.2 Overcoming the Challenges

To overcome the above challenges, businesses can take the following steps gradually:

- **Phased implementation**: Don't expect to do a lot of planning and action in a short time. Start with low-hanging fruit (e.g., energy efficiency) to build momentum.
- **Collaboration**: Sustainable business model policies and programs cannot be implemented alone. To implement them, you need to enlist the support of various groups that are concerned about social and environmental issues. Some of these groups include NGOs, governments, and industry groups.
- **Innovation**: Leverage the capabilities of various technologies to drive innovation. New technologies can provide various solutions that you should be aware of—For example, leverage technology (AI, IoT) to optimize sustainable practices.
- **Consumer Engagement: Educate the market** and align sustainability with brand value. Education and culture building in the market is one of the basic principles for implementing a sustainable business model. Therefore, you can introduce your business through comprehensive communication channels and media (e.g., Patagonia's "Worn Wear" program).

The findings of this study also have significant implications for policymakers. The study shows that promoting social sustainability in corporate business models is a powerful tool that can facilitate recovery and financial stability in times of crisis. Policymakers can support sustainable business models through various supporting programs. Additionally, they can implement education and awareness policies for all citizens. Ultimately, this study aims to enhance managers' understanding of the importance of social sustainability. Companies should continually view social sustainability as a core business goal, alongside investing in developing processes and tools that align businesses with society. There is also a need to implement advanced management systems within the framework of a sustainable business model to ensure the company's greater resilience against external pressures and improve its long-term performance.

6.3 Future Research Directions

This research focused on examining sustainable business models in leading companies. Therefore, based on the experiences and findings of this research, future researchers can conduct further research on the following topics:

- **Interdisciplinary Integration**: There is a need for more interdisciplinary research to foster sustainable development across various domains, such as construction and smart cities.
- **Case Studies and Practical Applications**: More in-depth case studies are needed to fully understand the practical implications and effectiveness of SBMs in various contexts.
- **Regulatory and Policy Impact**: Future research should explore the impact of regulatory policies and incentives on the adoption of SBMs.

References

Baumgartner, R. J. (2014). Managing corporate sustainability and CSR: A conceptual framework combining values, strategies and instruments contributing to sustainable development. *Corporate Social Responsibility and Environmental Management, 21*(5), 258–271.

Bocken, N. M., Short, S. W., Rana, P., & Evans, S. (2014). A literature and practice review to develop sustainable business model archetypes. *Journal of Cleaner Production, 65*, 42–56.

Dembek, K., & York, J. (2019). Sustainable business models: Rethinking value and impact. In *Rethinking strategic management: Sustainable strategizing for positive impact* (pp. 131–148).

Drimmelen, R. V. (2013). New business models for sustainable development. In *Handbook of sustainable engineering* (pp. 859–879). Springer.

Eriksson, H.-E., & Penker, M. (2000). Business modeling with UML. *New York, 12*.

Evans, S., Vladimirova, D., Holgado, M., Van Fossen, K., Yang, M., Silva, E. A., & Barlow, C. Y. (2017). Business model innovation for sustainability: Towards a unified perspective for creation of sustainable business models. *Business Strategy and the Environment, 26*(5), 597–608.

Goni, F. A., Gholamzadeh Chofreh, A., Estaki Orakani, Z., Klemeš, J. J., Davoudi, M., & Mardani, A. (2021). Sustainable business model: A review and framework development. *Clean Technologies and Environmental Policy, 23*, 889–897.

Gordijn, J., & Akkermans, H. (2001). Designing and evaluating e-business models. *IEEE Intelligent Systems, 16*(04), 11–17.

Hamel, G., & Trudel, J. D. (2001). *Leading the revolution*. Elsevier.

Hayes, J., & Finnegan, P. (2005). Assessing the of potential of e-business models: Towards a framework for assisting decision-makers. *European Journal of Operational Research, 160*(2), 365–379.

Hedman, J., & Kalling, T. (2003). The business model concept: Theoretical underpinnings and empirical illustrations. *European Journal of Information Systems, 12*(1), 49–59.

Johnson, M. W., Christensen, C. M., & Kagermann, H. (2008). Reinventing your business model. *Harvard Business Review, 86*(12), 50–59.

Mahadevan, R. (2002). A DEA approach to understanding the productivity growth of Malaysia's manufacturing industries. *Asia Pacific Journal of Management, 19*, 587–600.

Mazur, B., Walczyna, A., & Wilczewska, W. (2024). Sustainable business model implementation in polish enterprises. *Sustainability, 16*(15), 6412.

Pinzone, M., Guerci, M., Lettieri, E., & Huisingh, D. (2019). Effects of 'green' training on pro-environmental behaviors and job satisfaction: Evidence from the Italian healthcare sector. *Journal of Cleaner Production, 226*, 221–232.

Rana, P., Short, S. W., Evans, S., Granados, M. H., & Valkokari, K. (2017). Toolset for sustainable business modelling. In *Value networks in manufacturing: Sustainability and performance excellence* (pp. 123–153).

Rayport, J. F., Jaworski, B. J., & Kyung, E. J. (2005). Best face forward: Improving companies' service interfaces with customers. *Journal of Interactive Marketing, 19*(4), 67–80.

Schaltegger, S., Lüdeke-Freund, F., & Hansen, E. G. (2016). Business models for sustainability: A coevolutionary analysis of sustainable entrepreneurship, innovation, and transformation. *Organization & Environment, 29*(3), 264–289.

Schaltegger, S., & Synnestvedt, T. (2002). The link between 'green' and economic success: Environmental management as the crucial trigger between environmental and economic performance. *Journal of Environmental Management, 65*(4), 339–346.

Schaltegger, S., & Wagner, M. (2011). Sustainable entrepreneurship and sustainability innovation: Categories and interactions. *Business Strategy and the Environment, 20*(4), 222–237.

Sousa-Zomer, T. T., & Cauchick-Miguel, P. A. (2019). Exploring business model innovation for sustainability: An investigation of two product-service systems. *Total Quality Management & Business Excellence, 30*(5–6), 594–612.

Timmers, P. (1998). Business models for electronic markets. *Electronic Markets, 8*(2), 3–8.

Zott, C., Amit, R., & Massa, L. (2010). The business model: Theoretical roots, recent developments, and future research. *IESE Research Papers, 3*(4), 1–43.

Sustainable Product and Service Development

Sahar Dehyouri

Abstract

Sustainable product and service development is one of the key foundational elements of sustainable entrepreneurship, and it involves integrating environmental, social, and economic aspects into the center of business innovation. This chapter explores the principles, strategies, and frameworks necessary to design products and services that reduce their ecological footprint while enhancing their social impact. It examines essential concepts such as eco-design, circular economy, lifecycle assessment (LCA), and cradle-to-cradle (C2C) design. The role of stakeholders and their voices in shaping regulatory requirements and consumer demand for sustainable Innovation is also part of this discussion. The chapter also highlights case studies of successful sustainable ventures that align a company's profit with the planet through the use of green materials and resources, energy-efficient processes, and service-based business models while addressing challenges such as cost, scalability, and greenwashing. It offers a practical method for entrepreneurs to apply. By embedding the theories and principles into valuable applications and case studies, the chapter gives a practical guide for various audiences (entrepreneurs, policymakers, and teachers) to embrace sustainability thinking when developing products and services. Sustainable entrepreneurship is both ethical and moral, and it fosters a competitive dynamic in a changing society.

Keywords

Sustainable production · Service development · Lifecycle thinking · Sustainable entrepreneurship · Business innovation

Learning Objectives
By reading this chapter, readers will:

- Recognize fundamental principles of sustainable product and service design, such as eco-design, lifecycle thinking, and circular economy models.
- Assess the environmental and social impacts of products and services using tools such as Life Cycle Assessment (LCA) and Cradle-to-Grave (C2C) frameworks.
- Understand prominent drivers of sustainable Innovation, including regulations, consumer demand, and stakeholder expectations.
- Analyze real-world case studies to evaluate how businesses successfully integrate sustainability in product design, materials sourcing, and service delivery.
- Understand approaches for addressing common issues in sustainable entrepreneur-

S. Dehyouri (✉)
Department of Agriculture, Environmental Sciences Research Center, Isl.C., Islamic Azad University, Islamshahr, Iran
e-mail: Dehyouri.s@gmail.com

M. Akbari et al. (eds.), *Sustainable Entrepreneurship*, University of Tehran Science and Humanities Series, https://doi.org/10.1007/978-3-032-04909-4_5

ship, including cost limitations, scalability, and risks associated with greenwashing.

1 Introduction

1.1 The Urgent Need for Sustainable Development

The emphasis on sustainability shifted from an afterthought in the twenty-first century to a business necessity. Climate change, resource depletion, social inequalities, and other threats are forcing companies to integrate sustainability into their innovation plans to remain competitive and resilient. The Sustainable Development Goals (SDGs) of the United Nations emphasize the role the private sector must play in facilitating systemic change (Agreement, 2015). Those businesses that integrate Innovation and sustainability create long-term competitive advantages, such as lower costs, enhanced brand reputation, or influence with regulatory bodies (Nassis et al., 2023).

Developing a circular economy focusing on design reuse, recycling, and regeneration for sustainable Innovation is vital (Shehata Nakhil, 2024). Practices this in its take-back programs and new product designs (Gbadamasi et al., 2021). Reveal that the inference Co. has also reduced its carbon footprint by 96% through design efficiencies and recycled materials. Life Cycle Assessments (LCAs) are instrumental in identifying the environmental hotspots in a product's life cycle; these are the points in the life cycle when raw materials are lowered, which ultimately decreases waste and energy used from raw material extraction to the product's end of life (Ellen MacArthur Foundation, 2017).

Engaging stakeholders—such as customers, NGOs, and governments—favors inclusive Innovation. Hu and Zeng (2024) explains how the co-creation of Unilever's Sustainable Living Plan leads to scalable solutions. At the same time, technologies such as AI, IoT, and Blockchain promote sustainability by enabling supply chain transparency (AbuGhanem, 2025) and optimizing resource utilization. For example, Google has developed an AI system to reduce the energy used in data centers, demonstrating how digital Innovation can create environmental benefits (Rahsepar et al., 2023).

Corporations should align their innovation efforts with the UN Sustainable Development Goals (SDGs) so that their impact can be approximated and monitored (Enquist & Sebhatu, 2023). State IKEA integrates SDG 7 (Clean Energy) and SDG 12 (Responsible Consumption) into its business model through energy investments and the use of sustainable materials. Likewise, developing an organizational culture that values sustainability would be helpful. Training employees, offering incentive programs, and having leadership commitment, such as Tesla Co., can foster a sense of sustainability and encourage bottom-up Innovation (Chandratreya, 2025). Evaluating impacts transparently is critical to maintaining credibility and enabling continuous improvement. Mechanisms like the Global Reporting Initiative (GRI) and the Carbon Disclosure Project (CDP) standardize best practices in sustainability reporting. Tesla's Impact Report establishes an industry benchmark by articulating carbon-reduction initiatives and ethical imperatives in the supply chain (Silva, 2024). Companies that consistently track and communicate sustainability performance build stakeholder trust and reinforce long-term viability (Lyu et al., 2023).

The Existential Need for Sustainable Development: Environmental crises of an unprecedented scale, such as climate change and biodiversity loss, have arisen primarily from our unsustainable consumption and production systems. The Intergovernmental Panel on Climate Change (IPCC) reports that global average temperatures are likely to rise by 1.5 °C as soon as 2030, and as a result, immediate action is required across all areas of society (Lee et al., 2023). The United Nations Environment Programme (UNEP) also declared that 'material extraction and processing are responsible for over 90% of global water stress and biodiversity loss' (Bruyninckx et al., 2024). These alarming statistics provide an important context for the shift away from traditional linear production models to sustainability-informed models in the design of production and services. Sustainable entrepreneurship represents a paradigm shift in which environmental and social issues are not secondary to commercial viability but are instead primary drivers of business strategy. As Elkington (2020) states in Green Swans: The Coming Boom in Regenerative Capitalism, “The successful businesses of the future will be the ones that create value whilst solving ecological and social problems.” This chapter will examine how entrepreneurs can design products and services that satisfy the dual objectives of commercial profitability while also considering the limitations of planetary boundaries and human well-being.

1.2 Defining Sustainable Product and Service Development

Sustainable products and services must satisfy three key criteria: environmental integrity to minimize resource use, pollution, and carbon emissions throughout the lifecycle; social equity to ensure fair labor practices, accessibility, and community benefits; and economic viability to maintain competitiveness while internalizing externalities (Bocken & Short, 2021). The Ellen MacArthur Foundation emphasizes that genuinely sustainable design moves beyond “less bad” approaches to create regenerative systems that restore natural capital (MacArthur & Heading, 2019).

Also, the European Commission articulated the circular economy model as a systemic alternative to the take-make-waste linear economy (Rashid & Malik, 2023). Key principles include “Designing out waste and pollution (e.g., through modular construction),” Keeping products and materials in use (e.g., via repair and remanufacturing), and “Regenerating natural systems (e.g., through biodegradable materials).”

While the Triple Bottom Line (People, Planet, and Profit) remains relevant, scholars (Raworth, 2018) in Doughnut Economics argue for more ambitious frameworks that operate within planetary boundaries while meeting human needs. This requires businesses to consider Absolute environmental limits (e.g., carbon budgets), Social foundations (e.g., living wages), and Systems thinking in design processes.

1.3 Key Drivers of Sustainable Innovation

The key drivers of sustainable Innovation are categorized into three notions. These notions are Regulatory Landscape, Market Forces, and Technological Enablers. In the Regulatory Landscape, increased global policy changes are accelerating sustainable transitions. As a result, the European Union’s Corporate Sustainability Reporting Directive (Odobaša & Marošević, 2023) requires comprehensive disclosures of environmental, social, and governance (ESG) information. Meanwhile, California’s SB 253 is an expected law requiring significant firms to disclose Scope 1–3 emissions (Ecklund, 2023). More broadly, the UN Plastic Treaty will create binding international regulations around plastic production (Arora et al., 2024).

On the other hand, as Market Forces, Consumer preferences are shifting dramatically. Eighty-five percent of global consumers report changing their purchase behaviors toward sustainability (Purcărea et al., 2022), and the second hand apparel market is projected to double by 2027 (Michalowska et al., 2023). Additionally, 75% of institutional investors now incorporate ESG criteria into their investment decisions (GSIA, 2023).

In the end, as Technological Enablers, Breakthrough innovations are making sustainability feasible, such as AI-powered lifecycle assessment tools (El Merroun, 2024), Mycelium-based packaging replacing polystyrene (Verma et al., 2023), and Blockchain for supply chain transparency (Cordova & Aguirre, 2022). Despite progress, significant barriers remain. These barriers include Cost Premiums—Sustainable alternatives often carry 15–30% cost premiums (Bhattacharyya et al., 2023), Supply Chain Complexity—68% of companies struggle with Scope 3 emissions tracking (Ellram & Tate, 2025), and Greenwashing Risks—40% of sustainability claims may be misleading (Amatucci & Mollo, 2024). For example, we could explain the modular smartphone company Fairphone, which demonstrates Conflict-free mineral sourcing (Waseem, 2024), 60% reduced CO_2 emissions compared to the industry average, and a significantly increased 5+-year product lifespan (vs. the industry average of 2.5 years). Additionally, IKEA's furniture giant's initiatives, IKEA Circular Design, include 60% renewable or recycled materials (Lacy et al., 2020), Furniture leasing programs in 30 markets, and an AI-powered circular design toolkit.

1.4 Principles and Practices for Eco-Friendly Solutions

Sustainable solutions are necessary to mitigate pressing environmental problems, including climate change, depletion of natural resources, and biodiversity loss. The principles outlined above are sustainability (a development that provides for the needs of the present, absent negotiating the capability of future generations to meet their own needs) (Kamijo et al., 2017), resource efficiency (using less material and energy) (Allwood et al., 2013), and the definition of circular economy (reuse and recycle) (Minunno et al., 2020). Additional fundamental principles include the prevention of pollution, the preservation of biodiversity, and the application of systems thinking throughout the life cycle of products to design and dispose of them responsibly (Iacovidou et al., 2021). The public and private sectors, including businesses and individuals, can replicate these sustainability principles and reduce their environmental footprint by incorporating sustainable practices. Sustainable practices in the macro-catalytic programs for the built environment include eco-design (e.g., Fairphone's modular smartphones), green manufacturing (Tesla's renewable-powered facility), and sustainable sourcing (IKEA's wood sourcing practice, which utilizes FSC-certified wood).

Additionally, there are energy (e.g., Google investing in solar) (Cohen et al., 2020), waste resources (e.g., Apple recycling smartphones) (Lefebvre, 2023), water conservation (e.g., Singapore's NEWater initiative) (Kwa, 2023), sustainable transport (e.g., electric vehicles) (Graham, 2024), and green buildings (e.g., Bullitt Center) (Chen et al., 2024). While obstacles to implementation exist, such as tangible costs and regulations, eco-friendly solutions still represent considerable opportunities for companies. Corporations such as Patagonia (sustainable outdoor apparel) (Shehata Nakhil, 2024) and Interface (carbon-neutral flooring) (Amoruso et al., 2022) exemplify eco-friendly practices that lead to Innovation and increased market share. By applying these sustainable principles and practices, businesses and individuals can positively impact the transition to a suburban, rural, or urban ecological lifestyle.

1.5 From Idea to Market: Pathways to Sustainable Innovation

While sustainable Innovation is critical to addressing global problems like climate change, resource depletion, and social injustice, it also benefits the economy and the communities unique to its location. Sustainable Innovation is the Innovation of products and processes designed to meet the environmental, social, and economic dimensions (the three-legged stool). For example, the World Economic Forum assessed global sustainable innovation solutions and projected that \$12 trillion in market opportunities might be realized by 2030 (Zu, 2020). Businesses are shifting to sustainable Innovation for various reasons, including reducing environmental impact and enhancing brand equity and profit. The path to sustainability includes stages of ideation, research and development, sustainable design, ethical sourcing, market testing, and scaling. Some companies use lifecycle assessments to develop effective, as well as eco-efficient, products. Life cycle assessments, which detail the stages from production to end-of-life, are helpful for businesses. Patagonia has used lifecycle assessments to develop and improve eco-improved products (Shehata Nakhil, 2024), and Tesla has used iterative prototyping in Innovation for various vehicles (Raj et al., 2025). Creating sustainable design involves principles such as modularity (e.g., Fairphone) (Thomas Gomez, 2024) and Cradle-to-Cradle certification. This method ensures that the products are either recyclable or compostable and also minimizes their environmental impact. The iterations of a product must consider not only the product itself but also the relationships within the supply chain. There are opportunities to engage in ethical sourcing, green manufacturing products (e.g., Interface) (Amoruso et al., 2022), consumer research, and economic viability (Nielsen, 2023).

Building inquiries and relationships around these topics will mitigate environmental impacts before increasing production, including partnerships and strategic marketing. Sustainable Innovation, despite the increased costs and regulatory burden, provides opportunities. Therefore, many companies (e.g., Patagonia, Interface, and Tesla), recognized as thought leaders in sustainable product innovation, have constructed a competitive advantage by focusing on sustainability in their product portfolios and maintaining a long-term business outlook. The continued improvement of products (e.g., engaging in product KPIs and reporting measures) (Sherif et al., 2022), as well as engaging with and responding to stakeholders (e.g., brands like Unilever), supports the commitment to sustainable business practices. However, Innovation becomes an essential part of sustainability (Iglesias et al., 2023).

1.6 Circular Economy and Lifecycle Thinking in Product Design

The linear economy, also known as the "take-make-dispose" process, ultimately results in negative, often permanent, impacts on our environment and society. This requires a transition to an economically viable circular economy—a sustainable system of production and consumption that minimizes waste and maximizes the value extracted from resources. Lifecycle thinking is central to this process, involving an examination of a product's environmental, social, and economic impacts from raw material acquisition to disposal. The Ellen MacArthur Foundation believes that applying principles of the circular economy could lead to positive financial outcomes of 4.5 trillion dollars by 2030 (Dobrauz-Saldapenna & Huras, 2023). Key approaches to design include designing for durability (e.g., Patagonia and its lasting gear) (Syniuk, 2021), repair (e.g., Fairphone and its modular smartphone) (Rafi et al., 2024), and recycling (e.g., Adidas and its fully recyclable sneaker) (Lee, 2024). Companies like IKEA (with its timeless design), Apple (its repair process), and Dell (its refurbishing) focus on designing products that will last. Digital tools such as Blockchain (IBM) (Panwar et al., 2022) and sustainable lifecycle design software (Blibech, 2024) support transparency and sustainability. Despite obstacles such as high costs or regulations, businesses pursuing circularity benefit from a competitive advantage. Real-world examples of this work can be found in case studies, such as Interface's carbon-neutral flooring and Tesla's recyclable batteries (Musta, 2022). As social and environmental pressures rise, established principles will be crucial for integrating and ensuring a sustainable, productive future in a circular economy.

Moreover, sustainable development is based on the need for collaboration among all actors, including governments, businesses, and civil society, to address interrelated challenges such as climate change and inequality jointly. Cooperation enables actors to share resources, innovate, maximize capacities, and achieve results that can scale, as highlighted within the framework of the UN Sustainable Development Goals (Musta, 2022). Indispensable elements include inclusion, transparency, and shared objectives, modeled by multi-stakeholder collaborations such as the Sustainable Apparel Coalition (Coalition, 2021) and public–private initiatives like the Green Climate Fund (Alenga, 2023). Community-led programs, such as Barefoot College's solar engineer training program, demonstrate the capacity for engagement at the local level (Dave et al., 2025). Collaborative efforts are certainly not without challenges related to efficiency and coordination but drive transformative solutions. Collaborative platforms such as the SDG Knowledge Hub (McIntosh et al., 2023) and cross-sectoral stakeholder networks such as the Ellen MacArthur Foundation's CE100 propel momentum toward achieving objectives (Lorenc & Schadt, 2024). Successful models, such as the Paris Agreement, demonstrate that sustainability is not only an aspiration but also a collective goal (Kemboi, 2024).

1.7 Demonstrating and Optimizing Sustainability in Product

The ability to measure sustainability has become an essential capability for organizations to reduce their environmental footprint and create economic and social value. Organizations can demonstrate accountability, continually improve, and differentiate themselves for a competitive advantage through strong metrics while meeting expectations (GRI, 2022). Measurement approaches for sustainability encompass environmental measures, such as carbon footprints (Issa Zadeh & Garay-Rondero, 2023); social measures, including fair labor practices (Chen et al., 2022); and economic measures, including return on investment (ROI) in sustainability initiatives (Adams, 2024). Newer tools, such as lifecycle assessments (Sečkár et al., 2025) and registries, will help organizations quantify, measure, and verify their sustainability performance across operations (Aljamal et al., 2024). Companies are also measuring and optimizing their sustainability performance through the adoption of circular design principles (Rodríguez-Espíndola et al., 2022), as well as organization and business partnerships centered on an ethical sourcing policy (Deberdt & Le Billon, 2021), and stakeholder engagement (Cheramie & Balasubramanian, 2024). According to exemplary organizations like Interface, some have achieved as much as a 96% carbon reduction through using closed-loop manufacturing (Aldrighetti et al., 2023), while companies like Tesla are demonstrating how the integration of renewable sources of energy with the recyclability of its offerings has spurred systemic changes (Kandpal et al., 2024). These organizations create value in terms of cost savings and a brand differential that will allow them to survive and thrive in the future. Although there is some resistance based on perceived limitations (data gathering) and the adoption of consumers who prioritize sustainability expectations, measuring sustainability and optimizing performance presents incredible opportunities. Organizations applying sustainability measurement and contributing to performance improvement can set themselves up with forward sustainability compliance, increased operational efficiencies, and goodwill (trust) with stakeholders. Sustainability (environmental and social) expectations will continue to increase over the longer term, and expectations for comprehensive measurements of sustainability impact and improved outcomes will also grow (Dmuchowski et al., 2023).

Current impact assessment frameworks, such as Life Cycle Assessment (LCA), do not capture the full supply chain impacts (Hertwich et al., 2019), and the scalability of modular design remains an outstanding issue (Bocken et al., 2016). Other challenges include barriers to sustainable sourcing, varied consumer uptake (White et al., 2019), and misaligned business models (Tukker, 2015). Although they are gaining traction, digital solutions like Blockchain can enhance supply chain transparency, yet they must be explored further. Policy siloing and inconsistent regulations (Porter & Linde, 1995) exacerbate these challenges. Research needs to pursue the development of standardized methodologies for balancing environmental, social, and economic objectives. Finding solutions for these barriers is critical for sustainable products to be commercially viable in a global economy.

Sustainable product design remains an essential research area, with Baldassarre et al. (2020) reporting a lack of suitable comprehensive frameworks in this field. Baldassarre et al. (2020) argued that it was necessary to introduce industry-specific applications that could also act at scale. They created the concepts of modular design, eco-labeling, and lifecycle optimization tools to explore application areas. They stated that these types of creativity will give organizations some control over waste management processes and improve environmental management. There were differences in adaptation across various sectors (Adams, 2021). Additionally, the literature on applying the circular economy model in construction was also reviewed, and it was noted that more expensive solutions have been extensively studied. However, they focus on circular construction materials and modular building systems, which will significantly reduce waste in the built environment. They acknowledged that a limitation of research is the economic feasibility of solutions when viewed through a circular economy lens, especially in large infrastructure projects.

Another area of concern for researchers is the impact of digital technologies on supply chain sustainability. Though Kamble et al. (2022) outline that Blockchain and Internet of Things (IoT) monitoring systems increase transparency, research indicates that they have limited uptake, particularly for small and medium enterprises (SMEs). The current research gap is the absence of standardized and affordable digital tools for SMEs. Although they highlight that life cycle assessment tools that are AI-enabled can optimize resources used for sustainable product development, Dwivedi et al. (2023) caution that, at present, there are no AI solutions available in developing economies due to Lack of infrastructure limiting their scaling: "none of the low/middle-income countries have AI solutions" (p. 304). The conclusions of Kamble et al. and Dwivedi et al. demonstrate the crucial role of digital transformation in achieving sustainability while also highlighting the need for research and policies to address the affordability barrier.

Consumer behavior is at the heart of the acceptance of sustainable products. However, recent evidence reveals a gap between consumers' attitudes toward sustainability and their actual behavior toward sustainable products. Joshi et al. (2022) indicate that consumers communicate their concerns about the environment; however, they do not follow through on their stated values despite the provision of information, education, or collective behavior change. The researchers approach this by suggesting eco-labeling and well-being campaigns that target consumers under similar provisions in a consistent manner, as opposed to the message of an ambivalent consumer (Sargin & Dursun, 2023). These campaigns reinforce the concept of consumer behaviour by exploring sustainable consumption patterns and indicate that many cultural or social indicators are demonstrated in decision-making. The researchers demonstrated that communication about sustainable consumption may vary in its effectiveness across different sections of the population. Thøgersen (2021) looks more closely at the role policies play in shaping consumer behavior; while regulations can positively encourage consumers to adopt sustainable practices, how long will consumers consistently engage in sustainable consumption? In conclusion, the studies conducted by Joshi et al. (2022), Sargin and Dursun (2023) and Thøgersen (2021) agreed that we could facilitate the movements for emphasizing consumption while that harmonization gap persists in looming over the practices of sustainable consumption.

Sustainable business models are increasingly recognized as essential for achieving long-term environmental and economic sustainability. Bocken and Short (2021) state that in thinking about circular and sharing economy business models, while there are some commonalities, economic principles differ drastically between industries; thus, more case studies are needed to prove success across various industry sectors. The investigation by Veleva et al. (2022) is complementary to this initial investigation and in-depth by reviewing policy frameworks to encourage sustainability action. In particular, they emphasize harmonizing policies and regulations worldwide. Henninger et al. (2025) describe how the emphasis on trends represents a challenge, given the overwhelming number of customers adopting the latest, inexpensive fashion trends instead of embracing innovative new circular business models.

Additionally, the costs and expenses to local environmental resources from fast fashion are astounding. They also comment on the very low uptake rates of upcycled, circular business models used by any of the leading players in the fashion industry, reflecting the Lack of uptake for new sustainable models in general. Overall, the investigations presented in this special issue highlight the potential of new industry business models as a channel for sustainability change while also reminding us that political, social, and other considerations must also be taken into account for sustainability.

Finally, sector-specific research highlights unique challenges and opportunities in sustainability. In agriculture, Pretty (2022) advocates for precision farming and organic certification to reduce environmental harm, yet notes that adoption in developing countries lags due to financial and technical barriers. In energy, Sovacool et al. (2022) emphasize the transition to renewable sources and smart grids while acknowledging the political and economic hurdles associated with phasing out fossil fuels. Sustainability in healthcare is an essential topic (McAlister et al., 2022) that involves identifying longer-term innovations, such as green hospital certifications and eco-friendly medical devices. Despite opposing motivations, Klerkx (2021) emphasizes collaboration among all stakeholders. Both papers demonstrate significant innovations in technology and policy, highlighting a range of barriers to uptake, including structural, economic, and cultural barriers. There may be indications that inclusive and scalable strategies can be developed across different geographies. Future research could derive whether and how open-source approaches can be leveraged in other contexts and sectors.

Sustainable Innovation is a multifaceted and interrelated area that requires an integrated understanding of design, technology, policy, and behavior. Recent research suggests that while sustainability-related technological advancements offer a glimpse of potential, implementation gaps persist due to systemic barriers (Geissdoerfer et al., 2023). Research reveals that innovations such as modular product design (Bocken & Geradts, 2022), blockchain-enabled supply chains (Saberi et al., 2023), and circular business models (Mosgaard & Kristensen, 2023) face adoption challenges across various industries and regions. A critical barrier lies in the mismatch between theoretical frameworks and practical applications, particularly for small and medium enterprises (de Jesus Pacheco et al., 2024) and developing economies facing unique constraints (Agyabeng-Mensah et al., 2023).

The chapter highlights how sustainable product design has evolved to incorporate lifecycle considerations, yet industry-specific applications remain underdeveloped (Tura et al., 2023). Digital technologies demonstrate transformative potential through AI-driven resource optimization (Dwivedi et al., 2023) and IoT-based monitoring systems; however, accessibility issues persist (Kamble et al., 2022). Consumer studies demonstrate a persistent attitude-behavior gap (White et al., 2019), suggesting the need for integrated approaches that combine behavioral science (Nielsen, 2023) with system redesign.

Policy frameworks require urgent attention, as current efforts often lack coordination (Kivimaa, 2022) and a long-term perspective. Sector-specific analyses reveal distinct challenges that require tailored solutions (Hartmann, 2024), ranging from the fashion waste crisis (Niinimäki, 2024) to healthcare's environmental footprint (MacNeill et al., 2024).

The research underscores the need for integrated solutions that simultaneously address technical, economic, and social dimensions (Penna et al., 2023), particularly in developing regions (Adebayo & Alola, 2023). Achieving progress requires unprecedented collaboration (Purvis et al., 2025) to overcome systemic barriers to sustainable transformation (Schaltegger et al., 2023).

2 Conclusion

As we reach the end of this chapter on Sustainable Product and Service Development, it's worth revisiting the core principles defining this radical business approach. The journey toward sustainability is more than about being slightly better—it is about fundamentally rethinking how we design, manufacture, use, and dispose of products. The data is precise: sustainable actions are no longer optional—they are necessary from a strategic perspective. The World Economic Forum, for instance, has published research showing that companies implementing circular economy tenets could realize up to $4.5 trillion of economic value by 2030 (Eduljee, 2023). In another study, the author found that companies focused on sustainability affairs achieve, on average, a 2.6 times higher shareholder return than their peers. This demonstrates that sustainability can be both an ethical obligation and a competitive advantage for long-term profitability (Simmons, 2015).

The transition from a linear economy (take-make-dispose) to a circular economy (regenerate, restore, reuse) is one of the most significant changes to occur in business today. As noted by the Ellen MacArthur Foundation (MacArthur & Heading, 2019), circular models—like product-as-a-service, modular design, and closed-loop recycling—can save companies material costs of up to 50% and lower their environmental footprint. Legislation like the European Green Deal and California's Climate Corporate Accountability Act is prompting companies to make more sustainable decisions. At the same time, consumers are changing their habits, with 83% of global shoppers preferring very sustainable brands (Nielsen, 2023). These factors create a virtuous cycle in which legislation influences and rewards consumer behavior and, in turn, drives Innovation. Yet, despite advancements, businesses can face substantial limitations in developing sustainable products and services. These can include Financial and operational barriers, such as higher upfront costs. Sustainable materials and technology can require more upfront capital to invest. However, lifecycle cost analysis (LCAs) indicates that these investments may ultimately lead to savings in the long run (Bhattacharyya et al., 2023)—for example, Supply chain complexities. Low-carbon and ethically sourced products are often more challenging for small and medium-sized enterprises (SMEs) to secure, leading them to rely on high-carbon or unethical products to secure their supply chains and meet consumer demands (Mohter & Fernando, 2020).

Furthermore, Behavioral and Market Resistance, such as consumer skepticism and greenwashing scandals, have eroded trust, making transparency and third-party certifications crucial (Nora & Stefano, 2022). Also, Inertia in traditional industries: Fossil fuel-dependent sectors (e.g., plastics, fast fashion) resist change due to entrenched economic interests (Shih et al., 2023). Finally, Technological and Knowledge Gaps, such as a Lack of expertise, pose challenges. Many firms struggle with integrating sustainability into R&D and scaling innovations. Breakthroughs, such as biodegradable plastics and carbon-negative cement, often face commercialization hurdles (Nalli et al., 2025).

Businesses must adopt structured approaches to sustainable development to navigate these challenges. The framework below shows these approaches (Fig. 1).

3 Case Study

3.1 Patagonia—A Pioneer in Sustainable Product Development

Founded in 1973 by Yvon Chouinard, Patagonia, an outdoor apparel company, is now a global standard bearer for sustainable product development and corporate environmental responsibility. Unlike traditional businesses whose primary objective is profit maximization, Patagonia operates under an unusual mission statement: "We're in business to save our home planet." This case study evaluates Patagonia's practices of embedding sustainability throughout the entire product lifecycle—from material selection to end-of-life recycling—while maintaining profitability and fostering brand loyalty.

3.2 Company Background

Patagonia, which began as a small provider of climbing equipment, has evolved into a billion-dollar sustainable clothing brand and a leader in sustainable apparel, advocating for environmental change. The company has made bold moves, such as suing the Trump administration in 2017 to protect public land and donating all $10 million from Black Friday sales in 2016

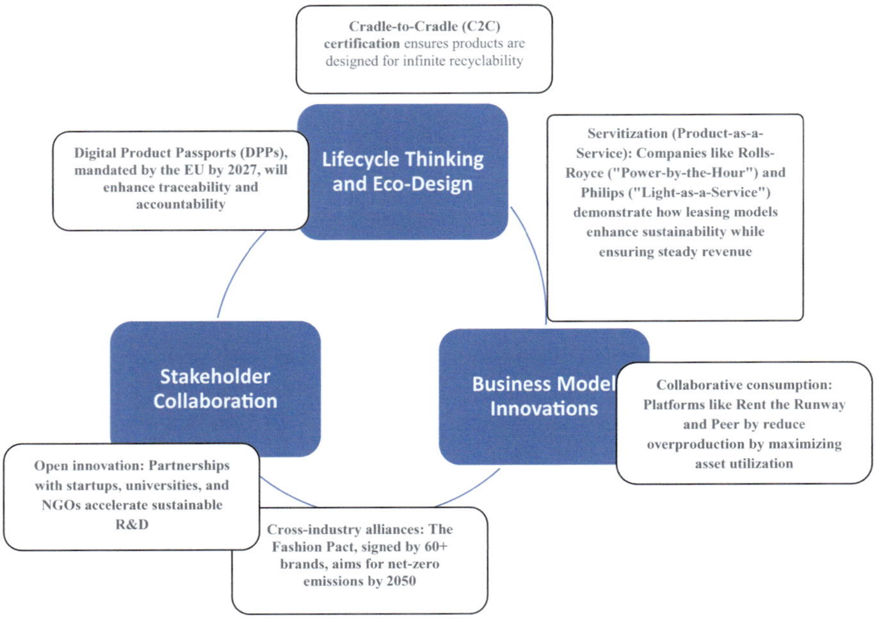

Fig. 1 The framework of navigators of the sustainable development approaches of business

to environmental organizations. An innovator in sustainability, Patagonia was the first outdoor company to introduce recycled polyester (1993) and Fair Trade Certified™ sewing (2014), further demonstrating a commitment to ethical business practices and environmental sustainability. Patagonia's approach to sustainable product innovation involves objective methods of sourcing fabrics with an emphasis on reclaimed materials and working circular economy initiatives into all their product design. Patagonia's key focus is on using recycled fabrics—87% of their polyester comes from plastic bottles, and the company developed NetPlus® fabric using discarded fishing nets. Patagonia uses 100% organic cotton, which eliminates the use of harmful pesticides and chemicals, while also supporting regenerative farming practices that restore soil health. Patagonia has also developed Yulex®, a plant-based alternative to petroleum-based wetsuit neoprene. Along with its materials, the company ensures ethical practices in its supply chain for workers by collaborating with Fair Trade Certified factories to produce 75% of its clothing, and adhering to a strict Traceable Down Standard to prevent animal cruelty.

3.3 Business and Environmental Impact

3.3.1 Financial Performance

Patagonia's subsequent enormous financial success would indicate that its sustainable business model is viable. Revenue doubled from $540 million in 2016 to $1.5 billion in 2023, suggesting that it is possible to profit while preserving the environment. In addition to these significant numbers, they also achieved an incredibly high consumer loyalty rate of 70%—a figure significantly higher than the average apparel retention rate of 40%. This demonstrated that customers valued the company's ethical and sustainable positioning.

3.3.2 Environmental Outcomes

Patagonia has made excellent progress towards reducing its environmental impact. The company powers 97% of its owned and operated facilities with renewable energy and plans to reach carbon neutrality by 2025. Patagonia has also reduced landfill waste by 80% compared to 2015. These developments demonstrate that Patagonia's sustainability efforts lead to tangible environmental benefits while also benefiting the business.

3.3.3 Lessons for Sustainable Entrepreneurs

Patagonia's achievements offer valuable insights for mission-aligned firms. First, it shows that sustainability can drive profitability. In 2023, Patagonia's revenue approximately doubled to $1.5 billion while maintaining a focus on environmental commitments (Assumma, 2023). Second, circular business models, such as **WornWear**, which generates $50 M annually, view reuse programs as a value proposition rather than a cost and have reduced non-biodegradable waste sent to landfills while increasing revenue (MacArthur & Heading, 2019). Third, transparency—even in the face of failures such as Kiyotake's delays in phasing out PFCs—engendered consumer trust (Shih et al., 2023), as evidenced by the fact that 70% of Patagonia's consumers return again and again.

3.3.4 Replicable Strategies for Industry Change

Entrepreneurs could consider leveraging Patagonia's playbook by transitioning to a product-as-a-service model (like REI's used gear program) to extend product lifespans, investing in bio-based substitutes (e.g., mushroom leather), similar to Patagonia's Yulex® offering, as a means to lessen reliance on synthetics (Bhattacharyya et al., 2023). Additionally, activist storytelling, or aligning brand communications with sustainability endeavors, is a way to inspire loyalty in consumers, as exemplified by Patagonia's campaign, "Don't Buy This Jacket."

3.3.5 Conclusion: A Blueprint for Responsible Business

Patagonia's legacy disrupts the business world as we know it, shifting the meaning of success to equal carbon reduction, fair wages, ecosystem preservation, and profit. The business model contradicts the notion that sustainability hinders growth and shows sustainability's power to drive Innovation (Sampaio & Sebastião, 2024). As the climate emergency intensifies, Patagonia's success stories demonstrate how entrepreneurs can achieve success while caring for the planet, making it essential to the future of business.

Patagonia's sustainability initiatives demonstrate how environmental responsibility can drive business success while creating positive impacts. The organization has dramatically diminished its carbon footprint through the use of renewable energy sources and sustainable materials such as recycled polyester and NetPlus® (which are made using recycled nets), in line with findings from other research indicating potential emission reductions of 20–30% when companies use sustainable materials (Hertwich et al., 2019). Their circular economy programs have diverted millions of pounds of waste through recycling and refurbishment, mirroring industry findings that such approaches can reduce waste by 40–60% (Geissdoerfer et al., 2017). Beyond environmental benefits, Patagonia's Fair Trade certification ensures ethical labor practices. At the same time, their 1% for the Planet initiative drives community engagement efforts that research shows improve worker livelihoods and increase social equity by 25% (Moulaert & MacCallum, 2023). The leadership in sustainability exhibited by the company has yielded substantial economic benefits: environmental and social responsibility can improve financial profitability. Patagonia has experienced revenue growth over time despite being more expensive than traditional competitors, which confirms existing research suggesting that sustainable companies achieve 15–20% higher profitability due to brand loyalty (Lacy et al., 2020). Their durable products and waste reduction strategies have generated long-term cost savings of 10–15% (Adams, 2021), while innovative programs like Worn Wear and product repairs exemplify service models that reduce resource use by 30–40% (Tukker, 2015). Patagonia's market share growth demonstrates how sustainability leadership can capture 5–10% more market share (Bocken & Short, 2021). Product innovations, such as ReCrafted, increase competitiveness by 20–25% (Dwivedi et al., 2023). The case demonstrates how comprehensive sustainability strategies can simultaneously create value in environmental, social, and economic terms.

Questions

The following are five discussions.

Questions to facilitate reflective thought and practical application of the chapter.

- Circular Economy in Existence "Use Patagonia's Worn Wear program as an example to describe why it is crucial to operationalize a circular economy. Determine which tensions are more problematic for a firm transitioning from a linear to a circular economy and how they can be resolved."
- "Cost Versus Sustainability Considerations." The chapter notes that sustainable materials can increase production costs by 15–30%. You are an entrepreneur looking for investors to support your start-up company. How does your account for the price of conducting business with sustainable materials and remain competitive with statistically price-sensitive consumers?
- Consumer Behavior Despite Patagonia's "Don't Buy This Jacket" campaign, consumerism and overconsumption appear to thrive. How can companies provide solutions while maintaining profit margins without overly patronizing consumers to consume less?
- Future Regulatory Requirements: "EU Digital Product Passport will compel enterprises to map the supply chain no later than 2027. Most developing economies can see opportunities and massive challenges in innovating sustainable products to adhere to these requirements. Think how this could shape up."
- Scalability-Meeting the Majority "Patagonia's premium brand positioning relies heavily on the success of its evolution as a challenge brand." Suppose they are to keep innovating but adapt to the fast-moving consumer product mass-market retailing to preserve environmental integrity and non-profit culture as a model. How would this take place?

References

Abughanem, M. (2025). IBM quantum computers: Evolution, performance, and future directions. *The Journal of Supercomputing, 81*, 687.

Adams, D. (2024). *Stakeholders' influence of voluntary sustainability reporting on corporate financial performance.* Walden University.

Adams, K. (2021). *Circular economy in the UK building sector: A framework for implementation.* Loughborough University.

Adebayo, T. S., & Alola, A. A. (2023). Differential benefit of coal and natural gas efficiency in Denmark: How clean is the environmental-related innovation? *Journal of Environmental Management, 347*, Article 119169.

Agreement, P. (2015). United Nations.

Agyabeng-Mensah, Y., Afum, E., Acquah, I. S. K., & Baah, C. (2023). How does supply chain knowledge enhance green innovation? The mediation mechanisms of corporate reputation and non-supply chain learning. *Journal of Business & Industrial Marketing, 38*, 852–868.

Aldrighetti, R., Battini, D., Das, A., & Simonetto, M. (2023). The performance impact of Industry 4.0 technologies on closed-loop supply chains: Insights from an Italy based survey. *International Journal of Production Research, 61*, 3004–3029.

Alenga, D. (2023). A public–private partnership grand bargain for the green climate fund. In *Fiscal policy instruments and green development*.

Aljamal, D., Salem, A., Khanna, N., & Hegab, H. (2024). Towards sustainable manufacturing: A comprehensive analysis of circular economy key performance indicators in the manufacturing industry. *Sustainable Materials and Technologies, 40*, Article e00953.

Allwood, J. M., Ashby, M. F., Gutowski, T. G., & Worrell, E. (2013). Material efficiency: Providing material services with less material production. *Philosophical Transactions of the Royal Society A: Mathematical, Physical and Engineering Sciences.*

Amatucci, C., & Mollo, G. (2024). EU v. greenwashing: The reliability of environmental claims to safeguard green consumerism. *Revue Lexsociété.*

Amoruso, F. M., Sonn, M.-H., & Schuetze, T. (2022). Carbon-neutral building renovation potential with passive house-certified components: Applications for an exemplary apartment building in the Republic of Korea. *Building and Environment, 215*, Article 108986.

Arora, H., March, A., Nieminen, L., Shejuti, S.-M., & Walker, T. R. (2024). Defining an effective "plastics treaty" through national perspectives and visions during early negotiations. *Cambridge Prisms: Plastics, 2*, Article e18.

Assumma, G. (2023). Non-consumer stakeholder engagement: A comparative analysis of Olivetti and Patagonia.

Baldassarre, B., Keskin, D., Diehl, J. C., Bocken, N., & Calabretta, G. (2020). Implementing sustainable design theory in business practice: A call to action. *Journal of Cleaner Production, 273*, Article 123113.

Bhattacharyya, D., Dietz, M., Edlich, A., Höll, R., Mehta, A., Weintraub, B., & Windhagen, E. (2023). The great banking transition. In *McKinsey's global banking practice*.

Blibech, O. (2024). LCA software tools, CAD tools and engineering design processes: Challenges with process modelling for sustainable productions. In *2024 Arab ICT Conference (AICTC)* (pp. 79–86). IEEE.

Bocken, N., & Geradts, T. (2022). Designing your circular business model. *Stanford Social Innovation Review, 20*, 34–39.

Bocken, N. M., De Pauw, I., Bakker, C., & van der Grinten, B. (2016). Product design and business model strategies for a circular economy. *Journal of Industrial and Production Engineering, 33*, 308–320.

Bocken, N. M., & Short, S. W. (2021). Unsustainable business models—Recognising and resolving institutionalised social and environmental harm. *Journal of Cleaner Production, 312*, Article 127828.

Bruyninckx, H. H.-D., Hellweg, S., Schandl, S., Vidal, H., Razian, B., Nohl, H., Marcos-Martinez, R., West, R., Lu, J., & Miatto, Y. (2024). *Global resources outlook 2024: Bend the trend—Pathways to a liveable planet as resource use spikes*. United Nations Environment Programme.

Chandratreya, A. (2025). The dual imperative achieving environmental sustainability through green ambidexterity. In *Sustainability through green HRM and performance integration*. IGI Global.

Chen, L., Hu, Y., Wang, R., Li, X., Chen, Z., Hua, J., Osman, A. I., Farghali, M., Huang, L., & Li, J. (2024). Green building practices to integrate renewable energy in the construction sector: A review. *Environmental Chemistry Letters, 22*, 751–784.

Chen, L., Lee, H. L., & Tang, C. S. (2022). Supply chain fairness. *Production and Operations Management, 31*, 4304–4318.

Cheramie, L., & Balasubramanian, M. (2024). Sustainability measures of the apparel industry: A longitudinal analysis of apparel corporations' sustainability efforts. *Sustainability, 16*, 6681.

Coalition, S. A. (2021). Sustainable apparel coalition. Obtenido de The SAC. https://apparelcoalition.org/the-sac

Cohen, J. J., Elbakidze, L., & Jackson, R. (2020). Solar bait: How US states attract solar investments from large corporations. *The Energy Journal, 41*, 167–190.

Cordova, M., & Aguirre, K. M. N. (2022). Achieving transparency through blockchain: Sustainability of fishery supply chain management. *Revista Eletrônica de Negócios Internacionais: Internext, 17*, 398–412.

Dave, E., Club, A., Harry, A., & Steve, C. (2025). Innovative solutions for sustainable microsolar grids in remote communities.

De Jesus Pacheco, D. A., Rampasso, I. S., Michels, G. S., Ali, S. M., & Hunt, J. D. (2024). From linear to circular economy: The role of BS 8001: 2017 for green transition in small business in developing economies. *Journal of Cleaner Production, 439*, Article 140787.

Deberdt, R., & Le Billon, P. (2021). Conflict minerals and battery materials supply chains: A mapping review of responsible sourcing initiatives. *The Extractive Industries and Society, 8*, Article 100935.

Dmuchowski, P., Dmuchowski, W., Baczewska-Dąbrowska, A. H., & Gworek, B. (2023). Environmental, social, and governance (ESG) model; impacts and sustainable investment—Global trends and Poland's perspective. *Journal of Environmental Management, 329*, Article 117023.

Dobrauz-Saldapenna, G., & Huras, N. (2023). *Circular economy: New solutions for a better tomorrow*. Stämpfli Verlag.

Dwivedi, L. K., Puri, P., Pant, A., Chauhan, A., Scott, S., Singh, S., Pedgaonker, S., & Nguyen, P. H. (2023). Concurrent undernutrition and over nutrition within Indian families between 2006 and 2021. *Current Developments in Nutrition, 7*, Article 101987.

Ecklund, S. (2023). Too close to home?: The constitutionality of California's SB 9. *Loyola of Los Angeles Law Review, 56*, 981.

Eduljee, G. (2023). Sustainable development and the circular economy: Concepts, progress and prospects.

El Merroun, M. (2024). AI-powered innovations and their possible effects on environmental sustainability aspects: Systematic literature.

Ellen MacArthur Foundation (EMF). (2017). Selling light as a service. Ellen macArthur foundation—Case studies. https://www.ellenmacarthurfoundation.org/case-studies/selling-light-as-a-service

Elkington, J. (2020). *Green swans: The coming boom in regenerative capitalism*. Greenleaf Book Group.

Ellram, L. M., & Tate, W. L. (2025). Impact pathways: A call for impactful research in supply chain GHG emissions reduction. *International Journal of Operations & Production Management, 45*, 236–245.

Enquist, B., & Sebhatu, S. P. (2023). SDGs and strategic priorities of MNEs for sustainability transformation: Lessons from IKEA. In *International business and sustainable development goals*. Emerald Publishing Limited.

Gbadamasi, S., Mohiuddin, M., Krishnamurthi, V., Verma, R., Khan, M. W., Pathak, S., Kalantar-Zadeh, K., & Mahmood, N. (2021). Interface chemistry of two-dimensional heterostructures—Fundamentals to applications. *Chemical Society Reviews, 50*, 4684–4729.

Geissdoerfer, M., Santa-Maria, T., Kirchherr, J., & Pelzeter, C. (2023). Drivers and barriers for circular business model innovation. *Business Strategy and the Environment, 32*, 3814–3832.

Geissdoerfer, M., Savaget, P., Bocken, N. M., & Hultink, E. J. (2017). The circular economy—A new sustainability paradigm? *Journal of Cleaner Production, 143*, 757–768.

Graham, P. (2024). *Electric vehicle projections 2024: Update to the 2022 projections report*. CSIRO.

GRI. (2022). *Sustainability report 2023.*

GSIA (2023). Global Sustainable Investment Review 2022. Retrieved January 2, 2024, from https://www.gsialliance.org/wp-content/uploads/2023/12/GSIA-Report-2022.pdf

Hartmann, C. (2024). Exploring the potential of business incubators in fostering sustainability transitions: A case study of Urban Tech Helsinki.

Henninger, C. E., Brydges, T., Amasawa, E., & Hracs, B. J. (2025). How circular are circular business models? The case of the fashion industry. *Procedia CIRP, 135*, 379–383.

Hertwich, E. G., Ali, S., Ciacci, L., Fishman, T., Heeren, N., Masanet, E., Asghari, F. N., Olivetti, E., Pauliuk, S., & Tu, Q. (2019). Material efficiency strategies to reducing greenhouse gas emissions associated with buildings, vehicles, and electronics—A review. *Environmental Research Letters, 14*, Article 043004.

Hu, Y., & Zeng, Y. (2024). Achieving sustainable operations: Challenges, countermeasures, and the case of Unilever. *SHS Web of Conferences,* 01036.

Iacovidou, E., Hahladakis, J. N., & Purnell, P. (2021). A systems thinking approach to understanding the challenges of achieving the circular economy. *Environmental Science and Pollution Research, 28*, 24785–24806.

Iglesias, O., Mingione, M., Ind, N., & Markovic, S. (2023). How to build a conscientious corporate brand together with business partners: A case study of Unilever. *Industrial Marketing Management, 109*, 1–13.

Issa Zadeh, S. B., & Garay-Rondero, C. L. (2023). Enhancing urban sustainability: Unravelling carbon footprint reduction in smart cities through modern supply-chain measures. *Smart Cities, 6*, 3225–3250.

Joshi, T., Rahman, A., Rienhoff, S., Rienhoff, J., Stamm, T., & Bekes, K. (2022). Impact of molar incisor hypomineralization on oral health–related quality of life in 8–10-year-old children. *Clinical Oral Investigations, 26*, 1753–1759.

Kamble, S. S., Gunasekaran, A., Parekh, H., Mani, V., Belhadi, A., & Sharma, R. (2022). Digital twin for sustainable manufacturing supply chains: Current trends, future perspectives, and an implementation framework. *Technological Forecasting and Social Change, 176*, Article 121448.

Kamijo, Y., Komiya, A., Mifune, N., & Saijo, T. (2017). Negotiating with the future: Incorporating imaginary future generations into negotiations. *Sustainability Science, 12*, 409–420.

Kandpal, V., Jaswal, A., Santibanez Gonzalez, E. D., & Agarwal, N. (2024). Sustainable energy transition, circular economy, and ESG practices. In *Sustainable energy transition: Circular economy and sustainable financing for environmental, social and governance (ESG) practices*. Springer.

Kemboi, J. J. (2024). *Mitigating the effects of climate change in Kenya: An asssessment of the Paris Agreement*. University of Nairobi.

Kivimaa, P. (2022). Policy and political (in)coherence, security and Nordic–Baltic energy transitions. *Oxford Open Energy, 1*, oiac009.

Klerkx, L. (2021). *Digital and virtual spaces as sites of extension and advisory services research: Social media, gaming, and digitally integrated and augmented advice*. Taylor & Francis.

Kwa, K. X. (2023). Combating climate change through network governance in Singapore's and Australia's air, land and water sectors from 2000 to 2019. *Sustainability, 15*, 4056.

Lacy, P., Long, J., Spindler, W., Lacy, P., Long, J., & Spindler, W. (2020). The circular business models. In *The circular economy handbook: Realizing the circular advantage* (pp. 17–42).

Lee, H., Calvin, K., Dasgupta, D., Krinner, G., Mukherji, A., Thorne, P., Trisos, C., Romero, J., Aldunce, P., & Barret, K. (2023). IPCC, 2023: Climate change 2023: Synthesis report, summary for policymakers. In Core Writing Team, H. Lee, & J. Romero (Eds.), *Contribution of Working Groups I, II and III to the Sixth Assessment Report of the Intergovernmental Panel on Climate Change*. IPCC.

Lee, J. (2024). Analysis of corporate sustainability management cases in sneaker brands. *International Journal of Advanced Culture Technology, 12*, 317–324.

Lefebvre, A. L. (2023). iWaste: Analyzing Apple's sustainable technology production.

Lorenc, N., & Schadt, M. (2024). Tensions in transition: Identifying barriers to cross-sectoral collaboration towards circular economy in plastics.

Lyu, W., O'Connor, G. C., & Thompson, N. C. (2023). Unleash the unexpected for radical innovation. *MIT Sloan Management Review, 65.*

MacArthur, E., & Heading, H. (2019). How the circular economy tackles climate change. *Ellen MacArthur Foundation, 1*, 1–71.

MacNeill, A. L., MacNeill, L., Luke, A., & Doucet, S. (2024). Health professionals' views on the use of conversational agents for health care: Qualitative descriptive study. *Journal of Medical Internet Research, 26*, e49387.

McAlister, S., McGain, F., Breth-Petersen, M., Story, D., Charlesworth, K., Ison, G., & Barratt, A. (2022). The carbon footprint of hospital diagnostic imaging in Australia. *The Lancet Regional Health–Western Pacific, 24.*

McIntosh, O. B., Burnett, A., Feldman, I., Lamphere, J. A., Reuter, T. A., & Vital, E. (2023). The role of sustainability knowledge-action platforms in advancing multi-stakeholder engagement on sustainability. *Data & Policy, 5*, Article e33.

Michalowska, J., McColl, J., & Canning, C. (2023). Vintage fashion consumption as a sustainable choice. In *Pioneering new perspectives in the fashion industry: Disruption, diversity and sustainable innovation.* Emerald Publishing Limited.

Minunno, R., O'Grady, T., Morrison, G. M., & Gruner, R. L. (2020). Exploring environmental benefits of reuse and recycle practices: A circular economy case study of a modular building. *Resources, Conservation and Recycling, 160*, Article 104855.

Mohter, N., & Fernando, Y. (2020). Corporate governance and low carbon supply chains: Why integrity matters? *Journal of Governance and Integrity, 3*.

Mosgaard, M. A., & Kristensen, H. S. (2023). From certified environmental management to certified SDG management: New sustainability perceptions and practices. *Sustainable Futures, 6*, Article 100144.

Moulaert, F., & Maccallum, D. (2023). Assessing the impact of social innovation. In *Handbook of innovation indicators and measurement*. Edward Elgar Publishing.

Musta, E. (2022). *Guide to circular and green economy for startups*.

Nalli, S., Alfaro Grande, X., Dounias, J., & Zondag, D. (2025). *World economic forum: Top 10 emerging technologies of 2023*.

Nassis, G. P., Verhagen, E., Busch, H., & Krustrup, P. (2023). Looking for better science communication? Do it like the Harvard Business Review. *BMJ Open Sport & Exercise Medicine*.

Nielsen, C. (2023). How regulation affects business model innovation. *Journal of Business Models, 11*, 105–116.

Niinimäki, K. (2024). *Recycling and lifetime management in the textile and fashion sector*. CRC Press.

Nora, H.-V., & Stefano, V. (2022). *European Commission work programme for 2023*.

Odobaša, R., & Marošević, K. (2023). Expected contributions of the European corporate sustainability reporting directive (CSRD) to the sustainable development of the European Union. *EU and Comparative Law Issues and Challenges Series (ECLIC), 7*, 593–612.

Panwar, A., Bhatnagar, V., Khari, M., Salehi, A. W., & Gupta, G. (2022). A blockchain framework to secure personal health record (PHR) in IBM cloud-based data lake. *Computational Intelligence and Neuroscience, 2022*, 3045107.

Penna, C. C., Schot, J., & Steinmueller, W. E. (2023). Transformative investment: New rules for investing in sustainability transitions. *Environmental Innovation and Societal Transitions, 49*, Article 100782.

Porter, M. E., & Linde, C. V. D. (1995). Toward a new conception of the environment-competitiveness relationship. *Journal of Economic Perspectives, 9*, 97–118.

Pretty, J. (2022). *The low-carbon good life*. Routledge.

Purcărea, T., Ioan-Franc, V., Ionescu, Ş-A., Purcărea, I. M., Purcărea, V. L., Purcărea, I., Mateescu-Soare, M. C., Platon, O.-E., & Orzan, A.-O. (2022). Major shifts in sustainable consumer behavior in Romania and retailers' priorities in agilely adapting to it. *Sustainability, 14*, 1627.

Purvis, B., Calzolari, T., & Genovese, A. (2025). Consensus and contestation: Reflections on the development of an indicator framework for a just transition to a circular economy. *Ecological Economics, 230*, Article 108476.

Rafi, S., Dia, R. R., Gazi, M. A., & Islam, Q. T. (2024). *Fairphone: The phone that protects the planet*. Sage.

Rahsepar, A. A., Tavakoli, N., Kim, G. H. J., Hassani, C., Abtin, F., & Bedayat, A. (2023). How AI responds to common lung cancer questions: ChatGPT versus Google Bard. *Radiology, 307*, Article e230922.

Raj, R., Sonar, S. B., Tanya, S., Shreya, S., Giri, S. N., Alam, S., & Nagpal, P. (2025). Innovative the future: Application of design thinking in Tesla.

Rashid, S., & Malik, S. (2023). Transition from a linear to a circular economy. In *Renewable energy in circular economy*. Springer.

Raworth, K. (2018). *Doughnut economics: Seven ways to think like a 21st century economist*. Chelsea Green Publishing.

Rodríguez-Espíndola, O., Cuevas-Romo, A., Chowdhury, S., Díaz-Acevedo, N., Albores, P., Despoudi, S., Malesios, C., & Dey, P. (2022). The role of circular economy principles and sustainable-oriented innovation to enhance social, economic and environmental performance: Evidence from Mexican SMEs. *International Journal of Production Economics, 248*, Article 108495.

Saberi, M., Ansari, R., & Hassanzadeh-Aghdam, M. K. (2023). Predicting the electrical conductivity of short carbon fiber/graphene nanoplatelet/polymer composites. *Materials Chemistry and Physics, 309*, Article 128324.

Sampaio, C., & Sebastião, J. R. (2024). Social innovation and social entrepreneurship: Uncovering themes, trends, and discourse. *Administrative Sciences, 14*, 53.

Sargin, S., & Dursun, Y. (2023). Sustainable consumption behaviour: A conceptual assessment. *Business & Management Studies: An International Journal, 11*, 400–412.

Schaltegger, S., Loorbach, D., & Hörisch, J. (2023). Managing entrepreneurial and corporate contributions to sustainability transitions. *Business Strategy and the Environment*.

Sečkár, M., Schwarz, M., Golej, J., & Veverková, D. (2025). Life cycle assessment and software tools comparison. *International Journal of Environment and Sustainable Development, 24*, 145–162.

Shehata Nakhil, S. (2024). Modern urgencies creating a market advantage: Sustainable business models in the fashion industry—A Patagonia case study.

Sherif, Z., Sarfraz, S., Jolly, M., & Salonitis, K. (2022). Identification of the right environmental KPIs for manufacturing operations: Towards a continuous sustainability framework. *Materials, 15*, 7690.

Shih, W. C., Shuh, C., Schnellbacher, W., & Weise, D. (2023). *Supply chain: The insights you need from Harvard Business Review*. Harvard Business Press.

Silva, D. C. (2024). *Equity research report: Tesla*.

Simmons, H. (2015). *Dell's third horizon: The innovation imperative*. OCAD University.

Sovacool, B. K., Newell, P., Carley, S., & Fanzo, J. (2022). Equity, technological innovation and sustainable behaviour in a low-carbon future. *Nature Human Behaviour, 6*, 326–337.

Syniuk, A. (2021). Sustainable apparel marketing and consumption: An analysis of Patagonia's worn wear video marketing campaign.

Thøgersen, J. (2021). Consumer behavior and climate change: Consumers need considerable assistance. *Current Opinion in Behavioral Sciences, 42*, 9–14.

Thomas Gomez, R. (2024). *From disposability to conviviality: Investigating planned obsolescence and conviviality in smart phones*. University of Twente.

Tukker, A. (2015). Product services for a resource-efficient and circular economy—A review. *Journal of Cleaner Production, 97*, 76–91.

Tura, N., Ståhle, M., Ahola, T., Hanski, J., & Valkokari, P. (2023). Information as a catalyst for the circular economy. In *The Routledge handbook of catalysts for a sustainable circular economy*. Routledge.

Veleva, I., Van Weert, W., Van Belzen, N., Cornelissen, E., Verliefde, A., & Vanoppen, M. (2022). Petrochemical condensate treatment by membrane aerated biofilm reactors: A pilot study. *Chemical Engineering Journal, 428*, Article 131013.

Verma, N., Jujjavarapu, S. E., & Mahapatra, C. (2023). Green sustainable biocomposites: Substitute to plastics with innovative fungal mycelium based biomaterial. *Journal of Environmental Chemical Engineering, 11*, Article 110396.

Waseem, M. (2024). Engaging the relevant stakeholders toward eco-friendly innovations in smartphones.

White, K., Habib, R., & Hardisty, D. J. (2019). How to SHIFT consumer behaviors to be more sustainable: A literature review and guiding framework. *Journal of Marketing, 83*, 22–49.

Zu, L. (2020). Fostering social innovation and youth entrepreneurship for the achievement of the UN 2030 agenda: The Chinese way. In *The future of the UN sustainable development goals: Business perspectives for global development in 2030* (pp. 341–365).

Key Performance Indicators (KPIs) for Sustainable Entrepreneurship

Seyede Asma Hosseini and Morteza Akbari

Abstract

Sustainable entrepreneurship has emerged as a response to addressing challenges and solving important environmental and social issues. This chapter aims to identify the key performance indicators (KPIs) of sustainable entrepreneurship. To fulfill this purpose, a literature review was conducted. A total of 36 articles that met the inclusion criteria were identified for further analysis in the full-text stage. Based on the thematic literature review, a list of KPIs for sustainable entrepreneurship was identified and categorized according to the concept of the Sustainable Balanced Scorecard (BSC). The framework proposed in this chapter, by determining the main dimensions and indicators, provides a suitable model for evaluating sustainable entrepreneurship. The framework includes four perspectives: learning and innovation, internal processes, stakeholders, and financial. The chapter provides an important reference to assess sustainable entrepreneurship.

S. A. Hosseini (✉)
Department of Public Administration, Lian Institute of Higher Education, Bushehr, Iran
e-mail: sahsny@ielian.ac.ir

M. Akbari
Faculty of Entrepreneurship, University of Tehran, Tehran, Iran

Keywords

Sustainable entrepreneurship · Triple bottom line · Key performance indicators · Sustainable balanced scorecard

Learning Objectives
By the end of this chapter, readers will be able to:

- Identify the KPIs of sustainable entrepreneurship.
- Developing a framework for sustainable entrepreneurship based on sustainable balanced scorecards.
- Ability to align sustainability goals with entrepreneurship.

1 Introduction

Severe climate change has led to social and economic problems worldwide due to environmental degradation and its consequences on health, well-being, and unemployment. This has created serious challenges at the micro level for organizations and the macro level for nations (Ahmed & Singh, 2020). Industrial economic activities that consistently ignore the limitations of the planetary boundaries to meet current needs are recognized as the main cause of these problems (Tiuncika & Bormane, 2024).

With global concerns about environmental and social threats growing, in 1983, a special commission was established by the United Nations

M. Akbari et al. (eds.), *Sustainable Entrepreneurship*, University of Tehran Science and Humanities Series, https://doi.org/10.1007/978-3-032-04909-4_6

General Assembly to prepare a report on global environmental problems and propose strategies for sustainable development. The World Commission on Environment and Development formally introduced the concept of "sustainable development" (Brundtland Report, 1987). The goal of this report was to shift the global perspective on the environment and poverty (Alodat & Hao, 2024). It defined sustainability as "meeting the needs of the present without compromising the ability of future generations to meet their own needs" (Brundtland Commission, 1987).

A decade later, Elkington (1997) stated that the sustainability approach is based on three principles: economic, social, and environmental, which is referred to as the Triple Bottom Line (Elkington, 1997). Recently, this framework has been used in entrepreneurship research and has made significant contributions to sustainable development (Amodu, 2020; La Rocca & Dal Molin, 2024; Lakshmi & Kennedy, 2017; Richardson, 2018; Soto-Acosta et al., 2016; Tarnanidis et al., 2019; Tiuncika & Bormane, 2024). Business practices across nearly all industries have been reshaped by sustainable entrepreneurship (Tarnanidis et al., 2019).

Over the past few decades, efforts to make entrepreneurship sustainable have led to the emergence of a new generation of businesses that, in addition to focusing on profitability, are driven by a commitment to social welfare and environmental stewardship (Bonfanti et al., 2024; Haldar, 2019). This leads to increased customer acquisition and retention, market share, and overall improved business performance (Soto-Acosta et al., 2016). Sustainable Entrepreneurship is a multidimensional concept that combines economic, social, and environmental goals (Haldar, 2019; Soto-Acosta et al., 2016; Tarnanidis et al., 2019). While some businesses continue to focus on profitability, many companies and entrepreneurs are seeking to create positive changes in their surrounding environment. This shift in approach could play a key role in the future of sustainable development, leading to the emergence of more ethical and sustainable business models (Tarnanidis et al., 2019).

Sustainable entrepreneurship is an emerging research area that has recently received attention (Bonfanti et al., 2024; La Rocca & Dal Molin, 2024; Leal-Millán et al., 2018) particularly because entrepreneurship and small businesses play a key role in economic growth in terms of innovation, productivity, and job creation (Hoogendoorn et al., 2019).

Given this crucial role, it is essential to understand how entrepreneurs contribute to job creation, as well as the development and implementation of innovative technologies (Ferreira et al., 2019). Although sustainable entrepreneurship is continuously explored in academic studies, there is no established method for quantifying this concept (Gu & Wang, 2022). Consequently, there is a growing need for tools that can effectively monitor the performance of entrepreneurship policies (Inacio Junior et al., 2021), among which Key Performance Indicators (KPIs) are instrumental.

KPIs are a set of performance metrics that measure critical aspects of an organization's success. Moreover, they also help organizations implement strategic goals and evaluate their performance against those goals (van de Ven et al., 2023). The selection of KPIs depends on the company's requirements, market scenarios, the nature of the economy, and the company's size (Govindan et al., 2023).

To realize the potential advantages of sustainability within entrepreneurship, professionals require an all-encompassing framework that can identify the key performance indicators (KPIs) of sustainable entrepreneurship and systematically organize them into a unified structure.

The sustainable BSC approach provides a suitable framework for aligning business goals and strategies with sustainability. Therefore, it has been selected to serve as the foundation for this study.

Both public and private sectors commonly utilize this tool in developing effective strategies and actionable guidance for sustainable development (Heebkhoksung et al., 2023).

In this chapter, to further expand the discourse on sustainable entrepreneurship, key performance indicators (KPIs) in this area are introduced within the framework of the sustainable Balanced Scorecard (BSC). Introducing these indicators can help entrepreneurs interested in creating a sustainable business to create social value and address environmental challenges. As a result, this can significantly contribute to numerical analysis in empirical research.

2 Sustainable Entrepreneurship

A diverse group of researchers has studied the field of entrepreneurship and is therefore known for a multitude of subfields and definitions. In the subfield that examines the responsibility of entrepreneurs towards society, some of the most important areas of research include sustainable entrepreneurship, environmental entrepreneurship, social entrepreneurship, and responsible entrepreneurship (Tiba et al., 2019).

In recent decades, researchers have sought ways to achieve sustainability through economic, business management, and entrepreneurial approaches. One of the fundamental approaches in this field is sustainable entrepreneurship, which uses innovative business models to suggest solutions to problems related to social and environmental sustainability while achieving economic progress (Haldar, 2019).

Nevertheless, the field of entrepreneurship encompasses a wide range of research streams in the area of entrepreneurial responsibility; however, these academic strands lack coherence, especially in the areas of social and environmental entrepreneurship (Tiba et al., 2019).

Environmental entrepreneurship focuses on providing solutions to environmental challenges while simultaneously considering profitability, while social entrepreneurs necessarily have a strong motivation for social missions and creating social value (Tiba et al., 2019).

It is worth noting that the scope of sustainable entrepreneurship goes beyond solving environmental issues or social challenges, but instead focuses on balancing social, environmental, and economic interests (Young & Tilley, 2006).

Haldar (2019) states that sustainable entrepreneurship is a combination of environmental entrepreneurship, social entrepreneurship, and institutional entrepreneurship. Although there are differences between these concepts, their underlying motivations are close (Haldar, 2019).

Sustainable entrepreneurship is not only a way to create successful businesses in a competitive market but also acts as a key factor in developing a sustainable future for society and the environment (Haldar, 2019).

According to the literature, researchers have considered different dimensions of sustainable entrepreneurship. By emphasizing each of the social and environmental dimensions or by considering all three in a balanced and simultaneous manner, a type of entrepreneurship emerges.

1. Environmental entrepreneurship
2. Social or responsible entrepreneurship
3. Triple Bottom Line-based entrepreneurship.

In the following section, we will describe each of these dimensions in different studies:

Social Entrepreneurship: In the last two decades, the concept of social entrepreneurship or social enterprise has attracted the attention of many academics (Bhardwaj et al., 2023; Imanuella et al., 2024; Littlewood & Holt, 2018; Tiba et al., 2019; Vanka et al., 2020). Social entrepreneurship is a fluid and context-dependent concept that manifests itself in a diverse set of activities and combinations, responding to social failures, creating new mixed partnerships, driving transformation, empowerment, and systemic social change, as well as generating new market opportunities at the local community level (Imanuella et al., 2024). Furthermore, social entrepreneurship is the process of creating new value with specific goals. In this context, innovative, beneficial, practical, and sustainable approaches can be considered examples of social entrepreneurship formed to address social issues (Kapusuz et al., 2018). As a result, social entrepreneurship has been recognized by policymakers, researchers, and practitioners as a valuable tool for addressing complex social problems, such as those outlined in the United Nations Sustainable Development Goals (Lorenzo-Afable et al., 2023). This type of entrepreneurship involves innovative approaches to achieving social goals and creating social value (Bhardwaj et al., 2023; Hoogendoorn et al., 2019; Lorenzo-Afable et al., 2023; Tiba et al., 2019). Moreover, the theoretical foundations of the success or failure of social entrepreneurship initiatives are rooted in human capital theory, resource dependency theory, social capital theory, institutional theory, and resource-based theory (Hoogendoorn et al., 2019). In addition to playing a traditional role in supporting disadvantaged individuals, social entrepreneurs also help policymakers address complex global challenges, such as environmental degradation, inequality, poverty, lack of access to primary healthcare, clean water, and sustainable energy (Bhardwaj et al., 2023).

Importantly, Social entrepreneurship is a unique field within the field of entrepreneurship studies. At the same time, the subfield of social entrepreneurship also shares many characteristics and elements with other larger research subfields, such as corporate social responsibility, responsible entrepreneurship, social innovation, and social impact investing. Thus, understanding the boundaries of this subfield of entrepreneurship and its relationship to other fields has become increasingly important as these subfields are constantly evolving (Lortie & Cox, 2018).

Social entrepreneurship is a type of entrepreneurship that aims to create social value innovatively and change social structures, characterized by risk-taking, initiative, and economic sustainability, to solve social problems. It may take the form of for-profit or non-profit organizations. The primary difference between social entrepreneurship and classical entrepreneurship lies in the prioritization of social goals over economic goals (Lortie & Cox, 2018).

Social entrepreneurship is currently undergoing rapid change, and emerging trends are shaping our responses to critical social and environmental issues. These recent developments reflect a growing understanding of the importance of combining business acumen with a sincere commitment to creating positive social and environmental impact. Looking ahead, it is clear that social entrepreneurship will continue to play a crucial role in addressing some of the world's most pressing challenges (Alka et al., 2023).

Environmental Entrepreneurship: Increasing attention to environmental protection has led to pressure from governments and the public on businesses to comply with environmental standards and laws (Yin et al., 2022). Green entrepreneurship, also referred to as environmental entrepreneurship or eco-entrepreneurship, involves the adoption of environmentally friendly innovations (Yin et al., 2022).

In response to these pressures, green entrepreneurship is a business model that proposes businesses should do as little harm to the environment as possible while generating profits. Specifically, it aims to produce eco-friendly products, utilize natural resources sustainably, and employ energy-saving techniques (Sri et al., 2024).

As a result, green entrepreneurship can lead to improved environmental performance by providing environmentally friendly services and products, reducing emissions, and committing to consumer safety and health. Moreover, in addition to reducing energy-saving costs, green entrepreneurship also brings economic benefits. Overall, green entrepreneurship is effective in enhancing both organizational and environmental performance (Momayez et al., 2023).

Triple Bottom Line Entrepreneurship: Meanwhile, the rapid depletion of natural resources due to growing customer needs has led to unexpected socio-economic concerns and environmental concerns (Momayez et al., 2023).

In light of this, Triple Bottom Line entrepreneurs strive to create optimal levels of economic, social, and environmental value. This type of entrepreneurship orientation encompasses all three dimensions of sustainability, thereby leading to sustainable development (Fischer et al., 2020). Accordingly, there has been an increasing interest in innovative business models such as market transformation models, sustainability-oriented models, hybrid models, and sustainable innovation models. However, the main challenge in this field is the potential tension between sustainability and profitability goals. Some research has explored how to organize sustainable decision-making to mitigate this tension and balance these two goals (Wesseh et al., 2024).

Furthermore, sustainable entrepreneurship brings numerous benefits, including implementing sustainability values at the company level, actively engaging with stakeholders, and promoting a healthier, more sustainable lifestyle in society. Additionally, social initiatives aimed at mitigating negative impacts can also lead to a competitive advantage and increased stakeholder satisfaction. Nevertheless, data indicates a limited understanding of how to start the path to sustainable entrepreneurship. Hence, significant progress is not being seen in the move towards sustainable entrepreneurship (Tiuncika & Bormane, 2024). In this regard, this chapter will identify the key indicators of sustainable entrepreneurship and their perspective within the Sustainable Balanced Scorecard framework.

2.1 KPIs

Today, it is crucial to have clear and precise metrics to measure a business's success. KPIs (Key Performance Indicators) are a set of metrics that quantify and measure the success of a specific part of a business. These metrics can assess the performance of a business, a department, or even an individual. In this regard, the primary purpose of these metrics is to determine whether a specific department within the business requires improvement or is moving in the right direction (Waleed et al., 2023). However, the selection of KPIs may vary depending on market conditions, company needs, organization size, and the nature of the economy. Moreover, each KPI can lead to a different ranking of the importance of the available options. Therefore, the appropriate selection of KPIs and their evaluation is of great importance (Govindan et al., 2023).

Key performance indicators are significant instruments for businesses for the following reasons:

- Ability to compare performance with industry standards
- Assess your progress and decide how to allocate resources best
- Evaluate the performance of departments, processes, and employees over time
- Measure customer satisfaction and loyalty.

Consequently, KPIs are essential tools for measuring performance, customer satisfaction, and organizational strategy. By continuously monitoring key performance indicators (KPIs) and utilizing data-driven analytics, businesses can enhance their performance, ultimately leading to increased profitability (Waleed et al., 2023).

2.2 Sustainable BSC

The BSC model, introduced by Kaplan and Norton in 1992, is a performance management tool that helps organizations translate their strategies into actionable plans and measurable outcomes (Figge et al., 2002; Mio et al., 2022; Nouri & Nikabadi, 2018; Raut et al., 2017). The BSC provides a structured approach to evaluating value creation by examining it from four distinct perspectives: Financial, Customer, Internal Business Processes, and Learning and Growth (Balaji et al., 2021; Nouri & Nikabadi, 2018).

Building on this, the concept of a sustainable Balanced Scorecard (BSC) is derived from the traditional BSC and combines its four perspectives with sustainability dimensions to incorporate sustainability-related objectives and indicators (Mio et al., 2022; Heebkhoksung et al., 2023). As such, the sustainable BSC can provide a suitable tool for monitoring performance and assessing progress toward sustainability objectives, both at the company and stakeholder levels (Mio et al., 2022). Figge et al. (2002) explored and discussed three key ways to integrate sustainability into the Balanced Scorecard (BSC) framework. The first approach involves embedding these aspects within the existing four standard perspectives. The second option is to add a separate perspective specifically focused on environmental and social concerns. Finally, a dedicated environmental and/or social scorecard can be developed as a standalone tool (Figge et al.,

Table 1 The search process

	Search in topic Refine	Date	Result
1	Article/Review article/Early access Language: English Publisher: Emerald/Springer/Elsevier/Taylor and Francis/Wiley	2019-03-28 to 2025-03-28	3285
2	WoS category: business/management		357
3	Final review		36

2002). In general, these architectures may differ depending on two criteria: the nature of the hierarchy between functional perspectives (and strategic objectives), which is influenced by the business's value system, and how the functional dimensions are designed, which refers to the approach taken to integrate sustainability into the sustainable BSC (Mio et al., 2022).

Accordingly, this chapter aims to establish key performance indicators (KPIs) for sustainable entrepreneurship using the Sustainable Balanced Scorecard (BSC) framework. To build this framework, the first of several available approaches was chosen. This approach enables us to give equal attention to the various key perspectives of the BSC while also facilitating the integration of core sustainability criteria into each perspective in a balanced manner (Nouri & Nikabadi, 2018).

In the next step, the literature review focused on identifying relevant sustainability criteria within the four traditional BSC perspectives—modified and renamed as needed to align with sustainability goals—as well as within the context of entrepreneurship.

3 Research Methodology

This chapter aims to identify key performance indicators (KPIs) of sustainable entrepreneurship. To fulfill the purpose of this chapter, A literature review was conducted. The search in the Core Collection section of the Web of Science (hereinafter referred to as WoS) was performed on March 29, 2025, covering the period from 2019 to 2025. The search formula is as follows: "Sustainable Entrepreneurship" OR "Green Entrepreneurship" OR "Ecopreneurship" OR "Social Entrepreneurship" OR "Environment * Entrepreneurship." The publishing field is a "Topic". As a result of searching for these keywords in the topic section, we retrieved 3285 records. Then, by applying the filter, the number of these records fell to 357. Only "Articles," "Article review," and "Early access" written in English were selected. In particular, we eliminated duplications and excluded papers that were published in journals not focused on the business and economics area. In the second step, we analyzed each paper's title, abstract, and keywords to determine whether to include it for further analysis. Finally, a total of 36 articles that met the inclusion criteria were identified for further analysis in the full-text stage. As shown in Table 1, the article selection process conducted in this study is documented. Based on the thematic literature review, a list of KPIs of sustainable entrepreneurship was identified and categorized under the four perspectives of the Sustainable BSC (as shown in Tables 2, 3, 4 and 5).

The resulting model in this study is built around the following key perspective.

3.1 Stakeholder

This perspective builds on the typical "customer perspective" found in the standard BSC but goes further by including the expectations of a broader group of stakeholders. Beyond customer satisfaction, it also considers the environmental and social concerns of various groups, including employees and communities. Essentially, it reflects how businesses should act to meet the sustainability expectations of these stakeholders (Nouri & Nikabadi, 2018).

Table 2 Stakeholder perspective

Stakeholder	Index	Source
Social and community-based participation	Participation in local/regional development (such as local community education)	Soto-Acosta et al. (2016), Moon and Moon (2018), Jiang et al. (2024), Fischer et al. (2020), Scartozzi et al. (2024), Bhardwaj et al. (2023) and Kaushik et al. (2023)
	Improving community quality of life and increasing public welfare	Haldar (2019), Bonfanti et al. (2024), Fischer et al. (2020), Silvestri and Veltri (2020) and Moon and Moon (2018)
	Transparency in providing information to stakeholders	Richardson (2018)
	Improving employees' quality of life	Ghag and Harshad (2024), Tarnanidis et al. (2019) and Tiba et al. (2019)
Economic empowerment	Job creation	Scartozzi et al. (2024), Bonfanti et al. (2024), Silvestri and Veltri (2020), Kimuli et al. (2023), La Rocca and Dal Molin (2024) and Tarnanidis et al. (2019)
	Reduction of inequalities (Gini index, poverty rate)	Scartozzi et al. (2024) and Bonfanti et al. (2024)
Social added value	Customers and social beneficiaries	Lortie and Cox (2018), Tiba et al. (2019), Nguyen et al. (2020) and van de Ven et al. (2023)
	Social progress	Scartozzi et al. (2024) and Rahayu (2024)
	Social legitimacy	Haldar (2019) and Derdabi and Dvouletý (2025)

This perspective encompasses three key components: social and community-based participation, economic empowerment, and Social added value. Understanding the preferences and needs of stakeholders (customers, employees, and the target community) can lead to the implementation of sustainable practices and guide the business's efforts toward advancing sustainable development goals.

3.2 Internal Processes

In the BSC model, this perspective is traditionally referred to as "internal processes," but here, the focus is on the performance of the entrepreneurship itself. This perspective emphasizes the key processes and activities that directly impact the sustainability of entrepreneurship (Nouri & Nikabadi, 2018). The entrepreneurship process encompasses the following dimensions: eco-friendly technology, Environmental Efficiency, Waste management, Pollution and emissions reduction, Compliance with Environmental Regulations, Sustainable Supply Chain, and Circular Economy.

3.3 Learning, Growth, and Innovation

The learning and growth perspective focuses on enhancing the organization's competitive advantage and building stakeholder capabilities to effectively address future challenges (Heebkhoksung et al., 2023). Dimensions of this perspective include sustainable Business Models, Green innovation, Entrepreneurial ecosystem development, and culture.

The launch of the value chain from the internal process perspective involves innovation and learning (Chung et al., 2016). This approach ultimately leads to the development of organizational processes that promote sustainability.

3.4 Financial Perspective

In the Sustainable BSC, the financial perspective follows the traditional BSC approach, with the difference that it is aligned with sustainability and social goals (Nouri & Nikabadi, 2018). A review of the existing literature led to the identification of five subcategories: Sustainable Profitability

Table 3 Internal processes perspective

Internal processes	Index	Source
Eco-friendly technology	Utilization of renewable energy	Moon and Moon (2018), Ghag and Harshad (2024), Jiang et al. (2024), Bonfanti et al. (2024), Tarnanidis et al. (2019), Fischer et al. (2020) and Kimuli et al. (2023)
	Use of energy-saving devices (such as efficient heating systems and pollution reduction)/use of tools to control water consumption (aerators, flow pressure regulators, timers, and photocells)	Ghag and Harshad (2024), Soto-Acosta et al. (2016), Labella-Fernández et al. (2024), Sri et al. (2024) and Jiang et al. (2024)
Environmental efficiency	Reducing energy/water consumption in the production process (reducing the volume of water used)	Labella-Fernández et al. (2024), Tarnanidis et al. (2019), Jiang et al. (2024), Bonfanti et al. (2025), Richardson (2018), La Rocca and Dal Molin (2024), Fischer et al. (2020) and Soto-Acosta et al. (2016)
	Use of domestic raw materials instead of imports	Kimuli et al. (2023) and Moon and Moon (2018)
Waste management	Percentage reduction in waste generation	Moon and Moon (2018), Musona et al. (2020), Kimuli et al. (2023), Jiang et al. (2024), Richardson (2018), Kimuli et al. (2023), Labella-Fernández et al. (2024), La Rocca and Dal Molin (2024) and Ghag and Harshad (2024)
	Waste segregation	Ghag and Harshad (2024) and Lupoae et al. (2024)
	Converting waste into marketable products or raw materials	Moon and Moon (2018), Musona et al. (2020), Kimuli et al. (2023), Jiang et al. (2024), Richardson (2018), Fischer et al. (2020) and Ghag and Harshad (2024)
Pollution and emissions reduction	Reduction of CO_2 consumption/ reduction of SO_2, NOx, and particulate matter emissions	Kimuli et al. (2023), Jiang et al. (2024), La Rocca and Dal Molin (2024), Fischer et al. (2020), Bonfanti et al. (2025), Gu and Wang (2022) and Scartozzi et al. (2024)
	Reducing dependence on natural resources	Jiang et al. (2024) and Labella-Fernández et al. (2024)
Compliance with environmental regulations	Compliance with environmental standards	Tiba et al. (2019), Tarnanidis et al. (2019), Bonfanti et al. (2025), Haldar (2019), Jiang et al. (2024), Gu and Wang (2022) and Labella-Fernández et al. (2024)
Sustainable supply chain	Green partners	Galkina (2021), Bonfanti et al. (2025) and Tiba et al. (2019)
	Avoiding collaboration with environmentally harmful companies	Galkina (2021) and Lupoae et al. (2024)
	Participation in ethical strategic alliances	Silvestri and Veltri (2020)
Circular economy	Access to resources during the implementation process	Haldar (2019)
	Creating environmental value by addressing environmental challenges	Bonfanti et al. (2024)

Table 4 Innovation and learning

Innovation and learning		
Sustainable/ business model	Business models based on sustainability criteria	Bonfanti et al. (2025), Hassan et al. (2025), Bonfanti et al. (2024) and Alberico et al. (2022)
	Integration of sustainability into organizational strategies	Galkina (2021), Moon and Moon (2018), Rahayu (2024) and Derdabi and Dvouletý (2025)
Green innovation	Green innovations (product/process)	Labella-Fernández et al. (2024), Hassan et al. (2025) and Jiang et al. (2024)
	Regional environmental knowledge (environmental awareness)	Jiang et al. (2024), Tiuncika and Bormane (2024) and Galkina (2021)
	Social entrepreneurship	Moon and Moon (2018), Silvestri and Veltri (2020), Hassan et al. (2025), Glasbeek et al. (2024) and Kapusuz et al. (2018)
	Employee training and development	Ghag and Harshad (2024), Galkina (2021), Tarnanidis et al. (2019), Tiba et al. (2019), La Rocca and Dal Molin (2024), Alberico et al. (2022) and Hassan et al. (2025)
	Green product design	Ghag and Harshad (2024), Silvestri and Veltri (2020), Soto-Acosta et al. (2016), Sri et al. (2024), Lupoae et al. (2024), Jiang et al. (2024), Labella-Fernández et al. (2024) and Tarnanidis et al. (2019)
	Product manufacturing process	Lupoae et al. (2024), Labella-Fernández et al. (2024), Tarnanidis et al. (2019), Haldar (2019), Fischer et al. (2020), Tiba et al. (2019) and Silvestri and Veltri (2020)
	Use of recyclable packaging	Tarnanidis et al. (2019), Ghag and Harshad (2024) and Lupoae et al. (2024)
Entrepreneurial ecosystem development	Attracted investors	Silvestri and Veltri (2020) and Hassan et al. (2025)
	Diversity of financing sources (public/private/ traditional/alternative)	Bonfanti et al. (2025) and Derdabi and Dvouletý (2025)
	Co-evolution with traditional market players	Bonfanti et al. (2024) and Silvestri and Veltri (2020)
Culture	Commitment to environmental values	Galkina (2021) and Tiuncika and Bormane (2024)
	Sustainability-oriented corporate culture	Alberico et al. (2022), Nguyen et al. (2020) and Ghag and Harshad (2024)

and Financial Returns, Financing and Investment Access, Green Operational Efficiency, Financial Legitimacy, and Alternative Instruments.

3.5 Practical Applications

In addition to contributing to the academic discourse on sustainable entrepreneurship, this chapter provides practical implications for enhancing operational and managerial effectiveness.

Specifically, in the early stages of a new business's development—where management is often informal and driven mainly by the founders—it can be beneficial to be aware of and pay attention to the environmental contexts to which one must adapt, as well as to recognize and exploit the power of influence that comes from the founder's passion. This awareness implies that sustainable entrepreneurs must carefully identify which stakeholders they intend to attract and which factors are most important to attract potential partners. In other words, sustainable entrepreneurs must first frame their strategy from the perspective of the target actors (La Rocca & Dal Molin, 2024).

Moreover, incubators can contribute to the development of a sustainable ecosystem through their services (such as consulting, training, networking, and access to financial resources), as well as by fostering a sustainability-oriented mindset and setting long-term sustainability goals. Therefore, entrepreneurs who aim to integrate sustainability with their business objectives can benefit from the capacities provided by these centers (Bonfanti et al., 2024).

Table 5 Financial perspective

Financial perspective	Index	Source
Sustainable profitability and financial returns	Net profit margin	Yin et al. (2022)
	Return on investment (ROI)	Derdabi and Dvouletý (2025)
	Compound annual growth rate (CAGR)	La Rocca and Dal Molin (2024)
	Long-term financial goals	Alberico et al. (2022)
	Sustainable income and revenue growth from green markets	Jiang et al. (2024), Labella-Fernández et al. (2024), Tarnanidis et al. (2019), Haldar (2019), Fischer et al. (2020), Derdabi and Dvouletý (2025), Bonfanti et al. (2024) and Moon and Moon (2018)
Financing and investment access	Access to traditional financial institutions (banks, loans)	Derdabi and Dvouletý (2025) and Haldar (2019)
	Participation of venture capital investors	Derdabi and Dvouletý (2025)
	Participation in crowdfunding campaigns	Derdabi and Dvouletý (2025) and La Rocca and Dal Molin (2024)
	Investment in green innovation and clean technologies	Jiang et al. (2024) and Bonfanti et al. (2025)
Green operational efficiency	Reduction in operational costs through environmental efficiency	Labella-Fernández et al. (2024), Kimuli et al. (2023) and La Rocca and Dal Molin (2024)
	Cost savings from waste recycling and reduced disposal	Kimuli et al. (2023)
Financial legitimacy and alternative instruments	Engagement in sustainable economic networks	Soto-Acosta et al. (2016), Haldar (2019) and La Rocca and Dal Molin (2024)
	Impact investing	Labella-Fernández et al. (2024), Moon and Moon (2018), Derdabi and Dvouletý (2025) and Bonfanti et al. (2025)
	Use of alternative financial instruments (ethical banking, impact investing)	Derdabi and Dvouletý (2025)

In addition, sustainable entrepreneurs can leverage the potential of digital technologies to drive sustainable business and develop sustainable business models. However, along with the use of digital technology, responsible behavior must be considered to prevent potential negative consequences and avoid unintended social and environmental outcomes. These negative consequences may result from a lack of knowledge or improper use of technology (Holzmann & Gregori, 2023).

Sustainable entrepreneurship has emerged as a sub-discipline of entrepreneurship that seeks to pursue social, environmental, and economic values simultaneously. Although sustainable entrepreneurship plays a vital role in sustainable development, our understanding of the factors that hinder or facilitate the initial development of sustainable businesses is limited.

4 Conclusions

Sustainability-focused entrepreneurship can contribute to sustainable development. This approach presents sustainable development as a business opportunity rather than a threat or obstacle to entrepreneurial activities (Tiuncika & Bormane, 2024).

In line with this perspective, sustainable entrepreneurship has gained growing recognition for its significant role in advancing sustainable development (Bhardwaj et al., 2023; Soto-Acosta et al., 2016). Indeed, adopting the sustainable entrepreneurship model can lead to substantial improvements in environmental and social outcomes while ensuring ongoing and sustainable economic growth (Tiuncika & Bormane, 2024).

To support this, this chapter reviewed relevant literature and identified 52 key performance indicators (KPIs) of sustainable entrepreneurship. Using the Sustainable BSC framework, the identified indicators were grouped into four primary categories:

(1) Stakeholder
(2) Process
(3) Learning, growth and innovation
(4) Financial.

Nevertheless, despite the efforts made, there is still a need for guidelines, standardized methods, and a unified framework that integrates the three sustainability goals with sustainable entrepreneurship (La Rocca & Dal Molin, 2024; Tiuncika & Bormane, 2024). Accordingly, this study presents a comprehensive list of sustainability indicators in entrepreneurship by reviewing the literature. It also considers the needs and demands of multiple stakeholders within the framework of sustainable business, taking into account the three dimensions of sustainability based on the Triple Bottom Line (economic, social, and environmental) model (Richardson, 2018; Soto-Acosta et al., 2016; Tarnanidis et al., 2019). This integration helps to balance the expectations and demands of entrepreneurs, employees, and the target community.

However, most studies of sustainable entrepreneurship focus on environmental or social dimensions. By contrast, by combining the three dimensions of sustainability, entrepreneurship, and the BSC, this model enables entrepreneurs and regulatory bodies to identify key indicators and utilize them to achieve sustainability.

Therefore, the framework proposed in this chapter, by determining the main dimensions and indicators, provides a suitable model for evaluating sustainable entrepreneurship. This framework also covers the research gap in studies such as La Rocca and Dal Molin (2024) and Bhardwaj et al. (2023), and consequently will be helpful for entrepreneurs in strategic and operational decision-making related to sustainability (Tiuncika & Bormane, 2024).

Looking ahead, future research could rank and moderate KPIs using qualitative or quantitative methods. Additionally, they could also identify the interrelationships between the identified dimensions and measures.

5 Case Study

This case study examines the creation and development of sustainable entrepreneurship in the village of Isfahak, located in South Khorasan, Iran. This village was devastated and severely damaged in the 1978 Tabas earthquake. In response, the villagers decided to build a new village adjacent to the semi-destroyed one. Nearly three decades later, Hossein Alizadeh, a resident, launched an eco-lodge project using his funds and the support of other villagers. He aimed to achieve profitability by renovating traditional adobe houses, while also providing a platform for job creation and enhancing the quality of life for the local community.

Simultaneously, along with the construction of this residence, a production unit was established to provide the materials needed to repair the remaining part of the village. Moreover, considering the geographical conditions and water shortage, in addition to saffron cultivation, they began to establish fields of drought-resistant plants, such as damask rose and medicinal plants. This was a significant help in reducing unemployment and preventing villagers from migrating to neighboring cities.

In 2017, the village was designated a national heritage site, and it began to receive increased support from the government. Following Isfahak's listing among the Best Tourism Villages by the United Nations World Tourism Organization (UNWTO) in 2023, there was a significant increase in both domestic and international tourist visits. The most important key indicators that contributed to this are as follows:

Innovation and Learning Perspective

This project not only led to long-term economic profitability but also contributed to environmental protection and the development of the local community. Therefore, it can be said that it has taken a crucial step in rural development and sustainable tourism through the adoption of sustainable business models.

The efforts of local activists to register this village in the list of national monuments in Iran led to the receipt of low-interest loans and government facilities. Additionally, technical support and consulting from the government, as well as environmental activists who voluntarily participated in this matter, were crucial steps in advancing and developing this business.

Internal Process

The traditional design of adobe houses, featuring domed roofs and thick walls made of adobe and mud, is not only beautiful and eye-catching but also leads to optimized energy consumption and enhanced material resistance to the region's hot and dry weather conditions. Additionally, installing solar systems to provide energy for household consumption and industrial production units is a crucial step toward reducing energy consumption.

Financial

Income from tourism, along with the production and packaging of perfumes, rose water, medicinal plants, handicrafts, and agricultural products, has led to economic growth and the creation of social value in this village, in addition to creating employment.

Stakeholders

The Isfahak Village Ecotourism Project, by emphasizing the local community's capacities, has led to the participation of village people in this process. Additionally, holding training courses and skill workshops by non-profit organizations and educated volunteers has contributed to both cultural and economic development. In addition, this project has helped empower households by creating job opportunities for locals and has led to a reduction in the village's poverty rate.

Sustainable entrepreneurship can be considered a strategy for the development and growth of businesses, mainly rural businesses. This case study can serve as a model for sustainable entrepreneurship in other rural areas while also highlighting key performance indicators (KPIs) for sustainable entrepreneurship.

Questions

- What is sustainable entrepreneurship, and how does it differ from traditional entrepreneurship?
- What is the sustainable BSC, and what is its purpose in this chapter?
- How can a startup integrate sustainability into its goals and strategies?
- What are the KPIs of sustainable entrepreneurship?
- How can sustainable entrepreneurship simultaneously contribute to environmental protection and profitability?
- In your opinion, what is the biggest challenge facing sustainable entrepreneurship?

References

Ahmed, S., & Singh, S. (2020). Examining the role of HR practice and employee engagement on employee's loyalty—The sustainability dimensions of textile industry in Bangladesh. In *Sustainable human resource management: Transforming organizations, societies and environment*. https://doi.org/10.1007/978-981-15-5656-2_16

Alberico, T. R., Ricardo, J. R., & Cruz, S. (2022). Sustainable entrepreneurship: A current review of literature. *International Journal of Business Research, 14*(5556), 1–25. https://doi.org/10.3390/su14095556

Alka, T. A., Sreenivasan, A., & Suresh, M. (2023). Shaping tomorrow: A bibliometric perspective through two decades of social entrepreneurship. *Journal of Social Entrepreneurship,* 1–29. https://doi.org/10.1080/19420676.2023.2266827

Alodat, A. Y., & Hao, Y. (2024). Environmental, social and governance (ESG) disclosure and firm performance: Moderating role of board gender diversity and sustainability committee. *Sustainable Development, 33*, 636–651. https://doi.org/10.1002/sd.3126

Amodu, N. (2020). Stakeholder protection and corporate social responsibility from a comparative company law perspective: Nigeria and South Africa. *Journal of African Law, 64*(3), 425–449. https://doi.org/10.1017/S0021855320000212

Balaji, M., et al. (2021). Balanced scorecard approach in deducing supply chain performance. *Materials*

Today: Proceedings [Preprint]. https://doi.org/10.1016/j.matpr.2021.05.541

Bhardwaj, R., Weerawardena, J., & Srivastava, S. (2023). Advancing social entrepreneurship research: A morphological analysis and future research agenda. *Journal of Social Entrepreneurship,* 1–35. https://doi.org/10.1080/19420676.2023.2199748

Bonfanti, A., et al. (2024). Convergences and divergences in sustainable entrepreneurship and social entrepreneurship research: A systematic review and research agenda. *Journal of Business Research, 170,* 114336. https://doi.org/10.1016/j.jbusres.2023.114336

Bonfanti, A., et al. (2025). Business incubators as a driver of sustainable entrepreneurship development: Evidence from the Italian experience. *International Journal of Entrepreneurial Behaviour and Research* [Preprint]. https://doi.org/10.1108/IJEBR-05-2024-0500

Brundtland, G. (1987). *Our common future: The world commission on environment and development.* Oxford University Press.

Chung, C. C., et al. (2016). A balanced scorecard of sustainable management in the Taiwanese bicycle industry: Development of performance indicators and importance analysis. *Sustainability (Switzerland), 8*(6). https://doi.org/10.3390/su8060518

Derdabi, A., & Dvouletý, O. (2025). Funding sustainable entrepreneurship: A hybrid systematic literature review and bibliometric analysis. *Nankai Business Review International* [Preprint]. https://doi.org/10.1108/NBRI-03-2024-0025

Elkington, J. (1997). *Cannibals with forks – Triple bottom line of 21st century business.* New Society Publishers.

Ferreira, J. J. M., Fernandes, C. I., & Kraus, S. (2019). Entrepreneurship research: Mapping intellectual structures and research trends. *Review of Managerial Science, 13*(1), 181–205. Available at: https://doi.org/10.1007/s11846-017-0242-3

Figge, F., et al. (2002). The sustainability balanced scorecard—Management to business strategy (Vol. 284, pp. 269–284).

Fischer, D., Brettel, M., & Mauer, R. (2020). The three dimensions of sustainability: A delicate balancing act for entrepreneurs made more complex by stakeholder expectations. *Journal of Business Ethics, 163*(1), 87–106. https://doi.org/10.1007/s10551-018-4012-1

Galkina, T. (2021). International ECOpreneurship: Environmental commitment and international partner selection of Finnish firms from the energy sector. *Journal of International Entrepreneurship,* 300–320. https://doi.org/10.1007/s10843-021-00286-8

Ghag, N., & Harshad, S. (2024). Sustainable entrepreneurship practices of Indian SMEs: A strategic approach using fuzzy Delphi and best worst method (pp. 1794–1809). https://doi.org/10.1002/bse.3572

Glasbeek, L., Wickert, C., & Schad, J. (2024). Evaluating definitions of social entrepreneurship: A rulebook from the philosophy of science (pp. 384–409). https://doi.org/10.1111/ijmr.12359

Govindan, K., et al. (2023). Analysis of supplier evaluation and selection strategies for sustainable collaboration: A combined approach of best–worst method and TOmada de Decisao Interativa Multicriterio. *Business Strategy and the Environment, 32*(7), 4426–4447. https://doi.org/10.1002/bse.3374

Gu, W., & Wang, J. (2022). Research on index construction of sustainable entrepreneurship and its impact on economic growth. *Journal of Business Research, 142,* 266–276. https://doi.org/10.1016/j.jbusres.2021.12.060

Haldar, S. (2019). Towards a conceptual understanding of sustainability-driven entrepreneurship. *Corporate Social Responsibility and Environmental Management, 26*(6), 1157–1170. https://doi.org/10.1002/csr.1763

Hassan, M., Mehmet, A., & Hassnian, B. (2025). Decoding sustainable entrepreneurship current research and future direction through application of machine learning-based structured topic modeling on intellectual corpus. *Journal of International Entrepreneurship.* https://doi.org/10.1007/s10843-025-00387-8

Heebkhoksung, K., Rattanawong, W., & Vongmanee, V. (2023). A new paradigm of a sustainability-balanced scorecard model for sport tourism.

Holzmann, P., & Gregori, P. (2023). The promise of digital technologies for sustainable entrepreneurship: A systematic literature review and research agenda. *International Journal of Information Management, 68,* 102593. https://doi.org/10.1016/j.ijinfomgt.2022.102593

Hoogendoorn, B., van der Zwan, P., & Thurik, R. (2019). Sustainable entrepreneurship: The role of perceived barriers and risk. *Journal of Business Ethics, 157*(4), 1133–1154. https://doi.org/10.1007/s10551-017-3646-8

Imanuella, S. F., Idris, A., & Kamaruddin, N. (2024). Social entrepreneurship and rural development in post-independence Indonesia. *Social Enterprise Journal, 21*(1), 46–66. https://doi.org/10.1108/SEJ-12-2023-0155

Inacio Junior, E., et al. (2021). The global entrepreneurship index as a benchmarking tool? Criticisms from an efficiency perspective. *Journal of Intellectual Capital, 22*(1), 190–212. https://doi.org/10.1108/JIC-09-2019-0218

Jiang, Z., et al. (2024). Intellectual capital and green entrepreneurship: A systematic literature review. *Journal of Intellectual Capital, 25*(4), 801–821. https://doi.org/10.1108/JIC-12-2023-0287

Kapusuz, A. G., Çavuş, M. F., & Pekkan, N. Ü. (2018). Perceived social support and social entrepreneurship: Gender perspective from Turkey. In V. Ratten, V. Braga, & C. S. Marques (Eds.), *Knowledge, learning and innovation: Research insights on cross-sector collaborations* (pp. 75–90). Springer International Publishing. https://doi.org/10.1007/978-3-319-59282-4_6

Kaushik, V., et al. (2023). Towards a precise understanding of social entrepreneurship: An integrated bibliometric–machine learning based review and research agenda. *Technological Forecasting and Social Change, 191.* https://doi.org/10.1016/j.techfore.2023.122516

Kimuli, S. N. L., et al. (2023). The potential of industrial symbiosis in fostering sustainable entrepreneurship among the micro and small enterprises (MSEs): Evidence from a developing country. *Journal of Work-Applied Management* [Preprint]. https://doi.org/10.1108/JWAM-07-2023-0073

La Rocca, A., & Dal Molin, S. (2024). Sustainable entrepreneurship: How do contextual factors play a role? *Industrial Marketing Management, 123*, 292–303. https://doi.org/10.1016/j.indmarman.2024.10.010

Labella-Fernández, A., et al. (2024). Environmental entrepreneurship orientation and duality of goals: The moderating roles of public and private knowledge. *Business Strategy and the Environment,* 648–669. https://doi.org/10.1002/bse.4014

Lakshmi, V. R., & Kennedy, H. (2017). The role of business sustainability in human resource management: A study on Indian manufacturing companies. *The South East Asian Journal of Management, 11*(1), 2017. https://doi.org/10.21002/seam.v11i1.7739

Leal-Millán, A., Peris-Ortiz, M., & Leal-Rodríguez, A. L. (2018). The route towards sustainable innovation and entrepreneurship: An overview, innovation, technology and knowledge management. https://doi.org/10.1007/978-3-319-57318-2_1

Littlewood, D., & Holt, D. (2018). How social enterprises can contribute to the sustainable development goals. In *Entrepreneurship and the sustainable development goals* (pp. 33–46). https://eprints.whiterose.ac.uk/129474/1/Chapter-Social-Entrepreneurship-and-the-SDGs.pdf

Lorenzo-Afable, D., Lips-Wiersma, M., & Singh, S. (2023). A narrative synthesis of the empirical literature on social value creation in social entrepreneurship: Gaps and opportunities for future research and action. *Journal of Social Entrepreneurship,* 1–32. https://doi.org/10.1080/19420676.2023.2195868

Lortie, J., & Cox, K. C. (2018). On the boundaries of social entrepreneurship: A review of relationships with related research domains. *International Entrepreneurship and Management Journal, 14*(3), 639–648. https://doi.org/10.1007/s11365-018-0514-7

Lupoae, O. D., Wilk, V., & Radu, R. I. (2024). Sustainable entrepreneurship in equine services. *Journal of Business Research, 170,* 114361. https://doi.org/10.1016/j.jbusres.2023.114361

Mio, C., Costantini, A., & Panfilo, S. (2022). Performance measurement tools for sustainable business: A systematic literature review on the sustainability balanced scorecard use (pp. 367–384). https://doi.org/10.1002/csr.2206

Moon, C. J., & Moon, C. J. (2018). Contributions to the SDGs through social and eco entrepreneurship: New mindsets for sustainable solutions. https://doi.org/10.1108/S2040-724620180000008008

Momayez, A., Rasouli, N., & Alimohammadirokni, M. (2023). Green entrepreneurship orientation, green innovation and hotel performance: The moderating role of managerial environmental concern. *Journal of Hospitality Marketing & Management, 32*(8), 981–1004. Available at: https://doi.org/10.1080/19368623.2023.2225495

Musona, J., et al. (2020). Bricolage in environmental entrepreneurship: How environmental innovators "make do" at the bottom of the pyramid. *Business Strategy and Development, 3*(4), 487–505. https://doi.org/10.1002/bsd2.112

Nguyen, H. T. T., Costanzo, L. A., & Karatas-Özkan, M. (2020). Stakeholders ' perceptions of sustainable entrepreneurship within the context of a developing economy. *Journal of Small Business Management,* 1–40. https://doi.org/10.1080/00472778.2020.1796465

Nouri, F. A., & Nikabadi, M. S. (2018). Developing the framework of sustainable service supply chain balanced scorecard (SSSC BSC). https://doi.org/10.1108/IJPPM-04-2018-0149

Rahayu, N. S. (2024). Assessing the determinant factors affecting green entrepreneurial intention among female entrepreneurs in Indonesia. *Cogent Business and Management, 11*(1). https://doi.org/10.1080/23311975.2024.2378919

Raut, R., Cheikhrouhou, N., & Kharat, M. (2017). Sustainability in the banking industry: A strategic multi-criterion analysis. *Business Strategy and the Environment, 26*(4), 550–568. Available at: https://doi.org/10.1002/bse.1946

Richardson, N. (2018). Entrepreneurial insights into sustainable marketing: A case study of U.K. music festivals. *Strategic Change, 27*(6), 559–570. https://doi.org/10.1002/jsc.2239

Scartozzi, G., et al. (2024). The social and environmental impact of entrepreneurship: A review and future research agenda. *Review of Managerial Science.* https://doi.org/10.1007/s11846-024-00783-9

Silvestri, A., & Veltri, S. (2020). Exploring the relationships between corporate social responsibility, leadership, and sustainable entrepreneurship theories: A conceptual framework. *Corporate Social Responsibility and Environmental Management, 27*(2), 585–594. https://doi.org/10.1002/csr.1822

Soto-Acosta, P., et al. (2016). Sustainable entrepreneurship in SMEs: A business performance perspective. *Sustainability (Switzerland), 8*(4), 1–12. https://doi.org/10.3390/su8040342

Sri, H., et al. (2024). Multiple pathways linking commitment to the environment and green entrepreneurship behavior: The mediating and moderating analysis. *Cogent Business & Management, 11*(1). https://doi.org/10.1080/23311975.2024.2402516

Tarnanidis, T., Papathanasiou, J., & Subeniotis, D. (2019). How far the TBL concept of sustainable entrepreneurship extends beyond the various sustainability regulations: Can Greek food manufacturing enterprises sustain their hybrid nature over time? *Journal of Business Ethics, 154*(3), 829–846. https://doi.org/10.1007/s10551-017-3443-4

Tiba, S., van Rijnsoever, F. J., & Hekkert, M. P. (2019). Firms with benefits: A systematic review of responsible entrepreneurship and corporate social responsibility literature. *Corporate Social Responsibility and Environmental Management, 26*(2), 265–284. https://doi.org/10.1002/csr.1682

Tiuncika, L., & Bormane, S. (2024). Sustainable management of manufacturing processes: A literature review. *Processes, 12*(6). https://doi.org/10.3390/pr12061222

van de Ven, M., et al. (2023). Key performance indicators for business models: A systematic review and catalog. *Information Systems and e-Business Management*. https://doi.org/10.1007/s10257-023-00650-2

Vanka, S., Singh, S., & Rao, M. B. (2020). From sustainability to sustainable HRM—An evolutionary perspective. In S. Vanka, et al. (Eds.), *Sustainable human resource management: Transforming organizations, societies and environment* (pp. 1–8). Springer Singapore. https://doi.org/10.1007/978-981-15-5656-2_1

Waleed, E., et al. (2023). "Key performance indicators (KPIs), key result indicator (KRIs) and objectives and key results (OKRs)" a new key incorporated results (KIRs) approach. *Information and Knowledge Management, 13*(2), 1–13. https://doi.org/10.7176/ikm/13-2-01

Wesseh, P. K., et al. (2024). Sustainable entrepreneurship: When does environmental compliance improve corporate performance? *Business Strategy and the Environment, 33*(4), 3203–3221. https://doi.org/10.1002/bse.3657

Yin, C., et al. (2022). Green entrepreneurship and SME performance: The moderating effect of firm age. *International Entrepreneurship and Management Journal,* 255–275. https://doi.org/10.1007/s11365-021-00757-3

Young, W., & Tilley, F. (2006). Can businesses move beyond efficiency? The shift toward effectiveness and equity in the corporate sustainability debate. *Business Strategy and the Environment, 15*(6), 402–415. https://doi.org/10.1002/bse.510

Technology, Innovation, and Digitalization

Role of Technology and Innovation in Sustainable Entrepreneurship

Atefeh Zolfaghari

Abstract

Sustainable entrepreneurship has evolved from a theoretical idea to an emerging necessity. Several key factors have contributed to this shift, including climate change, human resource scarcity, environmental degradation, and economic uncertainty. A key characteristic of sustainable entrepreneurs is their significant focus on ecological considerations within their business vision, contrasting with the traditional entrepreneurial objective of pursuing growth and generating profits. To examine the role of entrepreneurship (Bagheri et al., 2018) and innovation in sustainable entrepreneurship, this book chapter has gone to Sarma, 2024 model and modified this model by adding innovation and technology separately in three sections: society, environment, and sustainable economy. Integrating digital transformation with the circular economy—pairing renewable energy initiatives with social inclusion, utilizing artificial intelligence and data analytics for environmental monitoring, and applying digital tools to promote sustainable agriculture—highlights the intersection of diverse sustainability objectives. Furthermore, innovations in transportation, combined with urban planning and social equity, the role of waste management in job creation, educational technologies that foster environmental awareness, and advancements in healthcare powered by clean energy all illustrate how sustainable entrepreneurship thrives at the convergence of multiple goals, driven by technological innovation (Pereira et al., 2021). These interconnected strategies reflect the principles of Systems Thinking, which suggest that technological progress can facilitate comprehensive and practical solutions for sustainability (Meadows, 2008).

Keywords

Social sustainability · Environmental sustainability · Economic sustainability · Sustainable entrepreneurship · Innovation · Technology

Learning Objectives

- Role of Technology and Innovation in Environmental Sustainability.
- Role of Technology and Innovation in Social Sustainability.
- Role of Technology and Innovation in Economic Sustainability.
- The Influence of Technology and Innovation on Sustainable Entrepreneurship.

1 Introduction

In today's world, technological innovation plays a crucial role in enhancing a nation's efficiency,

A. Zolfaghari (✉)
ACECR, Tehran, Iran
e-mail: a_zolfaghari@ut.ac.ir

M. Akbari et al. (eds.), *Sustainable Entrepreneurship*, University of Tehran Science and Humanities Series, https://doi.org/10.1007/978-3-032-04909-4_7

productivity, and competitiveness. Simultaneously, there is a growing awareness of the importance of environmental sustainability, particularly in light of challenges such as climate change, resource depletion, and the detrimental effects of pollution (Mahardhani, 2023). Academic discourse has aimed to clarify the characteristics of entrepreneurial behavior that can be classified as "sustainable." The concept of "sustainability" was first introduced by the World Commission on Environment and Development (WCED, 1987). This concept underscores the importance of balancing three core objectives for the future advancement of humanity: social and ethical considerations, economic growth, and ecological health, commonly referred to as the "triple bottom line": People, Profit, and Planet (Schlange, 2006).

Sustainability refers to the ability to meet present needs without compromising the capacity of future generations to fulfill their own needs (Aguirre, 2024). Sustainable entrepreneurship has emerged as a transformative force in today's business landscape, increasingly emphasizing eco-friendly practices that align economic success with environmental responsibility (Odeyemi et al., 2024).

In the business context, the environment in which new ventures emerge is a significant area of research. Environmental factors not only create opportunities to capitalize on market inefficiencies but also influence whether various environments support (Hejazi et al., 2011, p. 100) or hinder the success (Hajiketabi et al., 2017, p. 650) of new ventures (Simón-Moya et al., 2014, p. 715).

Challenges such as environmental degradation, income inequality, and unequal access to resources and opportunities are on the rise. Consequently, there is an increasing demand for sustainable entrepreneurship, which involves adopting sustainable business practices. Entrepreneurship plays a vital role in advancing toward a more sustainable future by aligning social, economic, and environmental goals. Ecological entrepreneurs make significant contributions to this effort (Rosário et al., 2022).

Sustainability and economic growth, along with the integration and balance of social, environmental, and financial needs, are essential for sustainable development and overall social well-being. In the context of sustainable innovation, achieving sustainable entrepreneurship necessitates that entrepreneurial ecosystems evolve into sustainable ones. The interactions among entrepreneurial actors within this ecosystem are crucial (DiVito & Ingen-Housz, 2021).

2 Main Body

2.1 Sustainable Entrepreneurship

Sustainable entrepreneurship is a unique approach that combines the generation of environmental, social, and economic benefits, with a focus on enhancing the well-being of future generations (Sarma et al., 2024).

2.2 Innovation

Innovation as a factor in gaining competitive advantage has always been a focal point for organizations (Zolfaghari et al., 2022). This concept is crucial for achieving sustainable entrepreneurship. Emphasizing innovation in the development of business models (Lüdeke-Freund, 2020) is essential when launching new businesses, as it represents a significant step toward sustainability. As the population continues to grow, increasing pressure is placed on the planet's natural resources (Hummels & Argyrou, 2021). Sustainable innovations provide practical strategies to mitigate the environmental impact of human activities and safeguard the Earth from further degradation. Since all sectors play a role in minimizing environmental harm, these innovations are essential for annual investments across the board. Embracing and achieving expanding new sustainable solutions can create jobs, improve public health, and reduce greenhouse gas emissions (Jha & Pande, 2024).

2.2.1 Eco-Friendly Innovation

The world's growing population and rapid urbanization undoubtedly raise environmental concerns (Khandeparkar et al., 2024). Eco-friendly innovations enable production systems to create safe products and services that do not harm the environment. The purchase of eco-friendly products is encouraged by governments and promoted through educational systems due to their contributions to sustainable environmental development. Therefore, it is essential to examine the factors that influence the purchase of eco-friendly products, with consumers' beliefs in the effectiveness of these products playing a significant role in their purchasing decisions (Chen et al., 2021).

2.3 Technology

Technological applications in sustainable entrepreneurship include blockchain, the Internet of Things (IoT), and data analytics, which can significantly enhance transparency, traceability, and resource efficiency. Conventional technologies, such as energy-efficient machinery and processes, empower businesses to minimize resource consumption and reduce waste generation.

2.3.1 Digital Technologies

Recent studies highlight the transformative impact of digital technologies, including artificial intelligence (AI), the Internet of Things (IoT) (Abd El-Fatah et al., 2024), and blockchain, in promoting sustainable practices and enhancing operational efficiencies. For example, AI applications are being adopted to optimize resource management and energy usage, enabling businesses to lower their carbon emissions while simultaneously boosting productivity. One notable application is in smart grid technologies, which enhance energy distribution by integrating renewable energy sources and dynamically balancing supply and demand. The Internet of Things (IoT) enables businesses to gain detailed visibility into their supply chains, promoting more sustainable logistical operations and significantly reducing waste (Smith & Jones, 2023). These technologies facilitate the real-time collection and analysis of critical data, which, in turn, supports improved decision-making processes that are more closely aligned with environmental and social sustainability objectives (Chen et al., 2024). Furthermore, research highlights the potential of blockchain technology to enhance transparency and traceability in sustainable sourcing practices, thereby promoting ethical conduct and strengthening consumer confidence (Lee, 2022). By utilizing decentralized and immutable ledgers, organizations can effectively verify the provenance of their materials, ensuring compliance with stringent environmental and social responsibility standards (Garcia & Brown, 2025).

Ultimately, the synergistic integration of technological advancements and innovative approaches fosters a corporate culture focused on sustainability. This empowers entrepreneurs to create and implement scalable solutions that actively contribute to a more resilient and environmentally sustainable future (United Nations, 2023). As these technological frontiers continue to evolve and mature, their impact on the realm of sustainable entrepreneurship is expected to grow significantly (World Economic Forum, 2024).

2.3.2 Artificial Intelligence

Artificial intelligence (AI) significantly advances sustainable entrepreneurship by enhancing efficiency, optimizing resource management, and driving innovative solutions to environmental challenges. Additionally, AI algorithms analyze data to optimize the use of resources, such as raw materials and energy, across various business processes, thereby increasing resource optimization. AI technologies enhance supply chain efficiency by accurately predicting demand and optimizing inventory levels to maximize efficiency. This minimizes excess production and waste, fostering a more sustainable approach to logistics. Adopting and developing innovative supply chain management practices can enhance social sustainability in new projects. This approach may be worth exploring to bridge the gap in alignment and collaboration between public entities (such as government and civil society) and private interests (Bickley et al., 2024).

AI assists in designing eco-friendly products (Islam, 2024) by simulating various materials and processes to minimize environmental impact during the product development process. Additionally, AI-powered systems can reduce waste by analyzing waste patterns and recommending actionable strategies for reducing, recycling, or repurposing materials, thereby promoting a circular economy model (Avilés-Palacios & Rodríguez-Olalla, 2021).

AI enhances energy management and consumer engagement. AI-driven analytics provide valuable insights into consumer preferences for sustainable products, enabling entrepreneurs to tailor their offerings and marketing strategies to appeal to eco-conscious consumers (Uriarte-Gallastegi et al., 2024). AI tools assist entrepreneurs in making data-driven decisions that align with sustainability goals, ranging from investments in green technologies to assessing the social and environmental impacts of their business practices and making appropriate decisions (Malik, 2024).

By leveraging these capabilities, AI empowers sustainable entrepreneurs to innovate more effectively, make informed decisions, and contribute to a more sustainable economy.

2.3.3 Internet of Things (IoT)

The Internet of Things (IoT) serves as a powerful catalyst for sustainable entrepreneurship by facilitating more innovative resource management and enhancing operational efficiencies. Real-time data collected from IoT devices optimizes the consumption of energy, water, and materials, thereby minimizing waste. In supply chains, IoT improves transparency through real-time tracking, ensuring sustainable handling and ethical sourcing while also reducing food waste (Sharma et al., 2025). Additionally, predictive maintenance enabled by IoT sensors extends the lifespan of equipment and conserves valuable resources. IoT solutions identify waste patterns in production, enabling operational adjustments to be made. Smart agriculture leverages IoT data on soil conditions and weather patterns to optimize irrigation and minimize the use of pesticides (Gadanakis, 2024). Smart grids and connected devices enhance energy management and facilitate the integration of renewable energy sources, thereby promoting a more sustainable energy system. Insights into customer behavior derived from IoT data enable the tailoring of sustainable offerings. Environmental monitoring through IoT tracks air and water quality, thereby informing strategies for reducing environmental impact. Furthermore, IoT promotes collaboration and innovation among stakeholders by providing shared data platforms (Maione, 2024). By harnessing IoT technology, sustainable entrepreneurs can achieve significant efficiency gains, reduce their environmental footprint, and create value through responsible business models.

2.3.4 Blockchain

The integration of blockchain technology offers substantial benefits to sustainable entrepreneurship by promoting greater transparency, accountability, and streamlined processes throughout business operations (Nakamoto, 2008). Its immutable digital records enhance the clarity of supply chains, enabling consumers to verify the ethical and environmentally sound origins of products (Weber et al., 2016). Specifically, within the food and agricultural sectors, blockchain-enabled traceability strengthens safety protocols and reduces waste by providing a comprehensive history from production to consumption (Tian, 2017). Self-executing agreements, commonly referred to as smart contracts on the blockchain, automate compliance with regulations and facilitate transactions related to sustainably sourced resources (Buterin, 2014). Moreover, blockchain platforms facilitate the exchange of carbon credits and support the development of decentralized renewable energy markets, thereby optimizing energy allocation and promoting the adoption of cleaner energy sources (O'Grady et al., 2018). The inherent security features of blockchain protect critical sustainability-related data. This technology also fosters greater collaboration among various stakeholders and underpins social impact initiatives, such as fair trade programs (Kshetri, 2018). Additionally, blockchain enables innovative financing opportunities for sustainable startups, driving the transition towards a more accountable and environmentally conscious economic system (Tapscott & Tapscott, 2016).

2.4 The Role of Technology and Innovation in Sustainable Entrepreneurship

Public policies play a crucial role in promoting technological innovation and sustainability, both of which are essential for achieving a sustainable and advanced future. When effective, these policies can create a supportive ecosystem for innovation, advance research and development, and empower individuals to tackle evolving technological challenges (Mahardhani, 2023). The role of technology and innovation in sustainable entrepreneurship is becoming increasingly vital as businesses confront the dual challenges of economic growth and environmental sustainability. Research suggests that sustainable business models are not solely dependent on technologies or eco-friendly products and services. For instance, Evans et al. (2017) argue that such models can also emerge through innovation in business strategies. In this context, entrepreneurs design business models that focus on relationships and exchanges with stakeholders (Gray et al., 2014). Today's business landscape empowers consumers to exert significant influence over company operations, as they can share feedback and opinions across various platforms, including social media (Evans et al., 2017). The Internet facilitates two-way communication, transforming customers from passive recipients into active contributors in the value creation process (Schaltegger et al., 2016). Consequently, businesses increasingly utilize data from customers and other stakeholders to adapt their models to meet evolving expectations and demands. Thus, sustainable entrepreneurship fosters greater customer involvement in value creation and business operations, thereby strengthening the relationship between companies and consumers.

Innovation is a significant driver of sustainability. It accelerates changes in both environmental and social sustainability. The role of digital evolution has been instrumental in addressing climate change and promoting sustainable practices (Xiao & Su, 2022).

Sustainable entrepreneurship exists at the intersection of environmental sustainability, social sustainability, and economic sustainability (Sarma et al., 2024). To explore the role of innovation and technology in sustainable entrepreneurship, the impacts of these factors on environmental, social, and economic sustainability have been examined. Figure 1 illustrates the influence of technology and innovation on sustainable entrepreneurship.

Table 1 lists the scientific theoretical literature related to Environmental impact variation,

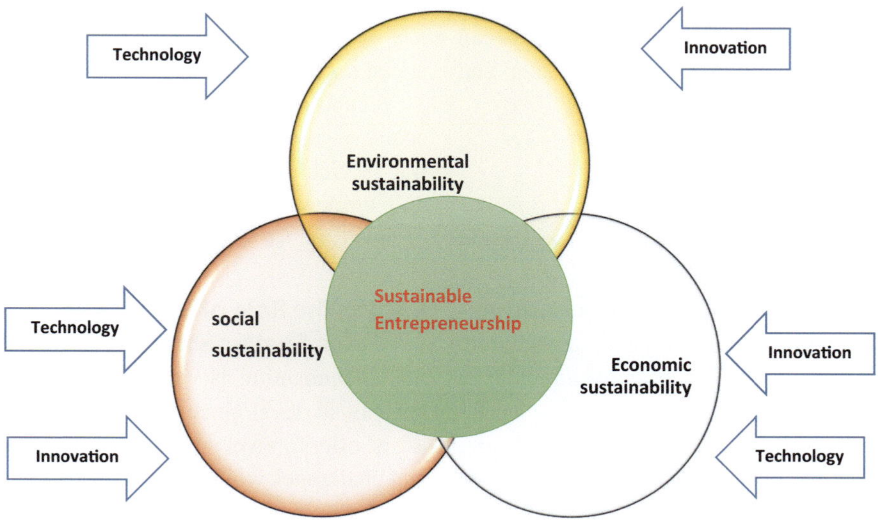

Fig. 1 Effect of technology and innovation on sustainable entrepreneurship. *Source* By writer and Sarma et al. (2024)

Table 1 Key funding in literature

Theme	Key funding	Supporting evidence	Policy implication
Environmental impact variation	The environmental impact of technological innovation varies widely, from positive to negative effects	Positive effects (Ahmed et al., 2022), negative effects (Mughal et al., 2022; Sinha et al., 2020)	There is a need for targeted policies to ensure that technological innovation contributes positively to environmental sustainability
Economic-environmental trade-offs	Some studies indicate a trade-off between economic growth and environmental sustainability in the context of technological innovation	Mixed effects in Morocco (Abir et al., 2024), adverse environmental effects alongside economic growth (Villanthenkodath & Mahalik, 2022)	Policies should aim to strike a balance between economic growth and environmental protection, potentially by promoting specific types of innovation
Renewable energy and resource efficiency	Some studies highlight the role of renewable energy consumption and resource efficiency in the relationship between innovation and sustainability	Inclusion of renewable energy consumption in sustainability indicators (Dam et al., 2024; Sinha et al., 2020)	Policies should promote innovations that enhance the adoption of renewable energy and improve resource efficiency

Economic-environmental trade-offs, and Renewable energy and resource efficiency.

2.4.1 The Role of Technology in Sustainable Economic

Technology plays a crucial role in promoting sustainable economic growth (Iqbal et al., 2023) by enhancing efficiency, reducing resource consumption, and encouraging innovation. It serves as a powerful enabler of sustainable economic practices. By enhancing efficiency, promoting innovation, and facilitating the transition to renewable resources, technology can contribute to creating a more sustainable and resilient economy (Fig. 2).

2.4.2 The Role of Innovation in Sustainable Economic

Innovation is a crucial driver of sustainable economic growth (Butt, 2024) as it fosters new ideas, processes, and solutions that enhance efficiency and promote environmental stewardship. Here are several key roles that innovation plays in sustainable economics.

In summary, innovation is crucial for driving sustainable economic growth by introducing new technologies and practices that enhance efficiency, minimize environmental impact, and promote resilience. By cultivating a culture of innovation, societies can tackle urgent challenges such as climate change, resource depletion, and social inequality, while paving the way for a sustainable future (Fig. 3).

2.4.3 The Role of Technology in Sustainable Social

Technology plays a significant role in promoting social sustainability (Al-Emran, 2023) by enhancing quality of life, fostering equity, and empowering communities. Here are some key ways in which technology contributes to social sustainability.

Technology plays a crucial role in promoting social sustainability by enhancing access to essential services, fostering community engagement, and empowering individuals and marginalized groups (Fig. 4).

2.4.4 The Role of Innovation in Sustainable Social

Social innovation is vital for achieving sustainable social development (Leal Filho et al., 2022), as it offers novel solutions to complex societal challenges while considering long-term environmental and economic well-being. It goes beyond traditional approaches, often involving

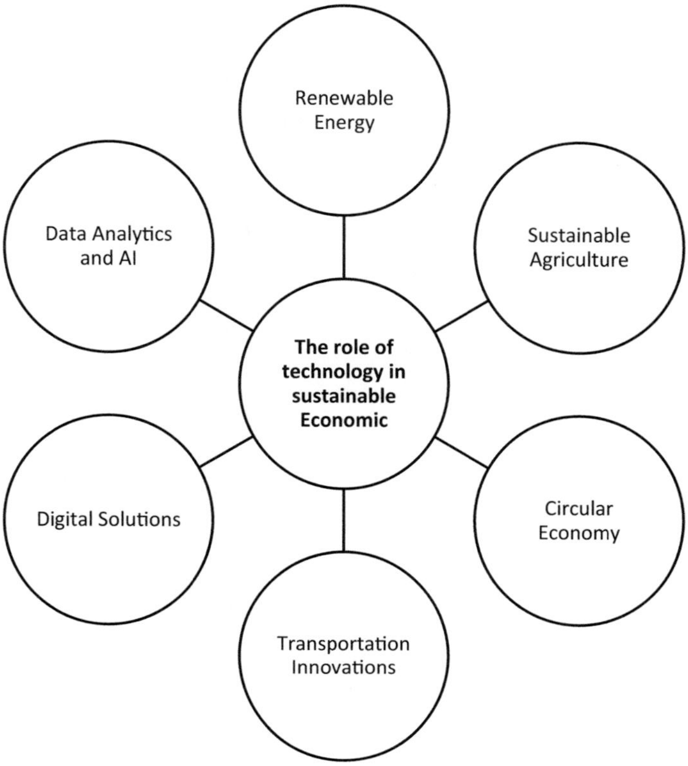

Fig. 2 The role of technology in sustainable economic development. *Source* Ali et al. (2024)

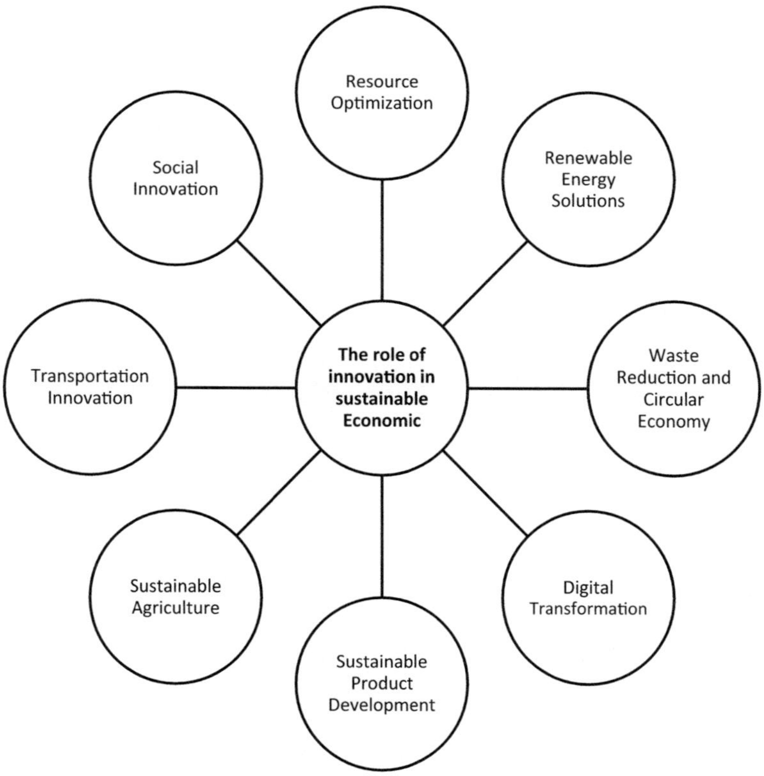

Fig. 3 The role of innovation in sustainable economic development. *Source* Reddy et al. (2023)

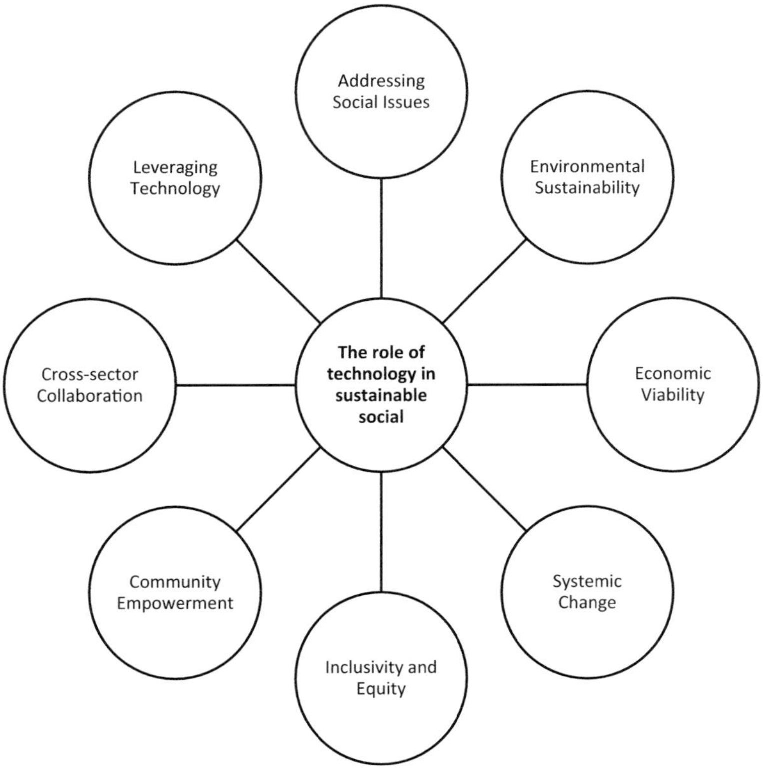

Fig. 4 The role of technology in sustainable social. *Source* Attah et al. (2024)

collaboration across sectors and the active participation of communities to create systemic change.

In essence, innovation in sustainable social development is about creating a future where social progress, environmental responsibility, and economic prosperity are interconnected and mutually reinforcing (Fig. 5).

2.4.5 The Role of Innovation in a Sustainable Environment

Innovation is indispensable for achieving sustainable environmental development (Ghobakhloo et al., 2021). It provides the means to address environmental challenges more effectively, efficiently, and in ways that foster long-term ecological health and resilience (Fig. 6).

Innovation is not merely a technological advancement but a fundamental driver for achieving environmental sustainability across all sectors. It provides the necessary tools, processes, and approaches to decouple economic growth from environmental degradation, thereby building a more resilient and healthy planet for future generations.

2.4.6 The Role of Technology in a Sustainable Environment

Technology plays a multifaceted and critical role in achieving sustainable environmental development (Islam, 2025). While the production and use of some technologies have historically contributed to environmental problems, innovation is now increasingly focused on creating solutions that mitigate these negative impacts and promote ecological health. Here's a breakdown of technology's role (Fig. 7).

It's essential to recognize that technology has an environmental impact, encompassing the energy consumption of data centers, the extraction of raw materials for electronics, and the issue of electronic waste. Therefore, a holistic approach to sustainable environmental development involves not only deploying green technologies but also ensuring the sustainable production, use, and disposal of all technologies.

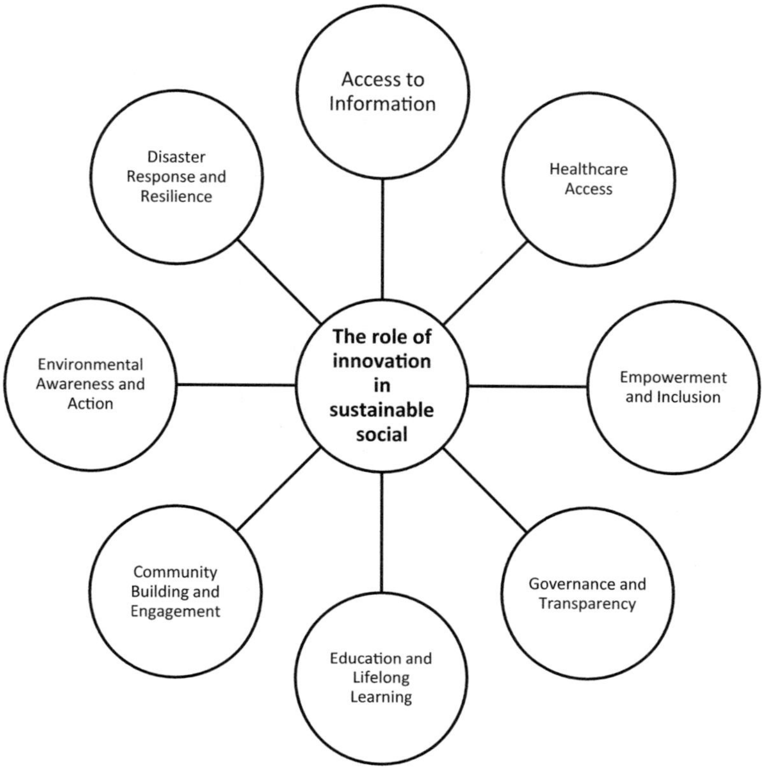

Fig. 5 The role of innovation in sustainable social. *Source* Kangana et al. (2024)

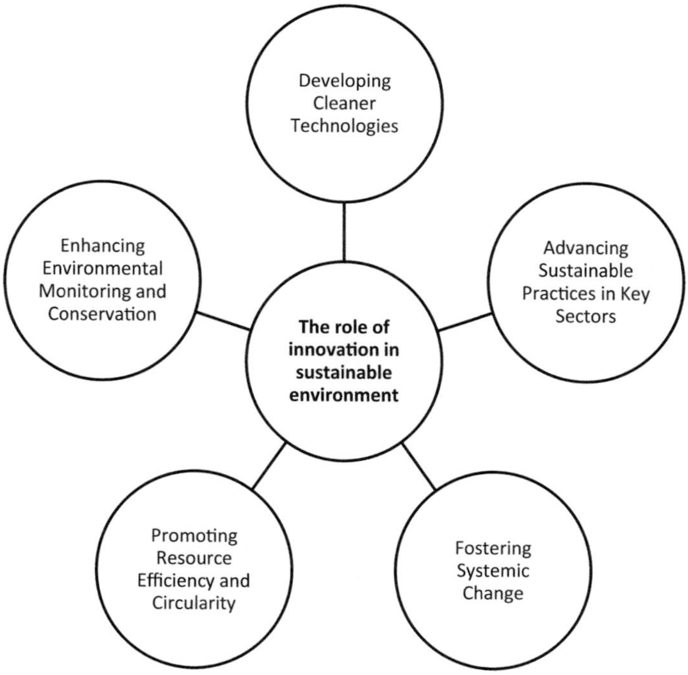

Fig. 6 The role of innovation in a sustainable environment. *Source* Kandpal et al. (2024)

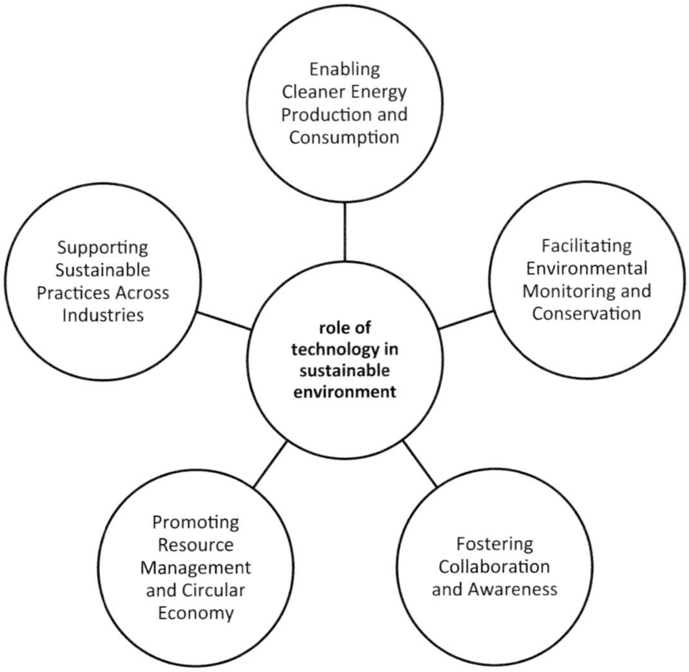

Fig. 7 The role of technology in a sustainable environment. *Source* Kandpal et al. (2024)

Ultimately, technology is a powerful tool that, when developed and applied responsibly, is essential for creating a more sustainable and resilient future for our planet.

2.4.7 Innovation and Technology in Economic Sustainability

The following is a review of the role of innovation and technology in a sustainable economy, society, and environment. Technological solutions or innovative practices simultaneously support more than one pillar of sustainability—i.e., it has economic value, environmental benefits, and/or social impact.

Innovation and Technology are central to all aspects of Sustainable Entrepreneurship. They enable scalable, efficient, and impactful solutions across the economy, environment, and society. Generally, new opportunities lie in combining digital transformation, green technologies, and social impact models.

Economic sustainability focuses on creating long-term value by balancing profitability (Khan, 2024) with environmental stewardship and social responsibility. Innovation and technology are central to this goal, enabling businesses to optimize resources, reduce costs, and align financial success with the well-being of the planet and society. Below, we explore key intersections of technology and economic sustainability, supported by the real world (Table 2).

Innovation and technology are reshaping economic sustainability by transforming how businesses generate value. From renewable energy adoption to circular economy models, these advancements prove that profitability and sustainability are not mutually exclusive. By leveraging cutting-edge tools, companies can reduce costs, mitigate risks, and capture emerging market opportunities. However, success requires systemic shifts—businesses must embed sustainability into core strategies, supported by policy incentives and cross-sector collaboration. As climate and resource challenges intensify, economic sustainability will no longer be optional but a competitive necessity (Ivanova, 2024) for thriving in the twenty-first century.

Table 2 Innovation and technology in economic sustainability (Makieła et al., 2022)

No.	Technology	How it work	Challenge	Innovation	Examples	Impact
1	Innovation in a sustainable economy	Profit meets responsibility	Traditional business models often prioritize short-term profits over long-term ecological and social health	Sustainable business models integrate triple-bottom-line principles (people, planet, profit) to create shared value	• Patagonia's Worn Wear • Unilever's sustainable living plan	Enhances brand loyalty, reduces regulatory risks, and future-proofs businesses against resource scarcity
2	Resource optimization	Efficient use of materials, energy, and labor	Inefficient resource use increases operational costs and environmental degradation	Technologies like AI, IoT, and automation streamline production, minimize waste, and boost productivity	• Siemens' digital enterprise • Precision agriculture	Lowers input costs, improves margins, and aligns with global sustainability targets
3	Sustainable product development	Eco-friendly products and services	Consumer demand for sustainable products outpaces supply	Companies design products with reduced environmental footprints using recycled materials, biodegradable packaging, or energy-efficient processes	• Tesla's electric vehicles (EVs) • Allbirds' carbon-neutral footwear	Opened new markets, attracts eco-conscious consumers, and differentiates brands in competitive industries
4	Renewable energy solutions	Clean energy for cost and emission reduction	Fossil fuel dependence exposes businesses to volatile energy prices and climate risks	Transitioning to renewable energy sources (such as solar, wind, and geothermal) reduces energy costs and carbon footprints	• Apple's 100% renewable energy commitment • Ørsted's wind farms	Stabilizes energy budgets, enhances corporate reputation, and aligns with net-zero pledges
5	Sustainable agriculture	Farming for long-term profitability	Conventional farming practices deplete soil health and water resources, while also facing the challenges of climate volatility	Agri-tech solutions improve yields, reduce inputs, and ensure resilience	• Hello Tractor (Africa) • Vertical farming	Boosts farmer incomes, secures food supply chains, and mitigates climate-related losses
6	Waste reduction and circular economy	Turning waste into value	Linear "take-make-waste" systems generate 2.01 billion tons of municipal solid waste annually (World Bank)	Circular economy models repurpose waste as raw material, creating revenue streams from discarded resources	• Renewlogy • IKEA's buy-back program	Reduces disposal costs, unlocks new income sources, and meets consumer demand for zero-waste practices

(continued)

Table 2 (continued)

No.	Technology	How it work	Challenge	Innovation	Examples	Impact
7	Transportation innovation	Fuel-efficient logistics and delivery systems	Freight and logistics account for 11% of global CO_2 emissions (International Energy Agency, IEA)	Electrification, route optimization software, and autonomous vehicles cut fuel use and emissions	• Amazon's electric delivery fleet • UPS's ORION algorithm	Lowers operational costs, improves delivery efficiency, and strengthens compliance with emissions regulations
8	Digital transformation	Enhancing business efficiency	Manual processes and outdated systems hinder agility and scalability	Cloud computing, blockchain, and AI automate workflows, reduce errors, and unlock data-driven insights	• Amazon's electric delivery fleet • UPS's ORION algorithm	Drives operational efficiency, reduces overhead, and supports agile decision-making in dynamic markets

2.4.8 Innovation and Technology in Environmental Sustainability

Environmental sustainability focuses on protecting ecosystems, reducing pollution, and ensuring responsible resource use. Innovation and technology are critical enablers of these goals, offering scalable solutions to address climate change, biodiversity loss, and resource depletion. Below, we explore key intersections of technology and environmental sustainability, supported by real-world examples (Table 3).

Innovation and technology are transforming environmental sustainability by addressing pollution, resource depletion, and ecosystem degradation. From cleaner production methods to AI-driven conservation tools, these advancements offer scalable solutions to global challenges. However, systemic change requires collaboration between governments, businesses, and individuals. By integrating policy frameworks with cutting-edge technology, societies can transition toward a regenerative future, one where economic growth harmonizes with planetary health (Čegar et al., 2024). The urgency of climate action demands that these innovations not only continue but accelerate, ensuring a livable world for future generations.

2.4.9 Innovation and Technology in Social Sustainability

Social sustainability focuses on fostering inclusive, equitable, and resilient societies where access to resources, opportunities, and participation in decision-making are guaranteed. Innovation and technology serve as catalysts for achieving these goals, addressing systemic challenges, and improving the quality of life. Below, we explore key intersections of technology and social sustainability, supported by real-world examples (Table 4).

Innovation and technology are pivotal in advancing social sustainability by dismantling barriers to equity, participation, and resilience. From democratizing education to revolutionizing disaster response, these tools empower marginalized groups, enhance governance, and foster global collaboration. However, ethical deployment is critical to avoid exacerbating inequalities. By prioritizing inclusivity and accessibility, societies can harness technology not just as a tool for progress but as a force for collective well-being.

Table 3 Innovation and technology in environmental sustainability (Caillol, 2025)

No.	Technology	How it work	Challenge	Innovation	Examples	Impact
1	Developing cleaner technologies	Low-emission production and green chemistry	Industrial processes and chemical manufacturing contribute significantly to greenhouse gas emissions and toxic waste	Cleaner technologies prioritize low-emission production and eco-friendly chemical processes	• Carbon capture and storage (CCS) • Green chemistry • Electric vehicles (EVs)	Reduces reliance on fossil fuels, lowers air pollution, and mitigates climate change risks
2	Promoting resource efficiency and circularity	Reduce, reuse, recycle	Linear "take-make-waste" models deplete finite resources and generate massive waste	Circular economy strategies and resource-efficient technologies extend product lifecycles and minimize waste	• Patagonia's Worn Wear TerraCycle • Smart water systems	Conserves natural resources, reduces landfill waste, and creates economic value from recycled materials
3	Advancing sustainable practices in key sectors	Green construction and manufacturing	Traditional construction and manufacturing are energy-intensive and polluting	Sustainable practices integrate renewable energy, energy-efficient design, and low-carbon materials	• Green buildings • Sustainable manufacturing • Low-carbon cement	Cuts energy consumption, reduces carbon footprints, and sets industry-wide sustainability standards
4	Enhancing environmental monitoring and conservation	IoT sensors and AI tracking	Limited data on ecosystem health hinders effective conservation efforts	IoT sensors and AI analyze environmental data in real-time to inform conservation strategies	• Smart forests, ocean health tracking • Air quality monitoring	Enables data-driven conservation, improves climate resilience, and empowers policymakers with actionable insights
5	Fostering systemic change	Policy and tech-enabled eco-conscious behavior	Individual and institutional behaviors often prioritize short-term gains over long-term sustainability	Technology and policy frameworks incentivize eco-friendly practices across sectors	• Carbon pricing • Smart grids • Behavioral nudges	Drives large-scale adoption of sustainable practices, aligns economic incentives with environmental goals, and fosters a culture of accountability

Table 4 Innovation and technology in social sustainability (Selvakumar, 2025)

No.	Technology	How it work	Challenge	Innovation	Examples	Impact
1	Access to information	Bridging the digital divide	Unequal access to information perpetuates economic and educational disparities	Technologies like broadband expansion, public Wi-Fi initiatives, and open platforms (e.g., free encyclopedias, open-source tools) democratize access to knowledge	• Internet.org by Meta aims • Wikipedia Zero	Empowers marginalized communities with critical resources (e.g., job listings, health guides) and fosters informed civic participation
2	Education and lifelong learning	E-learning platforms and upskilling tools	Traditional education systems often fail to meet evolving workforce demands or reach remote populations	Digital learning platforms and AI-driven tools enable personalized, scalable education	• Duolingo uses AI • LinkedIn learning • Coursera and Khan Academy	Reduces educational inequality, enhances employability, and bridges skill gaps in the labor market
3	Healthcare access	Telemedicine and wearable health tech	Geographic, financial, and infrastructural barriers limit healthcare access in low-income and rural areas	Telehealth services and wearable devices enable remote diagnostics and preventive care, allowing for more efficient and effective healthcare management	• Solar deploys solar-powered • Apple watch and Fitbit monitor	Lowers mortality rates, improves chronic disease management, and reduces healthcare costs
4	Community building and civic engagement	Platforms for participation	Disconnected communities struggle to advocate for their needs or hold institutions accountable	Digital tools connect citizens to local governance and amplify collective action	• Nextdoor • FixMyStreet	Strengthens social cohesion, increases transparency, and drives responsive policymaking
5	Empowerment and inclusion	Tech for marginalized groups	Women, disabled individuals, and other marginalized groups face exclusion from economic and social systems	Assistive technologies and gender-focused platforms promote equity and equality	• JAWS (screen-reading software) • Ava translates	Breaks down barriers to education, employment, and public life, fostering inclusive growth
6	Environmental awareness and action	Citizen science and activism apps	Limited public engagement in environmental issues hinders collective action	Apps engage citizens in data collection and advocacy	• iNaturalist • Climate reality project	Raises awareness, empowers communities to protect ecosystems, and informs climate policies

(continued)

Table 4 (continued)

No.	Technology	How it work	Challenge	Innovation	Examples	Impact
7	Governance and transparency	Blockchain for accountability	Corruption and opaque governance erode public trust in institutions	Blockchain ensures secure, tamper-proof record-keeping and transactions	• Estonia's e-governance • Follow the money tracks	Enhances accountability, reduces fraud, and strengthens democratic processes
8	Disaster response and resilience	AI and drone-assisted relief	Inefficient disaster response exacerbates human and economic losses	AI predicts disasters, while drones deliver aid swiftly	• IBM • Zipline	Saves lives, accelerates recovery, and builds community resilience to climate risks

3 Conclusion

Overlapping the role of tech and innovation in Social, economic, and environmental Sustainability will lead to Sustainable Entrepreneurship (Sarma et al., 2024). These overlaps demonstrate that sustainable entrepreneurship thrives at the intersection of multiple sustainability goals, supported by technology and innovation. Some cross-cutting innovations appear in more than one domain and drive sustainable entrepreneurship; they will be revealed in this chart (Table 5).

These Overlaps Enable Sustainable Entrepreneurship as Key Enablers. Technology provides scalable tools (such as IoT, AI, blockchain, and cloud), and innovation reimagines outdated systems and creates new markets.

Triple-bottom-line thinking ensures alignment with people, planet, and profit. New business models emerge that are not only profitable but also regenerative and socially inclusive.

Table 5 Innovation and technology in sustainability goals

No.	Sustainability goals	Feature	Examples
1	Digital transformation + circular economy (Ilić et al., 2022)	Digital platforms enable the sharing economy, product lifecycle tracking, and reverse logistics	Airbnb, Patagonia's Worn Wear program
2	Renewable energy + social inclusion (Shobande et al., 2023)	Off-grid solar solutions for rural communities empower local economies and reduce emissions	d.light, M-KOPA Solar
3	AI and data analytics, environmental monitoring (Harfouche et al., 2024)	Real-time pollution tracking, deforestation alerts, and climate modeling	IBM green horizon, rainforest connection
4	Sustainable agriculture + digital tools (Sarfraz et al., 2023)	Precision agriculture, utilizing drones, IoT, and mobile apps, enhances yield and minimizes waste	Hello Tractor (Africa), Taranis
5	Transportation innovation + urban planning + social equity (Kuzio, 2019)	Smart mobility solutions (e-bikes, shared transit) reduce emissions and improve accessibility	Lime scooters, the Moovit app
6	Waste management + job creation (Fatimah et al., 2020)	Circular economy startups create jobs through recycling, repair, and upcycling	TerraCycle, Renewlogy
7	Education technology + environmental awareness (Bhambri & Bajdor, 2024)	Gamified learning platforms teach sustainability concepts to youth	Eco-Schools, Classcraft
8	Healthcare + clean energy (Čegar et al., 2024)	Solar-powered clinics in remote areas provide sustainable access to essential services	We Care Solar

Entrepreneurs become agents of systemic change by addressing global challenges like climate change, inequality, and resource scarcity.

The integration of digital transformation with the circular economy-pairing renewable energy initiatives with social inclusion, utilizing artificial intelligence and data analytics for environmental monitoring, and applying digital tools to promote sustainable agriculture—highlights the intersection of diverse sustainability objectives. Furthermore, innovations in transportation, combined with urban planning and social equity, the role of waste management in job creation, educational technologies that foster environmental awareness, and advancements in healthcare powered by clean energy all illustrate how sustainable entrepreneurship thrives at the convergence of multiple goals, driven by technological innovation (Pereira et al., 2021). These interconnected strategies reflect the principles of Systems Thinking, which suggest that technological progress can facilitate comprehensive and effective solutions for sustainability (Meadows, 2008).

4 Case Study

The fashion industry is renowned for its creativity, innovation, and significant impact on global culture. However, its rapid expansion and unchecked production methods have raised serious concerns regarding its environmental and social consequences, making it one of the most polluting sectors in the world, second only to the petroleum industry (Woodside & Fine, 2019). Annually, this industry accounts for approximately 79 billion cubic meters of water consumption (around 20% of the world's total), emits about 1.7 billion tons of CO_2 (approximately 10% of global CO_2 emissions), and generates 92 million tons of textile waste (Abbate et al., 2023). These alarming statistics underscore the urgent need for more sustainable practices and actions to benefit current and future generations.

4.1 Patagonia

Patagonia, a renowned outdoor clothing brand, is widely acknowledged for its strong commitment to environmental sustainability and social responsibility. The company incorporates innovative strategies throughout its supply chain and overall business model, demonstrating that sustainability can be both effective and financially rewarding.

"Ethical, environmentally friendly, and outdoors" are the words that come to mind when Patagonia is mentioned. Founded in 1973 by Yvon Chouinard, Patagonia emerged from his deep love for sustainability and the great outdoors. Today, it has established itself as a prominent brand in the outdoor apparel market, dedicated to environmental conservation and activism. In the 2020 Fashion Transparency Index, Patagonia achieved a commendable 7th place out of 250 apparel brands. Despite operating only 37 stores, the company has a significant presence across the US, Europe, Japan, Argentina, Chile, and other regions. Patagonia has initiated progressive actions that have significantly altered perceptions of the fashion industry. Words like "ethical," "environmentally friendly," and "outdoors" are often associated with the brand (Yik, 2022).

Patagonia simultaneously pursues sustainability and quality objectives embedded within its mission statement: "Build the best product, cause no unnecessary harm, use business to inspire and implement solutions to the environmental crisis" (O'Rourke & Strand, 2017). This company created a market advantage by building a remarkable and unique sustainable business model in the fashion industry (Shehata Nakhil, 2024).

4.1.1 Technological Innovations

Patagonia tries to leverage driving sustainable innovation by embracing tensions (O'Rourke & Strand, 2017). The dimensions of technological Innovations in this company are:

I. **Material Innovation**

Patagonia prioritizes the use of recycled materials in its product line. For example, the company was a pioneer in creating fleece jackets made from recycled plastic bottles. This approach not only diminishes waste but also helps conserve natural resources and energy.

II. **Worn Wear Program**

The Worn Wear initiative by Patagonia promotes garment repair over disposal. This program enables customers to buy, sell, and trade pre-owned Patagonia items, leveraging technology to support a circular economy that reduces waste and extends product life.

III. **Supply Chain Transparency**

Patagonia is enhancing the transparency of its supply chain through blockchain technology, enabling consumers to trace the origins of their purchases and ensuring that sourcing practices are ethical and sustainable.

IV. **Environmental Impact Tracking**

The company leverages technology to track and report its environmental impact. This data-driven methodology provides Patagonia with the insights necessary to refine its sustainability efforts continually.

4.1.2 Sustainable Practices

In Sustainable Practices, activism and advocacy are essential. Beyond being a retailer, Patagonia actively champions various environmental causes. The company allocates a portion of its profits to support environmental initiatives and engages in efforts to conserve and protect natural resources.

As a certified B Corporation, Patagonia exemplifies its dedication to corporate responsibility through social and environmental performance, accountability, and transparency.

4.1.3 Results of the Case Study

Through its innovative approach to sustainable entrepreneurship, Patagonia has not only strengthened its reputation as a leader in corporate responsibility but has also shown that environmental commitment can lead to significant business success. The brand's growth in sales and international recognition underscores that sustainability can indeed be aligned with profitability.

Patagonia illustrates how technology and innovation can foster sustainable entrepreneurship. By weaving advanced technologies and eco-friendly practices into its business framework, Patagonia not only contributes positively to the environment but also connects with environmentally conscious consumers, cultivating brand loyalty and influencing the industry at large. This case study highlights the opportunities for businesses to pursue sustainable innovation and thrive in a market that is increasingly focused on environmental responsibility.

Questions

- Identify the key drivers of sustainable entrepreneurship.
- Analyze the impact of sustainability practices on business performance.
- Develop a sustainable business model using case examples.

References

Abbate, S., Centobelli, P., & Cerchione, R. (2023). From fast to slow: An exploratory analysis of circular business models in the Italian apparel industry. *International Journal of Production Economics, 260*, Article 108824.

Abd El-Fatah, S., Ahmed, M., & Gharib, I. (2024). The role of Internet of things in supporting products design and achieving sustainable entrepreneurship. *Journal of Design Sciences and Applied Arts, 5*(1), 259–265.

Abir, S. I., Al Shiam, S. A., Zakaria, R. M., Shimanto, A. H., Arefeen, S. S., Dolon, M. S. A., Sultana, N., & Shoha, S. (2024). Use of AI-Powered precision in machine learning models for real-time currency exchange rate forecasting in BRICS economies. *Journal of Economics, Finance and Accounting Studies, 6*(6), 66–83.

Aguirre, L. A. L. (2024). Pathways to sustainable entrepreneurship: Analysing drivers of sustainable entrepreneurial orientation. *Sustainable Technology and Entrepreneurship,* 100081.

Al-Emran, M. (2023). Beyond technology acceptance: Development and evaluation of technology-environmental, economic, and social sustainability theory. *Technology in Society, 75*, Article 102383.

Ali, Z. A., Zain, M., Hasan, R., Al Salman, H., Alkhamees, B. F., & Almisned, F. A. (2024). Circular economy advances with artificial intelligence and digital twin: Multiple-case study of Chinese industries in agriculture. *Journal of the Knowledge Economy,* 1–37.

Amatucci, F. M., Pascale, A. M., & Serluca, M. C. (2015). Green economy and social responsibility in the Italian agri-food sector: The focus on the wine sector. *International Journal of Globalisation and Small Business, 7*, 234–246.

Attah, R. U., Garba, B. M. P., Gil-Ozoudeh, I., & Iwuanyanwu, O. (2024). Strategic partnerships for urban sustainability: Developing a conceptual framework for integrating technology in community-focused initiatives. *GSC Advanced Research and Reviews, 21*(2), 409–418.

Avilés-Palacios, C., & Rodríguez-Olalla, A. (2021). The sustainability of waste management models in circular economies. *Sustainability, 13*(13), 7105.

Bagheri, A., Akbari, M., Zolfaghari, A., & Razi, S. (2018). Entrepreneurship research in Iran: Current trends and future agendas. In N. Faghih & M. Zali (Eds.), *Entrepreneurship education and research in the Middle East and North Africa (MENA). Contributions to Management Science*. Springer. https://doi.org/10.1007/978-3-319-90394-1_16

Bhambri, P., & Bajdor, P. (Eds.). (2024). *Handbook of technological sustainability: Innovation and environmental awareness*. CRC Press.

Bickley, S. J., Macintyre, A., & Torgler, B. (2024). Artificial intelligence and big data in sustainable entrepreneurship. *Journal of Economic Surveys*.

Buterin, V. (2014). *A next-generation smart contract and decentralized application platform*.

Butt, J. S. (2024). From concept to practice: Innovations driving sustainable economic development. *EIRP Proceedings, 19*(1), 181–190.

Caillol, S. (2025). Chemistry in 2025: Shaping the future of sustainable innovation and eco-friendly materials. *Green Materials, 13*(1), 1–2.

Čegar, S., Drezgić, S., & Čišić, D. (2024). Exploring the regenerative economy: A comprehensive review of literature. *Niš, 1*(1), 1–40.

Chen, L., Wang, Y., & Zhang, H. (2024). Real-time data analytics for sustainable supply chain management using IoT. *Journal of Sustainable Operations, 12*(1), 45–62.

Chen, S., Qiu, H., Xiao, H., He, W., Mou, J., & Siponen, M. (2021). Consumption behavior of eco-friendly products and applications of ICT innovation. *Journal of Cleaner Production, 287*, Article 125436.

Dam, M. R., & van der Laan, M. D. (2024). Techno-economic assessment of battery systems for PV-equipped households with dynamic contracts: A case study of the Netherlands. *Energies, 17*(12), 2991.

DiVito, L., & Ingen-Housz, Z. (2021). From individual sustainability orientations to collective sustainability innovation and sustainable entrepreneurial ecosystems. *Small Business Economics, 56*(3), 1057–1072.

Evans, S., Vladimirova, D., Holgado, M., Van Fossen, K., Yang, M., Silva, E. A., & Barlow, C. Y. (2017). Business model innovation for sustainability: Towards a unified perspective for creation of sustainable business models. *Business Strategy and the Environment, 26*(5), 597–608.

Fatimah, Y. A., Govindan, K., Murniningsih, R., & Setiawan, A. (2020). Industry 4.0 based sustainable circular economy approach for smart waste management system to achieve sustainable development goals: A case study of Indonesia. *Journal of Cleaner Production, 269*, 122263.

Gadanakis, Y. (2024). Advancing farm entrepreneurship and agribusiness management for sustainable agriculture. *Agriculture, 14*(8), 1288.

Garcia, M., & Brown, S. (2025). Blockchain for ethical sourcing: A case study analysis. *Sustainable Business Review, 15*(2), 112–130.

Ghobakhloo, M., Iranmanesh, M., Grybauskas, A., Vilkas, M., & Petraitė, M. (2021). Industry 4.0, innovation, and sustainable development: A systematic review and a roadmap to sustainable innovation. *Business Strategy and the Environment, 30*(8), 4237–4257.

Gray, B. J., Duncan, S., Kirkwood, J., & Walton, S. (2014). Encouraging sustainable entrepreneurship in climate-threatened communities: A Samoan case study. *Entrepreneurship & Regional Development, 26*(5–6), 401–430.

Harfouche, A., Merhi, M. I., Albizri, A., Dennehy, D., & Thatcher, J. B. (2024). Sustainable development through technological innovations and data analytics. *Information Systems Frontiers, 26*(6), 1989–1996.

Hajiketabi, A., Zolfaghari, A., Goodarzi, K., & Akhondi, A. R. (2017). Environmental factors and medical spin-offs: A case study. *Payesh (Health Monitor), 16*(5), 947–958.

Hejazi, S. R., Zolfaghari, A., & Farhoudi, A. (2011). Identifying environmental influencing factors on the growth of research-based spin-offs in Iran. In *ECEI2011-6th European Conference on Innovation and Entrepreneurship: ECEI 2011* (p. 446). Academic Conferences Limited.

Hummels, H., & Argyrou, A. (2021). Planetary demands: Redefining sustainable development and sustainable entrepreneurship. *Journal of Cleaner Production, 278*, Article 123804.

Ilić, M. P., Ranković, M., Dobrilović, M., Bucea-Manea-Ţoniş, R., Mihoreanu, L., Gheţa, M. I., & Simion, V. E. (2022). Challenging novelties within the circular economy concept under the digital transformation of society. *Sustainability, 14*(2), 702.

Iqbal, K., Sarfraz, M., & Khurshid. (2023). Exploring the role of information communication technology, trade, and foreign direct investment to promote sustainable economic growth: Evidence from Belt and Road Initiative economies. *Sustainable Development, 31*(3), 1526–1535.

Islam, H. (2025). Nexus of economic, social, and environmental factors on sustainable development goals: The moderating role of technological advancement and green innovation. *Innovation and Green Development, 4*(1), Article 100183.

Islam, R. (2024). *Sustainable entrepreneurship in the digital age: The role of AI in green business practices* [M.B.A. dissertation]. Novia UAS.

Ivanova, L. (2024). *Crafting cultures of sustainable agile* [Doctoral dissertation]. Technische Universität Wien.

Jha, V. K., & Pande, A. S. (2024). Making sustainable development happen: Does sustainable entrepreneurship make nations more sustainable? *Journal of Cleaner Production, 440*, Article 140849.

Kandpal, V., Jaswal, A., Santibanez Gonzalez, E. D., & Agarwal, N. (2024). Circular economy principles: Shifting towards sustainable prosperity. In *Sustainable energy transition: Circular economy and sustainable financing for environmental, social and governance (ESG) practices* (pp. 125–165). Springer Nature Switzerland.

Kangana, N., Kankanamge, N., De Silva, C., Goonetilleke, A., Mahamood, R., & Ranasinghe, D. (2024). Bridging community engagement and technological innovation for creating smart and resilient cities: A systematic literature review. *Smart Cities, 7*(6), 3823–3852.

Khan, F. A. (2024). Sustainable business management: Balancing profit and purpose. *Journal of Management Science Research Review, 2*(1), 24–32.

Khandeparkar, A. S., Paul, R., Sridhar, A., Lakshmaiah, V. V., & Nagella, P. (2024). Eco-friendly innovations in food packaging: A sustainable revolution. *Sustainable Chemistry and Pharmacy, 39*, Article 101579.

Kshetri, N. (2018). Blockchain and sustainable development: Opportunities and challenges. *Journal of Global Information Management (JGIM), 26*(4), 61–79.

Kuzio, J. (2019). Planning for social equity and emerging technologies. *Transportation Research Record, 2673*(11), 693–703.

Leal Filho, W., Fritzen, B., Ruiz Vargas, V., Paço, A., Zhang, Q., Doni, F., Azul, A. M., Vasconcelos, C. R. P., Nikolaou, I. E., Skouloudis, A., Weresa, M. A., Marczewska, M., Price, E., Anholon, R., Rampasso, I., Quelhas, O., Salvia, A. L., Ozuyar, P. G., Moggi, S., & Wu, Y. J. (2022). Social innovation for sustainable development: Assessing current trends. *International Journal of Sustainable Development & World Ecology, 29*(4), 311–322.

Lee, K. (2022). Enhancing consumer trust through blockchain-enabled traceability in sustainable products. *Journal of Consumer Ethics, 8*(4), 201–215.

Lüdeke-Freund, F. (2020). Sustainable entrepreneurship, innovation, and business models: Integrative framework and propositions for future research. *Business Strategy and the Environment, 29*(2), 665–681.

Mahardhani, A. J. (2023). The role of public policy in fostering technological innovation and sustainability. *Journal of Contemporary Administration and Management (ADMAN), 1*(2), 47–53.

Maione, G. (2024). The Internet of things and its impact on sustainable innovation reporting. In *Sustainable innovation reporting and emerging technologies: Promoting accountability through artificial intelligence, blockchain, and the Internet of things* (pp. 71–94). Emerald Publishing Limited.

Makieła, Z., Stuss, M. M., & Borowiecki, R. (2022). *Sustainability, technology and innovation 4.0*. Routledge.

Malik, S. (2024). Data-driven decision-making: Leveraging the IoT for real-time sustainability in organizational behavior. *Sustainability, 16*(15), 6302.

Meadows, D. (2008). *Thinking in systems: International bestseller*. Chelsea green publishing.

Mughal, N., Arif, A., Jain, V., Chupradit, S., Shabbir, M. S., Ramos-Meza, C. S., & Zhanbayev, R. (2022). The role of technological innovation in environmental pollution, energy consumption and sustainable economic growth: Evidence from South Asian economies. *Energy Strategy Reviews, 39*, 100745.

Nakamoto, S. (2008). *Bitcoin: A peer-to-peer electronic cash system.*

Odeyemi, O., Usman, F. O., Mhlongo, N. Z., Elufioye, O. A., & Ike, C. U. (2024). Sustainable entrepreneurship: A review of green business practices and environmental impact. *World Journal of Advanced Research and Reviews, 21*(2), 346–358.

O'Grady, W., Weise, T., & Lee, J. (2018). Blockchain for renewable energy: Status and outlook. *Energy Informatics, 1*(1), 1–11.

O'Rourke, D., & Strand, R. (2017). Patagonia: Driving sustainable innovation by embracing tensions. *California Management Review, 60*(1), 102–125.

Pereira, V., Silva, G. M., & Dias, Á. (2021). Sustainability practices in hospitality: Case study of a luxury hotel in Arrábida Natural Park. *Sustainability, 13*(6), 3164.

Reddy, V. V., Kumar, P., Rao, A. L. N., Kumar, R., Singh, S., Asha, V., & Kareem, S. H. (2023). Waste to wealth generation: Innovative methodologies in resource utilization and minimization in circular economy. *E3S Web of Conferences, 453*, 01035.

Rosário, A. T., Raimundo, R. J., & Cruz, S. P. (2022). Sustainable entrepreneurship: A literature review. *Sustainability, 14*(9), 5556. https://doi.org/10.3390/su14095556

Sarfraz, S., Ali, F., Hameed, A., Ahmad, Z., & Riaz, K. (2023). Sustainable agriculture through technological innovations. In *Sustainable agriculture in the era of the OMICs revolution* (pp. 223–239). Springer International Publishing.

Sarma, S., Attaran, S., & Attaran, M. (2024). Sustainable entrepreneurship: Factors influencing opportunity recognition and exploitation. *The International Journal of Entrepreneurship and Innovation, 25*(1), 56–69.

Schaltegger, S., Hansen, E. G., & Lüdeke-Freund, F. (2016). Business models for sustainability: Origins, present research, and future avenues. *Organization & Environment, 29*(1), 3–10.

Schlange, L. E. (2006, November). What drives sustainable entrepreneurs. In *Proceedings of the Applied Business and Entrepreneurship Association International (ABEAI) Conference* (pp. 1–11).

Selvakumar, P. (2025). The role of technology in enhancing social sustainability. In *Enhancing social sustainability in manufacturing supply chains* (pp. 275–302). IGI Global Scientific Publishing.

Sharma, M., Antony, R., Sharma, A., & Daim, T. (2025). Can smart supply chain bring agility and resilience for enhanced sustainable business performance? *The International Journal of Logistics Management, 36*(2), 501–555.

Shehata Nakhil, S. (2024). Modern urgencies creating a market advantage: Sustainable business models in the fashion industry—A Patagonia case study.

Shobande, O. A., Ogbeifun, L., & Tiwari, A. K. (2023). Re-evaluating the impacts of green innovations and renewable energy on carbon neutrality: Does social inclusiveness really matters? *Journal of Environmental Management, 336*, Article 117670.

Smith, A., & Jones, B. (2023). The role of IoT in optimizing logistics for environmental sustainability. *Environmental Management Journal, 21*(3), 305–320.

Simón-Moya, V., Revuelto-Taboada, L., & Guerrero, R. F. (2014). Institutional and economic drivers of entrepreneurship: An international perspective. *Journal of Business Research, 67*(5), 715–721.

Sinha, A., Shah, M. I., Sengupta, T., & Jiao, Z. (2020). Analyzing technology-emissions association in Top-10 polluted MENA countries: How to ascertain sustainable development by quantile modeling approach. *Journal of Environmental Management, 267*, 110602.

Tapscott, D., & Tapscott, A. (2016). *Blockchain revolution: How the technology behind bitcoin is changing money, business, and the world*. Penguin.

Terán-Yépez, E., Marín-Carrillo, G. M., del Pilar Casado-Belmonte, M., & de las Mercedes Capobianco-Uriarte, M. (2020). Sustainable entrepreneurship: Review of its evolution and new trends. *Journal of Cleaner Production, 252*, 119742.

Tian, F. (2017). A supply chain traceability system for food safety based on HACCP, blockchain & Internet of things. In *2017 International Conference on Service Systems and Service Management* (pp. 1–6). IEEE.

United Nations. (2023). *Transforming our world: The 2030 agenda for sustainable development*.

Uriarte-Gallastegi, N., Arana-Landín, G., Landeta-Manzano, B., & Laskurain-Iturbe, I. (2024). The role of AI in improving environmental sustainability: A focus on energy management. *Energies, 17*(3), 649.

Villanthenkodath, M. A., & Mahalik, M. K. (2022). Technological innovation and environmental quality nexus in India: Does inward remittance matter? *Journal of Public Affairs, 22*(1), e2291.

WCED. (1987). World commission on environment and development. *Our Common Future, 17*(1), 1–91.

Weber, I., Chen, S., & Bastiaens, Y. (2016). Traceability in supply chains using blockchain technology. In *Proceedings of the 2016 International Conference on Digital Transformation* (pp. 73–78).

Woodside, A. G., & Fine, M. B. (2019). Sustainable fashion themes in luxury brand storytelling: The sustainability fashion research grid. *Journal of Global Fashion Marketing, 10*(2), 111–128.

World Economic Forum. (2024). *The future of sustainable technology*.

Xiao, D., & Su, J. (2022). Role of technological innovation in achieving social and environmental sustainability: Mediating roles of organizational innovation and digital entrepreneurship. *Frontiers in Public Health, 10*, Article 850172.

Yik, S. (2022). Case 11: Patagonia–leader of a sustainable business. In *Business innovation* (pp. 267–274). Routledge.

Zolfaghari, A., Akbari, M., & Alizadeh, S. S. (2022). Open innovation development policy: Requirements for Iran. *Roshd-e-Fanavari, 3*(71), 33.

Digital Sustainable Entrepreneurship

Mahsa Kamalipoor and Morteza Akbari

Abstract

This chapter of the book focuses on the concept of sustainable digital entrepreneurship, a topic that has recently garnered the attention of academics. First, the idea of digital entrepreneurship was studied. Digital Entrepreneurship is defined as the pursuit of opportunities based on the use of digital media and other information and communication technologies. To further understand sustainable digital entrepreneurship, digital technologies were introduced. These technologies include artificial intelligence, Internet of Things, blockchain, big data, augmented reality, virtual reality, and cloud computing. Then, the role of digital technologies and digital business models in sustainable entrepreneurship was described. Sustainable digital entrepreneurship is defined as organizational activities that aim to enhance social and environmental value through the creative application and utilization of digital technology. To develop sustainable digital entrepreneurship, individuals, businesses, and governments are key players, and the socio-cultural and economic contexts of countries must be taken into account.

M. Kamalipoor · M. Akbari (✉)
Faculty of Entrepreneurship, University of Tehran, Tehran, Iran
e-mail: mortezaakbari@ut.ac.ir

Keywords

Digital sustainable entrepreneurship · Digital economy · Digital technologies · Digital business models

Learning Objectives

By the end of this chapter, readers will be able to

- Understand the concept of digital sustainable entrepreneurship.
- Analyze the impact of digital sustainable entrepreneurship on business, environment, and society.
- Describe how digital business models contribute to sustainability.
- Explain strategies for developing digital sustainable entrepreneurship in business.
- Develop a digital sustainable entrepreneurship business model using case examples.

1 Introduction

The topic of sustainability, which requires businesses to address social, environmental, and economic issues, is one of the major challenges they face today (Geissdoerfer et al., 2018). Because sustainable business models have been highly successful in implementing sustainable development strategies and concepts, they

M. Akbari et al. (eds.), *Sustainable Entrepreneurship*, University of Tehran Science and Humanities Series, https://doi.org/10.1007/978-3-032-04909-4_8

have gained popularity over the first and second decades of the twenty-first century (Szromek, 2021). The Sustainable Development Goals (SDGs) inspire many types of entrepreneurship that combine value creation with social protection and support. This opens up new avenues for us to create new investments and shape a phenomenon called sustainable entrepreneurship (Crecente et al., 2021). Among the most critical digital transformation technologies are mobile applications, knowledge work automation, the Internet of Things, cloud technology, advanced robotics, energy storage, social media, virtual and augmented reality, data analytics, artificial intelligence, and blockchain(Sebastian et al., 2017). Such technological advances have led to the digitalization of the economy, and the COVID-19 outbreak has accelerated the development of the digital economy, underscoring the need for its further development even more than before (Bloom et al., 2020). The expanding digital economy owes much to the entrepreneurial activities made possible by digital technologies (Zaheer et al., 2019), and digital entrepreneurship has been a particularly popular topic among manufacturing and service actors. Digital entrepreneurship has been proposed as a new approach to starting a business, based on innovative thinking and ideas for creating and developing areas that produce goods and services in society (Pergelova et al., 2019).

Sustainable development and digitalization are crucial factors that impact the performance of modern companies. The link between digitalization and sustainable entrepreneurship is essential for preventing and mitigating emerging environmental and social challenges (Smoląg & Chomiak-Orsa, 2025). Digital technologies are not simply tools that sustainable entrepreneurs can use to achieve specific predefined goals, often presented in other pieces, but instead, they are vital and lively actors that lead to unexpected and surprising results in sustainability efforts. Unfortunately, the transformative potential of digital technologies for sustainable entrepreneurship has not been thoroughly explored (George et al., 2021), and our understanding of the impact of digital technologies on sustainable entrepreneurial business models remains limited (Fuerst et al., 2023).

To address the question of how digital technologies can support sustainable entrepreneurial objectives, this chapter outlines the concept of digital entrepreneurship, examines digital technologies, and discusses the role of digital technologies and digital business models in promoting sustainability. Finally, we conclude by presenting a case study that sheds further light on digital sustainable entrepreneurship.

2 What Is Digital Entrepreneurship?

Entrepreneurship is a process in which an entrepreneur, with new, creative ideas and by identifying new opportunities, mobilizes resources to create new and innovative growing companies and organizations, and often leads to the introduction of a product or the provision of services to society (Thompson, 2020).

Entrepreneurship has undergone a global transformation in the last decade. One of these transformations is digital entrepreneurship (Zaheer et al., 2019). In recent years, nearly all organizations across various industries have taken steps to adopt and benefit from digital technologies (Matt et al., 2015). Digital entrepreneurship is a type of entrepreneurship in which new products and services are developed by leveraging the Internet, cloud technology, big data and analytics tools, and other emerging technologies (Recker & von Briel, 2019), thereby forming the concept of digital entrepreneurship. Research on digital entrepreneurship is located at the intersection of digital technologies and entrepreneurship (Nambisan et al., 2017). Several disciplines are involved in the digital entrepreneurship process: software development, which is rooted in information systems (IS), business model conceptualization and strategy formulation, which are rooted in management, and new venture creation, which is rooted in entrepreneurship (Zaheer et al., 2019). Thus, this type of entrepreneurship goes beyond technology and involves the interaction between technology, strategy, and business processes. It involves entrepreneurs utilizing information and communi-

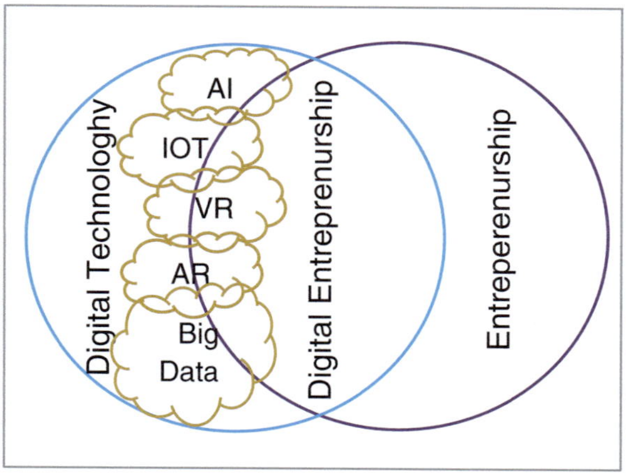

Fig. 1 The state of digital entrepreneurship

cation technologies to create value (Mahmood & Yu, 2005).

New goods and services are created through digital entrepreneurship, which leverages the Internet, cloud computing, big data and analytics tools, and other cutting-edge technologies (Giones & Brem, 2017). Digital entrepreneurship includes all companies (social, government, and private companies) that create services and goods for clients using digital technology (Sussan & Acs, 2017). Digital entrepreneurs seek to create value by creating and expanding economic activity by identifying and exploiting new or related products, processes, and markets dependent on information and communication technology (Li et al., 2017).

Figure 1 shows the position of digital entrepreneurship. Other researchers have provided some definitions of digital entrepreneurship in Table 1.

Three evolutionary trends have been proposed for digital entrepreneurship, which we summarize in Fig. 2 (Zaheer et al., 2019). As shown in the figure, businesses have moved from internet entrepreneurship to digital entrepreneurship.

Changes in big data analytics, the adoption of digital technologies, and their increasing utilization are commonly linked to digitalization (Parviainen et al., 2017). The rate of digitalization continues to increase, according to research. However, given the rapidly increasing rate of digitization in the contemporary era, it is imperative to consider digital entrepreneurship to comprehend its potential long-term effects on the transformation of such systems. (Satalkina & Steiner, 2020).

Academic research makes clear how important digitalization is as a catalyst for change in innovation systems. (Ardolino et al., 2018) and policy-relevant surveys from the European Commission, which indicate that 96% of business leaders consider digital technologies to be crucial for innovative development and sustained, qualitative economic growth (European Commission, 2014). Therefore, it is crucial to consider digital entrepreneurship in the innovation system to understand its potential impacts on the transformation and sustainable transition of such systems (Satalkina & Steiner, 2020).

3 Digital Technologies

Digital technologies (such as the Internet of Things, cloud computing, big data, blockchain, etc.) profoundly affect the activities and processes of companies, thus leading to changes in the value creation, value delivery, and value capture mechanisms of companies. The concept of “digital technology” has attracted considerable attention among professionals and researchers and is at the core of various research areas and disciplines, ranging from engineering to information systems to management. However, there is no agreed-upon definition of what constitutes a DT because its rapidly evolving nature has prevented researchers from delineating the boundaries of the concept and providing a sound conceptualization (Ancillai et al., 2023).

Digital technologies are transforming the way we live, learn, work, and generate ideas, and have become increasingly attractive to individuals, society, and companies (Bruno & Canina, 2019). For example, the hotel industry was severely disrupted by Booking.com and Airbnb, while Spotify revolutionized the music industry. Incumbent companies such as Starbucks were also transformed through digitalization. These disruptions are enabled by digital technologies such as the Internet of Things, big data analytics, cloud

Table 1 Definitions of digital entrepreneurship

Source	Digital entrepreneurship definitions
McMullen and Dimov (2013)	The process of starting a digital startup, whether as a new company or within an existing one, is known as digital entrepreneurship
Kollmann (2006)	E-entrepreneurship refers to establishing a new company with an innovative business idea within the net economy, which, using an electronic platform in data networks, offers its products and/or services based upon a purely electronic creation of value. It is essential that this value offer was only made possible through the development of information technology
Hull et al. (2007)	Technology development combined with innovative approaches to starting and running a firm is known as "digital entrepreneurship."
Hair et al. (2012, p. 2)	One definition of digital entrepreneurship is entrepreneurship where part or all of the business endeavor is conducted online rather than through more conventional means
Davidson and Vaast (2010)	Digital entrepreneurship involves identifying opportunities through the use of digital media and other information and communication technologies
Davidson and Vaast (2010)	It is a sub-branch of entrepreneurship that involves the use of digital media and technologies to digitize some or all of the physical elements present in traditional settings
Guthrie (2014, p. 116)	The establishment of a business to manufacture and make money from digital commodities over electronic networks is known as digital entrepreneurship
Sussan and Acs (2017)	Within both ecosystems, digital entrepreneurship is the convergence of digital infrastructure and entrepreneurial actors
Nambisan (2017)	Platforms, infrastructures, or objects that use computer power on public networks are referred to as digital technologies, and the nexus between digital technologies and entrepreneurship is known as digital entrepreneurship
Le Dinh et al. (2018)	Technology resources, such as the Internet and information and communication technologies (ICT), have been surpassed by digital entrepreneurship
Le Dinh et al. (2018)	The reconciliation of conventional entrepreneurship with the new methods of starting and doing business in the digital age is known as digital entrepreneurship
Davidson and Vaast (2010)	Digital entrepreneurship can be defined as the pursuit of opportunities based on the use of digital media and other information and communication technologies

computing, mobile technologies, and artificial intelligence (Feroz et al., 2021).

Table 2 introduces some of the most critical transformative digital technologies. As is clear, the term' digital technologies' is broader than information and communication technology. Digital technologies have several features that enable entrepreneurs, employees, and users to do things that could not previously be done effectively or at all using other, less sophisticated tools. Among the features of digital technologies are the following (Pinchot & Soltanifar, 2021):

- Automatic functions
- Capacity and range
- Evolvement
- Flexibility
- Interactivity
- Provisionality
- Speed
- Unpredictability/generativity.

Digital technologies create new options for individuals and organizations by lowering the cost of experimentation and iteration in creative activity. Stated differently, digital technologies encourage rapid iteration and experimentation along nonlinear paths in the entrepreneurial process through fluid and dynamic processes, facilitating rapid learning. Because of the generative nature of digital technologies, founders can build scalable products and services that create change and, consequently, drive growth. Digital technologies encourage disintermediation,

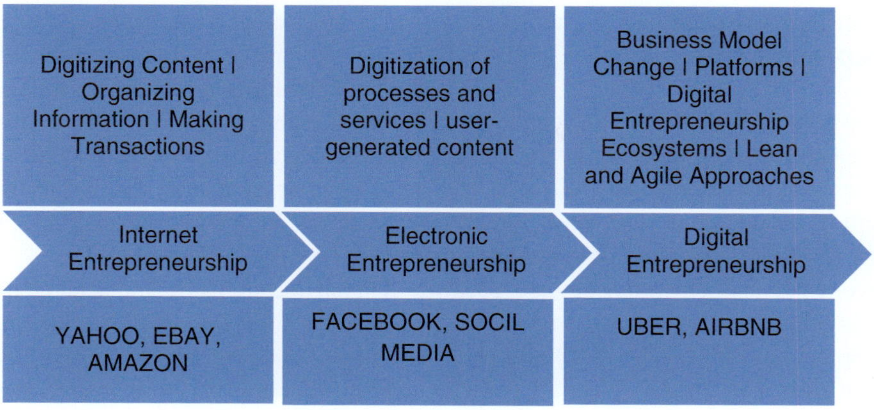

Fig. 2 Three evolutionary trends for digital entrepreneurship

Table 2 Transformative digital technologies

Transformative digital technologies	Definition
Artificial intelligence	The ability of a digital computer or computer-controlled robot to perform tasks typically associated with intelligent beings and having components of learning, reasoning, problem-solving, perception, and language (Copeland, 2020)
Augmented reality	Augmented reality adds a layer of digital information to the physical world, making it accessible to users. Through augmented reality, digital data—such as virtual, statistical, and aural content—is integrated into a real-world setting (Farshid et al., 2018)
Virtual reality	Virtual reality is a completely digital representation of the real world that incorporates virtual objects and information. It is a comprehensive substitute for the real world (Farshid et al., 2018)
Blockchain	Blockchain is a distributed, shared, encrypted, time-ordered, irreversible, immutable database and computing system with a consensus mechanism that creates added value by conducting direct interactions between users (Meijer, 2017, p. 39)
Cloud computing	Cloud computing is an information technology-based business model, provided as a service over the Internet, where both hardware and software computing services are delivered on-demand to customers in a self-service fashion, independent of device and location in which high levels of quality service, dynamically scalable, rapidly provisioned, shared, virtualized and released with minimal service provider interaction (Madhavaiah & Bashir, 2012)
Big data	Big Data is the Information asset characterised by such a High Volume, Velocity, and Variety to require specific Technology and Analytical Methods for its transformation into Value (De Mauro et al., 2016)
Internet of things	IoT technology connects physical objects around us with the Internet, enabling them to communicate with each other for business purposes (Cai et al., 2014)

leading to a greater emphasis on the ecosystem, not just the new venture team (NVT) (Zaheer et al., 2019).

Companies that have successfully adopted and exploited digital technologies use technologies to get closer to customers, empower employees, and transform their internal business processes. Such companies are better able than others to monetize their physical assets, access global markets for their goods and services, handle more demand per unit of existing physical capacity, and increase revenue by reaching new customers or engaging with existing customers in new ways (Pinchot & Soltanifar, 2021).

4 Digital Sustainable Entrepreneurship

Sustainable development is the foundation for the development of future generations, creating opportunities and challenges for managers to create socio-economic value. The three primary objectives of future human development—society, economy, and environment—are used to generally illustrate the concept of sustainable development as outlined by the Brundtland Commission (Stawicka, 2021).

One of the most important solutions for the sustainable growth of society, resources, and the environment is entrepreneurship. Research has shown that entrepreneurs do not neglect social and environmental issues in their entrepreneurial activities, and they consider personal interests and make rational decisions. This category of entrepreneurs, known as sustainable entrepreneurs, leverages entrepreneurial opportunities arising from neglected social and environmental issues to create value and establish the concept of sustainable entrepreneurship. In other words, sustainable entrepreneurship can be defined as an approach that focuses on preserving nature, supporting life, and society in the pursuit of perceived opportunities to create future products, processes, and services. This approach leads to benefits that collectively include individual, economic, and social benefits. One of the key factors influencing sustainable entrepreneurship is digital technologies (Xu et al., 2022). As an enabler of entrepreneurship and innovation, digital technologies are gradually unlocking their transformative potential to mitigate and address major environmental and social challenges, offering new opportunities for value creation (Li & Long, 2024). From this perspective, digital sustainable entrepreneurship has been considered. George et al. (2021) defined digital sustainable entrepreneurship as organizational activities that aim to enhance social and environmental value through the creative application and utilization of digital technology. Digital sustainable entrepreneurship represents the process of embedding social, environmental, and financial goals in digital products or e-value creation, as well as sustainable platforms. The challenges facing sustainable entrepreneurship often arise from information asymmetries, which manifest in the accuracy and timeliness of information transfer. Digital technologies address such problems by leveraging analytics, connectivity, intelligence, and ecosystem capabilities (Xu et al., 2022).

Digital technologies are leading to changes in people's lifestyles, enabling them to protect the environment and reduce pollution, while also creating equality and social justice by providing equal and transparent access to limited resources (Brenner, 2018). As it is clear, such technologies can impact the issue of sustainability in various ways. Let's examine the three pillars of sustainability and the role of technology.

I. For example, AI-based algorithms facilitate the optimization of resource use, energy efficiency, and waste management, thereby reducing carbon emissions and environmental footprints (Gupta & George, 2020). AI technology enables entrepreneurs to make informed decisions that reduce environmental impacts throughout the value chain by tracking and evaluating environmental data. Businesses can improve energy use and reduce carbon emissions by using AI-based models that can forecast patterns in energy demand (Islam, 2024). AI enables the creation of intelligent infrastructures that can autonomously adapt to changing conditions, thereby enhancing sustainability and resilience (Feng et al., 2020). According to Leng et al. (2020), blockchain is also seen as a tool with significant potential for attaining sustainability in business and industrial operations (Leng et al., 2020). By precisely tracking items and identifying work activities, blockchain can lower greenhouse gas emissions and resource usage (Saberi et al., 2018). BCG Group, in collaboration with the World Economic Forum, modeled autonomous vehicles in urban environments and showed that this emerging technology can reduce emissions, reduce traffic, shorten trips, and reduce deaths from accidents (improving both the environmental and social dimensions) (Schwab & Davis, 2018). Digital technologies, including the Internet of Things, cloud computing, and social media, have demonstrated effective applications in waste management (Ferrari et al., 2020). For example, an IoT-based system was designed and implemented to treat food waste generated by the Asian Institute of Technology (AIT) academic community, resulting in a significant reduction in food waste (Logan et al., 2019). It has been demonstrated that social media significantly influences people's recycling habits based on the notion of planned behavior (Sujata et al., 2019). Four broad categories can be used to categorize the effects of digital revolution on environmental sustainability: pollution control (Air pollution, CO_2 emission, Water treatment, Climate change/ Disaster management), waste management (solid waste, e-waste, food waste, and agri-waste), sustainable production (Sustainable manufacturing, Sustainable supply chain), or urban sustainability (Smart cities, Sustainable cities) (Feroz et al., 2021).

II. Digital technologies play a significant role in the social dimension of sustainability. Digital entrepreneurship offers the opportunity to harness creative forces at the lowest cost and maximum effectiveness in businesses, thereby leading to job creation (Richter et al., 2015) and providing a platform for economic empowerment and participation for low-income and marginalized individuals (Purnomo et al., 2020). Digital technology, such as blockchain, can potentially contribute to the social dimension of sustainable supply chains. Blockchain can be used in sustainable supply chains due to its security, traceability, transparency, and efficiency, as well as its ability to ensure the immutability and accessibility of data (Pan et al., 2020). Because information in blockchain does not change without the consent of the actors, it can block bad actors and hold corrupt individuals accountable for both social and personal crimes (Saberi et al., 2018). This technology enables people to register their properties from anywhere in the world, allowing them to access financial markets. Virtual and augmented reality can significantly enhance educational achievement and enable individuals to practice in safe, 3D environments, thereby improving their skills (Schwab & Davis, 2018). By analyzing vast data sets and understanding the needs of diverse stakeholders, entrepreneurs can develop solutions that create shared value (Mishra & Sinha, 2021). AI-powered platforms enable personalized services, accessibility, and inclusiveness, thereby increasing customer satisfaction and social impact (Cohen & Winn, 2007). In addition, through meaningful interactions with communities, stakeholders, and staff, AI technologies help entrepreneurs promote social cohesion, transparency, and trust. Including underrepresented groups as important partners in their business models is one advantage of employing digital technologies, which promotes social inclusion and community integration (Fuerst et al., 2023).

III. Digital technologies play a significant role in the economic sustainability dimension. Economic values in the sustainability dimension include price, quality, and cost. Digital technologies play a significant role in the success of innovation and the financial performance of companies (König et al., 2019). For example, using blockchain technology can benefit a company and its supply chain from various business dimensions that affect their economic performance. Blockchain can reduce business losses, transaction costs, and time by disintermediating the supply chain (Ward, 2017). Artificial intelligence enhances the operational effectiveness and competitiveness of sustainable investments by streamlining processes, enhancing decision-making, and optimizing resource allocation. Automation of routine tasks, predictive analytics, and intelligent algorithms enable entrepreneurs to increase productivity, reduce costs, and improve overall performance (Bapna et al., 2021). AI-based insights facilitate the real-time monitoring of supply chains, customer behavior, and market dynamics, enabling agile and data-driven decision-making processes (Liu et al., 2019). As a result, sustainable entrepreneurs can allocate resources more efficiently, increase scalability, and adapt to evolving market conditions, thereby enhancing long-term sustainability and growth, as well as improving business efficiency and profitability (Islam, 2024).

5 Digital Business Models and Sustainability

Entrepreneurs are not only utilizing digital technology-based innovations to address critical sustainability challenges, but also contributing to sustainability by developing business models that integrate innovations with new objectives (George et al., 2021). A business model is a simplified representation of the value proposition, value creation and delivery, and value capture elements, as well as the interaction between these elements within an organizational unit (Geissdoerfer et al., 2018). As digital technologies create new ways to generate and deliver value, they can render existing business models obsolete and give rise to new, digitally-centric business models (Palmié et al., 2022). A digital business model refers to a business model whose value proposition, value creation and delivery, and/or value capture are embodied in, or significantly enabled by, digital technologies (Soluk et al., 2021). Digital business models are diverse. For example, Wirtz (2019) identified four categories of B2C business models: content (e-information, e-entertainment, e-education, e-entertainment), commercial (e.attraction, e.negotiation, e.transaction, e.tailing), contextual (search engines, web directories, bookmarking), and communication (intraconnection, interconnection).

Each of these models can impact three aspects of sustainability. Content models, such as online learning platforms, have the potential to expand access to education, narrow the gaps between advantaged and disadvantaged communities, and thereby contribute to educational equity. Educational equity means that all learners, regardless of their field of study, ethnicity, gender, religion, culture, GPA, temperament, behavior, dress, and grooming style, demographic factors, and personality, have equal access to educational facilities (Heidari et al., 2023; Mehr et al., 2020). Such content business models can also enhance the customer experience and bring consumer satisfaction. For example, the use of adaptive learning technologies that utilize data analysis and machine learning to personalize education and respond to the diverse needs of individuals helps identify and address learning gaps in real-time (Darling-Hammond et al., 2020). E-commerce models, which are the process of buying and selling, transferring or exchanging products, services or information through computer networks connected to the Internet (Outland et al., 2018) lead to improved quality and faster access to customers, reduced costs of accessing customers (Fooladikia et al., 2021), expanded market coverage, improved productivity, significantly reduced transaction costs, and increased sales percentage (Adam et al., 2020). These are the economic effects of digital business models. Since e-commerce has two central elements—the buyer and the seller—they are inherently human in nature. The human nature of these two elements causes them to be governed by a set of social, cultural, and economic values. If an e-commerce system cannot be adaptable to the value spaces of these two elements, then it will not be used (Wu et al., 2018).

On the other hand, the basis of many e-commerce systems is the sharing economy (Mont et al., 2020). Consumers who give each other temporary access to shareable goods in exchange for money are called the sharing economy. In essence, shareable goods are goods that inherently provide additional capacity to their owners and give consumers the opportunity to lend or rent them to other consumers. The sharing economy through sharing platforms leads to the creation of new transactions, increased revenues, reduced transaction costs, consumer welfare, creating and facilitating less risky and more attractive connections between strangers through user ratings and reviews, reduced resource use, reusing unused assets, reducing CO_2 emissions (e.g. through car sharing), reducing the production of new goods or the construction of new facilities (e.g. hotels or co-working spaces), which is a response to the three dimensions of sustainability (Frenken & Schor, 2019).

Digital technologies can help create green awareness among IT professionals, businesses, and the general public by creating communities, engaging groups in participatory decision-making, and supporting education and environmental protection campaigns. Tools such as environmental web portals, blogs, wikis, and interactive simulations of an activity's environmental impact can be helpful in this regard (Murugesan, 2008).

The contextual model involves classifying and organizing information available on the Internet (e.g., Google, Indeed, Yahoo, etc.). Consider Indeed, a job search website. Indeed helps to reduce unemployment, improve workforce productivity, reduce the need for physical travel and printing resumes, and achieve net zero greenhouse gas emissions by increasing equal access to job opportunities, reducing the time it takes to find a job by 50%, helping to hire 30 million people who face barriers, attracting the best talent and fair hiring, providing transparency about organizational culture, employee treatment, and even corporate ethics through company reviews. Indeed, it contributes to sustainability by reducing unemployment, improving workforce productivity, reducing the need for physical travel and printing resumes, and achieving net-zero greenhouse gas emissions. About the Sustainable Development Goals, Indeed states:

> Our Environmental, Social, and Governance (ESG) commitments aim to create a future of work that is fair, inclusive, and sustainable. As the world's number one job site and a leading recruitment platform, we strive to make a positive impact on society by connecting people to better work for better lives. We are moving closer to achieving our mission by reducing our environmental footprint, streamlining job searches, eliminating bias and barriers to hiring, and fostering lasting equity for our employees. Our commitment to diversity, equity, inclusion, and belonging extends beyond our workforce. We have launched and will continue to expand our supplier diversity program, focusing on women, veterans, people with disabilities, and minority-owned businesses. We've achieved carbon neutrality since 2021 across our in-office, work-from-home, travel-related, and vendor emissions through the purchase of high-quality carbon offsets. 31,634 metric tons of carbon dioxide equivalent were saved through Indeed Interview, our virtual interview platform, between April 2022 and March 2023, the equivalent of keeping 7,529 gas vehicles off the road for one year. 90% of our energy consumption in the EMEA region is powered by renewable energy" (Indeed, N).

Communication models facilitate the exchange of information on networks, such as social networks (e.g., Facebook, Dropbox). Given the pervasiveness of communication and the ability to reach many people through a single post, the importance of using social media to develop a sustainable green lifestyle should not be underestimated. Sharing information through social media can increase public awareness of climate change, environmental and green information, and encourage people to participate in environmentally friendly activities. Encourage the environment. Social media influencers also drive the adoption of a green lifestyle by providing information-rich content and normalizing environmentally friendly behaviors. Social media enables businesses to innovate, enhance their reputation, increase success, and improve sustainability. Some companies have leveraged social media to develop new business opportunities and implement strategic management practices, while also enhancing organizational effectiveness by optimizing current business resources and procedures. Individuals can establish social connections through social networks using digital devices to fulfill their personal, professional, and recreational needs (Li et al., 2024). Dropbox has been focusing more on sustainability by maintaining the best power efficiency, optimizing overall power consumption, and sourcing renewable energy, stating this on its blog: "As a data center-based company, we understand our impact on global energy consumption and are proud to announce that all of our data center storage server power is covered by 100% renewable electricity. This means you leave no carbon footprint when you store your data with us" (Dropbox, 2021).

6 Strategy for Digital Sustainable Entrepreneurship

Intending to create a sustainable business model, organizations are shifting their focus from maximizing short-term profitability to creating long-term value for all stakeholders, encompassing not only economic issues but also social and environmental concerns (Geissdoerfer et al., 2018). Therefore, it is necessary to prepare themselves to move in this direction. We suggest that business owners always pay attention to strengthening the three dimensions (entrepreneurship, digital transformation, and sustainability) to move in this direction. We have outlined some basic actions focusing on digital entrepreneurship and sustainability below:

- **Strengthening the digital mindset and sustainability of the entrepreneur**: In general, the individual characteristics, motivation, and mindset of the entrepreneur are important for starting, running, and growing a business. Mindset represents a psychological process that is activated in response to a given task (French, 2016). One of the most important drivers of organizational change is a change in mindset (Gleeson, 2019). To be a sustainable digital entrepreneur, one needs to develop both a digital mindset and a sustainability mindset. A digital entrepreneurial mindset is defined as the willingness and ability to discover, evaluate, and exploit opportunities in a way that leverages digital technologies more effectively than a typical entrepreneur. Entrepreneurs with a digital mindset seek to create connections, utilize data, leverage cloud space to reduce costs, foster a learning culture, and be customer-centric by utilizing technology and an empowered workforce. Therefore, entrepreneurs, managers, and employees need to share a strategy to develop a digital entrepreneurial mindset at the business level (Pinchot & Soltanifar, 2021).
- On the other hand, in the sustainability dimension, understanding the manager and entrepreneur is a crucial reference for analyzing the company's decision-making and sustainable strategic initiatives. Companies pursuing digital entrepreneurship will only be able to exploit the digital capabilities and benefits of digital innovation that create value for society and the environment when managers and entrepreneurs have a positive attitude towards social and environmental issues (Xu et al., 2022). Therefore, managers and entrepreneurs need to adopt a mindset that fosters entrepreneurship.
- **Acquisition of digital knowledge and sustainability**: Most organizations face a lack of knowledge and capabilities needed to succeed in the digital aspects of business. Employees must acquire the knowledge required for digital transformation through a variety of methods (formal and informal). Leaders also need to stay informed about digital trends, understand their impact on their business, and learn how to leverage new technologies effectively. This does not mean that they know how IT works, but they need to have a good understanding of why it is important and how to use it (Whitehurst, 2015). Concerning the knowledge available in businesses in the field of digital technologies, it is necessary to (1) have a sufficient number of members of the organization who understand digital technology and can create and implement all the necessary innovations to drive the digital transformation required by the organization; (2) pay attention to the way the organization makes decisions and takes the necessary actions to demonstrate fluid knowledge of digital technology (Pinchot & Soltanifar, 2021). It is important to remember that building digital skills is a gradual process and does not happen overnight. Businesses should establish programs to enhance digital skills through hiring, participation, acquisition, and incubation (Schwab & Davis, 2018).
- On the other hand, gaining knowledge about environmental problems plays a crucial role in recognizing the opportunities that sustain the environment. For example, awareness of potential sources of air pollution in developing countries has led individuals to identify opportunities for technologies that maintain clean

air by significantly reducing particulate matter pollutants in households (Prahalad, 2010). The greater the knowledge of the natural and social environment, the more likely entrepreneurs are to recognize opportunities for sustainable development (Patzelt & Shepherd, 2011).

- **Having sponsors to support digital entrepreneurs**: Often, when business leaders call for more digital innovation, the results don't materialize because there are no sponsors for digital entrepreneurs within their organization or business. Successful digital and non-digital intrapreneurs often have several committed sponsors, who are positioned at different organizational levels. The facts indicate that every company requires numerous sponsors to support the large number of intrapreneurs in the era of rapid digital innovation. Sponsors spend time with digital innovators, coaching them on political and business issues and strategies, providing them with the necessary resources, and employing a coaching approach to help them develop their ideas (Pinchot & Soltanifar, 2021).
- **Quality management of digital entrepreneurs**: A business leader must consider the benefits of the innovations that people offer. Digital entrepreneurs within an organization need a greater degree of freedom to be effective. Therefore, managers should trust them based on competency criteria and provide more freedom of action to those who are qualified. A manager should also distinguish between real entrepreneurs who are pragmatic and those who talk a lot but lack the courage to pursue their ideas, and those who are merely seeking a position (Pinchot & Soltanifar, 2021).
- **Appropriate investment**: Economic factors play a crucial role in the formation of sustainable entrepreneurship. The economic life of sustainable businesses is related to the flow of money and investment. In other words, if there are no financial and economic assets, no activity for any type of entrepreneurship, including sustainable entrepreneurship, will take place (Aliabedi et al., 2022). On the other hand, with the link between digitalization and sustainability, the necessary investment in creating appropriate platforms for business digitalization and benefiting from the opportunities ahead in businesses with a sustainability focus, while taking into account social and environmental requirements, is crucial. A lack of financial resources prevents access to technology, particularly for production and manufacturing as part of the value creation process (Herman, 2022). As a result, businesses are forced to use more manual processes or low-level technologies (Fuerst et al., 2023). Of course, entrepreneurs also need to use entrepreneurial financing methods.
- **Strengthening social networks and engaging with stakeholders:** Sustainable entrepreneurs need more social networks and connections than typical entrepreneurs. These networks also include human, institutional, and tooling factors associated with creating business models. Creating sustainable business models, securing financing, identifying opportunities, and overcoming informal social barriers are shaped by the diversity of networks and social interactions between entrepreneurs and other actors (entrepreneurial ventures, government agencies, incubator/accelerator organizations, investors, higher education organizations). Successful sustainable businesses establish social connections that are tailored to the business type, age, model (organizational factors), and network actors, as well as demographic characteristics such as gender, race, and ethnicity (individual factors) (Neumeyer & Santos, 2018). These utilize networks that involve various actors to exchange knowledge, expand sustainable businesses, and promote sustainability (Chollet et al., 2014).
- **Adapting the digital business model to sustainable development goals**: It is evident that, with a focus on environmental and social issues, entrepreneurs and managers must modify their business model. Redefining business models in light of new digital technologies should be viewed as a continuous process that encompasses all aspects of core business operations, including environmental sustainability

(Feroz et al., 2021). The most significant changes are in the value proposition components, customer relationships, financial structure (including costs and revenues), infrastructure, and key partnerships. Understanding these four elements is crucial for effectively managing a business model and driving business model innovation with a focus on sustainability. By combining sustainability strategies, businesses can have different degrees of business model innovation (defensive, proactive, accommodative). Since digital businesses are byte-based and bit-end-based digital structures, they differ from physical businesses. They are editable, interactive, open, or reprogrammable and distributed (Kallinikos et al., 2013). Therefore, they can be easily changed and scaled, and their business model can be innovated.

NOTE: To prepare businesses for integrating digital entrepreneurship and sustainability, it is essential to recognize that the socio-cultural and economic contexts of countries differ, and sustainable entrepreneurship must be grounded in these contexts. On the other hand, 4e is not only businesses that are responsible for this preparation, but also governments and individuals.

7 Conclusion

The increasing problems of water scarcity, air pollution, low human development indicators, slow economic growth, and climate change have led to confusion and concern among policymakers, experts, and academics(Shakeel et al., 2020). Hence, the development of sustainable business models is currently the focus of attention in academic literature, among researchers, and policymakers. Support for sustainable business models is growing among proponents who aim to mitigate the negative environmental impacts of production and consumption systems, as well as address social challenges (Bergmann & Utikal, 2021). On the other hand, in the current era, the impacts of digital technologies are also widespread and go beyond consumer or organizational behavior (Zaheer et al., 2019). Some of these digital technologies include artificial intelligence, blockchain, Internet of Things, augmented reality, cloud computing, and big data. The use of digital technologies is one of the most promising developments in sustainability, and Sustainable entrepreneurship is considered key to providing solutions to significant social and environmental challenges (Feroz et al., 2021). In this regard, the concept of digital sustainable entrepreneurship has been considered. We introduce digital sustainable entrepreneurship as follows:

- ***Digital sustainable entrepreneurship is the process of creating and delivering economic, social, and environmental value through the creative combination of digital technologies to exploit sustainability-oriented opportunities.***

Considering the topics examined in the previous sections, the following conceptual framework (Fig. 3) is proposed for the concept of digital sustainable entrepreneurship. This framework shows the intersection of digital technologies, entrepreneurship, and sustainability. In this framework, we demonstrated that the digital entrepreneur explores and exploits opportunities based on digital technologies, thereby providing value. The product or service provided by a digital entrepreneur with a sustainability mindset creates value in all three dimensions: social, environmental, and economic. This value creation in the field of sustainability requires attention to effective policies in the dimension of digital entrepreneurship and sustainability by businesses, individuals, and the government. We suggest that other researchers and experts in this field examine the challenges of the digital sustainable entrepreneurship ecosystem, the paradox of digital technologies and sustainability, financing solutions for sustainable entrepreneurial businesses, and macro-policies towards digital sustainable entrepreneurship.

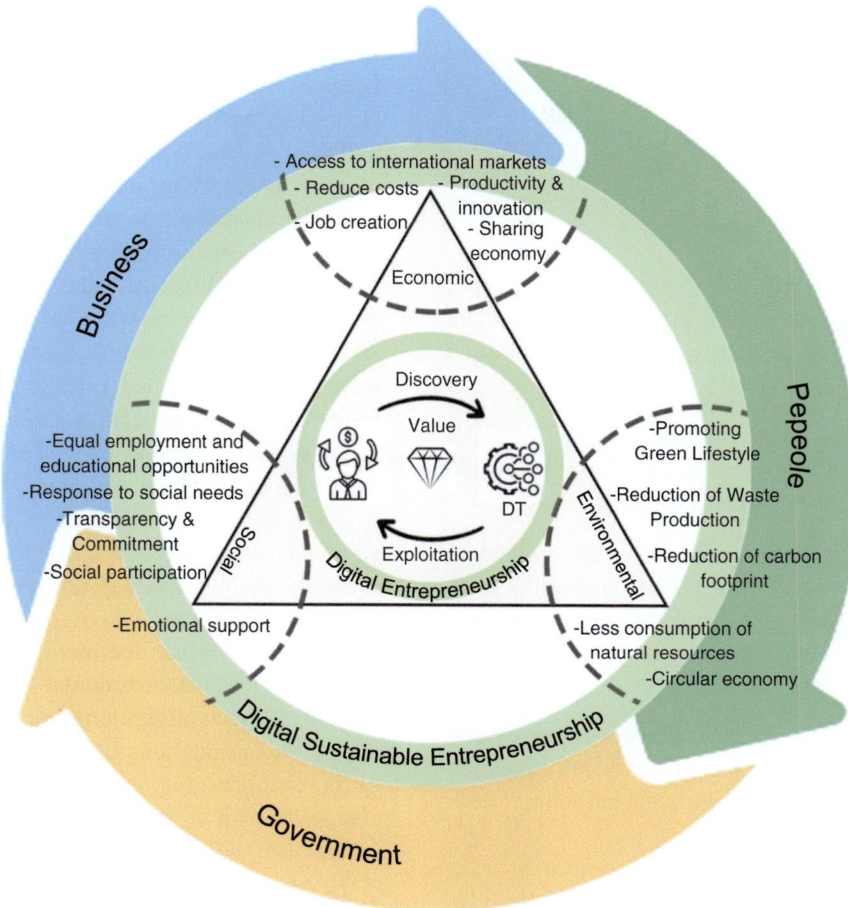

Fig. 3 Framework of digital sustainable entrepreneurship

8 Case Study

One of the most significant environmental issues of our day is waste recycling. In today's world, the excessive consumption of natural resources and the production of various waste products pose a serious threat to the environment.

ASPAK is a leading platform in the field of waste management and environmental protection. By focusing on promoting recycling and reducing environmental pollution, ASPAK enables citizens to manage their waste effectively and contribute to a healthier environment. ASPAK states on its platform that, instead of harming nature with our waste, we can turn it into valuable resources, thereby taking a positive step towards a sustainable future.

The ASPAK platform empowers individuals to make a meaningful contribution to environmental protection by participating in the collection and recycling of various waste materials. This platform is specifically designed for managing waste, including electronics, dry and wet recycling, household appliances, refrigerators and freezers, valves, and other dry waste. Each time the waste is delivered, its amount is recorded in the system. For every kilogram of waste collected, citizens will receive a coin as a reward that they can use to purchase services and special discounts on the platform.

How does Aspak contribute to sustainability?

Environmental protection: By collecting and recycling fruit mixture and peel, the release of organic and chemical substances into the environment is prevented, which helps maintain ecological balance.

Reducing air and soil pollution: Burning or burying organic waste leads to air and soil pollution. By recycling this waste, these problems are minimized.

Promoting a recycling culture: This project collects fruit mixtures and peels, turning them into recycled materials, and aims to promote a recycling culture in the Isfahan community.

Job creation: This collection and recycling process can lead to job opportunities in the region and contribute to the local economy.

Organic Compost Production: By collecting mixed waste and fruit peels and converting them into organic compost, inorganic compost sources are avoided, and farmers will have access to this natural resource.

Instead of harming nature with our waste, we can turn it into valuable resources, thereby taking a positive step towards a sustainable future.

Questions

- What is the concept of digital sustainable entrepreneurship?
- How can sustainability benefit from digital technologies?
- How do digital business models link to sustainability?
- How can you develop the concept of sustainability in your business using digital technologies?
- What challenges may businesses face in implementing sustainability dimensions with a focus on digital technologies?

9 Further Reading

Recommendations for additional resources for readers who wish to explore the topic in more depth:

- **Shaping the Future of the Fourth Industrial Revolution:** This book offers a clear explanation of the technologies driving the Fourth Industrial Revolution and presents a comprehensive vision for humanity's future. The connection between technology and sustainability is evident throughout the book, and it presents strategies for individuals, businesses, and governments to benefit more effectively.
- **Digital Sustainability: Navigating Entrepreneurship in the Information Age:** This book offers innovative strategies and solutions for building a more environmentally conscious and socially responsible future. Featuring a collection of case studies, it articulates areas such as blockchain, artificial intelligence, machine learning, cybersecurity, and the Internet of Things, focusing on sustainable resource use and entrepreneurial decision-making processes, and helps redefine innovative and sustainable business models.
- **Sustainable Digital Technologies: Trends, Impacts, and Assessments:** This book explains the role of digital technology and its impact on sustainability. It highlights the key factors that drive the development of sustainable goals, with a focus on digital technologies and Industry 4.0. The concepts of the Internet of Things, circular economy, and green cloud computing, along with their relationship to sustainability, are considered.
- **Digital Technologies for Sustainable Futures: Promises and Pitfalls (Routledge Studies in Sustainability):** This work examines the interaction between digitalization and sustainability, interrogating the hypothesis that increased digitalization automatically contributes to greater sustainability. It considers the dual role of digital technology: its potential to support socio-economic and environmental sustainability, while also creating contradictions and deadlocks that undermine these goals.

References

Adam, I. O., Alhassan, M. D., & Afriyie, Y. (2020). What drives global B2C e-commerce? An analysis of the effect of ICT access, human resource development and regulatory environment. *Technology Analysis & Strategic Management, 32*(7), 835–850. https://doi.org/10.1080/09537325.2020.1714579

Aliabedi, V., Ataei, P., & Gholamrezai, S. (2022). Identification of the relationships among the indicators of sustainable entrepreneurial ecosystems in agricultural startups. *Journal of Innovation & Knowledge, 7*(4), Article 100245. https://doi.org/10.1016/j.jik.2022.100245

Ancillai, C., Sabatini, A., Gatti, M., & Perna, A. (2023). Digital technology and business model innovation: A systematic literature review and future research agenda. *Technological Forecasting and Social Change, 188*, Article 122307. https://doi.org/10.1016/j.techfore.2022.122307

Ardolino, M., Rapaccini, M., Saccani, N., Gaiardelli, P., Crespi, G., & Ruggeri, C. (2018). The role of digital technologies for the service transformation of industrial companies. *International Journal of Production Research, 56*(5), 2116–2132. https://doi.org/10.1080/00207543.2017.1324224

Bapna, R., Gangadharan, G. R., & Joshi, Y. V. (2021). AI for social good: Unlocking new business opportunities. *MIS Quarterly, 45*(2), 533–541. https://doi.org/10.25300/MISQ/2021/16037

Bergmann, T., & Utikal, H. (2021). How to support startups in developing a sustainable business model: The case of a European social impact accelerator. *Sustainability, 13*(6), 3337. https://doi.org/10.3390/su13063337

Bloom, N., Bunn, P., Chen, S., Mizen, P., & Smietanka, P. (2020). The economic impact of coronavirus on UK businesses: Early evidence from the decision maker panel. *VOX CEPR Policy Portal*. https://voxeu.org/article/economic-impact-coronavirus-uk-businesses

Brenner, B. (2018). Transformative sustainable business models in the light of the digital imperative—A global business economics perspective. *Sustainability, 10*(12), 4428. https://doi.org/10.3390/su10124428

Bruno, C., & Canina, M. (2019, April). Creativity 4.0. Empowering creative process for digitally enhanced people. *The Design Journal, 22*(Suppl. 1), 2119–2131. https://doi.org/10.1080/14606925.2019.1594935

Cai, H., Da Xu, L., Xu, B., Xie, C., Qin, S., & Jiang, L. (2014). IoT-based configurable information service platform for product lifecycle management. *IEEE Transactions on Industrial Informatics, 10*(2), 1558–1567. https://doi.org/10.1109/TII.2014.2306391

Chollet, J. F., Couderchet, M., & Bonnemain, J. L. (2014). Crop protection: New strategies for sustainable development. *Environmental Science and Pollution Research, 21*, 4793–4796. https://doi.org/10.1007/s11356-014-2521-8

Cohen, B., & Winn, M. I. (2007). Market imperfections, opportunity and sustainable entrepreneurship. *Journal of Business Venturing, 22*(1), 29–49. https://doi.org/10.1016/j.jbusvent.2006.05.002

Copeland, B. J. (2020). Artificial intelligence. https://www.britannica.com/technology/artificial-intelligence/Reasoning

Crecente, F., Sarabia, M., & del Val, M. T. (2021). Sustainable entrepreneurship in the 2030 horizon. *Sustainability, 13*(2), 909. https://doi.org/10.3390/su13020909

Darling-Hammond, L., Flook, L., Cook-Harvey, C., Barron, B., & Osher, D. (2020). Implications for educational practice of the science of learning and development. *Applied Developmental Science, 24*(2), 97–140. https://doi.org/10.1080/10888691.2018.1537791

Davidson, E., & Vaast, E. (2010). Digital entrepreneurship and its sociomaterial enactment. In *Proceedings of the 43rd Hawaii International Conference on System Sciences*, Honolulu (pp. 1–10). IEEE Xplore. https://doi.org/10.1109/HICSS.2010.150

De Mauro, A., Greco, M., & Grimaldi, M. (2016). A formal definition of Big Data based on its essential features. *Library Review, 65*(3), 122–135. https://doi.org/10.1108/LR-06-2015-0061

Dropbox. (2021). *Reaching our sustainability goals through our data centers*. https://blog.dropbox.com/topics/technology/reaching-our-sustainability-goals-through-our-data-centers

Commission, E. (2014). *Digital entrepreneurship: The EU vision, strategy and actions*. European Commission.

Farshid, M., Paschen, J., Eriksson, T., & Kietzmann, J. (2018). Go boldly!: Explore augmented reality (AR), virtual reality (VR), and mixed reality (MR) for business. *Business Horizons, 61*(5), 657–663. https://doi.org/10.1016/j.bushor.2018.05.009

Feng, T., Li, Q., Zhou, H., & Shen, Y. (2020). Application of artificial intelligence in green supply chain management: A literature review. *Journal of Cleaner Production, 258*, Article 120687. https://doi.org/10.1016/j.jclepro.2020.120687

Feroz, A. K., Zo, H., & Chiravuri, A. (2021). Digital transformation and environmental sustainability: A review and research agenda. *Sustainability, 13*, 1530. https://doi.org/10.3390/su13031530

Ferrari, F., Striani, R., Minosi, S., De Fazio, R., Visconti, P., Patrono, L., Catarinucci, L., Corcione, C. E., & Greco, A. (2020). An innovative IoT-oriented prototype platform for the management and valorisation of the organic fraction of municipal solid waste. *Journal of Cleaner Production, 247*, Article 119618. https://doi.org/10.1016/j.jclepro.2019.119618

Fooladikia, E., Taghipour, F., & Naghsh, A. R. (2021). Importance and role of social networks in trusting e-commerce (B2C). *Karafan Quarterly Scientific Journal, 18*(2), 91–109. https://doi.org/10.48301/kssa.2021.130158

French, R. P. (2016). The fuzziness of mindsets: Divergent conceptualizations and characterizations of mindset theory and praxis. *International Journal of Organizational Analysis, 24*(4), 673–691. https://doi.org/10.1108/IJOA-09-2014-0797

Frenken, K., & Schor, J. (2019). Putting the sharing economy into perspective. In O. Mont (Ed.), *A research agenda for sustainable consumption governance* (pp. 121–135). Edward Elgar Publishing. https://doi.org/10.4337/9781788117814.00016

Fuerst, S., Sanchez-Dominguez, O., & Rodriguez-Montes, M. A. (2023). The role of digital technology within the business model of sustainable entrepreneurship. *Sustainability, 15*(14), 10923. https://doi.org/10.3390/su151410923

Geissdoerfer, M., Vladimirova, D., & Evans, S. (2018). Sustainable business model innovation: A review. *Journal of Cleaner Production, 198*, 401–416. https://doi.org/10.1016/j.jclepro.2018.06.240

George, G., Merrill, R. K., & Schillebeeckx, S. J. (2021). Digital sustainability and entrepreneurship: How digital innovations are helping tackle climate change and sustainable development. *Entrepreneurship Theory and Practice, 45*(5), 999–1027. https://doi.org/10.1177/1042258719899425

Giones, F., & Brem, A. (2017). Digital technology entrepreneurship: A definition and research agenda. *Technology Innovation Management Review, 7*(5), 44–51. https://timreview.ca/sites/default/files/article_PDF/GionesBrem_TIMReview_May2017.pdf

Gleeson, B. (2019, 27 December). Five key ingredients for successful organisational change. *Forbes*. https://www.forbes.com/sites/brentgleeson/2018/12/27/5-key-ingredients-for-successful-organisationalchange/#4b4225f476dd

Gupta, R., & George, S. (2020). *Sustainable development through artificial intelligence: Emerging research and opportunities*. IGI Global.

Guthrie, C. (2014). The digital factory: A hands-on learning project in digital entrepreneurship. *Journal of Entrepreneurship Education, 17*(1), 115–133. https://www.researchgate.net/publication/288234328_The_digital_factory_A_hands-on_learning_project_in_digital_entrepreneurship

Hair, N., Wetsch, L. R., Hull, C. E., Perotti, V., & Hung, Y. T. C. (2012). Market orientation in digital entrepreneurship: Advantages and challenges in a Web 2.0 networked world. *International Journal of Innovation and Technology Management, 9*(6), 1250045. https://doi.org/10.1142/S0219877012500459

Heidari, P., Zare, S., Sharifi, M. E., & Shaban Boloukat, H. (2023). A review of educational justice. *Rooyesh in Medical Sciences Education, 15*(25), 13–21. http://roy esh.tums.ac.ir/article-1-76-fa.html

Herman, E. (2022). The interplay between digital entrepreneurship and sustainable development in the context of the EU digital economy: A multivariate analysis. *Mathematics, 10*(10), 1682. https://doi.org/10.3390/math10101682

Hull, C. E. K., Hung, Y. T. C., Hair, N., Victor, P., & DeMartino, R. (2007). Taking advantage of digital opportunities: A typology of digital entrepreneurship. *International Journal of Networking and Virtual Organizations, 4*, 290–303. https://doi.org/10.1504/IJNVO.2007.015166

Islam, R. (2024). *Sustainable entrepreneurship in the digital age: The role of AI in green business practices*. Routledge. https://urn.fi/URN:NBN:fi:amk-2024060521089

Kallinikos, J., Aaltonen, A., & Marton, A. (2013). The ambivalent ontology of digital artifacts. *MIS Quarterly, 37*(2), 357–370. https://doi.org/10.25300/MISQ/2013/37.2.05

Kollmann, T. (2006). What is e-entrepreneurship? Fundamentals of company founding in the net economy. *International Journal of Technology Management, 33*, 322–340. https://doi.org/10.1504/IJTM.2006.009247

König, M., Ungerer, C., Baltes, G., & Terzidis, O. (2019). Different patterns in the evolution of digital and non-digital ventures' business models. *Technological Forecasting and Social Change, 146*, 844–852. https://doi.org/10.1016/j.techfore.2019.05.006

Le Dinh, T., Vu, M. C., & Ayayi, A. (2018). Towards a living lab for promoting the digital entrepreneurship process. *International Journal of Entrepreneurship, 22*(1), 1–17. https://www.researchgate.net/publication/326990005_Towards_a_living_lab_for_promoting_the_digital_entrepreneurship_process

Leng, J., Ruan, G., Jiang, P., Xu, K., Liu, Q., Zhou, X., & Liu, C. (2020). Blockchain-empowered sustainable manufacturing and product lifecycle management in industry 4.0: A survey. *Renewable and Sustainable Energy Reviews, 132*, 110112. https://doi.org/10.1016/j.rser.2020.110112

Li, F., & Long, J. (2024). Exploration and exploitation of multiple values: The dynamic evolution process of sustainable entrepreneurship in Chinese digital platform corporates. *Sustainable Development, 32*(3), 2342–2357. https://doi.org/10.3390/su16188069

Li, J., Chiu, D. K., Ho, K. K., & So, S. (2024). The use of social media in sustainable green lifestyle adoption: Social media influencers and value co-creation. *Sustainability, 16*(3), 1133. https://doi.org/10.3390/su16031133

Li, W., Du, W., & Yin, J. (2017). Digital entrepreneurship ecosystem as a new form of organizing: The case of Zhongguancun. *Frontiers of Business Research in China, 11*(1), 5. https://doi.org/10.1186/s11782-017-0004-8

Liu, D., Shi, Y., Li, J., & Jin, H. (2019). The role of artificial intelligence in achieving sustainable development goals. *Nature Communications, 10*(1), 1–3. https://doi.org/10.1038/s41467-019-14108-y

Logan, M., Safi, M., Lens, P., & Visvanathan, C. (2019). Investigating the performance of internet of things based anaerobic digestion of food waste. *Process Safety and Environmental Protection, 127*, 277–287. https://doi.org/10.1016/j.psep.2019.05.027

Madhavaiah, C., & Bashir, I. (2012). Defining cloud computing in business perspective: A review of research. *Metamorphosis, 11*(2), 50–65. https://doi.org/10.1177/0972622520120205

Mahmood, A., & Yu, C. M. (2005). E-entrepreneurship in knowledge economy: Implications for the Asia-Pacific economies. *The Business Review, Cambridge, 4*(1), 153–160.

Matt, C., Hess, T., & Benlian, A. (2015). Digital transformation strategies. *Business & Information Systems*

Engineering, 57(5), 339–343. https://doi.org/10.1007/s12599-015-0401-5

McMullen, J. S., & Dimov, D. (2013). Time and the entrepreneurial journey: The problems and promise of studying entrepreneurship as a process. *Journal of Management Studies, 50*(8), 1481–1512. https://doi.org/10.1111/joms.12049

Mehr, M., Khosravi, S. M., & Sheikhi (2020). Educational equality: The incomplete puzzle of education in the 21st century. *Nama Educational Studies Journal, 50*, 9–15. https://nama.ajaums.ac.ir/article-1-339-en.pdf

Meijer, D. B. (2017). *Consequences of the implementation of blockchain technology*. MSc thesis. Delft University of Technology. https://repository.tudelft.nl/record/uuid:da0b8d80-d19e-4149-bfbd-64b0ca79042a

Mishra, P., & Sinha, N. (2021). Sustainable entrepreneurship: A bibliometric review and research agenda. *International Entrepreneurship and Management Journal* [online]. https://doi.org/10.1007/s11365-021-00758-2

Mont, O., Palgan, Y. V., Bradley, K., & Zvolska, L. (2020). A decade of the sharing economy: Concepts, users, business and governance perspectives. *Journal of Cleaner Production, 269*, Article 122215. https://doi.org/10.1016/j.jclepro.2020.122215

Murugesan, S. (2008). Harnessing green IT: Principles and practices. *IT Professional, 10*(1), 24–33. https://doi.org/10.1109/MITP.2008.12

Nambisan, S. (2017). Digital entrepreneurship: Toward a digital technology perspective of entrepreneurship. *Entrepreneurship Theory and Practice, 41*(6), 1029–1055. https://doi.org/10.1111/etap.12254

Nambisan, S., Lyytinen, K., Majchrzak, A., & Song, M. (2017). Digital innovation management: Reinventing innovation management research in a digital world. *MIS Quarterly, 41*(1), 223–238. https://doi.org/10.25300/MISQ/2017/41:1.03

Neumeyer, X., & Santos, S. C. (2018). Sustainable business models, venture typologies, and entrepreneurial ecosystems: A social network perspective. *Journal of Cleaner Production, 172*, 4565–4579. https://doi.org/10.1016/j.jclepro.2017.10.167

Outland, E. T., King, D., Lee, J. K., Liang, T. P., & Turban, D. C. (2018). *Electronic commerce 2018: A managerial and social networks perspective* (5th ed.). Springer. https://doi.org/10.1007/978-3-319-58715-8

Palmié, M., Miehé, L., Oghazi, P., Parida, V., & Wincent, J. (2022). The evolution of the digital service ecosystem and digital business model innovation in retail: The emergence of meta-ecosystems and the value of physical interactions. *Technological Forecasting and Social Change, 177*, Article 121496. https://doi.org/10.1016/j.techfore.2022.121496

Pan, X., Pan, X., Song, M., Ai, B., & Ming, Y. (2020). Blockchain technology and enterprise operational capabilities: An empirical test. *International Journal of Information Management, 52*, Article 101946. https://doi.org/10.1016/j.ijinfomgt.2019.07.002

Parviainen, P., Kääriäinen, J., Tihinen, M., & Teppola, S. (2017). Tackling the digitalization challenge: How to benefit from digitalization in practice. *International Journal of Information Systems and Project Management, 5*(1), 63–77. https://doi.org/10.12821/ijispm050104

Patzelt, H., & Shepherd, D. A. (2011). Recognizing opportunities for sustainable development. *Entrepreneurship Theory and Practice, 35*(4), 631–652. https://doi.org/10.1111/j.1540-6520.2011.00457.x

Pergelova, A., Manolova, T., Simeonova-Ganeva, R., & Yordanova, D. (2019). Democratizing entrepreneurship? Digital technologies and the internationalization of female-led SMEs. *Journal of Small Business Management, 57*(1), 14–39. https://doi.org/10.1111/jsbm.12494

Pinchot, G., & Soltanifar, M. (2021). Digital intrapreneurship: The corporate solution to a rapid digitalisation. In Soltanifar, M., Hughes, M., & Göcke, L. (Eds.), *Digital entrepreneurship*. Future of business and finance. Springer. https://doi.org/10.1007/978-3-030-53914-6_12

Prahalad, C. K. (2010). *The fortune at the bottom of the Pyramid*. Wharton School Publishing. https://doi.org/10.19177/reen.v1e220081-23

Purnomo, A., Susanti, T., Sari, A. K., Firdaus, M., & Dewi, R. (2020). A study of digital entrepreneurship through bibliometric visualizing from 1993 to 2019. In *2020 International Conference on Information Management and Technology (ICIMTech)* (pp. 911–915). IEEE. https://doi.org/10.1109/ICIMTech50083.2020.9211237

Recker, J., & von Briel, F. (2019). The future of digital entrepreneurship research: Existing and emerging opportunities. In *Proceedings of the International Conference on Information Systems (ICIS 2019)*, Munich, Germany, 15–18. https://www.researchgate.net/publication/336085226_The_Future_of_Digital_Entrepreneurship_Research_Existing_and_Emerging_Opportunities

Richter, C., Kraus, S., & Syrjä, P. (2015). The shareconomy as a precursor for digital entrepreneurship business models. *International Journal of Entrepreneurship and Small Business, 25*(1), 18–35. https://doi.org/10.1504/IJESB.2015.068773

Saberi, S., Kouhizadeh, M., & Sarkis, J. (2018). Blockchain technology: A panacea or pariah for resources conservation and recycling? *Resources, Conservation and Recycling, 130*, 80–81. https://doi.org/10.1016/j.resconrec.2017.11.020

Satalkina, L., & Steiner, G. (2020). Digital entrepreneurship and its role in innovation systems: A systematic literature review as a basis for future research avenues for sustainable transitions. *Sustainability, 12*(7), 2764. https://doi.org/10.3390/su12072764

Schwab, K., & Davis, N. (2018). *Shaping the future of the fourth industrial revolution*. Crown Currency. https://rudyct.com/sm.vuca/Shaping%20the%20Fourth%20Industrial%20Revolution%20by%20Klaus%20Schwab%20et-al-2018.pdf

Sebastian, I. M., Ross, J. W., Beath, C. M., Mocker, M., Moloney, K. G., & Fonstad, N. O. (2017). How big old companies navigate digital transformation. *MIS Quarterly Executive, 16*(3), 197–213. https://aisel.aisnet.org/misqe/vol16/iss3/1/

Shakeel, J., Mardani, A., Chofreh, A. G., Goni, F. A., & Klemeš, J. J. (2020). Anatomy of sustainable business model innovation. *Journal of Cleaner Production, 261*, Article 121201. https://doi.org/10.1016/j.jclepro.2020.121201

Smoląg, K., & Chomiak-Orsa, I. (2025). Future prospects and trends in digital entrepreneurship based on sustainable development. In *Digital sustainability* [Online] (pp. 93–110). https://doi.org/10.1201/9781003484226-7

Soluk, J., Miroshnychenko, I., Kammerlander, N., & De Massis, A. (2021). Family influence and digital business model innovation: The enabling role of dynamic capabilities. *Entrepreneurship Theory and Practice*. https://doi.org/10.1177/1042258721998946

Stawicka, E. (2021). Sustainable development in the digital age of entrepreneurship. *Sustainability, 13*(8), 4429. https://doi.org/10.3390/su13084429

Sujata, M., Khor, K. S., Ramayah, T., & Teoh, A. P. (2019). The role of social media on recycling behaviour. *Sustainable Production and Consumption, 20*, 365–374. https://doi.org/10.1016/j.spc.2019.08.005

Sussan, F., & Acs, Z. J. (2017). The digital entrepreneurial ecosystem. *Small Business Economics, 49*(1), 55–73. https://doi.org/10.1007/s11187-017-9867-5

Szromek, A. R. (2021). The sustainable business model of spa tourism enterprise—Results of research carried out in Poland. *Journal of Open Innovation: Technology, Market, and Complexity, 7*(1), 73. https://doi.org/10.3390/joitmc7010073

Thompson, N. A. (2020). Book review: Practice theory in action: Empirical studies of interaction in innovation and entrepreneurship by Betsy Campbell. *International Small Business Journal, 38*(6), 512–514.

Ward, T. J. A. D. (2017). Blockchain could help us save the environment: Here's how. In *World Economic Forum Annual Meeting* [online]. https://www.weforum.org/agenda/2017/01/blockchain-could-help-us-save-the-environment-here-s-how/

Whitehurst, J. (2015). *Driving digital transformation: New skills for leaders, new role for the CIO* (pp. 1–16). Harvard Business Review Analytical Report. https://hbr.org/sponsored/2015/03/driving-digital-transformation-new-skills-for-leaders-new-role-for-the-cio

Wirtz, B. W. (2019). Digital business models. In *Digital business models: Concepts, models, and the alphabet case study* (pp. 137–152). Springer. https://doi.org/10.1007/978-3-030-13005-3

Wu, W., Huang, V., Chen, X., Davison, R. M., & Hua, Z. (2018). Social value and online social shopping intention: The moderating role of experience. *Information Technology & People, 31*(3), 688–711. https://doi.org/10.1108/ITP-10-2016-0236

Xu, G., Hou, G., & Zhang, J. (2022). Digital sustainable entrepreneurship: A digital capability perspective through digital innovation orientation for social and environmental value creation. *Sustainability, 14*(18), 11222. https://doi.org/10.3390/su141811222

Zaheer, H., Breyer, Y., & Dumay, J. (2019). Digital entrepreneurship: An interdisciplinary structured literature review and research agenda. *Technological Forecasting and Social Change, 148*, Article 119735. https://doi.org/10.1016/j.techfore.2019.119735

Policy, Marketing, and Human Resources

Entrepreneurship Policies and Strategies for Sustainable Development

Bahman Khodapanah, Hamid Padash, and Fariba Mesri Eisalou

Abstract

This chapter examines the critical role of entrepreneurship in promoting sustainable development, with a focus on policies and strategies that align business practices with environmental, social, and economic objectives. In this context, sustainable entrepreneurship is defined in terms of the United Nations' Sustainable Development Goals (SDGs), emphasizing the need to integrate sustainability into business models. Key themes covered in this chapter include green entrepreneurship, innovation, and social enterprises, all of which contribute to systemic change and the creation of continuous economic benefits. Additionally, digital infrastructure and regulatory frameworks are examined as fundamental factors in fostering an enabling environment for sustainable business practices. By analyzing policy measures that promote sustainable entrepreneurship, this chapter provides insights into how businesses can address global challenges while ensuring economic viability. Through case studies and practical applications, readers will gain a deeper understanding of the entrepreneurial role in developing sustainable solutions. This chapter will be particularly valuable for policymakers, entrepreneurs, and researchers seeking practical strategies to institutionalize sustainability within business frameworks.

Keywords

Sustainable entrepreneurship · Sustainable development · Entrepreneurship policies · Entrepreneurship strategies · Policy frameworks

Learning Objectives

By the end of this chapter, readers will be able to

- Sustainable Development Goals (SDGs).
- Identify key policies and strategies that promote sustainable business practices.
- Analyze the role of digital infrastructure in enhancing sustainable entrepreneurship.
- Evaluate real-world case studies to understand the practical application of sustainability principles.
- Develop practical strategies for integrating sustainability into entrepreneurial activities.

B. Khodapanah · F. Mesri Eisalou
Department of Business Management, Faculty of Social Sciences, University of Mohaghegh Ardabili, Ardabil, Iran
e-mail: khodapanah@uma.ac.ir

F. Mesri Eisalou
e-mail: mesri.fariba1997@gmail.com

H. Padash (✉)
Department of Corporate Entrepreneurship, Faculty of Entrepreneurship, University of Tehran, Tehran, Iran
e-mail: padash@ut.ac.ir

M. Akbari et al. (eds.), *Sustainable Entrepreneurship*, University of Tehran Science and Humanities Series, https://doi.org/10.1007/978-3-032-04909-4_9

1 Introduction

In the twenty-first century, economic development, social factors, and environmental concerns have become increasingly interdependent to the extent that economic growth is no longer feasible without accounting for its social and environmental ramifications (Apostu & Gigauri, 2023; Biehl et al., 2012). These factors are fundamental to sustainability, and neglecting them can have profound and lasting consequences. Therefore, all nations, regardless of their developmental status or economic system—whether market-oriented or planned—must formulate their economic and social development objectives in alignment with the principles of sustainability (Goel & Joshi, 2017). Entrepreneurship, as a driving force of economic development, plays a pivotal role in fostering social transformation and ensuring the sustainable production of goods and services (Apostu & Gigauri, 2023; Biehl et al., 2012). Sustainable entrepreneurship prioritizes voluntary and proactive initiatives that address critical challenges, including poverty, public health, recycling, global peace, justice, and human security, through innovative solutions. This approach not only contributes to achieving the Sustainable Development Goals (SDGs) but also helps mitigate environmental degradation. Conversely, sustainable development can also influence entrepreneurship, transforming it into an effective mechanism for advancing sustainability (Apostu & Gigauri, 2023). Recognizing its significance, the United Nations General Assembly has repeatedly highlighted entrepreneurship as a potential solution to sustainability challenges. In December 2012, during its sixty-seventh session, the General Assembly adopted its first resolution on Entrepreneurship for Development, recognizing that entrepreneurship is not only a key driver of economic growth but also a vital instrument for addressing global challenges, such as unemployment and poverty. The resolution further emphasizes the importance of expanding equal opportunities for all, with a particular focus on disadvantaged groups, women, and young people (Mehra, 2022).

Given this significance, entrepreneurship necessitates a comprehensive approach and well-structured policymaking to advance sustainability goals. Governments should adopt an inclusive strategy to foster entrepreneurship, emphasizing equal economic participation and empowering diverse societal groups (Mehra, 2022). At the same time, sustainability acknowledges that organizations operate within a complex and interdependent environment, where persistent unsustainable development can lead to irreversible consequences. Thus, urgent and coordinated action is essential to achieving a Systemic Equilibrium between economic productivity enhancement, Social Equity, and environmental sustainability (Goel & Joshi, 2017). Accordingly, the formulation and implementation of effective policies not only facilitate entrepreneurship in alignment with sustainability principles but also create opportunities for governments, businesses, and entrepreneurs to develop innovative strategies that promote sustainability. These strategies can help establish a sustainable equilibrium between economic growth, social justice, and environmental preservation. Committed to sustainable resource management and intergenerational environmental preservation.

2 Literature Review

2.1 Entrepreneurship

In recent decades, entrepreneurship—the process of creating and pursuing innovative opportunities to generate value in society—has been considered by writers and policymakers as a key tool and mechanism for economic and social development (Roundy & Dutch, 2019). Interest in the role of entrepreneurship in economic growth and development began in the late 1990s, driven by international organizations such as the Organization for Economic Co-operation and Development (OECD) and research organizations like the Global Entrepreneurship Monitor. However, until now, there has been limited knowledge about entrepreneurship as a policy area and how to design and implement policies. In other words, entrepreneurship development policymaking, especially at the regional level, is an emerging field that is currently in the early stages of its life cycle and is not yet well developed (González-López et al., 2019). There appears to be considerable ambiguity regarding the type of policies used to encourage and stimulate entrepreneurship development versus traditional business development policies. A review of the economies of different countries reveals that some regions perform better economically than others, with more people engaged in entrepreneurial activities and higher economic growth compared to the national average. This increased growth is attributed to the appropriate economic structure, the existence of comparative advantages in various activities, and effective regional policy and planning. The institutional context, including the cultural structure, influences the nature of entrepreneurial opportunities and the individuals who identify and exploit those opportunities (Alia et al., 2019).

Entrepreneurship is a multifaceted and ambiguous concept, with varying perspectives on its definition. While some scholars emphasize business development, others focus on entrepreneurial behavior, which extends beyond the business sector to include activities in non-business domains as well (Ahmed & McQuaid, 2005). Entrepreneurship is broadly understood as the process of establishing and sustaining a business. Though often associated with "self-employment," the two concepts are not necessarily synonymous (Agwunobi, 2020). The earliest formal definition of entrepreneurship was provided by Cantillon in 1734, who described it as "self-initiative with uncertain returns," highlighting its inherent risk dimension (Delacroix et al., 2019). While Smith (1950) metaphorically referred to the entrepreneur as an "invisible hand," his role was not explicitly elaborated. The significance of entrepreneurship was more clearly articulated when Schumpeter (1934) identified innovation as its defining characteristic. In contrast, Kirzner (1973) emphasized the entrepreneur's ability to recognize and capitalize on opportunities by reorganizing and recombining resources to meet market demands. Further refining the concept, Stevenson and Jarillo (1990) defined entrepreneurship as a dynamic process in which entrepreneurs generate novel ideas that create value and actively pursue emerging opportunities (Bhattarai, 2020).

Audretsch (1995) and Cunningham and Lischeron (1991) highlight the challenges of reaching a universally and comprehensively defined definition of entrepreneurship. They emphasize three distinct stages: at the first stage, the term refers to small businesses or enterprises; at the second stage, it involves the establishment of new firms; and at the third and highest level, it pertains to innovation and the coordination of complex production systems (Ahmed & McQuaid, 2005). In contrast, Sarasvathy (2000) and Timmons and Spinelli (2004) emphasize the outcome-oriented dimension of entrepreneurship, which includes the creation of new products, businesses, innovations, societal value, and organizations (Bhattarai, 2020). Consequently, many definitions of entrepreneurship converge on the notion of value creation that yields positive societal impacts (Bhattarai, 2020; Jones et al., 2020), ultimately contributing to wealth generation and enhancing social welfare (Mehmood et al., 2019). In recent years, this perspective has evolved, with growing recognition of the potential positive relationship between entrepreneurship and sustainability. This emerging paradigm views entrepreneurship as a catalyst for driving sustainable change and development (Baiocco et al., 2023). This approach, known as sustainable entrepreneurship, is explored in the following sections.

Despite some exceptions, most classical growth theories fail to explicitly acknowledge the role of entrepreneurship in economic development. These traditional theories instead concentrate on internal production factors—including land, infrastructure, fixed capital, and other static production elements—as well as external factors such as labor and circulating capital. Classical theories remain valid and reliable for explaining the "old economy," which was characterized by the dominance of large corporations, national markets, and mass production systems

(Fordism). In this economic paradigm, the manufacturing sector served as the cornerstone of both national and regional economies, while innovation was conceptualized as a linear, sequential process. Schumpeter (1942) notably developed an influential framework explaining the significance of entrepreneurship in driving innovation and economic growth. However, his analysis focused primarily on the national level and large-scale enterprises. He firmly believed that, particularly regarding technological change and innovation, "large-scale business is the most important engine of progress" (Schumpeter, 1942, p. 106). His empirical observations of major economic development waves, inspired by Kondratiev's pioneering work in primary sector analysis, were subsequently adapted to regional contexts by other scholars (Marshall and others).

Nevertheless, these researchers similarly overlooked the potential contributions of startups and small firms. Consequently, regional development was predominantly viewed as the outcome of large business activities. In essence, startups and small firms occupy a peripheral position in traditional theories of regional growth and development. While companies—rather than policies, natural resources, or infrastructure—are recognized as the primary drivers of regional development, these companies are typically large, well-established enterprises with significant market presence.

2.2 Sustainable Development and Its Dimensions

In 1984, the United Nations established an independent commission comprising 22 member states representing both developed and developing nations and tasked it with formulating long-term environmental strategies for the international community. This initiative led to the publication of the World Commission on Environment and Development report, *Our Common Future*, which is widely regarded as a pivotal milestone in integrating the concept of sustainable development into international development policy and discourse. The report has since been translated into more than 24 languages (Elliott, 2013). Sustainable development is inherently multidisciplinary, encompassing a wide range of issues, including science, innovation, technology, research and development, information technology and e-commerce, economic growth, public health, foreign direct investment, multinational corporations, international debt and aid, trade, politics, conflict, natural disasters, population dynamics, and global security concerns such as terrorism (Ahmed & McQuaid, 2005). While nearly all academic disciplines engage with sustainable development in some capacity, early contributions primarily emerged from environmental fields (Elliott, 2013). The vast scope of sustainable development underscores both its significance and the complexity of its implementation at global, national, and local levels.

Linguistically, sustainable development refers to the maintenance and continuity of development over time, representing a fusion of two fundamental concepts: *sustainability*, which denotes durability and persistence, and *development*, which signifies capacity expansion, gradual progress, and the enhancement of conditions. This concept encompasses the positive transformations societies have undergone and continue to experience throughout history (Elliott, 2013; Klarin, 2018). There are numerous definitions of this term in circulation, each with varying interpretations and countless practical applications. The significance of definitions lies in their role as the foundation for formulating strategies, policies, and mechanisms that aim to achieve sustainable development. Moreover, how we perceive, analyze, and comprehend human and environmental conditions serves as the cornerstone for subsequent interventions (Elliott, 2013).

Jacobs (1991) describes sustainable development as a "contested concept," similar to notions like "democracy" or "equality," which possess broadly accepted meanings but are subject to significant debate regarding their implementation. In this context, the World Commission on Environment and Development defines sustainable development as the ability of nations and economic enterprises to manage resources in a manner that fulfills present needs without compromising the capacity of future generations to meet

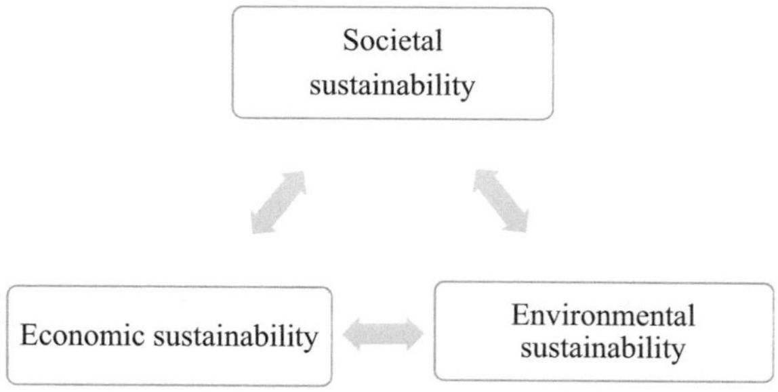

Fig. 1 Dimensions of sustainability

their own (Elliott, 2013; Hall et al., 2010). Sustainable development has emerged as both an influential and controversial concept in business and politics, advocating for a balanced integration of social, environmental, and economic objectives. It asserts that human well-being should not come at the cost of environmental degradation—a principle known as the "sustainability triad" (Apostu & Gigauri, 2023; Hall et al., 2010). Sustainability acknowledges that human activities exert negative impacts on environmental resources, potentially threatening human survival. Achieving sustainability necessitates a comprehensive rethinking of all aspects of human activity and should not be confined solely to the scientific and technological domains. Within this framework, the primary focus remains on humanity, aiming to foster economic development within the constraints of finite resources (Ahmed & McQuaid, 2005; Moggi et al., 2022).

The dimensions of sustainability encompass three fundamental areas: economic, environmental, and social sustainability. Economic sustainability serves as the foundation for a company's long-term viability, referring to its capacity to manage capital, reserves, and assets—both tangible and intangible—effectively. Environmental sustainability ensures that businesses operate without causing ecological harm or generating excessive pollutant emissions. Social sustainability requires companies to conduct their operations in alignment with stakeholder needs while adhering to ethical and value-based principles (İyigün, 2015). Achieving comprehensive sustainable development necessitates maintaining a balance among these three pillars. However, attaining this equilibrium is challenging, as each dimension must uphold its objectives without compromising the stability of the others (Klarin, 2018). Figure 1 illustrates the interconnected nature of sustainability dimensions as an integrated framework.

2.3 Entrepreneurship and Sustainable Development

Sustainable development, along with its associated concepts—sustainability and social responsibility—can only be effectively realized through sustainability-driven entrepreneurship (Yaşlıoğlu & Toplu Yaşlıoğlu, 2020). Many works that warn of environmental crises ultimately adopt an optimistic perspective, suggesting that the future of human civilization depends on the efforts of social and environmental entrepreneurs (Hall et al., 2010). In this context, early research on sustainable entrepreneurship introduced the idea of integrating entrepreneurship with sustainable development, emphasizing that sustainability presents a significant opportunity for entrepreneurs and innovators to drive meaningful change (Ribeiro & Leitão, 2024).

Sustainable entrepreneurship refers to the identification, development, and exploitation of entrepreneurial capabilities that provide a competitive advantage by generating environmental benefits within the communities where businesses operate. It serves as an umbrella term for both social and environmental entrepreneurship, continuously contributing to the well-being of both humanity and the environment (Hooi et al., 2016). While eco-entrepreneurship primarily focuses on protecting the natural environment, sustainable entrepreneurship aims to create economic, social, and environmental value simultaneously (Kimuli et al., 2020). As a key driver of innovation, entrepreneurship can significantly contribute to environmental, social, and economic innovations, all of which play a crucial role in advancing sustainable development. Technology and innovation, both shaped by and shaping entrepreneurship, are essential in the development of new products and services, as well as in improving existing ones. This process enables businesses to operate more efficiently and effectively, thus furthering sustainable development (Ahmed & McQuaid, 2005). Consequently, sustainable entrepreneurship (SE) is recognized as a distinct form of entrepreneurship, encompassing all entrepreneurs who pursue goals related to sustainability and environmental responsibility.

2.4 The Role of Entrepreneurship in Addressing Social, Economic, and Environmental Challenges

Systematically, Entrepreneurship has been recognized as a dynamic mechanism for addressing economic, social, and environmental challenges through opportunity identification and innovative value creation. As a result, it has become a topic of extensive research across multiple scientific disciplines, including business, management, economics, social sciences, and psychology (Filser et al., 2019). In this context, entrepreneurship plays a fundamental role in driving economic progress. Innovation and the creation of new opportunities facilitate the optimal allocation of resources and enhance economic flexibility in response to changing circumstances (Muhammad Muneeb et al., 2020). Furthermore, recent studies highlight that entrepreneurs not only contribute to economic growth but also serve as agents for reducing social inequality and promoting environmental protection. In this regard, Tietenberg and Lewis (2012) provide a comprehensive framework for understanding the economic, social, and environmental impacts of entrepreneurship within the broader context of well-being. They argue that "the real measure of development increases when we, as a nation or as a world, are better off and decreases when we are worse off." This perspective aligns with the views of numerous researchers who emphasize the crucial role of entrepreneurship in improving social and environmental well-being (Neumann, 2021).

Entrepreneurship is increasingly recognized as a powerful tool for social change, enhancing both the individual and collective well-being of people by fostering essential human capabilities, such as the ability to work, earn an income, and accumulate wealth. The United Nations Human Development Report (1998) highlights that as families move toward entrepreneurship and economic empowerment, they begin to experience greater self-respect, a sense of belonging to society, and a sense of self-fulfillment. Additionally, entrepreneurship has the potential to address the current healthcare crisis by developing products that improve health outcomes while simultaneously reducing costs (Dhahri & Omri, 2018).

Entrepreneurship also plays a crucial role in addressing environmental challenges. Specifically, entrepreneurial initiatives can make significant contributions to preserving ecosystems, combating climate change and deforestation, enhancing agricultural practices, improving freshwater supplies, and conserving biodiversity (Shepherd & Patzelt, 2011). Moreover, the goal of eco-entrepreneurship is to provide solutions to mitigate environmental degradation. Eco-entrepreneurs are actively engaged in driving transformative environmental innovations that aim to solve complex environmental issues. Given the complexity and uncertainty of achieving the Sustainable Development Goals, current companies are often insufficient on their own to accomplish these objectives. This gap presents a valuable opportunity for eco-entrepreneurs to tackle these challenges through the implementation of eco-innovations (Iqbal et al., 2020; Table 1).

Socially, entrepreneurial initiatives have a profound positive impact on gender equality and social inclusion by empowering women and facilitating the integration of young people into the labor market. Entrepreneurship transcends boundaries of gender, race, personal char-

Table 1 The role of entrepreneurship in addressing social, economic, and environmental challenges (Filser et al., 2019)

Row	Role
1	Economic dimension: entrepreneurship contributes to economic growth by creating job opportunities, improving decent work standards, promoting sustainable agriculture, and strengthening innovation
2	Social dimension: entrepreneurship can play a significant role in enhancing social cohesion, reducing inequalities, and providing opportunities for all groups, including women, youth, individuals with disabilities, and the most vulnerable populations
3	Environmental dimension: entrepreneurship can address environmental challenges by offering innovative technologies to reduce and adapt to climate change, adopting resilience measures, and promoting sustainable consumption and production practices

acteristics, height, and culture. Women worldwide contribute equally to economic development through entrepreneurship (Sajuyigbe & Fadeyibi, 2017). The glass ceiling that once constrained women's career advancement has now opened new avenues for business ownership, allowing women to leverage their business acumen while fostering stronger family relationships (Ambepitiya, 2013).

As a result, women are no longer excluded from economic activities and possess the necessary skills and capabilities to transition from domestic roles to professional environments, thereby making equal contributions to local economic development. Moreover, women entrepreneurs are increasingly assuming prominent roles in socio-economic development across both developed and developing economies, as they represent a significant portion of the workforce in both formal and informal sectors. The United Nations (2014) affirms that women entrepreneurs make significant contributions to national economies through their involvement in establishing and growing small and medium-sized enterprises (SMEs) (Sajuyigbe & Fadeyibi, 2017). Thus, a clear relationship exists between women's entrepreneurship and sustainable development. Entrepreneurs introduce new values to society, often through innovations or novel ideas in the form of business ventures. Women's entrepreneurship serves as a tool for empowering individuals, improving the status of rural women, and addressing pressing social challenges. Women entrepreneurs play a pivotal role in promoting sustainable practices across economic, social, and environmental domains, ultimately advancing sustainable development (Ambepitiya, 2013).

2.5 Entrepreneurship, Policymaking, and the Pathway to Sustainable Development

As mentioned earlier, achieving sustainable development requires a deep and accurate understanding of its goals because without a proper understanding of these goals, achieving sustainable development in any field will not be possible. Therefore, these goals must first be appropriately recognized and explained; then, appropriate policies and strategies must be formulated and implemented. In this regard, the United Nations determined a set of 17 Sustainable Development Goals (SDGs) in 2015, which are presented in Table 2. These goals are considered a comprehensive roadmap for achieving a better and more sustainable future by 2030.

Considering these goals and gaining a deeper understanding of the concept of entrepreneurship, it can be concluded that entrepreneurship plays a crucial role in achieving the Sustainable Development Goals (SDGs). By creating job opportunities, promoting innovation, enhancing productivity, and increasing the economic resilience of societies, entrepreneurship paves the way for achieving multiple SDGs, including poverty eradication (Goal 1), gender equality (Goal 5), sustainable economic growth (Goal 8), innovation and infrastructure (Goal 9), and responsible consumption and production (Goal 12). Furthermore, the formulation and implementation of supportive policies and strategic frameworks by governments and decision-making institutions can significantly bolster the capacity

Table 2 Sustainable development goals (Huđek & Hojnik, 2020)

Goal number	Type of goal	Goal description
1	Eradicating poverty	Ending all forms of poverty everywhere
2	Eliminating hunger	Ensuring comprehensive food security, enhancing nutritional outcomes, and advancing agricultural systems
3	Wellness and quality of life	Ensuring healthy lives and promoting well-being for all at all ages
4	Quality education systems	Providing inclusive and equitable quality education and promoting lifelong learning opportunities for all
5	Gender equality	Eliminating gender-based disparities and strengthening the role of women and girls in society
6	Clean water and sanitation	Expanding access to safe water and sanitation infrastructure for individuals across all communities
7	Affordable and clean energy	Guaranteeing that all individuals have access to energy sources that are economical, consistent, sustainable, and technologically advanced
8	Decent work and economic growth	Fostering long-term, inclusive, and environmentally responsible economic progress alongside comprehensive and meaningful employment for all
9	Industry, innovation, and infrastructure	Establishing durable infrastructure, facilitating inclusive and green industrialization, and promoting the generation of new ideas and technologies
10	Reduced inequalities	Mitigating disparities both within national borders and across international boundaries
11	Sustainable cities and communities	Developing urban areas and human communities that are accessible, secure, adaptable, and environmentally sustainable
12	Responsible consumption and production	Promoting responsible consumption and production behaviors that support environmental sustainability
13	Climate action	Undertaking swift initiatives aimed at countering climate change and its associated impacts
14	Life below water	Protecting and managing oceans, seas, and marine ecosystems responsibly to support sustainable development
15	Life on land	Protect, restore, and promote the sustainable use of terrestrial ecosystems
16	Peace, justice, and strong institutions	Promote peaceful and inclusive societies, provide access to justice for all, and build effective, accountable, and inclusive institutions at all levels
17	Partnerships for the goals	Strengthen the means of implementation and revitalize the global partnership for sustainable development

for sustainable entrepreneurship and guide it toward realizing these goals. Therefore, a nuanced understanding of the SDGs, along with their interpretation from both a policymaking and entrepreneurial perspective, is indispensable for achieving comprehensive and sustainable development. These goals should be viewed not only as a global roadmap but also as a strategic platform for informing policies, fostering institution-building, advancing education, driving investment, and enhancing human capacities in the field of entrepreneurship.

3 Sustainable Entrepreneurship and Government Goals

3.1 Policy Measures to Maximize Benefits

The Organization for Economic Co-operation and Development (OECD) emphasizes the need for entrepreneurship-related policies to be adaptable in both their design and implementation at the local level. Accordingly, governments are expected to adopt context-appropriate strategies that foster diverse forms of entrepreneurship across economic and social sectors, thereby not only advancing development but also minimizing and potentially reversing its adverse environmental impacts (Ahmed & McQuaid, 2005). In this context, countries and international organizations considering local conditions and overarching development goals have undertaken initiatives within the framework of specific strategies and programs aimed at promoting green growth, enhancing social inclusion, and reducing poverty. These efforts illustrate how entrepreneurial capacities can be leveraged to achieve inclusive and sustainable development. Table 3 presents selected examples of policy measures adopted by countries and international organizations to maximize economic, social, and environmental benefits through entrepreneurship-oriented policies.

As illustrated in Table 1, the entrepreneurship policies adopted by countries such as Colombia, Fiji, the Philippines, and Ethiopia emphasize gender equality and the enhancement of youth participation in entrepreneurial activities. These policies not only contribute to reducing social inequalities but also serve as effective drivers for accelerating economic growth. Likewise, the Croatian government has implemented a national strategy for Female-Led Entrepreneurship as part of its sustainable development agenda, while South Africa has adopted a youth entrepreneurship development strategy. Both initiatives aim to foster an entrepreneurial culture among young people and accelerate the growth of youth-led businesses, thereby creating more employment opportunities for this demographic group (Elmonshid & Sayed, 2024; Mehra, 2022; Raman et al., 2022).

Similarly, Italy, along with countries such as Poland, Thailand, and the United States, has implemented policies to support social entrepreneurs and innovative businesses with environmental and social objectives. These policies specifically encourage innovative startups engaged in the development and marketing of sustainable products and services. Due to their focus on replacing conventional physical products with innovative solutions, such startups are considered more environmentally and socially sustainable compared to traditional manufacturing firms. These policy measures have contributed to strengthening the resilience, dynamism, and adaptive capacity of entrepreneurial ecosystem elements, which are essential for fostering entrepreneurship aligned with sustainable development (Anuar et al., 2020). Moreover, the UK government has introduced programs aimed at stimulating social entrepreneurship through the establishment of dedicated support funds and tailored initiatives for voluntary and charitable organizations. Startups that focus on delivering innovative solutions play an active role in improving social conditions and enhancing public well-being (Jiménez, 2024).

Additionally, the British government actively supports social entrepreneurs in implementing public welfare projects. At the European level, various operational programs have been launched to strengthen social entrepreneurship in countries such as Bulgaria, France, Luxembourg, Slovenia, and Sweden. Furthermore, green entrepreneurship has received growing attention in many countries, as sustainable growth driven by the promotion of advanced technologies and the adoption of more efficient processes and products plays a significant role in fostering economic development (Jiménez, 2024; Mehra, 2022). Therefore, the international landscape increasingly reflects a policy environment in which governments simultaneously promote growth through entrepreneurship and the realization of sustainable development goals (such as the United Nations Sustainable Development Goals), seeking to align business interests with the broader needs and expectations of society (Criado-Gomis et al., 2017).

Policy measures that can serve as strategic levers for maximizing the economic, social, and environmental benefits of entrepreneurship typically include the following: (1) optimizing regulatory frameworks; (2) enhancing education and skills development with education recognized as a central element in policy frameworks for sustainable entrepreneurship, as it can foster awareness and strengthen responsible attitudes, thereby encouraging individuals to think and act toward the creation of envi-

Table 3 Selected policy measures for integrating sustainable development into national entrepreneurship strategies (DESA, 2020; Mehra, 2022)

Row	Country/ economy	Initiative/program	Description
1	Chile	Green Growth Strategy	This strategy was formulated in 2013 and remained in effect until 2020. Its objective was to reduce environmental impacts while simultaneously fostering employment and enhancing social equity. The strategy pursued economic growth within the framework of sustainable development
2	Global	Global Green Growth Institute	It is an intergovernmental organization established to accelerate the transition to green growth, grounded in social inclusion and environmental sustainability. The organization collaborates with countries to develop and implement strategies that simultaneously address poverty, inequality, and environmental challenges while promoting economic growth
		Green Growth Strategy (OECD)	This framework supports countries in pursuing economic growth while minimizing environmental degradation, climate change, and the inefficient exploitation of natural resources. The strategy focuses on job creation and economic development through the more sustainable use of natural resources and energy
3	Colombia	Department of Social Prosperity	This department operates within the framework of the 2011 Presidential Decree entitled 'Prosperity for All' and is responsible for coordinating all government actions related to poverty reduction and peacebuilding. Its key responsibilities include overseeing all programs related to poverty alleviation, promoting entrepreneurship—including impact investment and social entrepreneurship—and fostering innovative approaches to the delivery of welfare and peace-oriented services
		Youth Entrepreneurs strategy	The 'Young Entrepreneurs' strategy was part of the President's Youth Program in 2013, aimed at developing young people's entrepreneurship culture
4	Peru	Ministry of Social Development and Inclusion	The institution was established in 2011 to coordinate social policies and programs across various levels of government. In 2012, the ministry developed a policy for social development and inclusion aimed at eradicating poverty and extreme poverty
5	Poland	National Strategy for Social Entrepreneurship	In 2014, Poland introduced a comprehensive framework for fostering entrepreneurship as a social phenomenon. The goal of this strategy was to make social entrepreneurship a driver of employment, social cohesion, and development by 2020
6	Thailand	Office of Social Enterprises	In 2010, Thailand established the Office of Social Enterprises under the Thai Health Promotion Foundation Act to lead the planning and implementation of the Social Enterprise Strategy (2010–2014)
7	United States	Office of Social Innovation and Civic Participation	This institutional department serves as a strategic intermediary, systematically coordinating cross-sectoral interactions between public and private entities to advance social innovation ecosystems. Additionally, it functions as a critical advisory mechanism, providing comprehensive policy recommendations to the Presidential administration regarding social enterprise development and innovative social transformation strategies while simultaneously facilitating policy formulation and implementation processes

(continued)

Table 3 (continued)

Row	Country/ economy	Initiative/program	Description
8	Fiji	National Youth Policy	This policy considers youth entrepreneurship as one of its primary objectives and includes small enterprises as a key focus area in the country's 20-year Youth Development Strategic Plan
9	Ethiopia	National Growth and Transformation Program and National Gender Action Plan	This program identifies the micro, small, and medium-sized enterprise (MSME) sector as a key driver for increasing women's income through job creation. Additionally, the National Action Plan for Gender Equality seeks to eliminate gender and cultural biases that hinder women's participation in economic activities
10	Cambodia	Youth Entrepreneurs strategy	These centers were established in 2019 to provide empowerment and financial support to young founders and startups
11	Philippines	Youth Development Program 2017–2022	This program is a comprehensive document that addresses various aspects of the national youth agenda, encompassing nine areas of action. The objectives for youth economic empowerment within the program are summarized as follows: optimizing youth participation in the labor force, preventing youth involvement in vulnerable employment, promoting youth engagement in entrepreneurial activities, and strengthening youth participation in employment promotion support systems
		Youth Entrepreneurship Act	This law provides training in entrepreneurship and financial literacy to foster youth entrepreneurship. Enacted in 2015, the law states that the primary strategy was to promote the sustainability of Filipino youth by nurturing their talents and skills in the areas of finance and entrepreneurship through education and specialized programs
		Go Negosyo Act	This law was enacted in 2015 to support entrepreneurial activities. Under this law, 'Negosyo Centers' are to be established in all provinces to provide accessible services and business support

ronmentally friendly and sustainable businesses; (3) improving access to financial resources; (4) facilitating the exchange of technology and innovation; and (5) promoting awareness and networking. These measures should be integrated as part of a coherent and coordinated framework designed to foster entrepreneurship (Burchi et al., 2021; Mehra, 2022). Consequently, policy actions must be specifically tailored to the national contexts of each country, as different nations face distinct economic, political, and social challenges (Hojnik, 2024; Spencer et al., 2010).

3.2 Supportive Policies for Entrepreneurs

- **Strengthening Conditions for Sustainable Entrepreneurial Growth**

Entrepreneurship, in its pursuit of sustainable transformation, is inherently aligned with the overarching objectives of governments seeking to achieve sustainable development. To enhance the role and contribution of entrepreneurship to this form of development, it is essential for governments to adopt a set of targeted policies and action policies that move beyond a narrow focus on economic growth to incorporate social and environmental dimensions holistically. In this regard, governments must design and implement policies that facilitate effective support for entrepreneurs and contribute to their empowerment, thereby advancing national development goals and accelerating progress toward global sustainability objectives. Such policies can directly foster the necessary conditions for the emergence and strengthening of sustainable entrepreneurship. From this perspective, policymakers can create an enabling and incentive-driven environment for the growth of inclusive, innovative, and sustainable businesses by employing effective strategies consistent with sustainability principles (Jiménez, 2024; Mehra, 2022; Watson et al., 2023).

Thus, the effectiveness of entrepreneurship policies should not be assessed solely based on the level of entrepreneurial activity but also in terms of their contribution to achieving sustainability goals. Policies must aim to align economic growth with sustainable development, ensuring both the advancement of entrepreneurs and the pursuit of broader sustainability objectives (Biermann et al., 2022; Hojnik, 2024). Achieving sustainable development inevitably involves making difficult choices, a challenge rooted in the undeniable fact that the planet's fundamental limits, determined by the laws of physics and biology, cannot be ignored. Human activities cannot continue beyond these boundaries indefinitely. While it is sometimes possible to temporarily exceed these limits, remaining in such conditions without causing irreversible harm to the environment and society is ultimately unsustainable. As such, sustainable development requires sound political decision-making since no technological solution alone can resolve the complex economic, social, and environmental issues confronting humanity (Hummels & Argyrou, 2021).

Therefore, societies must manage their economic, social, and natural resources wisely, as the failure to take action for their preservation could result in irreversible damage (Criado-Gomis et al., 2017). In this regard, the transition toward sustainable business models is critical. This perspective was also highlighted in the Brundtland Report, which emphasized the need to design business models that are more attuned to the principles of sustainability (Criado-Gomis et al., 2017). In recent years, proactive efforts to encourage the adoption of sustainable business models have become a central priority in policy agendas, and governments worldwide have been introducing various policy measures to promote this shift (Paletta et al., 2021). Through the implementation of appropriate recommendations, policymakers can establish a supportive ecosystem in which inclusive and sustainable business models can flourish (Mehra, 2022).

Addressing the negative social and environmental externalities of business can no longer rely solely on conventional tools such as taxation, fines, or regulation. Instead, businesses must be encouraged through incentives, supportive government measures, effective public–private partnerships, innovation support, enhanced competitiveness, and strengthened stakeholder engagement to redesign their business models and create conditions conducive to sustainable entrepreneurship (Mehra, 2022; Paletta et al., 2021). Such businesses not only contribute to job creation and economic growth but also foster social inclusion and address pressing environmental challenges. Examples include renewable energy projects, circular economy initiatives, and social enterprises, all of which are closely aligned with the Sustainable Development Goals (SDGs), particularly those related to climate action and poverty reduction. Governments play a crucial role in creating the enabling environment necessary for these sustainable development pathways to emerge and thrive (Arejiogbe et al., 2023; Mehra, 2022; Zhu & Wang, 2024).

To achieve these objectives, governments must prioritize public education on climate change and the use of renewable energy, while also fostering an entrepreneurial culture that aligns with sustainable development. In this regard, the design of educational programs focused on environmentally friendly products and renewable energy resources is essential. Moreover, the provision of financial resources, business training, and advisory services, particularly in emerging economies, can create an environment that enables innovation and sustainability. Policymakers' support for financial institutions in creating favorable financing conditions and offering loans can further help entrepreneurs align their business activities with the Sustainable Development Goals (Iqbal et al., 2020).

In addition, policies should not be formulated in isolation from broader political priorities or corporate interests. Recent policy planning has increasingly focused on guiding the economy from a linear model toward a circular economy and on replacing fossil-based resources with bio-based alternatives (Paletta et al., 2021). The principles of the circular economy offer an effective strategy for promoting sustainable business models. By moving away from the traditional linear 'take-make-dispose' model and transitioning toward a closed-loop system in which waste is eliminated and resources are continuously recovered and reused, organizations can align their activities more closely with sustainability goals. Companies such as IKEA and Philips have successfully integrated these principles into their operational models, focusing on extending product lifecycles, recycling resources, and minimizing waste generation (Elsawy & Youssef, 2023).

Moreover, since 2015, the circular economy has been adopted as a strategic principle in the policies of many countries, and its implementation has been pursued through various approaches. For instance, some governments have facilitated the transition from traditional economic models to more sustainable ones by enacting laws and policies that promote this shift. Germany and Japan have been among the pioneering countries in formulating explicit policies related to the circular economy, followed by China and the European Union (De Melo et al., 2022; Lahti et al., 2018). In Europe, the 'Circular Economy Package' was adopted by the European Parliament in 2018, comprising a set of actions and policies aimed at reducing waste across the continent. Specific recycling targets were established for EU member states concerning materials such as packaging, plastics, wood, ferrous metals, aluminum, glass, paper, and cardboard (Lahti et al., 2018).

Similarly, in 2002, China's central government recognized the circular economy as a new national development strategy. The primary policy framework in this regard is the "Circular Economy Promotion Law," enacted in 2009, which emphasizes resource efficiency as a means of protecting and enhancing the environment. Subsequently, multiple action plans (for example, the State Council, 2013) were developed to provide more detailed guidelines for specific sectors and to clarify the mechanisms for implementing the law (McDowall et al., 2017).

In addition to the aforementioned measures, investing in digital infrastructure and enhancing technological skills are also critical policies that can accelerate digital entrepreneurship development and assist entrepreneurs in adopting sustainable practices, particularly in less digitally advanced regions (Herman, 2022). Alongside these actions, facilitating regulations, reducing barriers, and creating supportive ecosystems for startups and small and medium-sized enterprises can strengthen their role in achieving economic and social sustainability (Elmonshid & Sayed, 2024; Fig. 2).

4 Conclusion

Entrepreneurship policies and strategies serve as essential instruments for advancing sustainable development. In alignment with the Sustainable Development Goals (SDGs), Entrepreneurial Activity fosters economic growth and job creation, also generating positive social and environmental impacts. To achieve these objectives, it is essential for governments to formulate policies that holistically address the multidimensional aspects of entrepreneurship while promoting economic, social, and environmental sustainability simultaneously. These policies should prioritize fostering innovation, supporting sustainable business models, and establishing the necessary infrastructure for entrepreneurial activities. Additionally, enhancing access to education, facilitating financing, and reducing barriers to business creation can significantly streamline the entrepreneurial process. In this context,

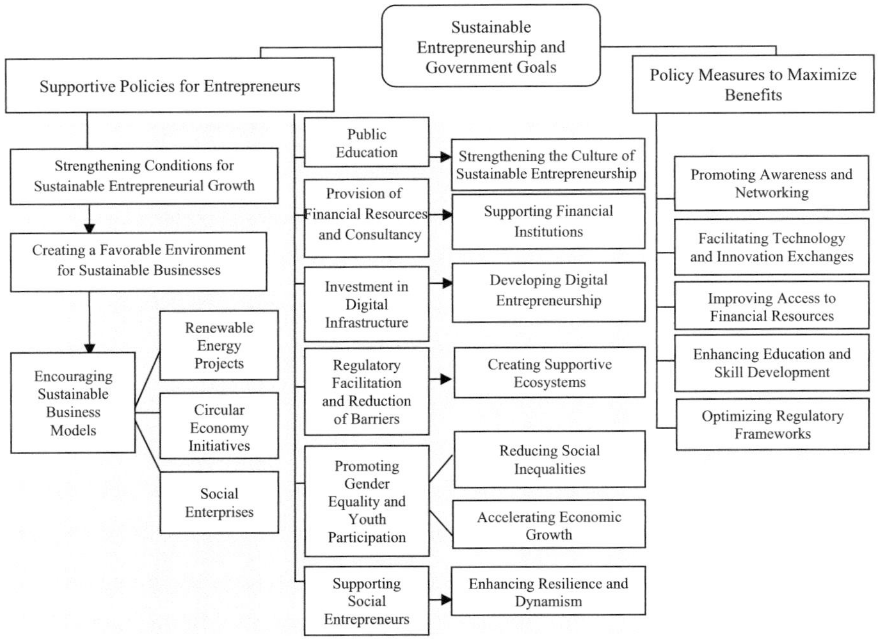

Fig. 2 Policy framework for sustainable entrepreneurship development (*source* Research Findings)

targeted policies—such as incentivizing green technologies, promoting social entrepreneurship, and cultivating an enabling environment for entrepreneurs—play a crucial role in sustainable development. Ultimately, achieving meaningful progress toward sustainability requires entrepreneurship policies that extend beyond economic growth to encompass social and environmental considerations, thereby contributing to the realization of global sustainable development goals.

5 Case Study

5.1 The Akshar Educational Model and Sustainable Entrepreneurship

Child labor remains a prevalent issue in India, particularly in Assam, where poverty and illiteracy are the primary driving forces. Many children are compelled to leave school due to financial constraints, forcing them into hazardous labor and perpetuating the cycle of poverty and illiteracy. The issue of child labor and illiteracy in Assam is deeply entrenched, having persisted for centuries. Widespread poverty continues to deprive these children of educational opportunities, reinforcing their socio-economic disadvantages. The Akshar Foundation presents an innovative and effective solution to this critical social issue by integrating education, environmental awareness, and employment opportunities. Its model merges traditional classroom learning with technical and vocational training, aligning with Mahatma Gandhi's vision of skill-based education. With the increasing accessibility of the internet, the emphasis in education is shifting from mere information retention to the acquisition of skills. Akshar's educational model enables students to earn an income while they learn. By engaging in teaching and other school-related activities, students receive financial incentives, allowing those who previously dropped out due to economic hardships to return to school. This approach not only supports their families but also facilitates their continued education, helping them progress through the curriculum without financial obstacles. By providing economic support alongside education, this model effectively prevents school dropouts and fosters long-term social transformation. The Akshar initiative serves as a crucial step toward breaking the cycle of poverty and illiteracy, empowering children with the knowledge and skills needed for a sustainable future (Hazarika et al., 2022).

Another innovative aspect of the Akshar Foundation's model is its unique tuition structure, in which students pay their fees with plastic waste. Each student is required to bring between 15 and 25 plastic bottles or other plastic waste from home each week. Additionally, students collect plastic waste from 15 to 20 nearby households and repurpose it into eco-bricks, which are then utilized for small-scale construction projects or as materials for garden and wall decorations. This policy exemplifies an innovative approach to sustainable entrepreneurship, offering a holistic model of sustainable development by integrating education, employment, and environmental sustainability. By enabling students to earn an income while studying, this initiative provides a practical solution to combating child labor and illiteracy. Moreover, by alleviating financial burdens on families, it creates pathways for children to return to school and complete their education. Beyond reducing educational disparities, such policies contribute to the economic empowerment of local communities, serving as an effective strategy for fostering sustainable social and economic development. Ultimately, models like Akshar's not only help break the cycle of poverty and illiteracy but also drive long-term, sustainable transformation in underprivileged societies by strengthening human capital and promoting environmental responsibility.

5.2 Procter & Gamble's (P&G) Water Purification Initiative in Developing Countries

Procter & Gamble, in collaboration with organizations such as the U.S. Centers for Disease Control and Prevention, has developed an innovative and cost-effective water purification product designed to address the critical issue of water pollution in developing countries, including Haiti, Kenya, Pakistan, and Uganda. This initiative aligns with the Sustainable Development Goals (SDGs), particularly those focused on improving health and well-being by reducing waterborne diseases (Altowaijri, 2016). By significantly lowering the cost of water purification, this solution helps reduce mortality rates associated with water-related illnesses. Beyond addressing immediate health concerns, Procter & Gamble's initiative contributes to broader social objectives, including reducing child mortality and controlling the spread of disease. Additionally, this effort supports local economic development by generating employment opportunities in the production and distribution of water purification products, further enhancing the initiative's long-term impact on sustainable development.

5.3 Questions

- What are the three main dimensions of sustainable development discussed in the chapter?
- How does the Akshar Educational Model integrate education with sustainability?
- How can women's entrepreneurship contribute to social empowerment and sustainable development?
- What policy measures are suggested for promoting sustainable entrepreneurship at the national level?
- According to the chapter, how does entrepreneurship contribute to achieving the Sustainable Development Goals (SDGs)?
- Imagine you are a policymaker in a developing country. How would you design a strategy to promote youth entrepreneurship aligned with SDG 8 (Decent Work and Economic Growth)?
- Given the examples from Chile and the Philippines, how could a local government integrate environmental protection into its entrepreneurship development programs?

References

Agwunobi, A. (2020). *Entrepreneurial health and well-being in the digital economy: an intersectional crit-*

ical realist investigation [Doctoral dissertation, Loughborough University]. https://doi.org/10.26174/thesis.lboro.12605453.v1

Ahmed, A., & McQuaid, R. W. (2005). Entrepreneurship, management, and sustainable development. *World Review of Entrepreneurship, Management and Sustainable Development, 1*(1), 6–30. https://doi.org/10.1504/WREMSD.2005.007750

Altowaijri, R. A. (2016). *Progressing sustainable development in Saudi Arabia: The role of the public and private sectors* [Doctoral dissertation, American University of Paris (France)].

Ambepitiya, K. R. (2013). The role of women entrepreneurs in establishing a sustainable development in Sri Lanka. *Kelaniya Journal of Human Resource Management, 8*(2), 149. https://doi.org/10.4038/kjhrm.v8i2.7

Anuar, M. M., Mohamad, O., & Omar, M. W. (2020). Millennials' response toward cause-related marketing advertisements via social media and magazines: Which is more effective? In *Eurasian Studies in Business and Economics* (Vol. 13, Issue 2, pp. 239–250). https://doi.org/10.1007/978-3-030-40160-3_17

Apostu, S. A., & Gigauri, I. (2023). Sustainable development and entrepreneurship in emerging countries: Are sustainable development and entrepreneurship reciprocally reinforcing? *Journal of Entrepreneurship, Management and Innovation, 19*(1), 41–77. https://doi.org/10.7341/20231912

Arejiogbe, O. E., Moses, C. L., Salau, O. P., Onayemi, O. O., Agada, S. A., Dada, A. E., & Obisesan, O. T. (2023). Bolstering the impact of social entrepreneurship and poverty alleviation for sustainable development in Nigeria. *Sustainability (Switzerland), 15*(8). https://doi.org/10.3390/su15086673

Baiocco, S., Leoni, L., & Paniccia, P. M. A. (2023). Entrepreneurship for sustainable development: Co-evolutionary evidence from the tourism sector. *Journal of Small Business and Enterprise Development, 30*(7), 1521–1546. https://doi.org/10.1108/JSBED-01-2023-0003

Bhattarai, N. (2020). *Migrant entrepreneurs and the digital economy in the UK* [Doctoral dissertation, Royal Holloway, University of London]. https://pure.royalholloway.ac.uk/ws/files/40200086/2020bhattarainphd.pdf

Biehl, C. F., Hoepner, A. G. F., & Liu, J. (2012). Social, environmental, and trust issues in business and finance. In *Socially responsible finance and investing* (pp. 110–141). Wiley. https://doi.org/10.1002/9781118524015.ch7

Biermann, F., Hickmann, T., Sénit, C. A., Beisheim, M., Bernstein, S., Chasek, P., Grob, L., Kim, R. E., Kotzé, L. J., Nilsson, M., Ordóñez Llanos, A., Okereke, C., Pradhan, P., Raven, R., Sun, Y., Vijge, M. J., van Vuuren, D., & Wicke, B. (2022). Scientific evidence on the political impact of the Sustainable Development Goals. *Nature Sustainability, 5*(9), 795–800. https://doi.org/10.1038/s41893-022-00909-5

Burchi, A., Włodarczyk, B., Szturo, M., & Martelli, D. (2021). The effects of financial literacy on sustainable entrepreneurship. *Sustainability (Switzerland), 13*(9), 1–21. https://doi.org/10.3390/su13095070

Cunningham, J. B., & Lischeron, J. (1991). Defining entrepreneurship. *Journal of Small Business Management, 29*(1).

Criado-Gomis, A., Cervera-Taulet, A., & Iniesta-Bonillo, M.-A. (2017). Sustainable entrepreneurial orientation: A business strategic approach for sustainable development. *Sustainability, 9*(9), 1667. https://doi.org/10.3390/su9091667

De Melo, T. A. C., De Oliveira, M. A., De Sousa, S. R. G., Vieira, R. K., & Amaral, T. S. (2022). Circular economy public policies: A systematic literature review. *Procedia Computer Science, 204*, 652–662. https://doi.org/10.1016/j.procs.2022.08.079

Delacroix, E., Parguel, B., & Benoit-Moreau, F. (2019). Digital subsistence entrepreneurs on Facebook. *Technological Forecasting and Social Change, 146*, 887–899. https://doi.org/10.1016/j.techfore.2018.06.018

DESA. (2020). *Exploring youth entrepreneurship.*

Dhahri, S., & Omri, A. (2018). Entrepreneurship contribution to the three pillars of sustainable development: What does the evidence really say? *World Development, 106*(84504), 64–77. https://doi.org/10.1016/j.worlddev.2018.01.008

Elliott, J. (2013). *An introduction to sustainable development* (4th ed.). Routledge.

Elia, G., Margherita, A., & Passiante, G. (2020). Digital entrepreneurship ecosystem: How digital technologies and collective intelligence are reshaping the entrepreneurial process. *Technological Forecasting and Social Change, 150,*119791.

Elmonshid, L. B. E., & Sayed, O. A. (2024). The relationship between entrepreneurship and sustainable development in Saudi Arabia: A comprehensive perspective. *Economies, 12*(8), 198. https://doi.org/10.3390/economies12080198

Elsawy, M., & Youssef, M. (2023). Economic sustainability: Meeting needs without compromising future generations. *International Journal of Economics and Finance, 15*(10), 23. https://doi.org/10.5539/ijef.v15n10p23

Filser, M., Kraus, S., Roig-Tierno, N., Kailer, N., & Fischer, U. (2019). Entrepreneurship as catalyst for sustainable development: Opening the black box. *Sustainability, 11*(16), 4503. https://doi.org/10.3390/su11164503

Goel, M., & Joshi, B. P. (2017). Entrepreneurship and sustainable development. *Journal of Entrepreneurship and Management, 6*(3), 43.

González-López, M. J., Pérez-López, M. C., & Rodríguez-Ariza, L. (2019). Clearing the hurdles in the entrepreneurial race: The role of resilience in entrepreneurship education. *Academy of Management Learning & Education, 18*(3), 457–483.

Hall, J. K., Daneke, G. A., & Lenox, M. J. (2010). Sustainable development and entrepreneurship: Past contributions and future directions. *Journal of Business*

Venturing, 25(5), 439–448. https://doi.org/10.1016/j.jbusvent.2010.01.002

Hazarika, G., Salunke, P., Wadhwani, K., & Sen, K. (2022). Sustainable entrepreneurship—A dyad case study on recycling. *SMS Journal of Entrepreneurship & Innovation, 8*(01). https://doi.org/10.21844/smsjei.v8i01.30010

Herman, E. (2022). The interplay between digital entrepreneurship and sustainable development in the context of the EU digital economy: A multivariate analysis. *Mathematics, 10*(10), 1682. https://doi.org/10.3390/math10101682

Hojnik, B. B. (2024). Global country classification of sustainable startups levels. *Problemy Ekorozwoju, 19*(1), 193–200. https://doi.org/10.35784/preko.5383

Hooi, H. C., Ahmad, N. H., Amran, A., & Rahman, S. A. (2016). The functional role of entrepreneurial orientation and entrepreneurial bricolage in ensuring sustainable entrepreneurship. *Management Research Review, 39*(12), 1616–1638. https://doi.org/10.1108/MRR-06-2015-0144

Huđek, I., & Hojnik, B. B. (2020). Impact of entrepreneurship activity sustainable development. *Problemy Ekorozwoju/problems of Sustainable Development, 15*(2), 175–183.

Hummels, H., & Argyrou, A. (2021). Planetary demands: Redefining sustainable development and sustainable entrepreneurship. *Journal of Cleaner Production, 278*, Article 123804. https://doi.org/10.1016/j.jclepro.2020.123804

Iqbal, N., Khan, A., Gill, A. S., & Abbas, Q. (2020). Nexus between sustainable entrepreneurship and environmental pollution: Evidence from developing economy. *Environmental Science and Pollution Research, 27*(29), 36242–36253. https://doi.org/10.1007/s11356-020-09642-y

İyigün, N. Ö. (2015). What could entrepreneurship do for sustainable development? A corporate social responsibility-based approach. *Procedia—Social and Behavioral Sciences, 195*, 1226–1231. https://doi.org/10.1016/j.sbspro.2015.06.253

Jacobs, M. (1991). *The Green Economy*. Pluto Press.

Jiménez, J. C. (2024). *The role of public policy in scaling up sustainable entrepreneurship—A multi-level analysis* [Doctoral dissertation, Dissertation, Erfurt, Universität Erfurt, 2024].

Jones, P., Ratten, V., & Hayduk, T. (2020). Sport, fitness, and lifestyle entrepreneurship. *International Entrepreneurship and Management Journal, 16*, 783–793. https://doi.org/10.1007/s11365-020-00666-x

Kimuli, S. N. L., Orobia, L., Sabi, H. M., & Tsuma, C. K. (2020). Sustainability intention: Mediator of sustainability behavioral control and sustainable entrepreneurship. *World Journal of Entrepreneurship, Management and Sustainable Development, 16*(2), 81–95. https://doi.org/10.1108/WJEMSD-12-2019-0096

Kirzner, I. (1973). *Competition and entrepreneurship*. University of Chicago Press.

Klarin, T. (2018). The concept of sustainable development: From its beginning to the contemporary issues. *Zagreb International Review of Economics and Business, 21*(1), 67–94. https://doi.org/10.2478/zireb-2018-0005

Lahti, T., Wincent, J., & Parida, V. (2018). A definition and theoretical review of the circular economy, value creation, and sustainable business models: Where are we now and where should research move in the future? *Sustainability (Switzerland), 10*(8). https://doi.org/10.3390/su10082799

McDowall, W., Geng, Y., Huang, B., Barteková, E., Bleischwitz, R., Türkeli, S., Kemp, R., & Doménech, T. (2017). Circular economy policies in China and Europe. *Journal of Industrial Ecology, 21*(3), 651–661. https://doi.org/10.1111/jiec.12597

Mehmood, T., Alzoubi, H. M., Alshurideh, M., Algasaymeh, A., & Ahmed, G. (2019). *Schumpeterian Entrepreneurship Theory, 25*(4), 1–10.

Mehra, S. (2022). Integrate sustainable development goals into national entrepreneurship strategies. In *Emerging trends in biosciences* (p. 195).

Moggi, S., Pierce, P., & Bernardi, N. (2022). From sustainability to thrivability: A novel framework for entrepreneurial ecosystems. *International Entrepreneurship and Management Journal, 18*(2), 829–853. https://doi.org/10.1007/s11365-021-00787-x

Muhammad Muneeb, F., Karbassi Yazdi, A., Wanke, P., Yiyin, C., & Chughtai, M. (2020). Critical success factors for sustainable entrepreneurship in Pakistani Telecommunications industry: A hybrid grey systems theory/ best-worst method approach. *Management Decision, 58*(11), 2565–2591. https://doi.org/10.1108/MD-08-2019-1133

Neumann, T. (2021). The impact of entrepreneurship on economic, social and environmental welfare and its determinants: A systematic review. In *Management review quarterly* (Vol. 71, Issue 3). Springer. https://doi.org/10.1007/s11301-020-00193-7

Paletta, A., Foschi, E., Alimehmeti, G., & Bonoli, A. (2021). A step-by-step process towards an evolutionary policy encouraging the adoption of sustainable business models. *Sustainability (Switzerland), 13*(3), 1–17. https://doi.org/10.3390/su13031176

Raman, R., Subramaniam, N., Nair, V. K., Shivdas, A., Achuthan, K., & Nedungadi, P. (2022). Women entrepreneurship and sustainable development: Bibliometric analysis and emerging research trends. *Sustainability, 14*(15), 9160. https://doi.org/10.3390/su14159160

Ribeiro, G. K. A., & Leitão, J. (2024). Sustainable leadership and sustainable entrepreneurship: A systematic literature review. In *Discover sustainability* (Vol. 5, Issue 1). Springer. https://doi.org/10.1007/s43621-024-00383-3

Roundy, P. T., & Fayard, D. (2019). Dynamic capabilities and entrepreneurial ecosystems: The microfoundations of regional entrepreneurship. *The Journal of Entrepreneurship, 28*(1), 94–120.

Sajuyigbe, A. S., & Fadeyibi, I. O. (2017). Women entrepreneurship and sustainable economic development: Evidence from South Western Nigeria. *Journal of Entrepreneurship, Business and Economics*, *5*(2), 19–46. www.scientificia.com

Sarasvathy, S. D. (2000). Seminar on research perspectives in entrepreneurship (1997). *Journal of Business Venturing, 15*(1), 1–57.

Schumpeter, J. A. (1934). *The theory of economic development*. Harvard University Press.

Schumpeter, J. A. (1942). *Capitalism, socialism and democracy* (Vol. 13). Harper and Brother.

Shepherd, D. A., & Patzelt, H. (2011). The new field of sustainable entrepreneurship: Studying entrepreneurial action linking "what is to be sustained" with "what is to be developed." *Entrepreneurship: Theory and Practice*, *35*(1), 137–163. https://doi.org/10.1111/j.1540-6520.2010.00426.x

Smith, A. (1950). *An inquiry into the nature and causes of the wealth of nations, (1776)*. Methuen Publication.

Spencer, G. M., Vinodrai, T., Gertler, M. S., & Wolfe, D. A. (2010). Do clusters make a difference? Defining and assessing their economic performance. *Regional Studies, 44*(6), 697–715. https://doi.org/10.1080/00343400903107736

Stevenson, H. H., & Jarillo, J. C. (1990). A paradigm of entrepreneurship: Entrepreneurial management. *Strategic Management Journal, 11*(special issue), 17–27.

Tietenberg, T. H., & Lewis, L. (2012). *Environmental & natural resource economics*. Pearson series in economics (9th ed.). Pearson Education.

Timmons, J. A., & Spinelli, S., Jr. (2004). New venture creation: Entrepreneurship for the 21st century. *Language, 30*(700p), 28cm.

United Nations. (2014). Resolutions: *Female entrepreneurs are key to sustainable global development*. United Nations Foundation. http://www.unfoundation.org/blog/female-entrepreneurs.html on March 2017

Watson, R., Nielsen, K. R., Wilson, H. N., Macdonald, E. K., Mera, C., & Reisch, L. (2023). Policy for sustainable entrepreneurship: A crowdsourced framework. *Journal of Cleaner Production, 383*, 135234. https://doi.org/10.1016/j.jclepro.2022.135234

Yaşlıoğlu, M. M., & Toplu Yaşlıoğlu, D. (2020). Entrepreneurship for the future: A conceptual look toward sustainability entrepreneurship. In *Eurasian studies in business and economics* (Vol. 13, Issue 2, pp. 277–290). https://doi.org/10.1007/978-3-030-40160-3_20

Zhu, L., & Wang, Y. (2024). Entrepreneurship and carbon footprints in sub-Saharan Africa. *Problemy Ekorozwoju, 19*(1), 221–231. https://doi.org/10.35784/preko.5759

Enhancing Sustainable Entrepreneurship with Sustainable Supply Chain Strategies

Mahtab Iltarabian

Abstract

Sustainable entrepreneurship is becoming a cornerstone of long-term success in today's rapidly evolving corporate landscape, fostering creativity and ethical development through responsible growth. Emphasizing the harmony between social responsibility, environmental stewardship, and economic success, this chapter examines how entrepreneurial operations can utilize sustainable supply chain methods. Beginning with an introduction to an ecological supply chain, the book defines sustainability within an entrepreneurial context and emphasizes the need to align corporate activities with ethical resource management. We review fundamental management concepts, including strategic material acquisition, ethical labor practices, and the application of open and responsible procedures. The chapter examines how to incorporate sustainability into current business operations, the significant impact of technology on resource flow analysis and emissions reduction, and the benefits of forming close relationships. It also provides methods grounded in consistent key performance indicators and external audits to evaluate and explain its influence. Through real-world case studies and exemplary practices, the chapter presents pragmatic applications, challenges, and success stories, providing actionable insights and strategies for businesses striving for innovation and sustainable development in a competitive global market.

Keywords

Sustainable entrepreneurship · Supply chain management · Environmental stewardship · Ethical business practices · Innovation and technology in sustainability · Triple bottom line

Learning Objectives

By the end of this chapter, readers will be able to:

- Identify the importance of sustainability in entrepreneurial settings.
- Analyze sustainable supply chain principles and their economic, environmental, and social impacts.
- Examine the concepts of sustainable supply chains in terms of their social, environmental, and financial impacts.
- Review plans for incorporating sustainability into current corporate operations.
- Utilize cooperation and technology to enhance sustainability.
- Create and assess the influence of key performance criteria.

M. Iltarabian (✉)
Department of Computer Engineering, N.T.C., Islamic Azad University, Tehran, Iran
e-mail: m.iltarabian@iau.ir

M. Akbari et al. (eds.), *Sustainable Entrepreneurship*, University of Tehran Science and Humanities Series, https://doi.org/10.1007/978-3-032-04909-4_10

1 Introduction

In an era of rapid technological development and growing environmental concerns, sustainable entrepreneurship has become increasingly important for companies seeking to create long-lasting value (Vares et al., 2024). Fundamentally, sustainable entrepreneurship combines social responsibility with economic success to surpass conventional profit-driven strategies, including environmental stewardship (Gerlich et al., 2025). This all-encompassing strategy not only solves urgent global issues such as social inequalities, resource constraints, and climate change but also provides creative means of competitive differentiation and long-term development (Frank & Lucky, 2025).

Strategic supply chain management is fundamental for this metamorphosis (Bokareva & Avdalyan, 2020). Designed with sustainability in mind, modern supply networks have become strong instruments that promote resilience, efficiency, and openness (Rashid et al., 2025). They help companies maximize resource usage, lower waste, and ensure moral behavior throughout the value chain (Garrido & Nunes 2024). Entrepreneurs can build closer ties with suppliers, create cooperative networks, and ultimately improve their capacity to respond to environmental (Wu et al., 2021) and market volatility by aligning supply chain operations with sustainable values (Boscari et al., 2024).

This chapter examines the symbiotic relationship between supply chains and sustainable entrepreneurship. It examines how incorporating sustainability into the supply chain not only supports company innovation but also generates a competitive edge in today's interconnected marketplace. We will explore fundamental ideas, key management techniques, and practical applications that demonstrate how intelligent procurement, ethical work standards, and cutting-edge technologies can transform conventional supply chains into vibrant engines of sustainable development. Furthermore, the chapter will present actual case studies illustrating how forward-looking companies have effectively navigated the challenges of this change (Ruhana et al., 2024).

The need for sustainable practices becomes increasingly important as global markets evolve. Those who use sustainable supply chain methods are not only better suited to control risks but also to capitalize on the opportunities presented by a consumer base growing in environmental consciousness and a regulatory environment changing in this regard (Onukwulu et al., 2025). This introduction invites readers to rethink how sustainability can be integrated into every aspect of entrepreneurial operations, thereby ensuring long-term success and a positive impact on society and the environment. This sets the scene for a thorough investigation of sustainable supply chains.

2 Sustainable Supply Chain Strategies: Integration, Management, and Innovation

2.1 Ecological Supply Chain Overview

The ecological supply chain is becoming increasingly important in today's ever-changing corporate environment, promoting sustainable entrepreneurship (Le, 2022). This section lays the groundwork for understanding how incorporating environmental factors into supply chain management can yield transformative results for companies, society, and the planet (Shekarian et al., 2022).

2.1.1 Sustainability's Meaning in the Entrepreneurial Context

In an entrepreneurial setting, sustainability extends beyond the conventional concept of profit maximization to encompass a broader vision, including long-term economic viability, social responsibility, and environmental conservation (Rosário et al., 2022). Today's entrepreneurs must be creative yet also aware of resource constraints, environmental changes, and social impacts (Veleva, 2021). From this comprehensive perspective, sustainability is not only a fundamental concept that influences corporate strategy and decision-making but also an operational benefit (Ayandibu, 2024).

In essence, sustainable entrepreneurship demands a balance between generating financial value and ensuring that company operations do not compromise the social or environmental fabric of nations (Elmonshid & Sayed, 2024). This means putting plans aimed at reducing waste, minimizing energy consumption, and promoting recycling and the use of renewable resources into effect. It also involves creating value propositions that appeal to a client base expanding in environmental awareness, thereby balancing market success with responsible environmental preservation (Gomes, 2024).

2.1.2 Value of Correspondent Social, Environmental, and Economic Objectives

Harmonic integration of social, environmental, and economic goals produces strong, sustainable supply chains—a trio commonly referred to as the "triple bottom line. Achieving congruence among these objectives is crucial, as it ensures that social injustice or environmental damage is not compromised to guarantee corporate growth (Suteu et al., 2017).

Economically, by better-utilizing resources and reducing waste, sustainability projects can generate cost savings; simultaneously, they create new markets and strengthen brand reputation (Odeyemi et al., 2024). Environmentally, the key goals are to preserve ecological balance, reduce the carbon footprint, and thereby safeguard natural resources (Yang & Solangi, 2024). Socially, sustainable living promotes equal access to advantages, fair employment conditions, and community involvement. When these three pillars align, businesses can cultivate resilience and creativity, enabling them to demonstrate ethical leadership in the modern market.

2.1.3 Differentiation of Sustainable Supply Chains: Characteristics of Contemporary Corporate Structures

Sustainable supply chains set themselves apart by incorporating environmentally friendly and socially conscious methods into every level of the value chain (Feng et al., 2024). Unlike conventional supply chains that prioritize speed and cost efficiencies, sustainable supply chains emphasize a comprehensive strategy that incorporates the environmental and social implications of sourcing, manufacturing, transportation, and waste management.

Given the competitive landscape of today, sustainable supply chains are increasingly important. Including sustainability in their supply chain policies not only helps businesses reduce regulatory demands and resource limits but also fosters closer, trusted relationships with their consumers (Hwang, 2024). Forward-looking companies that aim to strike a balance between profitability and a commitment to long-term sustainability typically identify themselves on this basis (Jindal, 2025). Therefore, sustainable supply chains—symbolizing ingenuity, social responsibility, and a proactive commitment to building a more resilient and equitable global economy—have come to reflect the face of modern corporate organizations.

These ideas, taken together, highlight the importance that ecological problems play in redefining supply chains for sustainable business, thereby laying the foundation for more in-depth discussions on management ideas, technical enablers, and cooperative strategies in future parts.

2.2 Principles of Management for the Sustainable Supply Chain

Good management of sustainable supply chains relies on a set of guiding concepts that incorporate social, economic, and environmental elements into every aspect of activity (Jaboob et al., 2024). Using these concepts enables businesses to ensure that their supply chains are ethical, robust, and compliant with their long-term sustainability goals.

2.2.1 Wise Material Choices and Source

Most of a sustainable supply chain is built on reasonable material purchases. This concept emphasizes the need to choose goods not only for their low cost but also for their ethical production and environmental friendliness (Rahman et al., 2024). Sustainable sourcing involves evaluating the environmental impact of raw materials; therefore, companies should prefer renewable resources and recycled inputs whenever possible to help reduce resource depletion and waste (Silva & Figueiredo, 2022).

Buying products from neighboring companies ensures that local environmental needs are met, reduces transportation emissions, and supports local businesses through ethical and local procurement.

Clearly defining standards for evaluating vendors, including their waste management, environmental preservation, and resource management practices, ensures that only those following sustainability goals participate (Oyeyemi et al., 2024).

By replacing greener alternatives with non-renewable or environmentally damaging resources, encouraging innovation in the production of alternative materials will help drive long-term sustainability.

This approach not only reduces environmental impact but also establishes a robust supply chain that can adapt to legal changes and shifting resource availability.

2.2.2 Ethical Policies and Fair Trade Laws

Not only in terms of resources but also regarding ethical and fair interactions throughout the entire trading process in sustainable supply chains (Busch & Mühlrath, 2023). Two main elements that ensure every actor in the supply chain is treated with respect and fairness are fair trade policies and ethical working standards, thereby supporting social sustainability. Important components include:

Ensuring that labor practices match or exceed local and international standards helps guarantee workers' well-being, thereby boosting their loyalty and motivation (Fritz & Silva, 2024).

Establishing contracts with open terms and clear expectations that protect the rights of all participants helps build confidence and reduces the likelihood of exploitation. Through collaborations between inclusive suppliers, establishing ties with suppliers who prioritize ethical behavior helps create a more equitable economic environment (Tian et al., 2024). Regular audits and assessments help ensure adherence to ethical norms, thereby guaranteeing continuous development in fair business operations and working conditions. Including ethical concerns in every stage of the supply chain helps businesses not only mitigate risk but also enhance their brand in increasingly ethical markets.

2.2.3 Traceability, Transparency, and Responsibility

Building trust among stakeholders in sustainable supply chains fundamentally depends on traceability, accountability, transparency, and integrity (Holloway, 2024). These concepts guarantee that every activity and decision is adequately transparent for thorough investigation, hence improving the overall integrity of the supply chain system. Transparent reporting systems, sophisticated tracking tools, and independent validation of sustainable practices are essential elements.

Transparent reporting systems require consistent disclosure of supply chain practices, enabling stakeholders to understand the social and environmental impacts of the company's actions. Each entity in the supply chain is accountable for its operations through clearly defined duties and performance metrics, hence promoting continuous development and enhancement (Budler et al., 2023).

Sophisticated tracking technologies, including blockchain, RFID, and data analytics, provide real-time insights into material movement and source, thereby authenticating each action carried out (Yang et al., 2024). Third-party validation ensures that certifying organizations or independent auditors evaluate sustainability claims, thereby strengthening credibility and confidence among stakeholders (Hastig & Sodhi, 2022).

Collectively, these components enable firms to maintain regulatory compliance, manage risks, and cultivate strong relationships with consumers and partners. Implementing these management concepts can enhance supply chains, transforming them into robust, transparent, and environmentally sustainable systems that promote long-term sustainable development in an era where consumers and investors demand ethical and sustainable practices.

2.3 Integration of Sustainability Within Current Activities

If sustainable entrepreneurship is to flourish, ordinary company operations must incorporate sustainability into their core (Kılıç & Atilla, 2024). Through a thorough review and modification of existing processes, this integration ensures that every aspect, from suppliers to distribution routes, aligns with broader sustainability goals. The sections that follow provide a rigorous approach to incorporating sustainability into current activities.

2.3.1 Analyzing Distribution Channels and Current Suppliers

Beginning with a thorough analysis of current suppliers and channels of distribution, the road towards a sustainable supply chain is one whereby Beginning with evaluative assessments, in which companies carefully review their present vendors to determine adherence to ethical and environmental standards, this process moves forward Using sustainability scorecards, audits, and self-assessment questionnaires helps one to have a whole view of supplier performance (Changalima, 2024). Concurrent with this is a review of distribution channels to ensure that packaging, shipping, and logistics are optimized for energy efficiency and have the least impact on the environment (Segura et al., 2020). This stage also entails identifying hazards by pointing out places where present practices fall short of accepted sustainability criteria, including the use of ecologically unfriendly manufacturing techniques or excessively high carbon emissions from transportation. Moreover, maintaining open lines of contact with distributors and suppliers generates opportunities for cooperative development. Through this thorough screening process, businesses can identify areas of concern and establish the foundation for a more environmentally responsible operating plan.

2.3.2 Examine the Levers and Create Needs

Following the initial screening, an analysis is conducted to identify the differences between current procedures and planned sustainability. Through this gap analysis, the current state of operations is mapped against industry best practices and sustainability criteria, revealing areas of deficiency in resource use, waste management, and labor practices (Galleli et al., 2021). Once these gaps are identified, it becomes crucial to prioritize requests by determining which ones are most critical to address to achieve strategic sustainability goals. This may require upgrading technology, enhancing supplier involvement, or supporting process innovations to achieve this goal. Including employees, suppliers, consumers, and community members in this study helps engage key stakeholders to provide insightful analysis of both current issues and possible development areas. Moreover, bridging the gap between current practices and sustainable goals depends on a precise understanding of the expenditures required for infrastructure, technology, or training. Through rigorous analysis of these gaps and the definition of developmental needs, businesses can establish a focused roadmap that prioritizes interventions and prepares the ground for long-term environmental changes.

2.3.3 Adoptive Criteria and Environmental Standards Establishment

The penultimate phase in incorporating sustainability into current operations is setting strong environmental standards and adaptable criteria that serve as benchmarks for continuous improvement. This phase aligns with industry best practices and legal requirements, starting with fully defined, quantified, and legally enforceable criteria governing every step of the supply chain—from material procurement to final delivery (Inyang et al., 2023). Establishing key performance indicators (KPIs) to track development toward sustainability goals is vitally crucial, as these programs enable companies to analyze improvements in areas such as ethical labor standards, waste reduction, and energy efficiency (Hristov & Chirico, 2019). As crucial is the building of a dynamic structure that enables regular reviews and adjustments, thereby guaranteeing the relevance of sustainability standards in the face of changing industry trends and technological improvements. Companies should establish incentive programs that reward suppliers and partners who consistently meet or exceed stated criteria, fostering a culture of responsibility and ongoing development. Including certifying bodies or independent auditors for outside certification of compliance with these criteria also provides stakeholders with external validation, which strengthens confidence and reputation. Establishing clear environmental standards and adopting criteria helps companies build a strong framework for sustainability that not only reduces environmental hazards but also enables them to achieve leadership in the sustainable growth of specific sectors.

Including sustainability in present operations is a dynamic and continuous process needing proactive analysis, evaluation, and improvement of every aspect of the supply chain. Businesses can embed sustainability into their operations by methodically assessing current processes, spotting areas of weakness, and defining clear standards, therefore opening the path for strong and ethical development (Nguyen & Kanbach, 2024).

2.4 Enabling Technology to Meet Advanced Sustainability

In the context of sustainable supply chains, technology plays a crucial role in transforming traditional corporate processes into systems that are not only efficient but also transparent and environmentally friendly. Modern technological solutions enable businesses to access the tools and data necessary to make informed decisions in line with environmental policies. Technology has been the primary driving force behind this transformation, as businesses strive to maximize operational efficiency, thereby reducing their environmental impact. In this section, we investigate how modern technologies enable a detailed assessment of resource flows, drastically reduce emissions and waste, and simplify logistics by utilizing energy-efficient transportation. These changes are laying the road toward a time when every component of the supply chain supports a robust, ecologically conscious business (Saberi et al., 2018).

2.4.1 Data Analysis Resource Flows

Providing an unparalleled understanding of resource flows, data analytics has rapidly become the core of sustainable practices in supply chain management. Companies today monitor, evaluate, and optimize resource use through advanced analytics technologies applied across all phases of their operations. Real-time data on waste generation, material use, and energy consumption enables businesses to identify inefficiencies they might have otherwise overlooked. Through rigorous analysis of energy consumption patterns, managers can identify locations where energy is being lost, thereby enabling focused improvements that generate significant cost savings and reduce environmental impact (Sumon Gazi, 2024).

Predictive analytics helps businesses to forecast future resource needs, thereby improving this process even further. This foresight ensures the best use of resources and helps reduce surplus waste by enabling adjustments in procurement and manufacturing techniques. Forecasting demand and trends not only helps to reduce running disruptions but also enhances the complete supply chain. Moreover, firms may receive valuable information to inform strategic decisions through advanced data analysis. Comprehensive environmental statistic reporting increases transparency, allowing stakeholders—who are increasingly seeking clear evidence of a company's sustainability commitment—to gain confidence in the company's commitment. Ultimately, the intentional application of data analytics transforms unprocessed data into a powerful tool for continuous improvement, therefore guiding companies toward long-term sustainable success.

2.4.2 Emission and Waste Reduction Due to Green Technologies

Green technologies are at the forefront of efforts to reduce the environmental impact of supply chains, thereby directly addressing issues of waste and emissions. By incorporating solar and wind energy sources into manufacturing and logistical processes, companies can significantly reduce their reliance on fossil fuels and thereby lower greenhouse gas emissions. Moving to renewable energy not only improves the environment but also offers long-term financial savings as energy markets evolve. Apart from using renewable energy, companies are also rapidly replacing existing machinery with energy-efficient versions. Modern equipment and lighting solutions help reduce running costs and a smaller carbon footprint by using less energy while maintaining high production levels (Tambuskar et al., 2024).

Waste management innovations highlight even further the role green technology plays. Waste-to-energy systems and advanced recycling techniques are transforming formerly discarded products into valuable resources. These technical fixes address the twin problems of reducing landfill reliance and opening new revenue-generating paths. One additional important advancement in this field is smart manufacturing. Automated control systems and Internet of Things (IoT) devices enable exact, real-time monitoring of manufacturing processes. This technical synergy ensures that resource use is maximized and that any deviation from established criteria is promptly remedied, thereby reducing waste formation. These green technologies, taken together, offer a strong framework that not only addresses urgent environmental issues but also stimulates operational efficiencies, thereby integrating sustainability into the corporate strategy.

2.4.3 Simplifying Logistics with Energy-Efficient Transportation

Logistics is a fundamental component of supply chain management and, therefore, crucial for sustainability projects, particularly in terms of embracing alternative transportation options. Modern technology developments have transformed traditional logistics methods by focusing on reducing the carbon footprint of transportation systems and enhancing operational effectiveness. Among the most important changes in this field are fleet optimization. Companies can now select the most effective travel routes by utilizing advanced route optimization software and telematics systems, thereby reducing fuel consumption and emissions. These systems constantly evaluate traffic patterns, road conditions, and vehicle performance to adjust routes rapidly, ensuring deliveries are made in the most fuel-efficient way possible.

Another essential element in simplifying transportation is switching to alternate-fuel vehicles. Since they significantly reduce carbon emissions and the environmental impact of transportation activities, electric and hybrid cars have become increasingly popular as viable alternatives to conventional fossil fuel-powered fleets. These vehicles help create a more sustainable logistics system due to their seamless operation. Apart from technical developments in vehicles, real-time tracking systems, such as IoT sensors and GPS, provide constant monitoring of fleet performance. Through dynamic changes to delivery plans and courses made possible by real-time data, fuel economy helps improve overall service reliability (Balinado et al., 2024).

Moreover, the concept of multimodal transportation is becoming increasingly important as businesses seek to integrate road, sea, and rail to provide the most eco-friendly and effective logistical solutions. Through a thorough investigation of the environmental impact and efficiency of every mode of transportation, companies can create a logistics plan that leverages the advantages of each method, ensuring the best possible combination of resources for product delivery. Taken together, these energy-efficient transportation options significantly reduce the carbon footprint of logistics operations, thereby allowing for more stringent sustainability standards to be met, as well as improved operational performance and cost savings. Taken collectively, these energy-efficient transportation choices significantly reduce the carbon footprint of logistics operations, thereby enabling more stringent sustainability standards to be met, as well as operational performance and cost savings.

Combining enabling technologies in resource analytics, green innovations, and efficient logistics is significantly altering the supply chain environment in general. These changes enable companies to reduce their environmental impact while also increasing operational agility, openness, and responsibility. Companies that continue to utilize these technologies pave the way for a sustainable future where environmental stewardship and efficiency coexist, benefiting both the world and the bottom line.

2.5 Developing Alliances Through Collaboration

The search for sustainable supply chains cannot be achieved in isolation within the interconnected global economy of today. Environmental damage, resource constraints, and changing regulatory environments present challenges that require companies to extend their reach beyond their own operations and establish strong relationships with various stakeholders. Through multiple capacities, cooperative projects enable companies to apply different skills, disseminate technical innovations, and integrate strategic objectives with broader sustainability issues. Through these relationships, businesses can build solid networks covering not just present operational difficulties but also encourage moral corporate behavior and long-term growth. In sustainability, cooperation is primarily about developing a shared vision, where suppliers, local communities, and industry peers collaborate to drive systemic transformation that benefits the entire value chain. This part examines several aspects of cooperation and demonstrates how industry-wide cooperation, close ties, and community involvement can transform supply chains into engines of sustainable development (Yakovleva et al., 2019).

2.5.1 More Specifically Associated Vendors Due to Shared Values

Developing a sustainable supply chain mostly depends on establishing strong, long-term partnerships with suppliers who promise ethical behavior and environmental responsibility. Common values unite companies and their suppliers, thereby enhancing the entire supply chain and enabling more open, flexible, and rapid responses to new challenges. Not simply transactional but also based on mutual trust and a tireless desire for continuous improvement, this cooperative link is built on. Under these kinds of partnerships, both sides are encouraged to be innovative and implement ecologically friendly solutions transcending mere legal compliance. Those who follow sustainable practices, for example, can invest in certificates verifying their waste-reducing techniques, cleaner production methods, or environmental commitments. These projects help the purchasing firm fulfill its objectives and improve the overall level of supply chain sustainability.

Moreover, sometimes, the collaboration of like-minded companies transcends joint efforts to solve problems when both sides work together on research and development projects aimed at producing new materials or technologies that lower environmental impact. These strategic relationships enable the pooling of resources, risk-sharing, and the development of solutions tailored to specific industry issues. Suppliers are more eager to provide important information about their operations as confidence increases and lines of communication get more simplified. This will help to forecast better, manage quality, and finally provide a more responsive supply chain. Such coordinated projects generate a robust network in which every participant not only answers for their sustainability performance but also actively supports the supply chain to be generally healthy and efficient (Nnaji et al., 2024).

2.5.2 Community Involvement and Representation

Not only within corporate boardrooms and production sites but also within local communities, sustainable supply chains influence the basic fabric of society. Interacting with community members and their representatives can help ensure that sustainability initiatives are not only technically successful but also socially fair and culturally appropriate. Engaging local stakeholders helps companies gain a deeper understanding of regional issues and opportunities that might otherwise be overlooked. Community involvement provides a platform for communication, allowing people to share traditional knowledge, express their concerns, and collaborate with enterprises to co-create solutions addressing local environmental issues.

More than just planned meetings, this type of involvement is an ongoing commitment to cultivating lasting relationships with local businesses, leaders, and community members. These kinds of encounters contribute to providing a "social license to operate," which is increasingly important in the context of today's growing social consciousness and government monitoring. On the road toward a brighter future, active participants in sustainability planning often become friends. Their participation ensures that sustainability projects begin with local reality, thereby enhancing their importance and impact. Moreover, businesses create social capital, which enables programs such as job creation, infrastructure development, or environmental preservation to commit themselves to the welfare of surrounding communities, thereby fostering greater community loyalty and support. The involvement of local communities ultimately transforms the supply chain into a dynamic ecosystem where social and environmental goals are pursued in tandem, thereby fostering a spirit of shared responsibility and commonwealth (Kamenopoulos & Agioutantis, 2021).

2.5.3 Industry-Wide Cooperation Proportional to Enhanced System Management

Effective, sustainable supply chains are those that evolve into a cooperative industry-wide movement spanning more than just particular corporate initiatives. Many businesses find that no single company can handle the complexities of sustainability on its own; the challenges are too great and multifaceted. Here, industry-wide collaboration is important since it provides a structure wherein businesses may pool resources, exchange best practices, and jointly manage the changing terrain of environmental management and market expectations.

When companies within a sector band together, they can develop best practices that benefit each other. Typically, this type of group effort yields shared technologies, harmonized regulations that level the playing field, and coordinated research and development projects. This partnership also aims to influence legislators and regulatory authorities by providing a consistent voice in support of sustainable practices. A coordinated industry effort is significantly strengthened when companies share not only their successes but also their challenges and lessons learned. This free flow of knowledge encourages a transparent and always developing culture whereby others may quickly welcome discoveries in one area of the company, therefore accelerating the general development (Ariputro et al., 2024).

Group projects also enable companies to manage risks better. By spreading the load, the distribution of research, development, and regulatory compliance enables industry players to reduce the financial and operational risks associated with transitioning to more sustainable techniques. This risk-sharing arrangement is particularly crucial in unstable markets, where rapid changes could otherwise bring about the collapse of some companies. Apart from improved operational savings, the benefits of such cooperation include a more robust supply chain and increased market competitiveness. Through industry-wide cooperation, companies can not only address current sustainability challenges but also generate long-term economic, environmental, and social benefits. This will guide the systematic transformation.

In the field of sustainable supply chain management, building alliances via cooperation is ultimately a must-have tactic. Robust partnerships founded on shared values, including meaningful communication with local communities and supporting industry-wide cooperation, help companies build a comprehensive and robust framework that propels sustainable development through suppliers. These cooperative efforts not only address the numerous challenges of today's global market but also pave the way for a future where sustainability is integrated into every aspect of supply chain activities. By means of strategic alliances, businesses can leverage collective strengths, achieve greater success, and attain operational excellence, benefiting not only the company but also society and the environment as a whole.

2.6 Evaluating and Reporting Influence

Not only is it an operational need, but also a basic pillar in converting supply chains into transparent and efficient systems is measuring and sharing the impact of sustainability projects. Companies starting this road must have a strong structure that closely monitors development using well-defined criteria and simultaneously presents these results in a way that inspires trust and confidence among all stakeholders. From an abstract concept to a series of measurable successes routinely observed, enhanced, and recognized, this all-encompassing strategy transforms sustainability.

The foundation of this strategy is the development of key performance indicators (KPIs) closely aligned with the objectives of corporate sustainability. These KPIs help identify areas of great importance, such as moral work ethics, energy efficiency, waste reduction, and water conservation. Every KPI serves as a measurable benchmark, providing a methodical strategy for tracking progress over time. A company may want to reduce energy use by a specific percentage over a given period or lower waste output through improved recycling and process optimization. By tracking these metrics, companies can identify areas of inefficiency, uncover patterns, and make data-driven decisions that direct ongoing development. Methodically, KPIs integrate lofty sustainability goals into measurable results that not only track development but also guide strategic changes and forthcoming projects (Guerrero & Hansen, 2018).

Once these performance criteria are in place, the next phase involves thorough sustainability reporting combined with external auditing. Businesses structure their sustainability data using accepted reporting systems, including the Sustainability Accounting Standards Board (SASB) rules or the Global Reporting Initiative (GRI). These techniques ensure that, over many years and in accordance with industry criteria, the given material remains consistent, transparent, and comparable. Third-party audits value these records still much more. Independent validation by outside auditors provides an independent assessment of the claimed facts and reveals probable internal team mistakes or inconsistencies. Apart from validating the company's claims on sustainability, this extensive outside review process enhances its dedication to responsibility and transparency. By applying such strategies, businesses can demonstrate that their efforts toward sustainability are ongoing and effective.

Open communication of these sustainability results helps build and preserve confidence among various stakeholders, including the public, consumers, and investors. Interactive dashboards, thorough sustainability reports, and frequent progress updates provide stakeholders with a clear understanding of the business's current state. Organizations foster an honest and continually improving atmosphere by openly sharing both successes and challenges. This degree of openness helps stakeholders understand the actual impact of environmental projects, thereby promoting positive feedback and collaborative solutions to problems. Furthermore, by consistently sharing their growth, companies not only demonstrate their commitment to ethical business practices but also help establish themselves as leaders in the broader movement for social and environmental responsibility (Zaunbrecher & O'Donnell, 2020).

Apart from basic reporting, the process of measuring and distributing sustainability impacts drives ongoing internal improvement. Regular performance assessments enable companies to identify areas where their policies are either functioning effectively or where further development is needed. This iterative method ensures that sustainability remains a fundamental factor in the strategic development of the business, thereby motivating ideas that reduce environmental impact and enhance overall operational effectiveness. Incorporating these ideas into the corporate culture enables companies to be more adaptable in addressing new challenges and seizing emerging opportunities in a world that is increasingly environmentally conscious (Nayal et al., 2022).

An efficient sustainability strategy is formed mainly by the combination of strong key performance indicators (KPIs), thorough reporting systems, and open channels of communication. These approaches ensure that sustainability initiatives are not only executed correctly but also validated and presented in a manner that fosters long-standing stakeholder confidence. Businesses that monitor their growth and freely share their triumphs help build long-term organizational resilience, operational excellence, and a positive environmental and social impact. Sustainable supply chain strategies become a dynamic agent for transformation through an all-encompassing approach, thereby promoting continuous development and setting new benchmarks for ethical and responsible business behavior.

2.7 Case Studies and Exemplars of Best Practices

Through a collection of case studies and examples that illustrate both the successes and challenges faced by companies in their pursuit of sustainability, this part exemplifies the pragmatic application of sustainable supply chain methods. Analyzing various organizational experiences reveals effective approaches, creative ideas implemented by corporate leaders, and typical challenges during the transformation.

2.7.1 Businesses Effectively Using Sustainable Supply Chains

Many companies have successfully integrated sustainability into their supply chain activities, employing strategies that reduce environmental impact and enhance operational performance and competitive edge. A notable example is a global consumer products corporation that has reformed its sourcing strategy to prioritize renewable materials and energy-efficient production techniques. The company significantly reduced waste and energy consumption through supplier cooperation, which helped synchronize environmental goals and the implementation of advanced monitoring systems, thereby strengthening its brand reputation among consumers who place great importance on the environment. A global electronics company optimized resource allocation and reduced transportation emissions through real-time monitoring and data analytics, thereby enhancing logistics. These examples highlight how a commitment to sustainable practices can transform conventional supply chain models into dynamic, robust systems that advance both company and environmental objectives.

2.7.2 Perspectives of Innovators and Industry Leaders

Creative corporate executives and industry pioneers offer essential insights into the strategic adjustments required to establish sustainable supply chains. These executives argue that, rather than implementing new technologies or processes, the search for sustainability demands a review of fundamental corporate values. From the top down, they advocate for a comprehensive approach that incorporates sustainability into corporate governance, thereby encouraging openness, innovation, and cooperation throughout the company (Jia et al., 2019). Their narratives demonstrate that success often depends on creating a culture that prioritizes long-term environmental sustainability alongside short-term financial performance. These leaders provide a foundation for other businesses trying to strike a balance between profitability and the pressing need for environmental stewardship by spreading their challenges, ideas, and best practices.

2.7.3 Handling Common Transition Difficulties

Adopting sustainable supply chain techniques means several challenges, including significant initial expenditures in new technologies and resistance from ingrained organizational cultures. Many businesses struggle to move beyond operational stagnation and align with the diverse interests of various stakeholders. Businesses may find it challenging to integrate contemporary monitoring tools into outdated systems or control supplier delays in adopting sustainable practices, which can cause additional delays. Still, the case studies indicate that delayed adoption and targeted training programs, coupled with strategic alliances, foster shared responsibility and inventiveness, helping to tackle these problems. Starting pilot projects typically means that companies possess the necessary knowledge to disseminate successful initiatives throughout their entire supply chain, thereby gradually building a more sustainable, open, and robust system. The lessons learned from these transient difficulties underscore the need for tenacity, adaptability, and the constant development of sustainable methods as essential components of long-term success.

These case studies and examples demonstrate that even if the road to sustainable supply chains is challenging, creative thinking, effective group leadership, and a commitment to ongoing improvement can help illustrate the growing possibilities of sustainable practices and provide a roadmap for other companies to initiate their environmental efforts.

2.8 Sustaining and Increasing Velocity

The norms of any one corporation have little bearing on moral business practices. If sustainability programs are to have a truly revolutionary impact, they must reach beyond to influence the broader entrepreneurial sector. Encouraging broader change involves enlisting a diverse range of participants—from startups and established companies to politicians and community organizations—to create an environment where sustainable practices become the standard rather than the exception.

Establishing cooperative forums with major participants from various areas is one of the most effective approaches to support ongoing development. Through cooperative projects, idea sharing, and co-creation of solutions to shared concerns, participants in these forums can be creative catalysts. Public–private alliances, for instance, can help fund projects such as the development of sustainable urban design or green energy networks. By way of information sharing and resource pooling, these alliances could have broad consequences that are detrimental to independent businesses on their own.

Apart from official alliances, advocating more general change calls for proactive involvement with legislators and regulatory bodies. Companies who closely collaborate with governments to set rules supporting sustainable practices contribute to creating an environment that stimulates more innovation and investment in sustainability. This process depends critically on advocacy campaigns, industry forums, and think tanks, as they offer a collective voice capable of influencing public policy and, therefore, guiding regulatory changes. By means of consistent communication and cooperation, companies can contribute to creating frameworks and standards that enable sustainability to be naturally included in the economic scene (Du et al., 2024).

The encouragement of institutional change also depends entirely on community involvement. Often, front-stage concerns in sustainability are local communities; their knowledge and experience can be quite helpful in creating context-specific solutions. Involving local leaders and community organizations helps businesses ensure that their environmental projects are not only successful but also inclusive and equitable. Among other community involvement activities, local sustainability seminars, public forums, and corporate social responsibility initiatives help to foster trust and safeguard the social license for business. Communities that feel appreciated and involved are more inclined to support ongoing environmental and social initiatives.

Moreover, incorporating sustainable practices into the entrepreneurial ecosystem requires a shift in perspective throughout the entire corporate community. Education and awareness promote this cultural change, as successful stories and best practices are widely shared to motivate others. Industry conferences, case study books, and sustainability awards are among the tools that can magnify one person's achievements and set off a chain reaction, inspiring other companies to do likewise. As more businesses implement sustainable practices and demonstrate their clear advantages—such as lower operating costs, a better brand reputation, and increased regulatory compliance—a larger movement emerges, thereby supporting the value of sustainability everywhere (Ahlawat & Yaseen, 2024).

Encouragement of more general change also depends on utilizing social media and digital platforms to disseminate knowledge and coordinate support. Digital communication tools can quickly amplify sustainable projects in the interconnected world of today, reaching a worldwide audience and facilitating group problem-solving. Virtual networks and online communities provide venues where business owners can share ideas, receive feedback, and collaborate on projects aimed at systemic change. This digital integration not only accelerates creativity but also democratizes knowledge about sustainability, therefore enabling even small and medium-sized companies to join the global sustainability movement.

Ultimately, encouraging broader transformation within the entrepreneurial ecosystem is about creating a future where every aspect of economic activity incorporates environmentally friendly solutions. Beyond a single project, this comprehensive approach fosters a lasting legacy of social responsibility, environmental protection, and financial sustainability. By grouping many stakeholders around a shared vision for sustainability, companies can promote a collective change that benefits not just their own operations but also the broader global economy (Zhang et al., 2020).

Maintaining and accelerating the speed of sustainable entrepreneurship is naturally a challenging task, necessitating a deliberate mix of global cooperation, ongoing education, and ecosystem-wide reform. Companies can significantly accelerate their sustainability activities and achieve long-lasting, major expansion through international networks, sponsoring ongoing human development, and integrating a diverse range of stakeholders. This all-encompassing approach ensures that sustainable projects are not only initiated with momentum but also maintained and developed over time, thereby establishing strong systems that enable long-term success in an increasingly competitive and ecologically sensitive global market. By means of these coupled strategies, businesses can encourage systematic transformation, thereby generating new criteria for sustainable business practices that not only enhance their own bottom line but also society and the planet generally.

3 Conclusion

Finally, this chapter presents a thorough investigation of how sustainable entrepreneurship is grounded in sustainable supply chain strategies. From careful material procurement and ethical trading practices to the use of innovative technologies and worldwide cooperative networks, we have discussed the vital need to include environmental, social, and financial factors in every aspect of supply chain management throughout. The main lesson is that adaptability is necessary for establishing resilience in a continuously changing organizational environment. Companies that consistently adapt to new trends and challenges can ensure their survival and prepare for the unknown of the future. This adaptability not only enables one to survive but also serves as a proactive tool for creating value and influence for future generations. Most importantly involved in fostering ongoing transformation is creativity. As we have seen, driving development calls for embracing new technologies, questioning conventional wisdom, and supporting always better cultures. Creativity generates dynamic responses that not only address current environmental problems but also provide new directions for efficiency and development. Companies dedicated to continuous growth stay ahead and remain relevant and competitive in a rapidly evolving market.

Ultimately, long-term success depends on incorporating sustainability into the core of a company. This requires a comprehensive integration of sustainability concepts into the fundamental business strategy, operations, and decision-making processes rather than merely following specific initiatives. Beginning to define a company's character, sustainability offers a solid framework for concurrently enjoying financial, social, and environmental benefits.

When one considers these revelations, it becomes clear that the path toward sustainable enterprise is complex and requires a combination of strategic adaptation, continuous innovation, and a strong commitment to upholding moral standards. The lessons from this chapter provide a roadmap for not just surviving but also flourishing in a future when sustainability is the top priority as companies develop.

4 Case Study

EcoStyle: Pioneering Sustainable Supply Chain Transformation

Responding to escalating environmental issues and rising consumer demand for sustainable products, EcoStyle, a mid-sized clothing company recognized for its commitment to eco-friendly products, initiated a comprehensive review of its supply chain. Historically, EcoStyle's supply chain relied on conventional sourcing techniques and linear production processes, which, although affordable, significantly contributed to environmental damage through excessive waste, high energy consumption, and inefficient logistics.

EcoStyle's leadership initiated a strategic shift based on three primary pillars: ethical material sourcing, technologically driven process optimization, and open stakeholder communication, recognizing that its operations had to be realigned with sustainable criteria immediately. The corporation began by closely reviewing its present suppliers to assess their social and environmental practices. Working with companies that have adopted strict environmental policies and fair labor standards, EcoStyle established a foundation of trust and shared values.

One crucial turning point in the development was the inclusion of advanced data analytics into their resource management systems. Using a real-time monitoring system, EcoStyle logs waste output, material movement, and energy use from all levels of manufacturing. The first year of installation resulted in a 20% reduction in energy usage and a 30% reduction in trash generation, facilitated by proactive decision-making and the speedy identification of inefficiencies made possible by this technology.

EcoStyle's dedication to sustainability extends beyond its internal operations. The company developed a network of support by forming cooperative relationships with industrial partners and other towns, therefore enabling resource pooling and shared learning. Overcoming frequent transition obstacles, including high initial expenses and aversion to change, proved to be much aided by this network. EcoStyle improved its own operations by participating in industry-wide projects and implementing best practices recommended by environmental leaders, thereby helping to raise sustainability standards across the board.

One of the main components of EcoStyle's methodology was its open attitude to assessing and presenting influence. Designed especially to track developments in environmental efficiency and social responsibility, the corporation created a set of key performance indicators (KPIs). Shared with investors, consumers, and community stakeholders, regular sustainability reports certified by outside auditors were this transparency developed a solid reputation for responsibility and helped all the engaged parties to develop enduring confidence.

Analyzing EcoStyle's approach reveals several quite significant success factors. Firstly, the identification and resolution of inefficiencies made significant use of technology, thereby demonstrating the major environmental and financial benefits resulting from data-driven policy. Second, the focus on moral behavior and supplier alignment ensured that everyone in the supply chain was committed to embedded sustainability. At last, EcoStyle's cooperative approach—both within and outside the organization—emphasized the need to establish solid networks to sustain ongoing development.

Focus Points

- How may the emphasis on ethical buying and supplier screening by EcoStyle assist in building a more generally sustainable supply chain?
- How specifically may data analytics support operational improvements made with technology?
- What challenges EcoStyle faced during the switchover, and how may cooperative coalitions have addressed these?

5 Questions

1. Comprehension: Explain the first difficulties EcoStyle encountered using its conventional supplier chain. These difficulties affected the organization's operating effectiveness as well as its surroundings.
2. Application: How did EcoStyle's approach to resource management change with the inclusion of data analytics? Provide examples of how real-time tracking has improved operations. EcoStyle decided to work with vendors that matched its dedication to sustainability by means of critical thinking. Discuss the potential hazards and advantages of this approach. How could such alliances affect the whole sector?
3. Analyze how EcoStyle's sustainability transformation benefits from open reporting and outside audits.
4. How can this approach increase stakeholder confidence and drive further development?
5. Reflecting on EcoStyle's path, what lessons could other businesses choose about the need to change their supply chain policies in the face of social and environmental issues?
6. Synthesis: Suggest further tactics that EcoStyle might consider going forward to help it reach its environmental objectives. How can these approaches be combined with current methods to produce a supply chain that is even more robust?
7. Evaluation: Determine whether other industries could adopt EcoStyle's method of encouraging cooperative alliances within the entrepreneurial ecosystem as a model. What elements might affect its effective reproduction in several sectors?

These questions aim to assess knowledge about EcoStyle's situation, promote the application of sustainability principles to practical scenarios, and foster critical thinking about the strategic decisions involved in transforming a supply chain for long-term success.

References

Ahlawat, D. R., & Yaseen, M. A. (2024). Challenges and benefits of work-from-home policies in the IT sector: An Indian perspective. *EPRA International Journal of Economics, Business and Management Studies, 11*(9). https://doi.org/10.36713/epra18219

Ariputro, A., Syahtaria, M., Trismadi, T., Suwarno, P., Widodo, P., & Purwanto, D. (2024). Collaboration and technology transfer in the defense industry as drivers of innovation and global security enhancement. *PSLSE Journal, 1*(3), 12. https://doi.org/10.47134/pslse.v1i3.247

Ayandibu, A. O. (2024). A literature review on sustainable innovation and entrepreneurship. *International Journal of Research in Business and Social Science, 13*(9), 33–42. https://doi.org/10.20525/ijrbs.v13i9.3767

Balinado, J. R., Andres, G. L., Macawile, R. M., Bautista, J. V., & Fernandez, J. R. (2024). Consumer perception and purchase intention on the transition to hybrid and electric vehicles for sustainable mobility and transportation in the Philippines. *Journal of Electrical Systems, 20*(5s). https://doi.org/10.52783/jes.2433

Bokareva, E. V., & Avdalyan, A. I. (2020). Supply chain management project on strategic development of enterprises. *International Journal of Supply Chain Management, 9*(3), 892–900. https://ojs.excelingtech.co.uk/index.php/IJSCM/article/view/4942

Boscari, S., Van Donk, D. P., Pullman, M., & Xiao, C. (2024). Different paths to improving together: A taxonomy of buyer–supplier collaborations for sustainability in food supply networks. *Supply Chain Management: An International Journal* (Advance online publication). https://doi.org/10.1108/SCM-02-2024-0119

Budler, M., Quiroga, B. F., & Trkman, P. (2023). A review of supply chain transparency research: Antecedents, technologies, types, and outcomes. *Journal of Business Logistics*. https://doi.org/10.1111/jbl.12368

Busch, M., & Mühlrath, D. (2023). Fairness and trust in organic food supply chains. *British Food Journal, 126*(2), 864–878. https://doi.org/10.1108/BFJ-05-2023-0394

Changalima, I. A. (2024). It's fine for them, but what about us? Exploring the role of supplier management practices on public buyer satisfaction. *Social Sciences & Humanities Open, 9*, Article 100849. https://doi.org/10.1016/j.ssaho.2024.100849

Du, L., Liu, X., & Sun, H. (2024). Corporate sustainable development strategies: Under the collaborative governance of government and the public. *Sustainable Development, 32*(4), 3055–3064. https://doi.org/10.1002/sd.2827

Elmonshid, L. B. E., & Sayed, O. A. (2024). The relationship between entrepreneurship and sustainable development in Saudi Arabia: A comprehensive perspective. *Economies, 12*(8), 198. https://doi.org/10.3390/economies12080198

Feng, T., Qamruzzaman, M., Sharmin, S. S., & Karim, S. (2024). Bridging environmental sustainability and organizational performance: The role of green supply chain management in the manufacturing industry. *Sustainability, 16*(14), 5918. https://doi.org/10.3390/su16145918

Frank, A., & Lucky, A. (2025). Sustainable business practices: The future of corporate responsibility. *IDOSR Journal of Arts and Management, 10*(1), 25–27. https://doi.org/10.59298/idosrjam/2025/101.202700

Fritz, M. M. C., & Silva, M. E. (2024). Social sustainability in supply chains: A framework for addressing labor conditions and worker well-being. *Business Strategy and the Environment, 33*(2), 567–583. https://doi.org/10.1002/bse.3490

Galleli, B., Teles, N., Santos, J., Freitas-Martins, M., & Hourneaux Junior, F. (2021). Sustainability university rankings: A comparative analysis of UI Green Metric and the Times Higher Education World University Rankings. *International Journal of Sustainability in Higher Education* (ahead-of-print). https://doi.org/10.1108/IJSHE-12-2020-0475

Garrido, S., & Nunes, C. S. (2024). Business ethics and ethical behavior. In *Reference module in social sciences*. Elsevier. https://doi.org/10.1016/B978-0-443-13701-3.00143-2

Gerlich, C., Kanbach, D. K., & Kraus, S. (2025). Sustainable corporate entrepreneurship as a catalyst for sustainability—A systematic literature review. *The International Journal of Entrepreneurship and Innovation*. https://doi.org/10.1177/14657503241307138

Gomes, A. (2024). Understanding the shapers of sustainability in small businesses. *International Journal of Science and Research Archive, 13*(2), 3040–3054. https://doi.org/10.30574/ijsra.2024.13.2.2500

Guerrero, J., & Hansen, E. (2018). Cross-sector collaboration in the forest products industry: A review of the literature. *Canadian Journal of Forest Research, 48*(10), 1269–1278. https://doi.org/10.1139/cjfr-2018-0032

Hastig, G. M., & Sodhi, M. S. (2022). Blockchain for supply chain sustainability: A review of applications and challenges. *Journal of Operations Management, 68*(6), 789–813. https://doi.org/10.1002/joom.1194

Holloway, S. (2024). Consumer perceptions of supply chain responsiveness and its impact on brand loyalty in the apparel industry. Preprints. https://doi.org/10.20944/preprints202406.1519.v1

Hristov, I., & Chirico, A. (2019). Performance indicators (KPIs) in implementing sustainable strategies. *Sustainability, 11*(20), 5742. https://doi.org/10.3390/su11205742

Hwang, J. (2024). Sustainability in supply chain management. *International Journal of Multidisciplinary Research Updates, 8*(2), 110–120. https://doi.org/10.53430/ijmru.2024.8.2.0064

Inyang, V., Kanakana, G. M., & Laseinde, O. T. (2023). Application of sustainable smart manufacturing technologies and toolkits in the automotive industry. *International Journal of Low-Carbon Technologies, 18*, 412–422.https://doi.org/10.1093/ijlct/ctad023

Jaboob, A. S., Awain, A. M. B., Ali, K. A. M., & Mohammed, A. M. (2024). Introduction to operation and supply chain management for entrepreneurship. In *Applying business intelligence and innovation to entrepreneurship* (pp. 29–52). IGI Global Scientific Publishing. https://doi.org/10.4018/979-8-3693-1846-1.ch004

Jia, F., Gong, Y., & Brown, S. (2019). Multi-tier sustainable supply chain management: The role of supply chain leadership. *International Journal of Production Economics, 217*, 44–63. https://doi.org/10.1016/j.ijpe.2018.07.022

Jindal, P. (2025). Economic, social, and environmental aspects of sustainable development—Direct and indirect effects on business practices. In S. Grima, K. Sood, E. Özen, & R. Dalli Gonzi (Eds.), *Science of sustainable systems: Greening our economy for a sustainable future* (pp. 227–240). Elsevier. https://doi.org/10.1016/B978-0-443-23603-7.00018-2

Kamenopoulos, S., & Agioutantis, Z. (2021). The importance of the social license to operate at the investment and operations stage of coal mining projects: Application using a decision support system. *The Extractive Industries and Society, 8*(2), Article 100740. https://doi.org/10.1016/j.exis.2020.05.019

Kılıç, C., & Atilla, G. (2024). Industry 4.0 and sustainable business models: An intercontinental sample. *Business Strategy and the Environment, 33*(4), 3142–3166. https://doi.org/10.1002/bse.3514

Le, T. T. (2022). How humane entrepreneurship fosters sustainable supply chain management for a circular economy moving towards sustainable corporate performance. *Journal of Cleaner Production, 368*, Article 133178. https://doi.org/10.1016/j.jclepro.2022.133178

Nayal, K., Raut, R. D., Yadav, V. S., Priyadarshinee, P., & Narkhede, B. E. (2022). The impact of sustainable development strategy on sustainable supply chain firm performance in the digital transformation era. *Business Strategy and the Environment, 31*(3), 845–859. https://doi.org/10.1002/bse.2921

Nguyen, H. L., & Kanbach, D. K. (2024). Toward a view of integrating corporate sustainability into strategy: A systematic literature review. *Corporate Social Responsibility and Environmental Management, 31*(2), 962–976. https://doi.org/10.1002/csr.2527

Nnaji, U., Benjamin, L., Eyo-Udo, N., & Etukudoh, E. (2024). Incorporating sustainable engineering practices into supply chain management for environmental impact reduction. *GSC Advanced Research and Reviews, 19*, 138–143. https://doi.org/10.30574/gscarr.2024.19.2.0177

Odeyemi, O., Mhlongo, N., Elufioye, O., Ugochukwu, C., & Usman, F. (2024). Sustainable entrepreneurship: A review of green business practices and environmental impact. *World Journal of Advanced Research and Reviews, 21*, 346–358. https://doi.org/10.30574/wjarr.2024.21.2.0461

Onukwulu, E. C., Dienagha, I. N., Digitemie, W. N., Egbumokei, P. I., & Oladipo, O. T. (2025). Integrating sustainability into procurement and supply chain processes in the energy sector. *Gulf Journal of Advance Business Research, 3*(1), 76–104. https://doi.org/10.51594/gjabr.v3i1.68

Oyeyemi, O. P., Anjorin, K. F., Ewim, S. E., Igwe, A. N., & Sam-Bulya, N. J. (2024). Marketing-driven supply chain innovation: A framework for FMCG SME sustainability. *International Journal of Management & Entrepreneurship Research, 6*(10), 3541–3558. https://doi.org/10.51594/Ijmer.V6i10.1660

Rahman, M. M., Razimi, M. S. A., Ariffin, A. S., & Hashim, N. (2024). Navigating moral landscape: Islamic ethical choices and sustainability in Halal meat production and consumption. *Discover Sustainability, 5*(1), 225. https://doi.org/10.1007/s43901-024-00025-x

Rashid, A., Rasheed, R., Rahi, S., & Mohd Noor, N. A. (2025). Impact of supplier trust and integrated technology on supply chain resilience for sustainable supply chain in the FMCG sector. *Journal of Science and Technology Policy Management*. https://doi.org/10.1080/JSTPM-04-2024-0134

Rosário, A. T., Raimundo, R. J., & Cruz, S. P. (2022). Sustainable entrepreneurship: A literature review. *Sustainability, 14*(9), 5556. https://doi.org/10.3390/su14095556

Ruhana, F., Nasution, M., Susniwan, S., Ardie, A., & Sutiapermana, A. (2024). Combating poverty and social inequality in 2023: A global imperative for equitable development and social justice. *International Journal of Science and Society, 6*(1), 899–912. https://doi.org/10.54783/ijsoc.v6i1.1081

Saberi, S., Kouhizadeh, M., Sarkis, J., & Shen, L. (2018). Blockchain technology and its relationships to sustainable supply chain management. *International Journal of Production Research, 57*(7), 2117–2135. https://doi.org/10.1080/00207543.2018.1533261

Segura, M., Maroto, C., García del Río, B., & del Rosal, J. C. (2020). Improving food supply chain management by a sustainable approach to supplier evaluation. *Mathematics, 8*(11), 1952. https://doi.org/10.3390/math8111952

Shekarian, E., Ijadi, B., Zare, A., & Majava, J. (2022). Sustainable supply chain management: A comprehensive systematic review of industrial practices. *Sustainability, 14*(13), 7892. https://doi.org/10.3390/su14137892

Silva, M. E., & Figueiredo, M. D. (2022). Sustainable supply chain management: A literature review on emerging economies. *Supply Chain Management: An International Journal, 27*(5), 623–641. https://doi.org/10.1108/SCM-03-2021-0130

Sumon Gazi, M. (2024). Optimizing regional business performance: Leveraging business and data analytics in logistics & supply chain management for USA's sustainable growth. *Journal of Business and Management Studies, 6*(2), 144–152. https://doi.org/10.32996/jbms.2024.6.2.14

Suteu, I., Brovelli, A., & Scalzi, M. (2017). The harmonic integration of culture and skills in design: An interdisciplinary experience in higher education. In *EDULEARN17 Proceedings* (pp. 2453–2461). IATED. https://doi.org/10.21125/edulearn.2017.1514

Tambuskar, D. P., Jain, P., & Narwane, V. S. (2024). An exploration into the factors influencing the implementation of big data analytics in sustainable supply chain management. *Kybernetes, 53*(5), 1710–1739. https://doi.org/10.1108/K-07-2022-1057

Tian, H., Pertheban, T. R., & Han, X. (2024). Navigating industrial performance: Impact of corporate social responsibility on sustainable development with moderation of reverse supply chain management. *Corporate Social Responsibility and Environmental Management*. https://doi.org/10.1002/csr.3000

Vares, H., Kargar Shouraki, M., & Yazdi, N. (2024). Sustainable entrepreneurship in the age of digital transformation. *Journal of Management and Sustainability, 14*(2), 155. https://doi.org/10.5539/jms.v14n2p155

Veleva, V. (2021). The role of entrepreneurs in advancing sustainable lifestyles: Challenges, impacts, and future opportunities. *Journal of Cleaner Production, 283*, Article 124658. https://doi.org/10.1016/j.jclepro.2020.124658

Wu, W., Wang, H., & Wu, Y. J. (2021). Internal and external networks, and incubatees' performance in dynamic environments: Entrepreneurial learning's mediating effect. *The Journal of Technology Transfer, 46*(6), 1707–1733. https://doi.org/10.1007/s10961-021-09872-w

Yakovleva, N., Frei, R., & Rama Murthy, S. (2019). Sustainable development goals and sustainable supply chains in the post-global economy. In *Sustainability and the global economy*. Springer. https://doi.org/10.1007/978-3-030-15066-2

Yang, Z., & Solangi, Y. A. (2024). Analyzing the relationship between natural resource management, environmental protection, and agricultural economics for sustainable development in China. *Journal of Cleaner Production, 450*, Article 141862. https://doi.org/10.1016/j.jclepro.2023.141862

Yang, Y., Du, Y., Gupta, V. K., Ahmad, F., Amiri, H., Pan, J., Aghbashio, M., Tabatabaei, M., & Rajaei, A. (2024). Exploring blockchain and artificial intelligence in intelligent packaging to combat food fraud: A comprehensive review. *Food Packaging and Shelf Life, 43*, Article 101287. https://doi.org/10.1016/j.fpsl.2024.101287

Zaunbrecher, M., & O'Donnell, S. (2020). Improving industry safety via real collaboration: A case study from the Qld CSG sector. *The APPEA Journal, 60*(2), 577–582. https://doi.org/10.1071/AJ19206

Zhang, Q., Loh, L., & Wu, W. (2020). How do environmental, social and governance initiatives affect innovative performance for corporate sustainability? *Sustainability, 12*(8), 3380. https://doi.org/10.3390/su12083380

Branding Challenges for Sustainable Entrepreneurship

Shahriar Azizi, Navid Khanjanzadeh, and Hesamedin Nemati

Abstract

Sustainable entrepreneurship has emerged as a pivotal solution to address environmental, social, and economic crises. However, developing sustainable ventures, particularly in the early stages, faces significant branding challenges. This chapter examines four core challenges: competition with established brands (due to resource disparities in finance and human capital), building trust (amid greenwashing risks and information asymmetry), shifting consumer mindsets (where cost and convenience often overshadow sustainability priorities), and low customer awareness of sustainability concepts. Sustainable branding not only serves as a tool to attract investors and customers but also reduces information gaps and enhances trust, driving the growth of sustainable markets. The case study of "Zeero," an Iranian smart transportation startup, illustrates practical challenges such as competing with dominant ride-hailing platforms (e.g., Snapp and Tapsi), addressing public unawareness of eco-friendly alternatives, and emphasizing data transparency to validate environmental impact. By integrating authenticity, transparency, and strategic stakeholder engagement, sustainable entrepreneurs can overcome resource constraints, differentiate themselves in competitive markets, and achieve long-term viability. The chapter emphasizes the importance of integrating sustainability into core branding strategies to navigate complexities and promote systemic change toward responsible business practices.

Keywords

Branding · Branding challenges · Sustainability · Sustainable entrepreneurship

Learning Objectives

By the end of this chapter, readers will be able to:

- Identify the primary branding challenges faced by sustainable entrepreneurs.
- Analyze the impact of competition, trust issues, consumer behavior, and awareness on sustainable brand development.
- Evaluate strategies to differentiate sustainable brands and build long-term consumer trust.
- Assess the role of authenticity, transparency, and stakeholder collaboration in sustainable branding success.

S. Azizi (✉) · N. Khanjanzadeh · H. Nemati
Department of Business Administration, Faculty of Management and Accounting, Shahid Beheshti University, Tehran, Iran
e-mail: s-azizi@sbu.ac.ir

N. Khanjanzadeh
e-mail: n_khanjanzadeh@sbu.ac.ir

H. Nemati
e-mail: he_nemati@sbu.ac.ir

M. Akbari et al. (eds.), *Sustainable Entrepreneurship*, University of Tehran Science and Humanities Series, https://doi.org/10.1007/978-3-032-04909-4_11

1 Introduction

In today's world, sustainable entrepreneurship has emerged as one of the primary solutions to address environmental, social, and economic crises. This type of entrepreneurship seeks to balance the three pillars of economic, social, and environmental dimensions, aiming not only for financial success but also for long-term positive impacts on society and the environment. In fact, sustainable entrepreneurship is no longer a choice but a necessity, as businesses can no longer afford to ignore their social and environmental responsibilities. Therefore, concepts such as the Triple Bottom Line (economic, social, and environmental) have gained significant importance (Cohen & Winn, 2007; Dyllick & Hockerts, 2002). However, establishing and growing sustainable businesses comes with numerous challenges, especially in the early stages. These challenges can arise at any stage of the entrepreneurial process, but they are particularly acute during the pre-launch and launch phases (Bhave, 1994).

In these early stages, entrepreneurs face multiple issues, including identifying and evaluating sustainable opportunities, designing innovative business models, securing initial funding, and building strategic partnerships. Often, these entrepreneurs lack sufficient experience in sustainable business practices and may struggle to balance profitability with social and environmental principles (Choi & Gray, 2008). These challenges can lead to project failures and even undermine entrepreneurs' motivation. In such situations, sustainable branding can serve as a key strategy to overcome these hurdles and facilitate the process of sustainable entrepreneurship.

Sustainable branding not only helps clarify a business's values but can also play a crucial role in attracting financial resources, gaining social support, and fostering business partnerships (Cohen & Winn, 2007; Patzelt & Shepherd, 2011). While most sustainable entrepreneurs may face financial limitations and a lack of market awareness in the early stages, sustainable branding can provide an opportunity to differentiate in competitive markets and attract investors and customers. Sustainable branding is built on identity, values, and social and environmental responsibilities, helping entrepreneurs build credibility and trust among various stakeholders, including customers, investors, and local communities (Hockerts & Wüstenhagen, 2010).

For sustainable entrepreneurs, who are often focused on securing funding and attracting customers in the early stages, branding serves as a crucial tool to help them achieve their goals. Furthermore, sustainable branding can provide an opportunity to establish strategic collaborations with other stakeholders and develop products and services that have positive social and environmental impacts (Mitchell et al., 1997). Especially in today's world, where consumers are increasingly focused on corporate social responsibility, creating sustainable brands can make a significant difference between success and failure for startups.

The challenges of sustainable branding in entrepreneurship, particularly in the early stages, include issues such as defining brand identity, gaining the trust of customers and investors, and using effective marketing strategies to communicate with target audiences (Kaplan & Haenlein, 2010). This article will specifically examine the various challenges sustainable entrepreneurs face during the branding process. These challenges may include identifying core brand values, aligning economic goals with social and environmental responsibilities, and effectively engaging with diverse stakeholders (Choi & Gray, 2008).

Sustainable entrepreneurs inevitably encounter multifaceted branding challenges, particularly when aligning their mission with environmental and social values. By systematically identifying and articulating these obstacles—such as competing with established brands, overcoming consumer skepticism, and addressing awareness gaps—entrepreneurs can proactively devise strategies to mitigate risks and enhance market positioning. This chapter aims to identify and introduce the challenges in sustainable entrepreneurial branding.

2 Sustainable Branding

2.1 Characteristics and Principles of Sustainable Branding

In recent years, the intersection of sustainability and branding has garnered significant attention from researchers, experts, and consumers alike. Sustainable branding refers to the process of developing a brand identity that reflects an organization's commitment to sustainability, typically characterized by transparency, ethical practices, and social responsibility. With the growing consumer awareness of environmental issues, consumers seek brands that align with their values and lifestyles, leading to shifts in how companies communicate their sustainability efforts. This literature review examines the role of authentic narratives in sustainable branding and communication, emphasizing their impact on consumer choices.

Sustainable branding has its roots in the concept of Corporate Social Responsibility (CSR), which asserts that organizations have responsibilities beyond maximizing profit and are also committed to environmental and social justice (Carroll, 1991). The rise of the ethical consumer, who prioritizes sustainability in their purchasing decisions, has led brands to adopt more responsible practices (Thøgersen, 2006). As a result, brands are increasingly using sustainability as a point of differentiation in competitive markets. However, simply adopting sustainable practices is not enough; brands must effectively communicate these efforts to align with consumers.

Authentic narratives play a crucial role in establishing a connection between sustainable branding and consumer perceptions. Authenticity in branding refers to the perception that a brand is genuine, trustworthy, and aligned with its stated values (Beverland, 2006). Authentic narratives showcase the brand's journey in sustainability, its values, and its commitments, establishing an emotional connection with consumers. When brands share stories about their sustainability initiatives, such as ethical sourcing or reducing carbon footprints, they enhance transparency, increasing credibility and trust (Bridoux & Stoelhorst, 2016).

The power of storytelling in branding has been well-documented in marketing literature. Stories engage consumers on an emotional level, making information relatable and memorable (Holt, 2004). This emotional engagement is particularly important for sustainable brands, as consumers are often skeptical of "greenwashing"—exaggerated or misleading sustainability claims made by companies (Delmas & Burbano, 2011). Through authentic narratives, brands can combat skepticism and reinforce their commitment to sustainability. Authentic storytelling allows brands to demonstrate their true impact, rather than making abstract claims, which strengthens consumer trust and loyalty.

Consumer behavior studies indicate that authenticity has a significant impact on purchasing decisions. Research by Beverland and Farrelly (2010) shows that consumers are more likely to choose brands they perceive as authentic, especially in the context of sustainability. This preference is linked to consumers' desire for social identity and alignment with their personal values. Sustainable brands that effectively communicate their authentic narratives create a sense of community and belonging among consumers, leading to brand loyalty and advocacy. Furthermore, studies indicate that consumers are willing to pay a premium for products from brands that have genuine sustainability efforts, underscoring the financial implications of effective sustainable branding (Nielsen, 2015).

Additionally, the digital landscape has transformed the way brands communicate their sustainability efforts. Social media platforms give brands real-time opportunities to share their narratives and engage directly with consumers (Kaplan & Haenlein, 2010). This democratization of communication fosters greater transparency, as consumers can view and share information about a brand's sustainability practices. However, this also presents challenges, as misinformation and negative reviews can spread quickly (Hennig-Thurau et al., 2010). As a result, brands must actively manage their online presence and engage with consumers to maintain their authenticity and credibility.

The co-creation role of consumers in sustainable branding has also attracted considerable attention. Co-creation involves consumers participating in the branding process, helping to shape the brand's narrative (Prahalad & Ramaswamy, 2004). This approach not only enhances the authenticity of the narrative but also empowers consumers, making them active participants in the brand's sustainability journey. By involving consumers in initiatives such as product design, feedback mechanisms, or social projects, brands can foster deeper connections with their audiences and nurture loyalty and advocacy (Frow & Payne, 2011).

Despite the positive outcomes of authentic narratives in sustainable branding, several challenges remain. One key issue is narrative fatigue, where consumers feel overwhelmed by the sheer volume of sustainability messages. To address this, brands must focus on clarity and coherence in their messaging to ensure their narratives are not only engaging but also easily understood. Moreover, brands must tread carefully between authenticity and performativity, as consumers can often tell when sustainability efforts are undertaken purely for marketing purposes, which can lead to negative reactions and diminished trust (Huang & Rust, 2021).

Future research should explore the long-term impact of authentic narratives on consumer behavior and brand equity. While existing studies emphasize the short-term benefits of sustainable branding, there is a need for longitudinal research to evaluate how these narratives shape consumer perceptions over time. Additionally, understanding the factors that influence consumer behavior in response to sustainability narratives is critical for developing targeted communication strategies. As sustainability continues to be a driving force in markets, brands must remain agile in their storytelling approaches to ensure their narratives reflect their values while resonating with their target audiences.

Sustainable branding and its communication have become an integral part of shaping consumer choices in environmentally conscious markets. Authentic narratives are a powerful tool for brands to convey their commitment to sustainability, fostering increased trust and emotional engagement with consumers. By leveraging transparent storytelling and involving consumers in the branding process, companies can strengthen their credibility and loyalty among their audiences. As sustainable branding evolves, brands must manage challenges and seize opportunities to create meaningful narratives that align with consumers. Ultimately, the sustainable connection between sustainability, branding, and consumer behavior represents a rich area for marketing research and development.

In examining the principles and characteristics of sustainable branding, several key points that contribute to the development and expansion of sustainable product markets can be highlighted. These characteristics are not only defined based on economic and marketing concepts but also specifically work towards promoting sustainability and transparency in sustainable product markets.

2.1.1 Reducing Information Asymmetry

One of the key features of brands in sustainable markets is their ability to reduce information asymmetry. Brands serve as tools to provide credible and reliable information to consumers, enabling them to make more informed choices about sustainable products. Akerlof (1978) emphasizes that brands can reduce information asymmetry, thereby developing sustainable markets and contributing to the growth and enhancement of sustainability.

2.1.2 Trust Building

Brands also serve as tools to build and strengthen trust between consumers and producers. This trust is particularly important in sustainable markets, as consumers need to be assured that the products they purchase are genuinely sustainable and meet environmental standards. Akerlof (1978) also underscores the importance of trust in markets, noting that brands can help build this trust by providing transparent and reliable information.

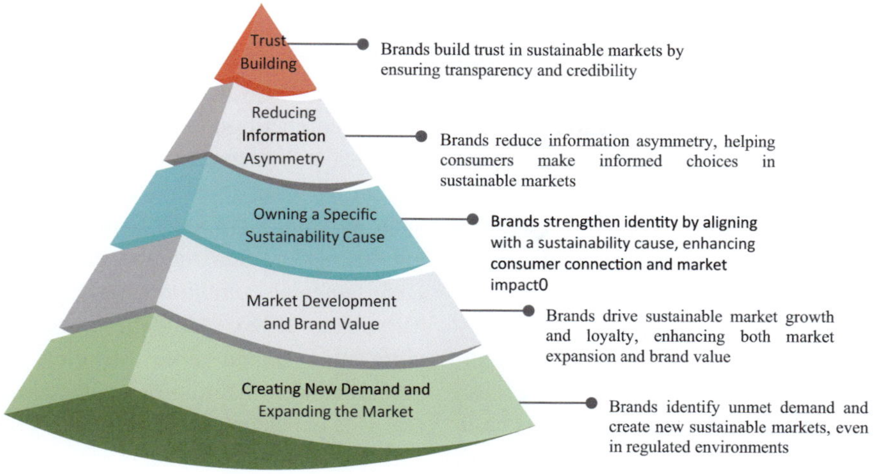

Fig. 1 Characteristics of sustainable branding

2.1.3 Market Development and Brand Value

Brands act as tools for developing sustainable product markets and increasing customer loyalty. By expanding the market and creating brand value, companies can leverage their investments in these markets and internalize the positive impacts of market development. Through this, brands not only help grow sustainable markets but also contribute to enhancing brand value (Lehner & Halliday, 2014).

2.1.4 Owning a Specific Sustainability Cause

Brands can give meaning to their identity by owning a specific sustainability cause and contributing to advancing sustainable markets. This specific sustainability goal can strengthen the connection between the brand and consumers and increase its positive impact in sustainable markets. Brands should integrate these goals into their strategy to effectively contribute to the development of sustainable markets (Lehner & Halliday, 2014).

2.1.5 Creating New Demand and Expanding the Market

Finally, brands help discover new ways to meet unmet demand and expand sustainable markets. This feature is significant in markets with formal institutions and restrictive sustainability regulations. Brands can serve as effective tools for identifying new demand and creating new markets for sustainable products (Lehner & Halliday, 2014; Fig. 1).

2.2 Sustainable Entrepreneurship Branding Challenges

In the past two decades, there has been a growing focus among scholars and practitioners on environmental issues related to entrepreneurial ventures, such as start-ups and incubators (Henry et al., 2020). These ventures are innovating products and processes to achieve cleaner production and sustainable development (Horne et al., 2019). Due to their flexible structures, they can quickly adapt to market needs, particularly during uncertain times, as seen with companies like Blue Origins and SpaceX (Di Vaio et al., 2022) This entrepreneurial behavior poses a challenge to traditional business models by promoting a collaborative and open innovation approach, ultimately benefiting social and environmental sustainability (Littlewood & Holt, 2018; Scuotto et al., 2017).

Alongside the growing demand for more sustainable business practices, sustainability-focused entrepreneurship has emerged as an important area of study within entrepreneurship research (Schaltegger & Wagner, 2011). Consequently, there has been a notable rise in the volume of published articles on sustainable entrepreneurship in recent years (Kraus et al., 2018). Nonetheless, fostering entrepreneurship within established companies presents a paradox, as much of the organizational focus is on maintaining control and predictability, whereas entrepreneurial endeavors are fundamentally uncontrollable and unpredictable (Morris & Kuratko, 2019). Moreover, since corporate entrepreneurship is fundamentally disruptive and brings about changes for individuals, organizations, markets, and industries, it can be perceived as a threat and is often met with resistance from both internal and external sources (Morris & Kuratko, 2019). Accordingly, branding challenges in sustainable entrepreneurship will be identified and introduced.

2.2.1 Competition with Big Brand

A major obstacle for sustainable entrepreneurs during the branding process is the competition with well-established brands that have significant financial resources, broad distribution channels, and strong customer loyalty. This disparity in market power puts sustainable entrepreneurs at a disadvantage, requiring them to develop innovative strategies to compete successfully (Mendes et al., 2022). Large brands take advantage of economies of scale and have access to substantial financial and human resources, allowing them to invest heavily in extensive marketing, technological advancements, and product development. In comparison, sustainable entrepreneurs frequently encounter limitations in financial and human resources (Belz & Binder, 2017). This disparity makes it difficult for them to establish brand awareness and draw in customers. For instance, the significant expenses associated with obtaining environmental certifications, such as Fair Trade or Carbon Neutral, can be prohibitively high for small businesses (Álvarez Jaramillo et al., 2019). In addition, large brands benefit from global distribution networks and strategic partnerships, while sustainable entrepreneurs are often limited to local or online markets. Research by Kumar and Christodoulopoulou (2014) highlights that penetrating major retail channels requires compliance with complex standards and entry costs that are unaffordable for many sustainable businesses (Bakos et al., 2020).

Customer loyalty acts as a crucial factor in the competition between large brands and sustainable entrepreneurs. While large brands benefit from historical advantages and resources such as complex loyalty programs, sustainable entrepreneurs often rely on ethical values and environmental commitments (Ong et al., 2015; Turyakira, 2018). Although large brands and companies benefit from historical credibility, greenwashing undermines trust. A study found that 45% of customers are skeptical of the sustainability claims made by large brands. In contrast, sustainable entrepreneurs enhance loyalty through transparency in their supply chains and open reporting on social and environmental impacts (Casidy & Lie, 2023).

2.2.2 Creating Trust

Trust is defined as consumers' confidence in a company's dependability, and its absence often results in reluctance to participate in business interactions (Nyamrunda & Freeman, 2021). Regarding trust's influence on transactions, the concept has been identified as a central element in forecasting consumers' anticipated outcomes and evaluations of business engagements (Ben Amor & Ben Yahia, 2022).

The difficulty of establishing trust can be viewed from two angles: consumers and stakeholders. Stakeholder theory is utilized to conceptualize the relationship between an organization and its stakeholders—those individuals who can influence or be influenced by the organization's actions. This theory emphasizes generating value without resorting to trade-offs, meaning that creating value should adhere to the principle of not reducing value for one group while enhancing it for another (Góes et al., 2023; Harrison et al., 2019). In other words, stakeholder theory aims to meet the needs of multiple stakeholders while still protecting the interests of the organization (de Mascena & Stocker, 2020).

Sustainable entrepreneurship incorporates environmental and social objectives into business models, requiring brands to harmonize profit with purpose. Nevertheless, trust continues to be a significant challenge, as stakeholders often doubt the authenticity and effectiveness of sustainability claims (Lyon & Montgomery, 2015). Sustainable brands face pressure to meet the diverse expectations of various stakeholders, including consumers, investors, and NGOs, each of whom emphasizes different aspects of sustainability. If these demands are not balanced effectively, it can undermine trust (Talbot et al., 2021; Yazici, 2020).

As the global demand for sustainable products rises, corporations are increasingly making claims about their environmental and social responsibility to attract mindful consumers. However, many resort to greenwashing, inflating or fabricating their sustainability efforts to gain a competitive edge (Lyon & Montgomery, 2015). This misleading practice takes advantage of information asymmetry, as consumers often lack the knowledge or resources to confirm these claims, resulting in a pervasive sense of distrust (Farbacher, 2023). By falsely portraying the environmental qualities of products or companies, greenwashing can deceive consumers, diminish trust, and ultimately obstruct authentic initiatives aimed at promoting sustainable consumption (Nisa et al., 2022). Deceptive assertions regarding sustainability undermine trust and create skepticism. Consumers are growing more wary of brands that do not provide verifiable evidence of their environmental or social impact (Schaltegger et al., 2019).

2.2.3 Changing the Mindset

Shifting customer mindsets to embrace sustainability values presents one of the most complex challenges for sustainable entrepreneurs. Despite increasing public awareness of environmental and social issues, many consumers continue to prioritize convenience, low prices, and traditional shopping habits. The deficiencies in consumer attitudes towards sustainability significantly hinder progress, affecting both individual behaviors and the strategic goals of various organizations (Urbańska et al., 2021). The lack of sustainability education creates reluctance to adopt environmentally friendly practices and innovations, highlighting a crucial area for intervention. Addressing this educational gap could lead to significant advancements in environmental stewardship and sustainable development. Moreover, this lack of education permeates the corporate sector, resulting in lost opportunities for sustainable innovation and growth. Many businesses fail to recognize the lasting benefits of integrating sustainability into their core operations, including cost savings from energy efficiency, opportunities for innovation in environmentally friendly technologies, and enhanced brand reputation through corporate social responsibility initiatives (Redclift, 2020). Furthermore, a limited understanding of the changing regulatory environment and the transition towards sustainable consumer preferences can lead to strategic misalignments and a weakened competitive position. Tackling this intricate challenge necessitates a coordinated effort across multiple sectors (Ciocnitu, 2024). Studies show that over 65% of consumers claim to value sustainability, yet only 25% adhere to it in their everyday purchases. This contradiction, known as the intention-action gap, is driven by factors such as cost priorities and lack of trust (Carrington et al., 2010). For instance, White et al. found that only 10% of customers are willing to pay more for sustainable products (White et al., 2019).

2.2.4 Lack of Customer Awareness

Customer awareness is an integral component of a company's marketing and communications strategy. It is a process that enables an entrepreneur to inform customers about their company, its achievements, and the products or services it offers (Raj & Rajan, 2017). However, the lack of customer awareness regarding sustainability concepts and related values is one of the main barriers to sustainable entrepreneurship branding. Many consumers do not have a clear understanding of sustainability-related terms or the actual impacts of their purchasing choices (Gong et al., 2019). This deficiency in awareness not only hinders the attraction of target audiences but also creates an opportunity for irresponsible brands to engage in greenwashing (Nguyen et al., 2019).

One of the underlying issues is the "awareness gap," where consumers feel overwhelmed by the sheer volume of sustainability messages they encounter (Mehta, 2024). This often results in confusion and skepticism toward sustainability claims.

A significant problem arises when companies exaggerate or misrepresent the environmental benefits of their products or practices. This ultimately erodes customer trust and undermines genuine environmental initiatives (Sudirjo et al., 2024). Researchers emphasize that recognizing the future impacts of purchasing choices is crucial for informed consumer decisions (Ferrer-Estévez & Chalmeta, 2023).

In addition to the main challenge of low consumer awareness of sustainability, companies are trying to overcome this lack of awareness. Sustainable entrepreneurs face two other challenges in this area. The high cost of educating consumers and high-quality educational content is often financially burdensome for small entrepreneurs. Also, competition with deceptive advertising from big brands that dominate messaging through high-budget marketing, and therefore, it will be difficult to compete with them (Mohammed, 2023).

To successfully market an eco-friendly product, it's essential to ensure that they are informed about and care about the issues your product seeks to tackle (Ottman, 2008). By focusing on education and transparency, sustainable entrepreneurs can bridge the awareness gap and foster a more informed consumer base that is equipped to make wise purchasing decisions (Fig. 2).

3 Conclusion

Sustainable entrepreneurship faces multifaceted branding challenges that require strategic navigation to balance economic viability with social and environmental responsibilities. This article highlights four primary challenges: competition with established brands, building trust, shifting consumer mindsets, and addressing low customer awareness. These hurdles are particularly pronounced in the early stages of venture development, where resource constraints and market skepticism amplify risks (Belz & Binder, 2017; Cohen & Winn, 2007).

First, sustainable entrepreneurs struggle to compete with large brands that dominate market share, distribution channels, and customer loyalty through economies of scale and extensive marketing budgets (Kumar & Christodoulopoulou, 2014; Mendes et al., 2022). To counteract this, differentiation through authentic sustainability narratives and transparency in supply chains can enhance credibility and attract ethically minded consumers (Casidy & Lie, 2023; Lyon & Montgomery, 2015). Second, trust remains a critical barrier, as stakeholders increasingly scrutinize greenwashing practices (Delmas & Burbano, 2011). Aligning profit with purpose and adopting verifiable sustainability certifications are essential to mitigate skepticism (Schaltegger et al., 2019).

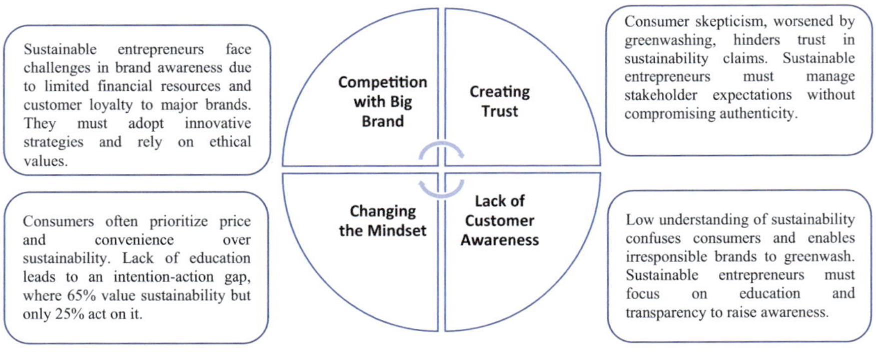

Fig. 2 Sustainable entrepreneurship branding challenges

Third, bridging the intention-action gap in consumer behavior demands innovative strategies to shift mindsets. Despite growing awareness, cost sensitivity and convenience often overshadow sustainability preferences (Carrington et al., 2010). Educating consumers through transparent storytelling and co-creation initiatives can foster deeper engagement (Ali & Cottle, 2021). Finally, low customer awareness perpetuates greenwashing risks and hinders market penetration (Gong et al., 2019). Leveraging digital platforms for targeted education and stakeholder collaboration can address this gap (Kaplan & Haenlein, 2010).

In conclusion, sustainable branding is not merely a marketing tool but a strategic imperative for long-term success. By embedding authenticity, transparency, and stakeholder engagement into their core strategies, sustainable entrepreneurs can overcome resource limitations, differentiate themselves in competitive markets, and drive systemic change. Future research should explore the longitudinal impacts of these strategies on brand equity and consumer loyalty to refine actionable frameworks for emerging ventures.

For future research, it is suggested that researchers study how branding rooted in authenticity shapes long-term consumer loyalty in sustainability-focused markets, with emphasis on strategies for startups to uphold credibility as greenwashing practices become more sophisticated. Also, evaluate the role of decentralized technologies, such as blockchain, in bridging information gaps and empowering small, eco-friendly businesses to rival larger, resource-dominant competitors. In addition to the proposed topics, analyze how cultural norms and socioeconomic conditions shape the uptake of sustainable transport innovations in developing economies, leveraging insights from real-world examples like Zeero's case study.

4 Case Study—Zeero

The knowledge-based company "Smart Clean Wheel," operating under the brand name "Zeero," is considered an innovative startup in the field of smart and clean logistics and transportation in Iran. Inspired by successful global examples and localizing and developing relevant technologies within the country, the company launched its pilot project in 2022 in collaboration with the Clean Transportation Headquarters of Tehran. This marks the first time in Iran that electric vehicles equipped with Internet of Things (IoT) technology are made available to the public in a shared format. By utilizing its own intelligent fleet management system, the company can effectively provide shared electric fleets to citizens in Tehran. Zeero aims to become one of the largest companies in Iran in the field of smart and clean transportation (for both passengers and freight) in the near future and has begun its commercial collaboration as the first electric freight fleet with various organizations.

Since its establishment, Zeero has made considerable progress in advancing sustainable transportation solutions in Iran. With an impressive total of 442,251.26 km covered by its electric vehicle fleet, the company has played a vital role in fostering a cleaner and greener environment. The initiatives undertaken by the company have led to the prevention of 105,786 kg of carbon dioxide emissions, underscoring its dedication to minimizing the carbon footprint linked to conventional transportation methods. Moreover, the company has saved 44,225 L of gasoline, further highlighting the effectiveness of its electric vehicles. This decrease in fuel consumption not only translates into cost savings but also supports energy sustainability.

As Zeero ventures into the Iranian transportation sector, it encounters intense competition from established rivals that command substantial market shares. The leading competitors include ride-hailing services such as Snapp and Tapsi, which have established themselves as key players in urban transportation through their distinctive business models and significant marketing investments. These platforms have altered consumer behaviors with considerable financial backing and widespread advertising, making it difficult for Zeero to change these habits without significant time and resources. This mindset is influenced by a reliance on private vehicles, a preference for lower costs associated with traditional services, and a lack of awareness regarding the long-term advantages of clean transportation. Furthermore, these major platforms utilize their extensive networks of drivers and vehicles to offer rapid access and competitive rates. Another challenge lies in building trust among users of these electric vehicles. Given that some businesses claim to base their operations on sustainability and sustainable development but behave in a manner inconsistent with these claims in practice, citizens have become increasingly skeptical of such enterprises. Another challenge facing ZiRo is the lack of public awareness regarding the availability of this electric transportation system. Many citizens are not only unaware of their access to these services but also lack sufficient understanding of their positive environmental impacts. Providing transparent data, such as the amount of fossil fuel savings, the prevention of thousands of tons of carbon dioxide emissions annually, and the reduction of noise pollution, can serve two key purposes: increasing public knowledge about the benefits of replacing traditional fleets with clean systems and encouraging citizens to choose ZiRo as a responsible, environmentally friendly option. This approach not only raises awareness but also motivates customers to participate in creating cleaner air for future generations.

4.1 Case Study Questions

- What approaches can Zeero adopt to effectively distinguish itself from dominant competitors such as Snapp and Tapsi in Iran's transportation sector, particularly in terms of influencing consumer preferences and capturing market share?
- How can Zeero improve public understanding of its electric transportation solutions and their ecological advantages to address the current knowledge gap among potential users?
- By what methods can Zeero establish credibility among consumers wary of corporate sustainability claims, and how does transparent reporting of environmental impact data contribute to fostering trust?

References

Akerlof, G. A. (1978). The market for "lemons": Quality uncertainty and the market mechanism. In *Uncertainty in economics* (pp. 235–251). Elsevier.

Ali, I., & Cottle, G. W. (2021). Reconceptualizing entrepreneurial performance: The creation and destruction of value from a stakeholder capabilities perspective. *Journal of Business Ethics, 170*(4), 781–796.

Álvarez Jaramillo, J., Zartha Sossa, J. W., & Orozco Mendoza, G. L. (2019). Barriers to sustainability for small and medium enterprises in the framework of sustainable development—Literature review. *Business Strategy and the Environment, 28*(4), 512–524.

Bakos, J., Siu, M., Orengo, A., & Kasiri, N. (2020). An analysis of environmental sustainability in small & medium-sized enterprises: Patterns and trends. *Business Strategy and the Environment, 29*(3), 1285–1296.

Belz, F. M., & Binder, J. K. (2017). Sustainable entrepreneurship: A convergent process model. *Business Strategy and the Environment, 26*(1), 1–17.

Ben Amor, N., & Ben Yahia, I. (2022). Investigating blockchain technology effects on online platforms transactions: Do risk aversion and technophilia matter? *Journal of Internet Commerce, 21*(3), 271–296.

Beverland, M. (2006). The 'real thing': Branding authenticity in the luxury wine trade. *Journal of Business Research, 59*(2), 251–258.

Beverland, M. B., & Farrelly, F. J. (2010). The quest for authenticity in consumption: Consumers' purposive choice of authentic cues to shape experienced outcomes. *Journal of Consumer Research, 36*(5), 838–856.

Bhave, M. P. (1994). A process model of entrepreneurial venture creation. *Journal of Business Venturing, 9*(3), 223–242.

Bridoux, F., & Stoelhorst, J.-W. (2016). Stakeholder relationships and social welfare: A behavioral theory of contributions to joint value creation. *Academy of Management Review, 41*(2), 229–251.

Carrington, M. J., Neville, B. A., & Whitwell, G. J. (2010). Why ethical consumers don't walk their talk: Towards a framework for understanding the gap between the ethical purchase intentions and actual buying behaviour of ethically minded consumers. *Journal of Business Ethics, 97*, 139–158.

Carroll, A. B. (1991). The pyramid of corporate social responsibility: Toward the moral management of organizational stakeholders. *Business Horizons, 34*(4), 39–48.

Casidy, R., & Lie, D. S. (2023). The effects of B2B sustainable brand positioning on relationship outcomes. *Industrial Marketing Management, 109*, 245–256.

Choi, D. Y., & Gray, E. R. (2008). The venture development processes of "sustainable" entrepreneurs. *Management Research News, 31*(8), 558–569. https://doi.org/10.1108/01409170810892127

Ciocnitu, C. R. (2024). Challenges and perspectives in sustainable entrepreneurship. In *19th International*

Conference on European Integration-Realities and Perspectives.

Cohen, B., & Winn, M. I. (2007). Market imperfections, opportunity and sustainable entrepreneurship. *Journal of Business Venturing, 22*(1), 29–49.

de Mascena, K. M. C., & Stocker, F. (2020). Gestao de Stakeholders: Estado da Arte e Perspectivas/ Stakeholder management: state of the art and perspectives. *Future Studies Research Journal: Trends and Strategy, 12*(1), 1–31.

Delmas, M. A., & Burbano, V. C. (2011). The drivers of greenwashing. *California Management Review, 54*(1), 64–87.

Di Vaio, A., Hassan, R., Chhabra, M., Arrigo, E., & Palladino, R. (2022). Sustainable entrepreneurship impact and entrepreneurial venture life cycle: A systematic literature review. *Journal of Cleaner Production, 378*, 134469. https://doi.org/10.1016/j.jclepro.2022.134469

Dyllick, T., & Hockerts, K. (2002). Beyond the business case for corporate sustainability. *Business Strategy and the Environment, 11*(2), 130–141.

Farbacher, T. (2023). The federal trade commission's green guides: Failing the American consumer and the planet. *Loyola Consumer Law Review, 36*, 136.

Ferrer-Estévez, M., & Chalmeta, R. (2023). Sustainable customer relationship management. *Marketing Intelligence & Planning, 41*(2), 244–262.

Frow, P., & Payne, A. (2011). A stakeholder perspective of the value proposition concept. *European Journal of Marketing, 45*(1/2), 223–240.

Góes, H. A. de A., Fatima, G., Santos Jhunior, R. de O., & Boaventura, J. M. G. (2023). Managing for stakeholders towards corporate environmental sustainability. *Corporate Social Responsibility and Environmental Management, 30*(4), 1561–1572.

Gong, M., Gao, Y., Koh, L., Sutcliffe, C., & Cullen, J. (2019). The role of customer awareness in promoting firm sustainability and sustainable supply chain management. *International Journal of Production Economics, 217*, 88–96.

Harrison, J. S., Barney, J. B., Freeman, R. E., & Phillips, R. A. (2019). *The Cambridge handbook of stakeholder theory.* Cambridge University Press.

Hennig-Thurau, T., Malthouse, E. C., Friege, C., Gensler, S., Lobschat, L., Rangaswamy, A., & Skiera, B. (2010). The impact of new media on customer relationships. *Journal of Service Research, 13*(3), 311–330.

Henry, M., Bauwens, T., Hekkert, M., & Kirchherr, J. (2020). A typology of circular start-ups: An Analysis of 128 circular business models. *Journal of Cleaner Production, 245*, 118528. https://doi.org/10.1016/j.jclepro.2019.118528

Hockerts, K., & Wüstenhagen, R. (2010). Greening Goliaths versus emerging Davids—Theorizing about the role of incumbents and new entrants in sustainable entrepreneurship. *Journal of Business Venturing, 25*(5), 481–492.

Holt, D. B. (2004). *How brands become icons: The principles of cultural branding.* Harvard Business Press.

Horne, A. C., Nathan, R., Poff, N. L., Bond, N. R., Webb, J. A., Wang, J., & John, A. (2019). Modeling flow-ecology responses in the Anthropocene: Challenges for sustainable riverine management. *BioScience, 69*(10), 789–799. https://doi.org/10.1093/biosci/biz087

Huang, M.-H., & Rust, R. T. (2021). A strategic framework for artificial intelligence in marketing. *Journal of the Academy of Marketing Science, 49*, 30–50.

Kaplan, A. M., & Haenlein, M. (2010). Users of the world, unite! The challenges and opportunities of Social Media. *Business Horizons, 53*(1), 59–68.

Kraus, S., Burtscher, J., Vallaster, C., & Angerer, M. (2018). Sustainable entrepreneurship orientation: Reflection on status-quo research on factors facilitating responsible managerial practices. In *Sustainable Entrepreneurship* (pp. 75–98). Routledge.

Kumar, V., & Christodoulopoulou, A. (2014). Sustainability and branding: An integrated perspective. *Industrial Marketing Management, 43*(1), 6–15. https://doi.org/10.1016/j.indmarman.2013.06.008

Lehner, M., & Halliday, S. (2014). Branding sustainability: Opportunity and risk behind a brand-based approach to sustainable markets. *Ephemera: Theory and Politics in Organization.*

Littlewood, D., & Holt, D. (2018). How social enterprises can contribute to the sustainable development goals (SDGs)—a conceptual framework. In *Entrepreneurship and the sustainable development goals* (Vol. 8, pp. 33–46). Emerald Publishing Limited. https://doi.org/10.1108/S2040-724620180000008007

Lyon, T. P., & Montgomery, A. W. (2015). The means and end of greenwash. *Organization & Environment, 28*(2), 223–249.

Mehta, D. (2024). Sustainable branding and communication: influencing consumer choices through authentic narratives. *Multidisciplinary Journal of Law for Future Security, 1*(2), 75–85.

Mendes, A. C. S., Ferreira, F. A. F., Kannan, D., Ferreira, N. C., & Correia, R. J. C. (2022). A BWM approach to determinants of sustainable entrepreneurship in small and medium-sized enterprises. *Journal of Cleaner Production, 371*, Article 133300.

Mitchell, R. K., Agle, B. R., & Wood, D. J. (1997). Toward a theory of stakeholder identification and salience: Defining the principle of who and what really counts. *Academy of Management Review, 22*(4), 853–886.

Mohammed, A. A. (2023). The influence of customer knowledge management on sustainable promotion. *Journal of Economics & Administrative Sciences, 29*(137).

Morris, M. H., & Kuratko, D. F. (2019). Addressing major challenges associated with sustainable entrepreneurship in established companies. *American Journal of Entrepreneurship, 12*(2), 5–34.

Nguyen, T. T. H., Yang, Z., Nguyen, N., Johnson, L. W., & Cao, T. K. (2019). Greenwash and green purchase intention: The mediating role of green skepticism. *Sustainability, 11*(9), 2653.

Nielsen, C. (2015). *The sustainability imperative: New insights on consumer expectations.* Nielsen Company.

Nisa, N. U., Mendoza, S. A. J., & Shamsuddinova, S. (2022). The concept of greenwashing and its impact on green trust, green risk, and green consumer confusion: A review-based study. *JABS, 8*(3), 1–18.

Nyamrunda, F. C., & Freeman, S. (2021). Strategic agility, dynamic relational capability and trust among SMEs in transitional economies. *Journal of World Business, 56*(3), Article 101175.

Ong, C. H., Salleh, S. M., & Yusoff, R. Z. (2015). Brand experience, trust components, and customer loyalty: Sustainable Malaysian SME brands study. *Asian Social Science.*

Ottman, J. A. (2008). The five simple rules of green marketing. *Design Management Review, 19*(4), 65–69.

Patzelt, H., & Shepherd, D. A. (2011). Negative emotions of an entrepreneurial career: Self-employment and regulatory coping behaviors. *Journal of Business Venturing, 26*(2), 226–238.

Prahalad, C. K., & Ramaswamy, V. (2004). Co-creation experiences: The next practice in value creation. *Journal of Interactive Marketing, 18*(3), 5–14.

Raj, G. P., & Rajan, A. P. (2017). A study on the customer awareness on green banking initiatives. *Intercontinental Journal of Finance Research Review, 5*(7), 54–65.

Redclift, M. (2020). *Wasted: Counting the costs of global consumption.* Routledge.

Schaltegger, S., Hörisch, J., & Freeman, R. E. (2019). Business cases for sustainability: A stakeholder theory perspective. *Organization & Environment, 32*(3), 191–212.

Schaltegger, S., & Wagner, M. (2011). Sustainable entrepreneurship and sustainability innovation: Categories and interactions. *Business Strategy and the Environment, 20*(4), 222–237.

Scuotto, V., Santoro, G., Bresciani, S., & Del Giudice, M. (2017). Shifting intra- and inter-organizational innovation processes towards digital business: An empirical analysis of SMEs. *Creativity and Innovation Management, 26*(3), 247–255. https://doi.org/10.1111/caim.12221

Sudirjo, F., Dizon, C. C. G., Rusmardiana, A., Deni, A., & Hantono, H. (2024). Sustainable marketing dilemmas: Moral, ethical, and practical in business communication perspectives. *Jurnal Studi Komunikasi, 8*(1), 11–20.

Talbot, D., Raineri, N., & Daou, A. (2021). Implementation of sustainability management tools: The contribution of awareness, external pressures, and stakeholder consultation. *Corporate Social Responsibility and Environmental Management, 28*(1), 71–81.

Thøgersen, J. (2006). Norms for environmentally responsible behaviour: An extended taxonomy. *Journal of Environmental Psychology, 26*(4), 247–261.

Turyakira, P. K. (2018). Ethical practices of small and medium-sized enterprises in developing countries: Literature analysis. *South African Journal of Economic and Management Sciences, 21*(1), 1–7.

Urbańska, M., Charzyński, P., Gadsby, H., Novák, T. J., Şahin, S., & Yilmaz, M. D. (2021). Environmental threats and geographical education: Students' sustainability awareness—Evaluation. *Education Sciences, 12*(1), 1.

White, K., Hardisty, D. J., & Habib, R. (2019). The elusive green consumer. *Harvard Business Review, 11*(1), 124–133.

Yazici, H. J. (2020). An exploratory analysis of the project management and corporate sustainability capabilities for organizational success. *International Journal of Managing Projects in Business, 13*(4), 793–817.

Sustainable Job Crafting: Introducing a Concept in Sustainable Human Resource Management

Elham Ebrahimi and Hamid Reza Irani

Abstract

This chapter explores the integration of job crafting and sustainability, introducing the concept of sustainable job crafting, which refers to employees' proactive efforts to redesign their jobs in alignment with sustainability principles. While job crafting traditionally focuses on modifying tasks, relationships, and perceptions to enhance meaning and engagement, this chapter extends its scope to incorporate environmental, social, and economic sustainability goals. Key discussions include different conceptualizations of job crafting, such as individual versus collaborative crafting and approach versus avoidance crafting, alongside their relevance to sustainability-oriented work practices. The chapter highlights how employees can actively shape their roles to enhance sustainability outcomes, emphasizing the importance of organizational support in fostering this process. Further, it identifies essential prerequisites for sustainable job crafting, including sustainability awareness, job flexibility, and employee authority. Additionally, structured interventions, such as job crafting training programs and sustainability-focused exercises, are examined to help employees align their work with long-term sustainability goals. By bridging job crafting with sustainability, this chapter presents a novel framework for organizations seeking to cultivate a sustainable workforce, highlighting that sustainability-driven job redesign is both an individual initiative and a strategic imperative.

Keywords

Job crafting · Sustainablility · Sustainable job crafting · Sustainable human resource management

Learning Objectives

By the end of this chapter, readers will be able to:

- Explain the concept of sustainable job crafting and its significance in integrating job redesign with sustainability principles.
- Differentiate between various forms of job crafting and analyze their relevance to sustainability-oriented work practices.
- Identify key organizational factors, such as sustainability awareness, job flexibility, and employee authority, that enable sustainable job crafting.

E. Ebrahimi
Department of Management, Institute for Humanities and Cultural Studies, Tehran, Iran
e-mail: e.ebrahimi@ihcs.ac.ir

H. R. Irani (✉)
Department of Entrepreneurship and Marketing, College of Farabi, University of Tehran, Qom, Iran
e-mail: hamidrezairani@ut.ac.ir

M. Akbari et al. (eds.), *Sustainable Entrepreneurship*, University of Tehran Science and Humanities Series, https://doi.org/10.1007/978-3-032-04909-4_12

- Evaluate structured interventions, including training programs and exercises, that facilitate sustainability-driven job redesign.
- Apply sustainable job crafting strategies to align individual roles with long-term environmental, social, and economic sustainability goals.

1 Introduction

Today, organizations must change rapidly and continually. Traditional jobs, once rigidly defined by management, are now becoming more dynamic as employees actively reshape their work experiences (Petrou et al., 2016). This shift highlights the growing relevance of job crafting, a concept that emphasizes an individual's proactive role in modifying tasks, relationships, and perceptions to enhance job engagement and performance (Wrzesniewski & Dutton, 2001). However, while job crafting has been widely explored in the context of employee engagement and well-being (van Wingerden et al., 2017), its integration with sustainability remains an emerging area of research.

Sustainability is a multidimensional concept that extends beyond environmental concerns to include economic stability, social responsibility, and long-term career viability (De Vos & Van der Heijden, 2017). Organizations striving for sustainability must not only adopt eco-friendly practices but also design work environments that foster continuous learning, adaptability, and meaningful job experiences. This raises a critical question: How can employees actively participate in shaping their roles to support both personal fulfillment and broader sustainability goals? The answer lies in sustainable job crafting, which combines the principles of job crafting with sustainability-oriented work design.

For sustainable job crafting to be effective, organizations must cultivate an environment where employees have the flexibility and authority to make meaningful changes in their roles (Wang, 2022). Job flexibility allows employees to modify their work structure in ways that align with sustainability principles, while employee authority ensures they have the necessary autonomy to implement these modifications. Additionally, raising sustainability awareness is essential for embedding sustainable job crafting within organizational culture. Employees must understand how their roles contribute to sustainability before they can actively reshape their work accordingly. Despite its potential, sustainable job crafting does not occur spontaneously. Organizations require structured interventions, such as training programs and guided exercises, to help employees identify opportunities for job redesign aligned with sustainability goals (Rai, 2018). These interventions empower employees to engage in purposeful modifications, ensuring that job crafting efforts align with both individual career aspirations and long-term organizational sustainability strategies.

This chapter explores the intersection of job crafting and sustainability, providing insights into how employees can actively shape their work while contributing to economic, social, and environmental well-being. It examines key enabling factors, structured interventions, and practical applications that support sustainable job crafting. By the end of this chapter, readers will have gained a deeper understanding of how job crafting can serve as a strategic tool for creating meaningful and sustainable work experiences that benefit both employees and organizations.

2 Sustainability Perspective on Human Resource Management

Despite being the subject of numerous studies, the concept of sustainability remains one of the most diverse and multifaceted notions, often appended as a suffix to various scientific disciplines. In the field of human resource management (HRM), sustainability signifies a shift away from a purely economic perspective, instead emphasizing a balance among three key dimensions: environmental, social, and economic. This approach is grounded in the Triple Bottom Line Model (Elkington, 1997). This emerging paradigm is referred to as Sustainable Human Resource Management (Sustainable HRM).

The intellectual origins of this paradigm can be traced back to the late 1990s and early 2000s, with early contributions from scholars such as Müller-Christ and Remer (1999), Gollan (2000), and Zaugg et al. (2001). These works represent some of the initial efforts to integrate sustainability into HRM. The breadth and complexity of topics discussed under the umbrella of sustainability in the HRM context are so extensive that, without a systematic classification, it would be difficult to provide a clear perspective on the subject. A structured categorization divides these diverse disciplines and research areas into three main domains: corporate sustainability, Corporate Social Responsibility (CSR), and sustainable work systems (Macke & Genari, 2019). Discussions within this framework occur at four levels: individual, process, organization, and society. Notably, as Ehnert (2009) suggested, sustainability-related topics within HRM often emphasize two key characteristics: environmental concerns and long-term orientation. Green HRM exemplifies the former (Tahir et al., 2024), while sustainable employability illustrates the latter (Courchesne et al., 2025). The distinctive feature of sustainability-oriented HRM approaches is their explicit focus on these two aspects—an area that has been largely overlooked in traditional HRM orientations. The intersection of sustainability and human resource management, conceptualized as Sustainable HRM, has been explored from multiple perspectives, further enriching this evolving field. In one research, Kramar (2014) categorized sustainable human resource management into three distinct research areas:

1. Capability Reproduction: focuses on the economic dimension of Sustainable HRM, emphasizing the development of sustainable competitive advantages. It also considers the internal impact of HRM policies in both economic and social/human terms.
2. Promotion of Social and Environmental Well-being: examines the broader external implications of Sustainable HRM, highlighting its effects on environmental, social, and human outcomes.
3. Interconnections: explores the relationships between managerial practices—particularly HRM—and organizational outcomes, with a focus on their influence on environmental and social sustainability. It also includes research on sustainable leadership and environmental sustainability.

In another work, Sustainable HRM is established through two perspectives: The role of HRM in promoting organizational sustainability and the sustainability of HRM processes (Macke & Genari, 2019). Hughes and Dundon (2025) examined sustainable human resource management across three levels: macro-, meso-, and micro-level. Drawing on stakeholder theory, systems theory, and strategic HRM, they identified three perspectives through which sustainable human resource management can be analyzed:

1. Macro-Level Systems: Encompassing labor markets, conditions, norms, values, and socio-economic-political governance.
2. Meso-Level Systems: Focusing on organizational HR subsystems such as recruitment, training, retention, and compensation.
3. Micro-Level Systems: Addressing the experiences, needs, and interests of managers, employees, and customers.

This classification provides a valuable framework for a more precise understanding of sustainable HRM and its diverse applications. For instance, when Kramar (2022) defines Sustainable HRM as policies related to job design, selection, training, performance management, and compensation that contribute to positive environmental outcomes, it becomes evident that this discussion aligns with the meso-level perspective. Furthermore, this categorization serves as a useful foundation for this book chapter, as it facilitates an exploration of the relationship between sustainability and job design—specifically, job crafting as a novel perspective—within the context of human resources subsystems. As noted by Siddiqui et al. (2022), this linkage remains underexplored in existing research, highlighting the need for further investigation into the integration of job crafting within sustainable HRM frameworks.

3 Theoretical Foundations and Evolution of Job Crafting

In today's dynamic work environments, employees' career paths resemble an adventure rather than a predetermined journey with fixed stations and a structured itinerary. Unlike a guided tour, career trajectories do not always conform to externally imposed frameworks. Instead, they often deviate from predefined guidelines and formal job descriptions, requiring employees to navigate uncertainties and actively shape their professional experiences. The concept of job crafting, introduced by Wrzesniewski and Dutton (2001), encapsulates this perspective through a term that conveys artistic and creative subjectivity. Their framework challenges the notion of jobs as static entities, instead emphasizing their inherent flexibility and dynamism. Job crafting enables employees to proactively modify aspects of their roles, aligning them with personal strengths, motivations, and evolving workplace demands. This perspective resonates with contemporary work environments, where adaptability and self-directed role modification have become essential for navigating continuous change and organizational transformations.

The evolution of job motivation theories has shifted from traditional, rigid frameworks to more modern, flexible approaches. This transformation marks a departure from top-down job design models toward bottom-up, employee-driven job crafting, which emphasizes creativity and autonomy. Moving beyond Scientific Management and Taylorism, which adopted fundamentally different and contrasting perspectives on motivation, traditional motivational job design theories—such as Hackman and Oldham's Job Characteristics Theory (1976) —advocated for the external addition of motivating factors, including variety, significance, identity, autonomy, and feedback, as determined by job designers. Later theories, such as the Self-Determination Theory (Ryan & Deci, 2000, 2023), acknowledged employees' intrinsic need for autonomy, choice in activities, and meaningful social connections. However, these frameworks did not explicitly emphasize employees' active role in shaping their jobs. A paradigm shift occurred with Wrzesniewski and Dutton (2001), who introduced the concept of job crafting based on the Job Identity Model. They defined job crafting as the physical, relational, and cognitive modifications individuals make within the boundaries of their jobs. Specifically, employees engage in job crafting through three primary mechanisms:

1. Task Crafting—Modifying job tasks by eliminating, altering, or enriching responsibilities to better align with personal skills and interests.
2. Relational Crafting—Expanding or reshaping work relationships to foster more meaningful and effective interactions with colleagues, supervisors, subordinates, customers, and other stakeholders.
3. Cognitive Crafting—Reframing the mental perception of one's job to enhance one's sense of meaning and motivation.

This shift marked a new era in job design, emphasizing employees' proactive role in shaping their work experiences rather than passively adapting to predefined structures. In this conceptualization, boundary changes encompass modifications in the number, quantity, quality, scope, nature, and perception of job-related tasks and relationships, provided that such changes yield positive outcomes for both the individual and the organization. Among the three forms of job crafting, cognitive crafting appears to occur more unconsciously, emerging naturally in response to intrinsic motivators. In contrast, task and relational crafting are more structured, making them teachable and transferable, which facilitates delegation and training. Notably, cognitive crafting—the process through which employees redefine their jobs in a more meaningful way—distinguishes Wrzesniewski and Dutton's (2001) definition of job crafting from other conceptualizations.

Over time, Wrzesniewski and Dutton's (2001) perspective on job crafting evolved as other scholars expanded upon the concept. Some researchers grounded their work in the Job Demands-Resources (JD-R) Theory (Bakker & Demerouti, 2007; Petrou et al., 2012; Tims et al., 2012), positing that job crafters actively balance job demands and resources to enhance work motivation and engagement. Within the JD-R framework, job resources refer to the tools, support, and opportunities employees utilize to perform their jobs effectively and manage job demands. Based on this theoretical perspective, job crafting manifests in four primary ways (Tims et al., 2012):

1. Increasing structural job resources
2. Increasing social job resources
3. Increasingly challenging job demands
4. Decreasing hindering job demands

In some instances, these four strategies are consolidated into broader classifications. For example, they are sometimes categorized into three groups: (1) seeking job resources, (2) seeking new challenges, and (3) reducing job demands. Alternatively, they may be condensed into two overarching categories: (1) seeking job resources and challenges and (2) reducing job demands (Petrou et al., 2012).

Various conceptualizations of job crafting extend beyond the original definitions proposed by Wrzesniewski and Dutton (2001). Some types of job crafting are shown in Table 1. For instance, Leana et al. (2009) differentiate between individual crafting and collaborative crafting. Individual crafting refers to job modifications undertaken independently by employees, whereas collaborative crafting involves teams or coworkers collectively engaging in job redesign to enhance their shared work experience. Another classification distinguishes between approach crafting and avoidance crafting, a framework derived from the integration of Job Identity Theory and the Job Demands-Resources (JD-R) Model (Bruning & Campion, 2018). Approach crafting is characterized by individuals who are motivated by positive stimuli, prompting them to increase structural or social job resources or enhance challenging job demands.

In contrast, avoidance crafting—also referred to as prevention-focused job crafting—occurs in individuals driven by the desire to remove negative stimuli, leading them to reduce hindering job demands. A similar distinction is made between promotion-focused and prevention-focused job crafting (Lichtenthaler & Fischbach, 2016, 2019). Promotion-focused job crafting reflects a proactive, opportunity-seeking approach in which employees actively reshape their roles to optimize personal and professional growth. Conversely, prevention-focused job crafting adopts a risk-averse, problem-avoidant strategy, wherein employees modify their work to minimize stressors and mitigate potential difficulties. These varying conceptualizations highlight the complexity of job crafting as a dynamic and multifaceted phenomenon that encompasses both proactive engagement and defensive adaptation strategies.

Table 1 Some types of job crafting

	Types of job crafting	References
1	• Task crafting • Relational crafting • Cognitive crafting	Wrzesniewski and Dutton (2001)
2	• Increasing structural job resources • Increasing social job resources • Increasing challenging job demands • Decreasing hindering job demands	Tims et al., (2012)
3	• Seeking job resources and challenges • Reducing job demands	Petrou et al., (2012)
4	• Individual crafting • Collaborative crafting	Leana et al. (2009)
5	• Promotion-focused job crafting • Prevention-focused job crafting	Lichtenthaler and Fischbach (2016, 2019)

4 Integrating Job Crafting and Sustainability: Establishing a Novel Framework

Despite the extensive body of literature on both job crafting and sustainability, the integration of these two concepts has received limited scholarly attention. However, job crafting offers a valuable framework for linking sustainability-related topics, such as green HRM, under the broader concept of green crafting. Both dominant approaches to job crafting—those proposed by Wrzesniewski and Dutton (2001) and Tims et al. (2012)—can be strategically aligned with an organization's sustainability objectives. For instance, when sustainability, environmental responsibility, or green human resource management (HRM) are embedded as organizational priorities, employees may internalize these values and integrate them into their daily work practices. This fosters a top-down approach in which sustainability considerations shape tasks, workplace relationships, and job perceptions. Employees, in turn, actively adjust their roles to contribute to broader environmental and social goals. Furthermore, within the Job Demands-Resources (JD-R) framework, employees may proactively modify their job resources or demands to incorporate sustainability-related elements. This can manifest in various ways, such as seeking additional resources to support eco-friendly initiatives, increasing collaboration with environmentally conscious colleagues, or reducing job demands that contribute to unsustainable practices. By embedding sustainability into job crafting, organizations can encourage employees to take ownership of green initiatives, fostering a culture of environmental responsibility while simultaneously enhancing job meaningfulness and engagement.

Research efforts aimed at establishing a connection between sustainability and job crafting can be categorized into two main groups. The first group comprises studies that focus on a specific dimension of sustainability—primarily the environmental aspect—leading to the introduction of concepts such as green crafting. These studies examine how employees actively modify their tasks, relationships, and job perceptions to align with environmental goals, often within the framework of Green Human Resource Management. By integrating sustainability into job crafting behaviors, these studies highlight the role of employees in fostering eco-friendly work practices and promoting organizational environmental initiatives. The second group consists of studies that have explored sustainability within the broader job design subsystem but have not yet explicitly addressed the job crafting approach. These studies typically focus on sustainable work design, emphasizing the creation of jobs that enhance long-term employee well-being, organizational resilience, and environmental or social sustainability. While these works contribute to understanding sustainable HRM, they do not necessarily consider the proactive role of employees in shaping their own work experiences through job crafting. The following sections provide examples of these two research streams, demonstrating how sustainability considerations are progressively being integrated into job design and job crafting literature.

Some researchers have applied the JD-R approach to job crafting as a means of linking green HRM with job crafting. For example, Tuan (2021) conceptualized green crafting as a crucial mediating mechanism that facilitates pro-environmental activities and enhances their meaningfulness. He proposed that green crafting functions as a mediating variable, translating the impact of a green organizational climate to individual green performance. Within this framework, green crafting comprises four dimensions. Increasing green-related structural resources involves an individual's effort to enhance environmental knowledge, skills, and personal development related to sustainability. Increasing green-related social resources refers to actively seeking interactions and feedback to support green initiatives. Increasing green-related challenging demands reflect an employee's proactive engagement in integrating environmental considerations into their tasks. Finally, decreasing hindering green task demands focuses on eliminating structural or relational barriers that prevent the incorporation of environmental concerns into job roles. In another study, Luu (2019) expanded on the concept of green crafting by introducing its collective dimension, referred to as collective green crafting. He argued that job crafting occurs at both individual and collective levels, with teams collaborating to optimize structural and relational resources, take on challenging tasks, and reduce hindering demands to achieve shared environmental sustainability objectives. When team members align their efforts toward sustainability, they engage in collective green crafting, a process that entails increasing job resources and challenges while minimizing obstacles to green behavior. Luu (2019) defined collective green crafting as the enhancement of structural green resources (e.g., developing green skills) and social green resources (e.g., seeking support and feedback). Additionally, teams can engage in new green initiatives as challenging tasks while reducing cognitive or emotional burdens related to sustainability through internal and external support, thereby decreasing hindrances to green demands.

As previously discussed, another significant research effort to integrate sustainability into human resource subsystems—particularly relevant to this book chapter—focuses on linking sustainability with job design. However, this integration follows a top-down approach rather than being directly shaped by employees. According to Siddiqui et al. (2022), since the emergence of the United Nations' Sustainable Development Goals (SDGs), sustainability principles have become increasingly embedded in all human resource management activities, and job design, as a fundamental HRM function, is no exception. However, they argue that traditional frameworks such as scientific management, Herzberg's dual-factor theory, and Job Characteristics Theory (Hackman & Oldham, 1976) have limited explanatory power in capturing the motivational and contextual influences on job design, making them unsuitable for linking job design to a sustainability-oriented approach. In contrast, the Job Demands-Control-Support (JDC/JDCS) Model (Karasek, 1979) and the Job Demands-Resources (JD-R) Model (Demerouti et al., 2001)—which classify job characteristics into demands and resources—have expanded the scope of job design to prioritize sustainability, paving the way for contemporary approaches that explore sustainable work environments and long-term organizational outcomes. We concur with Siddiqui et al. (2022) in their assertion that future research should focus on developing models that integrate environmental responsibilities and competencies into job roles to promote organizational sustainability. While we acknowledge the significance of the key emerging themes identified in their study—including the redefinition of decent work across cultures, the improvement of job quality, the integration of ergonomics, the conceptualization of green jobs, and the design of roles that enhance employee well-being—we argue that their framework remains incomplete. Notably, the critical concept of job crafting is absent from their discussion despite its potential to further enrich sustainable job design.

5 Sustainable Job Crafting

Sustainable job crafting can be defined by integrating the principles of sustainability and job crafting. Within the domain of sustainability, despite various discussions at both micro and macro levels, it is widely recognized that achieving sustainability requires a balanced focus on three key dimensions: economic, social, and environmental. Since the environmental dimension has received relatively less attention in human resource management research, studies related to sustainable HRM tend to place greater emphasis on this aspect. Additionally, sustainability in the context of HRM can be examined across three levels. At the macro level, it pertains to broad policy-making and strategic decisions. At the meso-level, it addresses organizational-level sustainability practices. At the micro level, it relates to individual behaviors and actions. On the other hand, job crafting refers to an individual's initiative, willingness, and ability to redesign their job by redefining tasks, relationships, and cognitive perceptions. Employees engage in job crafting by increasing structural, relational, and challenge resources while simultaneously reducing hindering demands, thereby making their work more meaningful and creative.

The integration of these two perspectives leads to the concept of sustainable job crafting, which refers to individuals' initiative, willingness, and ability to redesign their jobs in alignment with sustainability principles. Sustainability-related

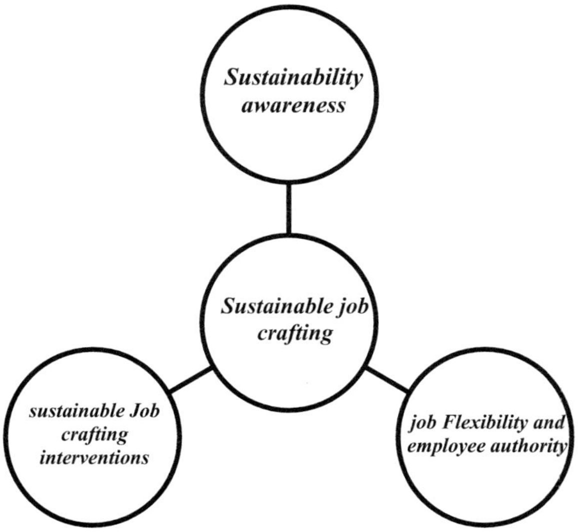

Fig. 1 Requirements for sustainability-oriented job crafting

concepts can be interpreted in multiple ways, ranging from incorporating environmental considerations to adopting a holistic approach that ensures success across the economic, social, and environmental dimensions. It can also involve career management strategies and fostering a long-term perspective on professional success. Overall, any goal that falls under the umbrella of sustainability can be considered an objective of job crafting. However, job redesign with a sustainability-oriented approach does not occur automatically. In the following sections, drawing on research such as Wang (2022) and Rai (2018), the requirements for sustainable job crafting, as illustrated in Fig. 1, will be discussed in detail.

5.1 Sustainability Awareness

Raising awareness and disseminating information about the concept of sustainability and its connection to organizational strategies and policies is a fundamental prerequisite for the formation of sustainable job crafting. It is unrealistic to expect this process to emerge and be implemented among employees without top-down awareness initiatives. To achieve this, organizations must first clearly define their intended approach to sustainability. Several critical questions must be addressed: Does the organization prioritize social and environmental sustainability and seek to integrate it into its operational strategies (Longoni & Cagliano, 2015)? Does it emphasize a long-term success approach through employees' career development (De Vos & Van der Heijden, 2017)? Is it committed to fulfilling the interests of various stakeholders, both direct and indirect, without compromising its ability to meet the needs of future stakeholders (Amrutha & Geetha, 2020)? Does it define sustainable human resource management as achieving current success without compromising future needs? Or does it conceptualize sustainability as balancing resource consumption with resource regeneration (Ehnert & Harry, 2012)? Does the organization define sustainable HRM as green human resource management (Amrutha & Geetha, 2020; Miah et al., 2024; Yin et al., 2025)? Does it equate sustainability with health, resilience, or diversity (Campos-García et al., 2024)? Does it aim to foster sustainable employability (Irfan et al., 2023) or other positive outcomes?

Addressing these and similar questions is essential to establishing a clear organizational understanding of sustainability. Moreover, it is crucial to identify and articulate specific applications of sustainability within the company. Only after employees have been properly informed about the concept and its relevance to the organization can they be expected to redesign their jobs and careers in alignment with these sustainability principles. Top-down communication plays a crucial role in embedding sustainability within HRM. Leadership must articulate the company's sustainability vision and integrate sustainable values into job expectations. When employees understand the significance of sustainability in their work, they are more likely to engage in sustainable job crafting, reshaping their tasks and responsibilities to support both personal and organizational sustainability goals.

5.2 Job Flexibility and Employee Authority

A fundamental prerequisite for sustainable job crafting is the presence of flexibility and authority within the workplace. Based on Cognitive Evaluation Theory (Deci et al., 1975), when individuals perceive their career-related behaviors as self-determined and autonomously controlled, they develop intrinsic motivation to engage in job crafting. Sustainable career development requires individuals to navigate dynamic work environments, adapt to uncertainties, and make timely adjustments. This necessitates that job roles be sufficiently adaptable, either through inherent job characteristics or through organizational policies that promote dynamism and flexibility. Without such adaptability, employees' efforts to modify their job structure or optimize their tasks become impractical. A sustainable career consists of self-regulatory processes that unfold within relatively stable career cycles. Throughout these cycles, individuals must redefine their career adaptation timeline to preserve and enhance, rather than deplete, human and career development. Every instance of successful adaptation consumes personal resources, highlighting that career sustainability is inherently tied to resource availability. Job crafting, in particular, is shaped by multiple environmental factors, including organizational policies, work culture, labor market conditions, and socio-cultural influences (Wang, 2022).

For employees to effectively engage in sustainable job crafting, they must possess both the flexibility to make modifications and the authority to implement meaningful changes in their roles. Job flexibility enables employees to adjust their work responsibilities, relationships, and cognitive perspectives to align with long-term sustainability objectives. In highly rigid job structures where roles are predefined and strictly maintained, employees may lack the capacity to introduce meaningful changes. This rigidity can stifle innovation, reduce job satisfaction, and ultimately hinder proactive efforts toward job optimization. Organizations that recognize the importance of adaptability can foster an environment where job crafting aligns with broader sustainability goals by embedding flexibility into their work design and policies. Beyond flexibility, authority is another critical factor that determines the extent to which employees can engage in sustainable job crafting. Employees must not only be allowed but also empowered to reshape their roles in ways that enhance both personal and organizational outcomes. This empowerment is contingent upon access to structural, relational, and challenging resources, which enable employees to make substantial, meaningful modifications to their work. Without the necessary authority, even in a flexible job structure, employees may find themselves constrained in their ability to act proactively.

5.3 Sustainable Job Crafting Interventions

Sustainable job-crafting interventions involve structured training programs and exercises designed to foster job-crafting behaviors among employees. Although job crafting is increasingly recognized as a mechanism for enhancing work engagement, relatively few studies have specifically explored interventions intended to promote this behavior. Rai (2018) synthesizes insights from a limited but significant body of research, including the work of Berg et al. (2013) and Van den Heuvel et al., (2012, 2015), to examine training methodologies and practical exercises that facilitate sustainable job crafting within organizations.

Van den Heuvel et al., (2012, 2015) developed a job crafting training program to enhance employees' awareness of how they could modify job characteristics, particularly job demands and job resources, to better align with their needs. The training program consists of three core components. The initial phase introduces employees to the Job Demands-Resources (JD-R) model and the concept of job crafting, helping them develop a Personal Crafting Plan (PCP) that outlines specific changes they intend to make in their work roles. During the implementation phase, employees put their plans into action over four weeks while maintaining a logbook to document their job crafting activities and reflect on their progress. In the final phase, a reflection meeting provides an opportunity to discuss achievements, challenges, and possible solutions, reinforcing job-crafting behaviors and supporting their long-term integration into daily work routines (Rai, 2018). This structured approach to job crafting interventions can be applied to sustainable job crafting by aligning employees' job modifications with sustainability objectives. Organizations can integrate sustainability-focused job crafting by educating employees on sustainability principles and their relevance to job roles, encouraging them to incorporate green initiatives into their Crafting Plans, and providing resources and support for sustainability-related job crafting. Access to mentorship, networking opportunities, or specialized training in green skills can further facilitate this process. Long-term reinforcement through reflection meetings and continuous learning programs can help sustain these efforts.

In addition to structured training programs, Berg et al. (2013) introduced a job crafting exercise designed to help individuals identify opportunities to reshape their work in alignment with their intrinsic motivations, strengths, and passions. When applying this exercise to sustainability, it should be customized to incorporate sustainability-related concepts and align with organizational sustainability goals. By encouraging employees to view their jobs as a dynamic set of elements rather than a rigid structure, this exercise fosters a proactive approach to sustainable job crafting. The sustainable job crafting exercise consists of several steps. Initially, participants create a "Before Sketch," a visual representation of how they currently allocate their time, energy, and attention across various tasks. These tasks are categorized into "Task Blocks" of different sizes, with the largest blocks representing tasks that demand the highest level of time and effort, the smallest blocks representing those requiring the least, and medium-sized blocks falling in between. This visualization enables employees to critically evaluate how their resources are currently allocated within their roles. Next, participants identify three key personal attributes—motives, strengths, and passions—while integrating sustainability goals that influence their sustainable work. These attributes guide the construction of an "After Diagram," which presents a redesigned version of their job. In this phase, employees create new task blocks that reflect how they would prefer to allocate their time and energy, aligning with sustainable work structures. By reshaping their responsibilities to include environmentally and socially responsible tasks, employees can enhance their sense of purpose while contributing to broader sustainability objectives. Following this, participants establish "Role Frames" by grouping tasks with similar functions to enhance the meaningfulness of their work. This categorization helps employees recognize how their tasks contribute to organizational sustainability goals, reinforcing the significance of their roles in this transformation. The final step of the exercise involves developing an "Action Plan," which outlines specific strategies and goals for integrating sustainability into their redesigned job structures in both the short and long term. By implementing this plan, employees take actionable steps toward crafting jobs that not only align with their strengths and interests but also support the company's sustainability initiatives. Through this process, employees gain the autonomy and awareness needed to integrate sustainability into their daily work, making sustainability-oriented job crafting a continuous and meaningful endeavor. Organizations that incorporate this exercise into their job-crafting interventions can create a workforce that is both engaged and committed to sustainability-driven innovation.

6 Conclusion

This chapter has examined the integration of sustainability into job crafting, emphasizing how employees can actively modify their roles to align with long-term environmental, social, and economic goals. Sustainable job crafting represents a proactive approach where individuals take ownership of their work by reshaping tasks, relationships, and cognitive perceptions in ways that support both personal fulfillment and broader sustainability objectives.

A key factor in enabling sustainable job crafting is job flexibility, as well as employee autonomy. When employees have the autonomy to make meaningful adjustments in their roles, they are more likely to engage in behaviors that contribute to sustainability. Additionally, sustainability awareness within organizations plays a crucial role in fostering a culture where employees recognize the importance of integrating sustainability into their daily tasks. Without clear communication from leadership and alignment with organizational goals, the potential for sustainable job crafting remains limited. Structured interventions, such as training programs and job crafting exercises, provide employees with practical tools to embed sustainability into their work. These initiatives not only enhance employee engagement and motivation but also contribute to long-term organizational resilience. By fostering a work environment that encourages sustainable job crafting, organizations can improve innovation, employee well-being, and overall performance while advancing sustainability efforts.

Despite its potential, sustainable job crafting remains an emerging area of research. Future studies should explore how different industries and job roles can best implement these practices, as well as the long-term impact on employees and organizations. By prioritizing sustainability within job design and empowering employees to shape their roles actively, organizations can create a workforce that is both adaptable and committed to sustainable development. Encouraging sustainable job crafting can lead to more meaningful work experiences, increased job satisfaction, and a stronger alignment between individual careers and global sustainability goals.

7 Case Study: Company X Utilizes Sustainability Job Crafting in Its Corporate Social Responsibility Plan

Company X operates in the production of feed additives and supplements for livestock and poultry and is recognized as a leading brand in this industry. Its products ensure the safety and health of feed for livestock, poultry, and aquaculture, playing a crucial role in the production of healthy milk and meat. The company's primary customers include animal feed manufacturers, poultry farms, livestock farms, veterinary pharmacies, and related industries. Currently, Company X produces additives for the livestock and poultry sector and is striving to diversify its products to meet all customer needs in this field.

As part of its corporate social responsibility initiatives, Company X has decided to redesign its job structures with a sustainability-oriented approach. The company has prioritized two key job categories—sales and research & development—for this redesign process. However, instead of adopting a purely top-down approach, it aims to leverage employee input to enrich job roles, enhance their meaningfulness from employees' perspectives, and, most importantly, align them with sustainable human resource management principles. After multiple brainstorming sessions, the company formulated a three-level implementation plan, as outlined in Fig. 2.

- **Macro-Level**: The company has placed sustainability awareness at the core of its strategy, revising its corporate strategies accordingly. After reassessing its operational strategies with a sustainability perspective, the company has organized orientation sessions for employees in sales and research & develop-

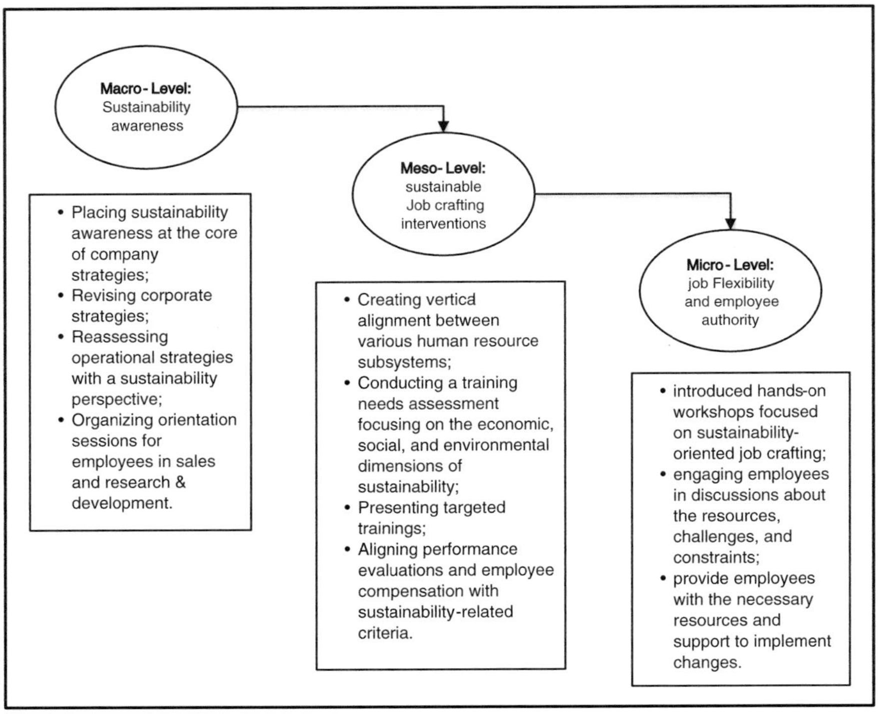

Fig. 2 A three-level plan for sustainable job crafting

ment. These sessions were not conducted separately; instead, they were designed to familiarize both frontline and support roles with each other's activities, fostering collaboration and shared learning.

- **Meso-Level**: The company has sought to create vertical alignment between various human resource subsystems—such as training and development, performance management, and compensation—with its broader business strategies. A training needs assessment was conducted, focusing on the economic, social, and environmental dimensions relevant to different job roles. Employees received targeted training based on these findings. Additionally, performance evaluations now incorporate sustainability-related criteria, which are assessed semiannually, with corresponding feedback provided to employees. A portion of employee compensation has also been linked to their performance scores in these sustainability metrics.
- **Micro-Level**: The company introduced hands-on workshops focused on sustainability-oriented job crafting. These workshops not only familiarize employees with job design principles and different job crafting approaches but also engage them in discussions about the resources, challenges, and constraints they face in crafting their roles with sustainability in mind. The goal is to provide employees with the necessary resources and support to implement these changes effectively.

After a year of implementation, the company plans to assess the effectiveness of the initiative. This evaluation will measure outcomes and, if necessary, revise the plan. They considered the following questions for the revision plan:

- What criteria should be used to assess the initiative's success in economic, social, and environmental dimensions?
- How do sales and research & development roles differ in their approach to sustainable job crafting?
- If the program were extended for another six months, what additional measures should be

introduced, and which elements should be modified or removed?

- What opportunities and challenges does the business context present for sustainable job crafting?
- Do different job crafting approaches yield varying results? At the micro, meso-, and macro levels, what complementary actions could enhance the effectiveness of the initiative?
- Which other human resource management systems can be aligned with this initiative, and how?

8 Questions

1. What are the key principles of sustainable job crafting, and how does it differ from traditional job crafting?
2. How do the Job Demands-Resources (JD-R) and Job Demands-Control-Support (JDCS) Models contribute to understanding sustainability-oriented job design?
3. If you were a manager implementing sustainability-oriented job crafting in your organization, what steps would you take to raise sustainability awareness among employees?
4. How can job flexibility and employee authority facilitate sustainable job crafting in an organization? Provide an example.
5. What potential challenges might organizations face when attempting to integrate sustainability principles into job crafting practices?
6. How might sustainable job crafting impact long-term employee well-being and organizational performance?
7. Based on the case study of Company X, how would you modify the initiative to better align with an organization operating in a different industry, such as healthcare or technology?
8. What role does leadership play in encouraging employees to engage in sustainable job crafting, and how can organizations ensure ongoing employee participation in these initiatives?

References

Amrutha, V. N., & Geetha, S. N. (2020). A systematic review on green human resource management: Implications for social sustainability. *Journal of Cleaner Production, 247*, 119131. https://doi.org/10.1016/j.jclepro.2019.119131

Bakker, A. B., & Demerouti, E. (2007). The Job Demands-Resources model: State of the art. *Journal of Managerial Psychology, 22*(3), 309–328. https://doi.org/10.1108/02683940710733115

Berg, J. M., Dutton, J. E., & Wrzesniewski, A. (2013). Job crafting and meaningful work. In B. J. Dik, Z. S. Byrne, & M. F. Steger (Eds.), *Purpose and meaning in the workplace* (pp. 81–104). American Psychological Association. https://doi.org/10.1037/14183-005

Bruning, P. F., & Campion, M. A. (2018). A role–resource approach–avoidance model of job crafting: a multimethod integration and extension of job crafting theory. *Academy of Management Journal, 61*(2), 499–522. https://doi.org/10.5465/amj.2015.0604

Campos-García, I., Alonso-Muñoz, S., González-Sánchez, R., & Medina-Salgado, M.-S. (2024). Human resource management and sustainability: Bridging the 2030 agenda. *Corporate Social Responsibility and Environmental Management, 31*(3), 2033–2053. https://doi.org/10.1002/csr.2680

Courchesne, S. A., Stynen, D., Semeijn, J. H., & Caniëls, M. C. J. (2025). Fostering sustainable employability in interorganizational networks: Which activities and conditions pay off? *Journal of Workplace Learning, 37*(9), 1–23. https://doi.org/10.1108/JWL-03-2024-0066

De Vos, A., & Van der Heijden, B. I. J. M. (2017). Current thinking on contemporary careers: The key roles of sustainable HRM and sustainability of careers. *Current Opinion in Environmental Sustainability, 28*, 41–50. https://doi.org/10.1016/j.cosust.2017.07.003

Deci, E. L., Cascio, W. F., & Krusell, J. (1975). Cognitive evaluation theory and some comments on the Calder and Staw critique. *Journal of Personality and Social Psychology, 31*(1), 81–85. https://doi.org/10.1037/h0076168

Demerouti, E., Bakker, A. B., Nachreiner, F., & Schaufeli, W. B. (2001). The job demands-resources model of burnout. *Journal of Applied Psychology, 86*(3), 499.

Ehnert, I. (2009). Introducing sustainability into HRM. In I. Ehnert (Ed.), *Sustainable human resource management: A conceptual and exploratory analysis from a paradox perspective* (pp. 1–31). Physica-Verlag HD. https://doi.org/10.1007/978-3-7908-2188-8_1

Ehnert, I., & Harry, W. (2012). Recent developments and future prospects on sustainable human resource management: Introduction to the special issue. *Management Revue, 23*(3), 221–238. http://www.jstor.org/stable/41783719

Elkington, J. (1997). *Cannibals with forks: The triple bottom line of 21st century*. Capstone Publishing.

Gollan, P. J. (2000). Human resources, capabilities and sustainability. In D. Dunphy, J. Benveniste, A. Griffiths, & P. Sutton (Eds.), *Sustainability: The corporate challenge of the 21st century*. Allen and Unwin.

Hackman, J. R., & Oldham, G. R. (1976). Motivation through the design of work: Test of a theory. *Organizational Behavior and Human Performance, 16*(2), 250–279. https://doi.org/10.1016/0030-5073(76)90016-7

Hughes, E., & Dundon, T. (2025). A multi-level systems perspective on (un)sustainable HRM in adult social care. *Human Resource Management*. https://doi.org/10.1002/hrm.22298

Irfan, S. M., Qadeer, F., Abdullah, M. I., & Sarfraz, M. (2023). Employer's investments in job crafting to promote knowledge worker's sustainable employability: A moderated mediation model. *Personnel Review, 52*(8), 2025–2047. https://doi.org/10.1108/PR-10-2021-0704

Karasek, J. R. A. (1979). Job demands, job decision latitude, and mental strain: Implications for job redesign. *Administrative Science Quarterly, 24*(2), 285–308.

Kramar, R. (2014). Beyond strategic human resource management: Is sustainable human resource management the next approach? *The International Journal of Human Resource Management, 25*(8), 1069–1089. https://doi.org/10.1080/09585192.2013.816863

Kramar, R. (2022). Sustainable human resource management: Six defining characteristics. *Asia Pacific Journal of Human Resources, 60*(1), 146–170. https://doi.org/10.1111/1744-7941.12321

Leana, C., Appelbaum, E., & Shevchuk, I. (2009). Work process and quality of care in early childhood education: The role of job crafting. *Academy of Management Journal, 52*(6), 1169–1192. https://doi.org/10.5465/amj.2009.47084651

Lichtenthaler, P. W., & Fischbach, A. (2016). Job crafting and motivation to continue working beyond retirement age. *Career Development International, 21*(5), 477–497. https://doi.org/10.1108/CDI-01-2016-0009

Lichtenthaler, P. W., & Fischbach, A. (2019). A meta-analysis on promotion- and prevention-focused job crafting. *European Journal of Work and Organizational Psychology, 28*(1), 30–50. https://doi.org/10.1080/1359432X.2018.1527767

Longoni, A., & Cagliano, R. (2015). Environmental and social sustainability priorities. *International Journal of Operations & Production Management, 35*(2), 216–245. https://doi.org/10.1108/IJOPM-04-2013-0182

Luu, T. T. (2019). Green human resource practices and organizational citizenship behavior for the environment: The roles of collective green crafting and environmentally specific servant leadership. *Journal of Sustainable Tourism, 27*(8), 1167–1196. https://doi.org/10.1080/09669582.2019.1601731

Macke, J., & Genari, D. (2019). Systematic literature review on sustainable human resource management. *Journal of Cleaner Production, 208*, 806–815. https://doi.org/10.1016/j.jclepro.2018.10.091

Miah, M., Gábor, S.-S., & Walter, V. (2024). A systematic literature review on green human resource management (GHRM): An organizational sustainability perspective. *Cogent Business & Management, 11*(1), 2371983. https://doi.org/10.1080/23311975.2024.2371983

Müller-Christ, G., & Remer, A. (1999). Umweltwirtschaft oder Wirtschaftsökologie? Vorüberlegungen zu einer Theorie des Ressourcenmanagements. In E. Seidel (Ed.), *Betriebliches Umweltmanagement im 21. Jahrhundert: Aspekte, Aufgaben, Perspektiven* (pp. 69–87). Springer. https://doi.org/10.1007/978-3-642-60245-0_6

Petrou, P., Demerouti, E., Peeters, M. C. W., Schaufeli, W. B., & Hetland, J. (2012). Crafting a job on a daily basis: Contextual correlates and the link to work engagement. *Journal of Organizational Behavior, 33*(8), 1120–1141. https://doi.org/10.1002/job.1783

Petrou, P., Demerouti, E., & Schaufeli, W. B. (2016). Crafting the change: The role of employee job crafting behaviors for successful organizational change. *Journal of Management, 44*(5), 1766–1792. https://doi.org/10.1177/0149206315624961

Rai, A. (2018). Job crafting intervention: Fostering individual job redesign for sustainable organisation. *Industrial and Commercial Training, 50*(4), 200–208. https://doi.org/10.1108/ICT-11-2017-0089

Ryan, R. M., & Deci, E. L. (2000). Self-determination theory and the facilitation of intrinsic motivation, social development, and well-being. *American Psychologist, 55*(1), 68–78.

Ryan, R. M., & Deci, E. L. (2023). Self-determination theory. In F. Maggino (Ed.), *Encyclopedia of quality of life and well-being research* (pp. 6229–6235). Springer. https://doi.org/10.1007/978-3-031-17299-1_2630

Siddiqui, H., Hummayun, G. K., & Mubeen, H. (2022). Job design and sustainability: Exploring uncharted link through literature. *RADS Journal of Business Management, 4*(2), 167–183.

Tahir, A. H., Umer, M., Nauman, S., Abbass, K., & Song, H. (2024). Sustainable development goals and green human resource management: A comprehensive review of environmental performance. *Journal of Environmental Management, 370*, 122495. https://doi.org/10.1016/j.jenvman.2024.122495

Tims, M., Bakker, A. B., & Derks, D. (2012). Development and validation of the job crafting scale. *Journal of Vocational Behavior, 80*(1), 173–186. https://doi.org/10.1016/j.jvb.2011.05.009

Tuan, L. T. (2021). Effects of environmentally-specific servant leadership on green performance via green climate and green crafting. *Asia Pacific Journal of Management, 38*(3), 925–953. https://doi.org/10.1007/s10490-019-09687-9

Van den Heuvel, M., Demerouti, E., & Peeters, M. C. W. (2012). Successful job crafting through group training. In J. De Jonge, M. C. W. Peeters, S. Sjollema, & H. De Zeeuw (Eds.), *Scherp in Werk* (Vol. 5, pp. 7–20). Van Gorcum Assen.

van den Heuvel, M., Demerouti, E., & Peeters, M. C. W. (2015). The job crafting intervention: Effects on job resources, self-efficacy, and affective well-being. *Journal of Occupational and Organizational*

Psychology, *88*(3), 511–532. https://doi.org/10.1111/joop.12128

van Wingerden, J., Bakker, A. B., & Derks, D. (2017). Fostering employee well-being via a job crafting intervention. *Journal of Vocational Behavior*, *100*, 164–174. https://doi.org/10.1016/j.jvb.2017.03.008

Wang, Q. (2022). Research on job crafting from the perspective of sustainable career: Motivation, paths and intervention mechanisms. *Advances in Psychological Science*, *30*(3), 499–510. https://doi.org/10.3724/SP.J.1042.2022.00499

Wrzesniewski, A., & Dutton, J. (2001). Crafting a job: Revisioning employees as active crafters of their work. *Academy of Management Review*, *26*(2), 179–201. https://doi.org/10.5465/amr.2001.4378011

Yin, X., Khan, A. J., Basheer, M. F., Iqbal, J., & Hameed, W. U. (2025). Green human resource management: A need of time and a sustainable solution for organizations and environment. *Environment, Development and Sustainability*, *27*(1), 1379–1400. https://doi.org/10.1007/s10668-023-03915-5

Zaugg, R. J., Blum, A., & Thom, N. (2001). *Sustainability in Human Resource Management* (51).

Sustainable Entrepreneurship in a National and International Context

Sustainable Entrepreneurship Development in Iran with an Emphasis on Environmental Issues

Hamideh Maleksaeidi

Abstract

Iran is facing a serious environmental crisis, the severity and scope of which have attracted global attention. Environmental crises in Iran have also had economic, social, health, and policy impacts. Experts and specialists believe that failure to address this issue can pose a significant threat to the country's security and stability. Therefore, some experts and policymakers are looking for solutions to reduce environmental crises as much as possible and minimize their negative consequences. In response to this concern, this chapter analyzes the necessity of sustainable entrepreneurship development in Iran to deal with environmental crises and identify challenges facing sustainable entrepreneurship development. The findings indicated that water scarcity, natural resource degradation, pollution, accumulation of waste, and climate change are Iran's most important environmental issues. The study showed how sustainable entrepreneurship using eco-friendly technologies and strategies can significantly solve environmental problems. However, sustainable entrepreneurship development in Iran faces various challenges, including high initial cost, high interest rates, and long-term payoff, changing the mindset of people about the environment and environmentally friendly goods, as well as technological, infrastructural, legal, and policy limitations that must be overcome.

Keywords

Sustainable entrepreneurship · Environmental crisis · Challenges · Iran

Learning Objectives

By the end of this chapter, readers will be able to:

- Identify the role of sustainable entrepreneurship in reducing and combating environmental problems.
- Analyze the necessity of sustainable entrepreneurship development in Iran.
- Recognize the challenges facing sustainable entrepreneurship development in Iran.

1 Introduction

While the efforts of governments to improve sustainability, whether in word or action, are outstanding, one new concept rising from sustainability is sustainable entrepreneurship, which combines social and environmental dimensions with entrepreneurial economic processes (Abereijo, 2016). The sustainable orientation of

H. Maleksaeidi (✉)
Faculty of Agriculture, University of Kurdistan, Sanandaj, Iran
e-mail: h.maleksaeidi@uok.ac.ir

M. Akbari et al. (eds.), *Sustainable Entrepreneurship*, University of Tehran Science and Humanities Series, https://doi.org/10.1007/978-3-032-04909-4_13

entrepreneurship contributes to addressing environmental degradation and achieving sustainable development goals in various countries, particularly in developing nations (Apostu & Gigauri, 2023; Salaheldeen et al., 2024). From an environmental point of view, both developed and developing countries face many environmental challenges. Still, the nature of these challenges and how they affect the livelihoods and lives of people in developing countries is deeper and more complex (Bowonder, 1987a). Environmental problems in developed countries are often caused by industrialization and growing consumerism. In developing countries, however, these problems are often caused by a lack of development and education (Bowonder, 1987b). In addition to environmental problems, developing countries are also grappling with various social challenges, according to the conducted research (e.g. Khuhawar & Shah, 2019; Pohlan, 2024; Rözer et al., 2020). Many of these problems are rooted in economic problems, especially unemployment. As stated, sustainable entrepreneurship can help solve social and environmental issues, particularly in developing countries. But before that, it is necessary to understand the nature of these problems and their dimensions. Since Iran is one of the developing countries that face intensified environmental challenges, this research intends to investigate the necessity of sustainable entrepreneurship in Iran with an emphasis on environmental issues in this country. In this chapter, we will discuss some of the environmental problems in Iran, the role sustainable entrepreneurship can play in addressing these issues, the importance of developing sustainable entrepreneurship in the country, and the challenges it faces.

2 Environmental Issues in Iran

Iran is facing unpredictable environmental crises that not only have challenged the sustainability of its various economic sectors but also threaten its stability if left unchecked. The intensity and scope of environmental pollution and degradation in Iran have increased unprecedentedly over the past few decades. In many cases, it is much greater than in other countries, to the point that it can no longer be ignored (Ebadi et al., 2020; Ranjbar Heidari & Jamshidi, 2016). This issue has caused concern and warning among many individuals and groups, non-governmental organizations (NGOs), and national and international environmental specialists (Tabbaz, 2016). The environmental crises in Iran result from the unfavorable performance of various economic sectors, including agriculture, industry, and services. Below, some of these crises and the role of various economic activities in their emergence are explained.

2.1 Water Scarcity

Iran, placed in the arid and semi-arid belt of the world, is among the countries facing water resource constraints. A review of the rainfall statistics in Iran over the past 50 years indicates that the long-term trend in rainfall has been downward, and the volume of rainfall has decreased in recent years compared to the past (Mokhtari Hashi & Moradi, 2021). Estimates indicate that per capita access to fresh water in Iran will fall below 1500 cubic meters per year by 2030 (Jamshidi et al., 2019); this is while the average annual water consumption in the country is estimated to be about 96 billion cubic meters, which is about 8% more than Iran's total renewable water resources (89 billion cubic meters) (Azadi & Mesgaran, 2018). There are various causes for water shortage in Iran. Some causes of water scarcity include population increase and dispersion, inefficient agriculture, mismanagement of water, and climate change (Khatibi & Arjjumend, 2019). Water scarcity in dry and low-rainfall areas causes water stress and excessive extraction of groundwater aquifers. Over-extraction of groundwater aquifers will have subsequent negative consequences and will lead to the gradual destruction of the ecosystem and the formation of numerous environmental crises, including pollution, dwindling groundwater reserves, land subsidence, desertification, and the destruction of habitats such as wetlands (Hakim & Alterman, 2023; Mokhtari Hashi & Moradi, 2021). All of this is exacerbated by the agricultural sector, which consumes about 90% of the water (Hakim & Alterman, 2023).

2.2 Natural Resource Degradation

Natural resources are the basis of sustainable life and development, which, in addition to creating balance in the ecosystem, play a fundamental role in creating food security, increasing gross domestic product, exports, earning foreign exchange, and strengthening the political position of countries (KianiRad & Noroozi, 2022). In Iran, natural resources have been the basis of life, production, and livelihood for centuries, and have provided the possibility of diverse and proportionate uses for the exploiters. During the period of exploitation, as long as the use of the land was within its ecological capacity, no form of destruction of natural resources was evident. However, throughout the current century and particularly in the last few decades, with increasing population and standard of living, the use has developed beyond the capacity of the ecosystems. The destruction of natural resources has become a concern for the exploiters, guardians, and researchers (Ansari et al., 2009). For example, according to official statistics from the Ministry of Agriculture-Jahad (2024), the average soil erosion rate in the country is about 16.5 tons per hectare, approximately 2–2.5 times the Asian average and 5–6 times the global average. In another example, a comparison of aerial photographs and satellite images shows that the area covered by forests in northern Iran between 1967 and 1994 has decreased by about 141,572 ha, or 1.7% (Velayati & Kadivar, 2009). Also, between 1969 and 2000, there were significant changes in the extent of Iran's forests, such that the area covered by Iran's natural forests decreased by about 11%. As a result, about 5 million hectares of land have been eroded (Velayati & Kadivar, 2009). Moreover, while rangelands, with an area of about 85 million hectares (Motamedi et al., 2020), constitute about 55% of Iran's total area and play a fundamental role in ecosystem sustainability, food security, and livelihoods, statistics show that during the past three decades, above 20% of Iran's rangelands have been destroyed in terms of quantity and quality. This degradation has occurred much faster than in many other countries worldwide (AlaAlD-invandi, 2021). Studies show that human interference, including uncontrolled and unprincipled livestock grazing, land use change, expansion of mining exploitation (Motamedi et al., 2020), and uncontrolled exploitation of wood and non-wood products (Safari et al., 2023) are the main factors in the destruction of forests, rangelands, and other natural resources in Iran.

2.3 Pollution

One of the main threats to Iran's environment is pollution from industries and agricultural projects. The lack of clean technologies and the use of old and unsustainable technologies for production cause the release of pollutants and dangerous chemicals, and since some industries do not adequately control and monitor pollutants and chemicals, these cases can lead to leaks of undesirable pollutants and their entry into soil, air and water resources (Gaur et al., 2024). Some large industries also lack access to proper waste disposal and recycling systems, failing to comply with health and environmental standards. As a result, their waste enters the environment, further polluting the soil and water (Borun, 2016). This issue also leads to a decline in biota and disruption of the aquatic environment's food chain, significantly impacting ecosystems and biological balance (Roy & Ahmaruzzaman, 2022). It should also be considered that pollution from unprincipled sewage disposal can lead to the transmission of diseases, human health disorders, and increasing medical costs, creating a heavy burden on the national economy (Jahandari & Abbasnejad, 2024). Chemicals and contaminants in wastewater can also penetrate the soil and change the physical and chemical qualities of the soil, reducing soil productivity, destroying soil organisms, and reducing agricultural production (Khamisabadi et al., 2019). Studies have also shown that excessive and untimely use of fertilizers and pesticides has several adverse environmental effects. These effects include leaching and displacement of soil compounds, pollution of groundwater and surface water, accumulation of heavy metals and nitrates, increased wetland feeding, air pollution and accumulation of nitrogen and sulfur gases, and production of greenhouse gases, global warming, depletion of the ozone layer, acidification of rain, and accumulation of chemicals in plants.

2.4 Accumulation of Urban, Industrial, and Agricultural Waste

In 2018, the World Bank reported that per capita municipal waste generation in Iran was 0.65 kg per person per day, almost equal to the average for the Middle East and the world (Kaza et al., 2018), though Iran ranks low in waste management and recycling. Despite the waste management law, the country performs poorly in this area, ranking 97th out of 180 countries and one of the few countries at the bottom of the list in the Middle East (Babaei et al., 2022). Also, according to the Waste Management Office of the Environmental Protection Organization of Iran in 2020, the volume of industrial waste production in Iran was estimated to be 32 million tons annually, which is the result of the activities of industrial, mining, refinery, petrochemical, and power plant centers (Babaei et al., 2022). Furthermore, In Iran, the annual production of waste in the agricultural sector is about 170 million tons (Safari Alamouti & Shams, 2016), and almost 30% of agricultural products are converted into waste without being consumed at various stages due to inadequacies in the storage and marketing systems (Khosravani et al., 2013). The accumulation of waste and its lack of proper recycling will have numerous environmental consequences. Unsanitary dispersion and disposal of waste will cause water, soil, and air pollution and will facilitate the spread of various diseases. There is no doubt about the microbial and chemical pollution of waste. Waste not only gives rise to disease, malodor, and disproportion in the environment but can also cause significant harm by polluting soil, water, and air. The accumulation of waste in the streets and its dispersion in the environment through leachate leakage leads to the contamination of surface and groundwater, and the penetration of the gases produced into the water causes its hardness. Waste leachate is highly toxic and contains chlorosulfate, magnesium, phosphorus, calcium, iron, sodium, potassium, nitrate, organic nitrogen, ammonia, and suspended solids. All urban and rural waste is contaminated with various microorganisms that can contaminate water resources. The contamination has been the reason for many cases of aquatic animal deaths and the drying of plants in the rivers. There are also various diseases caused by water pollution because of the presence of heavy metals, fluorine compounds, cadmium, nitrates, etc. (El-Fadel et al., 1997; Muhammad et al., 2019; Saffari, 2013; Siddiqua et al., 2022).

2.5 Climate Change and Disasters

Iran, with a total emission of greenhouse gases reaching nearly 741,616 million tons of CO_2, is one of the first countries to be blamed for causing climate change in the Middle East and the world. Iran's significant share of greenhouse gas emissions is attributed to its substantial oil and gas production, as well as rapid urbanization (Daneshvar et al., 2019). Iran is also one of the most vulnerable countries in the Middle East and North Africa to the impacts of climate change. Climate change in Iran is primarily characterized by a decline in rainfall and an increase in temperature. The average temperature in the hot seasons is around 50–54 °C. This is despite various reports indicating that Iran will experience an average temperature increase of 2.6 °C and a 35% decline in rainfall over the coming decades (Soleimani Morchekhorti, 2024). Even researchers have proven that over the last few decades, semi-arid areas in Iran have become increasingly dry (Gholami et al., 2017).

Climate change in Iran has brought disasters such as droughts and floods. Over the past two decades, floods have affected the lives of more than 11 million people and resulted in over 2600 fatalities (Daneshvar et al., 2019). In addition, floods can have many environmental effects.

Environmental effects of floods can include soil erosion, landslides, and damage to vegetation. Pollutants from floodwater can influence water quality, habitats, flora, and fauna (Aldardasawi & Eren, 2021). Also, climate change, along with human interventions, has led to fundamental changes in the severity, intensity, and duration of droughts in Iran (Shayeghi et al., 2024). Along with economic and social impacts such as reduced income and increased migration, drought will also have various environmental impacts such as drying up of springs, rivers, and habitats, desertification, deforestation, loss of plant and animal species, landslide, and soil erosion (Bahrami & Separi, 2021).

3 Sustainability Entrepreneurship and Tackling Environmental Challenges

In today's changing world, sustainable entrepreneurship is more important than ever. We currently face multiple and complex challenges, including climate change, environmental issues, and social inequalities. Therefore, it is necessary for entrepreneurs to play a more prominent role in building a sustainable world and a better future. One of the main reasons sustainable entrepreneurship matters is its positive influence on the environment (FasterCapital, 2024). According to Odeyemi et al. (2024), sustainable entrepreneurship, with its continued emphasis on environmental awareness and the use of ecological practices and technologies to achieve economic prosperity, is an important and fundamental concept that has emerged in today's business world. In other words, while traditional business practices focus on maximizing profits, which often leads to irreparable damage to natural resources and the environment (Odeyemi et al., 2024), sustainable entrepreneurs, by recognizing and understanding the importance of environmental protection, can manage their business in a way that minimizes harm to the environment (FasterCapital, 2024; Neumann, 2022). For example, imagine entrepreneurs using eco-friendly materials to produce global fashion and textiles or using waste to produce clean energy. The following are practical examples of the environmental achievements of sustainable entrepreneurship that will be effective for developing countries, including Iran, which is facing environmental problems.

3.1 Sustainable Entrepreneurship and Waste Recycling

Today, one of the environmental problems that countries worldwide, including Iran, face is the accumulation of urban, industrial, and agricultural waste (Zarghi & Rahmanizadeh, 2017). In this regard, waste recycling is one of the areas of entrepreneurship that not only helps protect the environment but also has a very high income. Many developed countries, aware of this issue, recycle and reuse a significant percentage of waste that is discarded and harmful to the environment using advanced and modern technologies. In addition to the positive environmental impacts of this activity, countries' income generation from this industry is sometimes so high that countries, including Sweden and Germany, import waste to recycle and export it to other countries. Additionally, studies indicate that entrepreneurship in the waste recycling sector has a high potential for generating employment. According to statistics published by the US Environmental Protection Agency in 2006, one job is created by burning 10,000 tons of waste, and six jobs are created by burying 10,000 tons of waste, while more than 36 jobs will be created by recycling the same amount of waste (Zarghi & Rahmanizadeh, 2017). Among the environmental benefits of waste recycling are savings in the use of natural resources and energy. For example, using recycled aluminum scrap to make aluminum cans requires 9% less energy than producing it from raw materials used in aluminum production. Similarly, producing one ton of paper from recycled resources saves about 17 trees and uses 50% less water. Also, using agricultural waste to produce productions such as renewable energy (Phiri et al., 2024; Sonu et al., 2023), cellulosic materials (Al Kamzari et al., 2023; Saberi Riseh et al., 2024), citric acid (Kumar et al., 2003), compost (Koul et al., 2022; Külcü & Yaldiz, 2014) and bioethanol (Jayakumaret al., 2023; Sarkar et al., 2012) will not only help reduce the negative environmental impacts of these wastes, but will also have significant economic, health and climatic positive consequences overall, while waste causes many diseases and damages to urban and rural settlements by polluting water, soil, and air, as well as its greenhouse effects. By utilizing sustainable entrepreneurship and creating recycling industries, such damages can be prevented, and the costs of treatment caused by diseases can be reduced. Among the countries that have been successful in recycling waste and reducing their environmental impacts are Japan, the United States, Australia, and Germany. By expanding entrepreneurial activities in the field of waste recycling, they have turned the threat of waste generation and its negative impacts on the environment into an opportunity to reduce these impacts and earn money.

3.2 Sustainable Entrepreneurship and Sustainable Water Management

In recent years, water scarcity and the necessity of proper management of this vital resource have become a suitable opportunity for entrepreneurs and entrepreneurial companies around the world. These individuals and companies are trying to find and provide innovative solutions, systems, and technologies in the field of management and optimal use of water in the industrial, agricultural, construction, and domestic sectors while helping to preserve this vital liquid and create the basis for increasing employment in society. Among the technologies that entrepreneurs and entrepreneurial companies have introduced in the field of optimal water management are sensors (Hamel et al., 2024), smart irrigation systems (Devendiran et al., 2023), water recycling systems, and water-saving methods (Chen et al., 2024), as well as the development of innovative technologies for water purification and improving water quality (Ahmad & Azam, 2019). These technologies include various processes such as reverse osmosis, nanofiltration, and the use of special filters to remove contaminants from water (Abdel-Fatah, 2018). In addition, the creation of technologies and systems for transferring water from water-rich areas to areas that need water, such as smart water distribution systems and new water transfer methods, such as drip irrigation lines, is also among these innovations (Mansingh et al., 2024). The development of water storage technologies and systems for better management of water resources, such as dam construction, the use of groundwater reservoirs, rainwater storage, and water collection systems, as well as the development of smart irrigation systems that automatically control irrigation based on the need and soil moisture level and improve water consumption in agriculture are also other examples of the use of sustainable entrepreneurship in water resource management (García-Ávila et al., 2023; Yihong et al., 2023).

3.3 Sustainable Entrepreneurship and Natural Resource Conservation

Sustainable entrepreneurship can play an essential role in managing natural resources through innovation and responsible practices. According to Regmi et al. (2023), entrepreneurs have a high capacity to find creative and sustainable solutions to complex challenges related to natural resource extraction, environmental protection, and achieving economic development. By adopting forward-looking perspectives through entrepreneurial activities, they are able to behave more equitably and responsibly in relation to natural resource management. For example, sustainable entrepreneurs help reduce the destruction of natural resources by employing innovative eco-friendly strategies and technologies such as more efficient use of energy through using energy-efficient and low-consumption machinery, producing and using renewable energy sources such as producing biofuels or using solar panels and wind turbines, and using smart monitoring solutions and systems (Odeyemi et al., 2024). One example that can be used to show the impact of sustainable entrepreneurship on natural resource conservation is the role of renewable energy production in reducing deforestation and rangeland degradation. For example, in some regions of Iran, including the north and west, the dispersion of villages in the forest and rangeland areas, their distance from urban centers, difficulty of communication routes and their closure in the cold seasons, the lack of appropriate transportation, insufficient distribution of oil and gas, and the high cost of obtaining fossil fuels have led residents of these regions to use forest and rangeland trees, shrubs, and bushes as fuel and as a result, destruction of forests and rangelands (Rahmani, 2023). In such circumstances, entrepreneurs can take effective steps to reduce the use of firewood and fossil fuels in various ways, such as helping to produce renewable energy and using innovative methods to educate people about the use of these resources.

3.4 Sustainable Entrepreneurship and Mitigating Climate Change

While climate change is one of the most important issues in the twenty-first century with destroying economic, social, and environmental impacts, sustainable entrepreneurship can play a vital role in confronting and adapting to these changes by providing innovative solutions that help decrease carbon emissions and encourage environmentally responsible practices (Pricopoaia et al., 2024). Among the strategies entrepreneurs consider to combat climate change is developing technologies that use solar, wind, and bioenergy sources to reduce dependence on fossil fuels and decrease greenhouse gas emissions. Malaysia's commitment to the 2030 Agenda and its use of entrepreneurial capacity to achieve sustainable development can be presented as a successful example. The country has significantly reduced greenhouse gas emissions by strongly engaging entrepreneurs and implementing technological and financial innovations in the field of renewable energy (Karaday et al., 2024). Another example of sustainable entrepreneurship strategies to face climate change is introducing and investing in green transportation solutions such as electric vehicles and smart mobility. Smart mobility not only includes the use of new transportation vehicles such as electric trains and trams or hydrogen-powered trams and buses instead of gas-powered vehicles but also includes innovative approaches in ride-sharing, car sharing, public transportation, etc. (Badamchizadeh et al., 2023; Pricopoaia et al., 2024; Rindone, 2022). Also, in the agricultural sector, innovative climate-smart farming strategies can help mitigate climate change and ensure food security. For example, by widely using laser land leveling, Uzbekistan has reduced water and seed losses and the need for fertilizers and pesticides while accurately leveling the land, leading to climate adaptation in various regions (Sutton et al., 2024). Overall, several studies have highlighted the role of sustainable entrepreneurship in addressing climate change. Some studies have also emphasized the role of sustainable entrepreneurship in addressing climate change. For example, George et al. (2021) focus on digital innovations undertaken by entrepreneurs. They indicated that startups directed by entrepreneurs considerably assist in combating climate change and other environmental challenges.

4 Necessity of Sustainable Entrepreneurship Development in Iran for Tackling Environmental Problems

As mentioned above, Iran faces numerous environmental problems that have even attracted global attention. Iran's ranking among the world's top countries in terms of environmental crises (ranked 67th in 2020) and the estimated damage of 8.43 billion dollars from environmental destruction in 2002 indicates that the country is far from the desired global environmental standards (Soleimani & Cheraghi, 2022). In addition to direct and indirect economic costs, environmental degradation has also led to negative social and individual consequences in society. For example, various studies (e.g. Farajollahi, 2021; Mehrabi & Vali, 2022; Mousavi Jahromi et al., 2023; Rezaei et al., 2013; Sojasi Qeidari et al., 2022) in Iran indicate that the occurrence of multiple droughts in the country has had direct impacts on spread of unemployment in rural areas, reduced income, increased poverty, and increased migration from these areas, increased tension and conflict, attitude and tendency towards inappropriate and false jobs, public dissatisfaction against governmental sector and reduced cohesion and cooperation in the society. Another example of the social consequences of environmental crises is the impact of deforestation on various aspects of people's social lives, as shown in multiple studies (John, 2020; Yadav, 2023). These studies show that deforestation leads to changes in the lifestyle of local people, migration and social instability, the loss of part of the cultural identity of local communities, social inequality, and a decline in public health.

Environmental degradation is also linked to numerous health consequences. For example, water pollution caused by industrial and agricultural activities, natural processes, and inadequate wastewater treatment facilities has numerous negative impacts on human health and other living organisms. According to UNESCO's World Water Development Report 2021, about 829,000 people die annually from diarrhea caused by unsafe drinking water, sanitation, and hand hygiene, including nearly 300,000 children under five years of age, which accounts for 5.3% of all deaths in this age group (Lin et al., 2022). Data from studies in developing countries show a clear relationship between cholera and contaminated water, while household water treatment and storage can reduce cholera (Gundry et al., 2004). In Iran, cholera is a disease that occurs in some years due to contaminated water. In the summer of 2005, the disease infected 1150 people, resulting in 11 deaths, increasing concern in society, and imposing economic costs on the country (Zahraei et al., 2007). In addition to disease, unsafe drinking water can lead to digestive diseases, impaired nutrient absorption, and malnutrition, particularly in children (Gundry et al., 2004). Air pollution in Iran, caused by greenhouse gas emissions from vehicles, industrial centers, deforestation, diesel fuel in power plants, etc., also burdens the country with many health-related costs every year, as well as leading to deaths. According to the report of the Iranian Ministry of Health, in 2024, there were about 30,692 deaths attributed to air pollution in Iran, which had an economic cost of about 12 billion dollars for the country (Iranian Ministry of Health, 2024).

Another point that should be noted is that Iran spends millions of dollars annually to import products such as alcohol, acetone, fuel ethanol, types of paper and its pulp, types of essential oils used in the food and health industries, animal and poultry feed and supplementary materials, as well as non-oil mulch, compost and biogas energy that obtained from the recycling and processing agricultural waste in developed countries. This is while, every year, a large amount of this waste is produced and accumulated in the country, but due to a lack of proper recycling, not only is its economic value not exploited, but improper disposal of these wastes also has environmental consequences for the country. This issue reveals the necessity of moving towards sustainable entrepreneurship in the field of waste recycling (Bani Imam & Nasiri, 2017; Daryabeigi Zand & Razaghi, 2014).

In summary, in the current situation where Iran is facing numerous environmental problems, and many of these problems are rooted in the type of activity and performance of various businesses in the industrial and agricultural sectors, the existence of drivers of sustainable development in business, adherence to the principles of sustainability, and the obligation to accept environmental and social responsibility in businesses are essential (Soleymani et al., 2021). Environmental protection helps to achieve sustainable development and reduce tensions within the country and with other countries, building trust and confidence at all levels and thus strengthening security (Faghih Hanini, 2021). Paying attention to environmental protection in businesses not only leads to the adjustment of environmental problems in the country and the reduction of economic and social costs resulting from environmental degradation but also helps to achieve sustainable development and reduce tensions within the country and with other countries, building trust and confidence at all levels, and thus strengthening security (Faghih Hanini, 2021). in this regard, sustainable entrepreneurship is a concept that, although it is very new in Iran and its development policies and tools have not reached the maturity they deserve and is in the early stages of its development in academic literature (Amousa et al., 2024), but can be considered as a solution to reduce environmental problems and deal with these problems. Sustainable entrepreneurs help to reduce greenhouse gas emissions, energy sustainability, protect natural resources, decrease pollution, and mitigate climate change through identifying new business opportunities and applying green and sustainable technologies and practices, such as using waste capacity to produce high-value-added materials, replacing fossil fuels with clean energy, or using technologies that have low energy consumption. Ultimately, as The Association for Entrepreneurship USA (2024) points out, "the rise of sustainable entrepreneurship heralds a promising era where economic prosperity is not at the expense of the environment, but rather in tandem with its preservation."

5 Challenges of Sustainable Entrepreneurship Development in Iran

Although sustainable entrepreneurship was emphasized as a way to deal with environmental problems and achieve sustainability, the development of this form of entrepreneurship in Iran will face challenges, some of which are discussed below.

5.1 High Initial Cost, High Interest Rate, and Long-Term Payoff

Businesses based on sustainability-related initiatives usually require a high initial investment cost, while the waiting period for their return is usually long (Ayarkwa et al., 2022). On the other hand, our country is currently facing a complex economic situation. High interest rates are considered one of the country's monetary policy tools to control inflation; theoretically, this increase in interest rates can reduce liquidity growth and control inflation by reducing demand and limiting credit, but in practice, more pressure is put on businesses and entrepreneurs face difficulties both as investors and in attracting investors. In such circumstances, the incentive to accept the risk of development and production is reduced, and people try to direct their capital towards low-risk businesses (Motiei, 2022; Vaez Barzani & Izadkhasti, 2011).

5.2 Changing the Mindset of People About the Environment and Environmentally Friendly Goods

In Iran, the environment and the importance of protecting it have not yet become a public issue, and it can be said that the root of many of the hardships and mistreatments towards nature and the environment in Iran is the lack of environmental consciousness and inattention to the countless benefits of nature (Salehi et al., 2018). Accordingly, another challenge facing sustainable entrepreneurship in Iran is the need to change the mindset of various groups, such as investors, policymakers, and consumers, regarding the importance of environmental issues. Gaining their support for plans, programs, or products aimed at protecting the environment depends on this shift in perspective. However, changing mindsets requires education and is a slow, time-consuming process.

5.3 Technological Limitation

Sustainable entrepreneurship requires continuous innovation and the use of primarily capital-intensive technologies. On the other hand, there are various challenges to their widespread adoption or integration into existing societal infrastructures (Ciocnitu, 2024). This challenge is particularly important in Iran, where the infrastructure for developing eco-friendly technologies is underdeveloped and needs to be expanded. In addition, some technologies are still in the early stages of their development and have limitations. For example, a major obstacle to the use of renewable energy sources such as biogas, wind, or solar energy is the limited capacity to store these energies on a large scale and in a long-term and cost-effective manner (Nnaji & Karakhan, 2020).

5.4 Infrastructural Limitation

Using sustainable and environmentally friendly technologies requires appropriate infrastructure such as power grids, transportation networks, and smart communication networks, while in developing countries, including Iran, at least in remote areas, these infrastructures are not well developed. Almost all sources (Anabestani & Jahantigh, 2018; Dorri Sedeh & Tavakoli, 2022; Hajarian, 2022; Maleksaeidi & Memarbashi, 2023; Rezaei et al., 2016) have cited weak infrastructure as one of the reasons for the lack of investment in various fields, including entrepreneurship, to achieve sustainability in Iran.

5.5 Legal and Policy Limitations

Compared to many countries, Iran has more complex and time-consuming bureaucratic processes, which can be considered one of the challenges of entrepreneurship in general and sustainable entrepreneurship in particular in Iran. In developed countries, the process of registering a company and obtaining licenses is much simpler and faster, and there are also more government incentives for startups in these countries. In Iran, the tax system is more complex, and government support is not as strong as in other countries. Additionally, constant and unpredictable changes in laws and regulations expose entrepreneurs to increased risks, and the necessary support for entrepreneurs and startup businesses in addressing legal and administrative issues is insufficient (Maleksaeidi & Memarbashi, 2023; Motiei, 2022).

6 Conclusion

Like many countries in the world, Iran has faced significant environmental problems and challenges in recent decades, which have directly impacted the economy and well-being of the country's population while also causing negative social and health consequences. The environmental crises in Iran are so severe and profound that, from the experts' perspective, failure to address them could threaten the country's future and stability. Experts and specialists are looking for a way to prevent further environmental crises as much as possible and mitigate the effects of existing crises. In response to this concern, while examining Iran's most important environmental problems, this chapter analyzed sustainable entrepreneurship as a relatively new concept in the entrepreneurship literature and its role in dealing with environmental changes. Then, based on these analyses, the need to pay attention to sustainable entrepreneurship in Iran and the challenges it faces for development were examined. The findings showed that Iran faces many environmental problems, the most important of which are water scarcity, natural resource degradation, air, water, and soil pollution, accumulation of urban, industrial, and agricultural waste, and climate change and disasters such as drought and flood. Based on the findings, these crises have not been limited to the environment, but have imposed significant economic, social, and health costs on society. According to the findings, sustainable entrepreneurship, by using environmentally friendly technologies and strategies such as using renewable energy instead of fossil energy, using energy-efficient technologies, recycling waste, especially agricultural waste, and producing high-value-added materials such as alcohol, acetone, fuel ethanol, types of paper and its pulp, types of essential oils used in the food and health industries, animal and poultry feed and supplementary materials, as well as non-oil mulch, compost and biogas energy etc., can not only help reduce environmental crises but also provide an opportunity to gain economic and social benefits from existing threats. However, the findings showed that the development of sustainable entrepreneurship in

Iran faces challenges, the most important of which are high initial costs, high interest rates, and long-term payoffs, as well as changing the mindset of people about the environment and environmentally friendly goods, technological, infrastructural, and legal and policy limitations. Therefore, the country must first overcome these challenges to develop this entrepreneurial system.

Questions

- What are the most important environmental issues in Iran?
- Given the environmental problems in Iran, what role can sustainable entrepreneurship play in solving and addressing these problems?
- What challenges will sustainable entrepreneurship development face in Iran?

References

Abdel-Fatah, M. A. (2018). Nanofiltration systems and applications in wastewater treatment: Review article. *Ain Shams Engineering Journal, 9*(4), 3077–3092.

Abereijo, I. O. (2016). Ensuring environmental sustainability through sustainable entrepreneurship. In A. Goswami, & A. Mishra (Eds.), *Economic modeling, analysis, and policy for sustainability* (pp. 234–249). IGI Global Scientific Publishing.

Ahmad, A., & Azam, T. (2019). Water purification technologies. In A. M. Grumezescu, & A. M. Holban (Eds.), *Bottled and packaged water* (pp. 83–120). Woodhead Publishing.

AlaAlDinvandi, A. (2021). The affective factors on pastures destruction and identifying its protection approaches. *Journal of Urban Studies on Space and Place, 5*(20), 71–90. (in Persian).

Aldardasawi, A. F., & Eren, B. (2021). Floods and their impacts on the environment. *Academic Perspective Procedia, 4*(2), 42–49.

Al Kamzari, S. M. A., Rao, L. N., Lakavat, M., Gandi, S., Reddy, S., & Sri, G. K. (2023). Extraction and characterization of cellulose from agricultural waste materials. *Materials Today: Proceedings, 80*(part 3), 2740–2743.

Amousa, F., Fazli, S., Arasti, Z., & Elahi, M. (2024). Designing a green entrepreneurship development model in the field of urban waste management. *Majlis and Rahbord, 31*(117), 233–272. (in Persian).

Anabestani, A., & Jahantigh, H. (2018). Investigating the challenges of lacustrine green entrepreneurship development in Sistan Area. *Spatial Planning, 8*(4), 1–26. (in Persian).

Ansari, N., SayedAghlaghi, J., & Ghasemi, M. H. (2009). Determination of socio-economic factors on natural resources degradation of Iran. *Multifunctional Grasslands in a Changing World, 1*, 733. (in Persian).

Apostu, S. A., & Gigauri, A. (2023). Sustainable development and entrepreneurship in emerging countries: Are sustainable development and entrepreneurship reciprocally reinforcing? *Journal of Entrepreneurship, Management and Innovation, 9*(1), 41–77.

Association for entrepreneurship USA. (2024). The rise of sustainable entrepreneurship and its impact on the environment. Available from: https://afeusa.org/articles/the-rise-of-sustainable-entrepreneurship-and-its-impact-on-the-environment/

Ayarkwa, J., Opoku, D. J., Antwi-Afari, P., & Li, R. Y. M. (2022). Sustainable building processes' challenges and strategies: The relative important index approach. *Cleaner Engineering and Technology, 7*, Article 100455.

Azadi, P., & Mesgaran, B. (2018). A national adaptation plan for water scarcity in Iran. Stanford Iran 2040 Project, Stanford University.

Babaei, A., Mojabi, S. M., Shayegan, S. J., Ghiyasi, S., Kuchakian, H., & Sadeghian, M. (2022). The state of waste management in Iran. *Environmental Research Journal, 2*, 3–11. (in Persian).

Badamchizadeh, P., Teymuri, I., Oskouee Aras, A., & Babaie Aghdam, F. (2023). The effect of smart mobility performance in mitigation of climate change, the experiences of European cities. In *7th Intercontinental Geoinformation Days (IGD)*, Peshawar, Pakistan, November 18–19, 2023.

Bahrami, R., & Separi, M. F. (2021). Investigating the socio-economic and environmental effects of drought on rural areas of Kurdistan Province of Iran. *Village and Development, 24*(1), 173–194. (in Persian).

Bani Imam, M., & Nasiri, S. L. (2017). Agricultural and food industry waste management. In *The Second National Conference on Agricultural Development*, Zamin-e-Salem, March 28, 2017, Karaj, Iran. (in Persian)

Borun, M. (2016). Study and management of industrial wastewater and their harmful effects on the environment. In *The Second International Congress on Earth, Space and Clean Energy with a Focus on Natural Resource Management, Agriculture and Sustainable Development*, Tehran, February 2016.

Bowonder, B. (1987a). Environmental problems in developing countries. *Progress in Physical Geography: Earth and Environment, 11*(2), 15–108.

Bowonder, B. (1987b). Management of environment in developing countries. *Environment Systems and Decisions, 7*, 111–122.

Chen, L., Chen, Z., Liu, Y., Lichtfouse, E., Jiang, J., Hua, J., Osman, A. I., Farghali, M., Huang, L., Zhang, Y., Rooney, S. W., & Yap, P. S. (2024). Benefits and limitations of recycled water systems in the building sector: A review. *Environmental Chemistry Letters, 22*, 785–814.

Ciocnitu, C. R. (2024). Challenges and perspectives in sustainable entrepreneurship. In *The 19th Edition of*

the International Conference, European Integration - Realities and Perspectives. Proceedings (pp. 531–540).
Daneshvar, M. R. M., Ebrahimi, M., & Nejadsoleymani, H. (2019). An overview of climate change in Iran: Facts and statistics. *Environmental Systems Research, 8*(1), 1–10.
Daryabeigi Zand, A., & Razaghi, N. (2014). Rural waste management with an emphasis on the status of agricultural waste in Iran: Challenges and solutions. In *The First National Conference of New Achievements in Biological and Agricultural Sciences*, May 31, 2014, Zabul University, Zabul, Iran. (in Persian)
Devendiran, R., Turukmane, A. V., Sathiyarai, A., & Srinivasa Rao, P. (2023). Smart irrigation: Revolutionizing water management in agriculture for sustainable practices and improved crop yield. In *6th International Conference on Recent Trends in Advance Computing*. Dec. 14-15 2023, Vellore Institute of Technology Chennai, Vandalur to Kelambakkam Road, Melakottaityur, Chennai, India.
Dorri Sedeh, S., & Tavakoli, M. (2022). Obstacles to green economy development in entrepreneurship in rural areas of Isfahan province. *Economic Geography Research, 2*(6), 24–39. (in Persian)
Ebadi, A. G. H., Toughani, M., Najafi, A., & Babaee, M. (2020). A brief overview on current environmental issues in Iran. *The Central Asian Journal of Environmental Science and Technology Innovation, 1*(1), 1–11.
El-Fadel, M., Findikakis, A. N., & Leckie, J. O. (1997). Environmental impacts of solid waste landfilling. *Journal of Environmental Management, 50*(1), 1–25.
Faghih Hanini, A. (2021). Good environmental governance and sustainable development. *Political Research in the Islamic World, 10*(3), 297–313. (In Persian).
Farajollahi, A. (2021). The socio-economic impacts of drought and water crisis on local communities of Iran. In *10th Regional Conference on Rainwater Catchment Systems*, September 2021.
FasterCapital. (2024). Why sustainable entrepreneurship matters and get started. Available from: https://fastercapital.com/content/Why-Sustainable-Entrepreneurship-Matters-and-Get-Started.html#Environmental-Impact--How-Sustainable-Entrepreneurs-Contribute
García-Ávila, F., Guanoquiza-Suárez, M., Guzmán-Galarza, J., Cabello-Torres, R., & Valdiviezo-Gonzales, L. (2023). Rainwater harvesting and storage systems for domestic supply: An overview of research for water scarcity management in rural areas. *Results in Engineering, 18*, Article 101153.
Gaur, N., Sharma, S., & Yadav, N. (2024). Environmental pollution. In V. K. Garg, A. Yadav, Ch. Mohan, S. Yadav, & N. Kumari (Eds.), *Advances in green and sustainable chemistry. Green chemistry approaches to environmental sustainability* (pp. 23–41). Elsevier.
George, G., Merrill, R. K., & Schillebeeckx, S. J. (2021). Digital Sustainability and entrepreneurship: How digital innovations are helping tackle climate change and sustainable development. *Entrepreneurship Theory and Practice, 45*(5), 999–1027.
Gholami, V., Jolandan, M. A., & Torkaman, J. (2017). Evaluation of climate change in northern Iran during the last four centuries by using dendroclimatology. *Natural Hazards, 85*(3), 1835–1850.
Gundry, S., Wright, J., & Conroy, R. (2004). A systematic review of the health outcomes related to household water quality in developing countries. *Journal of Water Health, 2*(1), 1–13.
Hajarian, A. (2022). Investigating the challenges and barriers to green entrepreneurship development from the perspective of experts and beneficiaries Case study: Rural communities of Macron coast. *Journal of Entrepreneurial Strategies Agriculture, 9*(18), 115–125. (in Persian).
Hakim, Sh., & Alterman, J. B. (2023). *Iran's environmental challenges*. Center for Strategic and International Studies.
Hamel, P., Ding, N., Cherqui, F., Zhu, O., Walcker, N., Bertrand-Krajewski, J. L., Champrasert, P., Fletcher, T. D., McCarthy, D. T., Navratil, O., & Shi, B. (2024). Low-cost monitoring systems for urban water management: Lessons from the field. *Water Research X, 22*, Article 100212.
Iranian Ministry of Health. (2024). Available from: behdasht.gov.ir (in Persian)
Jahandari, A., & Abbasnejad, B. (2024). Environmental pollution status and health risk assessment of selective heavy metal(oid)s in Iran's agricultural soils: A review. *Journal of Geochemical Exploration, 256*, Article 107330.
Jamshidi, M., Zand-Parsa, S., Pakparvar, M., & Niogi, N. (2019). Evaluation of evapotranspiration over a semi-arid region using multiresolution data sources. *Journal of Hydrometeorology, 20*, 947–964.
Jayakumar, M., Gindaba, G. T., Gebeyehu, K. B., Periyasamy, S., Jabesa, A., Baskar, G., John, B. I., & Pugazhendhi, A. (2023). Bioethanol production from agricultural residues as lignocellulosic biomass feedstock's waste valorization approach: A comprehensive review. *Science of the Total Environment, 879*, Article 163158.
John, W. (2020). Impacts of deforestation on socio-economic development and environment in Nigeria. *The International Journal of Social Sciences and Humanities Invention, 7*(03), 5852–5863.
Karaday, T., Yazıcı, S., & Ömür, G. A. (2024). Climate change and entrepreneurship: A review and research agenda. *Global Business and Organizational Excellence*, 1–28.
Kaza, S., Yao, L., Bhada-Tata, P., & Van Woerden, F. (2018). *What a waste 2.0: A global snapshot of solid waste management to 2050*. The World Bank.
Khamisabadi, A., Parvanak, K., & Nasrabadi, M. (2019). Effect of the use treated wastewater on microbial contamination of soils of urban landscapes. *Journal of Environmental Health Engineering, 7*(1), 42–52. (in Persian).
Khatibi, S., & Arjjumend, H. (2019). Water crisis in making in Iran. *Grassroots Journal of Natural Resources, 2*(3), 45–54.

Khosravani, F., Pezeshkirad, G., & farhadian, H. (2013). Investigate of the agricultural waste and waste management situation and present of solutions in order to achieve sustainable development goals. *Popularization of Science, 5*(1), 95–112. (In Persian)

Khuhawar, K. H., & Shah, H. (2019). Social problems due to unemployment. *Journal of Marketing and Information Systems, 1*(2), 8–9.

KianiRad, A., & Noroozi, H. (2022). Explanation of factors affecting the natural resources degradation with emphasis on poverty and food security. *Agricultural Economics and Development, 29*(4), 119–150. (in Persian).

Koul, B., Yakoob, M., & Shah, M. P. (2022). Agricultural waste management strategies for environmental sustainability. *Environmental Research, 206*, Article 112285.

Külcü, R., & Yaldiz, O. (2014). The composting of agricultural wastes and the new parameter for the assessment of the process. *Ecological Engineering, 69*, 220–225.

Kumar, D., Jain, V. K., Shanker, G., & Srivastava, A. (2003). Utilisation of fruits waste for citric acid production by solid state fermentation. *Process Biochemistry, 38*(12), 1725–1729.

Lin, L., Yang, H., & Xu, X. (2022). Effects of water pollution on human health and disease geterogeneity: A review. *Frontiers in Environmental Science, 10*, 1–16.

Maleksaeidi, H., & Memarbashi, P. (2023). Barriers of environmentally-friendly entrepreneurship development in Iran's agriculture. *Environmental Development, 46*, Article 100831.

Mansingh, H., Banik, R., Bag, S., Sunarative, A., & Das, T. (2024). Drip irrigation system: An effective approach for increasing water use efficiency. In S. Ranjan, S. Sow, P. Dayal, A. Kumar, & D. K. Roy (Eds.), *Green technologies for sustainable agriculture*. Biotechbooks.

Mehrabi, M., & Vali, A. A. (2022). Prioritizing the impacts of drought on economic, social and environmental dimensions in urban and rural areas (Case study of Yazd province). *Journal of Arid Regions Geographic Studies, 11*(41), 29–43. (in Persian)

Mokhtari Hashi, H., & Moradi, A. (2021). Environmental consequences of the water crisis in Iran. *Political Spatial Planning, 3*(2), 117–131. (in Persian).

Motamedi J., Jalili A., Arzani, H., & Khodagholi, M. (2020). Causes of rangeland degradation in the country and solutions to get out of the current situation. 5(4), 21–44. (in Persian)

Motiei, M. (2022). Legal-jurisdictional barriers to financing technological entrepreneurship with an institutional approach: A mixed method study. *Journal of Asset Management and Financing, 10*(4), 67–94. (in Persian).

Mousavi Jahromi, Y., Biabani, J., & Arzaghi, E. (2023). Spatial analysis of internal migration based on drought index in Iran. *Physical Development Planning, 10*(4), 73–88. (in Persian).

Muhammad, U. L., Shamsuddin, I. M., Yahaya, K. M., & Nuhu, M. (2019). Environmental impacts of waste disposal: An overview on the disposal of polyethylene bags in Gusau City Zamfara State. *Journal of Evolutionary Science, 1*(2), 31–39.

Neumann, T. (2022). Impact of green entrepreneurship on sustainable development: An ex-post empirical analysis. *Journal of Cleaner Production, 377*, Article 134317.

Nnaji, C., & Karakhan, A. A. (2020). Technologies for safety and health management in construction: Current use, implementation benefits and limitations, and adoption barriers. *Journal of Building Engineering, 29*, Article 101212.

Odeyemi, O., Usman, F. O., Mhlongo, N. Z., Elufioye, O. A., & Ike, Ch. U. (2024). Sustainable entrepreneurship: A review of green business practices and environmental impact. *World Journal of Advanced Research and Reviews, 21*(02), 346–358.

Phiri, R., Rangappa, S. M., & Siengchin, S. (2024). Agro-waste for renewable and sustainable green production: A review. *Journal of Cleaner Production, 434*, Article 139989.

Pohlan, L. (2024). Unemployment's long shadow: The persistent impact on social exclusion. *J Labour Market Research, 58*, 12.

Pricopoaia, O., Lupașc, A., & Mihai, I. O. (2024). Implications of innovative strategies for sustainable entrepreneurship—Solutions to combat climate change. *Sustainability, 16*(22), 9742.

Rahmani, R. (2023). The role of new energies in protecting natural resources. Office of Natural Resources Protection and Protection, Iranian Environmental Organization. (in Persian)

Rezaei, M., Partovi, L., & Mohammadpour, A. (2013). Evaluation of demographic and socio-economic consequences of environmental destruction in Urmie Lake. *Iran Population Studies, 1*(1), 129–150. (in Persian).

Rezaei, B., Kahrizi, D., & Najafpoor, H. (2016). An investigation on barriers and challenges of green entrepreneurship development in the agricultural sector. *Journal of Studies in Entrepreneurship and Sustainable Agricultural Development, 3*(1), 35–53. (in Persian).

Rindone, C. (2022). Sustainable mobility as a service: Supply analysis and test cases. *Information, 13*, 351.

Ranjbar Heidari, V., & Jamshidi, E. (2016). Studying the concept of environmental security with an overview of the challenges of environmental in Iran. *Journal of Socio-Cultural Strategy, 5*(4), 199–231. (in Farsi).

Regmi, R., Zhang, Z., & Zhang, H. (2023). Entrepreneurship strategy, natural resources management and sustainable performance: A study of an emerging market. *Resources Policy, 86*(Part B), 104202.

Roy, P., & Ahmaruzzaman, M. D. (2022). Industrial wastes as novel adsorbents for the removal of toxic impurities from wastewater. In Tyagi, I., Dehghani, M. H., Goscianska, G., & Karri, R. R. (Eds.), *Sustainable materials for sensing and remediation of noxious pollutants*. Elsevier.

Rözer, J. J., Hofstra, B., Brashears, M. E., & Volker, B. (2020). Does unemployment lead to isolation? The consequences of unemployment for social networks. *Social Networks, 63*, 100–111.

Saberi Riseh, R., Gholizadeh Vazvani, M., Hassanisaadi, M., & Thakur, V. K. (2024). Agricultural wastes: A practical and potential source for the isolation and preparation of cellulose and application in agriculture and different industries. *Industrial Crops and Products, 208*, Article 117904.

Safari, M., Asadi, A., & Dideh, H. (2023). Investigation and introduction of the most important factors of forest destruction, their effects and solutions. In *Fourth International Conference and Seventh National Conference on Protection of Natural Resources and Environment along with the Fifth National Conference on Forests of Iran, Ardabil*, https://civilica.com/doc/1956718 (in Persian)

Safari, A. P., & Shams, A. (2016). The factor analysis of problems of rural waste management in Qazving township. *Rural Development Strategies, 3*(1), 77–94. (In Persian)

Saffari A. (2013). Analysis of environmental hazards and rural waste managerial strategies Case study: Central Ojarood, Germy County. *Journal Space Economy and Rural Developemnt, 2*(3), 71–91. (in Persian)

Salaheldeen, M., Ateeq, A., Ali, S. A., Milhem, M., & Al Ani, Z. (2024). Green entrepreneurship and sustainability in developing countries: Opportunities, challenges, and future research directions.

Salehi, S., Emamgoli, L., & Mohammadi, J. (2018). An analysis of failure to popularize of the environment in Iran. *Sociological Review, 25*(1), 139–162. (in Persian).

Sarkar, N., Ghosh, S. K., Bannerjee, S., & Aikat, K. (2012). Bioethanol production from agricultural wastes: An overview. *Renewable Energy, 37*(1), 19–27.

Shayeghi, A., Rahmati Ziveh, A., Bakhtar, A., Teymoori, J., Hanel, M., Godoy, M. R. V., Markonis, Y., & AghaKouchak, A. (2024). Assessing drought impacts on groundwater and agriculture in Iran using high-resolution precipitation and evapotranspiration products. *Journal of Hydrology, 631*, Article 130828.

Siddiqua, A., Hahladakis, J. N., & Al-Attiya, W. A. K. A. (2022). An overview of the environmental pollution and health effects associated with waste landfilling and open dumping. *Environmental Science and Pollution Research, 29*, 58514–58536.

Sojasi Qeidari, H., Shyan, H., & Soleymani, Z. (2022). Analysis of the impact of environmental unsustainability on social unsustainability in Iran: Tensions and social damages caused by drought in rural areas. *Journal of Research and Rural Planning, 11*(4), 1–19.

Soleimani Morchekhorti, E. (2024). *The accelerating trend of climate change in Iran with a focus on the ultraviolet radiation index*. Research Center Research Center of Islamic Parliament of I.R. Iran. Strategic Report (in Persian).

Soleimani, E., & Cheraghi, M. (2022). *An introduction to good governance in the environmental sector*. Research Center Research Center of Islamic Parliament of I.R. Iran. Strategic Report (in Persian).

Soleymani, A., Yaghubi Frani, A., & Karimi, S. (2021). Space economy and rural development. *Designing a Model for Sustainable Entrepreneurship Development for Rural Businesses Hamedan Province, 10*(36), 49–78. (in Persian).

Sonu, A., Rani, G. M., Pathania, D., Umapathi, R., Rustagi, A., Huh, U. S., Gupta, V. K., Kaushik, A., & Chaudhary, V. (2023). Agro-waste to sustainable energy: A green strategy of converting agricultural waste to nano-enabled energy applications. *Science of the Total Environment, 875*, Article 162667.

Sutton, W. R., Srivastava, J. P., Neumann, J. E., Droogers, P., & Boehlert, B. B. (2024). *Reducing the vulnerability of Uzbekistan's agricultural systems to climate change impact assessment and adaptation options*. World Bank Report.

Tabbaz, M. (2016). Environmental challenges in today's Iran. *Iranian Studies, 49*(6), 943–961.

Velayati, S., & Kadivar, A. A. (2009). Environmental challenges of Iran's forests and rangelands and their consequences. *Journal of Geography and Regional Development, 7*(2), 53–72. (in Persian).

Vaez Barzani, M., & Izadkhasti, H. (2011). Analyzing the role of monetary interest rates in economic crises of the capitalist system: An Islamic approach. *Islamic Economics, 11*(44), 77–104.

Yadav, R. K. (2023). The impact of forest destruction: Ecological, social, and economic consequences. *Journal of Emerging Technologies and Innovation Research, 10*(7), 252–255.

Yihong, Z., Afzal, J., Afzal, M. A., Nishtar, Z., & Adullah, A. (2023). Water resource management and dams construction for sustainable development of Pakistan. In *1st International Conference on Women Development in Engineering, Science & Technology (WDEST'23)*, October 03–05, 2023.

Zahraei, S. M., Naghadeh, M. T., Najafabadi, M. S., Masoumi Asl, H., Javanmard, A., & Safari, M. H. (2007). Cholera epidemic in Iran in 2005. *Iranian Journal of Infectious Diseases and Tropical Medicine, 12*(36), 1–4. (in Persian).

Zarghi, A. A., & Rahmanizadeh, H. (2017). Economic evaluation of waste recycling industry. In *International Conference on Sustainable Development and Urban Development*. Iran. Available on: https://sid.ir/paper/898906/fa

International Agencies/ Institutions to Promote Sustainable Entrepreneurship

Sahar Dehyouri and Azia Zand

Abstract

The chapter discusses how international agencies and institutions can potentially promote sustainable entrepreneurship, covering the activities of the United Nations, the World Bank, and the IFC in terms of policy frameworks, funding, and capacity-building initiatives that can support sustainable business practices. The chapter also focuses on the roles of public–private partnerships in the achievement of Sustainable Development Goals (SDGs) by discussing many case studies on successfully implemented sustainable entrepreneurship initiatives through the support of international institutions. At the end of the chapter, readers will appreciate the critical role these agencies play in fostering sustainable economic growth and innovation.

S. Dehyouri (✉) · A. Zand
Department of Agriculture, Environmental Sciences Research Center, Isl.C., Islamic Azad University, Eslamshahr, Iran
e-mail: dehyouri.s@iau.ac.ir

A. Zand
e-mail: azitazand@gmail.com

Department of Agriculture, Integrated Crops Research Center, Isl.C., Islamic Azad University, Islamshahr, Iran

Keywords

Sustainable entrepreneurship · International agencies and institutions · United Nations · The World Bank · IFC

Learning Objectives

By the end of this chapter, learners will be able to do the following:

1. Identify major international agencies and institutions supporting sustainable entrepreneurship
2. Analyze the role of international institutions in shaping global sustainability policies and frameworks
3. Analyze what funding and capacity-building programs have done for practical, sustainable business
4. Understand the role of public–private partnerships and their importance for achieving the SDGs
5. Apply insights into how case studies develop strategies for sustainable entrepreneurship.

1 Introduction

The main developments have regarded sustainability entrepreneurship as one of the key mechanisms capable of tackling most twenty-first

M. Akbari et al. (eds.), *Sustainable Entrepreneurship*, University of Tehran Science and Humanities Series,
https://doi.org/10.1007/978-3-032-04909-4_14

century challenges, ranging from climate change and dwindling resources to social inequality and weakened economies. Against this backdrop, whereas the rationale behind entrepreneurship historically had targeted profit maximization alone, sustainable entrepreneurship effectively incorporates various considerations for the environment, social responsibility, and economic feasibility into business modeling. This holistic approach fosters innovation and supports the attainment of the United Nations SDGs, which provide a universal framework in efforts to develop a more sustainable and inclusive global economy (Nations, 2015). The growth in popularity of sustainable entrepreneurship shows increased recognition of the fact that business needs to take a key driver role for systemic change. Entrepreneurs and enterprises focused on sustainable practices are uniquely positioned to take a leading approach to innovative solutions, balancing profitability with positive environmental and social outcomes. For example, companies that invest in renewable energy, circular economy models, or inclusive supply chains have the potential to show how entrepreneurship can enable sustainability and simultaneously create economic value (Schaltegger & Wagner, 2011). Nevertheless, the road to sustainable entrepreneurship faces numerous challenges, relating but not limited to inadequate access to finance, weak Policy frameworks, and a general lack of technical expertise, most particularly in developing economies.

It is here that international agencies and institutions play an enabling role. Institutions such as the UN, the World Bank, and the IFC have been instrumental in promoting sustainable entrepreneurship through policy guidance, financial support, and capacity building. These institutions are catalytic agents of change that can help build an enabling environment for sustainable businesses to grow. In supporting these, the UN's SDGs provide a useful roadmap to orient entrepreneurial activity in support of global sustainability imperatives, while various others—the World Bank and IFC, among others—provide needed funding and technical competencies for sustainably oriented projects across the developing world (Farny & Binder, 2021; Gulakov & Vanclay, 2024). Agencies will make perhaps their most overarching contribution through facilitating Public–Private Partnerships(Hodge & Greve, 2007). These partnerships bring together governments, businesses, and civil society in a collective effort to resolve some of the complex sustainability challenges that no one sector can solve in isolation. Public–Private Partnerships tap into the strengths of each partner, combining public-sector resources and regulatory power with private-sector innovation and efficiency. For instance, the UN's Green Climate Fund is partnering with private investors to finance renewable energy projects in Africa; this shows how international institutions can mobilize resources and expertise in the scaling up of sustainable entrepreneurship (Basak & Karlsson-Vinkhuyzen, 2022).

Research on the role of international agencies in fostering sustainable entrepreneurship faces various challenges. Firstly, measuring and evaluating sustainable entrepreneurship is complex, as it deals with multidimensional aspects: economic, environmental, and social performance (Schaltegger & Wagner, 2011). Secondly, the availability and reliability of data are crucial as most of the initiatives that agencies like the UN or World Bank undertook are generally long-term and have impacts that are barely visible or well-documented (Gideon & Unterhalter, 2020). Third, there are differences in mandate and approaches among international agencies, making it challenging to generalize their effectiveness and best practices (Gulakov & Vanclay, 2024).

A key research gap lies in how Public–Private Partnerships can be best leveraged to achieve scale in sustainable entrepreneurship. While Public–Private Partnerships are important in mobilizing resources, their success often depends on the local contexts and stakeholder dynamics, which remain underexplored in the literature to date (Hodge & Greve, 2007). There is also limited research on how international agencies address equity and inclusion in their programs, particularly in supporting marginalized communities and small enterprises. This is according to the United Nations (2015). Finally, while the challenges of sustainability are dynamic and call for adaptive strategies, there is a lack of analysis of how agencies evolve their approaches to meet emerging needs.

Addressing these gaps may provide further insight into the role of international agencies in fostering sustainable entrepreneurship for the SDGs.

This chapter explores the role played by international agencies and institutions in promoting sustainable entrepreneurship. It starts with defining sustainable entrepreneurship and its relevance in view of global sustainability. It then gives an overview of key players on the international front in this field—the UN, World Bank, and IFC—and looks at their plans and strategies for enabling sustainable business. It describes cases that worked and examples that have been done with the help of such institutions, and how these interventions contributed to the growth of a sustainable economy and innovation.

By the end of the chapter, the reader will have a firm understanding of how international agencies and institutions contribute to the development of sustainable entrepreneurship. This chapter underlines the need for global cooperation in the sustainability challenges and puts into perspective the transformative potential of sustainable entrepreneurship in building a more equitable and resilient world.

Sustainable entrepreneurship is a new paradigm that combines economic growth, environmental preservation, and social equity within business practices. It involves a shift from the traditional profit-driven model to a more holistic framework that seeks to create long-term value for society and the planet. This concept increasingly resonates with global challenges such as climate change, resource depletion, and social inequality, making it a cornerstone of sustainable development efforts worldwide (Schaltegger & Wagner, 2011).

1.1 Definition and Key Characteristics

Sustainable entrepreneurship is the means of creating and managing businesses that address environmental, social, and economic challenges while remaining profitable. It focuses on the triple bottom line of people, planet, and profit to ensure that businesses positively contribute to society and the environment while remaining economically viable (Elkington & Rowlands, 1999).

Some of the key characteristics include:

- Social Equity: The entrepreneurs promote fair labor practices, community engagement, and social inclusion.
- Environmental Responsibility: By utilizing renewable energy sources, reducing waste, and promoting resource efficiency, businesses can minimize their ecological footprint.
- Economic Viability: Sustainable businesses attain profitability by aligning their operations with sustainability objectives.

1.2 Importance of Sustainable Entrepreneurship

Sustainable entrepreneurship addresses the major global challenges through the incorporation of environmental and social responsibility into core business models, thus promoting innovation, creating jobs, and developing the Sustainable Development Goals (SDGs) (Biswas et al., 2021). Entrepreneurs, by focusing on climate change, develop solutions that incorporate reducing carbon emissions and transitioning to renewable sources of energy. They address the issue of resource constraint by embracing innovations that optimize resource use and reduce waste, while simultaneously addressing social injustice by promoting inclusive growth and opening up opportunities for the marginalized. This strategy integrates economic goals with the health of the planet and society, making systemic transformation toward a more equitable and sustainable future possible.

1.3 The Triple Bottom Line (TBL)

The Triple Bottom Line (TBL) theory, founded by John Elkington in 1997, constitutes the basis of sustainable entrepreneurship with a balance between three interdependent pillars: People (offering fair treatment, diversity, and inclusion to workers and societies), Planet (seeking environmentally friendly operations to reduce damage), and Profit (having financial prosperity and improving social and environmental well-being) (Srivastava et al., 2022). This model emphasizes that businesses can thrive economically while fostering social equity and environmental stewardship, connecting long-term profitability with greater sustainability goals.

1.4 Role of International Agencies

International agencies like the United Nations, the World Bank, and the International Finance Corporation play a significant role in promoting sustainable entrepreneurship. They provide a policy framework, funding, and capacity-building programs that support business enterprises to go green. For example, the UN's SDGs and the Paris Agreement provide global frameworks for aligning entrepreneurial activities with sustainability goals (Latoszek & Latoszek, 2023; Nicholls & Ormiston, 2023).

1.4.1 Overview of International Agencies and Institutions in Promoting Sustainable Entrepreneurship

The policy frameworks, financial support, technical assistance, and capacity-building programs provided by international agencies and institutions stimulate sustainable entrepreneurship. These are the very agencies that act as agents of change, enabling entrepreneurs to respond to the most pressing challenges of our time: climate change, resource depletion, and social inequality. A summary of the roles of key international agencies in promoting sustainable entrepreneurship, the emerging trends, and challenges is presented herein, supported by academic literature and reports.

1.4.2 Key International Agencies and Their Roles

A. **United Nations (UN)**: The UN is on the frontline of championing sustainable development through initiatives such as the SDGs and the UN Global Compact. In 2015, the 17 SDGs were adopted as a universal framework for attaining sustainable development by 2030. Of particular relevance to sustainable entrepreneurship are Goal 8 (Decent Work and Economic Growth), Goal 9 (Industry, Innovation, and Infrastructure), and Goal 12 (Responsible Consumption and Production) (Nations, 2015). Also, this calls for the alignment of business with ten universal principles in human rights, labor, environment, and anti-corruption to join forces with the UN to achieve the SDGs (Podrecca et al., 2022).

B. **World Bank**: The World Bank fosters sustainable entrepreneurship by *mobilizing financial resources* (e.g., loans and grants for renewable energy, sustainable agriculture, and climate resilience initiatives such as the Climate Investment Funds [CIFs]), policy guidance (e.g., support for governments on *regulatory policy* and *green investment*), and technical support for advancing economic growth, poverty alleviation, and environmental sustainability (Organization, 2023). They aim to build enabling ecosystems where sustainable businesses can thrive, coupling economic development with planetary well-being and social justice (Abdulsalam et al., 2025).

C. **International Finance Corporation (IFC)**: As part of the World Bank Group, the IFC supports private sector development in emerging markets through the provision of financial resources, advisory services, and risk management tools. The IFC has invested in renewable energy, energy efficiency, and sustainable agriculture. It has been mobilizing capital through instruments like IFC Green Bonds by *Green Investments* (Aziz et al., 2021). It also provides technical support to facilitate the adoption of greener business practices, such as supply chain sustainability and carbon emission reduction, by enterprises through *Advisory Services* (Canton, 2021).

D. **The European Union (EU)**: The EU encourages sustainable entrepreneurship through various funding programs, policy schemes, and regulatory arrangements, including the Green Deal and the Circular Economy Action Plan. *Horizon Europe* is a funding program for research and innovation in clean energy, sustainable agriculture, and digital technologies (Kassim & Tholoniat, 2021). The *Circular Economy Action Plan* promotes recycling, reutilization, and resource efficiency with its policy measures and funding opportunities (Friant et al., 2021).

1.4.3 Contributions to Sustainable Entrepreneurship

A. **Policy Frameworks**: International agencies create frameworks like the SDGs and the Paris Agreement that guide businesses to work in tune with global sustainability goals. These frameworks address environmental, social, and economic challenges; hence, responsible business practices are promoted (Van Tulder & van Mil, 2022).

B. **Investment and Funding**: Agencies provide access to finance through grants, loans, and equity investments to sustain the project. For example, CIFs by the World Bank and Green Bonds by the IFC help mobilize capital for various climate-friendly initiatives (Vishnoi, 2022).

C. **Capacity Building**: Training, capacity building, and knowledge transfer are imparted through programs such as the United Nations Development Program and IFC Advisory Services to the entrepreneurs for implementing sustainable practices (Ejejigbe, 2021).

It is a transformative approach wherein economic success rhymes with environmental and social well-being. In addition to addressing global challenges, sustainable entrepreneurs contribute to a more sustainable and inclusive future. The support of international agencies and frameworks, such as the TBL, further enhances the impact of sustainable entrepreneurship. This introduction, together with key references, sets a basis for defining the concept of its importance; thus, this book will develop the concept of the involvement of international agencies in promoting sustainable business practices.

1.5 Models of Sustainable Entrepreneurship

Several models make up sustainable entrepreneurship, which integrates environmental and social objectives with economic objectives into the business processes. These represent pathways through which entrepreneurs can take up the challenges of climate change, resource depletion, and other social inequalities in their value-creating process. Some key models are summarized here, together with some academic references.

1.5.1 Triple Bottom Line Model

The TBL model was developed by John Elkington in 1997 and aims to explain how the three dimensions of people, planet, and profit are balanced. It urges businesses to consider the level of success based not only on the financial perspective but also on their social and environmental contribution (Homer et al., 2025).

1.5.2 Circular Economy Model

The Circular Economy Model focuses on waste reduction, resource reuse, and material recycling to achieve a closed-loop system. The key features are designing for longevity, high resource efficiency, and applying regenerative practices (Bocken et al., 2023; Corsini & Frey, 2024).

1.5.3 Social Entrepreneurship Model

Social entrepreneurship addresses social problems such as poverty and education through innovative business models. It focuses on social missions while being financially sustainable and scalable (Mohammadi et al., 2024; Muneer, 2023).

1.5.4 Green Entrepreneurship Model

In the Green Entrepreneurship Model (Kirkwood & Walton, 2010; Schaltegger, 2002), Green entrepreneurs develop such products and technologies that have a reduced environmental impact, for instance, renewable energy, eco-friendly products, etc. This model focuses on the environment, innovation, and usage of sustainable markets (Gupta & Dharwal, 2022; Trapp & Kanbach, 2021).

1.5.5 Shared Value Model

This model focuses, as advocated by Porter and Kramer (2011), on simultaneous economic value creation and solving social problems. It emphasizes collaboration between stakeholders, offering innovative solutions toward profitability and enhanced social impact (Korletey, 2024).

1.5.6 B Corporation Model

B Corps are businesses verified for high socio-economic and high environmental performance, focusing on the need for stakeholder well-being (Cao & Gehman, 2021; Vallet, 2021).

1.5.7 Bottom of the Pyramid (BoP) Model

The BoP business model focuses on low-income consumers by offering goods and services at very cheap prices to try to eradicate or reduce poverty. It focuses on inclusiveness, affordability, and scalability (Decker & Obeng Dankwah, 2023; Voros Fregolente & Carvalho, 2023).

1.5.8 Regenerative Business Model

The Regenerative Business Model goes beyond sustainability: it aims for the restoration of ecosystems and the making of net positive contributions holistically, integrating the social and environmental factors with economic systems in perspective (Hahn & Tampe, 2021; Konietzko et al., 2023). This model sets a framework within which entrepreneurs can integrate other sustainability models into their business models, ranging from the TBL and circular economy to social entrepreneurship and regenerative business, each with its unique strategy for addressing global challenges. Companies can use these approaches toward a more sustainable and equitable world and still achieve financial success.

1.5.9 Emerging Trends in Sustainable Entrepreneurship

Digital Technologies: Technologies such as AI, blockchain, and IoT are revolutionizing sustainable entrepreneurship. AI optimizes energy use and predicts environmental risks, while blockchain enhances supply chain transparency (Bocken & Geradts, 2022).

Social Entrepreneurship: Social entrepreneurship is gaining momentum as a model for addressing social and environmental challenges. Impact investing and inclusive business models are driving this trend, prioritizing social impact alongside profitability (Langer, 2024).

Green Innovation: Innovations in renewable energy, sustainable agriculture, and waste management are reducing environmental impact. Examples include advancements in solar and wind energy, precision farming, and vertical farming (Majeed et al., 2023).

International agencies and institutions, through policy frameworks, funding, and capacity-building programs, remain indispensable drivers in the pursuit of sustainable entrepreneurship. The new emerging trends in the field are shaped by digital technologies, social entrepreneurship, and green innovation. The challenges that will have to be faced involve breaking bureaucratic barriers and overcoming limitations in funding. The international agencies thus can help toward a more sustainable and equitable world by encouraging collaboration and innovation.

Table 1 provides an overview of how international agencies and institutions see the concepts of sustainability, profitability, importance, and efficiency of sustainable entrepreneurship with relevant references. This table emphasizes the strengths of each organization, along with its challenges, while promoting sustainable entrepreneurship.

Table 1 The perspectives of international agencies and institutions in terms of their sustainability efforts, profitability, importance, and efficiency

Agency/institution	Sustainability efforts	Profitability	Importance	Efficiency	References
United Nations (UN)	Promotes global sustainability through the Sustainable Development Goals (SDGs) and the UN Global Compact	Focuses on social and environmental impact rather than direct profitability	High importance is placed on setting global sustainability agendas and fostering international cooperation	Efficient in creating global frameworks but faces challenges in implementation at the local level	Nations (2015)
World Bank	Funds sustainable projects in areas such as renewable energy, climate resilience, and sustainable agriculture	Balances financial returns with social and environmental impact through loans and grants	Critical for providing financial resources and policy advice to developing countries	Efficient in mobilizing large-scale funding but criticized for bureaucratic delays	Abdulsalam et al.,(2025)
International Finance Corporation (IFC)	Invests in green projects and provides advisory services to promote sustainable business practices	Focuses on financial sustainability and profitability while supporting green investments	A key player in private sector development and sustainable entrepreneurship in emerging markets	High efficiency in leveraging private sector investments for sustainability	Canton (2021)
European Union (EU)	Promotes sustainability through the European Green Deal and Circular Economy Action Plan	Encourages profitable green businesses through funding programs like Horizon Europe	High importance is placed on driving sustainability policies and innovation in Europe	Efficient in policy implementation but faces challenges in aligning member states' interests	Bongardt and Torres (2022)
UNDP	Supports sustainable development through capacity-building programs and technical assistance	Focuses on social and environmental impact rather than direct profitability	Important for building local capacity and supporting sustainable entrepreneurship in developing countries	Efficient in delivering targeted programs but limited by funding constraints	UNDP (2021)
UNEP	Focuses on environmental sustainability through initiatives like the Green Economy Initiative	Primarily focused on environmental impact rather than profitability	Critical for promoting environmental sustainability and green business practices globally	Efficient in raising awareness but limited in direct implementation capacity	UNDP (2021)
OECD	Provides policy recommendations and data to promote sustainable economic growth	Focuses on economic policies that balance sustainability and profitability	Important for shaping global economic policies that support sustainability	Efficient in research and policy advocacy, but lacks direct implementation power	Lamichhane et al., (2021)

1.6 The Relationship Between International Agencies and Sustainable Entrepreneurship

International agencies like the United Nations, World Bank, and International Finance Corporation are key drivers of sustainable entrepreneurship. They provide policy frameworks such as the SDGs and the Paris Agreement, funding mechanisms such as IFC Green Bonds, and capacity-building programs such as UNDP training initiatives. These efforts support entrepreneurs in addressing global challenges like climate change, resource depletion, and social inequality. By encouraging innovation, inclusive growth, and green technologies, international agencies create conditions for making businesses sustainable in line with global sustainability goals (Canton, 2021; Nations, 2015; Abdulsalam et al, 2025).

1.6.1 The Role of International Agencies in Promoting Sustainable Entrepreneurship

1. **Policy Frameworks**

International agencies create policy frameworks guiding sustainable entrepreneurship, allowing businesses to meet the set global sustainability objectives.

- The SDGs are the 17 goals that provide an overall framework for attaining sustainable development by 2030. Among these, Goals 8: Decent Work and Economic Growth, 9: Industry, Innovation, and Infrastructure, and 12: Responsible Consumption and Production are particularly relevant to sustainable entrepreneurship (Nations, 2015).
- Paris Agreement: The agreement on climate change between nations encourages businesses to take action towards reducing carbon emissions and the use of sustainable practices (Hickmann et al., 2021)

2. **Funding and Investment**

International agencies provide financial resources for sustainable entrepreneurship, such as grants, loans, and equity investments.

- World Bank Funding: The World Bank offers funding to sustainable projects, with the CIFs aiming to help developing countries invest in low-carbon and climate-resilient initiatives. World Bank, 2020.
- IFC Green Bonds: The International Finance Corporation (IFC) mobilizes capital for climate-friendly projects through the issuance of green bonds (Aziz et al., 2021).

3. **Capacity Building**

Capacity-building programs help entrepreneurs develop the skills and knowledge needed to implement sustainable practices.

- UNDP Capacity Building: The United Nations Development Programme (UNDP) provides training and technical assistance to entrepreneurs in developing countries (UNDP, 2021).
- IFC Advisory Services: The IFC provides advisory services to enhance the sustainability performance of businesses, such as carbon emissions reduction and supply chain sustainability (Canton, 2021)

1.6.2 Contributions of International Agencies to Sustainable Entrepreneurship

International agencies also act as facilitators, enabling environments for sustainable entrepreneurship through promoting policy reforms and establishing public–private partnership building. The World Bank advises governments to develop enabling regulatory frameworks that favor green investments; these indeed contribute to building supportive ecosystems for sustainable business operations (Abdulsalam et al., 2025). Similarly, the UN Global Compact encourages the collaboration of business, government, and civil society in pursuit of the SDGs, for collective action toward sustainability, while (Podrecca et al., 2022) provide supportive innovations and technology. For instance, the European Union's Horizon Europe program funds research and innovation in clean energy, sustainable agriculture, and digital technologies that spur the development of innovative solutions for sustainability. According to Bürgin (2023), in addition, the Advisory Services of the International Finance Corporation have been supporting the transfer of green technologies to developing countries, thereby enabling the private sector to take up environmental concerns. International agencies also focus on inclusive growth through support for sustainable entrepreneurship initiatives that empower the poorest of the poor. The United Nations Development Program, through the BoP model, supports businesses serving low-income populations, creating opportunities for economic participation (UNDP, 2021). Furthermore, the IFC invests in social enterprises that generate both financial returns and social impact, demonstrating the potential of sustainable entrepreneurship to address social inequalities (Canton, 2021). By fostering policy reforms, supporting innovation, and promoting inclusive growth, international agencies empower entrepreneurs to tackle global challenges such as climate change, resource depletion, and social inequality. These efforts not only meet the Millennium Development Goals regarding sustainability but also set the grounds for long-term economic growth and care for the environment.

Indeed, international agencies face several obstacles to sustainable entrepreneurship, such as bureaucratic barriers, complicated frameworks of regulations, and lengthy approval processes that delay the projects, which in turn increase costs, according to OECD 2019. Secondly, a shortage of funds acts as one major challenge because, even after international funding initiatives, access to finance is always poor in the majority of countries that are still developing; this demand still exceeds supply, and the constraints place certain barriers to growth ventures for the sustainability of the World Bank 2020. In addition, local resistance has its roots in cultural and political skepticism, which hampers progress. Communities may distrust new technologies or business models and thus resist their adoption. According to Hall et al. (2016), these challenges signify the need for streamlined regulations, increased funding, and community engagement in order to foster sustainable entrepreneurship. Such findings have significant ramifications for how international agencies can support entrepreneurs in realizing global sustainability goals.

Also, International agencies influence sustainable entrepreneurship through policy frameworks, funding, capacity building, technical assistance, and networking (Di Vaio et al., 2022; Emon & Khan, 2023). They create the global guidelines for businesses to be environmentally responsible by setting goals like the SDGs and the Paris Agreement, as stated by Nations (2015). Financial support by World Bank Climate Investment Funds and IFC Green Bonds finances renewable energy and sustainable agriculture projects from 2020, according to World Bank and IFC from 2021 (Canton, 2021; Esteves et al., 2021). Capacity-building programs put on by UNDP and IFC provide training and knowledge transfer that empower entrepreneurs to apply sustainable practices. The technical assistance of technology transfer and advisory services offered by the IFC in 2021 helped companies adopt green innovations. Networking platforms, such as the UN Global Compact, provide opportunities for collaboration on the road to achieving sustainability goals among businesses, governments, and civil society. These mechanisms constitute an enabling environment for sustainable entrepreneurship that can address global challenges of climate change and inequality while promoting economic growth and environmental preservation.

Table 2 categorically shows how and where international agencies are influential in creating the desired situation in the realms of sustainable entrepreneurship by identifying their spheres of influence with their references. It gives a detailed insight into some important areas where international agencies have to apply efforts to facilitate business behavior change.

Table 2 The kinds of mechanisms used by international agencies to influence sustainable entrepreneurship

Mechanism	Description	Level of influence	References
Policy development			
Global Policy Frameworks	Development of global frameworks like the SDGs and the Paris Agreement	High	Nations (2015)
National Policy Reforms	Advising governments on policy reforms to create an enabling environment	Medium	Sachs et al., (2022)
Funding and investment			
Financial resources	Providing grants, loans, and equity investments for sustainable projects	High	Canton (2021)
Impact investing	Promoting investments that generate both financial returns and social impact	Medium	UNDP (2021)
Capacity building			
Training and education	Offering training programs to develop skills and knowledge for sustainability	Medium	UNDP (2021)
Knowledge transfer	Sharing best practices, research findings, and case studies	Medium	Sachs et al., (2022)
Technical assistance			
Technology transfer	Facilitating access to green technologies and innovations	High	Canton (2021)
Technical support	Providing consulting services, feasibility studies, and project management	Medium	Sachs et al., (2022)
Networking and collaboration			
Public–private partnerships	Facilitating partnerships between businesses, governments, and civil society	High	Liu et al., (2024)
Networking opportunities	Organizing events and platforms for entrepreneurs to connect and collaborate	Medium	Gupta and Dharwal (2022)

On the other hand, international agencies face significant challenges and criticisms despite their vital role in promoting sustainable entrepreneurship. Bureaucratic barriers often delay funding and project implementation, as evident in the World Bank's complex approval processes, which hinder timely support for entrepreneurs (Fidlerová et al., 2022). Funding limitations also pose a challenge, with demand exceeding available resources, particularly in developing countries. For instance, the International Finance Corporation (IFC) struggles to meet the high demand for green project funding (Rosário et al., 2022). Local resistance further complicates efforts, as communities may distrust new technologies or business models, slowing project adoption (Van Tulder et al., 2021). Additionally, dependency on external funding threatens the long-term sustainability of initiatives, as projects may falter once funding ends (Hummels & Argyrou, 2021).

To address these issues, agencies are streamlining bureaucratic processes, increasing funding through mechanisms like green bonds, and engaging local communities to ensure projects align with their needs (Canton, 2021; Latoszek & Latoszek, 2023; Nations, 2015; UNDP, 2021). Enhancing transparency and accountability in resource allocation is also crucial to building trust and improving outcomes (Canton, 2021). By addressing these challenges and criticisms, international agencies can better support sustainable entrepreneurship and contribute to global sustainability goals.

Table 3 provides a good overview of the challenges and criticisms of international agencies in promoting sustainable entrepreneurship, supported by references. It shows where the main

Table 3 The challenges and criticisms faced by international agencies in promoting sustainable entrepreneurship

Raw	Description	References
Challenges		
Bureaucratic barriers	Complex bureaucratic structures can slow down decision-making and implementation processes	Knill et al., (2024) and Abdulsalam et al., (2025)
Funding limitations	The demand for funding often exceeds the available resources, limiting the scale and scope of initiatives	Busch and Barkema (2021) and Canton (2021)
Local resistance	Cultural and political resistance to sustainable practices can slow down progress	Bobby Banerjee et al., (2023) and UNDP (2021)
Dependency on external funding	Reliance on external funding can create sustainability challenges once the funding ends	UNDP (2021)
Criticisms		
Lack of local context	Initiatives may not consider the specific needs and conditions of local communities	Canton (2021) Abdulsalam et al., (2025)
Inefficiency and bureaucracy	Lengthy approval processes can delay the execution of sustainable projects	Canton (2021) Abdulsalam et al., (2025)
Lack of accountability	Concerns about transparency and the effectiveness of resource allocation	Canton (2021)

areas of improvement lie that would enhance the effectiveness of international support for sustainable entrepreneurship.

1.7 Future Directions on the Relationship Between International Agencies and Sustainable Entrepreneurship

As global challenges related to climate change, resource depletion, and social inequality escalate, the potential role of international agencies in stimulating sustainable entrepreneurship also intensifies. The following essay debates emerging trends, policy reforms, and technological changes regarding future directions that are expected to impact the efficiency of the mechanisms employed by international agencies. With the use of references and a forecasting table, this summary sets out to delineate an overview of the shifting relationship between international agencies and sustainable entrepreneurship.

1.7.1 Emerging Trends

Digital Technologies: Digital technologies such as AI, blockchain, and IoT are changing the sustainable entrepreneurship landscape. These technologies make it possible for an enterprise to optimize resource utilization, reduce waste of materials, and bring more transparency to the value chain.

- AI for Sustainability: AI optimizes resource efficiency, predicts environmental risk, and provides efficient energy use; hence, Bocken and Geradts (2022).
- Blockchain for Transparency: Blockchain ensures supply chain transparency and allows a business to verify the sustainability of its products. This is according to the World Economic Forum, 2020.

Social Entrepreneurship: Social entrepreneurship is fast gaining momentum as a model for addressing social and environmental challenges.

- Impact Investing: Investors are increasingly supporting social enterprises that generate both financial returns and social impact, according to Dees (1998).
- Inclusive Business Models: Businesses are developing models serving low-income populations, hence unlocking economic opportunities for them (Mair & Martí, 2006).

Green Innovation: Green innovation focuses on technologies and practices that have minimal impact on the environment.

- Renewable Energy: Solar, wind, and battery technologies are driving innovation in the transition towards a low-carbon economy (Schaltegger, 2002).
- Sustainable Agriculture: Precision farming and vertical farming enhance resource efficiency while minimizing environmental impact (Kirkwood & Walton, 2010).

1.7.2 Forecasting Efficiency of Mechanisms

Table 4 estimates the efficiency percentage of mechanisms adopted by international agencies in promoting sustainable entrepreneurship for a decade, 2023–2033.

Efficiency Percentage and Formula Calculation

The explanation of the derivation of the efficiency percentage of the mechanisms adopted by international agencies towards the promotion of sustainable entrepreneurship requires an understanding of the formula and methodology adopted in the derivation of the percentage. The table forecasts the efficiency percentage for different mechanisms across varying periods running from 2023 through 2033. Estimate actual and potential outputs using evidence from reports, studies, and initiatives—UNDP, IFC, and UN Global Compact (Canton, 2021; Debrah et al., 2024; Pittaluga, 2021). Such forecasting is most likely to be based on a combined effect of qualitative and quantitative factors that include:

- Key Drivers: Factors driving the efficiency of each mechanism, such as the adoption of SDGs, the growth of green bonds, etc.
- Evidence: Trends or data observed that substantiate the predictions, for instance, improved alignment with SDGs and expansion of green bonds.
- References: Studies or reports on which the predictions have been based.

General Formula for Efficiency Percentage

The efficiency percentage is generally calculated as:

$$\text{Efficiency Percentage} = (\text{Actual Output or Impact}/\text{Potential Output or Impact}) \times 100$$

In this context:

- Actual Output or Impact: Measured results of the mechanism, such as the number of businesses aligned with SDGs, and the amount of green financing mobilized, among others.
- Potential Output or Impact: Best possible outcomes if the mechanism were fully optimized, such as total businesses in the target region, and total funding required for sustainability goals, among others.

Policy Frameworks

Global sustainability initiatives are propelled by the country's ratification of the SDGs and the 2015 Paris Agreement (Nations, 2015), and companies are increasingly adjusting their operations to conform to these frameworks (Kücükgül et al., 2022). Even while 100% corporate alignment with SDGs is the ultimate aim, current forecasts show gaps in implementation, with only 65% of businesses expected to comply by 2025. Closing this gap and attaining broader adherence to sustainability targets require accelerated policy action, cooperation, and resource mobilization.

Table 4 Prediction of the efficiency percentage

Mechanism	2023–2025 (%)	2026–2028 (%)	2029–2031 (%)	2032–2033 (%)	Key drivers	Evidence	References
Policy frameworks	65	75	85	90	Adoption of SDGs, Paris Agreement, and national policy reforms	Increased alignment of business practices with SDGs and climate goals	Kücükgül et al., (2022) and Nations (2015)
Funding and investment	60	70	80	85	Growth of green bonds, impact investing, and blended finance	Expansion of IFC Green Bonds and World Bank Climate Investment Funds	Canton (2021) and Debrah et al., (2024)
Capacity building	55	65	75	80	Expansion of training programs and knowledge-sharing platforms	UNDP and IFC capacity-building initiatives in developing countries	Pittaluga (2021)
Technical assistance	50	65	75	85	Increased adoption of green technologies and advisory services	IFC's technology transfer programs and UNDP's green technology initiatives	Shabir et al., (2022) and Sikder et al., (2022)
Networking and collaboration	60	75	85	90	Growth of public–private partnerships and global sustainability networks	UN Global Compact's partnerships and the World Bank's collaborative platforms	Kolmodin (2023)

$$\text{Efficiency Percentage } (2023 - 2025) = (65/100) \times 100 = 65\%$$

Financing and Investment

The global shift to sustainability is being propelled by the expansion of impact investing and green bonds (Debrah et al., 2024). Organizations such as the World Bank, Climate Investment Funds, and IFC Green Bonds are growing to fund low-carbon infrastructure and renewable energy (Canton, 2021). However, because of issues including disjointed legal frameworks and low investor knowledge, only 60% of the required investment is anticipated by 2025.

The future of sustainable entrepreneurship is certainly pegged onto the various emergent trends the international agencies The global shift to sustainability is being propelled by the expansion of impact investing and green bonds. Organizations such as the World Bank, Climate Investment Funds, and IFC Green Bonds are growing to fund low-carbon infrastructure and renewable energy. However, because of issues including disjointed legal frameworks and low investor knowledge, only 60% of the required investment is anticipated by 2025. They are capable of adopting policy reforms and improved technologies. These agencies make the world a better place, sustainable, and more equitable place by addressing challenges and gaps that reduce the effectiveness of their mechanisms.

2 Conclusion

The chapter has highlighted that international agencies and institutions indeed play a very important role in the promotion of sustainable entrepreneurship through policy frameworks, funding, capacity building, technical assistance, and networking. Yet, there is a lack of structured frameworks that address the existing challenges toward value addition with a realization of maximum impact. This calls for a framework addressing the alignment of goals, accountability, transparency, inclusiveness, and adaptability for international agencies in supporting sustainable entrepreneurship toward contributing to the achievement of the SDGs (Fig. 1).

The proposed framework shows the way international agencies can better support sustainable entrepreneurship. In aligning goals, accountability, promoting transparency, upholding ethical standards, and a focus on impact, international agencies will overcome challenges and maximize their contributions to sustainable development.

International agencies and institutions have much relevance for fostering sustainable entrepreneurship against the background of global challenges such as climate change, resource depletion, and social inequality. These agencies will need to address the issues of sustainable business from various dimensions, including the strengthening of policy frameworks, increased funding, enhanced capacity building, fostering innovation, and promoting public–private

Fig. 1 The framework of international agencies/institutions that promote sustainable entrepreneurship

partnerships. This would also provide an enabling environment for sustainable practices whereby the agency aligns national policies with global goals like the SDGs and the Paris Agreement. Secondly, scaling up the mechanism of green financing: IFC Green Bonds and World Bank Climate Investment Funds are critical in providing access to finance that entrepreneurs, in particular, need in developing countries. These capacity-building programs entail training and technical assistance to be provided to the entrepreneur in order to prepare him or her for applying concepts of sustainability by making use of green technologies.

Sustainable entrepreneurship initiatives require a scaling-up process through collaboration. A PPP can be an instrumental vehicle that makes effective utilization of comparative advantages of governments, businesses, and civil societies to address complicated challenges of sustainability more holistically. While initiatives like the UN Global Compact spur business strategy changes at platforms with regard to sustainability principles, involving local communities makes the initiatives contextual and inclusive. Issues of equity and inclusion equally need to be addressed at an international agency level through specific funding and training programs that would support the most marginalized groups. Equitable provision for low-income groups through inclusive business models—in this case, workshop access to clean energy—can simultaneously realize socio-economic development. Third, an effective program in essence requires high-ranking monitoring and evaluation so that any achievement thereof can be measured and, importantly, afterward, engender accountability that might help attain trust from major stakeholders in such interventions.

Effectiveness, in this regard, for international agencies evolves towards emergent contexts, emerging trends, and evolving challenges of sustainable entrepreneurship. AI, blockchain, and IoT are among the digital technologies that can bring increased efficiency in resource utilization and greater transparency to the supply chain. Climate-resilient practices, circular economy principles of recycling, and resource efficiency will go a long way in enabling companies to adapt to environmental changes. International agencies should foster innovation, collaboration, and inclusivity to empower entrepreneurs as agents of systemic change toward a sustainable and equitable world. These would mitigate some of the global challenges while unlocking economic opportunities; hence, they are able to show the transformative potential of sustainable entrepreneurship.

3 Case Study: United Nations-Business Partnerships—Good Intentions and Contradictory Agendas

3.1 Introduction

The paper "United Nations-Business Partnerships: Good Intentions and Contradictory Agendas" by Jem Bendell in 2009 deals with the dynamics existing in the partnerships between the United Nations and business. These partnership attempts to address some global challenges, such as poverty, environmental degradation, and social inequality, through private sector resources and expertise. But these partnerships also result in a lot of problems emerging, from the difference in agendas to priorities. In this case study, the key themes, findings, and implications arising from the chapter are critically summarized against the UN-business partnerships.

3.2 Background

The UN has increasingly turned to partnerships with businesses in pursuit of its sustainable development goals. These partnerships are seen as one avenue through which resources, innovation, and expertise can be mobilized from the private sector. The collaboration between a multilateral organization like the UN and profit-driven corporations raises questions of alignment of goals, accountability, and effectiveness.

Bendell (2017) examines these issues through a critical lens, highlighting both the potential and pitfalls of such partnerships.

3.3 Key Themes and Findings

1. **Good Intentions and Shared Goals**

Most UN-business partnerships are essentially driven by good intentions, with both parties seeking to address pressing global challenges. The UN and businesses share a common goal of attaining sustainable development, poverty reduction, and protection of the environment. For example, the partnerships under the UN Global Compact aim to have businesses align their operations to conform to ten universal principles in areas related to human rights, labor, and the environment, as outlined by Utting and Zammit (2009). Businesses bring in financial resources, technical expertise, and innovation that could help magnify the UN's efforts in achieving better results for its initiatives.

2. **Conflicting Agendas**

The study highlights the stress unions of UN-business collaborations underpinned by diverging agendas: if companies are in the business of profits and shareholder value, the UN is about social/environmental impact, inevitably leading to friction, which can water down effectiveness as a partnership. In addition, "blue washing"—the use of UN branding to improve a company's perception (Utting & Zammit, 2009)—can be a publicist's lie, concealing accurate contributions to sustainability for the sake of a more pristine reputation than actual development outcomes. The misfit motivations could potentially disincentivize the partnership to the level of frivolous partnerships that benefit corporations rather than go global for sustainable development goals.

3. **Accountability and Transparency**

An important issue linked to accountability in UN-business partnerships is the concept of accountability itself. The chapter underlines a lack of strong mechanisms to hold businesses accountable for their commitments in partnerships due to a Lack of Accountability. For instance, the UN Global Compact uses voluntary reporting, which cannot ensure absolute compliance with its principles. Making Gaps in Transparency, Decision-making in partnerships is often not transparent, raising questions over how priorities are set and resources allocated.

4. **Power Imbalances**

Power imbalances between the UN and businesses can lead to differential outcomes from partnerships. Large multinational corporations often have a major influence on partnerships, which can marginalize the voices of smaller stakeholders, including local communities and civil society organizations, through *Corporate Dominance*. Also, the UN's reliance on corporate funding could compromise its independence and ability to hold businesses accountable.

5. **Case Examples**

Case examples show various challenges and contradictions within the UN-business partnerships. Although the UN Global Compact drew thousands of businesses, it still relies on volunteers, with businesses often self-reporting their efforts against the criteria in question. Most of these partnerships operating within the conflict zone encounter ethical dilemmas where profit supersedes human rights and social stability by Conflict Zone Partnerships.

3.4 Challenges

Among the four main challenges facing international agencies' partnerships are alignment difficulties (misalignment of incentives between UN long-term sustainability goals with the short-term profit motives of businesses), lack of enforcement allowing companies to shy away from liability if they fail on commitments and therefore eroding the credibility of partnership; Ethical Ambiguity in working with peer industries that are morally repugnant (e.g. fossil fuels, tobacco), these agencies might have a tarnished reputation; missing local voices, in particular, small businesses and communities are absent from decision-making dominated by large corporations who frequently ignore what is wanted on the ground at an often grassroots level. These issues undermine the effectiveness and inclusiveness of partnership and erode trust.

3.5 Implications and Recommendations

The report gives a set of recommendations to enhance UN-business partnerships, calling for fortified accountability by way of *mandatory reporting, independent audits and redress to non-compliance*; *more openness in decision-making/resource-allocation for trust building*; a fair power relationship by enhancing local communities' voices through participatory platforms; ethical vetting to filter out UN values adverse sectors (e.g. fossil fuels) and concrete and measurable outcomes over PR mantra with explicit objectives and performance measures for partnerships to result in actual sustainable development impact. These moves were intended to align partnerships with equity, integrity, and effectiveness.

3.6 Conclusion

UN-business partnerships have the potential to address global challenges by leveraging the resources and expertise of the private sector. However, as Bendell (2017) highlights, these partnerships are often fraught with contradictions and challenges, including misaligned agendas, lack of accountability, and power imbalances. To realize their full potential, UN-business partnerships must prioritize accountability, transparency, and ethical considerations. By addressing these issues, the UN can build more effective and equitable partnerships that contribute to sustainable development.

4 Questions

Examples of application and critical thinking questions on sustainable entrepreneurship and the role of international agencies.

Application Questions

- How would you apply eco-design principles to create a sustainable product?
- How can a small business incorporate a circular economy model into its operations?
- What steps would you take to secure funding for your sustainable agriculture project from an international agency?

Critical Thinking Questions

- What challenges could a startup find in implementing a circular supply chain model?
- How might international agencies balance the need for profitability, intending to promote sustainable entrepreneurship?"
- What might be some risks to relying heavily on public–private partnerships (PPPs) to achieve sustainable development goals?"
- How might international agencies ensure that support for sustainable entrepreneurship benefits marginalized communities?

References

Abdulsalam, T. A., Tajudeen, R. B., Subair, M. L., Ogherohwo, T. F., & Raji, Y. M. (2025). Global environmental challenges in the current business scenario: A roadblock for sustainable development. In *Green economy and sustainable development* (pp. 253–282). Springer Nature Switzerland.

Aziz, H. M., Sorguli, S., Hamza, P. A., Sabir, B., Qader, K., Ismeal, B. A., & Gardi, B. (2021). Factors affecting international finance corporation. *International Journal of Humanities and Education Development (IJHED), 3*(3), 148–157.

Basak, R., & Karlsson-Vinkhuyzen, S. (2022). The Green Climate Fund: History, status and legitimacy. In *Handbook of international climate finance* (pp. 135–166). Edward Elgar Publishing.

Bendell, J. (2017). *Evolving partnerships: A guide to working with business for greater social change*. Routledge.

Biswas, S. S., Ahad, M. A., Nafis, M. T., Alam, M. A., & Biswas, R. (2021). Introducing "α-Sustainable Development" for transforming our world: A proposal for the 2030 agenda. *Journal of Cleaner Production, 321*, Article 129030.

Bobby Banerjee, S., Maher, R., & Krämer, R. (2023). Resistance is fertile: Toward a political ecology of translocal resistance. *Organization, 30*(2), 264–287.

Bocken, N., Baldassarre, B., Keskin, D., & Diehl, J. C. (2023). Design thinking tools to catalyse sustainable circular innovation. In *The Routledge handbook of catalysts for a sustainable circular economy* (pp. 359–387). Routledge.

Bocken, N., & Geradts, T. (2022). Designing your circular business model. *Stanford Social Innovation Review, 20*(2), 34–39.

Bongardt, A., & Torres, F. (2022). The European green deal: More than an exit strategy to the pandemic crisis, a building block of a sustainable European economic model. *JCMS: Journal of Common Market Studies, 60*(1), 170–185.

Bürgin, A. (2023). The European Commission: A climate policy entrepreneur. In *Handbook on European Union climate change policy and politics* (pp. 23–37). Edward Elgar Publishing.

Busch, C., & Barkema, H. (2021). From necessity to opportunity: Scaling bricolage across resource-constrained environments. *Strategic Management Journal, 42*(4), 741–773.

Canton, H. (2021). International Finance Corporation—IFC. In *The Europa Directory of International Organizations 2021* (pp. 323–324). Routledge.

Cao, K., & Gehman, J. (2021). Certified B corporations and benefit corporations.

Corsini, F., & Frey, M. (2024). The paradigm of the circular economy: Barriers and enabling factors for companies. In *Crowdfunding for environmental sustainability and the circular economy: Empowered strategies for sustainable growth* (pp. 13–36). Springer.

Debrah, C., Chan, A. P., Darko, A., Owusu-Manu, D.-G., & Ohene, E. (2024). Green 13 finance. *Rethinking pathways to a sustainable built environment* (p. 277).

Decker, S., & Obeng Dankwah, G. (2023). Co-opting business models at the base of the pyramid (BOP): Microentrepreneurs and multinational enterprises in Ghana. *Business & Society, 62*(1), 151–191.

Dees, J. G. (1998). Enterprising nonprofits: What do you do when traditional sources of funding fall short. *Harvard Business Review, 76*(1), 55–67.

Di Vaio, A., Hassan, R., Chhabra, M., Arrigo, E., & Palladino, R. (2022). Sustainable entrepreneurship impact and entrepreneurial venture life cycle: A systematic literature review. *Journal of Cleaner Production, 378*, Article 134469.

Ejejigbe, M. S. (2021). *Exploratory study of interventions and challenges faced by indigenous NGOs supporting startup businesses (established by young entrepreneurs) in Nigeria's Niger-Delta Region*. University of Wales Trinity Saint David (United Kingdom).

Elkington, J., & Rowlands, I. H. (1999). Cannibals with forks: The triple bottom line of 21st century business. *Alternatives Journal, 25*(4), 42.

Emon, M. M. H., & Khan, T. (2023). The impact of cultural norms on sustainable entrepreneurship practices in SMEs of Bangladesh. *Indonesian Journal of Innovation and Applied Sciences (IJIAS), 3*(3), 201–209.

Esteves, A. M., Genus, A., Henfrey, T., Penha-Lopes, G., & East, M. (2021). Sustainable entrepreneurship and the Sustainable Development Goals: Community-led initiatives, the social solidarity economy and commons ecologies. *Business Strategy and the Environment, 30*(3), 1423–1435.

Farny, S., & Binder, J. (2021). Sustainable entrepreneurship. In *World encyclopedia of entrepreneurship* (pp. 605–611). Edward Elgar Publishing.

Fidlerová, H., Stareček, A., Vraňaková, N., Bulut, C., & Keaney, M. (2022). Sustainable entrepreneurship for business opportunity recognition: Analysis of an awareness questionnaire among organisations. *Energies, 15*(3), 849.

Friant, M. C., Vermeulen, W. J., & Salomone, R. (2021). Analysing European Union circular economy policies: Words versus actions. *Sustainable Production and Consumption, 27*, 337–353.

Gideon, J., & Unterhalter, E. (2020). Reflections on public private partnerships: An introduction. In *Critical reflections on public private partnerships* (pp. 1–25). Routledge.

Gulakov, I., & Vanclay, F. (2024). Continuing to Put People First: Embedding community investment in the sustainability standards of international financial institutions. *Sustainable Development*.

Gupta, M., & Dharwal, M. (2022). Green entrepreneurship and sustainable development: A conceptual framework. *Materials Today: Proceedings, 49*, 3603–3606.

Hahn, T., & Tampe, M. (2021). Strategies for regenerative business. *Strategic Organization, 19*(3), 456–477.

Hall, J. C., Lacombe, D. J., & Pokharel, S. B. (2016). Freedom and entrepreneurship: A spatial econometric approach. *Journal of Entrepreneurship and Public Policy, 5*(3), 404–411.

Hickmann, T., Widerberg, O., Lederer, M., & Pattberg, P. (2021). The United Nations Framework Convention on Climate Change Secretariat as an orchestrator in global climate policymaking. *International Review of Administrative Sciences, 87*(1), 21–38.

Hodge, G. A., & Greve, C. (2007). Public–private partnerships: An international performance review. *Public Administration Review, 67*(3), 545–558.

Homer, S. T., Lim, W. M., & Poon, W. C. (2025). A concentric locus of control and triple bottom line model for responsible management: Theory development inspired by students from the Global South. *The International Journal of Management Education, 23*(1), Article 101083.

Hummels, H., & Argyrou, A. (2021). Planetary demands: Redefining sustainable development and sustainable entrepreneurship. *Journal of Cleaner Production, 278*, Article 123804.

Kassim, H., & Tholoniat, L. (2021). The European Commission: Crisis and crisis management in the European Union: The European Council and European Commission revisited. *The Palgrave handbook of EU crises* (pp. 297–313).

Kirkwood, J., & Walton, S. (2010). How ecopreneurs' green values affect their international engagement in supply chain management. *Journal of International Entrepreneurship, 8*, 200–217.

Knill, C., Steinebach, Y., & Zink, D. (2024). How policy growth affects policy implementation: Bureaucratic overload and policy triage. *Journal of European Public Policy, 31*(2), 324–351.

Kolmodin, S. (2023). A platform for collaboration: The views of civil society organisations on a local compact. *Central European Journal of Public Policy, 17*(1), 14–26.

Konietzko, J., Das, A., & Bocken, N. (2023). Towards regenerative business models: A necessary shift? *Sustainable Production and Consumption, 38*, 372–388.

Korletey, V. (2024). Creating shared value for stakeholders: A strategic approach to building sustainable. *Business Ethics in Africa: Values, Profits and Responsibility, 1*, 147.

Kücükgül, E., Cerin, P., & Liu, Y. (2022). Enhancing the value of corporate sustainability: An approach for aligning multiple SDGs guides on reporting. *Journal of Cleaner Production, 333*, Article 130005.

Kramer, M. R., & Porter, M. (2011). *Creating shared value* (Vol. 17). FSG.

Lamichhane, S., Eğilmez, G., Gedik, R., Bhutta, M. K. S., & Erenay, B. (2021). Benchmarking OECD countries' sustainable development performance: A goal-specific principal component analysis approach. *Journal of Cleaner Production, 287*, Article 125040.

Langer, C. (2024). Social entrepreneurship as enactment of social innovation in tourism and hospitality: An analysis of its implementation and functioning.

Latoszek, E., & Latoszek, A. (2023). The World Bank: A new narrative for new challenges? *Global public goods and sustainable development in the practice of international organizations* (p. 195).

Liu, B., Li, J., Wang, D., Liu, H., Wu, G., & Yuan, J. (2024). Public–private partnerships: A collaborative framework for ensuring project sustainable operations. *Engineering, Construction and Architectural Management, 31*(1), 264–289.

Mair, J., & Martí, I. (2006). Social entrepreneurship research: A source of explanation, prediction, and delight. *Journal of World Business, 41*(1), 36–44.

Majeed, Y., Khan, M. U., Waseem, M., Zahid, U., Mahmood, F., Majeed, F., Sultan, M., & Raza, A. (2023). Renewable energy as an alternative source for energy management in agriculture. *Energy Reports, 10*, 344–359.

Mohammadi, P., AlHattali, K. S. S., & Alghatrifi, I. N. S. (2024). Rethinking the boundaries and definition of social entrepreneurship: A critical literature review. *Business Management and Strategy, 15*(1), 226–240.

Muneer, A. (2023). Social entrepreneurship and sustainable development: A synthesis of business and social goals. *Liberal Journal of Management & Social Science, 1*(1), 11–18.

Nations, U. (2015). *Transforming our world: The 2030 agenda for sustainable development* (Vol. 1, p. 41). United Nations, Department of Economic and Social Affairs.

Nicholls, A., & Ormiston, J. (2023). Financial market transformations for. *Social economy science: Transforming the economy and making society more resilient* (p. 106).

Organization, W. H. (2023). *Trends in maternal mortality 2000 to 2020: Estimates by WHO, UNICEF, UNFPA*. World Health Organization.

Regional Economic Outlook: International Monetary Fund eBooks. (2022).

Pittaluga, F. (2021). Project Information Document (PID)-Monitoring and Evaluation Capacity Building in Anglophone Africa-P177221.

Podrecca, M., Sartor, M., & Nassimbeni, G. (2022). United Nations Global Compact: Where are we going? *Social Responsibility Journal, 18*(5), 984–1003.

Rosário, A. T., Raimundo, R. J., & Cruz, S. P. (2022). Sustainable entrepreneurship: A literature review. *Sustainability, 14*(9), 5556.

Sachs, J., Kroll, C., Lafortune, G., Fuller, G., & Woelm, F. (2022). *Sustainable development report 2022*. Cambridge University Press.

Schaltegger, S. (2002). A framework for ecopreneurship: Leading bioneers and environmental managers to ecopreneurship. *Greener Management International* (38), 45–58.

Schaltegger, S., & Wagner, M. (2011). Sustainable entrepreneurship and sustainability innovation: Categories and interactions. *Business Strategy and the Environment, 20*(4), 222–237.

Shabir, M., Ali, M., Hashmi, S. H., & Bakhsh, S. (2022). Heterogeneous effects of economic policy uncertainty and foreign direct investment on environmental quality: Cross-country evidence. *Environmental Science and Pollution Research, 29*(2), 2737–2752.

Sikder, M., Wang, C., Yao, X., Huai, X., Wu, L., KwameYeboah, F., Wood, J., Zhao, Y., & Dou, X. (2022). The integrated impact of GDP growth, industrialization, energy use, and urbanization on CO_2 emissions in developing countries: Evidence from the panel ARDL approach. *Science of the Total Environment, 837*, Article 155795.

Srivastava, A. K., Dixit, S., & Srivastava, A. A. (2022). Criticism of triple bottom line: TBL (with special reference to sustainability). *Corporate reputation review* (pp. 1–12).

Trapp, C. T., & Kanbach, D. K. (2021). Green entrepreneurship and business models: Deriving green technology business model archetypes. *Journal of Cleaner Production, 297*, Article 126694.

UNDP, S. (2021). UNEP-WCMC (2021). *Creating a nature-positive future: The contribution of protected areas and other effective area-based conservation measures*. UNDP.

Utting, P., & Zammit, A. (2009). United Nations-business partnerships: Good intentions and contradictory agendas. *Journal of Business Ethics, 90*, 39–56.

Vallet, A. (2021). Impact and recurring factors of Business Model Canvas in B Corp Certification.

Van Tulder, R., Rodrigues, S. B., Mirza, H., & Sexsmith, K. (2021). The UN's sustainable development goals: Can multinational enterprises lead the decade of action? *Journal of International Business Policy, 4*(1), 1.

Van Tulder, R., & van Mil, E. (2022). *Principles of sustainable business: Frameworks for corporate action on the SDGs*. Routledge.

Vishnoi, J. N. (2022). *The role of the financial institutions in transition to low carbon economy*. Wien.

Voros Fregolente, M., & Carvalho, M. M. (2023). Exploring BoP generations through business model innovation lens: A review and framing. *Sustainability, 15*(17), 12817.

Special Issues

Sustainable Entrepreneurship in Small and Medium Enterprises (SMEs)

Hamed Hosseini Zarrabi and MohamadReza Amini

Abstract

In the context of a global economy, Small and Medium Enterprises (SMEs) play a pivotal role in driving sustainable entrepreneurship. This paper examines the definition, core characteristics, theoretical underpinnings, and critical success factors of sustainable entrepreneurship within SMEs. Central to this discourse are key success factors such as effective leadership, managerial competence, strategic planning, and innovation capacity. These elements collectively enhance SMEs' resilience and adaptability, fostering long-term viability. Organizational culture, particularly visionary leadership and a learning-oriented environment, further amplifies sustainability outcomes. Managerial capabilities—specifically strategic decision-making, financial discipline, and resource efficiency—are emphasized as critical enablers of sustainable performance. However, SMEs often encounter multifaceted barriers that hinder sustainable initiatives, which may stem from internal constraints or external institutional and market pressures. Understanding the interplay between these barriers is essential for achieving strategic alignment with sustainability goals. The study also highlights key drivers of sustainable entrepreneurship and presents case study reviews that illustrate practical applications and successful models across different sectors. This integrated approach contributes to a more comprehensive framework for understanding and advancing sustainability in SMEs, offering valuable insights for practitioners, policymakers, and researchers alike.

Keywords

SME · Sustainable entrepreneurship · Barrier · Emerging economies

Learning Objectives

- **Define** sustainable entrepreneurship within the context of Small and Medium Enterprises (SMEs) and identify its core characteristics and critical success factors.
- **Distinguish** between internal and external barriers to sustainable entrepreneurship and explain their dynamic interactions.
- **Explore** the primary drivers of sustainable entrepreneurship and how they can be leveraged within SMEs to overcome contextual challenges.
- **Review** case studies that demonstrate the practical implementation of sustainable entrepreneurship strategies across various sectors.

H. H. Zarrabi (✉) · M. Amini
Faculty of Entrepreneurship, University of Tehran, Tehran, Iran
e-mail: hamed.zarrabi@ut.ac.ir

M. Akbari et al. (eds.), *Sustainable Entrepreneurship*, University of Tehran Science and Humanities Series, https://doi.org/10.1007/978-3-032-04909-4_15

1 Introduction

Sustainable entrepreneurship in small and medium-sized enterprises (SMEs) can be defined as the practice of conducting business in a manner that is economically viable, socially responsible, and environmentally friendly. This approach integrates sustainability principles into the core business operations and decision-making processes, ensuring that the enterprise makes a positive contribution to economic development, social well-being, and environmental preservation. Sustainable entrepreneurship involves creating value across three dimensions: economic, social, and environmental (Fong et al., 2023; Reiche, 2014; Tazhibekova & Shametova, 2024). This means that SMEs not only focus on profitability but also on their impact on society and the environment.

Sustainable entrepreneurship in SMEs refers to the ongoing commitment by businesses to behave ethically and contribute to economic development while improving the quality of life for various stakeholders, including the workforce, local communities, and future generations (Crals & Vereeck, 2005). It involves creating economic, social, and environmental value across enterprises (Fong et al., 2023).

Sustainable practices play a critical mediating role in enhancing business performance in SMEs, with entrepreneurial orientation positively influencing business performance indirectly through engagement in sustainable development practices (Crals & Vereeck, 2005). The study of SMEs in the food industry found that social and environmental factors, including customer orientation and recycling, significantly influence sustainable performance (Štverková & Pohludka, 2021).

The most frequently cited definition of sustainability is the one used in a report by the Brundtland Commission to the 1987 United Nations conference. It defined sustainable development as "meeting present needs without compromising the ability of future generations to meet their needs." A reasonable working definition of a sustainable system might now be:

A sustainable system fulfills present and future needs while using, and not harming, renewable resources and unique human environmental resources of a site, including air, land, water, energy, mineral resources, and human ecology, as well as those of other (off-site) sustainable systems (Andreas et al., 2011).

The concept of sustainable entrepreneurship in SMEs centers on a transformative business model where leaders are not only profit-oriented but also committed to fostering community well-being, ecological conservation, and ethical behavior. This paradigm shift is evident in community-based social enterprises (CBSEs), which highlight the dual role of business and social innovation in addressing local challenges (Suriyankietkaew et al., 2022). Sustainable entrepreneurship integrates the principles of economic viability, social equity, and environmental stewardship, transcending the conventional profit-maximization framework in favor of more holistic value creation (Suriyankietkaew et al., 2022).

Sustainability can be understood through various lenses, including tangible products and intangible processes, and assessed in terms of absolute or relative success. It also encompasses static and dynamic perspectives, depending on whether the artifact interacts with a fixed or evolving ecological and social environment. The evolution of sustainability research and definitions can be categorized into four major transitional phases. These stages reflect the field's growing complexity and integration into various domains over time.

1. **Sustainability Awareness**—This phase centered on fundamental research and definitions that shaped the understanding of sustainability, with environmental reporting by practitioners serving as a key milestone.
2. **Sustainability Management**—Definitions and research during this phase emphasized the components of sustainability—social, economic, and environmental—reflecting a practitioner's focus on ISO 9001, ISO 14001, and carbon footprint methodologies.
3. **Sustainability Adoption**—Research explored the implementation of sustainability within the framework of ESG as businesses began utilizing voluntary reporting to demonstrate

their commitment to these principles among stakeholders.

4. **Sustainability Strategic Integration**—In this phase, sustainability concepts were integrated into broader strategic frameworks, including economic, finance, and risk management theories. A key development was the introduction of the ISO 26000 standard, signaling the shift from best practices to core corporate strategy (James, 2014).

Despite the upsurge of interest in and commitment to more sustainable operations among larger organizations, A Simple Path to Sustainability, the vast majority of smaller business leaders have yet to embrace this type of thinking. There are many reasons for this:

1. Leaders of small organizations are often consumed by immediate concerns such as cash flow and staffing, leaving little room for long-term planning. This short-term focus hinders their ability to integrate sustainability into strategic decision-making.
2. Sustainable practices are often viewed as costly investments, perceived as feasible only for large corporations. This belief leads to concerns that sustainability will lower profit margins and increase operational expenses.
3. Some business leaders believe that addressing global sustainability is the responsibility of others. This mindset fosters complacency, allowing them to prioritize short-term gains over long-term environmental impact.

Since small and medium-sized enterprises significantly contribute to GDP and employment, their engagement is essential for achieving global sustainability. Ultimately, all businesses must recognize that long-term success depends on integrating sustainable practices into their core strategies (Andreas et al., 2011).

2 Theoretical Frameworks

The evolution of sustainable entrepreneurship in SMEs is underpinned by several interrelated theoretical frameworks. These frameworks provide a robust foundation for understanding how small and medium-sized enterprises (SMEs) can continuously innovate and adapt while maintaining stability and resilience. Two primary frameworks warrant discussion: the emerging models for sustainability in emerging economies and the role of entrepreneurial orientation in facilitating flexible and adaptive business practices.

2.1 Framework for Emerging Economies

In emerging economies, sustainable entrepreneurship often intersects with community-based practices to form alternative business models that extend beyond traditional market mechanisms. The concept of a community-based social enterprise (CBSE) illustrates how localized leadership practices nurture human capital and foster ethical values, social capital, and pro-environmental behaviors (Suriyankietkaew et al., 2022). This model is characterized by the active involvement of local stakeholders who collectively manage community resources and support the local economy. In the context of Thailand, for example, an award-winning social enterprise in a coastal fishing village demonstrated how value-oriented leadership practices can contribute to economic resilience even in the aftermath of a severe global crisis such as COVID-19 (Suriyankietkaew et al., 2022).

Furthermore, these community-centered business models have led to the development of alternative sustainable business models that align with broader sustainability and resilience goals. The proposed models incorporate strategic, systems thinking, and interpersonal competencies alongside a robust ethical framework to ensure that businesses not only survive but also thrive in an uncertain global environment. These frameworks emphasize the integration of internal and external stakeholder interests, thereby encouraging inclusive growth and long-term sustainable development.

Another critical framework is the Quintuple Helix Model (5HM), which extends traditional innovation models by incorporating environmental dimensions alongside the roles of academia, industry, government, and civil society (Borrero & Yousafzai, 2024). The 5HM framework argues that a collaborative, multi-stakeholder approach is essential for addressing the multifaceted challenges of sustainable development. In this model, universities and research institutions act as institutional entrepreneurs, bridging knowledge gaps and fostering innovation in regional sustainable ecosystems. By leveraging these frameworks, SMEs in emerging economies can develop strategies that simultaneously address market needs and broader socio-environmental challenges.

3 Characteristics of Sustainable Entrepreneurship in SMEs

3.1 Dimensions

The increasing emphasis on sustainability in business practice has positioned small and medium-sized enterprises (SMEs) as pivotal actors in achieving long-term environmental, social, and economic objectives. Sustainable entrepreneurship within SMEs is multidimensional, typically categorized into four interconnected dimensions: environmental, social, economic, and strategic. The environmental dimension focuses on eco-friendly practices, green innovations, and resource efficiency, though implementation is often constrained by SMEs' limited access to capital and knowledge, as highlighted by Anaman et al. (2025). Challenges such as resource shortages and difficulties in adapting product designs for global exports further hinder sustainable practices (Babgohari et al., 2023). Studies of Malaysian manufacturing SMEs reveal that firms' dynamic capabilities fully mediate the relationship between environmental dimensions and sustainable business performance, emphasizing the role of adaptive strategies in overcoming barriers (Yacob et al., 2022). Additionally, digitalization has been identified as a potential driver for improving the environmental aspects of sustainable development, particularly in addressing institutional and cultural dimensions (Belyaeva & Lopatkova, 2020).

The social dimension encompasses entrepreneurial values, corporate social responsibility, and stakeholder engagement, which are deeply influenced by socio-cultural contexts (Hosseininia & Ramezani, 2016). Research indicates that while SMEs demonstrate significant efforts in implementing economic sustainability measures, the managerial and ecological dimensions hold the highest potential for improvement (Jakobs et al., 2024). In Malaysian SMEs, dynamic capabilities also mediate the link between social dimensions and sustainable outcomes, underscoring the importance of organizational adaptability in fostering social responsibility (Yacob et al., 2022). Furthermore, the adoption of sustainable practices in SMEs is heavily shaped by entrepreneurs' individual values and socio-cultural specificities, as observed in international studies (Spence et al., 2011).

From an economic perspective, sustainable entrepreneurship enhances competitiveness through innovation, resilience, and long-term profitability. Bajdor et al. (2021) emphasize the importance of integrating sustainability into core operational frameworks to align SMEs with global competitiveness and environmental responsibility (Kapoor et al., 2023). Key indicators of economic sustainability include productivity improvement, financial stability, and profitability, which are crucial for the survival and growth of SMEs (Alfi et al., 2024). The mediating role of dynamic capabilities is again evident in Malaysian SMEs, where economic dimensions are closely tied to sustainable business outcomes, reinforcing the interdependence of financial health and adaptive strategies (Yacob et al., 2022).

The strategic dimension integrates leadership orientation, business planning, and capability development to align sustainability goals with organizational growth and development. Empirical studies demonstrate that strategic orientation mediates the relationship between environmental, social, and economic dimensions and overall sustainable performance (Soto-Acosta et al., 2016). Strategic drivers of sustainable entrepreneurship encompass sustainable entrepreneurial orientation, organizational structures, performance metrics, and network collaborations, whereas barriers often stem from contextual challenges, such as regulatory environments (Ferreira & Ferreira, 2024). In Malaysian SMEs, dynamic capabilities strategically bridge economic, social, and environmental practices with sustainable outcomes (Yacob et al., 2022). Additionally, research on Iranian SMEs in the food industry highlights the influence of entrepreneurs' characteristics, such as work experience and education, on their adoption of sustainable strategies, emphasizing the human-centric nature of strategic decision-making (Hosseininia & Ramezani, 2016).

By holistically addressing these four dimensions, SMEs can not only meet evolving sustainability expectations but also cultivate competitive advantages in dynamic global markets. The integration of environmental stewardship, social responsibility, economic resilience, and strategic agility enables SMEs to navigate resource constraints, leverage digitalization, and align entrepreneurial values with long-term sustainability goals.

3.2 Key Success Factors

In today's volatile economic landscape, small and medium-sized enterprises (SMEs) must strategically leverage key success factors to achieve sustainable entrepreneurship. Effective leadership, managerial competence, and strategic planning are foundational determinants of organizational success, enabling SMEs to navigate uncertainty and capitalize on opportunities (Abdullahi et al., 2018; Ng & Kee, 2012). These entrepreneurial characteristics, including resilience, adaptability, and innovation, are further amplified by organizational cultures rooted in visionary leadership and learning orientation, which drive sustainable performance (Sukma, 2024). Managerial capabilities, encompassing strategic decision-making, financial prudence, and resource efficiency, are equally critical, as highlighted across multiple studies (Rodrigues et al., 2021; Utami et al., 2024; Velmurugan et al., 2024).

Human resources and intellectual capital form the backbone of SME competitiveness. Employee education, skills development, and motivational practices directly enhance productivity and innovation (Chiun Lo et al., 2016; Rodrigues et al., 2021). Strategic human capital management fosters dynamic capabilities and internal knowledge flows, while a culture of empowerment and trust encourages employee contributions to organizational learning (Balić & Vuka, 2023; Rodrigues et al., 2021; Yuik et al., 2020). Education and training for entrepreneurs and staff are emphasized as vital for building sustainability competencies (Meshram & Rawani, 2019; Rodrigues et al., 2021; Yuik et al., 2020).

Organizational innovation, encompassing technological adoption, process redesign, and novel solutions, is crucial for maintaining market relevance and driving growth (Ng & Kee, 2012; Othman et al., 2022). SMEs that embrace innovation and calculated risk-taking position themselves to adapt and seize emerging opportunities (Velmurugan et al., 2023, 2024). A strong market orientation—proactively aligning products with customer needs—enhances global competitiveness, while external networks and partnerships strengthen resilience and access to resources (Rodrigues et al., 2021).

Financial capacity remains a persistent challenge, with limited access to capital hindering growth (Al-Tit et al., 2019). However, public support mechanisms, external financing, and prudent financial management can bridge these gaps (Aktar & Kabir, 2024; Ng & Kee, 2012). Concurrently, embedding sustainability into organizational missions through environmental and social responsibility bolsters legitimacy, stakeholder trust, and competitive advantage (Álvarez-García et al., 2022). Firms that integrate sustainability into their strategic decisions outperform those that treat it peripherally, with green innovations improving both environmental and economic outcomes (Mondal et al., 2024; Ta'Amnha et al., 2024).

Government support, including policies, financial aid, and infrastructure, is essential for overcoming resource constraints (Aktar & Kabir, 2024; Meshram & Rawani, 2019). Entrepreneurial resilience—the ability to withstand crises—is another critical factor that enables SMEs to thrive in uncertainty (Enwereji et al., 2024; Fatoki, 2018). Finally, opportunity identification and a focus on economic sustainability, such as efficient resource management, underpin long-term success (Soto-Acosta et al., 2016; Utami et al., 2024; Velmurugan et al., 2023, 2024). By holistically addressing these interconnected factors—leadership, human capital, innovation, financial sustainability, and strategic partnerships—SMEs can secure competitive advantages and achieve enduring sustainability in dynamic markets. These factors collectively contribute to the sustainable entrepreneurship of SMEs, ensuring their growth and long-term success in a dynamic business environment.

3.3 Barriers

Sustainable development has become an increasingly dominant paradigm in global economic, political, and environmental discourse. While large enterprises have attracted the bulk of research and policy focus, small and medium-sized enterprises (SMEs) represent the vast majority of businesses worldwide and play a central role in achieving sustainability transitions. However, SMEs face numerous barriers that hinder their ability to engage with sustainability goals effectively. These barriers may be internal (arising from within the organization) or external (stemming from the broader institutional, legal, or market environment), and they often interact in complex ways.

The comprehensive literature review by Álvarez Jaramillo et al. (2019) analyzed 46 highly cited articles indexed in Scopus from 2013 to 2017. From this analysis, the authors identified 175 unique barriers to sustainability in SMEs, which were then categorized by sector, sustainability tools, and whether they were internal or external in nature.

3.3.1 Internal Barriers

Lack of Resources

The lack of resources was the most frequently cited barrier, appearing in 13 studies. This includes not only financial capital but also technical infrastructure and human capacity. SMEs often lack the budgetary flexibility to invest in sustainability tools or hire the necessary expertise for implementation (e.g., Ghazilla et al., 2015; Heidrich & Tiwary, 2013). Outdated equipment (e.g., Trianni & Cagno, 2015) and time constraints (e.g., Heidrich & Tiwary, 2013; Meath et al., 2016) further exacerbate this issue.

Lack of Expertise and Awareness

Another widely documented internal challenge is the lack of technical expertise and awareness about sustainability practices. For instance, internal resistance to change (Ghazilla et al., 2015; Oliveira Neto et al., 2017), ignorance of regulations (e.g., Hjorth & Brem, 2016), and lack of training or R&D capacity (e.g., Auer & Jarmai, 2018) prevent SMEs from adopting or adapting relevant tools. Johnson and Schaltegger (2016) emphasized that the complexity of sustainability standards often alienates SMEs without environmental or CSR backgrounds.

Organizational Structure and Management Resistance

Several studies highlight weak organizational structures and management resistance to sustainability as barriers (e.g., Auer & Jarmai, 2018; Lewis et al., 2015). When top management lacks commitment or sees sustainability as non-essential, integration fails. In many cases, company policies are too rigid or slow to adapt, delaying sustainability initiatives (e.g., Bocken, 2015; Meath et al., 2016).

3.3.2 External Barriers

High Capital Costs and Financial Constraints

High initial capital costs were the second most cited barrier (in 11 studies), particularly relevant for the implementation of green logistics, energy efficiency, and certification processes (e.g., Hjorth & Brem, 2016; Malá et al., 2017). Even when SMEs express interest in sustainability, financial barriers—especially in developing economies—can be prohibitive (e.g. Dayaratne & Gunawardana, 2015; Oliveira Neto et al., 2017).

Regulatory and Policy Barriers

A significant number of SMEs operate in environments where regulations are either insufficiently enforced or poorly designed. Several authors note the lack of supportive policies or the presence of corrupt regulatory systems (e.g., Hasan, 2016; Zhou et al., 2015). Moreover, tools designed for large businesses often fail to translate effectively to SME contexts due to differences in scale, complexity, and operational capacity (e.g., Conway, 2015; Johnson & Schaltegger, 2016).

Market and Infrastructure Limitations

Lack of customer demand for sustainable products and weak market infrastructure also pose external challenges. Some SMEs report low societal awareness and engagement with sustainability, which in turn reduces the incentive to change practices (e.g., Ghadge et al., 2017). Moreover, distribution limitations and poor logistics infrastructure in sectors like dairy and wood processing hinder the implementation of green supply chains (e.g., Malá et al., 2017).

3.3.3 Sector-Specific Challenges

The barriers are categorized by sectors, including automotive, construction, dairy, medical devices, food processing, and ICT. For instance, in the automotive sector, barriers to implementing Material Flow Cost Accounting (MFCA) include a lack of compatibility with existing equipment and skepticism about the tool's benefits (Sulong et al., 2015). In the ICT sector, high costs, low digital literacy, and lack of tailored applications were key obstacles (Mbuyisa & Leonard, 2015). Similarly, in recycling and manufacturing, the lack of reliable life cycle data and the absence of internal certification processes were frequently mentioned (Heidrich & Tiwary, 2013).

3.3.4 The Role of Innovation and Its Barriers

Innovation, although vital to sustainability, is hindered by several barriers. The lack of absorptive capacity—a firm's ability to understand and apply external knowledge—was cited as a key issue (Lewis et al., 2015). In some cases, innovation was stalled by inadequate R&D, low pricing expertise, and customer resistance to new technologies (e.g., Mourtzis et al., 2016). Auer and Jarmai (2018) and Bocken (2015) emphasized that even when SMEs are motivated, the innovation tools available are often ill-suited to their scale, or they lack the required external validation to secure investment.

3.3.5 Intersections of Internal and External Barriers

Many barriers reflect a hybrid nature, where internal limitations are worsened by external inefficiencies. For example, a lack of external stakeholder engagement (e.g., customers, suppliers, government agencies) can demotivate internal efforts (Auer & Jarmai, 2018; Ghadge et al., 2017). In other cases, poor regulation compounds internal ignorance or resistance, creating a feedback loop of inertia (Hasan, 2016; Lewis et al., 2015). This interplay underscores the importance of multi-actor collaboration, where businesses, policymakers, and civil society organizations work together in tandem.

3.4 Drivers of Sustainable Entrepreneurship

Sustainable entrepreneurship has emerged as a multifaceted response to the increasing global demand for businesses that strike a balance between economic performance and environmental and social accountability. However, despite its conceptual appeal, translating sustainable entrepreneurship into widespread practice remains uneven and context-dependent. Understanding the conditions that enable (drivers) the emergence of sustainable entrepreneurship is crucial for researchers, educators, and policymakers aiming to design effective support structures. This section is based exclusively on the findings of a comprehensive literature review conducted by Almeida and Garrod (2024), which analyzes drivers of entrepreneurship as discussed across 49 academic publications. Without focusing on a specific regional case, this synthesis highlights generalizable patterns from the literature that inform how sustainable entrepreneurship can be fostered.

3.4.1 Drivers of Sustainable Entrepreneurship

Family Background and Entrepreneurial Role Models

One of the most prominent drivers identified is **entrepreneurial family background**. The reviewed literature consistently shows that individuals with close family members who have operated a business are significantly more inclined to develop entrepreneurial intentions. This influence operates through various pathways: access to financial and social resources, early exposure to business activities, and the normalization of entrepreneurship as a viable career path. The presence of role models within the family enhances self-efficacy and reduces the fear of failure (Altinay et al., 2012; Liu & Zhao, 2021; Mohan, 2022).

Age and Maturity

Another internal driver is the **age of the prospective entrepreneur**. The review indicates that older individuals tend to show stronger entrepreneurial intentions than younger ones. This correlation is attributed to higher levels of personal maturity, accumulated life and work experience, and greater readiness to assume risk. These factors collectively enhance the ability to perceive opportunities and navigate uncertainty (Santos et al., 2017; Vegetti & Adăscăliţei, 2017).

Entrepreneurship Education and Experiential Learning

Entrepreneurship education (EE) is a frequently cited driver across the literature. However, its effectiveness is strongly linked to how education is delivered. **Experiential learning**, which includes activities such as business plan development, startup simulations, networking exercises, and mentorship, is shown to have a significantly greater impact than purely theoretical instruction. These interactive components improve the psychological readiness of learners and equip them with practical tools to take entrepreneurial action (Guerrero & Peña-Legazkue, 2013; Santos et al., 2017; Zhao et al., 2005).

Academic Field and Curriculum Alignment

The academic background of individuals also plays a role. Programs with a stronger emphasis on business management, economics, and innovation are more likely to foster entrepreneurial intent. In contrast, students in disciplines with limited entrepreneurial integration in the curriculum demonstrate weaker inclinations. This suggests that entrepreneurship is not equally promoted across fields of study, and the design of academic curricula plays a substantial role in shaping attitudes (Carr & Sequeira, 2007; Kolvereid, 1996).

Social Networks and Peer Support

Social and peer networks emerge as another key driver. These networks serve as sources of encouragement, knowledge, and even material support, such as unpaid labor or early customers. Entrepreneurs embedded in **supportive communities** tend to have better chances of overcoming early-stage barriers. The presence of like-minded individuals also contributes to a culture of entrepreneurship, reinforcing its perceived legitimacy and achievability (Baldacchino et al., 2008).

4 Case Studies

Empirical examples from around the world offer critical insights into the implementation of sustainable entrepreneurship in diverse contexts. The following case studies illustrate the practical application of the theoretical frameworks and characteristics discussed above, demonstrating both the successes and challenges encountered by SMEs in various sectors and regions.

4.1 Case Study 1: Sustainable Entrepreneurship in Iran's Food SMEs

Small and medium-sized enterprises (SMEs) form the backbone of Iran's industrial and economic landscape, contributing to over 75% of all enterprises in the country (Hosseininia & Ramezani, 2016). Among various sectors, the food industry stands out due to its extensive network of production units—numbering over 7370—of which an overwhelming 94% fall within the SME classification. Despite this economic prominence, the sustainability dimensions of SMEs in Iran, especially in the food sector, have been underexplored. Previous research has largely overlooked how small and medium-sized enterprises (SMEs) conceptualize and apply sustainable entrepreneurship (SE), particularly concerning environmental and social factors. This omission is critical, given that the sustainability of these businesses directly impacts local employment, food security, and environmental outcomes.

The study by Hosseininia and Ramezani (2016) seeks to address this gap by identifying the most influential social and environmental factors affecting SE in Iran's food sector. The authors adopt a structured approach to isolate and rank eight key factors—four social and four environmental—that collectively shape sustainable behavior among SMEs. Their goal is not only to quantify the relevance of these factors but also to understand the role of entrepreneurial characteristics, such as gender, education, and work experience, in mediating sustainable outcomes. This case study thus offers valuable insight into how sustainability is understood, internalized, and operationalized in a key segment of Iran's SME ecosystem.

The findings reveal that, among all variables assessed, social factors have the most significant influence on **sustainable entrepreneurship**. Specifically, **customer orientation** was ranked as the most influential factor, with a mean score of 5.57, suggesting that entrepreneurs in the food sector view satisfying and retaining customers as central to both economic viability and social responsibility. Entrepreneurs associated sustainability with building trust, responding to client preferences, and offering high-quality, health-conscious food products. According to the authors, this orientation toward customer needs translates into tangible actions, such as modifying product composition, packaging, and labeling in line with sustainability principles. These measures, while customer-focused, also generate indirect environmental benefits, reinforcing the interlinkage between the social and ecological domains of SE.

The second most influential factor was **human resources**, with a mean score of 4.95. Entrepreneurs demonstrated a strong commitment to the well-being of their employees, which included paying fair wages, providing insurance, and supporting transport needs. Importantly, staff development and training were not perceived merely as economic investments but as **ethical imperatives** aligned with the broader goals of sustainable business practice. The authors note that well-treated and well-trained employees exhibit higher loyalty, reduced turnover, and improved productivity—benefits that create a reinforcing loop between sustainability and operational efficiency. These findings confirm that in the Iranian context, **internal stakeholder welfare is deeply embedded in entrepreneurs' understanding of what constitutes a sustainable business**.

In the environmental dimension, **recycling** emerged as the most crucial factor, with a mean score of 4.49 and a regression coefficient (β) of 0.444, making it the most statistically significant environmental predictor of SE. Entrepreneurs widely acknowledged that waste management and recycling could reduce production costs, comply with governmental regulations, and appeal to environmentally conscious consumers. The interviews further indicated that **many SMEs had established informal or semi-structured recycling systems**, ranging from fundamental waste separation to more advanced reuse practices. These environmental measures, though driven partly by economic motives, reflect a growing awareness of ecological responsibility.

Closely following recycling in importance was the "**future of the Earth and environ-**

ment" factor (mean = 4.45), which measures entrepreneurs' expressed concern for environmental preservation and intergenerational equity. This concern reflects a **value-driven dimension of SE**, wherein entrepreneurs demonstrate a willingness to consider the long-term ecological consequences of their operations, even if such concern does not always translate into immediate action. The study reports that several respondents discussed the moral responsibility of passing on a healthy environment to future generations, which suggests that environmental consciousness in Iranian food SMEs is often **rooted in personal and cultural ethics** rather than compliance or market pressure alone.

The factor of **staff training**—while ranked fifth overall—was also statistically significant in predicting sustainable entrepreneurship. SMEs that invested in employee learning and professional development were more likely to implement socially and environmentally sustainable practices. Training in areas such as waste management, customer communication, and food safety not only improved technical competence but also fostered a shared organizational culture centered on **sustainability**. These findings suggest that in SMEs, where roles are often overlapping and informal, human capital development becomes a lever for cross-functional sustainability adoption.

Interestingly, the **least influential factors** were "social support" and "physical standards of the workplace," which had relatively lower scores and explanatory power. While entrepreneurs valued social cohesion and workplace safety, these dimensions appeared to be perceived as **baseline operational standards** rather than differentiators of sustainability performance. Their low rankings suggest that such factors, though important, are not primary motivators or enablers in the strategic sustainability decisions of food SMEs in Iran.

Beyond individual factor analysis, the study offers compelling insight into the **demographic characteristics of sustainable entrepreneurs**. It finds no significant relationship between **gender or age** and sustainable orientation, contradicting assumptions in some global literature that younger or female entrepreneurs may be more inclined toward sustainability. Instead, **educational attainment** and **work experience** showed significant positive relationships with both social and environmental sustainability pillars. Entrepreneurs with university degrees and greater exposure to business operations were more likely to embed sustainable thinking into their strategies. However, an unexpected finding was that the **number of previously launched enterprises** had a **negative correlation** with SE outcomes, suggesting that serial entrepreneurs in this context may prioritize growth or short-term profitability over sustainability.

From a thematic perspective, the authors argue that sustainable entrepreneurship in Iranian food SMEs often follows a **social-first trajectory**. That is, businesses tend to initiate sustainability through **social dimensions**—such as employee care and customer relations—before advancing into environmental practices, including waste reduction and energy efficiency. This order reflects **the availability of resources and the perceived return on investment: entrepreneurs can more easily and immediately impact social dimensions without requiring significant** financial or technical inputs. Over time, these social commitments may create a stronger foundation and culture from which more complex environmental initiatives can emerge.

The implications of these findings are multilayered. First, the **dominance of social factors** highlights the need for sustainability frameworks and policies that resonate with local business culture and values. Programs aiming to promote SE in Iran or similar contexts must first **engage SMEs on social grounds**—through workforce development, customer trust-building, and employee health—before expecting widespread adoption of environmental technologies. Second, the statistically significant role of education and experience underlines the importance of **targeted training and capacity-building initiatives** to foster sustainability-minded entrepreneurs. Finally, the observed gap between concern for the environmental future and actual environmental action signals a **need for bridging mechanisms**, such as accessible recycling systems, incentive structures, or peer learning platforms.

In conclusion, the study by Hosseininia and Ramezani (2016) provides a nuanced, empirically grounded portrait of sustainable entrepreneurship in Iranian food SMEs. It shows that **customer orientation and human resources are core drivers**, while **recycling and environmental concerns** also play significant but secondary roles. Entrepreneurial success in sustainability depends not on age or gender but on **education and experience**, with social engagement often paving the way for environmental responsibility. As the authors note, fostering SE in such contexts requires a **bottom-up approach**—starting with the people inside the business and expanding outward toward broader ecological impacts. These insights provide valuable guidance for researchers, policymakers, and practitioners seeking to support sustainability in SMEs operating in similar socio-economic contexts.

4.2 Case Study 2: Sustainable Entrepreneurship in Small and Medium Enterprises: Insights from Indonesia's Creative Industries

The concept of sustainable entrepreneurship has emerged as a critical paradigm for addressing environmental degradation and social inequities while maintaining economic viability. Unlike conventional business models, it integrates ecological and societal well-being into the core of entrepreneurial activities. Small and medium enterprises (SMEs), particularly in developing regions like Indonesia, face unique challenges and opportunities in adopting these practices due to their resource constraints and deep community ties.

4.2.1 The Intersection of Culture and Sustainability

In Jambi, a province rich in cultural heritage, creative-industry SMEs—such as batik producers, woodcraft artisans, and traditional textile weavers—exemplify how sustainability can be embedded in local economies. These businesses often rely on indigenous knowledge and natural materials, inherently aligning with principles of environmental stewardship. For instance, many artisans use plant-based dyes and sustainably sourced timber, practices that have been passed down through generations. However, modernization and market pressures threaten these traditions, creating tensions between preserving cultural identity and meeting contemporary demands.

4.2.2 Motivational Drivers for Sustainable Practices

The adoption of sustainable entrepreneurship in this context is influenced by a combination of external and internal factors:

1. **Community Influence (Social Norms)**

In collectivist societies, such as Indonesia, entrepreneurial behavior is heavily shaped by communal expectations. SMEs in Jambi often prioritize practices that garner social approval, such as reducing waste or supporting local livelihoods. A batik workshop owner, for example, might switch to organic dyes not solely for environmental reasons but to align with a growing community-wide movement toward eco-conscious production. This reflects how societal values can drive change even in the absence of formal regulations.

2. **Economic Perceptions (Perceived Desirability)**

Entrepreneurs are more likely to embrace sustainability when they perceive tangible benefits, such as access to premium markets or enhanced brand loyalty. In Jambi's tourism-dependent creative sector, SMEs that leverage sustainable practices report an increased interest from ethically minded consumers. A woodcarving cooperative, for instance, gained international buyers after certifying its use of reclaimed timber, demonstrating how market incentives can catalyze green transitions.

3. **Structural Barriers and Enablers**

While personal attitudes toward sustainability are essential, their impact is often moderated by systemic challenges. Limited access to financing, lack of green technology, and fragmented supply chains hinder SMEs from fully committing to sustainable practices. Conversely, initiatives like government grants for eco-friendly materials or NGO-led training programs have proven effective in lowering these barriers.

4.2.3 The Role of Policy and Institutional Support

Sustainable entrepreneurship in SMEs cannot thrive solely through individual effort. Institutional frameworks play a pivotal role in creating an enabling environment. In Indonesia, programs like the **Green Business Certification** for SMEs and partnerships with local universities to disseminate sustainable techniques have shown promise. However, gaps remain in scaling these efforts, particularly in rural areas where infrastructure is weak.

4.2.4 Toward a Culturally Grounded Approach

The case of Jambi underscores that sustainability in SMEs must be context-sensitive. Preserving traditional techniques, such as handloom weaving or natural dyeing, can simultaneously safeguard cultural heritage and reduce environmental footprints. Policymakers and support organizations are urged to:

- Design interventions that align with local values and practices.
- Foster collaborations among artisans, researchers, and marketers to develop innovative, sustainable solutions.
- Amplify consumer awareness to drive demand and create shifts.

4.2.5 Conclusion

Sustainable entrepreneurship in SMEs, as illustrated by Jambi's creative industries, is a dynamic interplay of cultural preservation, economic pragmatism, and community-driven change. While global sustainability frameworks provide guidance, their implementation must be adapted to local realities. The experiences of these SMEs offer a microcosm of how grassroots entrepreneurship can contribute to broader ecological and social resilience, provided systemic barriers are addressed and cultural assets are leveraged.

4.3 Case Study 3: Sustainable Entrepreneurship in Polish SMEs

Poland's small and medium enterprises demonstrate a unique approach to balancing business objectives with sustainability commitments. The sector reveals a strong inclination toward social responsibility, particularly in employee welfare and community engagement, while environmental initiatives lag behind. This pattern suggests that cultural and economic factors prioritize immediate social impact over ecological concerns.

Five distinct organizational profiles emerge based on the implementation of sustainability. A small but significant minority of firms successfully integrate both social and environmental dimensions, serving as benchmarks for the industry. These leaders tend to be more innovative, adopting advanced technologies and streamlined processes that enhance both sustainability and competitiveness. Their success stories reveal potential synergies between responsible business practices and financial performance.

The majority of Polish SMEs exhibit selective adoption, focusing either on social or environmental aspects, but rarely both. Socially-oriented firms emphasize fair labor practices, workplace safety, and community involvement. Environmentally conscious businesses concentrate on energy efficiency and waste reduction, although renewable energy adoption remains limited across the sector. A concerning segment exhibits minimal engagement with sustainability principles, often due to resource constraints or a perceived lack of relevance to their operations.

Key challenges include an inconsistent understanding of sustainability concepts, financial barriers to adopting green technologies, and a lack of institutional support. The research highlights how traditional business values in transitional economies can both enable and constrain sustainable transformation. Successful cases demonstrate that sustainability integration often begins with pragmatic, cost-saving measures that gradually evolve into comprehensive strategies.

4.4 Case Study 4: Thailand—Community-Based Social Enterprises

In Thailand, a notable case study involves a community-based social enterprise located in a coastal fishing village. This enterprise has successfully incorporated sustainable leadership practices to manage resources and foster local socio-economic development (Suriyankietkaew et al., 2022). The leadership of this enterprise has focused on developing key competencies, including strategic management, systems thinking, anticipatory planning, interpersonal communication, and ethical decision-making, to navigate the adverse effects of the COVID-19 pandemic (Suriyankietkaew et al., 2022). These competencies have enabled the enterprise to promote pro-environmental behaviors, safeguard cultural heritage, and stimulate local economic renewal.

The business model adopted by this enterprise is deeply rooted in community engagement. Local stakeholders are actively involved in decision-making processes, ensuring that the enterprise remains responsive to the community's evolving needs. By prioritizing holistic growth over short-term financial gains, the enterprise has achieved a balance between economic viability and social responsibility. This case exemplifies how sustainable leadership practices and entrepreneurial orientation can be leveraged to build resilience in SMEs, especially in regions where traditional economic structures are undergoing significant transformation.

4.5 Case Study 5: Canada—ICT SMEs and IT-Enabled Innovation

Another illustrative case study can be found in the Canadian Information and Communication Technology (ICT) sector, where SMEs have successfully harnessed IT capabilities to drive product innovation performance (PIP) (Lyver & Lu, 2018). In these firms, strategic entrepreneurship plays a mediating role by balancing the exploration of new technological opportunities with the exploitation of existing competencies. The integration of IT resources has not only enhanced the innovative capacity of these SMEs but also provided them with a competitive edge in a globalized market.

The empirical analysis of Canadian ICT SMEs reveals that a substantial proportion of these enterprises have sustained operations for over a decade, highlighting the importance of technological adaptability and entrepreneurial resilience (Lyver & Lu, 2018). These firms demonstrate that the combination of IT capabilities and strategic orientation is instrumental in fostering a robust pipeline of innovative products and services. The success of these SMEs provides a model for how other sectors can leverage technology and entrepreneurial agility to achieve sustainable outcomes.

4.6 Case Study 6: Green Transport Corridors in Europe

The Green Transport Corridor initiative, spearheaded by the European Commission, offers a compelling example of how sustainability frameworks can be operationalized in the logistics and transportation sectors (Prause & Hunke, 2014). This initiative aims to shift the dominant mode of road traffic to more efficient and environmentally friendly modes of transport, such as short-sea shipping, rail, and inland waterways. The framework for Green Transport Corridors is based on a set of stringent requirements and key performance indicators (KPIs) that ensure the realization of economic, environmental, and social benefits.

SMEs operating within these corridors are integrated into a complex network of stakeholders that include manufacturers, forwarders, and governmental agencies. Collaborative models and harmonized regulations are essential components of this ecosystem, ensuring that sustainable logistics solutions are not only technically feasible but also economically viable (Prause & Hunke, 2014). The KPIs—ranging from emission reductions to enhanced safety metrics and improved logistical efficiency—provide clear benchmarks for success. This case study illustrates the potential for cross-sector collaboration in driving sustainability and underscores the importance of cohesive regulatory frameworks in promoting environmentally friendly business practices.

4.7 Case Study 7: Spain—Circular Economy and the Quintuple Helix Model

In Spain, particularly in the Huelva region, a transformative shift toward a circular economy (CE) has been documented. The transition to a Circular Economy (CE) in this region draws heavily on the Quintuple Helix Model, which emphasizes multi-stakeholder collaboration as a critical driver of sustainable development (Borrero & Yousafzai, 2024). Here, SMEs are positioned as vital actors in the shift from a linear to a circular model of economic activity, where waste is minimized and resource efficiency is maximized.

The Spanish case study highlights several challenges inherent in the transition to a circular economy, including low levels of current circularity (only 8.6% globally) and significant obstacles such as institutional inertia and insufficient innovation frameworks (Borrero & Yousafzai, 2024). Despite these challenges, regional initiatives have successfully mobilized stakeholders from government, academia, industry, and civil society to create a more resilient and sustainable economic ecosystem. The role of institutional entrepreneurs—particularly universities acting as centers of innovation—is emphasized as a key factor in overcoming these challenges and facilitating the broader adoption of CE principles.

References

Abdullahi, M. A., Mohamed, Z., Shamsudin, M. N., Sharifuddin, J., & Ali, F. (2018). Effects of top leadership culture and strategic sustainability orientation on sustainable development among Malaysian herbal-based SMEs. *Business Strategy and Development, 1*(2), 128–139. https://doi.org/10.1002/bsd2.17

Aktar, S., & Kabir, A. (2024). A comprehensive study on sustainable entrepreneurship for cottage, micro, small, and medium enterprises in Bangladesh. *European Journal of Sustainable Development Research, 8*(2), em0254.

Alfi, S., Taleghani, M., Vatanparast, M. R., & Homayounfar, M. (2024). A dynamic model for assessing the sustainability of small and medium sized enterprises. *Journal of Optimization in Industrial Engineering, 17*(2), 1–13. https://doi.org/10.22094/QJIE.2024.1184391

Almeida, A., & Garrod, B. (2024). Drivers and inhibitors of entrepreneurship in Europe's Outermost Regions: Implications for entrepreneurship education. *The International Journal of Management Education, 22*(2), Article 100975.

Altinay, L., Madanoglu, M., Daniele, R., & Lashley, C. (2012). The influence of family tradition and psychological traits on entrepreneurial intention. *International Journal of Hospitality Management, 31*(2), 489–499. https://doi.org/10.1016/j.ijhm.2011.07.007

Al-Tit, A., Omri, A., & Euchi, J. (2019). Critical success factors of small and medium-sized enterprises in Saudi Arabia: Insights from sustainability perspective. *Administrative Sciences, 9*(2), 32.

Álvarez Jaramillo, J., Zartha Sossa, J. W., & Orozco Mendoza, G. L. (2019). Barriers to sustainability for small and medium enterprises in the framework of sustainable development—Literature review. *Business Strategy and the Environment, 28*(4), 512–524. https://doi.org/10.1002/bse.2261

Álvarez-García, J., Hormiga-Pérez, E., Sarango-Lalangui, P. O., & del Río-Rama, M. de la C. (2022). Leaders' sustainability competences and small and medium-sized enterprises outcomes: The role of social entrepreneurial orientation. *Sustainable Development, 30*(5), 927–943. https://doi.org/10.1002/sd.2291

Anaman, P. D., Ahmed, I. A., Suleman, A. R., & Dzakah, G. A. (2025). Environmentally sustainable business practices in micro, small, and medium enterprises: A sub-saharan African country perspective. *Business Perspectives and Research, 13*(2), 175–190. https://doi.org/10.1177/22785337231162740

Andreas, F., Cooperman, E. S., Gifford, B., & Russell, G. (2011). A simple path to sustainability: Green business strategies for small and medium-sized businesses. In *A*

Simple Path to Sustainability: Green Business Strategies for Small and Medium-Sized Businesses. https://doi.org/10.5860/choice.49-3354

Auer, A., & Jarmai, K. (2018). Implementing responsible research and innovation practices in SMEs: Insights into drivers and barriers from the Austrian medical device sector. *Sustainability (Switzerland), 10*(1), 17. https://doi.org/10.3390/SU10010017

Babgohari, A. Z., Esmaelnezhad, D., & Taghizadeh-Yazdi, M. (2023). Prioritising SMEs internationalisation practices considering their various interrelating barriers: A sustainability and resiliency approach. In *Decision-making in international entrepreneurship: Unveiling cognitive implications towards entrepreneurial internationalisation*. https://doi.org/10.1108/978-1-80382-233-420231011

Bajdor, P., Pawełoszek, I., & Fidlerova, H. (2021). Analysis and assessment of sustainable entrepreneurship practices in polish small and medium enterprises. *Sustainability (Switzerland), 13*(7), 3595. https://doi.org/10.3390/SU13073595

Baldacchino, L., Cassar, V., & Caruana, A. (2008). Startup success in a small island state: A study among entrepreneurs in Malta. *Island Studies Journal, 3*(1), 73–96. https://doi.org/10.24043/ISJ.216

Balić, N., & Vuka, I. (2023). State of sustainable entrepreneurship in SMEs: Development of a cross-country sustainability pulse check. In *Lecture notes in networks and systems LNNS* (Vol. 741). https://doi.org/10.1007/978-3-031-38318-2_7

Belyaeva, Z., & Lopatkova, Y. (2020). The impact of digitalization and sustainable development goals in SMEs' strategy: A multi-country european study. In *Palgrave studies of cross-disciplinary business research, in Association with EuroMed Academy of Business*. https://doi.org/10.1007/978-3-030-45835-5_2

Bocken, N. M. P. (2015). Sustainable venture capital—Catalyst for sustainable startup success? *Journal of Cleaner Production, 108*, 647–658. https://doi.org/10.1016/J.JCLEPRO.2015.05.079

Borrero, J. D., & Yousafzai, S. (2024). Circular entrepreneurial ecosystems: A Quintuple Helix Model approach. *Management Decision, 62*(13), 141–177. https://doi.org/10.1108/MD-08-2023-1361

Carr, J. C., & Sequeira, J. M. (2007). Prior family business exposure as intergenerational influence and entrepreneurial intent: A Theory of Planned Behavior approach. *Journal of Business Research, 60*(10), 1090–1098. https://doi.org/10.1016/J.JBUSRES.2006.12.016

Chiun Lo, M., Chai Wang, Y., Justin Wah Constance, R., & Ramayah, T. (2016). The critical success factors for organizational performance of SMEs in Malaysia: A partial least squares approach. *Review of Business Management, 18*(61). https://doi.org/10.7819/rbgn.v18i61.3058

Conway, E. (2015). Engaging small and medium-sized enterprises (SMEs) in the low carbon agenda. *Energy, Sustainability and Society, 5*(1). https://doi.org/10.1186/S13705-015-0060-X

Crals, E., & Vereeck, L. (2005). The affordability of sustainable entrepreneurship certification for SMEs. *International Journal of Sustainable Development and World Ecology, 12*(2), 173–183. https://doi.org/10.1080/13504500509469628

Dayaratne, S. P., & Gunawardana, K. D. (2015). Carbon footprint reduction: A critical study of rubber production in small and medium scale enterprises in Sri Lanka. *Journal of Cleaner Production, 103*, 87–103. https://doi.org/10.1016/J.JCLEPRO.2014.09.101

Enwereji, P. C., van Rooyen, A. A., & Joshi, M. (2024). Entrepreneurial resilience strategies for startups: A view from the Vuca Paradigm. *Corporate and Business Strategy Review, 5*(4), 140–151.

Fatoki, O. (2018). The impact of entrepreneurial resilience on the success of small and medium enterprises in South Africa. *Sustainability (Switzerland), 10*(7), 2527. https://doi.org/10.3390/su10072527

Ferreira, N. C. M. Q. F., & Ferreira, J. J. M. (2024). Quo Vadis sustainable entrepreneurship? A systematic literature review of related drivers and inhibitors in SMEs. *IEEE Transactions on Engineering Management, 71*, 9644–9660. https://doi.org/10.1109/TEM.2023.3305475

Fong, C. W., Yen, Y. Y., & Ramasamy, S. (2023). Sustainable entrepreneurship in Malaysian companies. *Asian Economic and Financial Review, 13*(2), 98–107. https://doi.org/10.55493/5002.v13i2.4708

Ghadge, A., Kaklamanou, M., Choudhary, S., & Bourlakis, M. (2017). Implementing environmental practices within the Greek dairy supply chain drivers and barriers for SMEs. *Industrial Management and Data Systems, 117*(9), 1995–2014. https://doi.org/10.1108/IMDS-07-2016-0270

Ghazilla, R. A. R., Sakundarini, N., Abdul-Rashid, S. H., Ayub, N. S., Olugu, E. U., & Musa, S. N. (2015). Drivers and barriers analysis for green manufacturing practices in Malaysian smes: A preliminary findings. *Procedia CIRP, 26*, 658–663. https://doi.org/10.1016/J.PROCIR.2015.02.085

Guerrero, M., & Peña-Legazkue, I. (2013). The effect of intrapreneurial experience on corporate venturing: Evidence from developed economies. *International Entrepreneurship and Management Journal, 9*(3), 397–416. https://doi.org/10.1007/S11365-013-0260-9

Hasan, M. N. (2016). Measuring and understanding the engagement of Bangladeshi SMEs with sustainable and socially responsible business practices: An ISO 26000 perspective. *Social Responsibility Journal, 12*(3), 584–610. https://doi.org/10.1108/SRJ-08-2015-0125

Heidrich, O., & Tiwary, A. (2013). Environmental appraisal of green production systems: Challenges faced by small companies using life cycle assessment. *International Journal of Production Research, 51*(19), 5884–5896. https://doi.org/10.1080/00207543.2013.807372

Hjorth, S. S., & Brem, A. M. (2016). How to assess market readiness for an innovative solution: The case of heat recovery technologies for SMEs. *Sustainability*

(Switzerland), 8(11), 1152. https://doi.org/10.3390/su8111152

Hosseininia, G., & Ramezani, A. (2016). Factors influencing sustainable entrepreneurship in small and medium-sized enterprises in Iran: A case study of food industry. *Sustainability (Switzerland), 8*(10), 1010. https://doi.org/10.3390/su8101010

Jakobs, L., Salamanca Cano, A. K., & Rauch, E. (2024). Sustainability in SMEs: A comparative study to assess current status and setting targets for action. *IFAC-PapersOnLine, 58*(19), 836–841. https://doi.org/10.1016/j.ifacol.2024.09.188

James, L. (2014). Sustainability footprints in SMEs: Strategy and case studies for entrepreneurs and small business. In *Sustainability footprints in SMEs: Strategy and case studies for entrepreneurs and small business.* https://doi.org/10.1002/9781118779477

Johnson, M. P., & Schaltegger, S. (2016). Two decades of sustainability management tool for SMEs: How far have we come? *Journal of Small Business Management, 54*(2), 481–505. https://doi.org/10.1111/JSBM.12154

Kapoor, P., Goel, S., & Nagar, N. (2023). Delineating a trajectory for progressive SMEs towards a sustainable future: A cluster-based approach. *International Journal of Productivity and Quality Management, 40*(3), 340–363. https://doi.org/10.1504/IJPQM.2023.134814

Kolvereid, L. (1996). Prediction of employment status choice intentions. *Entrepreneurship Theory and Practice, 21*(1), 47–58. https://doi.org/10.1177/104225879602100104

Lewis, K. V., Cassells, S., & Roxas, H. (2015). SMEs and the potential for a collaborative path to environmental responsibility. *Business Strategy and the Environment, 24*(8), 750–764. https://doi.org/10.1002/BSE.1843

Liu, X., & Zhao, W. G. W. (2021). Family education? Unpacking parental factors for tourism and hospitality students' entrepreneurial intention. *Journal of Hospitality, Leisure, Sport and Tourism Education.* https://doi.org/10.1016/J.JHLSTE.2020.100284

Lyver, M. J., & Lu, T. J. (2018). Sustaining innovation performance in SMEs: Exploring the roles of strategic entrepreneurship and IT capabilities. *Sustainability (Switzerland), 10*(2), 442. https://doi.org/10.3390/su10020442

Malá, D., Sedliačiková, M., Kaščáková, A., Benčiková, D., Vavrová, K., & Bikár, M. (2017). Green logistics in Slovak small and medium wood-processing enterprises. *BioResources, 12*(3), 5155–5173. https://doi.org/10.15376/BIORES.12.3.5155-5173

Mbuyisa, B., & Leonard, A. (2015). ICT adoption in SMES for the alleviation of poverty. In *IAMOT 2015—24th International Association for Management of Technology Conference: Technology, Innovation and Management for Sustainable Growth, Proceedings.*

Meath, C., Linnenluecke, M., & Griffiths, A. (2016). Barriers and motivators to the adoption of energy savings measures for small- and medium-sized enterprises (SMEs): The case of the ClimateSmart Business Cluster program. *Journal of Cleaner Production, 112,* 3597–3604. https://doi.org/10.1016/J.JCLEPRO.2015.08.085

Meshram, S., & Rawani, A. M. (2019). Analyzing success factors of small and medium enterprises (SMEs): A study in Indian context. In *Lecture Notes in Mechanical Engineering.* https://doi.org/10.1007/978-981-13-6412-9_78

Mohan, P. S. (2022). An investigation into entrepreneurial intentions in Caribbean Small Island Developing States. *Journal of Innovation and Entrepreneurship, 11*(1). https://doi.org/10.1186/S13731-022-00253-0

Mondal, S., Singh, S., & Gupta, H. (2024). Examining the impact of green entrepreneurship drivers on sustainable development in the context of emerging country. *Benchmarking.* https://doi.org/10.1108/BIJ-11-2023-0835

Mourtzis, D., Fotia, S., Gamito, M., Neves-Silva, R., Correia, A., Spindler, P., Pezzotta, G., & Rossi, M. (2016). PSS design considering feedback from the entire product-service lifecycle and social media. *Procedia CIRP, 47,* 156–161. https://doi.org/10.1016/J.PROCIR.2016.03.092

Ng, H. S., & Kee, D. M. H. (2012). The issues and development of critical success factors for the SME success in a developing country. *International Business Management, 6*(6), 680–691.

Oliveira Neto, G. C., Leite, R. R., Shibao, F. Y., & Lucato, W. C. (2017). Framework to overcome barriers in the implementation of cleaner production in small and medium-sized enterprises: Multiple case studies in Brazil. *Journal of Cleaner Production, 142,* 50–62. https://doi.org/10.1016/J.JCLEPRO.2016.08.150

Othman, A. K., Mahmud, A. A., Mustafa, M. J., & Abujarad, I. (2022). Organisational success factors of medium-sized enterprises (SMEs) in Malaysia: A qualitative focus group study. *Academy of Entrepreneurship Journal, 28*(3).

Prause, G., & Hunke, K. (2014). Sustainable entrepreneurship along green corridor. *Entrepreneurship and Sustainability Issues, 1*(3), 124–133. https://doi.org/10.9770/jesi.2014.1.3(1)

Reiche, K. (2014). Responsible entrepreneurship. In *CSR, sustainability, ethics and governance* (pp. 279–280). https://doi.org/10.1007/978-3-642-38753-1_24

Rodrigues, M., Franco, M., Silva, R., & Oliveira, C. (2021). Success factors of smes: Empirical study guided by dynamic capabilities and resources-based view. *Sustainability (Switzerland), 13*(21), 12301. https://doi.org/10.3390/su132112301

Santos, S. C., Caetano, A., Spagnoli, P., Costa, S. F., & Neumeyer, X. (2017). Predictors of entrepreneurial activity before and during the European economic crisis. *International Entrepreneurship and Management Journal, 13*(4), 1263–1288. https://doi.org/10.1007/S11365-017-0453-8

Soto-Acosta, P., Cismaru, D.-M., Vătămănescu, E.-M., & Ciochină, R. S. (2016). Sustainable entrepreneurship in SMEs: A business performance perspective. *Sustainability (Switzerland), 8*(4), 342. https://doi.org/10.3390/su8040342

Spence, M., Ben Boubaker Gherib, J., & Biwolé, V. O. (2011). Sustainable entrepreneurship: Is entrepreneurial will enough? A North–South comparison. *Journal of Business Ethics, 99*(3), 335–367. https://doi.org/10.1007/s10551-010-0656-1

Štverková, H., & Pohludka, M. (2021). Sustainable entrepreneurship in small and medium-sized enterprises in the Czech Republic. In *Proceedings of the 14th International Conference on Strategic Management and Its Support by Information Systems 2021, SMSIS 2021* (pp. 303–311).

Sukma, A. (2024). Exploring factors influencing SMEs' success: The perspective of entrepreneurial and learning orientations. *Journal of Innovation in Business and Economics, 8*(01), 33–46.

Sulong, F., Sulaiman, M., & Norhayati, M. A. (2015). Material Flow Cost Accounting (MFCA) enablers and barriers: The case of a Malaysian small and medium-sized enterprise (SME). *Journal of Cleaner Production, 108*, 1365–1374. https://doi.org/10.1016/J.JCLEPRO.2014.08.038

Suriyankietkaew, S., Krittayaruangroj, K., & Iamsawan, N. (2022). Sustainable leadership practices and competencies of SMEs for sustainability and resilience: a community-based social enterprise study. *Sustainability (Switzerland), 14*(10), 5762. https://doi.org/10.3390/su14105762

Ta'Amnha, M. A., Al-Qudah, S., Asad, M., Magableh, I. K., & Riyadh, H. A. (2024). Moderating role of technological turbulence between green product innovation, green process innovation and performance of SMEs. *Discover Sustainability, 5*(1), 324

Tazhibekova, K., & Shametova, A. (2024). Ecological initiatives and their influence on the competitiveness and sustainability of companies: "Green" strategies of SMEs. *Journal of the Knowledge Economy*. https://doi.org/10.1007/s13132-024-02062-0

Trianni, A., & Cagno, E. (2015). Diffusion of motor systems energy efficiency measures: An empirical study within Italian manufacturing SMEs. *Energy Procedia, 75*, 2569–2574. https://doi.org/10.1016/J.EGYPRO.2015.07.302

Utami, I. D., Rizal, A., Novianti, T., & Winarso, K. (2024). Factors influencing sustainability of Batik SMEs entrepreneurs. *BIO Web of Conferences*. https://doi.org/10.1051/bioconf/202414601011

Vegetti, F., & Adăscăliţei, D. (2017). The impact of the economic crisis on latent and early entrepreneurship in Europe. *International Entrepreneurship and Management Journal, 13*(4), 1289–1314. https://doi.org/10.1007/S11365-017-0456-5

Velmurugan, R., Jegadeeswari, S., Manikandan, R., Hemalatha, T. M., & Amutha, K. (2023). Determinants of MSMES entrepreneurs sustainability. *E3S Web of Conferences, 449*, 05007.

Velmurugan, R., Thirumalaisamy, R., Paquibut, R., & Abouraia, M. (2024). An integrated framework for the financial sustainability and entrepreneurial success of micro, small, and medium enterprises. In *The AI revolution: Driving business innovation and research* (Vol. 2, pp. 233–246). Springer.

Yacob, P., Peter, D., & Chin, K. S. (2022). Sustainable business practices in manufacturing SMEs: The mediating effect of dynamic capabilities. *International Social Science Journal, 72*(243), 73–89. https://doi.org/10.1111/issj.12303

Yuik, C. J., Perumal, P. A., & Feng, C. J. (2020). Exploring critical success factors for the implementation of lean manufacturing in machinery and equipment SMEs. *Engineering Management in Production and Services, 12*(4), 77–91. https://doi.org/10.2478/emj-2020-0029

Zhao, H., Hills, G. E., & Seibert, S. E. (2005). The mediating role of self-efficacy in the development of entrepreneurial intentions. *Journal of Applied Psychology, 90*(6), 1265–1272. https://doi.org/10.1037/0021-9010.90.6.1265

Zhou, Y., Xu, G., Minshall, T., & Liu, P. (2015). How do public demonstration projects promote green-manufacturing technologies? A case study from China. *Sustainable Development, 23*(4), 217–231. https://doi.org/10.1002/SD.1589

Financing Sustainable Entrepreneurship: Models, Challenges, and Strategic Innovations

Mohsen Motiei and Amir Mohammad Khani

Abstract

As a new approach to economic development, sustainable entrepreneurship—also known as entrepreneurial modeling for sustainable development—aims to practice the triple bottom line, encompassing economic profitability, social responsibility, and environmental protection. The financing of sustainable businesses in a world marked by crises such as climate change, poverty, and social inequalities is a significant challenge. The purpose of this chapter is to identify and analyze the existing mechanisms of finance for sustainable entrepreneurship, including traditional finance, venture capital, microfinance, and social investment, as well as their limitations in supporting the needs of sustainable entrepreneurs. Key challenges include cultural, economic, and political issues, as well as poor access to financial resources and a lack of transparent data. It then proposes ways to take the path, for instance, through the use of blockchain technologies, the crowdfunding model, and the roles of governments and international institutions in pursuing this path. The Grameen Bank and Patagonia case studies demonstrate how equally innovative financial models can contribute to social empowerment, build stronger economic resilience, and protect the environment. In this chapter, the recommendations to policymakers, researchers, and investors stress the desirability of more humane, transparent, and responsible financial instruments for development to be more widely sustainable globally.

Keywords

Sustainable entrepreneurship · Green finance · Crowdfunding · Government policy · Environmental sustainability

1 Introduction

Sustainable entrepreneurship involves creating and developing businesses that not only strive for financial profitability but also prioritize their social and environmental responsibilities. This type of entrepreneurship specifically seeks solutions that meet the needs of society today while also preparing the ground for future generations without harming natural and environmental resources. In other words, sustainable entrepreneurship is recognized as a tool for achieving sustainable development and enhancing the quality of life in various societies (Gutterman, 2022; Schaltegger, 2013). In today's world, where

M. Motiei (✉)
Department of Management and Accounting, Allameh Tabataba'i University, Tehran, Iran
e-mail: motiei@atu.ac.ir

A. M. Khani
Department of Industrial Management, Faculty of Management, University of Tehran, Tehran, Iran
e-mail: amir.mo.khani@ut.ac.ir

M. Akbari et al. (eds.), *Sustainable Entrepreneurship*, University of Tehran Science and Humanities Series, https://doi.org/10.1007/978-3-032-04909-4_16

we face global crises such as climate change, environmental pollution, poverty, and social inequalities, sustainable entrepreneurship plays a crucial role in mitigating these challenges and creating sustainable solutions. Instead of only considering short-term economic profitability, sustainable entrepreneurs simultaneously emphasize their environmental and social impacts, seeking to create businesses that, in addition to generating wealth, improve people's living conditions and protect natural resources (Kirby & Healey-Benson, 2024; Scartozzi et al., 2024).In this context, sustainable entrepreneurship is not only an economic path but also a fundamental solution to deal with increasing environmental and social problems. This type of entrepreneurship can help address social and environmental crises at both local and global levels, particularly in societies plagued by poverty, unemployment, and environmental issues (Azamat et al., 2023; Salimi & Ton Vrauwdeunt, 2024).

Financial challenges are one of the most significant obstacles for sustainable entrepreneurs. Many entrepreneurs working in sustainable fields aim to bring about long-lasting, positive change in society and the environment, but traditional financing methods are primarily geared toward short-term profit and quick returns. The conflict between short-term financial goals and the longer-term goals of sustainable entrepreneurship makes it difficult for many sustainable entrepreneurs to access financing easily, especially in the early stages of their venture (Yiadom et al., 2025). The other major challenge lies in the investors being unaware and not understanding. However, investors are often reluctant to invest in this type of entrepreneurship because they lack knowledge on how to invest in such a business model, as it is complex and unclear to them. The second part, sustainable entrepreneurship, is an ecology of much longer timescales than the timescales of the first, faster ones, and sustainable entrepreneurs are looking for gross potentials for returns: the things that they produce have no relevance for investor interests; the first part of making locus is looking for gross positives (Hoogendoorn et al., 2019; Wöhler & Haase, 2022). Moreover, problems of access to finance, complex bureaucracies, and a lack of transparent systems to evaluate sustainable projects are other major barriers to financing in this sector. Nevertheless, many financial institutions still lack clear and credible criteria for evaluating and supporting sustainable startups, which prevents sustainable entrepreneurs from accessing financing efficiently (Mukherjee et al., 2024; Nunes et al., 2022).

This chapter examines the mechanisms of financing for sustainable entrepreneurship and the barriers to its implementation. This chapter aims to identify various sources of financing for sustainable entrepreneurs and their associated challenges and barriers to accessing these resources. Next, the main concepts of sustainable entrepreneurship and the primary financing modes—venture capital, traditional finance, microfinance, and social investment (also known as impact investing)—are reviewed. The following section will discuss the main challenges faced by sustainable entrepreneurs in financing, including cultural, economic, and political barriers, as well as access issues to financial resources and data availability. Secondly, the remaining part provides a closer examination of strategic and solution-oriented approaches to addressing these challenges. Strategies utilizing new technologies, such as Blockchain, will be employed to enhance transparency in financing; additionally, models like crowdfunding will be implemented. The chapter ends with a case study of one or more projects that have employed sustainable finance practices. They can be practical models for entrepreneurs and investors in this field. It will also provide suggestions and recommendations for future research and policy actions to ensure access to sustainable finance.

2 Main Body

2.1 Core Concepts

2.1.1 Sustainable Entrepreneurship: Definition and Characteristics

This is an entrepreneurship that goes beyond being economically profitable and includes social and environmental impacts, which is generally referred to as sustainable entrepreneurship. This is also known as an entrepreneur who strives to create businesses that can contribute to social and environmental development, as well as economic growth. In reality, sustainable entrepreneurship focuses not solely on short-term profitability but instead on long-term and relatively balanced development in the economic, social, and environmental areas. Its core purpose is to generate sustainable innovations for global problems, which can be particularly significant in addressing climate change, social inequalities, and natural resource crises (Karadayı et al., 2024; Motiei et al., 2021).

2.1.2 Characteristics of Sustainable Entrepreneurship

Sustainable entrepreneurship, as a new approach to the business world, possesses distinct characteristics that differentiate it from traditional entrepreneurship. Here, we will examine the six main characteristics of sustainable entrepreneurship:

1. **Triple Bottom Line**: A key characteristic of sustainable entrepreneurship is that it consistently integrates economic, social, and environmental aspects. The triple bottom line concept is based on the idea that a successful business should be evaluated based on three key dimensions: profit (financial), people (social), and planet (environmental). This model aims for businesses to be not only financially profitable but also to have a positive impact on society and the environment (Nica et al., 2025; Nogueira et al., 2025).
2. **Focus on innovation and green technologies**: Another key characteristic of sustainable entrepreneurship is the emphasis on innovation in both technology and business processes. Those innovations are typically designed to reduce the use of natural resources and minimize environmental pollution. For instance, green technologies and renewable energy are integral components in the production process of sustainable entrepreneurship. These technologies are utilized by sustainable companies to make financial processes more transparent and trustworthy, or clean energy technology to reduce carbon emissions (Pricopoaia et al., 2024; Sultan, 2025).
3. **Commitment to social responsibility and transparency**: Sustainable entrepreneurs embody this social responsibility and incorporate it into all aspects of their business. The social responsibility of companies includes providing features of products that directly and sometimes indirectly support the growth of local community life, eliminating social and economic discrimination, and respecting human rights and employee welfare. Apart from that, transparency is one of the key pillars of sustainable entrepreneurship. Most sustainable businesses employ models that facilitate the easy assessment of a business's social and environmental impacts (TaghaviFard et al., 2022; Taghavifard et al., 2023).
4. **Circular Economy**: This is perhaps one of the main things of sustainable entrepreneurship. In this model, a concerted effort is made to design the production, consumption, and recycling cycle to minimize the use of natural resources and eliminate production waste. In this model, sustainable entrepreneurs aim to develop processes that convert raw materials into recycled materials, ultimately eliminating the emission of harmful waste. For instance, companies like Patagonia have successfully implemented an economic model that utilizes recycled materials and designs repairable and recyclable products (Corvellec et al., 2021; Morseletto, 2020).
5. **Focus on creating long-term value**: The second characteristic of sustainable entrepreneurship is focusing on creating long-term value instead of immediate returns. Among sustainable entrepreneurs, long-term goals are common, and their profitability is expected to be sustained over the long term rather than in the short term. This approach sells off these kinds of businesses in the context that investing in them is riskier, but once that is done, it can bring benefits that are sustainable to all stakeholders, such as the local communities, employees, and envi-

ronment (Kurznack et al., 2021; Serio et al., 2020)

6. **Sustainable Supply Chains**: In sustainable supply chains, entrepreneurs focus on both their supply chain and internal business processes. This entails making choices of suppliers and business practices that are sociable and ecological as far as possible, with sustainable standards. Sustainable entrepreneurs are eager to choose supplier companies that actively strive to reduce waste, improve working conditions, and utilize sustainable raw materials in their supply chain. It not only has a positive effect on the environment but also reinforces consumers' perception of the brand (Abbasi et al., 2025; Koberg & Longoni, 2019).

2.1.3 The Importance of Sustainable Entrepreneurship

This is because sustainable entrepreneurship is crucial, as it provides solutions to the challenges of global crises. Today, the world's most pressing challenges include climate change, depletion of natural resources, and social inequalities. These crises can be viewed as an opportunity for sustainable entrepreneurship to present one of its solutions. The principles of sustainability can help businesses become more effective in mitigating the negative impacts of the crisis. This type of entrepreneurship, which aims to improve both social and environmental situations while generating economic value, directly contributes to tackling these issues at a global level (Rosário et al., 2022; Shahid et al., 2023).

Moreover, the minds of people in the modern world are more concerned about brands that not only provide good quality products but also fulfill their obligation towards protecting the earth, alongside caring about other vital matters, such as social issues. Not only is this an ethical trend, but it also offers significant economic potential as it shifts this attitude. Companies or businesses that follow these principles will gain more trust from consumers and investors and also exhibit better progress (Neumann, 2022). Secondly, sustainable entrepreneurship focuses on creating long-term and sustainable value rather than prioritizing short-term profitability. On the one hand, this leads businesses down the road to greater profits in the future and helps ensure the social and economic stability of societies. Since sustainable entrepreneurship has the potential to continuously contribute to sustainable economic development for the long-term benefit of people, it can serve as an effective means of guaranteeing social welfare and justice (Di Vaio et al., 2022).

2.2 Available Financing Models

In today's world, financing is one of the fundamental challenges for sustainable entrepreneurs. These businesses need financial resources that, in addition to supporting economic growth, also address social and environmental impacts. Different financing models help entrepreneurs secure the resources they need to implement sustainable projects. In this section, we will examine some of the common financing models used in sustainable entrepreneurship.

2.2.1 Traditional Financing Classically

The traditional methods of financing that the entrepreneurs take recourse to include funds from banks, financial institutions, and other lenders to initiate and develop their businesses. The entrepreneur receives this model if he meets some conditions specified; for example, for a bank loan or credit, he has to provide collateral that interests the bank or credit company, is formed for a specific period, and has to repay the money (Nippani et al., 2024; Worakantak et al., 2024).

Features and Challenges:

- Lack of Accessibility: Many sustainable entrepreneurs struggle to access bank loans as easily as other types of entrepreneurs. The reason is that the banks generally look for businesses that can generate immediate and consistent returns, which sustainable entrepreneurship may require more time to achieve due to its long-term goal focus on social and environmental sustainability (Ključnikov et al., 2024).

- High Interest Rates: Banks and financial institutions also charge their customers very high interest rates, which are deemed stringent due to the high rates being charged. However, traditional startups and sustainable businesses often need more time to mature, and these terms are not easy or acceptable (Danisman & Tarazi, 2024).
- Banks have limited access to data: the data that banks have access to is not always satisfactory enough to predict risk. In reality, most sustainable businesses are still in the early stages and have no financial results or data yet to qualify for a loan (Dinh et al., 2024).

Meanwhile, since traditional financing is more accessible to entrepreneurs who are financially sound and can repay loans, it remains a first choice for such entrepreneurs seeking to obtain capital.

2.2.2 Venture Capital

Venture capital is financing used especially for startups and businesses with high growth potential. Investors provide money for a fraction of the business's shares and will earn money from the company's growth if it's successful in this model (Fu et al., 2024).

Features and Challenges:

- High Growth Potential: There is a high growth potential for businesses that will be invested in venture capital. This model appears to be suitable for sustainable entrepreneurs with sustainability-based ideas. Many venture capitalists are eager to invest in new startups with the potential for high returns in the long run (Gucciardi, 2024).
- Rigorous Valuation: Often, Venture capitalists are looking for a thorough valuation of their prospective investment company. For this assessment, they should take into consideration the business's social and environmental impacts, as well as its financial profitability. This is a complex assessment for sustainable entrepreneurs whose business model focuses on social and ecological implications (Shuwaikh et al., 2024).
- High Risk: Venture capitalists seek companies with high development capabilities; however, this type of speculation also accompanies high risks. And if one doesn't, fortunately, it happens to many startups and young businesses as well. Therefore, the business model must be fully implemented or tested, because many of them may face challenges along the way, leading to failure (Ruhnka & Young, 1991).
- More investment support: Venture capitalists offer advice, business relationships, and executive support. Besides funding, venture capitalists can provide additional investment support. Such support can significantly contribute to the development and growth of sustainable startups (Jeong et al., 2020).

2.2.3 Microfinance

It is about small-scale financing for small and medium-sized businesses, particularly in developing countries or communities with limited financial resources. This model is ideal for entrepreneurs who require only a minimal amount of financing to start or expand their business (Sarfo et al., 2024).

Features and Challenges:

- Easier Access: Microfinance Institutions usually have easier access to loans compared to banks. Small businesses and entrepreneurs often struggle to use traditional financial systems; therefore, these institutions are beneficial (Gyimah et al., 2024).
- Small Size and Quick Payback: Microloans often have a short repayment term and are typically used for small financial needs. This is a great feature for small and newly emerging business entities that don't require such high capital to introduce or broaden (Sangwan, 2023).
- Monitoring and Repayment Challenges: Another major challenge in providing microfinance is monitoring the utilization of resources and collecting repayment for loans. Many of

these resources are used to finance work by entrepreneurs who may not be able to repay on time (Hamid Uddin et al., 2022).

2.2.4 Social Investment (Impact Investing)

Social investment can be defined as investments that aim not only to achieve financial profitability but also to induce positive social and environmental change. In the majority of cases, such investments are in projects that address social and environmental issues while also yielding a suitable financial return (Popescu et al., 2024).

Features and Challenges:

- Create positive social and environmental impacts: A distinctive feature of social investment is its focus on creating positive social and environmental effects. This is a suitable model for sustainable entrepreneurs with social and ecological objectives.
- The biggest challenge for social investors is to identify impactful projects that genuinely and permanently benefit society and the environment. Precise criteria and transparent data are necessary for this assessment, but they may be challenging for many sustainable projects.
- Financial and social returns: You are seeing financial and social returns from investors in this space. Therefore, this means that entrepreneurs are expected to demonstrate success in both economic and social aspects.
- Long-term time horizon: Social and environmental impacts are typically longer-term, and the investment time may be significantly longer.

3 Challenges and Barriers

3.1 Cultural, Economic, and Political Barriers

Sustainable entrepreneurship, due to its multidimensional nature, faces numerous challenges and barriers to implementation and growth. There are no financial problems; these are the barriers that originate not just from financial but also from cultural, economic, and political structures. It is essential to understand these barriers to develop effective policies and strategies that support sustainable entrepreneurs (Hoogendoorn et al., 2019).

3.1.1 Cultural Barriers

Traditional attitudes towards the idea of economic success are one of the main cultural barriers to developing sustainable entrepreneurship (Hasan & Khan, 2023). In many societies, financial profitability and rapid growth are often considered the key indicators of success. This perspective undermines the entire value of sustainable entrepreneurship by highlighting the complexities between profit, society, and the environment. In most cultures, awareness of the necessity of sustainability and social responsibility in the public is very low (Hasan & Khan, 2023). It may happen that individuals in society would rather prefer extreme consumerism, a negligent attitude towards the origin of products, and a disregard for their environmental impacts. The latter arises because this type of consumer culture leads to a reduction in market incentives for entrepreneurs to produce value-oriented and sustainable products. Additionally, educational systems often fail to educate students about sustainable entrepreneurship, resulting in a lack of basic skills and knowledge among the new generation (Fülöp & Cifuentes-Faura, 2024).

3.1.2 Economic Barriers

The biggest hindrance from an economic point of view is the absence of suitable financial 'infrastructure' and long-term financial support tools (Kato et al., 2024a, 2024b). Only businesses capable of generating quick returns receive the majority of the available financial resources. However, there are sustainable businesses that emphasize long-term and positive outcomes, prioritizing them over financial gain. Other essential barriers include economic fluctuations, high bank interest rates, the absence of credit guarantees commensurate with the nature of sustainable businesses, and the lack of government budget allocation for green innovations

(Moolkham, 2024). Furthermore, the lack of specific criteria to assess the social and environmental returns of investments leads many investors to doubt this area. However, the fact that many entrepreneurs are excluded from using advanced technologies, as well as the fact that they will be forced to bear the burdens of technology-based infrastructure, such as renewable energy or pollution-reducing technologies, are severe obstacles (Jonsdottir et al., 2022).

3.1.3 Political and Institutional

The first obstacle to entrepreneurship development for sustainable development is the lack of a legal framework supporting policy-making and institutional structure. Despite this, existing laws and regulations are still based on traditional economic models and fail to meet the requirements of sustainable businesses (Watson et al., 2023). Some other institutional barriers include a lack of coordination between government institutions, inadequate incentive policies for green businesses, instability of environmental regulations, and insufficient tax support, which create an unfavorable environment for the growth of sustainable entrepreneurship (Sîrb, 2017). Policies often overlook sustainability, and it is generally considered that policymakers still prioritize short-term economic growth. Additionally, the bureaucracy is heavy, the processes of obtaining permits are complex, and securing approval for innovative plans takes valuable time and resources from sustainable entrepreneurs, forcing them to fall behind in competition with traditional competitors (Fathoni et al., 2024).

Collectively, these cultural, economic, and political barriers have created a multi-layered structure that poses serious challenges to the development of sustainable entrepreneurship. Overcoming these challenges requires changing social attitudes, reforming economic and institutional policies, and public education to promote a culture of sustainability. Only with such a transformation can a conducive environment be created for the real growth of sustainable entrepreneurship.

3.2 Access to Finances and Transparent Data

Limited access to finance and a lack of transparent data are among the most serious obstacles that sustainable entrepreneurs face in their journey. While sustainable entrepreneurship requires targeted, diverse, and sustainable financial support, traditional financing structures and inefficient information systems have created significant obstacles in this path.

3.2.1 Limited Access to Finance

The key issue for sustainable businesses to grow is the availability of finance, which is readily available but not as easily accessible as it should be. Banks and credit institutions, as well as most other traditional financial structures, are based on traditional financial perspectives—short-term profitability and immediate cash flow (Mansouri & Momtaz, 2022). Sustainable businesses are typically early in their operations, and their growing financial returns will be realized over time. Without the existence of specific financial instruments to invest in sustainable projects, such as green bonds or ESG funds, entrepreneurs have quite a limited range of options for raising capital (Maltais & Nykvist, 2020). Venture capitalists, despite their generally risk-averse nature, predominantly prioritize investment in startups that demonstrate rapid financial growth. They exhibit comparatively limited interest in ventures where their capital might be at risk, particularly those emphasizing social or environmental returns. Furthermore, institutional support—such as that provided by development banks, government financing initiatives, or specialized facilities aimed at fostering sustainable enterprises—is frequently inadequate, absent, or hindered by bureaucratic inefficiencies. Firstly, it deprives them of financial resources not only to support them while launching sustainable startups but also to grow them, producing thriving startups that can penetrate the market as well (Makki et al., 2020).

3.2.2 Lack of Transparent Data and Accurate Assessment Tools

Another major obstacle to financing sustainable entrepreneurship is the scarcity of accurate, transparent, and reliable data (Peng & Walid, 2022). To be financed, investment and financial institutions require information about the financial performance, social and environmental impacts, risks, and opportunities of projects in a clear and documented form (Mansouri & Momtaz, 2022). Uncertainty in the investment process increases due to the lack of such data. At the same time, sustainable entrepreneurs themselves often lack standard assessment tools that can quantitatively and qualitatively demonstrate the non-financial impacts of their work (Lee et al., 2021). Most projects considered 'sustainable' are at the lower end of the decision-making spectrum when investors must choose between purely economic projects due to the lack of clear frameworks for measuring environmental or social effectiveness (Popescu et al., 2024). Moreover, many countries lack open and widely accessible databases for entrepreneurs and investors. If information is available, it is often divided between economic, environmental, or target market considerations and is rarely available publicly, being either outdated or fragmentary. This ensures that entrepreneurs cannot make data-driven decisions and present transparent, defensible financial plans (Jonsdottir et al., 2022).

3.2.3 Lack of Digital and Information Infrastructure

Another barrier related to this is the weakness of information and digital infrastructure. The information technology infrastructure that entrepreneurs can use to evaluate their market, manage resources, and connect with suppliers and investors simply does not exist or is ineffective in many regions, particularly in developing countries (Yuwono et al., 2024). For example, the activities of sustainable entrepreneurs are characterized by a lack of transparency and credibility, as modern technologies like Blockchain for recording transparent data and digital platforms for evaluating performance are beyond the reach of most sustainable entrepreneurs. This also erodes public trust and investor confidence, making it less likely for one to obtain finance (Soluk et al., 2021).

In general, the lack of financial resources and the shortage of transparent data lead to a vicious cycle. In layperson's terms, sustainable entrepreneurs cannot adequately provide and defend their projects, and investors avoid investment in this area due to a lack of information. Reliable information systems to address these challenges, the development of financial instruments tailored to the business's sustainability characteristics, and policies that clarify and facilitate sustainable projects.

4 Research Background

The paper provides an integrated and comprehensive overview of key research conducted in the field of sustainable entrepreneurship financing through Table 1, titled "Research Background." Over the past two decades, this table illustrates the shift in the scientific literature from theoretical concepts to practical solutions and creative tools for sustainable project financing. In this series of studies, researchers sought to investigate the role played by venture capital, crowdfunding, Fintech, supportive legislation, and environmental, social, and governance (ESG) indices in promoting sustainable entrepreneurship.

However, a critical review of Table 1 (Research Background) studies reveals that, although these studies provide a relatively comprehensive picture of research developments in the field of sustainable entrepreneurial finance, they also lack certain studies or statements that would enable a deeper analysis of the research gaps. The following reviews these studies in an integrated and analytical way. First, many early studies, such as those by Choi and Gray (2008), employed the case study method. This provides a helpful way to go in-depth into a particular case, but suffers from a lack of generalizability of the results. Having examined these studies, they have helped identify the role of green investment and nontraditional sources of finance; however, they do not provide a broader theoretical framework or analyt-

Table 1 Research background

Authors	Title	Purpose	Tools	Conclusion	Case study
Choi and Gray (2008)	Investment development processes	Examining the process of developing sustainable businesses	Case study	Sustainable entrepreneurs leverage non-conventional sources of finance	Sustainable businesses
Hall et al. (2010)	Sustainable development and entrepreneurship	Examination of entrepreneurship's contribution to sustainable development	Resource review	Entrepreneurship is a platform for developing sustainable products	Startups
Haigh (2012)	Connecting sustainability goals to finance activities	Exploring the relationship between sustainability goals and finance	Resource analysis	Sustainable finance helps economic and social sustainability Resource analysis	Startups
Mbetwa and Silumbe (2016)	Entrepreneurship and business management	Examining the links between business management and sustainable entrepreneurship with financing	Review of sources and analysis of secondary data	Direct communication between government, financial institutions, and the public for sustainable development	Entrepreneurs in the business environment
Bento et al. (2019)	Crowdfunding for sustainability investments	Crowdfunding for sustainable businesses review	Data analysis and regression	High survival rates and fundraising success for sustainable businesses	Crowdfunding platforms
Arshad et al. (2019)	Sustainable crowdfunding for livelihood entrepreneurship	Analysis of crowdfunding for sustainable entrepreneurship	Conceptual framework	Crowdfunding plays a key role in developing sustainable entrepreneurship	Kiva platform example
Lagoarde-Segot (2020)	Developing countries	Need for innovation in financing mechanisms	Review of resources and reports	Analysis of SDG financing mechanisms	Financing the sustainable development goals
Ayandibu and Vezi-Magigaba (2021)	Climate change and entrepreneurship financing	Examining the relationship between climate change and entrepreneurship financing	Analysis of data from existing sources	Climate change provides new financial opportunities for entrepreneurs	Developing countries
Burchi et al. (2021)	The impact of financial literacy on sustainable entrepreneurship	Analyzing sustainable entrepreneurship and financial literacy	OLS model	Financial literacy helps entrepreneurs sustain success	European countries

(continued)

Table 1 (continued)

Authors	Title	Purpose	Tools	Conclusion	Case study
Mansouri and Momtaz (2021)	Financing sustainable entrepreneurship	Examining the role of ESG criteria in financing sustainable startups	Machine learning	Sustainable startups receive more favorable valuations but underperform	Tech startups
Lin (2021)	Venture capital in increasing sustainable investment	Development of sustainable venture funds	Legal analysis	Government laws and regulations should support sustainable investment	Venture capital
Bhatnagar et al. (2022)	India's Green startup boom	Analyzing the impact of green funding on sustainable startups in India	Case study	Startup success increased with green funding	Green startups in India
Imamah and Nafisa (2023)	FinTech results in sustainable business creation	Examining the function of financial technology in ecological entrepreneurship	Digital technology, FinTech	FinTech promotes economic growth and expands underprivileged groups' access to financial services	Financial technology and entrepreneurship
Sunarni (2023)	Analyzing how financing methods and environmental sustainability affect social enterprise success	Environmental sustainability, financing models, and risk management	Case study analysis	Identifies positive relationships between sustainability factors and the success of social entrepreneurship ventures	CV. Pangalusna Sukabumi
Mehrabi (2023)	A comprehensive analysis of the literature on sustainable finance	Systematic review of sustainable finance and its impact on innovation	Systematic review	The role of venture capital in the development of innovation	Literature review
Zamfirache et al. (2023)	Interest in financing for sustainable education	Survey of student interest in sustainable entrepreneurship funding	Questionnaire	High trust in European funding among students	Romanian students
Rink (2024)	Sustainable lending for small businesses	Analyzing the role of banks in the sustainable financing of small businesses	Survey of German banks	Identifying sustainable lending approaches and financial values in this context	German banks and SMEs
Valerievna and Igorevna (2024)	The effective role of sustainable finance in green business transformation: current issues and possible solutions	Analysis of green financing challenges and solutions and its impact on green businesses	Literature review and modeling of solutions	Proposing regulatory frameworks and KPIs to increase access to green loans	"Brown" and green businesses

(continued)

Table 1 (continued)

Authors	Title	Purpose	Tools	Conclusion	Case study
Kato et al. (2024a, 2024b)	The effect of sustainable finance options on small and medium businesses' long-term viability	Examining how sustainable finance affects SMEs	Multiple regression model	Sustainable finance increases SME productivity	East African SMEs
Yang et al. (2024)	Charting the signaling landscape that connects investment inputs and entrepreneurship with a sustainability focus	Signals of entrepreneurship and sustainability-focused investing	Modeling	Highlights the importance of organizational signals for investing in sustainable ventures and helps them overcome the constraints of conventional entrepreneurship	Self-descriptions of some small sustainable technologies

ical model that can be applied in macro policy-making. Other studies, such as those by Hall et al. (2010) and Haigh (2012), have explored the relationship between sustainable development goals and entrepreneurship by reviewing the relevant literature. The approach used in this review is certainly valuable; however, it is primarily descriptive and empirical in nature. Since there is a lack of numerical or comparative data that permits these studies to specify the direct and indirect financing effects on the accomplishment of sustainable projects with satisfactory precision, it is difficult to clarify whether and how the financing affects the rate of success of sustainable projects. For instance, mid-level studies by Bento et al. (2019) and Arshad et al. (2019) employ regression analysis or other conceptual frameworks to quantify their approaches, which still maintain a localized focus on specific countries. However, their generalizability to different geographic areas or technological platforms remains questionable. In addition, most of these studies pay too much attention to crowdfunding without thoroughly investigating its limitations (e.g., low scalability or short-term sustainability of the financial resources). Recent studies, such as those by Mansouri and Momtaz (2021) and Lin (2021), propose innovative approaches within the context of sustainable finance that leverage advanced technologies, including machine learning and legal analytics. These are more methodologically sophisticated studies, but in some cases, they suffer from a form of 'technological reductionism' in which the role of these technologies in the financial decisions of large entrepreneurs is overstated, without regard for cultural, institutional, or social barriers to the use of these technologies in different countries.

Of the studies conducted in 2023 and 2024, including Imamah and Nafisa (2023), Mehrabi (2023), and Rink (2024), there has been a greater diversity of research methods, and studies have become stronger as a result of data analysis, systematic review, and modeling. However, most of this research lacks a concentration on the regional, infrastructural, and cultural differences. For example, in the case of including a low-income group in the fintech products portfolio of a company, not having the necessary digital infrastructure can lead to incomplete or misleading findings. The studies by Sunarni (2023), Zamfirache et al. (2023), and Valerievna and Igorevna (2024) are somewhat scattered and address only the aspects of education and students' views on sustainable finance or certain financial model analyses, without connecting these topics. This research lacks theoretical coherence across the economic, social, psychological, and institutional dimensions of these studies, thereby rendering it difficult to gain a holistic understanding of the sustainable finance ecosystem.

Additionally, business success and financial capabilities are typically the focus of most studies, while little attention has been paid to the long-term social and environmental impacts of sustainable businesses. Such a lack has resulted in the fact that, in too many cases, finance has been approached independently of social justice or environmental sustainability as just a matter of necessity (survival) or economic growth. Key weaknesses in this body of research relate to the final three points: theoretically, studies in this area are weak in terms of theoretical engagement, lack integrated theoretical models, and exhibit a paucity of cross-country or cross-regional comparative studies. Future research should, therefore, develop comprehensive analytical models for comparing this situation at an international level, make comparative analyses, and examine the interplay of financial, cultural, technological, and institutional factors simultaneously.

5 Strategies and Solutions

5.1 Using Technology—Focusing on Blockchain for Transparency

Most importantly, the use of new technologies is proposed by many of the strategies recently outlined as one of the best techniques to achieve sustainable entrepreneurship financing. In this sense, blockchain technology has been progressively adopted for tasks as important as one of the most advanced and efficient interactive mechanisms to enhance financial transparency, public confidence, and optimize transaction costs (Khalegi et al., 2024). It is due to the multidimensional nature of sustainable entrepreneurship that systems capable of recording and controlling financial flows with high accuracy, transparency, and traceability are required; this is where Blockchain enters. Blockchain is a decentralized and immutable database, which means that Transactions can be recorded regardless of the need for intermediaries, by which all types of parties (investors, entrepreneurs, governments, and regulatory bodies) can access financial information (Bhandari et al., 2025). Such a level of transparency is necessary for reducing corruption, building trust in the capital-raising process, and ensuring the proper use of financial resources. As an example, from the perspective of funders in crowdfunding projects, one of the main concerns is the path of spending the funds. Blockchain can automatically demonstrate the path of spending funds, and any violation becomes impossible or at least detectable.

Smart contracts based on blockchain technology can automate payment processes, resource allocation, performance evaluation, and financial reporting, eliminating intermediaries and reducing costs while also ensuring transparency (Dashkevich et al., 2024). Especially in cases where the project has phased financing, that is, when the funding is made in stages according to the demonstrated progress record of the expected project results or goals, these types of contracts are particularly effective. Moreover, the use of blockchain technology is not limited to this; far more can be achieved in terms of tracking sustainable supply chains (Yadav et al., 2023). This technology can uniquely demonstrate the sourcing path of raw materials and manufacturing conditions, thereby ensuring compliance with environmental and human rights standards. This allows sustainable entrepreneurs to clearly show how the raw materials they use were obtained, how people were treated during the manufacturing process, and so forth (Biswas et al., 2022). Building this not only adds trust to the consumer but also differentiates the sustainable brand from its unsustainable competitors, becoming one of the reasons for both capital calls and market recognition (Valencia-Payan et al., 2024).

Although blockchain adoption has its own set of challenges, a robust technology infrastructure, initial implementation costs, the absence of technical knowledge among many entrepreneurs, and the lack of a clear legal and regulatory framework in many countries can prevent this technology from being widely adopted (Kaur et al., 2022). In addition, some blockchain algorithms, specifically those that consume high amounts of energy, such as Proof of Work, may have an opposing impact on environmental sustainability principles unless consistent use of more environmentally friendly algorithms, such as Proof of Stake, is undertaken (Magdalena et al., 2025). In general, Blockchain's use in the context of the target country's economic, cultural, and legal environment should be technological, and concerning these, should be accompanied by appropriate training and infrastructure and should be through a technological approach so that it can be leveraged as a tool for strengthening the transparent, sustainable, trust-building financing structures in sustainable entrepreneurship (Kaur et al., 2022). Entrepreneurs are not the only ones who can benefit from this technology, as it can pave the way for investors with a responsible and long-term perspective.

5.2 Innovative Financing Models: Crowdfunding

Crowdfunding is one of the innovative models that has also contributed much to democratizing financing in the context of sustainable entrepreneurship (Tenner & Hörisch, 2021). With the adoption of digital technology, this model attracts capital from numerous members of society, serving as an effective instrument for financing projects in social, environmental, and cultural aspects, and is even recognized for its efficiency (Dinh et al., 2024). Crowdfunding provides sustainable entrepreneurs with a means to reduce their reliance on official sources of financing by attracting capital from the general public, allowing them to reach their sustainable goals independently of affiliation with large financial institutions or particular investors. Forms of crowdfunding include donation-based, reward-based, lending-based, and equity-based investments (Hoque, 2024). Backers are rewarded in the case of the reward-based model with non-financial benefits, such as products or services. It is called an equity-based model because investors become partners in the project and get a share of ownership in the business. Because crowdfunding offers flexibility in its structure, this format is applicable not only to small startups but also to medium and small projects in fields such as renewable energy, organic farming, local crafts, and educational and medical initiatives (Zhao & Ryu, 2020).

Crowdfunding for sustainable entrepreneurship is a key benefit (Futre, 2025):

- Helping micro and local entrepreneurs gain easier access to finance, which they may otherwise have difficulty obtaining from banks or traditional investors.
- Direct connection with the target community, enabling the achievement of support from consumers who consider sustainability values as a key criterion.
- A large number of backers can build social credibility for the projects.
- Increasing transparency and accountability through continuous updates on the project status on crowdfunding platforms.

Although crowdfunding offers benefits, it also has its limitations. They encompass intense rivalry for the investor's attention, demands of digital marketing and efficient advertising, as well as some constraints on the amount of funds that can be raised (Chen, 2021). Some countries do not have the necessary legal frameworks for protecting small investors or regulating entrepreneurs' rights, which can reduce investment security. Then, the success of a sustainable crowdfunding project greatly depends on the power of storytelling, clarity of social/environmental goals, and transparency about the use of funds (Goethner et al., 2021). The more honest and compelling the narrative the entrepreneur can create for their audience, the more likely they are to succeed. The crowdfunding models can also create an ecosystem of sustainable entrepreneurship with other funding methods from a behind-the-scenes perspective (Dinh et al., 2024). In this aspect, other successful examples like Kiva (in the field of microfinance entrepreneurship with social eligibility in developing countries), Kickstarter, and Indiegogo (in the field of creative and sustainable products) proved the existence of crowdfunding as an appropriate channel to strengthen this ecosystem alongside other financing programs (Fareed & Viotto, 2024). Therefore, crowdfunding is not merely a financial tool but also a scheme for the formation of social interaction, increasing public awareness, and recycling the process of sustainable development. If properly supported by education and institutional policy, this model may serve as one of the principal sustainable finance mechanisms (Motylska-Kuzma, 2018).

5.3 The Role of Governments and International Institutions in Financing Sustainable Entrepreneurship

In the context of sustainable entrepreneurship financing structures, governments and international institutions play a critical and decisive role as regulators and macro policymakers, as well as facilitators, investors, and observers. Recent advances in the private sector and innovative technologies are significant, but without institutional and policy support, the necessary infrastructure is not in place to make sustainable development a reality, and capital streams are not funded for projects with a sustainable development approach (Watson et al., 2023).

1. Governments as facilitators of sustainable finance

Governments can utilize various policy instruments to facilitate the financing of sustainable entrepreneurs. These are the most important of these instruments (Abbasi et al., 2025; Watson et al., 2023):

- Incentive tax policies: Tax incentives include exemptions or reductions in taxes for businesses operating in environmentally or socially responsible areas, which are significant incentives to attract investment in these areas.
- Subsidies and direct financial assistance: Reducing risk and attracting private capital are among the most effective ways to mitigate risk by subsidizing, or at the very least, helping to fund green projects and social entrepreneurship initiatives in their early stages.
- Government support for the initiation of public–private investment funds: One way a government can help is by creating funds that combine public-sector investment with private-sector participation for investments in sustainable businesses.
- Credit guarantees and guarantees: The government's ability to provide guarantees and risk coverage for loans and investments in sustainable projects can provide an incentive for new investors to enter.

2. Role of the government institution in the reform of the legal and regulatory structure

Clear, transparent, and sustainable rules are required for green finance. Governments can (Watson et al., 2023):

- Create legal frameworks for green bonds, ESG, and impact investing.
- Provide regulatory bodies with the ability to evaluate the social and environmental performance of projects.
- It becomes mandatory for businesses to report their sustainability. They will not only establish transparency and trust in the market but will also attract foreign capital.

3. Identification of the role of international institutions in strengthening local capacities and global frameworks.

International institutions, such as the World Bank and the International Monetary Fund (IMF), as well as the United Nations Development Program (UNDP) and the United Nations Environment Program (UNEP), or regional organizations like the European Union, play an indispensable role in supporting sustainable finance. The following are these institutions (Endo et al., 2022; Wagenhofer, 2023):

- Financing large-scale infrastructure projects and sustainable enterprises in developing countries
- Contribution in the field of reporting, assessment, and monitoring sustainable performance in developing international standards.
- It was about creating a space for international cooperation, establishing platforms where countries could collaborate and share successful experiences with other countries, and translating this knowledge into technical expertise.
- Allotting the budgets of development aid for small and medium-sized businesses with a sustainability thought.

For instance, examples of such programs include the UN SDG Impact and the Green Climate Fund (GCF) projects, which utilize their platforms to attract international resources for local projects in the fields of environment and sustainable development.

4. As a monitor of financial justice and social sustainability by governments and international institutions

The social justice dimension is often a neglected aspect of many financial models. By providing oversight, legal support, and ethical frameworks, as well as fostering entrepreneurship among vulnerable groups, deprived areas, and marginalized communities, governments and international institutions can ensure that financing is not solely for the benefit of large investors. Monitoring these ESG criteria can also help prevent 'greenwashing' by them. Finally, the support of governments and international institutions for the success and expansion of sustainable entrepreneurial finance is structural and a backbone. These actors are provided with financial, legal, educational, and institutional support as a framework to implement sustainability projects with greater motivation, confidence, and security, ultimately achieving economic development alongside social and environmental justice (Ben Mahjoub, 2024).

6 Case Study

6.1 Grameen Bank—A Successful Example of Sustainable Finance in Social Entrepreneurship

The Grameen Bank in Bangladesh is one of the best-known and most influential global examples of successful sustainable finance on an institutional level. Established to support people experiencing poverty, particularly women in rural areas, it has become a model for global microfinance and sustainable social entrepreneurship. In the 1970s, Muhammad Yunus founded Grameen Bank with the idea that people are not poor because they lack the ability; rather, it is because they do not have access to sustainable financial resources and economic opportunities. Unlike traditional banks, the bank has developed innovative financing programs offering small loans to individuals who lack collateral and bank credit, enabling them to start or expand micro-businesses (Yunus et al., 2010).

6.1.1 Grameen Bank's Sustainable Financial Model

This model operates based on trust, responsibility, and, of course, without collateral. These loans go out without collateral, and repayments are made in small communities (group borrowers) weekly or monthly (Inayatul Faizah & Ahmad Husaeni, 2019).

- Women and social empowerment: More than 90% of Grameen Bank's borrowers are women. It is not focusing only on women that helped raise household incomes but also contributed to enhancing women's social and economic status.
- Business from below: loans are invested in sustainable agriculture, handicrafts, livestock, local retail, and small-scale service activities.

Such activities are environmentally friendly and contribute to regional economic growth.

- High in repayment and trust-building: The repayment rate at Grameen Bank averages over 95%, indicating that this model is highly efficient in markets where traditional banks have little chance of success.

6.1.2 Economic, Social, and Environmental Impacts

- Economic impacts: As a model, it increases the rate of growth of self-employment, decreases dependence on government assistance, generates cash flow within the local economy, and reduces poverty rates. However, many borrowers have been able to establish small family businesses and even expand on them.
- Social impacts: The Grameen model has increased social participation, reduced early marriages, improved children's education, and given women self-confidence. Support networks in villages have gradually evolved into borrower groups, thereby enhancing social solidarity within villages.
- Environmental impacts: Although Grameen Bank's thrust has been on social issues, the projects it supports continue to be guided by the touchstones of environmental sustainability. Especially in the fields of organic farming, traditional livestock farming, and the use of renewable resources.

6.1.3 Complementary Grameen Innovations

Recently, affiliates have been formed and expanded services beyond microlending. However, at the same time, they have made a tremendous contribution to increasing the bank's social and economic impact, transforming the Grameen model into a more inclusive and adaptable framework for the sustainable development of developing countries. Grameen Shakti, which has existed since 1996 and focuses on developing renewable energy in rural Bangladesh, is one example of such an entity (Yunus et al., 2010). In unelectrified areas, it has distributed more than 1.8 million solar home systems, which have contributed to providing access to clean energy, decreasing reliance on fossil fuels, and improving the quality of life for rural households (Jaap Bouma et al., 2017). Local technicians have indeed been trained by Grameen Shakti to install and maintain these systems, thus offering tremendous employment opportunities.

On the other hand, Grameen Telecom was founded to provide communication technologies to the people in rural areas who were less privileged. The Village Phone Program was one of its most prominent projects, in which loans were made to rural women to purchase a mobile phone and provide service to the community as a local operator. However, this program, while increasing communication levels and access to information, has also provided women with further economic and income opportunities. As a global non-profit organization, the Grameen Foundation has effectively developed digital technologies for financial empowerment, digital financial literacy, and business consulting while collaborating with other banks and organizations to deploy the Grameen model in various countries (Alom et al., 2025). With the advent of Big Data, mobile phone systems, and digital financial platforms, the foundation adopted the social banking model and integrated it with new technologies, further spreading it globally. Furthermore, in the internationalization of the model, Grameen America and Grameen Foundation India are instances from the United States and India. In this regard, the Grameen model has even been demonstrated to be adaptable to diverse needs under various cultural and economic conditions (Mousa & Ozili, 2022).

The Grameen Bank model demonstrates that innovative financial approaches can promote social justice, economic growth for local communities, and individuals' ability to cope with structural poverty, all within the principles of sustainability. Indeed, it is an excellent example of sustainable social entrepreneurship founded on inclusive microfinance, which has had a profound and far-reaching impact, both theoretically and practically. As a special feature, Grameen does not regard access to finance as an end in itself but rather as a tool for social empowerment, greater self-reliance, and cultural and economic transformation at the community level. This differs from purely economic approaches in that its focus is on women, the use of social trust rather than bank collateral, the formation of repayment cooperative groups, and the development of complementary services in areas such as energy, communications, and education simultaneously. But this model has shown in communities where resources are few that if you do financing properly, transparently, and locally, you can get economic independence rather than economic dependency. In addition, its replication and localization in countries around the world, including the US, India, Uganda, and the Philippines, demonstrate that it is indeed culturally and economically flexible. While simultaneously achieving financial sustainability, Grameen has shown that financial institutions can take on a social and humanitarian mission—the mission of increasing profitability in line with promoting positive social and environmental change. In this sense, Grameen is more than just a bank; it marks the beginning of a movement to establish financial justice and sustainable development worldwide. When considered from a local level, the work of the Grameen Bank is a testament not only to the ingenuity of social entrepreneurs everywhere but also a model for policymakers, social investors, international institutions, and entrepreneurs alike to pursue the nature of finance not to serve only as an economic instrument but as the purveyor of lasting social change.

7 Conclusion

This review of financing mechanisms for sustainable entrepreneurship considered how to balance economic growth, social justice, and environmental sustainability, and it was a comprehensive and multi-layered review. The key concepts were analyzed, which show that sustainable entrepreneurship departs from the traditional business view based on profit-making, and new models of financing must be built on principles of responsibility and long-term. The four main financing models explored were traditional financing, venture capital, microfinance, and impact investing. These models can fund sustainable entrepreneurs, but at the same time, they are not compatible with sustainable goals; they require collateral, and there are also short-term return issues and valuation complexities. The data relevant to challenges were analyzed, and it was found that the main cultural barriers (provision of services with a traditional attitude to profits), economic barriers (insufficient economic resources corresponding to goals of sustainability), and political and institutional barriers (there are inadequate incentive policies in this area; inadequate legal structures) hinder the development of sustainable finance. The reason behind this is a lack of access to transparent and reliable data, the absence of global standards for assessing social and environmental returns, and the absence of digital infrastructure in many developing countries, which has hindered many sustainable projects from accessing proper financial resources. In the solution section of the paper, the technologies used were Blockchain, crowdfunding tools, and fintech approaches, as these are innovative and transformative technologies. Not only do these tools serve as a means to improve transparency and facilitate transactions, but they also increase access to finance for emerging and micro-entrepreneurs. At the same time, the roles of governments and international institutions as policymakers and public investors were stressed. These institutions also provide financial, legal, and educational support that builds investor confidence, sustains infrastructure projects, and expands the viability of successful models. Grameen Bank is a real case study of how sustainable finance models that combine financial, social, and environmental goals can achieve success in the proper design of economic systems.

7.1 Future Research

On this basis, clear directions for future research and policy-making are:

- Hybrid models of sustainable financing: models that leverage public, private, technological, and social resources. Such research is capable of offering more flexible solutions that are more readily adapted to diverse economic and cultural conditions.

- How sustainable financing supports and sustains social and environmental structures: Rather than evaluating sustainable financing solely on financial terms, this paper seeks to consider the systemic impacts that financing has and the potential to create, as well as mitigate, inequality, increase community resilience, and protect natural resources.
- Social and environmental performance measurement tools can be developed through a data-driven approach, leveraging big data, artificial intelligence, and predictive analytics to create reliable indicators that measure the success of sustainable projects.
- International comparative study of models to assess the transferability of models: Many models have been developed in different locations, and they can be studied for potential transfer to other countries.
- Evaluation of government financial policies and international laws in the field of supporting sustainable investment: A comparative study of how government financial policies and international laws support sustainable investment, based on tax policies, incentives, disclosure requirements, and legal frameworks, will provide insight into more effective policies.
- Fintech in service to the reduction of inequality in access to finance: In a lot of underserved places, Fintech can bear a considerable role in financial justice development if partnered with financial education and digital infrastructure.

The future will be built on sustainability now, and that sustainability is sustainable entrepreneurship. The path is locked by sustainable finance, not a simple economic tool, but a driver of social, cultural, and environmental change. It is true, however, that financial returns are as significant as social and environmental returns and that the latter deserve their proper place in financial decision-making. This chapter illustrates that future financial models will have to be more humane, transparent, and responsible, and this is part of the global discourse on sustainable development.

References

Abbasi, S., Khoshkenar, A., Devrim İçtenbaş, B., Etemadifar, S., & Sheykhian, A. (2025). Designing a sustainable supply chain network with a financial approach in a catastrophic situation. *Discrete Dynamics in Nature and Society, 2025*(1). https://doi.org/10.1155/ddns/9269307

Alom, K., Rahman, Md. Z., Khan, A. I., Akbar, D., Hossain, Md. M., Ali, Md. A., & Mallick, A. (2025). Digital finance leads women entrepreneurship and poverty mitigation for sustainable development in Bangladesh. *Journal of Innovation and Entrepreneurship, 14*(1). https://doi.org/10.1186/s13731-025-00471-2

Arshad, N., Ramírez-Pasillas, M., & Hollebeek, L. D. (2019). Sustainable crowdfunding for subsistence entrepreneurship. In *Contributions to management science* (pp. 49–62). https://doi.org/10.1007/978-3-030-23604-5_4

Ayandibu, A. O., & Vezi-Magigaba, M. F. (2021). Climate change and entrepreneurial financing. In *Advances in finance, accounting, and economics book series* (pp. 399–410). https://doi.org/10.4018/978-1-7998-7967-1.ch024

Azamat, O., Fayzullokh, S., & Nilufar, A. (2023). The impact of entrepreneurship on poverty reduction. *International Journal of Professional Business Review, 8*(3), e01012–e01012. https://doi.org/10.26668/businessreview/2023.v8i3.1012

Ben Mahjoub, L. (2024). Greenwashing practices and ESG reporting: An international review. *International Journal of Sociology and Social Policy*. https://doi.org/10.1108/ijssp-08-2024-0365

Bento, N., Gianfrate, G., & Thoni, M. H. (2019). Crowdfunding for sustainability ventures. *Journal of Cleaner Production, 237*, Article 117751. https://doi.org/10.1016/j.jclepro.2019.117751

Bhandari, M., Tiwari, G., & Dhakal, M. (2025). Enhancing transparency and accountability in sustainable finance through blockchain technology: A systematic review of the literature. *Journal of Intelligent Management Decision, 4*(1), 23–43. https://doi.org/10.56578/jimd040102

Bhatnagar, M., Taneja, S., & Özen, E. (2022). A wave of green startups in India—The study of green finance as a support system for sustainable entrepreneurship. *Green Finance, 4*(2), 253–273. https://doi.org/10.3934/gf.2022012

Biswas, D., Jalali, H., Ansaripoor, A. H., & De Giovanni, P. (2022). Traceability vs. sustainability in supply chains: The implications of Blockchain. *European Journal of Operational Research, 305*(1), 128–147. https://doi.org/10.1016/j.ejor.2022.05.034

Burchi, A., Włodarczyk, B., Szturo, M., & Martelli, D. (2021). The effects of financial literacy on sustainable entrepreneurship. *Sustainability, 13*(9), 5070. https://doi.org/10.3390/su13095070

Chen, L. (2021). Investigating the impact of competition and incentive design on performance of crowdfunding projects: A case of independent movies. *Journal of Theoretical and Applied Electronic Commerce Research, 16*(4), 791–810. https://doi.org/10.3390/jtaer16040045

Choi, D. Y., & Gray, E. R. (2008). The venture development processes of "sustainable" entrepreneurs. *Management Research News, 31*(8), 558–569. https://doi.org/10.1108/01409170810892127

Corvellec, H., Stowell, A. F., & Johansson, N. (2021). Critiques of the circular economy. *Journal of Industrial Ecology, 26*(2), 421–432. https://doi.org/10.1111/jiec.13187

Danisman, G. O., & Tarazi, A. (2024). ESG activity and bank lending during financial crises. *Journal of Financial Stability, 70*, Article 101206. https://doi.org/10.1016/j.jfs.2023.101206

Dashkevich, N., Counsell, S., & Destefanis, G. (2024). Blockchain financial statements: Innovating financial reporting, accounting, and liquidity management. *Future Internet, 16*(7), 244. https://doi.org/10.3390/fi16070244

Di Vaio, A., Hassan, R., Chhabra, M., Arrigo, E., & Palladino, R. (2022). Sustainable entrepreneurship impact and entrepreneurial venture life cycle: A systematic literature review. *Journal of Cleaner Production, 378*(1), Article 134469.

Dinh, J., Isaak, A. J., & Wehner, M. C. (2024). Sustainability-oriented crowdfunding: An integrative literature review. *Journal of Cleaner Production.* https://doi.org/10.1016/j.jclepro.2024.141579

Endo, K., Edelenbos, J., & Gianoli, A. (2022). Sustainable infrastructure: A systematic literature review on finance arrangements and governance modes. *Public Works Management & Policy*, 1087724X2211302. https://doi.org/10.1177/1087724x221130284

Fareed, F., & Viotto, J. (2024). Crowdfunding platforms and financial inclusion: Fulfilled promise or disillusion? *Finance Research Letters, 73*, Article 106642. https://doi.org/10.1016/j.frl.2024.106642

Fathoni, F., Sulastuti, S., Yudhi, R., & Soerjatisnanta, H. (2024). Striking a balance: Streamlining environmental licensing for business growth and environmental protection in Indonesia. *Scholars International Journal of Law, Crime and Justice*, *7*(06), 240–248. https://doi.org/10.36348/sijlcj.2024.v07i06.005

Fu, H., Qi, H., & An, Y. (2024). When do venture capital and startups team up? Matching matters. *Pacific-Basin Finance Journal, 85*, Article 102361. https://doi.org/10.1016/j.pacfin.2024.102361

Fülöp, M. T., & Cifuentes-Faura, J. (2024). Sustainable entrepreneurship education: Impacts of the circular economy, health literacy, and well-being. *Sustainable Development*. https://doi.org/10.1002/sd.3262

Futre, B. (2025). Unveiling the route to sustainable entrepreneurship: A comprehensive examination of sustainable new ventures financing through systematic literature review. *Vision: The Journal of Business Perspective.* https://doi.org/10.1177/09722629251325089

Goethner, M., Hornuf, L., & Regner, T. (2021). Protecting investors in equity crowdfunding: An empirical analysis of the small investor protection act. *Technological Forecasting and Social Change, 162*, Article 120352. https://doi.org/10.1016/j.techfore.2020.120352

Gucciardi, G. (2024). Do venture capital investments contribute to the achievement of the sustainable development goals? *Business Strategy and the Environment.* https://doi.org/10.1002/bse.3942

Gutterman, A. (2022). Sustainable entrepreneurship. *SSRN Electronic Journal.* https://doi.org/10.2139/ssrn.4192086

Gyimah, P., Owusu-Afriyie, R., & Lussier, R. N. (2024). Advancing rural entrepreneurship: Does microfinancing matter? *Journal of Small Business Strategy, 34*(2). https://doi.org/10.53703/001c.117667

Haigh, M. (2012). Connecting sustainability goals to financing activity. *Journal of Sustainable Finance & Investment, 2*, 85–87. https://doi.org/10.1080/20430795.2012.711973

Hall, J. K., Daneke, G. A., & Lenox, M. J. (2010). Sustainable development and entrepreneurship: Past contributions and future directions. *Journal of Business Venturing, 25*(5), 439–448. https://doi.org/10.1016/j.jbusvent.2010.01.002

Hamid Uddin, Md., Akhter, S., Mollah, S., & Mahi, M. (2022). Differences in bank and microfinance business models: An analysis of the loan monitoring systems and funding sources. *Journal of International Financial Markets, Institutions and Money*, *80*, 101644–101644. https://doi.org/10.1016/j.intfin.2022.101644

Hasan, M., & Khan, T. (2023). The impact of cultural norms on sustainable entrepreneurship practices in SMEs of Bangladesh. *Indonesian Journal of Innovation and Applied Sciences*, *3*(3), 201–209. https://doi.org/10.47540/ijias.v3i3.962

Hoogendoorn, B., van der Zwan, P., & Thurik, R. (2019). Sustainable entrepreneurship: The role of perceived barriers and risk. *Journal of Business Ethics, 157*(4), 1133–1154. https://doi.org/10.1007/s10551-017-3646-8

Hoque, Md. M. (2024). Crowdfunding for innovation: a comprehensive empirical review. *Future Business Journal, 10*(1). https://doi.org/10.1186/s43093-024-00387-5

Imamah, N., & Nafisa, L. (2023). Fintech attaining sustainable entrepreneurship. In *Fintech attaining sustainable entrepreneurship* (pp. 62–75). https://doi.org/10.11594/futscipress15

Inayatul Faizah, S., & Ahmad Husaeni, U. (2019). Economic empowerment for poor women using grameen bank model in Indonesia. *KnE Social Sciences, 3*(13), 880. https://doi.org/10.18502/kss.v3i13.4255

Jaap Bouma, J., Jeucken, M., & Klinkers, L. (2017). *Sustainable banking the greening of finance.* Routledge.

Jeong, J., Kim, J., Son, H., & Nam, D. (2020). The role of venture capital investment in startups' sustainable growth and performance: Focusing on absorptive capacity and venture capitalists' reputation. *Sustainability, 12*(8), 3447. https://doi.org/10.3390/su12083447

Jonsdottir, B., Sigurjonsson, T. O., Johannsdottir, L., & Wendt, S. (2022). Barriers to using ESG data for investment decisions. *Sustainability, 14*(9), 5157. https://doi.org/10.3390/su14095157

Karadayı, T., Yazıcı, S., & Akdemir Ömür, G. (2024). Climate change and entrepreneurship: A review and research agenda. *Global Business and Organizational Excellence*. https://doi.org/10.1002/joe.22281

Kato, A. I., Chiloane-Tsoka, E. G., & Mugambe, P. (2024b). Unlocking the potential: the influence of sustainable finance solutions on the long-term sustainability of small and medium-sized enterprises. *Cogent Business & Management, 11*(1). https://doi.org/10.1080/23311975.2024.2391122

Kato, A. I., Chiloane-Tsoka, E. G., & Mugambe, P. (2024). Unlocking the potential: the influence of sustainable finance solutions on the long-term sustainability of small and medium-sized enterprises. *Cogent Business & Management, 11*(1). https://doi.org/10.1080/23311975.2024.2391122

Kaur, J., Kumar, S., Narkhede, B. E., Dabić, M., Rathore, A. P. S., & Joshi, R. (2022). Barriers to blockchain adoption for supply chain finance: The case of Indian SMEs. *Electronic Commerce Research*. https://doi.org/10.1007/s10660-022-09566-4

Khalegi, F., Kadyraliev, A., Tursunalieva, D., Orozbekov, A., & Bigali, A. (2024). Blockchain and sustainable finance: Enhancing transparency and efficiency in green investments. *Scientific Bulletin of Mukachevo State University Series "Economics," 11*(3), 125–137. https://doi.org/10.52566/msu-econ3.2024.125

Kirby, D. A., & Healey-Benson, F. (2024). Harmonious entrepreneurship: A sustainable model for addressing global inequality. *Journal of Entrepreneurship in Emerging Economies*. https://doi.org/10.1108/jeee-12-2023-0497

Ključnikov, A., Civelek, M., Kupec, V., & Badie, N. B. (2024). Breaking down entrepreneurial barriers: exploring the nexus of entrepreneurial behavior, innovation, and bank credit access through the lens of the theory of planned behavior. *International Entrepreneurship and Management Journal, 21*(1). https://doi.org/10.1007/s11365-024-01025-w

Koberg, E., & Longoni, A. (2019). A systematic review of sustainable supply chain management in global supply chains. *Journal of Cleaner Production, 207*(1), 1084–1098. https://doi.org/10.1016/j.jclepro.2018.10.033

Kurznack, L., Schoenmaker, D., & Schramade, W. (2021). A model of long-term value creation. *Journal of Sustainable Finance & Investment*, 1–19.

Lagoarde-Segot, T. (2020). Financing the sustainable development goals. *Sustainability, 12*(7), 2775. https://doi.org/10.3390/su12072775

Lee, B. H., Georgallis, P., & Struben, J. (2021). Sustainable entrepreneurship under market uncertainty: Opportunities, challenges, and impact. *SSRN Electronic Journal*. https://doi.org/10.2139/ssrn.3800795

Lin, L. (2021). Venture capital in the rise of sustainable investment. *SSRN Electronic Journal*. https://doi.org/10.2139/ssrn.3880285

Magdalena, R., Mohammed, K. S., Nassani, A. A., & Nicoleta Dascalu. (2025). Evaluating the environmental effects of bitcoin mining on energy and water use in the context of energy transition. *Scientific Reports, 15*(1). https://doi.org/10.1038/s41598-025-92314-z

Makki, A. A., Alidrisi, H., Iqbal, A., & Al-Sasi, B. O. (2020). Barriers to green entrepreneurship: An ISM-based investigation. *Journal of Risk and Financial Management, 13*(11), 249. https://doi.org/10.3390/jrfm13110249

Maltais, A., & Nykvist, B. (2020). Understanding the role of green bonds in advancing sustainability. *Journal of Sustainable Finance & Investment*. https://doi.org/10.1080/20430795.2020.1724864

Mansouri, S., & Momtaz, P. (2021). Financing sustainable entrepreneurship: ESG measurement, valuation, and performance in token offerings. *ERPN: Labor Economics (Topic)*. https://doi.org/10.2139/ssrn.3844259

Mansouri, S., & Momtaz, P. P. (2022). Financing sustainable entrepreneurship: ESG measurement, valuation, and performance. *Journal of Business Venturing, 37*(6), Article 106258. https://doi.org/10.1016/j.jbusvent.2022.106258

Mbetwa, S., & Silumbe, R. (2016). Entrepreneurship and business management—Exploring linkages for sustainable development. *International Journal of Scientific & Technology Research, 5*, 183–188.

Mehrabi, R. (2023). Systematic literature review of sustainable finance: Implications for entrepreneurial finance through venture capitals (VC). *Economic Analysis, 56*(2), 68–83. https://doi.org/10.28934/ea.23.56.2.pp68-83

Moolkham, M. (2024). The impact of sustainable development on the relevance of accounting information and financial activities: Evidence from Thailand. *PLoS ONE, 19*(11), e0313299–e0313299. https://doi.org/10.1371/journal.pone.0313299

Morseletto, P. (2020). Targets for a circular economy. *Resources, Conservation and Recycling, 153*, Article 104553. https://doi.org/10.1016/j.resconrec.2019.104553

Motylska-Kuzma, A. (2018). Crowdfunding and sustainable development. *Sustainability, 10*(12), 4650. https://doi.org/10.3390/su10124650

Motiei, M., Khani, A. M., & Beyrami, S. (2021). The effect of green supply chain and green human resource management on environmental performance: The mediating role of green innovation. *Logistics Thought, 20*(77), 165–197. https://doi.org/10.22034/lot.2021.96691

Mousa, R., & Ozili, P. K. (2022). Reimagining financial inclusion in the post COVID-19 world: The case of

Grameen America. *International Journal of Ethics and Systems*. https://doi.org/10.1108/ijoes-12-2021-0230
Mukherjee, A., Owen, R., Scott, J. M., & Lyon, F. (2024). Financing green innovation startups: A systematic literature review on early-stage SME funding. *Venture Capital*. https://doi.org/10.1080/13691066.2024.2410730
Neumann, T. (2022). Impact of green entrepreneurship on sustainable development: An ex-post empirical analysis. *Journal of Cleaner Production, 377*, Article 134317. https://doi.org/10.1016/j.jclepro.2022.134317
Nica, I., Chiriță, N., & Georgescu, I. (2025). Triple bottom line in sustainable development: A comprehensive bibliometric analysis. *Sustainability, 17*(5), 1932. https://doi.org/10.3390/su17051932
Nippani, S., Muldoon, J., Yonai, D., & Matricano, D. (2024). Small-business owners and their choice of banks: Some lessons from the banking literature and game theory in the aftermath of the COVID-19 pandemic. *Journal of Small Business Management*. https://doi.org/10.1080/00472778.2024.2399191
Nogueira, E., Gomes, S., & Lopes, J. M. (2025). Unveiling triple bottom line's influence on business performance. *Discover Sustainability, 6*(1). https://doi.org/10.1007/s43621-025-00804-x
Nunes, A. K. da S., Morioka, S. N., & Bolis, I. (2022). Challenges of business models for sustainability in startups. *RAUSP Management Journal, 57*(4), 382–400. https://doi.org/10.1108/rausp-10-2021-0216
Peng, H., & Walid, L. (2022). The effects of entrepreneurs' perceived risks and perceived barriers on sustainable entrepreneurship in Algeria's SMEs: The mediating role of government support. *Sustainability, 14*(17), 11067. https://doi.org/10.3390/su141711067
Popescu, I.-S., Schaubroeck, T., Gibon, T., Petucco, C., & Benetto, E. (2024). Investment funds are responsible for substantial environmental and social impacts. *Communications Earth & Environment, 5*(1). https://doi.org/10.1038/s43247-024-01479-4
Pricopoaia, O., Lupașc, A., & Mihai, I. O. (2024). Implications of innovative strategies for sustainable entrepreneurship—Solutions to combat climate change. *Sustainability, 16*(22), 9742–9742. https://doi.org/10.3390/su16229742
Rink, S. (2024). Sustainable small business lending. *Sustainable Futures, 8*, Article 100292. https://doi.org/10.1016/j.sftr.2024.100292
Rosário, A. T., Raimundos, R. J., & Cruz, S. P. (2022). Sustainable entrepreneurship: A literature review. *Sustainability, 14*(9), 5556. https://doi.org/10.3390/su14095556
Ruhnka, J. C., & Young, J. E. (1991). Some hypotheses about risk in venture capital investing. *Journal of Business Venturing, 6*(2), 115–133. https://doi.org/10.1016/0883-9026(91)90014-5
Salimi, N., & Vrauwdeunt, T. (2024). Sustainability: Does it contribute to the survival of entrepreneurship? *Journal of Environmental Management, 373*, 123454–123454. https://doi.org/10.1016/j.jenvman.2024.123454
Sangwan, N. (2023). Make microfinance great again: Can flexibility in repayments improve business outcomes in North India? *The Indian Economic Journal*. https://doi.org/10.1177/00194662231199232
Sarfo, C., Zhang, J. A., O'Kane, C., & O'Kane, P. (2024). Perceived value of microfinance and SME performance: The role of exploratory innovation. *International Journal of Innovation Studies, 8*(2). https://doi.org/10.1016/j.ijis.2024.02.003
Scartozzi, G., Delladio, S., Rosati, F., Nikiforou, A. I., & Caputo, A. (2024). The social and environmental impact of entrepreneurship: A review and future research agenda. *Review of Managerial Science*. https://doi.org/10.1007/s11846-024-00783-9
Schaltegger, S. (2013). Sustainable entrepreneurship. In *Encyclopedia of corporate social responsibility* (pp. 2458–2462). https://doi.org/10.1007/978-3-642-28036-8_742
Serio, R. G., Dickson, M. M., Giuliani, D., & Espa, G. (2020). Green production as a factor of survival for innovative startups. Evidence from Italy. *ArXiv (Cornell University)*. https://doi.org/10.48550/arxiv.2005.12102
Shahid, M. S., Hossain, M., Shahid, S., & Anwar, T. (2023). Frugal innovation as a source of sustainable entrepreneurship to tackle social and environmental challenges. *Journal of Cleaner Production, 406*, Article 137050. https://doi.org/10.1016/j.jclepro.2023.137050
Shuwaikh, F., Tanguy, A., Dubocage, E., & Alolah, O. (2024). Insights for sustainable business practices: Comparative impact of independent and corporate venture capital funding on financial and environmental performance. *Research in International Business and Finance, 73*, 102632. https://doi.org/10.1016/j.ribaf.2024.102632
Sîrb, S. E. (2017). *Barriers and incentives to green entrepreneurship in transition economies—Case study of SMEs in Romania.*
Soluk, J., Kammerlander, N., & Darwin, S. (2021). Digital entrepreneurship in developing countries: The role of institutional voids. *Technological Forecasting and Social Change, 170*, Article 120876. https://doi.org/10.1016/j.techfore.2021.120876
Sultan, M. (2025). Innovation, and green entrepreneurship are catalysts for sustainability in Oman. *Cleaner Waste Systems*. https://doi.org/10.1016/j.clwas.2025.100281
Sunarni, S. (2023). The effect of environmental sustainability factors, funding models, and risk management on social entrepreneurship success: A case study on CV. Pangalusna Sukabumi. *West Science Journal Economic and Entrepreneurship, 1*(03), 120–128. https://doi.org/10.58812/wsjee.v1i03.408
Taghavifard, M. T., Khani, A. M., & Beyrami, S. (2022). The social responsibility role of corporate in accepting green supply chain management with respect to the mediating role of big data analysis. *Logistics Thought, 20*(79), 69–106. https://doi.org/10.22034/lot.2022.210582.1149
Taghavifard, M. T., Khani, A. M., & Beyrami, S. (2023). Green HR model design in small and medium indus-

tries using Interpretive Structural Modeling (ISM). *Management Studies in Development and Evolution, 32*(107), 75–108. https://doi.org/10.22054/jmsd.2022.66417.4111

Tenner, I., & Hörisch, J. (2021). Crowdfunding sustainable entrepreneurship: What are the characteristics of crowdfunding investors? *Journal of Cleaner Production, 290*, Article 125667. https://doi.org/10.1016/j.jclepro.2020.125667

Valencia-Payan, C., Griol, D., & Corrales, J. C. (2024). Blockchain self-update smart contract for supply chain traceability with data validation. *Logic Journal of IGPL*. https://doi.org/10.1093/jigpal/jzae047

Valerievna, T. O., & Igorevna, G. I. (2024). Sustainable financing as a factor influencing eco-friendly business transformation: Existing challenges and potential solutions. *The American Journal of Interdisciplinary Innovations and Research, 6*(10), 8–20. https://doi.org/10.37547/tajiir/volume06issue10-02

Wagenhofer, A. (2023). Sustainability reporting: A financial reporting perspective. *Accounting in Europe, 21*(1), 1–13. https://doi.org/10.1080/17449480.2023.2218398

Watson, R., Nielsen, K. R., Wilson, H. N., Macdonald, E. K., Mera, C., & Reisch, L. (2023). Policy for sustainable entrepreneurship: A crowdsourced framework. *Journal of Cleaner Production, 383*, Article 135234. https://doi.org/10.1016/j.jclepro.2022.135234

Wöhler, J., & Haase, E. (2022). Exploring investment processes between traditional venture capital investors and sustainable startups. *Journal of Cleaner Production, 377*, Article 134318. https://doi.org/10.1016/j.jclepro.2022.134318

Worakantak, J., Newbery, R., & Kimmitt, J. (2024). Entrepreneurial finance and institutional logics in an emerging economy. *International Small Business Journal*. https://doi.org/10.1177/02662426241240136

Yadav, N., Luthra, S., & Garg, D. (2023). Blockchain technology for sustainable supply chains: A network cluster analysis and future research propositions. *Environmental Science and Pollution Research*. https://doi.org/10.1007/s11356-023-27049-3

Yang, N., Dousios, D., Korfiatis, N., & Chalvatzis, K. (2024). Mapping the signaling environment between sustainability-focused entrepreneurship and investment inputs: A topic modeling approach. *Business Strategy and the Environment, 33*(6), 5405–5422. https://doi.org/10.1002/bse.3748

Yiadom, E. B., Indome, P. N., Mawutor, J. K. M., & Domfe, G. (2025). Balancing finance and sustainability: The impact of financial access on carbon emissions through innovation and entrepreneurship in a global study. *International Review of Economics and Finance, 98*, Article 103878. https://doi.org/10.1016/j.iref.2025.103878

Yunus, M., Moingeon, B., & Lehmann-Ortega, L. (2010). Building social business models: Lessons from the Grameen experience. *Long Range Planning, 43*(2–3), 308–325. https://doi.org/10.1016/j.lrp.2009.12.005

Yuwono, T., Suroso, A., & Novandari, W. (2024). Information and communication technology in SMEs: a systematic literature review. *Journal of Innovation and Entrepreneurship, 13*(1). https://doi.org/10.1186/s13731-024-00392-6

Zamfirache, A., Suciu, T., Anton, C., Albu, R., & Ivasciuc, I. (2023). The interest shown by potential young entrepreneurs in Romania regarding feasible funding sources, in the context of a sustainable entrepreneurial education. *Sustainability*. https://doi.org/10.3390/su15064823

Zhao, L., & Ryu, S. (2020). Reward-based crowdfunding research and practice. *Advances in Crowdfunding*, 119–143. https://doi.org/10.1007/978-3-030-46309-0_6

Women and Sustainable Entrepreneurship

Nayereh Eftekhar

Abstract

The core theme of this chapter is the critical contribution of women to sustainable entrepreneurship, leading to economic, social, and environmental sustainability. It begins with the fact that women disproportionately suffer from climate change, poverty, and resource shortages, and are underrepresented in the mainstream business and decision-making systems. While women are often affected by dominant systems, they can also emerge as key drivers of change, using sustainable entrepreneurship to generate impactful and lasting solutions. The chapter demonstrates that female entrepreneurs can thrive by incorporating long-term vision, compassion, and value-driven leadership into their business practices. The Kenyan, Nigerian, and Indonesian case studies demonstrate how women's businesses respond to pressing global concerns, such as food security, waste management, and the environment. Although female entrepreneurs are economically competent, they face numerous daunting challenges. These problems include not having adequate access to finance, being left out of valuable networks, gender discrimination, and difficulty in balancing life and work. The chapter offers solutions to these problems, such as gender-sensitive money policies, mentoring programs, and conducive business cultures. Finally, the chapter concludes that supporting women in sustainable entrepreneurship is not merely an issue of gender equality but also a step towards creating a better, stronger, and inclusive future.

Learning Objectives

At the end of this chapter, readers will be able to:

- Understand the role of women in sustainable entrepreneurship.
- Analyze the impact of women-led sustainability entrepreneurship.
- Recognize barriers and solutions for women in sustainable entrepreneurship.
- Develop strategies for promoting gender-inclusivity in sustainable entrepreneurship.
- Analyze how empowering women contributes strategically to sustainable development.

1 Introduction

In Davos 2025, held in January, Anna Bjerde, Managing Director of Operations at the World Bank, stated during the "Adding Trillions with Gender Parity" session: "If you had women participating as actively as men in the economy, the GDP of the world would be 20% higher." Globally,

N. Eftekhar (✉)
Department of Entrepreneurship Development, Faculty of Entrepreneurship, University of Tehran, Tehran, Iran
e-mail: neftekhar@ut.ac.ir

M. Akbari et al. (eds.), *Sustainable Entrepreneurship*, University of Tehran Science and Humanities Series, https://doi.org/10.1007/978-3-032-04909-4_17

women disproportionately bear the consequences of social and environmental hardships, often in overlooked ways (Matharu & Juneja, 2021). For instance, climate change significantly impacts agriculturalists, especially women, who comprise nearly half of the global agricultural workforce (FAO, 2023). However, many women lack equitable access to agricultural assets, credit, and land (Agarwal, 2010). During droughts, floods, and extreme weather events, women face heightened challenges in recovery due to their disadvantaged positions. Beyond tending to farms, they often shoulder the additional responsibility of providing for their families. As a result, climate-related agricultural failures disproportionately devastate them. Without access to adaptive resources such as improved irrigation or climate-resilient seeds, women are trapped in deepening poverty (Pörtner et al., 2022).

Women are often more vulnerable to the challenges posed by inadequate access to clean water. In many rural settings, it is women and girls who are responsible for hauling water several miles for domestic use, and this role is made even more arduous by climate change (UNICEF, 2016). Climate change is exacerbating prolonged droughts and the contamination of water supplies, ultimately increasing the burden of acquiring potable water, often requiring girls to drop out of school and restricting women's ability to work and participate in the community (Sorenson et al., 2011). Poor sanitation infrastructure also aggravates these situations across all stages of life. It particularly makes life for females more challenging during the menstrual and childbirth phases, which contribute to health conditions and health inequities (Sommer et al., 2016). These challenges around other environmental issues are also offset by the inequities that women face in social and economic contexts. For example, natural disasters have contributed to many more deaths of females than males, and the deaths occur not naturally due to biological factors but through social inequity and socially constructed gender roles (Neumayer & Plümper, 2007). Often, women do not have early warnings, the ability to evacuate, or the economic means to rebuild their lives after disasters. Women are also more likely to work in lower-wage informal economic jobs that offer little in the way of job security, access to healthcare, or savings plans, which additionally place them in precarious situations when disasters occur (ILO, 2020).

However, women are not merely victims of unsustainable practices—they are also powerful agents of change, capable of driving impact through sustainable entrepreneurship. Sustainable entrepreneurship integrates environmental, social, and economic considerations into business practices to create value for communities and address global challenges (Rosário et al., 2022). Thanks to their unique perspective on fostering community-focused solutions, women across the globe are increasingly taking leadership roles in this field (Dhatt et al., 2017). In rural Kenya, women entrepreneurs have launched solar-powered, community-driven initiatives to deliver affordable clean energy, significantly decreasing reliance on fossil fuels (Alstone et al., 2015). These initiatives address environmental issues while creating financial opportunities for women, breaking cycles of inequality and poverty.

The rise of women in sustainable entrepreneurship highlights their adaptability and ingenuity (Ertac & Tanova, 2020). In India, women-led self-help groups are engaging in climate-resilient agricultural practices, such as vermicomposting and water harvesting, which enhance soil fertility, increase yields, and empower women economically and socially (Baraskar et al., 2018; Basak & Roy Chowdhury, 2024). Similarly, in Latin America, women entrepreneurs are transforming waste management. In Colombia, the firm Ingerecuperar, led by women, focuses on recycling waste into useful materials, simultaneously addressing environmental and housing challenges (Aidis, 2012). These examples demonstrate that women respond to environmental challenges and pioneer innovative, eco-friendly solutions that benefit their communities and the planet (Mansour et al., 2024).

Female entrepreneurship's impact extends beyond ecological benefits (Bullough et al., 2015); it also drives social transformation (Jeong & Yoo, 2022). Women in leadership positions within sustainable ventures challenge traditional gender norms, serving as role models who encourage the next generation of girls to pursue careers in STEM fields (Bannò et al., 2023). Female-owned businesses often prioritize social integration by creating employment for underserved communities and reinvesting profits into community development (Terjesen et al., 2016). Together, these initiatives underscore the transformational potential of female entrepreneurship in addressing ecological and social challenges.

2 Why This Subject Matters

Emphasizing women's perspectives and contributions to sustainable entrepreneurship and environmental resilience is vital (Raman et al., 2022). Despite their highly recognized contributions to natural resource management, household sustenance, and civic well-being, women's exclusion from mainstream business discourses and decision-making processes has been perpetuated (Milazzo & Goldstein, 2019). Such exclusion has perpetuated gender inequalities and restricted the development of holistic solutions to global challenges (Santos Silva & Klasen, 2021). The lived experiences of women, shaped by environmental and social burdens, are often overlooked by conventional approaches that fail to acknowledge their perspectives (Moreno & Shaw, 2018). It ensures that the most vulnerable are addressed by making solutions truly inclusive and effective by including women's voices in sustainability.

The reason why women's inclusion is imperative is that they are often on the front line in most environmental crises around the world (Datta et al., 2024). Many women fetch water, firewood, and food across the world. Such roles place them on the front line to observe changes in climate, deforestation, and water scarcity (Hlahla, 2022). Yet, excluding them from access to resources, education, and a platform for contribution to solutions represents a significant lost opportunity. Studies have demonstrated that the greater the inclusion of women in environmental decision-making and entrepreneurship, the more the outcomes tend to be equitable, sustainable, and beneficial to the entire community (Foster, 2017). Women leaders' focus is on long-term resilience rather than short-term gains; this, in turn, creates inclusiveness and sustainability.

In this way, increasing women's voices in sustainable entrepreneurship challenges entrenched gender roles and eventually leads to societal change (Raman et al., 2022). Women's work holds a model regarding how sustainability could be embedded with social impact avenues to be provided to marginalized people for economic empowerment (Barrachina Fernández et al., 2021). Recognizing and supporting women as agents of change is not about equality; rather, it's a strategic imperative (Markham, 2013). This will enable us to build a resilient, equitable, and sustainable future for all through an inclusive approach (Thomas, 2024).

3 Why Women Are Appropriate Leaders in Sustainable Entrepreneurship

There is increased recognition of the leadership roles played by women as a force that drives forward innovative, inclusive, and effective solutions to social and environmental issues. Sustainability indeed has to do with everyone's affairs, yet specific qualities in a woman are peculiarly matched with this undertaking. Research shows that companies and initiatives with higher gender diversity are more likely to innovate and successfully engage in long-term solutions around sustainability challenges (Glass et al., 2016). Many leaders who rise to leadership help propagate a governing organization and village values, attributed to women's leadership, based on cohesive forms, empathy, community-supported development, and intergenerational equity—key pillars and tenets of sustainable practices themselves (Eagly & Carli, 2007). This has equipped us to not only make better decisions but also formulate solutions that are equitable and inclusive.

Furthermore, women's societal roles and cultural influences play a critical part in shaping their leadership in sustainability. Historically, women have navigated systemic barriers and marginalization, developing resourcefulness and adaptability that are crucial for solving complex environmental and social problems (Agarwal, 2010). Their lived experiences as caregivers, resource managers, and community builders often provide unique insights into the impacts of environmental degradation and the needs of vulnerable populations (Leach et al., 2010). This perspective equips women with the ability to identify systemic flaws and develop innovative, grassroots-level solutions that can drive transformational change.

As the world gears toward escalated efforts to combat climate change for sustainable development, understanding the special motives and capabilities women bring to leadership on sustainability issues is quite instrumental. This calls for intrinsic motivation, including empathy and long-term vision, to address societal and cultural factors that shape the problem-solving capability. Recognizing and fostering women's leadership in sustainability is not only an ethical necessity but also a strategic necessity for building a more resilient and equitable world (UN Woman, 2020).

3.1 Intrinsic Motivations: Empathy, Vision, and Values

3.1.1 Capacity of Women for Empathy

Much research asserts the capacity of women for empathy (Pang et al., 2023) and their strong community ties as key drivers for engaging in sustainable initiatives (Barrachina Fernández et al., 2021). Different studies indicate that women prioritize social and environmental well-being and are more sensitive to community deficiencies and the influences of environmental challenges (Glass et al., 2016). The emphasis on relationality and interconnectedness, explored in Gergen's work on relational being (Gergen, 2011), suggests that women's understanding of their place within the community and the environment might inherently promote a more sustainable approach. This is further supported by the fact that women's leadership is often collaborative, building, and strongly relational in process, a feature noted by a number of researchers (Kezar & Wheaton, 2017; Selzer et al., 2017). This makes it inclusive and, hence, more effective at problem-solving (Oliver et al., 2022). Again, of interest is community well-being—a feature of the growing literature on social entrepreneurship—where women's ventures often include positive social impact alongside financial imperatives (Haugh & Talwar, 2016; Kimbu & Ngoasong, 2016).

3.1.2 Women's Long-Term View

The literature consistently highlights a longer-term perspective and a focus on intergenerational equity as key motivators for women in sustainability (Spijkers, 2018). Unlike traditional business models that often prioritize short-term profits, women frequently demonstrate a greater concern for the long-term consequences of actions on both the environment and society (Pierli et al., 2022). Their long-term concerns are intrinsically linked to their historical roles as caretakers and their inherent concern for the well-being of their families and communities across generations (Kaplan, 1994; Mussida & Patimo, 2021). The concept of intergenerational equity, central to sustainable development (Spijkers, 2018), resonates deeply with women's inherent concern for the future. Robinson's work on microfinance (Robinson, 2001) underlines further that when women gain economic empowerment, the long-term effects flow down through improved health and education in children and communities. This long-term focus seems at cross purposes with the often short-sighted nature of many more traditional business models. Furthermore, the focus on values-driven social entrepreneurship by Haskell et al. (2009) underlines the fact that for many women, sustainability is not just a business opportunity but an integral part of their personal and professional ethos.

3.1.3 Value Driven Entrepreneurship

A substantial body of research indicates that many women entrepreneurs in the sustainability sector are driven by deeply held values such as compassion, collaboration, and a sense of purpose (Haskell et al., 2009). Their ventures are often guided by a desire to create positive social and environmental impact, exceeding the mere pursuit of financial gain (Scartozzi et al., 2024). This values-driven approach is a key differentiator in their entrepreneurial endeavors, aligning their personal ethics with their business goals (Eagly & Carli, 2007). This contrasts with the often more narrowly defined goals of profit maximization found in traditional business models (Grossman & Stiglitz, 1977). This emphasis on intrinsic motivation, explored in the context of spirituality and business by Bouckaert and Zsolnai (2012), suggests that women's entry into sustainability could be deeply placed in a larger sense of purpose and social responsibility. The notion of the "triple bottom line," identified by Robinson (2001), which encompasses social, environmental, and economic aspects, is often at the core of their businesses, reflecting an expanded view of success beyond that which is realized through pure economic return.

3.2 Societal and Cultural Influences

3.2.1 Resourcefulness and Adaptability

Historically, women have had limited access to resources and opportunities (Collins, 1986). This has led to a very high degree of resourcefulness and flexibility, providing them with problem-solving skills and resilience that are particularly relevant in the face of complex sustainability issues (Leach et al., 2010). When facing systemic barriers with limited resources, new solutions to environmental and social challenges are sometimes required (Agarwal, 2010). Not only is this resilience not an individual characteristic, but it is also a result of social structures that have superficially limited women's access to power and resources (Robinson, 2001). Additionally, research on women entrepreneurs in developing countries highlights how they frequently devise ingenious strategies to overcome financial and logistical challenges, demonstrating a remarkable capacity for innovation in resource-limited contexts (UN Woman, 2020). This adaptability is critical for tackling the unpredictable and multifaceted problems associated with sustainability.

3.2.2 Marginalized Voices: A Critical Perspective

The process of marginalization itself has fostered a critical lens on systems that perpetuate inequity and environmental degradation (Abate, 2019). Women from marginalized communities often endure the direct consequences of unsustainable practices and social injustices (Matharu & Juneja, 2021). Their lived experiences provide them with unique insights into the intersections of social and environmental challenges, enabling them to frame sustainability solutions with a deep understanding of these interconnections (Leach et al., 2010). The "outsider within" perspective (Collins, 1986) provides evidence that occupying a marginal position in dominant systems enables the identification of systemic flaws and the creation of solutions that target the root causes of inequality and environmental harm, rather than addressing symptoms alone (ADPC, 2021). Furthermore, literature on social innovation emphasizes the critical role of diverse perspectives in crafting effective solutions, underlining the importance of incorporating the experiences and insights of marginalized communities (Mulgan, 2023).

3.3 The Interaction Between Intrinsic and Societal Factors

Finally, it is about investigating and scrutinizing the intersections of the intrinsic motivations we presented earlier, which are not only intrinsic values or life experiences but rather the interplay between the two intrinsic motivations, which matters to women's leadership in sustainability. A woman's increased tendency toward empathy may have originated from her own lived experience of marginalization (Collins, 1986), which enables her to develop particularly innovative and community-driven solutions. In addition, her long-term perspective (Spijkers, 2018) is typically drawn from the historical challenges she has experienced, as it propels her to create systems that are sustainable and promote intergenerational equity (Eagly & Carli, 2007). A strong value orientation, combined with ingenuity and adaptiveness (Haskell et al., 2009), which has been guided and shaped in the context of her experiences with systemic barriers (Setyaningrum et al., 2022), provides women with a very valuable capacity to lead in the area of sustainability. Understanding this connection is critical to building and enabling an environment that supports and empowers women as sustainability leaders.

4 Successful Cases in Female Sustainable Entrepreneurship

There are different sustainable female entrepreneurs around the world who are trying to make a difference in their environment and the world, considering different stakeholders currently or in the future. In this section, a few cases are introduced.

4.1 Joyce Kamande

Joyce Kamande, a Kenyan entrepreneur, introduces herself on LinkedIn as follows:

> A Mother|CO-Founder|Agribusiness|Rice Farmer| Leadership development|Public Narrative| Circular Economy advocate Biochar Producer| Youth empowerment|Climate Change| Sustainability|Biochar Storyteller

She is a dedicated social entrepreneur passionate about enhancing food security and improving the livelihoods of small-scale farmers throughout Africa. She is the co-founder and COO of Safi Organics in Kenya, exemplifying female leadership in waste entrepreneurship. Safi partners with over 3500 smallholder farms led by women to transform agricultural waste into organic fertilizers. Their flagship product, biochar, captures carbon and improves soil quality. Safi's fertilizer has led to a 30% increase in crop yields and a 50% rise in income for over 12,000 farmers, while simultaneously recycling 60,000 tons of waste and restoring 7000 acres of farmland (Khalid, 2023; Walsh, 2024).

4.2 Dupe Killa-Kafidipe

Dupe Killa-Kafidipe is the founder, owner, and chief executive officer of Platinum Fisheries, Nigeria, a company dedicated to promoting a healthy seafood industry through sustainable, responsible, and productive farming practices. With extensive C-Suite experience, she is an exemplary, award-winning business professional focused on business consulting, change management, and process re-engineering. Dupe has also demonstrated academic excellence as an exceptional graduate of Harvard Business School, IESE New York, the China–Europe International Business School, and INSEAD. In addition to her business accomplishments, Dupe plays a role in humanitarian efforts, working on strategic planning for a project that supports children affected by the conflict in Nigeria's northeast. The goal is to have a lasting impact on both business and society through influential leadership and transformational projects (Clarkon, 2023; Lionesses of Africa, 2023).

The establishment of Platinum Fisheries was inspired by personal experiences combined with external circumstances. Coming from an area reliant on fishing, she witnessed the damage overfishing had on fragile aquatic ecosystems; at the same time, she witnessed the increasing unsustainability of Nigeria's reliance on imports of frozen seafood; the constant rise in costs of foreign-exchange value caused prices to exceed the reach of ordinary citizens. This economic stress affected health and education, worsening the difficulty of vulnerable communities. Having been hungry as a youth, she became determined that no child would ever go through what she had been through. Realizing a need for action, she put herself forward to take action with her passion, thought process, knowledge, experience, and expertise to develop new solutions to combat these global challenges (Changtoek, 2024; Clarkon, 2023; Lionesses of Africa, 2023).

4.3 Astrid Paramita

A few years ago, Astrid and her husband founded Food Cycle. It was both an intervention into the economic and social context in Indonesia. Her food recovery program, A Blessing to Share, developed a practical intervention in this country's quickly growing food waste. According to the Economist Intelligence Unit, Indonesia became the second most food-wasteful country in the world, a staggering fact that motivated Astrid to take action (Leba, 2020).

But Astrid realized the extent of food waste, especially at weddings, and visualized how she could make a difference. Wedding celebrations in Indonesia often involve a significant waste of food, much of which ends up in the trash or landfills to decompose naturally. To tackle this problem, she has collaborated with Bride Story, a wedding vendor online platform, by motivating couples to donate excess food instead of just wasting it. A Blessing to Share distributes surplus food to orphanages and food banks in Jakarta through such partnerships (Ahluwalia, 2018; Leba, 2020).

In Astrid's Words, she believes that their program creates a win–win situation. Not only are they helping underprivileged communities, but they are also offering couples a sustainable way to manage their event surplus. In essence, they provide a structured food management solution that works for all the parties involved (Ahluwalia, 2018).

On her LinkedIn profile, Astrid introduces herself as a stay-at-home mother, entrepreneur, and food banker—titles that reflect her deep commitment to both family and social impact.

The stories of Joyce Kamande, Dupe Killa-Kafidipe, and Astrid Paramita illuminate the potential of women entrepreneurs in addressing pressing global challenges. These women epitomize passion, innovation, and commitment to sustainability within their fields of specialization. What they have in common is the ability to take any experience or challenge and turn it into solutions that have an impact on social and environmental concerns. Through the power of diverse perspectives, problem-solving skills, and values-based methodologies, they have reshaped the notion of entrepreneurship as a force for good. These stories will inspire and show how essential women are in ensuring progress on sustainability and inclusion. In the future, much more of the work and voices of such leaders will need to be amplified in a bid to solve global challenges for a more equitable and resilient world.

5 Unique Challenges Women Face in Sustainable Entrepreneurship

Women face multifaceted challenges in sustainable entrepreneurship, focusing on funding gaps, networking and mentorship limitations, gender bias, and work-life balance conflicts. The analysis draws upon various scholarly works exploring women's experiences across diverse geographical contexts and entrepreneurial sectors.

5.1 The Funding Gap: Access to Capital and Investment

A continuous and major challenge facing women entrepreneurs is the funding gap—an imbalance in access to capital and investment relative to men (Ahluwalia, 2018; Bosse & Taylor, 2012; Kuschel et al., 2017). The gap may take a number of forms. Studies of evidence demonstrate that women-led firms receive much less funding than those led by men, particularly early-stage funding (Edelman et al., 2018). For instance, one study revealed that women-led businesses receive less than 5% of early-stage equity investment, highlighting a significant disparity in access to growth capital (Edelman et al., 2018). The funding gap is particularly severe in certain sectors, such as the technology sector, another area where men are overrepresented (Kuschel et al., 2017). Women's underrepresentation in leadership roles in both the venture capital and angel investor communities may influence the funding gap, leading to a bias in their funding decisions (Brush et al., 2004). The Great Recession highlighted the vulnerability of women-led firms to funding difficulties compared to male-led firms during periods of economic contraction. Specifically, women-led firms are more likely to face funding issues than male-led firms when small-business lending contracts (Thebaud & Sharkey, 2016). All this is part of the systemic nature of the funding gap and sensitivity to economic conditions. In addition to a gap in equity funding, women in emerging market economies often deal with biases, norms and collateral forms when trying to access formal financial resources (De Vita et al., 2014; Mashapure et al., 2022). The lack of tailored training and support programs further compounds these difficulties (De Vita et al., 2014).

5.2 Networking and Supporting Relationships

Networking and strong mentorship remain some of the core aspects of successful entrepreneurship. The barriers for women, however, are quite common in these aspects, according to (Ertac & Tanova, 2020; McGowan et al., 2012). Thus, the Lack of access to supportive networks, mentors, and advisors is a great challenge, according to Ertac and Tanova (2020). These assure much-needed access to information, resources, and opportunities, according to Montgomery (2017). However, the predominance of male-dominated networks within many industries may exclude women from these very support systems (Edelman et al., 2018). A qualitative study involving 14 women entrepreneurs in Northern Ireland examined the need for deeper insight into their entrepreneurial journeys, shedding light on the challenges they face in balancing personal lives with the demands of running a business (McGowan et al., 2012). Some women are indeed motivated by factors similar to those of their male counterparts, such as independence and financial gain; however, for many, the move to entrepreneurship was motivated by finding a balance between work and family (McGowan et al., 2012). In this respect, it seems essential for a networking opportunity to be inclusive and truly relevant to women entrepreneurs, given their specific challenges and needs. The absence of role models and mentors who assist women in addressing these challenges might add to female entrepreneurs' frustrations as they seek guidance from people who understand their experiences (Ojediran & Anderson, 2020). In the context of academic medicine, research on women's access to mentoring is contradictory, with some studies affirming sex differences and others negating them (Laine & Turner, 2004). This means further research into the effectiveness and accessibility of the mentoring programs that women in different fields need.

5.3 Gender Bias and Stereotypes: How to Overcome Implicit and Explicit Discrimination

Gender bias and stereotypes are among the major development and growth constraints that women face in business (Ahl & Marlow, 2012; Edelman et al., 2018; Malmström et al., 2017; Ruth Eikhof et al., 2013). It has been indicated that angel investors may unconsciously perceive women-led ventures as less legitimate than male-led ones, even when the actual legitimacy of the ventures is comparable. This is due to perceived stereotypes that link femininity with less legitimacy in entrepreneurial ventures (Edelman et al., 2018). Most media representations of female entrepreneurs serve to reinforce traditional gender roles, constraining women's entrepreneurial aspirations (Ruth Eikhof et al., 2013) and further engendering the stereotype.

Similarly, in equity crowdfunding, signals of quality appear to be more or less effective depending on whether they are issued by male or female entrepreneurs and reflect gender stereotypes in the way those signals are interpreted (Kleinert & Mochkabadi, 2022). The dominant heteronormative assumptions underlying entrepreneurial discourse typically render women inferior or incomplete compared to men (Ahl & Marlow, 2012). This perpetuates existing power imbalances and narrows the epistemological boundaries of studies, which restrains deeper insight into women's experiences (Ahl & Marlow, 2012).

A discursive analysis of governmental venture capitalists' decisions showed how gender stereotypes are socially created and mobilized in decisions on the potential of entrepreneurs and how this results in female entrepreneurs receiving considerably less capital than their male counterparts (Malmström et al., 2017). Language and rhetoric in such decisions reflect deep-seated bias in funding allocation (Malmström et al., 2017). In Pakistan, discriminatory socio-cultural values and traditions are deeply engraved and contribute to unequal opportunities for women entrepreneurs to emerge (Ruth Eikhof et al., 2013).

5.4 Work-Life Balance: Juggling Entrepreneurial Pursuits and Family Responsibilities

Women, in general, face a particularly challenging task in balancing entrepreneurial endeavors with familial and caregiving responsibilities. The reality is that much of the burden associated with launching and operating a business is, at almost every turn, counterbalanced by parenting and domestic concerns (Emon & Nipa, 2024; McGowan et al., 2012). This is applicable to a host of present-day cultures where women are largely encumbered with greater levels of responsibility for domestic and child-rearing obligations, and the issue is compounded by minimal options for childcare that are affordable and available (Emon & Nipa, 2024). Women entrepreneurs need to make painful or challenging choices, often sacrificing their aspirations for their roles as wives and mothers or compromising and limiting their ability to advance in their career paths (Laine & Turner, 2004). In medicine, for example, women make career decisions that affect their earning power, thereby exercising their prerogatives around flexible work arrangements to balance family and career (Laine & Turner, 2004). This work and family imbalance can also inhibit women's access to networks and mentorship, such as less connection to and participation in professional development activities due to time constraints, limiting relationship building. The absence of supportive policy and infrastructure can deepen these difficulties in some situations (Emon & Nipa, 2024).

5.5 Overcoming Barriers: Strategies and Initiatives to Promote Gender Equality

Despite huge challenges, various strategies and initiatives are emerging that break barriers in promoting gender equality within the ecosystem of sustainable entrepreneurship (Edelman et al., 2018; Emon & Nipa, 2024; Mashapure et al., 2022; Roomi & Parrott, 2008). Some policies and programs that will help overcome such challenges include government initiatives, women entrepreneur development associations, reserved quotas, and microcredit programs, thus fostering gender-inclusive entrepreneurship (Emon & Nipa, 2024). These interventions often focus on improving access to funding, providing training and skills development, and fostering networking opportunities (Kuschel et al., 2017; Roomi & Parrott, 2008). Public awareness campaigns targeting gender stereotypes (Edelman et al., 2018; Kleinert & Mochkabadi, 2022; Malmström et al., 2017) are also crucial in creating a more supportive environment.

In the context of academic medicine, interventions such as mentorship programs and transparent compensation practices have shown promise in addressing gender inequity. However, these have to be further researched and tested for efficiency (Laine & Turner, 2004). In Pakistan, coordinated action among media, education policymakers, and government bodies has been recommended to advance women's entrepreneurship by improving service access and building inclusive economic networks (Roomi & Parrott, 2008). Moreover, targeted programs aimed at improving the legitimacy of women and unconscious biases in investment processes are also essential to bridge the funding gap (Edelman et al., 2018). Beyond such specific interventions, changes in the broader society are also necessary to dismantle deep-rooted gender stereotypes and to create a more level and inclusive playing field for entrepreneurship (Ahl & Marlow, 2012; Ruth Eikhof et al., 2013).

6 Lessons and Best Practices for Supporting Women in Sustainable Entrepreneurship

Support for women in sustainable entrepreneurship is an issue pertaining not only to equality between genders but also to the driver for reaching sustainable development. Women bring special perspectives, creativity, and resilience into the entrepreneurial landscape, and their participation in sustainable businesses stands to meet some of the urgent environmental and social challenges. This section underlines the critical lessons learned from and best practices of various approaches adopted toward women's empowerment in the field of sustainable entrepreneurship, ranging from policy recommendations and enabling ecosystems to the role of investors, education, and awareness to holistic strategies on supporting women entrepreneurs for greener and more inclusive growth (Barrachina Fernández et al., 2021; Raman et al., 2022).

6.1 Policy Recommendations

Setting policies supportive of female entrepreneurship is instrumental in achieving equality in gender development for sustainability (Foss et al., 2019). These are among such policy actions to support females to engage in entrepreneurship sustainably, including, among others.

6.1.1 Finance Accessibility

Access to finance remains crucial for starting and scaling their businesses, and thus, this is one issue that most women entrepreneurs must address (Malmström et al., 2017). The requirement now needed involves gender-responsive financial policies by policymakers, which would afford better access to women in terms of funding that would be made available through grants, low-interest loans, and equity investment opportunities in women-led businesses (OECD, 2021).

6.1.2 Mentorship Programs

It is crucial for women entrepreneurs in terms of guidance, support, and even valuable industry connections (Cardella et al., 2020). Mentorship programs can be initiated through the government or an organization that matches established entrepreneurs with aspiring women entrepreneurs (Almheiri et al., 2024). This will help them build confidence and skills to meet challenges arising in entrepreneurship (Stavropoulou & Protopapa, 2013).

6.1.3 Target-Oriented Training

In the area of Entrepreneurship training, target-oriented programs shall be imparted to cater to the needs of entrepreneurship. Training programs shall be imparted on various aspects of Entrepreneurship, such as business planning, marketing, finance management, and sustainable practices (Petridou & Glaveli, 2008). The policymaker may collaborate with educational institutions and industry experts for the development and delivery of such training programs (Bullough et al., 2015; Petridou & Glaveli, 2008).

6.1.4 Inclusive Procurement Policy

Inclusive procurement policies by governments and large corporations will ensure purchase preference for women-owned businesses, hence providing huge market opportunities for women entrepreneurs to sustain their businesses (Basheka Benon, 2018; Orser et al., 2021).

6.2 Building Supportive Ecosystems

Inclusive and supportive ecosystems go a long way in creating opportunities for collaboration and mentorship for women in sustainable entrepreneurship (de Bruin & Swail, 2024; Henry et al., 2017). Some of the most important constitutive aspects of a supportive ecosystem include:

6.2.1 Providing Networking Opportunities

The initiative is rather important in bringing on board avenues where women entrepreneurs engage with their peers, mentors, and experts in their industry. Events on sustainable entrepreneurship, including networking, workshops, and conferences, will bring them into valuable contact and help them learn from the experiences of successful entrepreneurs (Manello et al., 2020; Moletta et al., 2023).

6.2.2 Co-working and Innovation Spaces

Co-working and innovation spaces are other avenues of collaboration that could be afforded to women entrepreneurs through resources, support, and community. Indeed, this would provide an open space for innovation, collaboration, and sharing experiences (Mitev et al., 2019; Sperindé & Nguyen-Duc, 2020).

6.2.3 Supportive Communities

When women's communities are supportive, offering emotional and practical support in various forms (online, social media groups, local associations), they can provide spaces where women can share experiences, seek advice, and celebrate their successes (Meinhart et al., 2022; Schmitz & Gabel, 2023).

6.2.4 Recognition and Awards

Women's achievements should be recognized by the provision of awards and public recognition that would inspire others. Indeed, governments and organizations can create award and recognition programs that recognize the contribution of women in sustainable entrepreneurship (Barrachina Fernández et al., 2021; Raman et al., 2022).

6.3 Role of Investors

Investors also should support women-led sustainable businesses, including impact investors and venture capitalists (Ubfal, 2024). They can do the following:

6.3.1 Gender-Lens Investing

This might include bringing gender into their investment decisions, as in gender-lens investing. In this, for example, investment is advocated for companies that are woman-led and that are supportive of the advancement of gender equality and women's empowerment (Thorarinsdottir et al., 2024).

6.3.2 Investment in Women-Led Businesses

Investors can actively seek out and invest in women-led sustainable businesses. The potential of the investors to bridge the funding gap for women entrepreneurs is the second advantage accruable to investors of such businesses (Ubfal, 2024).

6.3.3 Capacity Building

Besides financial support, investors could extend capacity-building support to women entrepreneurs through mentorship, business development services, and access to networks and markets (Mishra, 2015).

Impact measurement can serve as a tool to demand greater accountability and transparency, demonstrating that investments in women-led sustainable businesses truly make a difference, not only in financial terms, but also as a recognition of women's ability and competence to lead.

6.4 Education and Awareness

Increasing interest in the field of STEM and entrepreneurship among girls and women will help in creating a pipeline for future women entrepreneurs (Kuschel et al., 2020). Some of the key strategies include:

6.4.1 Early Education

Girls need to be encouraged right from the nascent stage in the education sector to pursue subjects related to STEM (Msambwa et al., 2024) in order to improve their problem-solving attitude (Yabas et al., 2022). Schools can introduce curricula and extracurricular activities related to STEM that inspire and engage girls in these subjects (Merayo & Ayuso, 2023).

6.4.2 Role Models

Highlighting successful female entrepreneurs and professionals in sustainability-related ventures can provide meaningful role models for girls and young women, inspiring them to pursue similar paths (Barrachina Fernández et al., 2021). Connecting these women and girls to mentorship programs led by accomplished female leaders from relevant fields can add significant value, offering guidance, encouragement, and support for engaging in sustainability initiatives (Walsh, 2024).

6.4.3 Entrepreneurship Programs

The entrepreneurship programs may be undertaken by institutes of learning, which will in turn equip girls and women with knowledge and skills essential in the entrepreneurial process (Cardella et al., 2020). Training should also touch on business sustainability, including involvement in social responsibility and care for the environment (Yadav & Unni, 2016).

6.4.4 Public Awareness

People can be made aware of the advantages accruing from women's participation in the sustainability field, and entrepreneurship should be highlighted to break stereotypes and pave the way for more to enter these fields (Abdelwahed et al., 2022).

7 Conclusion: The Future Is Female and Will Be Sustainable

This chapter's discussion illustrates the central role played by women in sustainable entrepreneurship in mobilizing their particular competencies toward creating innovative and leading transformational solutions related to environmental and social problems, from their more empathetic style of leading to overcoming the issues of systemic barriers and resource deficiencies, pioneering inclusive, sustainable growth led by them. Entrepreneurs like Joyce Kamande, Dupe Killa-Kafidipe, and Astrid Paramita illustrate how women's ventures can be used to combat some of the pressing global issues, such as food insecurity, environmental degradation, and social inequality, while fueling economic empowerment in marginalized communities.

7.1 Outlook: Women as Catalysts for a Sustainable Future

The world faces growing environmental crises and deep-seated social inequities. In these respects, women bear a unique position of influence in leading the transition toward a sustainable future. Their intrinsic motivations—empathy, long-term vision, and value-driven approaches—combined with their lived experiences, equip them to create innovative and inclusive solutions. Research underscores that organizations with greater gender diversity are more likely to achieve long-term success in sustainability initiatives (Glass et al., 2016). By addressing intergenerational equity and prioritizing community well-being, women entrepreneurs are not just contributing to economic development but are also reshaping societal norms.

7.2 Call to Action: Empower Women to Lead

Addressing systemic barriers to unlocking the full potential of women in sustainable entrepreneurship requires concerted efforts. It is about supportive ecosystems for governments, organizations, and investors, a fair share in access to finance, and the promotion of gender-inclusive policies. Mentorship programs, networking, and STEM-focused education are all so crucial for the pipeline of future women leaders. Society must also change the narrative and create awareness of recognition and appreciation for women's inputs toward sustainability in order to keep the next generations coming.

7.3 Final Inspiring Thought

Malala Yousafzai, a laureate, once said that, "We cannot all succeed when half of us are held back."

Precisely so, on the way to sustainability. It is not only a step toward achieving equity, but it is also a strategic necessity toward the attainment of a resilient, equitable world. We must amplify women's voices and arm them with tools for success if we want a future that will be not just female but also sustainable, inclusive, and transformational.

8 Case Study: Dr. Lamees Hamdan and Shiffa—A Pioneer in Sustainable Beauty in the UAE

8.1 Background and Context

When Lamees Hamdan was pregnant in 2004, she wondered if there were organic beauty products that a pregnant woman could use. It was the beginning of the ideation of Shiffa, an organic skincare product in the UAE.

Dr. Lamees Hamdana, a medical doctor with an Irish education from the Royal College of Surgeons in Ireland, decided to combine science with nature to create beauty products that are both environmentally friendly and free from non-organic ingredients. In 2004, when she was disappointed due to not being able to locate chemical-free, safe, and effective skincare products, she decided to create her own. This led to the formulation of Shiffa, the luxury organic skincare line that brings together potent natural actives and scientific technology with the ideals of sustainability and fair trade. Using the local language, she chose the name Shiffa, which means "healing" in Arabic. The pillars of Shiffa are sustainability, wellness, and organic beauty.

Dr. Hamdan's background in medicine greatly inspired her vision for skincare, along with a strong belief in nature's healing. Her products are made from organic, sustainably sourced materials all over the world, with healing minerals, special flora, and bioactive ingredients that are kind to the skin but not disastrous for the planet.

8.2 Sustainability in the Heart of Shiffa Business

Sustainability is at the heart of Shiffa's business model. Dr. Hamdan ensures that every stage of production is environmentally responsible, from sourcing to packaging and transportation. Some key sustainability initiatives include:

1. Ethical and Sustainable Sourcing
    - Shiffa is committed to sustainability in two ways. First, its ingredients are sourced from sustainable suppliers and organic farms; everything from rose oil to frankincense is organic. Second, Shiffa follows fair trade practices, which ensure the use of small-scale farmers and communities, which makes it committed to sustaining local businesses as well.
2. Eco-Friendly Formulations
    - Shiffa avoids using harmful ingredients to lessen the chance of harm to users and the environment. They entirely avoid products like parabens, sulfates, and other non-eco-friendly products. In addition, Shiffa uses naturally derived ingredients and biodegradable, eco-friendly ingredients to reduce ecosystem harm to the water system.
3. Sustainable Packaging
    - Shiffa exemplifies its commitment to sustainable packaging by using recyclable bottles and biodegradable packaging materials, eliminating the use of plastic. They are looking into and attempting to find zero-waste packaging options as a company to limit the carbon footprint even further.
4. Carbon Footprint Reduction
    - Shiffa's Dr. Hamdan proactively encourages local manufacturing to cut down on emissions from shipping where possible. Shiffa also provides sustainable shipping options and encourages recycling their packaging once used to minimize carbon footprints.
5. Raising awareness of clean beauty products in the Middle East

- Shiffa has played a part in raising awareness about people making eco-conscious consumer decisions in the Middle East. Dr. Hamdan often educates audiences on the value of ethical entrepreneurship, and her company encourages businesses to do the same, even if they are not involved with the beauty industry.

8.3 Questions About the Case Study

1. In what ways does Shiffa guarantee that its ingredients are ethically- and sustainably-sourced?
2. If you were to launch a sustainable skincare brand, what aspects of Shiffa's brand could you adopt?
3. How could the beauty industry in the Middle East incorporate more sustainable practices, like those of Shiffa?
4. What are the biggest challenges sustainable beauty brands (like Shiffa) face when growing into international markets, and how could they overcome them?
5. Do you believe that the awareness of sustainability in skincare is increasing for consumers in the Middle East? What more can be done to educate consumers about ethical beauty? If you were launching a sustainable skincare brand, what lessons could you take from Shiffa's approach?

9 Final Questions

9.1 Comprehension Questions

1. How do women specifically participate in sustainable entrepreneurship?
2. How is sustainable entrepreneurship defined considering environmental, social, and economic elements?
3. What are the main barriers facing women entrepreneurs who want to start and expand sustainable businesses?

9.2 Application Questions

4. According to the chapter's case studies, how would you approach a sustainable business model that puts social and environmental impact at the forefront?
5. Imagine you become an advisor to a woman entrepreneur starting a sustainability-based business. What strategies will overcome funding and networking barriers?
6. How can government and financial institutions formulate more inclusive policies and practices to include women in sustainable entrepreneurship?

9.3 Critical Thinking Questions

7. How does the exclusion of women from mainstream decision-making inhibit global sustainability efforts?
8. Do you believe women-led sustainable businesses offer a competitive advantage compared to traditional business models? Why or why not?

10 Further Reading

1. Suresh, V., SS, S. Enabling and inhibiting factors of sustainability of Farmers' Producers Organisations in India. *Discov Sustain* **5**, 333 (2024). https://doi.org/10.1007/s43621-024-00527-5

This paper provides valuable insights for readers who want to explore the subject of sustainability in agricultural cooperatives further. It offers a comprehensive analysis of the factors that contribute to or hinder the long-term success of Farmer Producer Organizations (FPOs) in India, categorized under economic, social, and environmental dimensions.

2. Aikhuomogbe, S. (2022). Raising Sustainability Awareness at the Grassroots. In: Ogun-

yemi, K., Atanya, O., Burgal, V. (eds) Management and Leadership for a Sustainable Africa, Volume 3. Palgrave Studies in African Leadership. Palgrave Macmillan, Cham. https://doi.org/10.1007/978-3-031-05284-2_5

This book chapter is very useful to readers who want to know more about sustainability education and participation, particularly at the community level. It gives examples like Unilever's sustainability practices that can be replicated to raise awareness about sustainability. It also contains doable tips from employing visual narrative and community learning practices that enhance scholarship and policy-oriented approaches to education for sustainability.

3. Gbajumo-Sheriff, M. (2022). Gender Equity in Nigeria: Sustainability, Diversity, and Inclusion. In: Ogunyemi, K., Atanya, O., Burgal, V. (eds) Management and Leadership for a Sustainable Africa, Volume 1. Palgrave Studies in African Leadership. Palgrave Macmillan, Cham. https://doi.org/10.1007/978-3-031-04911-8_4

This source would be particularly helpful to readers who want to learn more about how gender equity contributes to corporate sustainability models. The source demonstrates how firms, particularly in emerging economies, can implement policies and practices that foster inclusive workplaces where women have an equal opportunity to contribute to and benefit from economic growth.

References

Abate, R. S. (2019). *Climate change and the voiceless: Protecting future generations, wildlife, and natural resources*. Cambridge University Press.

Abdelwahed, N. A. A., Bastian, B. L., & Wood, B. P. (2022). Women, entrepreneurship, and sustainability: The case of Saudi Arabia. *Sustainability, 14*(18).

ADPC. (2021). *Applying a gender lens to climate actions: Why it matters* (Climate Talks Series: CARE for South Asia Project, Issue.

Agarwal, B. (2010). *Gender and green governance: The Political economy of women's presence within and beyond community forestry*. Oxford University Press. https://doi.org/10.1093/acprof:oso/9780199569687.001.0001

Ahl, H., & Marlow, S. (2012). Exploring the dynamics of gender, feminism and entrepreneurship: Advancing debate to escape a dead end? *Organization, 19*(5), 543–562. https://doi.org/10.1177/1350508412448695

Ahluwalia, S. (2018). *New program tackles food waste in Indonesia*. CGTN. https://news.cgtn.com/news/3d3d674d3251444d79457a6333566d54/share_p.html

Aidis, R. (2012). *Innovative women entrepreneurs in Latin America: The case of ingerecuperar in Colombia*.

Almheiri, A., Chopra, A., & Haddad, A. (2024). The importance of mentorship for women entrepreneurs in United Arab Emirates (UAE). In R. Khamis & A. Buallay (Eds.), *AI in business: Opportunities and limitations: Volume 1* (pp. 277–299). Springer Nature Switzerland. https://doi.org/10.1007/978-3-031-48479-7_24

Alstone, P., Gershenson, D., & Kammen, D. (2015). Decentralized energy systems for clean electricity access. *Nature Climate Change, 5*, 305–314. https://doi.org/10.1038/nclimate2512

Bannò, M., Filippi, E., & Trento, S. (2023). Women in top echelon positions and their effects on sustainability: A review, synthesis and future research agenda. *Journal of Management and Governance, 27*(1), 181–251. https://doi.org/10.1007/s10997-021-09604-7

Baraskar, M., Choudhary, S., Wankhede, A., & Jain, S. K. (2018). Impact of vermicompost production in terms of income and employment generation. *Global Journal of Applied Agricultural Research, 8*(2), 111–116.

Barrachina Fernández, M., García-Centeno, M. D., & Calderón Patier, C. (2021). Women sustainable entrepreneurship: Review and research agenda. *Sustainability, 13*(21).

Basak, D., & Roy Chowdhury, I. (2024). Role of self-help groups on socioeconomic development and the achievement of Sustainable Development Goals (SDGs) among rural women in Cooch Behar District, India. *Regional Sustainability, 5*(2), 100140. https://doi.org/10.1016/j.regsus.2024.100140

Basheka Benon, C. (2018). Inclusive public procurement. *Administratio Publica, 26*(1), 86–106. https://doi.org/10.10520/ejc-adminpub-v26-n1-a5

Bosse, D., & Taylor, P. (2012). The second glass ceiling impedes women entrepreneurs. *Journal of Applied Management and Entrepreneurship, 17*, 52–68.

Bouckaert, L., & Zsolnai, L. (2012). Spirituality and business: An interdisciplinary overview. *Society and Economy, 34*(3), 489–514. https://doi.org/10.1556/socec.34.2012.3.8

Brush, C. G., Carter, N. M., Gatewood, E. J., Greene, P. G., & Hart, M. (2004). *Gatekeepers of venture growth: A diana project report on the role and participation of women in the venture capital industry*.

Bullough, A., de Luque, M. S., Abdelzaher, D., & Heim, W. (2015). Developing women leaders through entrepreneurship education and training. *Academy of Management Perspectives, 29*(2), 250–270. https://doi.org/10.5465/amp.2012.0169

Cardella, G. M., Hernández-Sánchez, B. R., & Sánchez-García, J. C. (2020). Women entrepreneurship: A systematic review to outline the boundaries of scientific literature. *Frontiers in Psychology, 11*. https://doi.org/10.3389/fpsyg.2020.01557

Changtoek, P. (2024). *Taking Nigerian aquaculture to the next level*. The Fish Site. Retrieved 19 March 2025 from https://thefishsite.com/articles/dupe-killa-kafidipe-platinum-fisheries-taking-nigerian-aquaculture-to-the-next-level

Clarkon, J. (2023). *Nigeria: Entrepreneur taps into the country's vast demand for fish*. How We Made it in Africa. Retrieved 19 March 2025 from https://www.howwemadeitinafrica.com/nigeria-entrepreneur-taps-into-the-countrys-vast-demand-for-fish/156067/

Collins, P. H. (1986). Learning from the outsider within: The sociological significance of Black feminist thought. *Social Problems, 33*(6), s14–s32.

Datta, R., Chapola, J., Waucaush-Warn, J., Subroto, S., & Hurlbert, M. (2024). Decolonizing meanings climate crisis and land-based adaptions: From Indigenous women's perspectives in Western Canada. *Women's Studies International Forum, 104*, 102913. https://doi.org/10.1016/j.wsif.2024.102913

de Bruin, A., & Swail, J. (2024). Advancing gender inclusivity: moving entrepreneurial ecosystems onto new paths. *International Journal of Gender and Entrepreneurship, ahead-of-print*(ahead-of-print). https://doi.org/10.1108/IJGE-02-2024-0070

De Vita, L., Mari, M., & Poggesi, S. (2014). Women entrepreneurs in and from developing countries: Evidences from the literature. *European Management Journal, 32*(3), 451–460. https://doi.org/10.1016/j.emj.2013.07.009

Dhatt, R., Theobald, S., Buzuzi, S., Ros, B., Vong, S., Muraya, K., Molyneux, S., Hawkins, K., González-Beiras, C., Ronsin, K., Lichtenstein, D., Wilkins, K., Thompson, K., Davis, K., & Jackson, C. (2017). The role of women's leadership and gender equity in leadership and health system strengthening. *Global Health, Epidemiology and Genomics, 2*, e8, Article e8. https://doi.org/10.1017/gheg.2016.22

Eagly, A. H., & Carli, L. L. (2007). *Through the labyrinth: The truth about how women become leaders*. Harvard Business School Press.

Edelman, L. F., Donnelly, R., Manolova, T., & Brush, C. G. (2018). Gender stereotypes in the angel investment process. *International Journal of Gender and Entrepreneurship, 10*(2), 134–157. https://doi.org/10.1108/IJGE-12-2017-0078

Emon, M. H., & Nipa, M. N. (2024). Exploring the gender dimension in entrepreneurship development: A systematic literature review in the context of Bangladesh. *Westcliff International Journal of Applied Research, 8*(1), 34–49.

Ertac, M., & Tanova, C. (2020). Flourishing women through sustainable tourism entrepreneurship. *Sustainability, 12*(14).

FAO. (2023). *The status of women in agrifood systems*.

Foss, L., Henry, C., Ahl, H., & Mikalsen, G. H. (2019). Women's entrepreneurship policy research: A 30-year review of the evidence. *Small Business Economics, 53*(2), 409–429. https://doi.org/10.1007/s11187-018-9993-8

Foster, E. A. (2017). Gender, environmental governmentality, and the discourses of sustainable development. In *Routledge handbook of gender and environment* (pp. 216–228). Routledge.

Gergen, K. J. (2011). *Relational being: Beyond self and community*. Oxford University Press. https://doi.org/10.1093/acprof:osobl/9780199846269.001.0001

Glass, C., Cook, A., & Ingersoll, A. R. (2016). Do women leaders promote sustainability? Analyzing the effect of corporate governance composition on environmental performance. *Business Strategy and the Environment, 25*(7), 495–511. https://doi.org/10.1002/bse.1879

Grossman, S. J., & Stiglitz, J. E. (1977). On value maximization and alternative objectives of the firm. *The Journal of Finance, 32*(2), 389–402.

Haskell, D., Haskell, J., & Kwong, J. (2009). Spiritual resources for change in hard places: A values-driven social entrepreneurship theory of change. In *Complexity science and social entrepreneurship: Adding social value through systems thinking*.

Haugh, H. M., & Talwar, A. (2016). Linking social entrepreneurship and social change: The mediating role of empowerment. *Journal of Business Ethics, 133*(4), 643–658. https://doi.org/10.1007/s10551-014-2449-4

Henry, C., Orser, B., Coleman, S., & Foss, L. (2017). Women's entrepreneurship policy: A 13 nation cross-country comparison. *International Journal of Gender and Entrepreneurship, 9*(3), 206–228. https://doi.org/10.1108/IJGE-07-2017-0036

Hlahla, S. (2022). Gender perspectives of the water, energy, land, and food security nexus in sub-Saharan Africa. *Frontiers in Sustainable Food Systems*. https://doi.org/10.3389/fsufs.2022.719913

ILO. (2020). *World employment and social outlook: Trends 2020* (9789220314081, 9789220314098).

Jeong, E., & Yoo, H. (2022). A systematic literature review of women in social entrepreneurship. *Service Business, 16*(4), 935–970. https://doi.org/10.1007/s11628-022-00512-w

Kaplan, L. D. (1994). Woman as caretaker: An archetype that supports patriarchal militarism. *Hypatia, 9*(2), 123–133. http://www.jstor.org/stable/3810173

Kezar, A., & Wheaton, M. M. (2017). The value of connective leadership: Benefiting from women's approach to leadership while contending with traditional views. *About Campus, 21*(6), 19–26. https://doi.org/10.1002/abc.21274

Khalid, D. (2023). *The Women-Waste-Climate Nexus: Unlocking the potential of womenentrepreneurs to combat the global waste crisis and accelerate the race to net zero*. I. M. I. o. T. M. S. S. o. Management.

Kimbu, A. N., & Ngoasong, M. Z. (2016). Women as vectors of social entrepreneurship. *Annals of Tourism Research, 60*, 63–79. https://doi.org/10.1016/j.annals.2016.06.002

Kleinert, S., & Mochkabadi, K. (2022). Gender stereotypes in equity crowdfunding: The effect of gender bias on the interpretation of quality signals. *The Journal of Technology Transfer, 47*(6), 1640–1661. https://doi.org/10.1007/s10961-021-09892-z

Kuschel, K., Ettl, K., Díaz-García, C., & Alsos, G. A. (2020). Stemming the gender gap in STEM entrepreneurship—Insights into women's entrepreneurship in science, technology, engineering and mathematics. *International Entrepreneurship and Management Journal, 16*(1), 1–15. https://doi.org/10.1007/s11365-020-00642-5

Kuschel, K., Lepeley, M.-T., Espinosa, F., & Gutiérrez, S. (2017). Funding challenges of Latin American women start-up founders in the technology industry. *Cross Cultural & Strategic Management, 24*(2), 310–331. https://doi.org/10.1108/CCSM-03-2016-0072

Laine, C., & Turner, B. J. (2004). Unequal pay for equal work: The gender gap in academic medicine. *Annals of Internal Medicine, 141*(3), 238–240. https://doi.org/10.7326/0003-4819-141-3-200408030-00019

Leach, M., Scoones, I., & Stirling, A. (2010). *Dynamic sustainabilities: Technology, environment, social justice*. Earthscan.

Leba, E. E. (2020). *Distributing food to others*. Kompas. Retrieved 19 March 2025 from https://www.kompas.id/baca/english/2020/12/22/distributing-food-to-others/

Lionesses of Africa. (2023). *Dupe Killa-Kafidipe, an entrepreneur finding solutions to global food insecurity*. Retrieved 19 March 2025 from https://www.lionessesofafrica.com/blog/2023/5/21/dupe-killa-kafidipe-an-entrepreneur-finding-solutions-to-global-food-insecurity

Malmström, M., Johansson, J., & Wincent, J. (2017). Gender stereotypes and venture support decisions: How governmental venture capitalists socially construct entrepreneurs' potential. *Entrepreneurship Theory and Practice, 41*(5), 833–860. https://doi.org/10.1111/etap.12275

Manello, A., Cisi, M., Devicienti, F., & Vannoni, D. (2020). Networking: A business for women. *Small Business Economics, 55*(2), 329–348. https://doi.org/10.1007/s11187-019-00300-3

Mansour, M., Al Zobi, M. t., Altawalbeh, M., Abu Alim, S., Lutfi, A., Marashdeh, Z., Al-Nohood, S., & Al Barrak, T. (2024). Female leadership and environmental innovation: do gender boards make a difference? *Discover Sustainability, 5*(1), 331. https://doi.org/10.1007/s43621-024-00545-3

Markham, S. A. (2013). *Women as agents of change: Having voice in society and influencing policy* (Women's voice, agency, and participation research series, Issue). http://documents.worldbank.org/curated/en/773451468150287912/Women-as-agents-of-change-having-voice-in-society-and-influencing-policy

Mashapure, R., Nyagadza, B., Chikazhe, L., Msipa, N., Ngorora, G. K. P., & Gwiza, A. (2022). Challenges hindering women entrepreneurship sustainability in rural livelihoods: Case of Manicaland province. *Cogent Social Sciences, 8*(1), 2132675. https://doi.org/10.1080/23311886.2022.2132675

Matharu, S. K., & Juneja, D. (2021). Factors impacting resilience of women entrepreneurs in India in the face of COVID-19. *Vision, 28*(4), 448–458. https://doi.org/10.1177/09722629211043299

McGowan, P., Redeker, C. L., Cooper, S. Y., & Greenan, K. (2012). Female entrepreneurship and the management of business and domestic roles: Motivations, expectations and realities. *Entrepreneurship & Regional Development, 24*(1–2), 53–72. https://doi.org/10.1080/08985626.2012.637351

Meinhart, M., Seff, I., Lukow, N., Tothezan, I., Poulton, C., & Stark, L. (2022). Examining the linkage between social support and gender-based violence among women and girls in humanitarian settings: A systematic review of the evidence. *Current Epidemiology Reports, 9*(4), 245–262. https://doi.org/10.1007/s40471-022-00310-y

Merayo, N., & Ayuso, A. (2023). Analysis of barriers, supports and gender gap in the choice of STEM studies in secondary education. *International Journal of Technology and Design Education, 33*(4), 1471–1498. https://doi.org/10.1007/s10798-022-09776-9

Milazzo, A., & Goldstein, M. (2019). Governance and women's economic and political participation: Power inequalities, formal constraints and norms. *The World Bank Research Observer, 34*(1), 34–64. https://doi.org/10.1093/wbro/lky006

Mishra, A. K. (2015). Impact of capacity building on women entrepreneurs—A literature analysis. *International Journal Multidisciplinary Research and Development, 2*, 344–348.

Mitev, N., de Vaujany, F.-X., Laniray, P., Bohas, A., & Fabbri, J. (2019). Co-working spaces, collaborative practices and entrepreneurship. In K. Riemer, S. Schellhammer, & M. Meinert (Eds.), *Collaboration in the digital age: How technology enables individuals, teams and businesses* (pp. 15–43). Springer International Publishing. https://doi.org/10.1007/978-3-319-94487-6_2

Moletta, J., de Carvalho, G. D. G., do Nascimento, R. F., Barboza, B. M. L., Resende, L. M., & Pontes, J. (2023). Business networks of women entrepreneurs: an analysis of the expectation and reality of factors that affect trust in a business network. *Journal of Intelligent Manufacturing, 34*(3), 1021–1036. https://doi.org/10.1007/s10845-021-01834-8

Montgomery, B. L. (2017). Mapping a mentoring roadmap and developing a supportive network for strategic career advancement. *SAGE Open, 7*(2), 2158244017710288. https://doi.org/10.1177/2158244017710288

Moreno, J., & Shaw, D. (2018). Women's empowerment following disaster: A longitudinal study of social change. *Natural Hazards, 92*(1), 205–224. https://doi.org/10.1007/s11069-018-3204-4

Msambwa, M. M., Daniel, K., Lianyu, C., & Antony, F. (2024). A systematic review using feminist perspectives on the factors affecting girls' participation in

STEM subjects. *Science & Education*. https://doi.org/10.1007/s11191-024-00524-0

Mulgan, G. (2023). The social economy and the Fourth Industrial Revolution: The risks of marginalization and how to avoid them. In G. Krlev, D. Wruk, G. Pasi, & M. Bernhard (Eds.), *Social economy science: Transforming the economy and making society more resilient* (pp. 90–105). Oxford University Press. https://doi.org/10.1093/oso/9780192868343.003.0005

Mussida, C., & Patimo, R. (2021). Women's family care responsibilities, employment and health: A tale of two countries. *Journal of Family and Economic Issues, 42*(3), 489–507. https://doi.org/10.1007/s10834-020-09742-4

Neumayer, E., & Plümper, T. (2007). The Gendered nature of natural disasters: The impact of catastrophic events on the gender gap in life expectancy, 1981–2002. *Annals of the Association of American Geographers*. https://doi.org/10.1111/j.1467-8306.2007.00563.x

OECD. (2021). Entrepreneurship policies through a gender lens. In *OECD studies on SMEs and entrepreneurship*. https://doi.org/10.1787/71c8f9c9-en

Ojediran, F., & Anderson, A. (2020). Women's entrepreneurship in the global south: Empowering and emancipating? *Administrative Sciences, 10*(4).

Oliver, A. G., Pfarrer, M. D., & Neville, F. (2022). Grand challenges and female leaders: An exploration of relational leadership during the COVID-19 pandemic. *Business & Society, 63*(4), 954–987. https://doi.org/10.1177/00076503221141880

Orser, B., Liao, X., Riding, A. L., Duong, Q., & Catimel, J. (2021). Gender-responsive public procurement: Strategies to support women-owned enterprises. *Journal of Public Procurement, 21*(3), 260–284. https://doi.org/10.1108/JOPP-11-2019-0078

Pang, C., Li, W., Zhou, Y., Gao, T., & Han, S. (2023). Are women more empathetic than men? Questionnaire and EEG estimations of sex/gender differences in empathic ability. *Social Cognitive and Affective Neuroscience, 18*(1), nsad008. https://doi.org/10.1093/scan/nsad008

Petridou, E., & Glaveli, N. (2008). Rural women entrepreneurship within co-operatives: Training support. *Gender in Management: An International Journal, 23*(4), 262–277. https://doi.org/10.1108/17542410810878077

Pierli, G., Murmura, F., & Palazzi, F. (2022). Women and leadership: How do women leaders contribute to companies' sustainable choices? *Frontiers in Sustainability*. https://doi.org/10.3389/frsus.2022.930116

Pörtner, H.-O., Roberts, D. C., Mintenbeck, M. M. B. T. E. P. K., Alegría, A., Craig, M., Langsdorf, S., Löschke, S., Möller, V., Okem, A., & Rama, B. (2022). *Climate change 2022*.

Raman, R., Subramaniam, N., Nair, V. K., Shivdas, A., Achuthan, K., & Nedungadi, P. (2022). Women entrepreneurship and sustainable development: Bibliometric analysis and emerging research trends. *Sustainability, 14*(15).

Robinson, M. (2001). *The microfinance revolution: Sustainable finance for the poor*. World Bank Publications.

Roomi, M. A., & Parrott, G. (2008). Barriers to development and progression of women entrepreneurs in Pakistan. *The Journal of Entrepreneurship, 17*(1), 59–72. https://doi.org/10.1177/097135570701700105

Rosário, A. T., Raimundo, R. J., & Cruz, S. P. (2022). Sustainable entrepreneurship: A literature review. *Sustainability, 14*(9).

Ruth Eikhof, D., Summers, J., & Carter, S. (2013). Women doing their own thing": Media representations of female entrepreneurship. *International Journal of Entrepreneurial Behaviour & Research, 19*(5), 547–564. https://doi.org/10.1108/IJEBR-09-2011-0107

Santos Silva, M., & Klasen, S. (2021). Gender inequality as a barrier to economic growth: A review of the theoretical literature. *Review of Economics of the Household, 19*(3), 581–614. https://doi.org/10.1007/s11150-020-09535-6

Scartozzi, G., Delladio, S., Rosati, F., Nikiforou, A. I., & Caputo, A. (2024). The social and environmental impact of entrepreneurship: A review and future research agenda. *Review of Managerial Science*. https://doi.org/10.1007/s11846-024-00783-9

Schmitz, C. L., & Gabel, S. G. (2023). Women, human rights, and gender equality. *Journal of Human Rights and Social Work, 8*(4), 359–360. https://doi.org/10.1007/s41134-023-00288-7

Selzer, R., Howton, A., & Wallace, F. (2017). Rethinking women's leadership development: Voices from the trenches. *Administrative Sciences, 7*(2).

Setyaningrum, R. P., Norisanti, N., Fahlevi, M., Aljuaid, M., & Grabowska, S. (2022). Women and entrepreneurship for economic growth in Indonesia. *Frontiers in Psychology, 13*, Article 975709. https://doi.org/10.3389/fpsyg.2022.975709

Sommer, M., Caruso, B. A., Sahin, M., Calderon, T., Cavill, S., Mahon, T., & Phillips-Howard, P. A. (2016). A time for global action: Addressing girls' menstrual hygiene management needs in schools. *PLoS Medicine, 13*(2), Article e1001962.

Sorenson, S. B., Morssink, C., & Campos, P. A. (2011). Safe access to safe water in low income countries: Water fetching in current times. *Social Science & Medicine, 72*(9), 1522–1526. https://doi.org/10.1016/j.socscimed.2011.03.010

Sperindé, S., & Nguyen-Duc, A. (2020). Fostering open innovation in coworking spaces: A study of Norwegian startups. In A. Nguyen-Duc, J. Münch, R. Prikladnicki, X. Wang, & P. Abrahamsson (Eds.), *Fundamentals of software startups: Essential engineering and business aspects* (pp. 161–178). Springer International Publishing. https://doi.org/10.1007/978-3-030-35983-6_10

Spijkers, O. (2018). Intergenerational equity and the sustainable development goals. *Sustainability, 10*(11).

Stavropoulou, O., & Protopapa, S. (2013). A strengths-based approach to mentoring women entrepreneurs:

How to free the strengths within them. *Journal of Innovation and Entrepreneurship, 2*(1), 13. https://doi.org/10.1186/2192-5372-2-13

Terjesen, S., Couto, E. B., & Francisco, P. M. (2016). Does the presence of independent and female directors impact firm performance? A multi-country study of board diversity. *The Journal of Management and Governance (JMG), 20*, 447–483 (2016). https://doi.org/10.1007/s10997-014-9307-8

Thebaud, S., & Sharkey, A. (2016). Unequal hard times: The influence of the great recession on gender bias in entrepreneurial financing. *Sociological Science, 3*, 1–31. https://doi.org/10.15195/v3.a1

Thomas, A. (2024). The role of women's entrepreneurship in achieving sustainable development goals (SDGs): A comprehensive review. *Journal of Biotechnology & Bioinformatics Research, 6*, 1–11. https://doi.org/10.47363/JBBR/2024(6)174

Thorarinsdottir, F. V., Óladóttir, Á. D., Darmstadt, G. L., & Benediktsdóttir, S. (2024). Gender lens investing: A scoping review from theoretical foundations to public equity applications. *Sustainability, 16*(20).

Ubfal, D. (2024). Supporting women-led businesses: Narrative review of recent causal evidence. *The World Bank Research Observer*, lkae012. https://doi.org/10.1093/wbro/lkae012

UNICEF. (2016). *UNICEF: Collecting water is often a colossal waste of time for women and girls.* Unicef. https://www.unicef.org/press-releases/unicef-collecting-water-often-colossal-waste-time-women-and-girls

Walsh, D. (2024). *Building the case for women waste entrepreneurs.* MIT Management Sloan School. Retrieved 2 Feb 2025 from https://mitsloan.mit.edu/ideas-made-to-matter/building-case-women-waste-entrepreneurs

UN Woman. (2020). *Woman and sustainable development goals.*

Yabas, D., Kurutas, B. S., & Corlu, M. S. (2022). Empowering girls in STEM: Impact of the girls meet science project. *School Science and Mathematics, 122*(5), 247–258. https://doi.org/10.1111/ssm.12540

Yadav, V., & Unni, J. (2016). Women entrepreneurship: Research review and future directions. *Journal of Global Entrepreneurship Research, 6*(1), 12. https://doi.org/10.1186/s40497-016-0055-x

Sustainable Social Entrepreneurship: A Multifaceted Approach to Addressing Complex Global Challenges

Shokooh Sadat Alizadeh Moghadam and Mohammad Shariati

Abstract

Sustainable social entrepreneurship emerges as a multidimensional paradigm at the confluence of social and sustainable entrepreneurship, addressing complex global challenges through integrated value creation across social, environmental, and economic domains. This chapter examines the conceptual foundations, convergences, and divergences of sustainable social entrepreneurship in relation to its parent fields, highlighting its capacity to promote systemic innovation, long-term impact, and intergenerational equity. Social entrepreneurship prioritizes social impact, leveraging hybrid models to address societal issues like poverty and inequality, while sustainable entrepreneurship focuses on environmental preservation and resource efficiency within a triple-bottom-line framework. Sustainable social entrepreneurship synthesizes these objectives, employing innovative, hybrid revenue models to ensure financial autonomy while advancing social and environmental goals. Drawing on a comparative structural analysis, the study delineates distinctions in financial mechanisms, impact scope, business models, and stakeholder engagement across these domains. Key challenges include institutional barriers, such as inadequate legal frameworks, scalability constraints due to context-specific complexities, and financial obstacles stemming from limited investment appeal. Opportunities for synergy arise through multi-stakeholder collaborations involving governments, the private sector, universities, and international organizations, alongside local capacity-building efforts to enhance sustainability. The chapter also forecasts future trends, highlighting the transformative potential of digital technologies, crowdfunding, and Generation Z's social consciousness in reshaping investment landscapes, necessitating enhanced transparency and real-time impact reporting. Practical recommendations advocate for clear legislative frameworks, tax incentives, and robust digital infrastructure to support sustainable social enterprises alongside entrepreneurial skill development and university-led research to foster innovation. A coherent conceptual framework is proposed to clarify interactions among these entrepreneurship models, guiding policymakers and practitioners in designing effective strategies. By analyzing successful cases and addressing tensions, such as mission drift and cultural resistance, this study highlights the need for tailored impact assessment tools and scalable models to enhance the role of sustainable social entrepreneurship in achieving the United Nations Sustainable Development Goals.

S. S. A. Moghadam (✉) · M. Shariati
Faculty of Entrepreneurship, University of Tehran,
Tehran, Iran
e-mail: sh.s.alizadeh@ut.ac.ir

M. Akbari et al. (eds.), *Sustainable Entrepreneurship*, University of Tehran Science and Humanities Series, https://doi.org/10.1007/978-3-032-04909-4_18

Learning Objective

Upon completing this chapter, readers will be able to:

1. Define and distinguish social, sustainable, and sustainable social entrepreneurship.
2. Identify core characteristics of sustainable social entrepreneurship.
3. Assess hybrid revenue models and their challenges in sustainable social ventures.
4. Analyze institutional, financial, and scalability challenges, as well as synergy opportunities.
5. Understand integrated impact measurement tools across entrepreneurship types.
6. Anticipate trends such as digital technologies and the influence of Generation Z.
7. Propose strategies for policymakers, entrepreneurs, and universities to advance the field of study.

1 Introduction

Social entrepreneurship, as an academic field, centers on generating social value through innovative activities that leverage opportunities and resources available within society (Talić et al., 2020). Addressing multifaceted and complex social challenges often relies on the social capital of entrepreneurs, encompassing networks, knowledge, trust, and critical resources at their disposal. Such endeavors may incur higher initial costs and, at times, limited economic justification. Although social entrepreneurs strive to mobilize constrained and accessible resources as catalysts for social change through strategies like bricolage or entrepreneurial shortcuts, these approaches are fraught with uncertainties and risks. These challenges manifest as barriers at various stages of a venture's lifecycle, impacting the pace and quality of its growth. In such contexts, financial sustainability and attention to economic profitability become priorities for social entrepreneurs, prompting them to seek innovative solutions that balance economic and social objectives while ensuring operational profitability and addressing societal concerns (Dixit et al., 2025). Achieving economic sustainability alongside social value creation remains an ongoing challenge for social entrepreneurs.

Emerging within the entrepreneurship literature is a concept that emphasizes the simultaneous creation of innovative environmental, social, and economic value. This concept seeks to develop business models that embed these values from inception, integrating them into the core DNA of the venture rather than incorporating sustainability considerations at later stages. Termed *sustainable entrepreneurship*, this approach prioritizes economic profitability while ensuring environmental sustainability, stakeholder community resilience, and sustainable resource use in the establishment, growth, and development of ventures, reflected in their objectives, processes, outcomes, and impacts (Savitz, 2013).

While social entrepreneurship and sustainable entrepreneurship share significant conceptual overlaps, distinct differences and tensions exist in their approaches and definitions. Social entrepreneurship primarily focuses on resolving social issues, prioritizing social impact while considering economic performance. In contrast, sustainable entrepreneurship places the most significant emphasis on resource and environmental sustainability, striving to safeguard the interests of both internal and external stakeholders while maintaining economic viability. At the intersection of these domains, both practically and theoretically, a new concept emerges: *sustainable social entrepreneurship*. This subfield of social entrepreneurship addresses social challenges through market-based mechanisms while also prioritizing environmental preservation and sustainable resource use. Sustainable social entrepreneurship is thus defined as a subset of social entrepreneurship that focuses on creating environmental and social value through economically viable business models (Bonfanti et al., 2024).

Sustainable social entrepreneurship introduces a comprehensive and integrated approach wherein innovative opportunities are identified, evaluated, and exploited to generate social, environmental, and economic value simultaneously. This approach ensures social impact while safeguarding resources for future generations. Key characteristics of sustainable social entrepreneurship include balancing social, environmental, and economic objectives; fostering long-term impact; promoting systemic innovation; developing multifaceted business models; adopting a multi-stakeholder perspective; implementing integrated impact measurement; and employing innovative financial structures (Bonfanti et al., 2024). Sustainable social entrepreneurship can lead to the emergence of integrated business models capable of addressing complex global challenges, offering solutions to market and government failures in areas such as universal access to education or clean energy, thereby bridging critical gaps (Shepherd & Patzelt, 2011).

The multifaceted nature of these objectives and values has given rise to organizations with hybrid characteristics. The emergence of hybrid organizations that simultaneously pursue financial profit, social impact, and environmental sustainability underscores the need to revisit traditional management concepts. Concepts such as social benefit enterprises, impact investing, impact assessment, and impact reporting have become integral to sustainable social ventures, necessitating the development of models tailored to this form of entrepreneurship. While numerous studies have separately explored social entrepreneurship (Sampaio & Sebastião, 2024) and sustainable entrepreneurship (Abbas & Bulut, 2024), the domain of sustainable social entrepreneurship remains underexplored (Incio Chavesta et al., 2025). This conceptual ambiguity risks inconsistency in definitions and boundaries, hindering cohesive theoretical development, rigorous empirical research, and the formulation of relevant policies. For instance, the broad notion of "social" in social entrepreneurship may overlap with environmental objectives in sustainable entrepreneurship, complicating the identification of fundamental distinctions.

Formulating a coherent conceptual framework to elucidate the interactions and tensions among economic, social, and sustainable entrepreneurship models can organize fragmented knowledge across these domains. Such a framework would clarify areas of synergy and conflict, propose solutions for managing discrepancies, and provide valuable guidance for policymakers in designing effective support policies for sustainable social entrepreneurship. Furthermore, the development of metrics and tools to measure and evaluate impact and performance in sustainable social ventures would be a critical outcome of this framework.

To this end, the present chapter undertakes a comparative analysis of social entrepreneurship and sustainable entrepreneurship, reviews key literature to identify existing gaps, and proposes a framework for sustainable social entrepreneurship.

2 Social Entrepreneurship Versus Sustainable Entrepreneurship

Social Entrepreneurship

Social entrepreneurship is a process in which individuals or organizations employ entrepreneurial thinking to create, organize, and manage a venture with the primary aim of generating social value rather than solely pursuing financial profit. Social entrepreneurs aim to identify and address social, cultural, or environmental issues, employing innovative and market-oriented approaches to achieve these objectives, with the ultimate goal of fostering positive and sustainable societal change (Elkington & Hartigan, 2008). Distinctive characteristics of social entrepreneurship, compared to general entrepreneurship, include impact orientation, prioritizing social change over financial gain; institutional innovation, creating novel structures to address longstanding issues; hybrid structuring, blending market and social logics; emphasis on financial sustainability, generating revenue without reliance on charitable donations; and stakeholder focus, simultaneously addressing the needs of target communities and broader society (Praszkier & Munnik, 2023).

Numerous well-documented examples of social entrepreneurship models have been studied. In each case, a social mission is defined to address a specific social issue or harm, implemented through innovative solutions and tailored financing models. A prominent example is Grameen Bank in Bangladesh, which embarked on a mission to provide microcredit to rural women through small, collateral-free loans with group-based repayment mechanisms, significantly improving their financial status and employment opportunities (Yunus, 2007). Innovative revenue and financing models are hallmark features of these ventures. For instance, TOMS Shoes' "one-for-one" model addresses the social issue of footwear scarcity by donating a pair of shoes to a child in need for every pair purchased by a customer. Another example is D-Rev, a startup focused on equipping hospitals in low-income regions by designing cost-effective medical devices for emerging markets, thereby making affordable technologies accessible to hospitals in developing countries (Battilana & Lee, 2014).

Research on social entrepreneurship highlights individual-level motivations, including intrinsic drivers such as finding personal meaning and direct engagement with social issues, as well as extrinsic drivers like social recognition and role modeling. At the organizational level, efficient use of limited resources, institutional innovation, and networking are identified as key drivers. At the macro level, social entrepreneurship is propelled by market and government failures in addressing social issues, cultural shifts, the rise of enabling technologies, and the presence of constraining limitations (Ledi, 2024).

Challenges in social entrepreneurship encompass financial and internal process inefficiencies, market structure constraints, mission drift, scalability risks, governance issues, and cultural barriers (Simsek, 2024). Scalability and growth pose significant challenges for ventures that are primarily committed to resolving social issues rather than optimizing economic performance. As social entrepreneurship often targets the base of the economic pyramid, it relies on local stakeholder participation, adaptive technologies, and creative financial models to address social problems. Success typically requires a blend of creativity, technology, a deep understanding of local communities, flexible governance models, and financial support (Prahalad, 2005).

Securing resources and attracting investment remain primary challenges, with traditional and venture capital investors often hesitant to fund social entrepreneurship projects. The mission-revenue duality creates persistent pressure for profitability, potentially leading to deviations from social objectives (Battilana & Lee, 2014). Additional challenges include managing diverse stakeholders, the absence of legal frameworks for hybrid organizations, and difficulties in measuring social impact (Collavo, 2022).

The social entrepreneurship process is classically delineated into seven stages. It begins with identifying a social issue through methods such as field research and stakeholder analysis. This is followed by designing innovative solutions that integrate social and economic value; planning resource mobilization through hybrid funding models, including impact investments or grants; implementing and testing solutions via limited trials or rapid impact assessments to mitigate risks; measuring impact using varied social impact metrics; scaling through one of three primary strategies (direct expansion, knowledge dissemination, or systemic influence); and finally, institutionalizing the model into a sustainable structure, often through hybrid frameworks or partnerships with government entities (Zahra et al., 2009).

Social entrepreneurship has given rise to organizations with nonprofit, for-profit, and social mission-driven structures, as well as hybrid models. Nonprofit organizations focus exclusively on creating social value without pursuing financial profit. For-profit organizations with social missions pursue both financial and social objectives, leveraging profitable business models to address social and environmental issues. Profitability in these ventures is essential not only for financial sustainability but also for expanding their social impact. Examples include B Corps and social-purpose limited liability companies. Hybrid organizations combine elements of both nonprofit and for-profit models, offering greater flexibility in resource mobilization and achieving social goals. For instance, an organization might maintain a nonprofit arm for charitable activities and a for-profit arm for selling products or services aligned with its social mission. The choice of an organizational structure depends on the venture's objectives, mission, funding strategy, and the legal and social environment in which it operates (Nicolopoulou et al., 2017).

Sustainable Entrepreneurship

Sustainable entrepreneurship is defined as the creation of innovative value by simultaneously pursuing economic, social, and environmental objectives to meet the needs of present and future stakeholders—namely, organizations, people, and the planet (Newth, 2016). Like social entrepreneurship, sustainable entrepreneurship is conceptualized as a response to various market failures and can be regarded as institutional entrepreneurship, emphasizing the external impacts of entrepreneurial activities on society and the environment. The concept hinges on the relative importance of market-driven (economic) goals compared to non-market (social and environmental) goals. In this perspective, sustainable entrepreneurship is pursued by businesses that prioritize social and environmental objectives at least as much as economic ones. It involves identifying and exploiting entrepreneurial opportunities within existing incentive systems while also creating new opportunities that necessitate changes in these systems to achieve sustainability (Nicolopoulou et al., 2017).

Key focus areas of sustainable entrepreneurship include preserving the natural environment, mitigating environmental degradation, promoting the efficient use of natural resources, and advocating eco-friendly practices. Additionally, it encompasses supporting human and ecological well-being, generating social value, strengthening communities through job creation, advancing social justice, and enhancing quality of life. Sustainable entrepreneurs typically adopt a long-term perspective, considering the implications of their decisions for future generations (Odeyemi et al., 2024).

The process of sustainable entrepreneurship, as delineated in prior studies, begins with identifying sustainable opportunities, which involves understanding social and environmental needs and devising innovative ways to address them through sustainable business models. This is followed by developing solutions with a triple-bottom-line approach—designing products, services, and processes that concurrently achieve economic, social, and environmental goals. Securing financing and establishing a sustainable venture often involves drawing on diverse funding sources, including loans, social investments, and crowdfunding. Finally, entering or creating a sustainable market involves introducing sustainable products and services, as well as engaging customers who value sustainability principles (Belz & Binder, 2015).

3 Convergences and Overlaps Between Social and Sustainable Entrepreneurship

An examination of social entrepreneurship and sustainable entrepreneurship, along with their respective examples, reveals significant convergences and overlaps that are critical for understanding these conceptual domains. The primary points of alignment include:

Shared Aspirations for Positive Impact Toward a Good Society: Both forms of entrepreneurship aim to effect transformative change beyond organizational profitability, though their focal points may differ. Their ultimate goal aligns with the normative vision of a "good society"—a just and equitable society that maximizes collective well-being. In contemporary discourse, a good society is inherently sustainable (Sparviero, 2024). Social entrepreneurship traditionally focuses on addressing social challenges and the needs of marginalized groups (e.g., poverty, inequality, environmental degradation, access to education, and healthcare), while sustainable entrepreneurship specifically targets environmental issues and promotes sustainable development. However, these objectives frequently intersect, as many social problems are rooted in environmental challenges and vice versa. Sustainable entrepreneurs establishing ventures based on sustainability principles implicitly pursue social goals as well (Belz & Binder, 2015). Job creation, economic development, improved quality of life, and natural resource preservation are common outcomes of both types of entrepreneurship.

Creation of Non-economic Value Aligned with Market Logic: Both sustainable and social entrepreneurs leverage market mechanisms to achieve their higher-order goals. For social entrepreneurship, identifying a pressing social or environmental issue that aligns with personal passion and values is a critical first step, with economic sustainability through stable revenue streams being emphasized as essential for continuity (Salgueiro, 2024). Hybrid organizations pursuing social and sustainability missions exhibit characteristics of social welfare logic (e.g., fulfilling social missions, collaboration, care ethics, altruism) alongside market logic (e.g., wealth creation, competition, financial discipline, and self-promotion) (Salgueiro, 2024).

Pursuit of Combined Social, Environmental, and Economic Goals: While social entrepreneurship directly seeks social value creation, sustainable entrepreneurship prioritizes environmental value; both often generate multifaceted value. For instance, a sustainable entrepreneurship venture using recycled materials not only contributes to environmental preservation but also creates local employment opportunities, fostering economic improvement—a form of social value creation. Similarly, a social entrepreneurship organization focused on empowering local communities through organic farming generates social value while promoting sustainable agricultural practices and environmental conservation (Maaßen et al., 2023; Newth, 2016; Raufflet et al., 2016).

Innovation in Business Models, Products, and Services: To achieve social and environmental objectives, both social and sustainable entrepreneurs frequently innovate in business models, processes, products, and services (Ballesteros-Sola & Raible, 2024). In sustainable entrepreneurship, adopting energy-efficient technologies exemplifies such innovation. Social entrepreneurs may employ innovative models, such as dual-mission organizations (for-profit with a social purpose), to achieve financial sustainability while meeting social goals. Innovations in integrating market mechanisms into nonprofit revenue models to advance sustainability objectives are also prevalent. However, low risk tolerance and rigid organizational structures are reported as challenges to implementing such innovations (Newth, 2016).

Flexibility in Organizational Structure: Neither social nor sustainable entrepreneurship is confined to a specific organizational type, as it manifests in both for-profit and nonprofit entities. Nonprofits may adopt entrepreneurial approaches to address social or environmental issues, while for-profit firms may embed social and environmental goals within their business models, operating as social-purpose or sustainable enterprises. Examples in the literature include sustainable ventures such as Fairnopoly (a transparent online marketplace with social and environmental value) and Polarstern (providing renewable energy services). These ventures, while profit-oriented, emphasize creating "social value" and may operate as nonprofits or specialized social enterprises (Raufflet et al., 2016).

Focus on Opportunity Recognition, Evaluation, and Exploitation: Social entrepreneurs identify opportunities to address unmet social needs or resolve societal issues, while sustainable entrepreneurs seek opportunities to develop sustainable products and services, enhance resource efficiency, and reduce environmental impacts. This process requires creativity, risk-taking, and perseverance to transform ideas into viable and sustainable solutions (Newth, 2016).

Dependence on Similar Enablers for Success: Individual factors, such as the entrepreneur's passion and social/sustainability-oriented (O'Steen et al., 2024; Sparviero, 2024), alongside knowledge, experience, and expertise (Summers & Dyck, 2011), are critical. Organizational factors, including supportive structures, collaborative networks, participatory governance models, and organizational learning, also play a role (Summers & Dyck, 2011). Environmental factors—such as access to financing, education, public awareness, governmental and local support, and social norms—further enable both social and sustainable entrepreneurship (Summers & Dyck, 2011; Salgueiro, 2024).

Stakeholder-Centric Approach: Both entrepreneurship types prioritize addressing the expectations and preferences of direct and indirect stakeholders in their design and operations (Sparviero, 2024).

Unique Impact Measurement and Evaluation Methods: Due to the qualitative nature of social and sustainable entrepreneurship outcomes, measuring impact using conventional profitability metrics is challenging, necessitating specialized tools and approaches (Ballesteros-Sola & Raible, 2024).

4 Divergences and Distinctions Between Social and Sustainable Entrepreneurship

While the concepts of social entrepreneurship and sustainable entrepreneurship are closely related and, at times, complementary, they exhibit distinct divergences in specific areas. The primary distinction lies in their differing focal points, which can give rise to additional differences. The literature identifies the following key points of divergence between these two concepts:

Distinct Missions and Primary Focus: The mission, vision, and values of social enterprises often reflect a blend of the social entrepreneur's personal values and a vision for a better future grounded in societal needs (O'Steen et al., 2024). Social entrepreneurship prioritizes a social mission over generating returns for shareholders, with profitability valued as a means to ensure self-sustainability and expand investments to address broader social challenges (Salgueiro, 2024). In contrast, sustainable entrepreneurship emphasizes environmental concerns and the development of green products, with a focus on long-term impacts on both the environment and society (Talić et al., 2020).

Different Motivations and Drivers: In social entrepreneurship, the entrepreneur's passion serves as a powerful driver for initiating a social venture, with personal values integrating with societal needs to shape the venture's mission and values (O'Steen et al., 2024). Sustainable entrepreneurship, however, is driven by deep-seated ethical and environmental values, idealistic visions for positive change, personal commitment to sustainability principles, or a strategic recognition of long-term economic opportunities tied to sustainability (Raufflet et al., 2016).

Varying Scope and Temporal Horizons: Social entrepreneurship often addresses issues pertinent to the current generation (intra-generational equity), whereas sustainable entrepreneurship incorporates considerations for future generations (inter-generational equity) (Nicolopoulou et al., 2017).

Differing Scales of Impact: Social entrepreneurship may operate at a local level, focusing on specific communities (Salgueiro, 2024), while sustainable entrepreneurship typically adopts a broader perspective, aiming for systemic and long-term impacts on sustainability issues at a larger scale (Maaßen et al., 2023).

Distinct Metrics for Measuring Impact and Success: Although both forms of entrepreneurship may pursue financial objectives, social entrepreneurship is primarily evaluated by the extent of social impact achieved (Sparviero, 2024). Sustainable entrepreneurship, in addition to economic considerations, is evaluated based on its ability to mitigate adverse environmental impacts and promote sustainability goals (Talić et al., 2020).

Variation in Number and Prioritization of Objectives: While both approaches seek financial sustainability, their prioritization of objectives differs. Social entrepreneurship typically adopts a dual focus (social and economic), whereas sustainable entrepreneurship embraces a triple-bottom-line approach (economic, social, and environmental). In social entrepreneurship, the social mission may take precedence in decisions regarding business models and resource allocation. In sustainable entrepreneurship, efforts to balance and integrate economic, social, and environmental goals influence the design of the business model and financial metrics (Nicolopoulou et al., 2017).

Diverse Sources of Funding: Although both domains may draw on similar funding sources, the emphasis on specific types of financing varies. Social entrepreneurship often prioritizes attracting impact investments, while sustainable entrepreneurship may also seek funding tied to environmental sustainability, such as investments in clean technologies or carbon markets (Nicolopoulou et al., 2017).

Given these dimensions of convergence and divergence, it is noteworthy that some studies position sustainable entrepreneurship as a subset of social entrepreneurship or as an evolved concept derived from it, while others view social entrepreneurship as a tool or approach for achieving sustainable development goals (Kamaludin, 2023) (Fig. 1).

5 The Emergence of Sustainable Social Entrepreneurship: A Future-Oriented Approach to Social Value Creation

The concept of sustainable social entrepreneurship is frequently defined in the literature as the integration of economic sustainability into the creation of social value (Nicolopoulou et al., 2017), while other studies frame it as adherence to sustainable development principles in business design, with respect for environmental preservation (Belz & Binder, 2015). However, social entrepreneurship inherently encompasses economic profitability alongside social value creation, supported by credible revenue models. A precise delineation of how sustainable entrepreneurship differs from environmental or green entrepreneurship is also of critical importance (Angelo et al., 2024).

It appears that sustainability, when paired with social entrepreneurship, extends beyond these

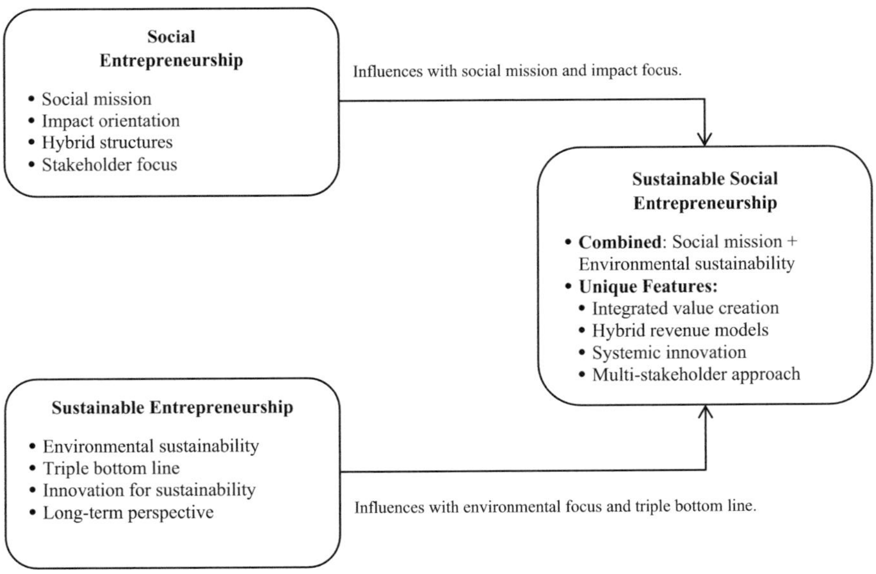

Fig. 1 Characteristics of and relationship among social, sustainable, and sustainable social entrepreneurship

dimensions. Sustainable social entrepreneurship is a form of entrepreneurship that meets present needs without compromising the ability, efficiency, or values of future generations to fulfill their requirements. Addressing a social issue or resolving a societal challenge with a vision that transcends current conditions is made possible through empowering target communities and building forward-looking infrastructures. Sustainable social entrepreneurship fosters sustainable development. Through well-designed and targeted educational programs, sustainable social entrepreneurs have advanced key domains, thereby expanding the theoretical and practical boundaries of sustainable social entrepreneurship (Angelo et al., 2024).

If the process of sustainable social entrepreneurship is categorized into three stages—opportunity recognition, business development, and sustainability, opportunity recognition emphasizes creating social value in identified opportunities while considering future scenarios. This approach not only addresses the immediate roots of social issues but also accounts for their future complexities and impacts. To this end, a matrix of possible social, environmental, and economic futures must be constructed, with solutions designed for the most desirable envisioned future. At the intersection of these three imagined futures, sustainable social entrepreneurship emerges, addressing current social problems while generating positive, long-term impacts across various dimensions. With these characteristics, sustainable social entrepreneurship forms the foundation for multidimensional, sustainable social business models characterized by flexible and innovative revenue streams and networked structures capable of rapid adaptation and correction. Further research is needed to explore how these three dimensions can be integrated while maintaining their individual significance and addressing both present and future social challenges (Singathurai, 2025).

In the first stage—identifying and evaluating the issue and designing a solution—the social dimension often dominates, influenced by the social entrepreneur's motivations and personality, which shape other dimensions. In the second stage, the development of a sustainable social business, integrating the three sustainability dimensions, becomes more feasible. Here, various facets of the business model are developed, and networked, multifaceted structures emerge to meet multidimensional objectives. In the sustainability stage, greater emphasis is placed on the economic dimension, with diverse and adaptive financial strategies leveraging realized potentials across different timeframes (Singathurai, 2025).

Given its profound and forward-looking perspective, sustainable social entrepreneurship raises challenges that promote the long-term protection of society, the environment, and its inhabitants. These challenges span critical social domains, including health and environmental changes, resource and energy conservation, waste management, and social impacts such as public welfare, unemployment, and crime, as well as the effects of emerging technologies on communities, necessitating innovative solutions (Brijmohan & Indravesh, 2023).

Sustainable social entrepreneurship represents a business model that balances economic, social, and environmental needs. Assessing social and environmental impacts is a critical component of project planning and decision-making, requiring creative evaluation methods. Integrating market and non-market impacts in project assessments is essential. The adoption of low-carbon and energy-efficient technologies plays a significant role in mitigating the adverse effects of economic growth. Targeted education can further support the development of social entrepreneurs, enabling more accurate assessments of sustainable social entrepreneurship's impact (Brijmohan & Indravesh, 2023).

The design of a sustainable social entrepreneurship model is grounded in four fundamental principles. The first principle, *social impact*, focuses on addressing social challenges such as poverty, inequality, inadequate education, or access to healthcare, offering innovative solutions to eradicate these issues in the present while shaping future outcomes. This social impact considers inter-generational equity, designing solutions to preserve the interests of future generations.

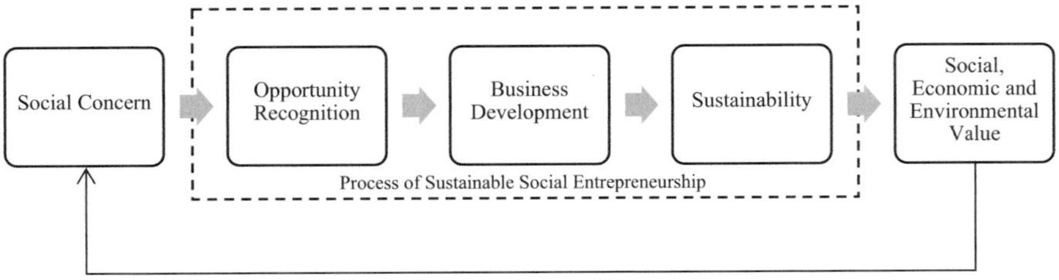

Fig. 2 Process of sustainable social entrepreneurship

The second principle, *environmental sustainability*, emphasizes preserving natural capital for future generations by reducing environmental harm, protecting available natural resources, and promoting the use of renewable resources, waste reduction, circular economy practices, and low-carbon cycles. The third principle, *a sustainable revenue model, entails generating income through the sale of products or services* independent of government or charitable aid. The fourth principle, *scalability*, involves expanding solutions to additional regions or communities without escalating environmental costs (Nicolopoulou et al., 2017) (Fig. 2).

6 The Necessity of Designing Robust Revenue Models in Sustainable Social Entrepreneurship

Sustainable social entrepreneurship strives to balance two seemingly divergent objectives: creating positive social impact and achieving sustainable revenue and profitability. In practice, this domain not only seeks to address social challenges but also aims to develop models that are economically self-sufficient and sustainable, capable of resolving social issues in the long term without reliance on charitable donations or government support. Exclusive dependence on philanthropy and public funding renders organizations vulnerable to political, economic, and social fluctuations, thereby compromising their autonomy in strategic decision-making. Sustainable revenue generation enables these organizations to pursue their activities independently of such volatilities (Singathurai, 2025).

Beyond ensuring organizational survival and autonomy, a stable revenue stream empowers sustainable social entrepreneurship organizations to expand their programs and projects, serve broader societal segments, and extend their operational scope. It also facilitates greater accountability to stakeholders and simplifies the attraction of social capital (Aswathy & Suresh, 2023).

To address the challenge of balancing social missions with profitability, hybrid revenue models have been designed for social organizations. These models simultaneously pursue social objectives and generate sustainable revenue. In hybrid models, organizations strive to maintain a focus on social issues while engaging in activities that yield consistent economic returns. Approaches include integrating commercial activities with social goals, launching social enterprises, establishing for-profit entities with social missions, and leveraging social and impact investment models (Battilana & Lee, 2014).

A critical consideration in designing revenue models for sustainable social entrepreneurship is the transformative potential of emerging technologies in enhancing impact and accessing financial resources. Technology and innovation serve as powerful tools for creating novel revenue models, reducing costs, and amplifying organizations' social effectiveness. Digital technologies and online platforms enable social organizations to achieve objectives such as expanding market access, attracting social and crowdfunding capital, enhancing financial and operational transparency and effectiveness, building trust among social investors, accessing innovative revenue models (e.g., subscription-based or platform-driven), reducing costs, and improving scalability (Aswathy & Suresh, 2023).

Given that sustainable and hybrid revenue models play a pivotal role in ensuring the long-term success of social organizations, their intelligent design, coupled with the strategic use of technology and innovation, not only fosters economic sustainability but also significantly enhances social impact. This approach paves the way for a brighter future for sustainable social entrepreneurship.

Implementing hybrid revenue models in social organizations and sustainable social enterprises while offering substantial benefits and potential is fraught with notable challenges. These challenges can be broadly categorized into three main areas:

Managerial and Organizational Challenges: One of the most prevalent challenges is mission drift, where organizations risk shifting focus from their core social objectives to economic profitability. Social organizations typically have a clear social mission, and engaging in commercial activities may divert their attention toward financial gains. Simultaneously, managing dual objectives (social and economic) demands sophisticated and flexible managerial skills, capacities, and structures. This often necessitates hiring managers experienced in both commercial business and social domains. Recruiting and retaining staff and managers who are committed to both social and commercial values is challenging, as it requires diverse skills, experiences, and mindsets (Ignas, 2025).

Cultural and Value-Based Challenges: Social organizations often operate within a culture that prioritizes human and social values, sometimes viewing revenue generation and profitability as secondary or even negative. Overcoming this mindset and fostering a culture that recognizes income generation and economic activity as legitimate and essential is a time-consuming and complex process. Specific stakeholders—such as prior donors, sponsors, or local communities—may perceive engagement in revenue-generating activities as "commercialization" or a deviation from the organization's primary mission, leading to resistance (Ignas, 2025).

Economic and Market Challenges: Social organizations may struggle to access stable and competitive markets for their products and services. Social products often face challenges, including perceived lower quality, higher prices, or limitations in distribution and marketing. Competition poses additional difficulties, as rivals typically possess greater financial resources and superior marketing and advertising capabilities. Moreover, securing initial funding and investments for launching hybrid social enterprises is a challenging task. Traditional investors may be reluctant to engage due to lower returns or higher risks associated with these ventures (Tiffany Hui-Kuang & Kun-Huang, 2024).

7 Structural Comparative Analysis of Sustainable Social Entrepreneurship with Social and Sustainable Entrepreneurship

Sustainable social entrepreneurship represents a novel hybrid concept emerging from the integration of *social entrepreneurship* and *sustainable entrepreneurship*. The primary distinctions among these domains lie in the degree of emphasis entrepreneurs place on economic and environmental dimensions in business design. Sustainable social entrepreneurship seeks to leverage the strengths of both domains in a balanced manner. This field has garnered significant attention in recent scholarly literature, with a rapidly growing body of publications. Table 1 presents a comparative analysis of sustainable social entrepreneurship, social entrepreneurship, and sustainable entrepreneurship across four dimensions: financial and economic aspects, social objectives and impact scope, business models, and stakeholder roles (Bansal et al., 2019). The financial and economic dimension examines funding mechanisms and preferred investment approaches for each model. The social objectives and impact scope dimension address the prioritization of social or environmental interventions. The busi-

Table 1 Structural comparative analysis of sustainable social entrepreneurship with social and sustainable entrepreneurship

Structural dimension	Social entrepreneurship	Sustainable entrepreneurship	Sustainable social entrepreneurship
Financial and economic aspects	The primary focus is on securing charitable donations, social support, government grants, and occasionally limited hybrid revenue models. Initial investments typically come from philanthropic and supportive sources	Emphasis on private investments, green financing, responsible investing, and market-based, profit-driven funding models reflecting a stronger economic orientation	Utilizes hybrid approaches: impact investing, responsible investing, government support, hybrid models (blending philanthropy and market-driven strategies), and financing rooted in social responsibility
Social objectives and impact scope	Absolute priority on social interventions, such as poverty alleviation, empowerment, support for vulnerable groups, and quality-of-life improvements. Environmental concerns are secondary	Simultaneous prioritization of economic, social, and environmental dimensions (triple bottom line), with particular emphasis on environmental issues and balancing economic and socio-environmental benefits	Prioritizes social issues alongside significant attention to environmental sustainability. The primary goal is addressing social challenges within a framework that fully adheres to economic and environmental sustainability considerations
Business model	Predominantly nonprofit or hybrid models. Revenue generation is limited, with a primary focus on creating social value. Economic sustainability often depends on external support	Sustainable and profitable business models (triple bottom line) emphasizing shared economic, social, and environmental value creation, with a strong focus on revenue generation and economic sustainability	Hybrid integrated models emphasize reasonable and sustainable revenue generation to ensure long-term operations and shared social-environmental value creation. Economic sustainability is essential, but not at the expense of social impact
Stakeholder roles	Strong focus on target social groups and extensive collaboration with NGOs, governments, and charities. Primary responsibility lies in addressing societal challenges	Broad collaboration with diverse stakeholders: governments, customers, responsible investors, environmental organizations, and society. Balanced responsibility toward both society and the environment	Integrated and cohesive approach: simultaneous collaboration with social stakeholders (NGOs, governments), environmental stakeholders (environmental organizations), and social and responsible investors. Full responsibility toward both society and the environment

ness model dimension evaluates revenue strategies, shared value creation, and the importance of sustainability. The stakeholder roles dimension considers the stakeholders involved and their engagement approaches, including responsibilities toward society and the environment, as well as collaboration with governments and NGOs (Boons & Lüdeke-Freund, 2013).

It is evident that while these three domains share commonalities, they exhibit significant structural and strategic differences in financial-economic aspects, social and environmental objectives, business models, and stakeholder roles. Understanding these distinctions enables managers, policymakers, and researchers to select and implement the most appropriate entrepreneurship model based on specific contexts, needs, and objectives.

8 Major Challenges Facing Sustainable Social Entrepreneurship

Sustainable social entrepreneurship is a multidimensional phenomenon shaped within specific social, economic, cultural, and institutional contexts, with its success or failure contingent on structural and institutional factors. Structural and institutional challenges, alongside financial and investment obstacles, are among the most critical barriers, and their direct and indirect impacts on the effectiveness and sustainability of social organizations are analyzed below.

Institutional and Structural Challenges

Institutional frameworks, policies, and regulations play a decisive role in the success or failure of sustainable social entrepreneurship. In many countries, clear legal and regulatory frameworks to support social enterprises are absent, forcing these businesses to operate under regulations designed for for-profit companies or nonprofits, which often misalign with their hybrid missions, leading to significant tax and financial complications (Ignas, 2025).

Additional challenges include the lack of tax incentives, limited governmental support for social investments, and the absence of defined support mechanisms, all of which hinder achieving economic sustainability. Bureaucratic complexities, inadequate communication and technological infrastructure, insufficient transportation and logistics, and limited access to markets and basic facilities further undermine the sustainability and success of social enterprises.

Beyond external institutional factors, internal structural and organizational challenges also pose barriers. For instance, engaging in innovative social and economic domains requires continuous, sometimes radical, adaptation by managers and employees, which may not always be realized. Many social organizations operate with traditional, risk-averse mindsets, impeding the adoption of innovative ideas and new business models. In organizations with a purely social or charitable legacy, engaging in revenue-generating economic activities may be perceived as "commercialization," which conflicts with their ethical values.

Scalability, Financial, and Investment Challenges

Scalability is a fundamental challenge for sustainable social entrepreneurship, defined as a social enterprise's ability to expand its activities, broaden its impact, and reach more communities without compromising social, economic, or environmental effectiveness. Sustainable social entrepreneurship often addresses complex, multifaceted issues lacking standardized solutions, as each region and community presents unique conditions. A model that is effective in one area may not perform well elsewhere, necessitating a deep understanding of local communities, cultural alignment, and adaptation to varying economic and environmental contexts (Battilana & Lee, 2014).

Moreover, sustainable social entrepreneurship relies heavily on active community participation, social capital, and public trust, which are time-consuming and complex to establish and difficult to replicate in new regions. Unlike scaling a commercial product or service, social scalability requires building social networks and trust-based relationships in each new context. Other factors complicating scalability include securing sufficient resources for rapid expansion, maintaining balance among multiple objectives (social, economic, and environmental), and the lack of adequate institutional infrastructure and support.

A significant challenge lies in standardizing and replicating innovative models. Sustainable social entrepreneurship models are often tailored to specific local conditions, making large-scale standardization and replication difficult. This demands flexibility, continuous reassessment, and adaptation to new contexts.

Financial and investment issues constitute another critical challenge. The social nature of this domain and the imperative for sustainable resource use create a paradox. On the one hand, there is a drive for resource efficiency driven by sustainability principles and limited availability; on the other hand, long-term, transformative goals necessitate substantial investments. This tension between short-term cost-saving and long-term strategic, inter-generational impact is a core financial management challenge (Brijmohan and Indravesh, 2023).

Attracting investment for social and environmental ventures is hindered by lower and longer-term returns, higher investment risks, and the lack of suitable financing tools. Social and environmental enterprises typically offer lower or delayed economic returns and struggle to provide sufficient guarantees, reducing their appeal to traditional investors. In many countries, suitable financial instruments—such as social investment funds, social bonds, or supportive financing facilities—are not available.

While responsible and social investors play a vital role in funding social enterprises, the number and volume of available investments remain limited relative to needs. Additionally, standardized global methods for assessing the impacts of social investment are insufficient, complicating decision-making for investors. Social investors require accurate, transparent, and reliable data on social and environmental impacts, which are often unavailable. Many social organizations also struggle to clearly define and report their social objectives and impacts, which undermines investor confidence (Rosyadi et al., 2024).

9 Opportunities for Synergy and Network Effects in Sustainable Social Entrepreneurship

One of the standout advantages of sustainable social entrepreneurship is its high capacity for generating synergy and network effects through effective interaction and collaboration among diverse stakeholders. These opportunities often arise from partnerships involving governments, the private sector, international organizations, universities, and local communities. This section explores these opportunities and the mechanisms for their realization.

Collaboration Among Governments, Private Sector, International Organizations, and Universities

Multilateral collaboration across societal sectors is a key strategy for enhancing the effectiveness and sustainability of sustainable social entrepreneurship. Each sector brings unique resources, capabilities, and strengths, and their cooperation can significantly amplify social and environmental impact.

a. **Role of Governments**: Governments can create a conducive environment for launching and scaling social enterprises by enacting clear, supportive legislation, offering tax incentives, and providing financial facilities. Through support funds, low-interest loans, subsidized programs, and public–private partnership contracts, governments can supply initial financial resources. Additionally, by backing incubators and accelerators, they can provide the infrastructure necessary for social innovation (Singathurai, 2025).
b. **Role of the Private Sector**: The private sector can bolster the managerial and operational capacities of social organizations by transferring expertise in management, marketing, and technology. Commercial enterprises can contribute to addressing social and environmental challenges through responsible and social investments, generating economic returns while fostering hybrid business models. Collaboration between social organizations and commercial firms can enhance the production and marketing capacity of social products (Singathurai, 2025).
c. **Role of International Organizations**: International organizations, such as the United Nations, the World Bank, global foundations, and development agencies, can transfer successful experiences from other countries to local communities. They support sustainable social projects through funding programs, technical assistance, and capacity-building initiatives. By promoting social-environmental evaluation standards, these organizations enhance transparency and build trust among social investors (Singathurai, 2025).
d. **Role of Universities and Research Institutions**: Universities contribute specialized, evidence-based knowledge through applied research to design innovative social entrepreneurship models. They can strengthen entrepreneurs' skills through educational programs, workshops, and mentorship initiatives. By fostering knowledge-exchange networks, seminars, and conferences, universities facilitate connections among researchers, social activists, and the private sector (Singathurai, 2025).

Development of Local Capacities

Building local capacities is another valuable opportunity in sustainable social entrepreneurship, contributing to long-term sustainability and sustained project effectiveness:

a. **Empowering Local Communities**: Training and upskilling local individuals in business management, technical expertise, and marketing fosters sustainable development and self-reliance. Creating new employment and economic opportunities for local communities improves the quality of life and reduces poverty (Angelo et al., 2024).
b. **Leveraging Local Resources and Knowledge**: Incorporating indigenous knowledge and experiences into business model design enhances project acceptance, effectiveness, and success. Utilizing local resources—human capital, raw materials, and traditional knowledge—reduces costs and strengthens economic sustainability (Angelo et al., 2024).
c. **Building Local and Regional Networks**: Establishing networks among social entrepreneurs and stakeholders (e.g., local governments, businesses, universities, and communities) facilitates the sharing of resources and knowledge, thereby strengthening local capacities. Creating social entrepreneurship clusters in specific geographic areas enhances effectiveness and fosters synergistic activities (Angelo et al., 2024).

10 Impact Assessment in Social, Sustainable, and Sustainable Social Entrepreneurship

In assessing the impact of *social entrepreneurship*, the primary focus is on social outcomes, employing predominantly social and qualitative-quantitative indicators. For *sustainable entrepreneurship*, impact assessment balances all three dimensions (economic, social, and environmental), utilizing broader, more comprehensive standards. In *sustainable social entrepreneurship*, a holistic and integrated approach is adopted, prioritizing both social and environmental impacts while ensuring economic sustainability and drawing on a combination of international standards (Kamaludin, 2023) (Table 2).

11 Forecasting Future Trends in Sustainable Social Entrepreneurship

Given the pressing need to address global challenges such as climate change and pervasive social inequalities in the coming years, coupled with advancements in digital transformation and artificial intelligence in business and sustainable development, the future of sustainable social entrepreneurship will be shaped by the emergence of new technologies and evolving business models that rely on them. The ease and scale of crowdfunding from global sources to address social issues, alongside the entry of Generation Z into investment markets with their inherent inclination toward responsibility and social impact, will invigorate the market for socially and responsibly invested funds. However, a critical trend will be the growing demand for deeper transparency and real-time, accurate reporting from sustainable social organizations (Angelo et al., 2024).

Hybrid and integrated organizational structures will gain increasing traction, with a shift from traditional models toward hybrid entities that balance economic profitability, social value creation, and environmental impact. This transformation will also serve green marketing strategies to attract customers and retain socially conscious talent. Consequently, managerial models, leadership approaches, and evaluation and valuation frameworks will undergo significant evolution.

This trajectory will amplify the role of social enterprises in achieving the United Nations' Sustainable Development Goals (SDGs), fostering expanded international support and the development of national policy frameworks. Multilateral networks and collaborations among societal sectors—governments, private entities, international organizations, universities, and local communities—will strengthen, aiming to create

Table 2 Impact assessment in social, sustainable, and sustainable social entrepreneurship

Evaluation criterion	Social entrepreneurship	Sustainable entrepreneurship	Sustainable social entrepreneurship
Success metrics and indicators	Primarily social indicators (e.g., number of beneficiaries, poverty reduction rates, improved access to education and healthcare, empowerment of marginalized groups)	A mix of economic (revenue, profitability, market growth), social (job creation, customer satisfaction), and environmental (emission reductions, resource optimization) indicators	Balanced integration of social (empowerment, poverty alleviation), environmental (reduced ecological footprint), and economic (sustainable revenue, social return on investment) indicators
Sustainability metrics (triple bottom line)	Primary focus on the social dimension, with limited attention to economic and environmental aspects	Comprehensive focus on all three economic, social, and environmental dimensions simultaneously (Balanced Triple Bottom Line)	Integrated focus on social and environmental dimensions, with economic sustainability as a prerequisite (Integrated Triple Bottom Line)
Social impact assessment	Highly emphasized; employs methods like SROI (Social Return on Investment), stakeholder analysis, and qualitative-quantitative social evaluations	Moderate emphasis: social impact is assessed alongside environmental and economic impacts using multidimensional impact measurement	Strongly emphasized extensive use of combined tools like SROI, stakeholder analysis, and quantitative–qualitative metrics, assessing social impact alongside environmental and economic outcomes
Tools and international standards	Primarily uses tools such as IRIS (Impact Reporting and Investment Standards), GIIRS (Global Impact Investing Rating System), SROI, and B-Corp standards to demonstrate social impact effectiveness to investors	Broader use of comprehensive sustainability standards, such as GRI (Global Reporting Initiative), ISO 14001, B-Corp Certification, and CSR (Corporate Social Responsibility) for multidimensional reporting	Extensive use of hybrid standards: IRIS, GIIRS, B-Corp, GRI, SROI, integrating social, environmental, and economic reporting

synergy and build capacity to tackle complex global challenges. This movement toward a probable future will introduce new roles and responsibilities for stakeholders in this domain (Angelo et al., 2024).

12 Strategies and Practical Recommendations for Simultaneously Advancing Social, Environmental, and Economic Value in Sustainable Social Entrepreneurship

Achieving concurrent social, environmental, and economic value in sustainable social entrepreneurship requires collaboration and synergy among governments (through policymaking and infrastructural support), entrepreneurs (via hybrid model design and skill development), and universities (through education, research, and networking). The coordinated implementation of these strategies can yield optimal social, environmental, and economic impacts.

Policy Recommendations and Macro-Level Strategies

Governments and policymakers play a crucial role in creating an environment that fosters the growth of sustainable social enterprises. To achieve this, they can enact clear and comprehensive legislation that includes legal definitions for social enterprises and sustainable social enterprises, thereby providing a solid framework for their operation. Additionally, offering tax incentives and exemptions to businesses with defined social and environmental missions can encourage their growth and expansion. Developing and strengthening both physical and digital infrastructure—such as incubators, social technology parks, and digital platforms—is essential to support these enterprises. Streamlining bureaucratic processes for registering and launching social and environmental businesses can further reduce barriers to entry. Governments should also offer supportive financing options, such as low-interest loans, and introduce financial support packages, including grants, for initiatives demonstrating clear social and environmental value. Establishing tax incentives for social investors and companies investing in sustainable social enterprises can attract additional funding. Moreover, creating awards and incentives, such as a National Sustainable Social Innovation Award, can recognize and motivate successful enterprises. Finally, prioritizing sustainable social enterprises in government tenders and public procurement contracts can ensure a steady demand for their services and products, embedding them within the economic fabric (Ballesteros-Sola & Raible, 2024).

Guidelines for Entrepreneurs

Sustainable social entrepreneurs are the primary drivers in advancing this field. To develop successful and sustainable models, they should adopt hybrid and innovative revenue models that balance the achievement of social and environmental goals with economic sustainability. Designing market-driven business models that maintain effectiveness in addressing social and environmental challenges is equally critical. Leveraging digital technologies and online platforms can enhance scalability while reducing operational costs, making these enterprises more competitive. Entrepreneurs must also develop managerial, financial, and marketing skills to balance multiple mission objectives effectively. Strengthening sustainable leadership competencies—such as systems thinking, long-term vision, and social-environmental responsibility—is vital for navigating complex challenges. Participation in training programs and courses focused on managing hybrid and integrated organizations can further equip entrepreneurs with the necessary tools to succeed in this dynamic landscape (Incio Chavesta et al., 2025).

Role of Universities and Research Institutions

Universities and research institutions play a crucial role in advancing sustainable social entrepreneurship by generating applied knowledge and developing human capacity. They can promote hybrid concepts that integrate innovation, social value, and revenue generation, fostering a new generation of socially conscious innovators. Designing and delivering specialized courses and workshops on sustainable social entrepreneurship at undergraduate and graduate levels can provide students with practical skills. Establishing dedicated academic programs, such as degrees in sustainable management, social innovation, and social enterprise, can institutionalize this field of study. Fostering a culture of social innovation and hybrid business models among students, researchers, and faculty can inspire meaningful change. Organizing startup competitions and social innovation events for students and alumni can stimulate entrepreneurial activity while creating research networks and supporting novel ideas, pushing the boundaries of knowledge. Establishing specialized research centers and institutes focused on social innovation and sustainable entrepreneurship, alongside hosting conferences, seminars, and workshops, facilitates knowledge exchange among researchers, policymakers, and social activists. Building collaborative networks for knowledge transfer among universities, social organizations, governments, and the private sector enhances local and regional capacities, thereby strengthening the overall capabilities of these entities. Ultimately, supporting applied research and developing case studies on successful international and domestic experiences in sustainable social entrepreneurship can provide valuable insights for practitioners and policymakers alike (Ledi, 2024).

13 Conclusion

Sustainable social entrepreneurship has been introduced as a novel and multidimensional approach situated at the intersection of *social entrepreneurship* and *sustainable entrepreneurship*. This concept emphasizes the simultaneous creation of social, environmental, and economic value, aiming to establish businesses that are both impactful and sustainable in the long term through innovative and hybrid revenue models. Key characteristics of sustainable social entrepreneurship include the creation of social value in balance with triple-bottom-line objectives, a focus on long-term impact and resource preservation for future generations, a multi-stakeholder and intergenerational approach, systemic innovation, extensive intra- and inter-systemic networking, and the prioritization of impact assessment over financial performance evaluation. Additionally, it involves hybrid, integrated business models and structures. Given the evident challenges in investment and scalability for sustainable social business models, designing hybrid revenue models that ensure financial sustainability, preserve organizational autonomy, and enhance social impact is imperative. Social investors, equipped with distinct personal attributes and skills, play a pivotal role in providing financial resources, capacity-building, knowledge transfer, and establishing transparent standards for assessing social and environmental impact.

The primary challenges facing sustainable social entrepreneurship include structural and institutional barriers (e.g., a lack of clear, supportive legislation), scalability, financial, and investment obstacles (e.g., difficulties in securing initial capital), and cultural and organizational resistances (e.g., reluctance to embrace change and the risk of mission drift). Opportunities for synergy and network effects emerge through effective collaboration among governments, the private sector, universities, and international organizations, accompanied by a focus on developing local capacities.

A well-defined conceptual understanding of sustainable social entrepreneurship—as a method that simultaneously pursues economic, social, and environmental goals while generating shared value among these areas—is crucial. Equally important is the creation of conceptual and practical models and frameworks to support such efforts across various geographic and cultural settings. Additionally, examining successful examples and suggesting strategies to overcome challenges and resolve potential conflicts are also recommended.

14 Case Study: SELCO

Introduction

This case study examines *SELCO (Solar Electric Light Company)*, a social enterprise in India, as a real-world example of sustainable social entrepreneurship. Drawing on publicly available information,[1] this report details SELCO's background, operations, impact, and alignment with the chapter's conceptual framework, highlighting its contributions to social, environmental, and economic objectives.

Background

SELCO was founded in 1995 by Harish Hande in Karnataka, India, with the mission of providing solar electricity to rural households that lacked access to the national grid. Recognizing energy poverty as a barrier to development, Hande established SELCO as a for-profit enterprise to ensure financial sustainability while pursuing social and environmental goals. Initially focused on selling and servicing photovoltaic (PV) systems in Karnataka, SELCO expanded its reach to serve over 100,000 households across India by 2009, as noted in the Yale SOM case study.

Operations

SELCO's operations are characterized by systemic innovation and multi-stakeholder engagement, key features of sustainable social entrepreneurship:

[1] You can refer to https://cases.som.yale.edu/selco-foundation for more information.

Tailored Solar Solutions: SELCO redesigns off-the-shelf solar components to make them affordable and suitable for low-income rural and urban households. This innovation ensures that solar systems meet the specific needs of their customers, such as lighting for homes or power for small businesses.
Financing Partnerships: To overcome the financial barriers faced by its target market, SELCO collaborates with banks and microfinance institutions to provide access to affordable financing. These partnerships enable customers to access solar systems through affordable financing options, such as microloans, making the technology accessible to those with limited upfront capital.
Service and Maintenance: SELCO provides ongoing maintenance and support for its PV systems, ensuring long-term reliability and customer satisfaction.

These operational strategies reflect SELCO's commitment to creating integrated value across social, environmental, and economic domains, as emphasized in the chapter.

Impact

SELCO's impact spans social, environmental, and economic dimensions, aligning with the triple-bottom-line approach of sustainable entrepreneurship:

Social Impact: By providing reliable electricity, SELCO enhances living conditions, supports education (e.g., enabling children to study after dark), and promotes economic activities, including small-scale businesses in rural areas. This addresses energy poverty, a significant social challenge in India, where over 400 million people live in poverty.
Environmental Impact: SELCO's use of solar energy reduces reliance on fossil fuels, lowering carbon emissions and contributing to environmental sustainability. This aligns with global efforts to combat climate change and preserve resources for future generations.
Economic Impact: As a for-profit enterprise, SELCO generates revenue through the sale and servicing of solar systems. Its partnerships with financial institutions create a sustainable business model that supports long-term operations without sole reliance on grants or donations.

According to the NextBillion article,[2] SELCO's urban model in Bangalore has electrified slum-dwelling households, demonstrating its ability to scale impact while maintaining its mission.

Alignment with Chapter Concepts

The chapter outlines several key characteristics and considerations of sustainable social entrepreneurship, which SELCO exemplifies:

1. **Integrated Value Creation**

SELCO creates social value by alleviating energy poverty, environmental value by promoting renewable energy, and economic value through its for-profit model. This integration of the triple bottom line (people, planet, profit) is a hallmark of sustainable social entrepreneurship, as defined in the chapter.

2. **Hybrid Revenue Models**

While specific details of SELCO's revenue streams are not fully disclosed in the sources, its for-profit status and partnerships with financial institutions suggest a hybrid revenue model. This likely includes direct sales, financing schemes, and possibly subsidies or grants, striking a balance between profitability and accessibility for low-income customers.

3. **Systemic Innovation**

SELCO's redesign of solar components to suit the needs of people with low incomes demonstrates systemic innovation. By addressing affordability and functionality, SELCO overcomes barriers to

[2] https://nextbillion.net/selco-A-case-study-in-solar-powered-social-impact/

adoption, making solar energy a viable solution for underserved communities.

4. **Multi-stakeholder Engagement**

SELCO collaborates with banks, microfinance institutions, and potentially government or NGO partners to achieve its mission. These partnerships facilitate financing, distribution, and policy support, reflecting the multi-stakeholder approach emphasized in the chapter.

5. **Challenges**

SELCO has faced significant challenges, as noted in the Yale SOM case study:

Financial Sustainability: Early attempts to scale through a franchised dealer network were detrimental, requiring restructuring with support from the International Finance Corporation (IFC).
Scalability Dilemma: SELCO faces the strategic decision of whether to deepen its impact among the poorest segments or expand geographically beyond Karnataka and Gujarat. This reflects the chapter's discussion of scalability constraints and the risk of mission drift.

6. **Opportunities for Synergy**

SELCO's partnerships with financial institutions and its urban model in Bangalore highlight opportunities for synergy through collaboration. The chapter suggests that multi-stakeholder collaborations can enhance impact, and SELCO's work with microfinance institutions exemplifies this potential.

7. **Impact Measurement**

SELCO's impact is quantifiable through metrics such as the number of households served (over 100,000) and the reduction in carbon emissions from displaced fossil fuel use. This aligns with the chapter's emphasis on integrated impact measurement, which balances both qualitative and quantitative outcomes.

8. **Future Trends**

The chapter identifies trends such as digital technologies, crowdfunding, and the social consciousness of Generation Z as opportunities for sustainable social entrepreneurship. SELCO could leverage digital tools for system monitoring or customer engagement, explore crowdfunding for new projects, or tap into the interest of younger generations in sustainability to expand its reach.

Challenges and Strategic Considerations

SELCO's journey has not been without obstacles, which mirror the challenges outlined in the chapter:

Institutional Barriers: Aligning investor interests with SELCO's social mission has been challenging, as noted in the Design Observer synopsis. Pressure from investors during supply shortages led to a "cramdown" of existing investors, highlighting the tension between profit and purpose.
Financial Sustainability: Early financial struggles required external support from the IFC, underscoring the difficulty of maintaining economic viability in a low-income market.
Scalability: The decision to deepen impact or expand geographically poses a strategic dilemma, as scaling too quickly risks diluting SELCO's mission or straining resources.

These challenges underscore the complexities of sustainable social entrepreneurship, where striking a balance between social impact and economic sustainability requires careful navigation.

Opportunities for Growth

SELCO's success in serving over 100,000 households positions it to explore new opportunities:

Urban Expansion: The NextBillion article notes SELCO's urban model in Bangalore, suggesting potential for further urban outreach.

Technological Innovation: Digital technologies could enhance SELCO's operations, such as remote monitoring of solar systems or mobile apps for customer support.

Policy Advocacy: Collaborating with governments to promote renewable energy policies could amplify SELCO's impact, aligning with the chapter's call for legislative support.

15 Conclusion

SELCO is a compelling example of sustainable social entrepreneurship, demonstrating how a for-profit enterprise can address energy poverty and environmental sustainability while maintaining economic viability. Its tailored solar solutions, financing partnerships, and multi-stakeholder engagement reflect the chapter's conceptual framework, particularly the emphasis on integrated value creation and systemic innovation. Despite challenges like financial sustainability and scalability, SELCO's impact on over 100,000 households underscores its role as a leader in the field. This case study offers valuable insights for policymakers, entrepreneurs, and researchers seeking to advance sustainable development through innovative business models.

Questions for Reflection

1. **Defining and Distinguishing Entrepreneurship Types**: How does SELCO's approach exemplify the distinctions between social, sustainable, and sustainable social entrepreneurship? Reflect on specific aspects of SELCO's operations that align with each type and explain why the integration of all three is crucial for addressing complex global challenges, such as energy poverty.
2. **Core Characteristics of Sustainable Social Entrepreneurship:** Identify and analyze two or three core characteristics of sustainable social entrepreneurship (e.g., integrated value creation, systemic innovation, multi-stakeholder engagement) demonstrated by SELCO. How do these characteristics contribute to its success, and what challenges might arise if one of these elements were absent?
3. **Hybrid Revenue Models and Challenges:** Considering SELCO's for-profit model and partnerships with financial institutions, what are the advantages and potential drawbacks of its hybrid revenue model? Reflect on how SELCO balances profitability with accessibility for low-income customers, considering the risks associated with financial dependency on external partners, such as banks or the IFC.
4. **Scalability and Synergy Opportunities:** SELCO faces a scalability dilemma between deepening impact among the poorest segments or expanding geographically. Reflect on the trade-offs of each option and propose how SELCO could leverage multi-stakeholder synergies (e.g., with governments, NGOs, or digital platforms) to address scalability challenges while maintaining its mission.
5. **Impact Measurement and Future Trends:** How does SELCO's approach to impact measurement (e.g., households served, carbon emissions reduced) align with the chapter's emphasis on integrated impact measurement tools? Reflect on how SELCO could incorporate emerging trends, such as digital technologies or Generation Z's social consciousness, to enhance its impact measurement and future growth.

References

Abbas, M. H., & Bulut, M. (2024). Navigating the landscape of sustainable entrepreneurship research: A systematic literature review. *Discover Sustainability, 5*(1), 171.

Angelo, B., et al. (2024). Convergences and divergences in sustainable entrepreneurship and social entrepreneurship research: A systematic review and research agenda. *Journal of Business Research, 170*, Article 114336.

Aswathy, S., & Suresh, M. (2023). Exploring the contribution of sustainable entrepreneurship toward sustainable development goals: A bibliometric analysis. *Green Technologies and Sustainability, 1*(3), Article 100038.

Ballesteros-Sola, M., & Raible, S. (2024). 21. Scaling social ventures. In B. C. Boots, L. G. Perry, & B. J. Williams (Eds.), *De Gruyter handbook of social entrepreneurship* (pp. 531–544). De Gruyter.

Bansal, S., et al. (2019). Social entrepreneurship as a path for social change and driver of sustainable development: A systematic review and research agenda. *Sustainability*. https://doi.org/10.3390/su11041091

Battilana, J., & Lee, M. (2014). Advancing research on hybrid organizing–Insights from the study of social enterprises. *Academy of Management Annals, 8*(1), 397–441.

Belz, F. M., & Binder, J. K. (2015). Sustainable entrepreneurship: A convergent process model. *Business Strategy and the Environment, 26*(1), 1–17.

Bonfanti, A., et al. (2024). Convergences and divergences in sustainable entrepreneurship and social entrepreneurship research: A systematic review and research agenda. *Journal of Business Research, 170*, 114336.

Boons, F., & Lüdeke-Freund, F. (2013). Business models for sustainable innovation: State-of-the-art and steps towards a research agenda. *Journal of Cleaner Production, 45*, 9–19.

Brijmohan, S., & Indravesh (2023). The impact of cultural and economic factors on social entrepreneurship and sustainable development goals. *Journal of Law and Sustainable Development, 11*(6).

Collavo, T. (2022). *Foundations of social entrepreneurship: Theory, practical tools and skills*. Routledge.

Dixit, K., et al. (2025). Leveraging social and intellectual capital for social entrepreneurship: a model for sustainable business practices in an uncertain environment. *Journal of Risk and Financial Management, 18*(2), 46.

Elkington, J., & Hartigan, P. (2008). *The power of unreasonable people: How social entrepreneurs create markets that change the world*. Harvard Business Press.

Ignas, M. B. (2025). From Mission Drift to Practice Drift: Theorizing drift processes in social enterprises and beyond. *Organization Studies, 46*(3), 385–407.

Incio Chavesta, J. E., et al. (2025). Exploring the challenges of social and sustainable entrepreneurship strategy: A review of the literature. *Corporate and Business Strategy Review, 6*(1, special issue), 349–360.

Kamaludin, M. F. (2023). Social sustainability within social entrepreneurship. *Technological Forecasting and Social Change, 192*, Article 122541.

Ledi, K. K. (2024). SDG motivation and social entrepreneurial intentions. The moderating roles of institutional and social support. *Journal of Social Entrepreneurship* 1–24.

Maaßen, C., et al. (2023). A process model for sustainable entrepreneurship: Evidence from a highly entrepreneurial European region. *Journal of Social Entrepreneurship* 1–32.

Newth, J. (2016). *Social entrepreneurship in an INGO: Exploring the challenges of innovation and hybridization* (pp. 389–410). Routledge.

Nicolopoulou, K., et al. (2017). *Sustainable entrepreneurship and social innovation*. Routledge Taylor & Francis Group Earthscan from Routledge.

O'Steen, B., et al. (2024). 13. It takes a seed: The importance of passion to a social venture's mission, vision, and culture. In *De Gruyter Handbook of Social Entrepreneurship* (p. 335).

Odeyemi, O., et al. (2024). Sustainable entrepreneurship: A review of green business practices and environmental impact. *World Journal of Advanced Research and Reviews, 21*(2), 34356–34358.

Prahalad, C. K. (2005). *The fortune at the bottom of the pyramid*. Financial Times/Prentice Hall.

Praszkier, R., & Munnik, P. (2023). *Social entrepreneurship* (pp. 1508–1517). Springer.

Raufflet, E., et al. (2016). *Sustainable development and entrepreneurship: Mapping definitions, determinants, actors and processes* (pp. 309–325). Routledge.

Rosyadi, R., et al. (2024). Impact investing: Aligning financial goals with social and environmental objectives. *Global International Journal of Innovative Research, 2*(3), 721–727.

Salgueiro, T. (2024). 17. Social entrepreneurship and CSR: Navigating hybridity with B lab's tools. In B. C. Boots, L. G. Perry, & B. J. Williams (Eds.), *De Gruyter handbook of social entrepreneurship* (pp. 433–454). De Gruyter.

Sampaio, C., & Sebastião, J. R. (2024). Social innovation and social entrepreneurship: Uncovering themes, trends, and discourse. *Administrative Sciences, 14*(3), 53.

Savitz, A. (2013). *The triple bottom line: How today's best-run companies are achieving economic, social and environmental success-and how you can too*. John Wiley & Sons.

Shepherd, D. A., & Patzelt, H. (2011). The new field of sustainable entrepreneurship: Studying entrepreneurial action linking "what is to be sustained" with "what is to be developed." *Entrepreneurship Theory and Practice, 35*(1), 137–163.

Simsek, K. D. (2024). *Impact investing* (pp. 48–57). Routledge.

Singathurai, S. (2025). Sustainable social entrepreneurship: Balancing profitability with environmental and social impact. *Journal of Information Systems Engineering and Management, 10*, 927–937.

Sparviero, S. (2024). 12. Business models in the social venture. In B. C. Boots, L. G. Perry, & B. J. Williams (Eds.), *De Gruyter handbook of social entrepreneurship* (pp. 307–334). De Gruyter.

Summers, D. B., & Dyck, B. (2011). A process model of social intrapreneurship within a for-profit company: First community bank. *Social and Sustainable Entrepreneurship* 139–174.

Talić, M., et al. (2020). Sustainable entrepreneurship: Creating opportunities for green products development. *Economics of Sustainable Development, 4*(2), 1–13.

Tiffany Hui-Kuang, Y., & Kun-Huang, H. (2024). Causal analysis of SDG achievements. *Technological Forecasting and Social Change, 198*, Article 122977.

Yunus, M. (2007). Creating a world without poverty: Social business and the future of capitalism. *Public Affairs.*

Zahra, S. A., et al. (2009). A typology of social entrepreneurs: Motives, search processes and ethical challenges. *Journal of Business Venturing, 24*(5), 519–532.

Ethics and Sustainable Entrepreneurship

Naser Valizadeh, Khadijeh Bazrafkan, and Hajar Zarei

Abstract

This chapter attempts to highlight the importance of ethics and its principles in entrepreneurship and sustainable business activities. To this end, the first section of the present chapter compares ethics in classical entrepreneurial theories with ethics in new or sustainable entrepreneurial theories. In the next step, an attempt was made to examine the most important indicators of entrepreneurial ethics. Also, in the third step of this chapter, social entrepreneurship was proposed and justified as a suitable tool and approach for sustainable and ethical entrepreneurship. In the fourth phase, an attempt was made to examine and discuss greenwashing as an unethical competitive strategy and to identify and discuss its negative consequences in both the short and long term. Finally, practical strategies for developing sustainable entrepreneurial ethics were presented. Additionally, a case study was presented on the topic of greenwashing and consumer perceptions of organic food products among Iranian consumers.

Keywords

Sustainable entrepreneurship · Ethics · Ethical business · Marketing · Greenwashing

Learning Objectives

- Readers will be able to critically evaluate the contrasting perspectives of business ethics and market-driven mechanisms in sustainable entrepreneurship.
- Readers will be able to identify and explain the key domains and indicators of ethics in sustainable entrepreneurship, including financial, marketing, production, social, and stakeholder ethics, and critically evaluate how cultural, social, and moral competencies contribute to ethical decision-making in business practices.
- Readers will be able to analyze the role of social entrepreneurship in addressing socioeconomic inequalities and evaluate the challenges and benefits of integrating the social, economic, environmental, and ethical dimensions of sustainability into entrepreneurial strategies.
- Readers will be able to identify the ethical dilemmas associated with greenwashing in entrepreneurial activities, understand the tactics companies use to create false envi-

N. Valizadeh (✉)
Department of Agricultural Extension and Education, School of Agriculture, Shiraz University, Shiraz, Iran
e-mail: n.valizadeh@shirazu.ac.ir

K. Bazrafkan
Agricultural Research, Education and Extension Organization (AREEO), Tehran, Iran

H. Zarei
Department of Agricultural Extension and Education, College of Agriculture, University of Tehran, Tehran, Iran

M. Akbari et al. (eds.), *Sustainable Entrepreneurship*, University of Tehran Science and Humanities Series, https://doi.org/10.1007/978-3-032-04909-4_19

ronmental claims, and evaluate the long-term impact of greenwashing on business reputation and consumer trust.
- Readers will be able to explore strategies for preventing greenwashing, including promoting transparency, accountability, and responsible marketing practices.

1 Introduction

Leading corporate management has been a key driver of sustainable development for many years. Sustainable development is defined as a kind of development that meets the needs of the present without compromising the ability of future generations to meet their own needs (Barbeito-Caamaño & Chalmeta, 2020; Kardos, 2012; Satpathy et al., 2024). Achieving sustainable development requires innovation in the field of sustainability. This research also requires entrepreneurs who can achieve environmental or social goals through the development of superior and successful products or processes in the general consumer market (Ghobakhloo et al., 2021; Silvestre & Ţîrcă, 2019; Szopik-Depczyńska et al., 2018). Sustainable entrepreneurs and sustainability managers significantly shape markets and society through their actions and behaviors. Therefore, actors and companies that achieve environmental improvements through their core activities can be considered sustainable entrepreneurs, and the actions and behaviors of such individuals in the business field can be referred to as sustainable entrepreneurship (Konys, 2019; Racelis, 2014; Rosário et al., 2022; Schaltegger & Wagner, 2011). When entrepreneurs undertake sustainability initiatives, they naturally face many opportunities and challenges. They may face a range of economic, socio-cultural, and even ethical issues and challenges, such as fundamental fairness, employee and customer relations, distributional issues, and other challenges (Arya et al., 2023; Hannefey, 2003; Lau, 2019; Malik & Abdallah, 2019; Nwabekee et al., 2024). Although many of these issues are primarily and purely related to legal and economic problems, they also involve ethical concerns. These ethical concerns include "human dignity and human rights issues," "a harmonious lifestyle based on justice, peace, and friendly relations," "corruption, especially in the supply chain and with customers," "financial and operational pressures that increase the incentive for expedient behavior (including dishonesty)," and "whether business and individual professions are moving toward becoming more ethical" (Racelis, 2014).

In the context of global business, organizations that wish to remain competitive in international markets must adopt appropriate ethical and social responsibility standards, given the cross-cultural and global nature of their business (Böhm et al., 2022; Brenkert, 2019; Crane et al., 2019; Racelis, 2014). Engagement in business activities plays a vital role in the socialization process of individuals in entire societies (Hamari et al., 2016; Mandrysz, 2020). Some experts believe that the level of ethical principles in businesses can influence the process of moral development and, consequently, the ethical behavior of employees, as well as the ethical behavior of other businesses and future competitors (DeTienne et al., 2021; Forsha, 2017; Giacalone & Thompson, 2006; Mitroff, 2004). Experts believe that some entrepreneurial and business activities foster a mindset that emphasizes profit maximization and materialism, resulting in the neglect of aspects such as social responsibility and ethics (Camilleri, 2016; Ghoshal, 2005; Sroka & Szántó, 2018). Such training may lead to the promotion of unethical behaviors among future entrepreneurs or business competitors (DeTienne et al., 2021; Forsha, 2017; Mitroff, 2004). In this regard, this chapter attempted to analyze and interpret the ethical dimensions of sustainable entrepreneurship.

2 Main Body

2.1 Business Ethics Versus the Invisible Hand of the Market in Sustainable Entrepreneurship

The focus on sustainable businesses with transparent processes and adherence to ethics and the

requirement to adhere to social responsibility and ethical standards related to a healthy environment and the protection of natural capital, on the one hand, and the unfavorable environmental conditions such as the reduction of raw material sources, increased waste, increased pollution levels, economic problems caused by pollution and social harms, on the other hand, have led to the greater importance of sustainable business ethics (Caldera et al., 2019; Lathabhavan, 2022). Additionally, recipients of services and consumers of products from organizations are also showing a greater tendency to prefer green services and products (Milfont & Markowitz, 2016; Trudel, 2019). Business ethics are a set of dos and don'ts that employees or owners of a business must follow to create the conditions they have committed to achieving, and ignoring them will lead to a decline in the standard of living of humanity (Becker, 2024; Crane et al., 2019; Weiss, 2021). From a perspective outside of entrepreneurial businesses, entrepreneurs who fulfill their moral and social responsibilities to society will enjoy greater public attention and a better position in society (Chenavaz et al., 2023; Tiba et al., 2019). Business or entrepreneurial ethics is rooted in the perspective of social responsibility, in contrast to the invisible hand of the market perspective (Chell et al., 2016; Donaldson & Fafaliou, 2003; Quarshie et al., 2016). This perspective, in its critique of entrepreneurial ethics, claims that the motivation for profit-seeking causes entrepreneurs and business owners to exert all their efforts and enter the arena of competition. For example, if an individual produces a substandard product, the invisible hand of the market will eliminate him. So, the market mechanism is the main regulator, not morality (Ahsan, 2020; Fuentes & Valenzuela-Garcia, 2019). Entrepreneurs are obligated to comply with the law, and since their primary mission is to generate a profit, they should not be expected to possess entrepreneurial ethics; therefore, social responsibility is deemed insignificant, and the law alone is paramount (McMullen & Bergman, 2017). In other words, social responsibility is often viewed as a means to generate and increase profits and to engage in activities that lead to profit; hence, discussing social responsibility is a mistake (Falck & Heblich, 2007; p. 1). From this perspective, if enterprises provide profitable goods and services to citizens, they have fulfilled their social responsibility (Hategan et al., 2018; Heald, 2018). Accordingly, industries may continue their activities without regard for ethics, resulting in environmental destruction, humanitarian disasters, climate pollution, and economic injustices. After all, the life of these industries depends on natural resources and capital that must be protected (Kolk, 2016; Singh et al., 2019).

This is in line with the perspective of social responsibility, where the responsible (ethical) entrepreneur strikes a balance between the three interests of biology, society, and economy, and the sustainable organization, balancing the needs of the earth, people, and profitability (Caldera et al., 2019; Comin et al., 2020). In other words, entrepreneurs should not only work for profit but also work completely disinterestedly for the collective interests of society, as the main goal is to achieve social welfare, which should be included in the business plan. If social responsibility is appropriately implemented, it also has a direct impact on financial outcomes. In other words, if the goal of social responsibility is not to make a profit, an entrepreneurial business that does not cover at least its expenses has violated social responsibility by wasting community resources (Alam et al., 2018; Neumann, 2021; Singgalen et al., 2022). Social responsibility is not merely a mechanical and profit-oriented process, but something more profound. Accordingly, the ethical principles of entrepreneurial business encompass social and humanitarian values that are not only related to human issues, ethics, and social relations but also refer to the welfare and happiness of society's people and other beings. These ethical principles should include rules for doing business, professional accountability for job-related duties, and personal accountability (Manasakis, 2018; Sroka & Szántó, 2018). Separating these principles from business creates injustice, and in other words, the two are complementary. From one perspective, the characteristics of business ethics encompass comfort at work, focus on work, optimal use of time and effort, self-reliance, and fairness (Rendtorff, 2020).

According to the World Commission on Environment and Development, the term "sustainability" in business ethics and sustainable entrepreneurial work means that while meeting the needs of the present, it does not compromise the ability of future generations to meet their own needs. Accordingly, sustainability in entrepreneurial business ethics depends on observing social obligations, embracing social responsibility, implementing green management, and protecting natural and environmental resources, which leads to development and enables the organization to achieve long-term economic viability (Ortiz-de-Mandojana & Bansal, 2016). Sustainability has dimensions: environmental sustainability (protection of the environment and natural resources, etc.); social sustainability (respect for the values, beliefs, convictions, basic principles, beliefs, and standards of society and efforts to achieve them); and economic sustainability (better management of limited resources and efficiency through process improvement) (Tennakoon et al., 2024). With this description, a sustainable entrepreneurial business is a business that, using the ethical principles of entrepreneurship, can advance commerce and trade both for-profit and for the collective benefit of society and public interests, and takes into account considerations related to social, economic, and environmental benefits, and considerations related to the protection of natural resources and capital, the environment, and the balance of the natural cycle (Azar et al., 2016; Lüdeke-Freund, 2020; Schaltegger et al., 2016).

2.2 Indicators of Ethics in Sustainable Entrepreneurial

The most important areas of entrepreneurial and business ethics are summarized as follows (Azar et al., 2016; Mahajan, 2015):

1. *Financial ethics*: Ethical issues and discussions in the financial field that companies and employees face include account manipulation, misleading analyses, transactions with related parties, special bonus programs for managers, bribery, and providing unreliable financial information.
2. *Marketing ethics*: Marketing ethics is a branch of entrepreneurial business ethics that addresses ethical principles in marketing operations and regulations. Ethical issues in this area include pricing, anti-competitive practices, fair competition, publishing misleading advertisements, and advertising content, among others.
3. *Production ethics*: This area of entrepreneurial and business ethics addresses the responsibilities of a company to ensure that its products and the production process do not have adverse effects on society. One of the acute problems in this area is that there is usually a degree of loss in the production process, and it is always difficult to define an acceptable degree of loss.
4. *Social sciences ethics*: In this field of entrepreneurial and business ethics, issues related to the environment, community rights, government and public enterprises, money laundering, embezzlement, tax evasion, use of public resources and facilities for personal gain, disregard for the law, and lack of attention to social responsibility are discussed.
5. *Business owners and stakeholders' ethics*: In this field of entrepreneurial and business ethics, the rights of shareholders and investors, banks, lenders, employees, and other stakeholders are studied and examined.

Different studies have proposed various indicators for measuring and evaluating ethics in the field of sustainable entrepreneurship. However, ethical indicators in sustainable entrepreneurial businesses can be reconceptualized as follows (Shoul et al., 2018):

(a) *Cultural competence*: This indicator is the result of three components: awareness, ability, and readiness. A person is culturally competent when they are aware of the importance of professional activities and can analyze social issues, possessing the personal knowledge and skills necessary to learn new teachings and techniques (awareness and ability).

The readiness component refers to whether the individual's activities in their field of work are consistently performed confidently. Cultural competence enables individuals to analyze their activities in light of social issues, understand their social significance, and behave under socially accepted ethical and legal standards. In a study, it was found that, according to research participants, the competence and ability to acquire up-to-date knowledge and skills related to environmental, social, economic, and business sustainability are the most essential indicators among other indicators of business ethics (Gibbs & Culleiton, 2016; Lin et al., 2015; Miloradova & Ishkov, 2015).

(b) *Social competence*: This indicator refers to the individual's ability to perform in a personal manner, find solutions in unusual situations, work in a system, be innovative, take care of oneself, and help and care for others (Abolghasemi et al., 2011; Usher et al., 2015).

(c) *Social responsibility*: The commitment of organizations active in business to maximize social benefits and minimize social costs, without individual intentions and only in the direction of collective interests, is called social responsibility, which itself has characteristics such as social accountability and organizational commitment. According to some management thinkers, when a business needs to generate a profit to justify its economic existence, it may act tyrannically and overlook its social responsibility. Economic performance is the first accountability and social responsibility of business. Therefore, a business that does not show profit at least equal to the cost incurred is invalid and wastes resources (Friedman, 2007; Luo & Bhattacharya, 2006).

(d) *Moral ideals*: This index has five branches: loyalty, trust, justice, fairness, love, and kindness (Adler, 1910; Silver, 1989).

(e) *Spiritual ideals*: Spiritual ideals are, in fact, the 10 levels of faith, perfection, sincerity of intention, and the crystallization of the truth of existence and humanity in man (Ashokkumar, 2014).

(f) *Social values*: These values are the principles, beliefs, norms, convictions, and standards of society and the factual bases for decisions in social areas that are made concerning humanitarian values and do not conflict with professional and specialized values. This indicator also includes mutual trust and hard work (Chen, 2000; Kim & Lee, 2015)

(g) *Moral integrity*: This indicator has two aspects: the continuity of individual morality and social morality (work ethic). Personal or individual ethics refer to the behavior, performance, actions, reactions, beliefs, and internal values of each individual in their personal life. It is explained that individual-level beliefs differ from societal-level beliefs. Additionally, social or business ethics refers to the behavior and performance of an individual in relation to other members of society within business contexts (Tavakoli, 2010).

(h) *Internal purification and introspection*: It is essential to recognize that life encompasses not only worldly and material aspects but also the spiritual dimension, which has a profound impact on an individual's morals (Ashokkumar, 2014).

2.3 Social Entrepreneurship as a Tool for Ethical, Innovative Business

Social entrepreneurship has been an active field in practice over the past few decades and has attracted considerable attention from academics (Choi & Majumdar, 2014). One of the most important reasons for paying attention to this form of entrepreneurship in businesses has been its ability to fill the socioeconomic gaps in societies that traditional entrepreneurship and business frameworks have created. Social entrepreneurship helps reduce social inequalities within societies by providing innovative, systematic, and cost-effective solutions (Choi & Majumdar, 2014; Huybrechts & Nicholls, 2012; Singh & Awasthy, 2025). These innovative forms of entrepreneurship and business offer robust, sustainable solutions to current and future global challenges, inspiring hope for a brighter future among entrepreneurs, business owners, and communities worldwide. Moreover, social entrepreneurial initiatives play a crucial role in advancing the United Nations Sustainable Development Goals on a global scale. (Littlewood & Holt, 2018; Singh & Awasthy, 2025). These characteristics make social entrepreneurial activities an ideal form of entrepreneurship and a promising alternative to the capitalist system (Battilana, 2018). However, it should not be forgotten that achieving social impact while maintaining financial sustainability simultaneously is a challenging task (Eiselein & Dentchev, 2020). This creates tensions that can lead to instability and, ultimately, mission deviation and failure of social entrepreneurial companies. Therefore, all efforts should be devoted to achieving sustainability so that their scope and impact increase and more people benefit from their services (Singh & Awasthy, 2025).

Sustainability is a multifaceted concept that is defined and conceptualized in different ways by different researchers and institutions. Some researchers and institutions consider it to mean the survival and continuity of entrepreneurial businesses. However, some people and thinkers refer to it as a complete adoption of sustainable practices and a fundamental rethinking of the entrepreneurial business model (Singh & Awasthy, 2025; Zhang & Swanson, 2014). Sustainability essentially has three interconnected and interrelated dimensions: economic, social, and environmental (Wallace, 2005). Organizations can benefit from a sustainability model that encompasses the three social, economic, and environmental dimensions to inform their entrepreneurial strategies. Of course, it should not be forgotten that translating these dimensions into entrepreneurial strategies is a complex process, and transformative approaches must be used in this context. The sustainability model, based on the three social, economic, and environmental dimensions, abandons the traditional and purely profit-oriented method of entrepreneurial business, aiming to balance the economic, social, and environmental dimensions of sustainability (Elkington, 2013; Singh & Awasthy, 2025). Ramanan and Taback (2016) expanded this three-dimensional model by adding a fourth dimension, ethics. In doing so, they attempted to highlight the importance of ethics as a balancing force and to prevent the disastrous consequences of an unethical economy. Singh and Awasthy (2025) argue that incorporating ethics as a fourth dimension to the three-dimensional approach to sustainability is essential, as without ethics, entrepreneurial businesses ultimately fail. Socially entrepreneurial companies are often more ethical (Chell et al., 2016; Dacin et al., 2011; Singh & Awasthy, 2025).

2.4 Greenwashing in Entrepreneurial Activities and Ethical Dilemmas

Today, many people have become increasingly aware of environmental issues. This change has prompted companies and entrepreneurs to present themselves as green, i.e., to introduce themselves as pro-environmental and sustainable (Hasan & Ali, 2015; Dangelico & Vocalelli, 2017). However, there is always a gap between reality and the claim, and sometimes, this gap is referred to as greenwashing. Greenwashing occurs when a brand or business attempts to present itself as more environmentally responsible than it is through the use of environmental slogans (Katait, 2017; Szabo & Webster, 2021). In other words, it presents itself as pro-environment without actually doing anything for the environment (Schmuck et al., 2018). In the world of entrepreneurship, especially during the early stages of starting a business, this temptation is particularly strong. An entrepreneur may want to demonstrate that their business is highly sustainable and environmentally friendly to attract customers or investors; however, in practice, there is often little substance behind these claims. In other words, greenwashing is a standard marketing ploy designed to make products appear more sustainable than they are (Katait, 2017; Szabo & Webster, 2021). It's essentially a way to convince customers that the company is making positive environmental choices. This is often done by portraying the product as more natural, healthier, or less toxic than its competitors.

This is where ethics comes into play. When a business uses environmental slogans just for show, it's playing with people's trust. Not only is it unethical, but it can also damage the brand in the long run (Wang & Walker, 2023). Customers may feel cheated, and investors may lose trust. In contrast, entrepreneurs who are genuinely committed to their social and environmental responsibilities will gain more credibility in the long run and receive more support from society and the market (Mendes et al., 2024; Wang et al., 2025). The right approach is to create a high level of transparency, present the facts, and back up environmental claims with tangible actions. Ultimately, to prevent greenwashing, training must be developed, transparency must be increased, and a culture of accountability must be institutionalized among businesses. Because trust is not something that is easily gained, but it is also easily lost (Álvarez-García & Sureda-Negre, 2023; Nygaard, 2023). In this way, companies use greenwashing to attract environmentally conscious customers without having to make significant changes to their business practices (Pimonenko et al., 2020). It is common for companies that greenwash to spend more time and money marketing the environmentally friendly nature of their products than they do to ensure the sustainability of their operations.

Greenwashing is so popular today because it works. Studies (Biswas & Roy, 2016; González-Rodríguez et al., 2020; Kucher et al., 2019) show that most shoppers are willing to pay more for environmentally friendly goods, and half of them consider product sustainability before making a purchasing decision. Greenwashing may seem harmless, but the reality is much worse. One problem is that it confuses consumers and distracts them from real eco-friendly programs (Katait, 2017; Szabo & Webster, 2021). Not all companies engage in greenwashing in a destructive manner. Often, the concept is as much a misunderstanding for marketers as it is for consumers (Caldas et al., 2021). Even so, unintentional greenwashing still spreads misinformation about what it takes to be sustainable and can convince well-intentioned consumers to make poor choices (Szabo & Webster, 2021; Wang et al., 2022).

Greenwashing has evolved since the 1980s, and it has become increasingly difficult for the uninitiated to spot. Here are some signs to look for when determining whether a brand is guilty of greenwashing.

- Selective disclosure: Companies often highlight positive environmental aspects of their products while deliberately avoiding mention of the negatives (Marquis et al., 2016). For example, an automobile manufacturer might praise the fuel efficiency of a vehicle while ignoring the environmentally destructive mining practices involved in producing its lithium battery.
- Symbolic actions: This is a standard practice for brands to draw attention to a small positive action that has little impact on changing their overall environmental footprint (Verma & Bharti, 2023). Donating Dawn dish soap to clean up contaminated animals after the company's product was dumped in the ocean is one example.
- Hidden exchanges: Brands might promote a new shift toward environmental awareness while ignoring their negative impacts (Sajid et al., 2024). For example, Starbucks introduced straw-free lids to reduce plastic waste; however, these new lids used more plastic than the previous ones.
- Lack of evidence: A company might make claims about environmental friendliness (e.g., 'Made with organic materials!') without providing any documentation or other evidence to support them (Parguel & Johnson, 2021).
- Ambiguity: Brands can present themselves as environmentally conscious by making broad statements about their sustainability that are overly vague (Urbański & Ul Haque, 2020). Examples include "new and improved," "non-toxic," and "made with biodegradable materials." Or the packaging of a plastic toy might be labeled "recyclable" without specifying whether it refers to the packaging, the toy itself, or its components.

- Bad and worse choices: This occurs when companies promote one of the beneficial aspects of a harmful product (Szabo & Webster, 2021). Examples include fuel-efficient SUVs or suing Walmart for misrepresenting the benefits of "biodegradable" plastic bottles.
- Overstated statements: Companies that engage in greenwashing may use statements that, while technically correct, give consumers a false sense of the products they are purchasing (Schmuck et al., 2018). For example, a clothing company might announce that its shirts are "now made with 50% more recycled fibers" when that amount has increased from 2 to 3% of the total garment. The company's claim is partly valid, but it is exaggerated as a benefit.

3 Conclusion

This chapter first attempts to discuss the different perspectives on the relationship between entrepreneurship and ethics. To this end, it presents two major approaches: ethical entrepreneurship and the traditional market-based perspective. The analysis and results of this chapter show that the market-based perspective strives to maximize profitability; however, from an ethical entrepreneurship perspective, the carbon footprint and the environmental consequences of entrepreneurial businesses are given new attention. This second approach has led to the formation of sustainable entrepreneurial businesses. The results also showed that consumers of goods produced by entrepreneurial businesses have become more sensitive to the reality of ethical principles in recent years; therefore, these businesses should make greater efforts to observe ethical principles in their products. The results of this chapter also showed that the lack of observance of ethical principles in entrepreneurial activities and businesses, although it can guarantee profits for business owners in the short term, can lead to bankruptcy and the liquidation of activities in the long term, which means a lack of sustainability in entrepreneurial activity.

Given that, in many cases, there is insufficient knowledge of the principles and indicators of ethics in sustainable entrepreneurial activities, this chapter also attempts to introduce some of the most important principles of sustainable entrepreneurial ethics. The results showed that sustainable entrepreneurial ethics encompasses various areas, including finance, marketing, production, social sciences, financial transparency, fair advertising, responsible production, and respect for community rights. The results also showed that in sustainable entrepreneurial ethics, great emphasis should be placed on the ethical responsibilities of business owners and stakeholders, including compliance with the law and fair treatment of employees, investors, and the general public. In summarizing the results related to the introduction of sustainable entrepreneurial indicators, cultural and social competence, social responsibility, moral and spiritual ideals, social values, moral integrity, and inner cultivation were proposed as criteria for the ethical evaluation of sustainable entrepreneurial activities, which are recommended to be used by entrepreneurs and business owners.

Additionally, this chapter identifies greenwashing as one of the most significant ethical challenges in the sustainable entrepreneurial process and then examines the evidence and areas where greenwashing can occur in entrepreneurial activities. Overall, this chapter recommends that entrepreneurs should try to prohibit greenwashing to ensure the sustainability of their business. Finally, this study examines one of the ethical dilemmas in the sustainable entrepreneurial business process, specifically greenwashing, in the context of organic food consumption, as a case study.

4 Case Study: Organic Food Consumers' Perceptions of "Greenwashing" as an Ethical Issue

Sustainable food choices are among the most essential concerns for food consumers. Trying to buy more sustainable foods, such as organic foods, or being curious about how they are produced, is one of the most important signs of this concern. Organic food products have many advantages. For example, these products can have positive effects and benefits on consumer health. Organic foods are also environmentally sustainable. Although the greenness of products can be a positive factor for food and many other products, it can also be used as a tool to deceive consumers. For example, greenwashing is one of the deceptive techniques that many manufacturers use to deceive the public, particularly consumers. Greenwashing has various consequences in the food industry. For example, this misleading practice can make companies genuinely committed to sustainable production suspect, which can reduce the long-term income of entrepreneurs, businesses, and companies active in this field. In addition, greenwashing undermines public trust, causing food consumers to lose faith in the producers of these products. Organic food consumers are highly sensitive to greenwashing, and accordingly, understanding their perceptions of greenwashing in organic products can help organic product producers communicate effectively with their customers. The results of this study can also help producers effectively gain the trust of their customers.

In this study, an in-depth interview research method was employed to collect the required data from 32 consumers of organic products in Fars Province, Iran. The research tool was an open-ended questionnaire, which, in addition to some personal and professional characteristics of the respondents, posed a general question about greenwashing in the production of organic food products: How do you perceive greenwashing in the production of organic food products? To collect data, the most important sellers of organic products in Fars Province, Iran, were first identified. Then, the data collection team went to these organic product sales centers and randomly distributed questionnaires to buyers of these products. First, the purpose of the research was fully explained to each participant, and they were asked to complete the questionnaire and participate in the interview if they wished. Each interview lasted about 3 h on average. The basis for the number and size of the sample was to reach theoretical saturation. In other words, the selection of the sample to complete the questionnaire and participate in the in-depth interview continued until theoretical saturation was achieved. Theoretical saturation is a point in qualitative research at which, as sampling continues, no new results are added to those previously obtained. From interviews 28 to 32, the researchers determined that no new results were being obtained, and therefore, sampling was terminated after the interview with the 32nd respondent. Word-for-word and line-by-line analysis of the speakers' statements was used to analyze the results. A total of 76 phrases or concepts were extracted in the context of organic food consumers' perception of "greenwashing." However, after refining and combining concepts that referred to the same theme, a total of 35 phrases or concepts were extracted. In the next step, these concepts were classified based on their proximity in meaning or logical connection, and finally, consumers' perceptions were divided into seven categories. This division and the frequency of each concept in the respondents' responses are presented in Table 1.

Questions

- How can ethical principles in entrepreneurship challenge or complement the concept of the "invisible hand" in market-driven systems?
- In what ways does social responsibility contribute to both the financial success and the sustainability of entrepreneurial ventures?
- What role do cultural and social competencies play in shaping ethical behavior within sustainable entrepreneurial businesses?
- How can the integration of spiritual and moral ideals enhance the ethical standards and social impact of sustainable enterprises?
- How can social entrepreneurship effectively balance the dual goals of social impact and financial sustainability?
- What is the significance of incorporating ethics as a fourth dimension in the sustainability model of entrepreneurial businesses?
- What ethical challenges do entrepreneurs face when attempting to appear environmen-

Table 1 Perceptions of the consumers regarding greenwashing as an ethical issue in organic food production

Category	Concept	Freq.
Trust and credibility	Brand trust	15
	Perceived authenticity	18
	Perceived organizational honesty	19
	Skepticism of marketing	16
	Source Credibility	7
	Perceived transparency	12
	Past brand behavior	13
Environmental awareness and knowledge	Environmental literacy	7
	Awareness of greenwashing	6
	Understanding of certified labels	9
	Perceived knowledge about sustainability	8
	The perceived complexity of environmental issues	10
Buying behavior and motivations	Motivation for Ethical Consumption	8
	Intention to buy green	24
	Perceived health benefits	26
	Price sensitivity	11
	Brand loyalty	16
	Willingness to pay more	18
Understanding marketing	Perceived honesty in green advertising	7
	Transparency in environmental claims	18
	Perceived exaggeration in marketing	12
	Trustworthiness of labels	11
	Perception of excessive abundance of environmental labels	15
Critical and intellectual evaluation	Critical examination of claims	15
	Need for mental processing	9
	Information-seeking behavior	7
	Intuitive processing versus systematic processing	11
	Perceived risk of deception	12
Emotional and psychological reactions	Moral outrage	8
	Brand disappointment	11
	Cognitive dissonance	14
	Perceived betrayal	9
Values and personal identity	Personal environmental identity	15
	Moral commitment	11
	Alignment with personal values	5

tally responsible without making substantial changes to their business practices?

- How can businesses develop strategies to ensure transparency and prevent greenwashing in their sustainability communications?

References

Abolghasemi, A., Rezaee, H., Narimani, M., & Zahed Babolan, A. (2011). A comparison of social competence and its components in students with learning disability and students with low, average and high

academic achievement. *Journal of Learning Disabilities, 1*(1), 6–23.
Adler, F. (1910). The moral ideal. *The International Journal of Ethics, 20*(4), 387–394.
Ahsan, M. (2020). Entrepreneurship and ethics in the sharing economy: A critical perspective. *Journal of Business Ethics, 161*(1), 19–33.
Alam, M. M., Shahriar, S. M., Said, J., & E Elahi, M. M. (2018). Waqf as a tool for rendering social welfare services in the social entrepreneurship context. *Global Journal Al-Thaqafah, 2018*, 87–98.
Álvarez-García, O., & Sureda-Negre, J. (2023). Greenwashing and education: An evidence-based approach. *The Journal of Environmental Education, 54*(4), 265–277.
Arya, B., Horak, S., Bacouel-Jentjens, S., & Ismail, K. (2023). Leading entrepreneurial sustainability initiatives in emerging economies. *International Journal of Emerging Markets, 18*(1), 64–85.
Ashokkumar, S. (2014). Thoughts on business ethics and corporate social responsibility from vedic literature. *Procedia Economics and Finance, 11*, 15–22.
Azar, A., Ramezanian, M. R., Farrokhi, H., & Dolatkhahi, K. (2016). Designing a model for evaluation of business ethics implication.
Barbeito-Caamaño, A., & Chalmeta, R. (2020). Using big data to evaluate corporate social responsibility and sustainable development practices. *Corporate Social Responsibility and Environmental Management, 27*(6), 2831–2848.
Battilana, J. (2018). Cracking the organizational challenge of pursuing joint social and financial goals: Social enterprise as a laboratory to understand hybrid organizing. *Management, 21*(4), 1278–1305.
Becker, C. U. (2024). *Business ethics: Methods, theories, and application.* Routledge.
Biswas, A., & Roy, M. (2016). A study of consumers' willingness to pay for green products. *Journal of Advanced Management Science, 4*(3).
Böhm, S., Carrington, M., Cornelius, N., de Bruin, B., Greenwood, M., Hassan, L., Jain, T., Karam, C., Kourula, A., Romani, L., Riaz, S., & Shaw, D. (2022). Ethics at the centre of global and local challenges: Thoughts on the future of business ethics. *Journal of Business Ethics, 180*(3), 835–861.
Brenkert, G. G. (2019). Mind the gap! The challenges and limits of (global) business ethics. *Journal of Business Ethics, 155*, 917–930.
Caldas, M. V. D. A., Veiga-Neto, A. R., Guimarães, L. G. D. A., Castro, A. B. C. D., & Pereira, G. R. B. (2021). Greenwashing in environmental marketing strategy in the Brazilian furniture market. *Revista De Economia e Sociologia Rural, 59*(3), Article e225336.
Caldera, H. T. S., Desha, C., & Dawes, L. (2019). Evaluating the enablers and barriers for successful implementation of sustainable business practice in 'lean' SMEs. *Journal of Cleaner Production, 218*, 575–590.
Camilleri, M. A. (2016). Reconceiving corporate social responsibility for business and educational outcomes. *Cogent Business & Management, 3*(1), 1142044.
Chell, E., Spence, L. J., Perrini, F., & Harris, J. D. (2016). Social entrepreneurship and business ethics: Does social equal ethical? *Journal of Business Ethics, 133*, 619–625.
Chen, C. T. (2000). Extensions of the TOPSIS for group decision-making under fuzzy environment. *Fuzzy Sets and Systems, 114*(1), 1–9.
Chenavaz, R. Y., Couston, A., Heichelbech, S., Pignatel, I., & Dimitrov, S. (2023). Corporate social responsibility and entrepreneurial ventures: A conceptual framework and research agenda. *Sustainability, 15*(11), 8849.
Choi, N., & Majumdar, S. (2014). Social entrepreneurship as an essentially contested concept: Opening a new avenue for systematic future research. *Journal of Business Venturing, 29*(3), 363–376.
Comin, L. C., Aguiar, C. C., Sehnem, S., Yusliza, M. Y., Cazella, C. F., & Julkovski, D. J. (2020). Sustainable business models: A literature review. *Benchmarking: An International Journal, 27*(7), 2028–2047.
Crane, A., Matten, D., Glozer, S., & Spence, L. J. (2019). *Business ethics: Managing corporate citizenship and sustainability in the age of globalization.* Oxford University Press.
Dacin, M. T., Dacin, P. A., & Tracey, P. (2011). Social entrepreneurship: A critique and future directions. *Organization Science, 22*(5), 1203–1213.
Dangelico, R. M., & Vocalelli, D. (2017). "Green Marketing": An analysis of definitions, strategy steps, and tools through a systematic review of the literature. *Journal of Cleaner Production, 165*, 1263–1279.
DeTienne, K. B., Ellertson, C. F., Ingerson, M. C., & Dudley, W. R. (2021). Moral development in business ethics: An examination and critique. *Journal of Business Ethics, 170*, 429–448.
Donaldson, J., & Fafaliou, I. (2003). Business ethics, corporate social responsibility and corporate governance: A review and summary critique.
Eiselein, P., & Dentchev, N. A. (2020). Managing conflicting objectives of social enterprises. *Social Enterprise Journal, 16*(4), 431–451.
Elkington, J. (2013). Enter the triple bottom line. In *The triple bottom line* (pp. 1–16). Routledge.
Falck, O., & Heblich, S. (2007). Corporate social responsibility: Doing well by doing good. *Business Horizons, 50*(3), 247–254.
Forsha, S. K. (2017). Virtue and moral development, changing ethics instruction in business school education. *College Student Journal, 51*(3), 429–443.
Friedman, M. (2007). The social responsibility of business is to increase its profits. In *Corporate ethics and corporate governance* (pp. 173–178). Springer.
Fuentes, S., & Valenzuela-Garcia, H. (2019). A crossroads for social entrepreneurship: Profits versus ethics. *Open Journal of Business and Management, 7*(2), 848–860.
Ghobakhloo, M., Iranmanesh, M., Grybauskas, A., Vilkas, M., & Petraitė, M. (2021). Industry 4.0, innovation, and sustainable development: A systematic review and a roadmap to sustainable innovation. *Business Strategy and the Environment, 30*(8), 4237–4257.

Ghoshal, S. (2005). Bad management theories are destroying good management practices. *Academy of Management Learning & Education, 4*(1), 75–91.

Giacalone, R. A., & Thompson, K. R. (2006). Business ethics and social responsibility education: Shifting the worldview. *Academy of Management Learning & Education, 5*(3), 266–277.

Gibbs, D. K., & Culleiton, A. L. (2016). A project to increase educator cultural competence in mentoring at-risk nursing students. *Teaching and Learning in Nursing, 11*(3), 118–125.

González-Rodríguez, M. R., Díaz-Fernández, M. C., & Font, X. (2020). Factors influencing willingness of customers of environmentally friendly hotels to pay a price premium. *International Journal of Contemporary Hospitality Management, 32*(1), 60–80.

Hamari, J., Sjöklint, M., & Ukkonen, A. (2016). The sharing economy: Why people participate in collaborative consumption. *Journal of the Association for Information Science and Technology, 67*(9), 2047–2059.

Hannafey, F. T. (2003). Entrepreneurship and ethics: A literature review. *Journal of Business Ethics, 46*, 99–110.

Hasan, Z., & Ali, N. A. (2015). The impact of green marketing strategy on the firm's performance in Malaysia. *Procedia-Social and Behavioral Sciences, 172*, 463–470.

Hategan, C. D., Sirghi, N., Curea-Pitorac, R. I., & Hategan, V. P. (2018). Doing well or doing good: The relationship between corporate social responsibility and profit in Romanian companies. *Sustainability, 10*(4), 1041.

Heald, M. (2018). *The social responsibilities of business: Company and community, 1900–1960*. Routledge.

Huybrechts, B., & Nicholls, A. (2012). Social entrepreneurship: Definitions, drivers and challenges. *Social entrepreneurship and social business: An introduction and discussion with case studies* (pp. 31–48).

Kardos, M. (2012). The reflection of good governance in sustainable development strategies. *Procedia-Social and Behavioral Sciences, 58*, 1166–1173.

Katait, S. K. (2017). Green washing in India an alarming issue: Misleading and deceptive environmental claims in advertising. *International Journal of Commerce and Management Research, 3*(2), 91–97.

Kim, Y. J., & Lee, C. I. (2015). Social values and economic dynamics. *Journal of Economic Dynamics and Control, 53*, 69–84.

Kolk, A. (2016). The social responsibility of international business: From ethics and the environment to CSR and sustainable development. *Journal of World Business, 51*(1), 23–34.

Konys, A. (2019). Towards sustainable entrepreneurship holistic construct. *Sustainability, 11*, 6749.

Kucher, A., Heldak, M., Kucher, L., Fedorchenko, O., & Yurchenko, Y. (2019). Consumer willingness to pay a price premium for ecological goods: A case study from Ukraine. *Environmental & Socioeconomic Studies, 7*(1), 38–49.

Lathabhavan, R. (2022). Sustainable business practices and challenges in Asia: A systematic review. *International Journal of Organizational Analysis, 30*(3), 778–794.

Lau, C. K. (2019). The economic consequences of business sustainability initiatives. *Asia Pacific Journal of Management, 36*(4), 937–970.

Lin, C. J., Chang, P. R., Wang, L. H., & Huang, M. C. (2015). Cultural competence course for nursing students in Taiwan: A longitudinal study. *Nurse Education Today, 35*(12), 1268–1274.

Littlewood, D., & Holt, D. (2018). How social enterprises can contribute to the Sustainable Development Goals (SDGs)—A conceptual framework. In *Entrepreneurship and the sustainable development goals* (pp. 33–46). Emerald Publishing Limited.

Lüdeke-Freund, F. (2020). Sustainable entrepreneurship, innovation, and business models: Integrative framework and propositions for future research. *Business Strategy and the Environment, 29*(2), 665–681.

Luo, X., & Bhattacharya, C. B. (2006). Corporate social responsibility, customer satisfaction, and market value. *Journal of Marketing, 70*(4), 1–18.

Mahajan, P. (2015). Business ethics in the scenario of corporate governance. Available at SSRN 2594116.

Malik, M., & Abdallah, S. (2019). Sustainability initiatives in emerging economies: A socio-cultural perspective. *Sustainability, 11*(18), 4893.

Manasakis, C. (2018). Business ethics and corporate social responsibility. *Managerial and Decision Economics, 39*(4), 486–497.

Mandrysz, W. (2020). Community-based social economy–social capital and civic participation in social entrepreneurship and community development. *Management Dynamics in the Knowledge Economy, 8*(1), 81–93.

Marquis, C., Toffel, M. W., & Zhou, Y. (2016). Scrutiny, norms, and selective disclosure: A global study of greenwashing. *Organization Science, 27*(2), 483–504.

McMullen, J. S., & Bergman, B. J., Jr. (2017). Social entrepreneurship and the development paradox of prosocial motivation: A cautionary tale. *Strategic Entrepreneurship Journal, 11*(3), 243–270.

Mendes, J. A. J., Oliveira, A. Y., Santos, L. S., Gerolamo, M. C., & Zeidler, V. G. Z. (2024). A theoretical framework to support green agripreneurship avoiding greenwashing. *Environment, Development and Sustainability*, 1–57.

Milfont, T. L., & Markowitz, E. (2016). Sustainable consumer behavior: A multilevel perspective. *Current Opinion in Psychology, 10*, 112–117.

Miloradova, N., & Ishkov, A. (2015). Environmental ethics as a social, professional and personal value of the students of civil engineering university. *Procedia Engineering, 117*, 246–251.

Mitroff, I. I. (2004). An open letter to the deans and the faculties of American business schools. *Journal of Business Ethics*, 185–189.

Neumann, T. (2021). The impact of entrepreneurship on economic, social and environmental welfare and its determinants: A systematic review. *Management Review Quarterly, 71*(3), 553–584.

Nwabekee, U. S., Abdul-Azeez, O. Y., Agu, E. E., & Ijomah, T. I. (2024). Innovative sustainability initiatives in the FMCG industry: A review of challenges and successes.

Nygaard, A. (2023). Is sustainable certification's ability to combat greenwashing trustworthy? *Frontiers in Sustainability, 4*, 1188069.

Ortiz-de-Mandojana, N., & Bansal, P. (2016). The long-term benefits of organizational resilience through sustainable business practices. *Strategic Management Journal, 37*(8), 1615–1631.

Parguel, B., & Johnson, G. (2021). Beyond greenwashing: Addressing 'the great illusion' of green advertising. *Responsible Organization Review, 16*(2), 59–66.

Pimonenko, T., Bilan, Y., Horák, J., Starchenko, L., & Gajda, W. (2020). Green brand of companies and greenwashing under Sustainable Development Goals. *Sustainability, 12*(4), 1679.

Quarshie, A. M., Salmi, A., & Leuschner, R. (2016). Sustainability and corporate social responsibility in supply chains: The state of research in supply chain management and business ethics journals. *Journal of Purchasing and Supply Management, 22*(2), 82–97.

Racelis, A. D. (2014). Sustainable entrepreneurship in Asia: A proposed theoretical framework based on literature review. *Journal of Management for Global Sustainability, 2*(1), 4.

Ramanan, R., & Taback, H. (2016). Ethics and corporate social responsibility for spirituality and sustainability. *Spirituality and sustainability: New horizons and exemplary approaches* (pp. 73–88). Springer.

Rendtorff, J. D. (2020). Principles of business ethics and business legitimacy. *Handbook of business legitimacy: Responsibility, ethics and society* (pp. 783–797). Springer.

Rosário, A. T., Raimundo, R. J., & Cruz, S. P. (2022). Sustainable entrepreneurship: A literature review. *Sustainability, 14*(9), 5556.

Sajid, M., Zakkariya, K. A., Suki, N. M., & Islam, J. U. (2024). When going green goes wrong: The effects of greenwashing on brand avoidance and negative word-of-mouth. *Journal of Retailing and Consumer Services, 78*, Article 103773.

Satpathy, A., Samal, A., Gupta, S., Kumar, S., Sharma, S., Manoharan, G., Karthikeyan, M., & Sharma, S. (2024). To study the sustainable development practices in business and food industry. *Migration Letters, 21*(S1), 743–747.

Schaltegger, S., & Wagner, M. (2011). Sustainable entrepreneurship and sustainability innovation: Categories and interactions. *Business Strategy and the Environment, 20*(4), 222–237.

Schaltegger, S., Lüdeke-Freund, F., & Hansen, E. G. (2016). Business models for sustainability: A co-evolutionary analysis of sustainable entrepreneurship, innovation, and transformation. *Organization & Environment, 29*(3), 264–289.

Schmuck, D., Matthes, J., & Naderer, B. (2018). Misleading consumers with green advertising? An affect–reason–involvement account of greenwashing effects in environmental advertising. *Journal of Advertising, 47*(2), 127–145.

Shoul, A., Hadavinejad, M., & Siahpor, A. (2018). Designing a fuzzy DEMATEL model of sustainable business ethics.

Silver, A. (1989). Friendship and trust as moral ideals: An historical approach. *European Journal of Sociology/archives Européennes De Sociologie, 30*(2), 274–297.

Silvestre, B. S., & Ţîrcă, D. M. (2019). Innovations for sustainable development: Moving toward a sustainable future. *Journal of Cleaner Production, 208*, 325–332.

Singgalen, Y. A., Sijabat, R., Widyastuti, P., & Harnadi, A. (2022). Community empowerment and social welfare development through social entrepreneurship. *Jurnal Studi Sosial Dan Politik, 6*(2), 217–231.

Singh, D., & Awasthy, R. (2025). Spirituality in social entrepreneurship: An ethical approach to sustainability. *Social Enterprise Journal, 21*(3), 492–518.

Singh, S. K., Chen, J., Del Giudice, M., & El-Kassar, A. N. (2019). Environmental ethics, environmental performance, and competitive advantage: Role of environmental training. *Technological Forecasting and Social Change, 146*, 203–211.

Sroka, W., & Szántó, R. (2018). Corporate social responsibility and business ethics in controversial sectors: Analysis of research results. *Journal of Entrepreneurship, Management and Innovation, 14*(3), 111–126.

Szabo, S., & Webster, J. (2021). Perceived greenwashing: The effects of green marketing on environmental and product perceptions. *Journal of Business Ethics, 171*, 719–739.

Szopik-Depczyńska, K., Kędzierska-Szczepaniak, A., Szczepaniak, K., Cheba, K., Gajda, W., & Ioppolo, G. (2018). Innovation in sustainable development: An investigation of the EU context using 2030 agenda indicators. *Land Use Policy, 79*, 251–262.

Tavakoli, M. J. (2010). Introduction to business ethics based on Islamic approach. *The Journal of Knowledge of Islamic Economics, 1*(1), 7–32.

Tennakoon, W. D. N. M. S., Janadari, M. P. N., & Wattuhewa, I. D. (2024). Environmental sustainability practices: A systematic literature review. *European Journal of Sustainable Development Research, 8*(3), 18.

Tiba, S., van Rijnsoever, F. J., & Hekkert, M. P. (2019). Firms with benefits: A systematic review of responsible entrepreneurship and corporate social responsibility literature. *Corporate Social Responsibility and Environmental Management, 26*(2), 265–284.

Trudel, R. (2019). Sustainable consumer behavior. *Consumer Psychology Review, 2*(1), 85–96.

Urbański, M., & Ul Haque, A. (2020). Are you environmentally conscious enough to differentiate between greenwashed and sustainable items? A global consumers perspective. *Sustainability, 12*(5), 1786.

Usher, L. V., Burrows, C. A., Schwartz, C. B., & Henderson, H. A. (2015). Social competence with an unfamiliar peer in children and adolescents with high functioning autism: Measurement and individual differences. *Research in Autism Spectrum Disorders, 17*, 25–39.

Verma, M., & Bharti, U. (2023). Combating greenwashing tactics and embracing the economic success of sustainability. *VEETHIKA—an International Interdisciplinary Research Journal, 9*(3), 6–12.

Wallace, B. (2005). Exploring the meaning (s) of sustainability for community-based social entrepreneurs. *Social Enterprise Journal, 1*(1), 78–89.

Wang, D., & Walker, T. (2023). How to regain green consumer trust after greenwashing: Experimental evidence from China. *Sustainability, 15*(19), 14436.

Wang, X., Zhao, M., & Cheng, L. (2025). Lies behind the green: Can business strategy lead to greenwashing? *Business Process Management Journal.*

Wang, Z., Ni, D., & Zheng, K. (2022). The role of false-claims ban regulation in greenwashing of firms with imprecise greenness information. *Sustainability, 14*(20), 13655.

Weiss, J. W. (2021). *Business ethics: A stakeholder and issues management approach.* Berrett-Koehler Publishers.

Zhang, D. D., & Swanson, L. A. (2014). Linking social entrepreneurship and sustainability. *Journal of Social Entrepreneurship, 5*(2), 175–191.

The Role of Leadership in Sustainable Entrepreneurship

Naser Valizadeh, Pouria Ataei, and Hamid Karimi

Abstract

This study examines the impact of leadership and future trends in sustainable entrepreneurship, drawing on extensive archival research and theoretical frameworks. It highlights the multidimensional nature of sustainable leadership, emphasizing economic, social, and environmental performance beyond traditional financial metrics. Using the Triple Bottom Line framework and various sustainability indicators, the study examines successful examples, including Interface, Nestlé, and Tesla, illustrating how visionary leadership drives organizational transformation toward sustainability. Additionally, the research explores emerging leadership paradigms—including systemic, adaptive, and collective leadership—that are essential to navigating the complexity and uncertainty of today's sustainability challenges. Additionally, this study investigates the role of local leadership in promoting farmer participation in sustainable agricultural value chains in rural Iran, utilizing the Theory of Planned Behavior as a framework. Results from this case study confirm that local leadership significantly enhances farmers' behavioral intentions by fostering trust, coordination, and access to resources. The main contribution of the study lies in integrating theoretical insights with practical examples to provide a comprehensive understanding of the evolving role of sustainable leadership. It offers guidance for policymakers and business leaders to develop leadership models that align profitability with social and environmental responsibility, preparing organizations to thrive in the next era of sustainable entrepreneurship.

Keywords

Sustainable leadership · Triple bottom line · Value chain development · Local leadership · Sustainable entrepreneurship

N. Valizadeh (✉)
Department of Agricultural Extension and Education, College of Agriculture, Shiraz University, Shiraz, Iran
e-mail: n.valizadeh@shirazu.ac.ir

P. Ataei
Department of Socio-Economic and Agricultural Extension Research, Fars Agricultural and Natural Resources Research and Education Center, Agricultural Research, Education & Extension Organization (AREEO), Shiraz, Iran

H. Karimi
Department of Agricultural Extension and Education, Faculty of Agriculture, University of Zabol, Zabol, Iran

M. Akbari et al. (eds.), *Sustainable Entrepreneurship*, University of Tehran Science and Humanities Series,
https://doi.org/10.1007/978-3-032-04909-4_20

Learning Objectives

1. Understanding the multidimensional nature of sustainable leadership and its impact on economic, social, and environmental performance.
2. Analyzing the role of local leadership in facilitating farmers' participation in sustainable value chain businesses.
3. Exploring key performance indicators used to measure leadership effectiveness in sustainable entrepreneurship.
4. Evaluating future trends and evolving paradigms in sustainable leadership, including climate leadership and purpose-driven entrepreneurship.
5. Developing strategies to empower local leaders and integrate sustainability principles into rural development programs.

1 Introduction

In recent years, the issue of sustainability has attracted the attention of various groups in society, especially organizations. The most common definition of sustainability is the practice of meeting the needs of the present without compromising the ability of future generations to meet their own needs, so that everyone has access to the necessary resources. In other words, meeting the needs of the present generation without undermining the ability of future generations to meet their own needs (Adelowo & Henrico, 2025). The ongoing challenge is to develop mechanisms and strategies that optimize processes, ultimately leading to a more sustainable world (Dote-Pardo et al., 2025). Given the intensification of global challenges such as climate change, resource scarcity, and social inequalities, the requirements for leadership in sustainable entrepreneurship are likely to intensify. Leaders who can anticipate these trends and position their organizations accordingly will gain a competitive advantage. This requires not only technical knowledge about sustainability issues but also the ability to think strategically about how industries and markets will change under the influence of these issues. The importance of sustainable leadership and sustainable entrepreneurship and their impact on organizations is increasingly evident. Organizational success and maximum performance depend on the leadership and entrepreneurial roles of leaders and entrepreneurs. In addition to serving as role models for other employees, leaders and entrepreneurs are responsible for realizing the company's vision and goals (Belgibayeva et al., 2023).

The concept of sustainable leadership has emerged from the combination of sustainable development and leadership (Liao, 2022). A sustainable leader must create an appropriate organizational environment for achieving sustainable development (Ahmad, 2025) and focus on relationships within and outside the company (Cheng et al., 2025). In other words, the leader plays multiple roles simultaneously. Sustainable leadership involves behaviors and actions that create lasting value for all stakeholders, including society, the environment, and future generations (Iqbal et al., 2025). Sustainable leadership goes beyond green transformational leadership and responsible leadership (Liao, 2022). It means having the capacity to make long-term decisions, promoting systematic innovation, cultivating a loyal team of employees, and delivering high-quality products, services, and solutions (Ahmad, 2025; Irviani, 2025).

The earliest studies on sustainable entrepreneurship highlight the importance of integrating the concepts of entrepreneurship and sustainable development, arguing that the concept of sustainable development presents a significant opportunity for entrepreneurs and innovators (Anand et al., 2021). From a sustainability perspective, sustainable entrepreneurship is seen as a means to achieve a more sustainable future (Ansari, 2025) and to solve complex social and environmental problems through the implementation of innovative solutions (Eynolghozat et al., 2025; Jamil et al., 2025). In this framework, the sustainable entrepreneur should be open to sustainability-related opportunities and be able to bridge the gap between sustainability issues and market needs (Naguib & Barbar, 2025). Although there is continuous growth in the literature on sustainable leadership (Suriyankietkaew et al., 2022) and sustainable entrepreneurship (Ansari, 2025; Purwana et al., 2025; Singh & Awasthy, 2025), there are significant gaps in understanding the relationship between leadership and entrepreneurship from a sustainability perspective. This gap highlights the need for in-depth research to clarify the interrelationship between these two fundamental concepts.

2 Main Body

2.1 Defining Sustainable Entrepreneurship in the Leadership Context

The concept of sustainable entrepreneurship has emerged as a vital paradigm in modern business, combining traditional entrepreneurial goals with the urgent need for environmental protection and social responsibility (Tajpour et al., 2025). At its core lies a fundamental shift in the way businesses operate, a move beyond profit maximization to a triple-bottom-line approach that balances economic, social, and environmental outcomes. This transformation is not simply about adopting greener practices but also requires redefining business models, value creation processes, and organizational cultures (Yiadom et al., 2025). Leadership plays a pivotal role in this transition, acting as a bridge between the abstract principles of sustainability and operational business strategies (Gracia-Zomeño et al., 2025).

In recent decades, the concept of entrepreneurship has shifted from traditional models that focus solely on financial profitability to one that emphasizes sustainable entrepreneurship. This shift has been driven by increasing awareness of environmental issues, social inequalities, and the limitations of natural resources. In the past, businesses were primarily evaluated based on short-term metrics, such as financial returns and market growth. However, today, the role of companies in creating value for society and preserving the environment has become equally important. The shift from traditional entrepreneurship to sustainable entrepreneurship reflects broader shifts in societal values and priorities. While businesses may have previously focused solely on financial metrics, today's entrepreneurs operate in an ecosystem where consumers, investors, and regulators demand responsible business practices (Adelowo & Henrico, 2025). This shift has been driven by growing awareness of climate change, resource depletion, and social inequalities, which have transformed sustainability from a marginal concern to a business imperative.

Leaders who understand this shift understand that sustainable entrepreneurship does not mean sacrificing profits for ethical principles but rather finding innovative ways to align the two. Sustainable entrepreneurship not only seeks economic profit but also prioritizes social and environmental responsibility (Purwana et al., 2025; Tajpour et al., 2025). For example, companies that use recycled materials or employ renewable energy are examples of this shift. This new approach necessitates adjustments in management structures, supply chains, and even marketing strategies. Today's entrepreneurs must balance the three main pillars of sustainability: Profit, People, and Planet.

A key characteristic of sustainable entrepreneurship lies in its long-term orientation. Traditional business models often prioritize short-term gains over long-term sustainability. Sustainable entrepreneurs, guided by forward-thinking leadership, take a more holistic view that considers intergenerational impacts and systemic implications (Dote-Pardo et al., 2025). This requires leaders to cultivate systems thinking—the ability to understand how different elements in an ecosystem interact and influence each other. For example, decisions about sourcing raw materials affect not only production costs but also environmental degradation, community relations, and brand reputation. Sustainability in business is based on principles that, without which, long-term goals cannot be achieved. The first principle is corporate social responsibility, which emphasizes a company's commitment to the well-being of its employees, customers, and society (Hou et al., 2025). This responsibility can take the form of providing fair wages, creating a safe working environment, or participating in social projects. The second principle is resource efficiency, which focuses on reducing waste and using raw materials more efficiently (Bejarano-Auqui et al., 2025). Leading companies in this field use new technologies to recycle materials or minimize energy consumption. This not only reduces costs but also attracts customers who trust environmentally friendly brands. The third principle is transparency and accountability. Sustainable businesses provide detailed reports on their social and environmental performance to gain the trust of stakeholders (Karlibaeva et al., 2024). This transparency makes companies feel more responsible for their decisions and encourages them to avoid mere show-offs. Finally, sustainable innovation is proposed as the fourth principle, which involves developing new products and services with minimal negative environmental impact (Ansari, 2025).

Measurement and accountability systems in sustainable entrepreneurship also reflect the distinct requirements of leadership. Unlike traditional businesses that can rely primarily on financial metrics, sustainable companies require more comprehensive performance indicators (Liu et al., 2022a, 2022b; Mamani et al., 2022). Leaders must develop and monitor metrics that measure social and environmental impacts alongside financial results. This triple bottom line approach requires sophisticated data collection and analysis capabilities, as well as a willingness to be transparent about successes and shortcomings. Business leaders play a critical role in realizing sustainable entrepreneurship. They must have a long-term perspective and be able to balance financial interests, social needs, and environmental preservation. A successful leader in this area thinks not only about the company's profitability but also about the impact of their decisions on employees, customers, and society at large. One of the most important tasks of leaders is to promote an organizational culture based on sustainability (Pandey & Pattnaik, 2020; Rausser & Strielkowski, 2023). This is possible by educating employees, setting clear goals, and encouraging green innovations. For example, some companies design programs to reduce their carbon footprint or support social projects in collaboration with non-governmental organizations (NGOs). In addition, leaders must be able to convince various stakeholders that sustainability is not only a moral imperative but also an economic opportunity. Investors today are more inclined than ever to invest in companies that comply with environmental, social, and governance criteria (Ribeiro & Leitão, 2024).

Therefore, a leader who can manage this balance effectively not only ensures the company's future but also sets a model for others. Ultimately, sustainable entrepreneurship is not a choice but a necessity for the survival of businesses in the twenty-first century. Companies that start thinking about integrating sustainability principles into their strategies today will enjoy a long-term competitive advantage and play a more effective role in building a better future. The definition of sustainable entrepreneurship within the leadership framework also demonstrates that this concept is both a fundamental departure from traditional business practices and a natural evolution of them. This concept reflects the maturity of capitalism, in which success is measured not only by what is achieved but also by what is offered to society and preserved for future generations.

2.2 Sustainable Leadership Competencies and Skills

In today's era, complex challenges such as climate change, social inequality, and natural resource constraints have compelled businesses to consider long-term sustainability. Systems thinking, as a holistic framework, enables leaders to view their organizations not as isolated entities but as part of an interconnected network of economic, social, and environmental factors. This perspective is a fundamental foundation for achieving long-term sustainability. Sustainable leadership necessitates a profound understanding of the intricacies of economic, social, and environmental systems. Systems thinking enables leaders to understand the interrelationships between the various components of a business and their broader impacts on society and the environment (Silvestri & Veltri, 2020). Unlike the traditional view that analyzes problems in isolation, this holistic approach provides solutions that simultaneously respond to financial, social, and environmental goals (Irviani, 2025). For example, a manufacturing company that focuses solely on reducing operating costs may use inexpensive but polluting raw materials. However, a leader with systems thinking considers, in addition to financial considerations, the impact of these materials on worker health, environmental pollution, and even long-term brand reputation. Such a leader understands that today's decisions shape tomorrow's consequences and, therefore, designs strategies that are not only profitable but also ensure the sustainability of the ecosystem.

Systems thinking is based on the principle that all components of a system are interconnected and that changes in one part can have unforeseen effects in other parts (Luhas & Mikkilä, 2024). For example, a company may reduce production costs by substituting cheaper raw materials; however, if these materials are harmful to employee health or the environment, they will ultimately face medical expenses, legal fines, or a loss of brand reputation. Leaders who view issues from this perspective can make decisions that not only increase financial profits but also have positive social and environmental impacts. One of the limitations of traditional leadership is that it focuses on solving surface problems without addressing the underlying factors. Systems thinking teaches leaders to identify and address the root causes of challenges rather than just addressing symptoms (Suriyankietkaew, 2019). For example, if an organization's employees are underperforming, the quick fix might be to increase the workload or lay off employees. However, a systems-oriented leader considers factors such as organizational culture, employee motivation, and work-life balance and implements solutions like improving the workplace or developing training programs (Luhas & Mikkilä, 2024). This approach not only addresses productivity issues but also enhances job satisfaction and organizational commitment. Many business decisions that seem profitable in the short term may cause irreparable damage in the long term. Systems thinking enables leaders to predict the potential consequences of their actions by modeling various scenarios (Cruz-Sandoval et al., 2023). For example, an agricultural company that uses traditional irrigation methods may incur lower costs in the short term, but in the long term, it will face a water shortage crisis and a decline in soil quality. In contrast, investing in smart irrigation systems not only reduces water consumption but also ensures long-term profitability.

In systems thinking, leaders must collaborate with various stakeholders as one of the essential actions that lead to long-term sustainability. No organization can achieve sustainability alone. Systems thinking emphasizes the importance of collaboration with various stakeholders, including the government, local community, non-governmental organizations, and even competitors (Luhas & Mikkilä, 2024). For example, a manufacturing company can develop environmentally friendly technologies in partnership with universities and research centers or implement training programs to empower the workforce in collaboration with local communities. Such collaborations not only solve common challenges but also create new opportunities for innovation and growth. The role of technology and data-driven thinking is also very important in systems thinking. Today, digital technologies such as artificial intelligence and the Internet of Things have provided leaders with powerful tools to implement systems thinking. Using these technologies, organizations can collect accurate data on energy consumption, waste generation, or the social impact of their activities and make more informed decisions based on them. For example, a retailer can optimize its supply chain to reduce food waste by analyzing customer purchase data, or a manufacturing company can minimize energy consumption by monitoring its equipment in real time.

Systems thinking is not just an analytical tool but a way of thinking that can fundamentally transform the way organizations lead. Leaders who adopt this approach not only prevent potential crises but also create opportunities to create shared value for business, society, and the environment. In today's world, where issues are highly interconnected, systems thinking is not an option but a necessity for organizations to survive and grow sustainably. One of the primary challenges in sustainable leadership is aligning the interests of various stakeholder groups. Shareholders consider financial returns, employees prioritize fair working conditions, customers value the quality and social responsibility of the company, and the local community may be concerned about the environmental impact of the business. In these situations, emotional intelligence helps leaders communicate effectively by understanding the emotions and motivations of these groups.

In today's complex world, where businesses face diverse and sometimes conflicting stakeholder demands, emotional intelligence has become one of the most essential skills for sustainable leaders. This concept, first introduced by Daniel Goleman, refers to the ability to understand, control, and appropriately express one's own and others' emotions (Mortan et al., 2014). In the context of organizational sustainability, emotional intelligence plays a vital role in balancing economic, social, and environmental interests. Leaders with high emotional intelligence can put themselves in the shoes of different stakeholders and understand their perspectives. This empathy enables them to better identify the expectations of employees, customers, investors, local communities, and even future generations (Gao & Tsai, 2024). For example, when deciding to change a production line to more sustainable methods, such a leader can understand both the financial concerns of shareholders and the environmental concerns of community activists.

In many cases, the interests of different stakeholder groups conflict with each other. Emotional intelligence enables leaders to manage these conflicts constructively. By controlling their own and others' emotions, they can create a safe space for discussion and reach win–win solutions (Esteves et al., 2024). A leader with high emotional intelligence knows how to remain calm in tense meetings, listen to all sides, and make suggestions that meet the basic needs of all groups as much as possible.

Shifting towards sustainability often requires fundamental changes in organizational culture. Leaders with high emotional intelligence can drive these changes with minimal resistance. They can articulate the organization's sustainability vision in an inspiring way that not only empowers employees to accept it but also encourages them to enthusiastically participate in its realization (Nakhaei et al., 2013). These leaders know how to connect people's values with organizational goals and create a sense of belonging (Jayaweera et al., 2025). One important aspect of emotional intelligence in sustainable leadership is the ability to communicate effectively with a wide range of stakeholders. Successful leaders in this area can articulate complex sustainability messages in simple, understandable language for different groups. They can create realistic hope while being transparent about the challenges (Sharma & Kumra, 2025). For example, when introducing a new sustainable initiative, such a leader can simultaneously explain their long-term benefits to investors and describe their positive social impacts on the local community. True sustainability is not possible without stakeholder trust. Leaders with high emotional intelligence know how to build trust through honesty, dependability, and accountability. They keep their promises, are transparent in times of crisis, and take responsibility for mistakes (Nakhaei et al., 2013). These traits are especially critical in the face of sustainability crises, where maintaining public trust can mean the difference between an organization's success and failure.

Sustainable leaders with high emotional intelligence not only pay attention to quantitative data but also closely monitor the emotions and attitudes of stakeholders (Zadorozhny et al., 2025). They can measure the emotional temperature of the organization and society through various tools, such as regular surveys, discussion forums, and social networks, and make informed decisions based on those results. This ability is especially valuable in times of rapid social and environmental change. In today's rapidly changing world of stakeholder expectations, emotional intelligence has become a key factor in the success of sustainable leadership. This skill enables leaders to build effective human relationships, manage conflict, and adapt to change amidst the complexities of sustainability. Emotional intelligence serves as a bridge between sustainability theories and their practical application in organizations. Leaders who develop this skill not only build more sustainable organizations but also inspire positive change in society as a whole. As businesses and organizations face complex economic, social, and environmental challenges, purpose- and values-based decision-making has become a fundamental pillar of sustainable leadership. This approach extends beyond traditional measures of short-term profitability, aiming to create lasting value for all stakeholders. Higher purpose refers to a deeper and more meaningful mission that goes beyond mere financial goals (Mawardi et al., 2025). Leading organizations in the field of sustainability define their purpose in terms of creating positive change in the world (Wang, 2024). For example, a clothing company might define its higher purpose as "respecting human dignity throughout the supply chain." This type of purpose serves as the organization's moral compass, guiding all decision-making.

The core values of sustainable organizations are usually shaped around concepts such as transparency, fairness, responsibility, and respect for the environment (Ünal, 2023). These values are not just slogans but are practically institutionalized in decision-making processes. A sustainable leader, when faced with difficult decisions, always asks himself: "Is this choice consistent with our core values?" This approach allows decisions to remain coherent and integrated even in challenging situations. Sustainable decision-making requires consideration of multiple dimensions. An effective model in this regard is the sustainable triple-bottom-line framework, which considers the economic, social, and environmental dimensions simultaneously (Muafi et al., 2025; Oduro Owusu & Gil, 2025). Successful leaders know how to balance these dimensions and make decisions that benefit all stakeholders in the long term. For example, when choosing a new supplier, in addition to price and quality, workers' working conditions and the environmental impact of production are also evaluated. Sustainable organizations adopt a participatory approach rather than a centralized, top-down decision-making process. This means involving a wide range of stakeholders in the decision-making process. For example, some companies are using innovative methods such as "stakeholder forums," in which representatives of employees, customers, local communities, and even future generations are involved in strategic discussions. This approach not only improves the quality of decisions but also facilitates their acceptance and implementation.

Purpose-based decision-making requires accurate impact measurement systems Fontana (Liu et al., 2022a, 2022b). Leading organizations utilize frameworks such as the Theory of Change, which enable them to map the relationship between actions and desired outcomes clearly. These organizations accurately measure and monitor not only traditional financial indicators but also metrics such as reducing inequality, improving working conditions, or reducing carbon footprints (Le Grange et al., 2024). Despite the apparent benefits, values-based decision-making comes with particular challenges. The most important of these challenges include competitive pressures, short-term stakeholder expectations, and the complexities of measuring non-financial impacts (Ahmad, 2025). To overcome these challenges, sustainable leaders need moral courage, perseverance, and the ability to convince stakeholders of the long-term benefits of this approach (Javed & Al-Mulali, 2025). Purpose- and values-based decision-making is the beating heart of sustainable leadership. This approach enables businesses to stay on track in today's turbulent world while helping to create positive change in society and the environment. Leaders who put this philosophy into practice not only build more resilient organizations but also inspire transformation throughout their businesses. Ultimately, value-based decision-making is not a choice but a necessity for survival and growth in the twenty-first century.

With rapid environmental change, economic volatility, and unexpected crises becoming the norm, resilience has become a key competency of sustainable organizations. This concept extends beyond the ability to survive in challenging conditions and encompasses the organization's capacity to anticipate crises, adapt swiftly to changes, and even emerge stronger from the crisis (Teniola et al., 2025). Sustainable organizations and businesses must develop resilience at three levels: strategic, operational, and cultural. At the strategic level, these organizations anticipate and plan for different future scenarios. At the operational level, they create resilient systems such as agile and distributed supply chains. At the cultural level, they institutionalize a spirit of learning and adaptation throughout the organization. This multi-level approach enables organizations to address various crises while upholding the principles of sustainability. Crisis management in sustainable businesses has unique characteristics. Instead of sectional and superficial reactions, these businesses seek radical solutions that simultaneously consider economic, social, and environmental dimensions. For example, in the face of a raw material shortage crisis, a sustainable business may simultaneously adopt several strategies: diversifying suppliers, investing in sustainable alternative materials, and partnering with the local community to develop sustainable resources. This comprehensive approach not only manages the current crisis but also prepares the organization for future challenges.

A distinctive feature of resilient, sustainable businesses is their ability to learn from crises and turn them into opportunities for improvement. After each crisis, these businesses systematically analyze events rather than blaming and denying them. They use tools such as after-action reviews to document lessons learned and incorporate them into future strategies (Abdelwahed et al., 2025; Zamfir et al., 2018). For example, many organizations have implemented more flexible work systems and stronger digital infrastructures in the wake of the COVID-19 pandemic, which has both increased productivity and reduced the business's carbon footprint. Sustainable businesses understand that how they respond to a crisis can be as crucial as the crisis itself. Rather than hiding, these businesses communicate honestly with their stakeholders during difficult times. They not only acknowledge the problems but also clearly articulate the actions they will take to solve them and prevent them from happening again. This transparency proactively fosters stakeholder trust and enhances the organization's long-term social capital.

Leading businesses in the field of sustainability do not consider resilience a cost but rather a strategic investment. These businesses continuously invest in developing resilience capabilities, including:

- Creating strategic reserves for critical raw materials
- Developing integrated information systems to monitor risks
- Training employees in multiple skills and systems thinking
- Creating networks of collaboration with other businesses and institutions

Resilience and crisis management in sustainable businesses is a proactive and forward-looking approach, not a reactive and short-term one (Alzamel, 2024). These businesses understand that crises are inevitable, but by planning strategically, investing in resilience capabilities, and maintaining a commitment to sustainability principles, they can use these challenges as opportunities for learning and transformation. In today's turbulent world, sustainable resilience is not an option but a necessity for the survival and long-term growth of businesses.

2.3 Visionary Leadership and Sustainability-Driven Innovation

In this era of increasingly complex environmental and social challenges, forward-looking leadership with a focus on sustainability has become a key factor in business success. Therefore, forward-looking leadership with a focus on sustainability is no longer a choice but an inevitable necessity. This type of leadership extends beyond day-to-day business management and, by creating an inspiring vision, guides the organization and its stakeholders toward a sustainable future. Forward-looking leaders understand that sustainability is not just a legal requirement or a tool to enhance the brand image but a strategic imperative for survival and growth in today's rapidly changing world. Organizations that adopt this type of leadership not only strengthen their competitive position but also make a valuable contribution to building a better future for everyone. These organizations have demonstrated that it is possible to be both profitable and have a positive impact on the world simultaneously (Li et al., 2021). Forward-looking leadership for sustainability demonstrates that business and sustainability are not incompatible but can reinforce each other and help create a more sustainable and prosperous future for all.

Building a sustainable vision and mission requires a deep understanding of the complex relationships between business, society, and the environment. A compelling vision must be both ambitious to inspire fundamental change and realistic to be actionable (Richardson & Burrell, 2023). For example, the vision of the Danish company Ørsted, which transformed itself from a fossil fuel-based energy company to a pioneer in renewable energy, is a prime example of this approach. By defining a global vision that relies entirely on green energy, the company transformed its strategy and operations, becoming a model for the energy industry (Dzhengiz et al., 2023). The mission of sustainable organizations should also clearly articulate how they combine creating value for shareholders with creating value for society and the environment. Integrating sustainable innovation into an organization's strategy is essential for any business, necessitating a systematic and holistic approach. This means integrating sustainability considerations into all strategic decision-making processes, from strategy formulation to implementation and evaluation (Khan et al., 2025). Leading companies in this field use tools such as stakeholder analysis, product life cycle assessment, and environmental and social impact modeling to ensure that their innovations are truly sustainable (Ekanem, 2024). For example, the Dutch company Philips, by implementing the "Innovation with Positive Impact" program, guides all its research and development activities based on sustainability criteria, designing products that meet customer needs while having a positive impact on society and the environment (van der Duin et al., 2014).

On the other hand, the role of leaders in fostering a culture of sustainability in businesses cannot be ignored. Organizational culture, as a set of shared values, beliefs, and norms, determines how employees behave and make decisions (Iddrisu, 2025). Forward-looking leaders institutionalize a culture of sustainability within their organizations through various measures, including modeling sustainable behaviors, establishing effective incentive systems, and developing comprehensive training programs (Hu et al., 2025). They understand that real change happens when sustainability becomes a natural part of the everyday thinking and practice of all employees rather than being confined to the CSR department.

One of the main challenges in leading a forward-looking sustainability journey is internal and external resistance (Pett et al., 2025). Many stakeholders may see the fundamental changes required for sustainable transformation as a threat to their short-term interests. Effective leaders manage this resistance through a combination of strong communication skills, transparency in decision-making, and stakeholder engagement in the change process. They know how to articulate the long-term benefits of sustainability in a way that is appealing and understandable to different stakeholder groups (Rojas Casarrubias et al., 2025). Measuring and evaluating the impact of sustainable innovations is also a critical part of this process. Forward-thinking leaders use advanced measurement systems that go beyond traditional financial metrics. These systems also accurately measure the social and environmental impacts of decisions and innovations, enabling continuous improvement and refinement (Fedajev et al., 2025). As noted earlier, some companies utilize frameworks such as the Theory of Change to map the connections between their actions and the desired social and environmental outcomes.

2.4 Stakeholder Engagement and Collaborative Leadership

Given the complexity of sustainability challenges, the need to rethink traditional leadership styles is increasingly felt. Old approaches that emphasized centralized and hierarchical decision-making no longer meet the needs of the present era. In contrast, participatory leadership has emerged as a new paradigm that emphasizes the active and meaningful participation of all stakeholders in the process of sustainable transformation. This style of leadership is rooted in the deep belief that isolated efforts cannot solve complex contemporary social and environmental issues; instead, they require comprehensive coordination and cooperation among all social actors. Engaging employees in the process of sustainable transformation is a fundamental pillar of this approach. Employees, often recognized as an organization's most important asset, play a dual role as both change agents and a unique source of ideas and innovations (Cleveland & Cleveland, 2025). Leading organizations in this field employ a diverse range of participatory mechanisms that extend beyond traditional methods. These mechanisms include interactive sustainability-focused training programs, collaborative green solution design workshops, sustainability-focused e-suggestion systems, and even internal entrepreneurship programs focused on social innovation (Grøn, 2025). For example, a leading company like Unilever has designed its "sustainable innovation program" in such a way that all employees at all levels of the organization can submit their ideas for more sustainable products and processes. The company allocates special funds and forms interdisciplinary teams to implement the best ideas (Nhu Laursen & Houman Andersen, 2023). This level of engagement not only leads to creative solutions but also significantly increases employees' sense of ownership and responsibility for the organization's sustainability goals. Extensive research in this area shows that organizations that effectively engage their employees in the sustainable transformation process not only have higher success rates in implementing sustainability initiatives but also see significant improvements in key organizational performance indicators such as productivity, employee retention, and customer satisfaction (Cleveland & Cleveland, 2025; Grøn, 2025).

Cross-functional collaborations to create effective social and environmental impact are another dimension of participatory leadership that has gained increasing importance in recent years. In an increasingly interconnected world, complex issues such as climate change, social inequalities, or global supply chain challenges cannot be solved within the limited framework of a single organization or even a specific industry. This is where strategic partnerships between different sectors of society, including businesses, governments at various levels, civil society organizations, educational and research institutions, and even commercial competitors, become a necessity. Successful examples of such partnerships can be seen in global initiatives such as the Alliance to End Plastic Waste, in which more than 80 companies from across the plastic value chain, including chemical manufacturers, consumer goods companies, retailers, and waste management companies, have come together to jointly invest more than $1.5 billion to address the plastic waste crisis in the environment (Kumar et al., 2025). Although these partnerships come with specific challenges, such as differing priorities, diverse working methods, and sometimes conflicting interests, they can achieve results that are beyond the capabilities of either organization alone if they are appropriately managed and utilize effective governance frameworks. The key to success in these partnerships is building mutual trust, clearly defining shared goals, and designing transparent accountability mechanisms. The role of leaders in shaping policy and advocacy activities also becomes particularly important in this context. Leading sustainability leaders have come to understand that real and lasting impact extends beyond their organizational boundaries and that to create systemic change, they must also actively participate at the macro-level policy level (Cheng et al., 2025). These leaders help shape the policy environment in various ways, including participating in the development of sustainable industry standards, collaborating with regulatory bodies to design progressive regulatory frameworks, supporting public awareness campaigns, and engaging in international forums (Irviani, 2025). For example, many CEOs of leading renewable energy companies not only actively participate in international climate change conferences, such as the COP, but also contribute their scientific expertise and practical experience to shaping clean energy policies at both national and international levels. This level of involvement requires a unique combination of skills, including the ability to work in complex multistakeholder environments, a deep understanding of public policy processes, the ability to build strategic alliances across organizations, and advanced communication and negotiation skills. These leaders know how to translate complex technical messages into the language of policymakers while also protecting the long-term interests of their businesses.

One of the main challenges in participatory leadership is managing the expectations and sometimes conflicting interests of different stakeholders. Effective leaders in this area utilize advanced tools, such as "dynamic stakeholder mapping," which not only analyzes the current positions of various stakeholders but also anticipates potential changes in their positions and priorities (Subrahmanyam, 2025). These leaders have a deep understanding that balancing short-term and long-term interests, between local focus and global perspective, and between financial and social/environmental goals requires a nuanced approach based on a thorough knowledge of stakeholder dynamics. They possess advanced communication skills, the ability to listen actively and deeply, and the capacity to build trust among groups with sometimes conflicting interests. These leaders know how to facilitate dialogue between different stakeholders in crises and offer creative solutions that meet the basic needs of all parties to the greatest extent possible. Measuring and evaluating the impact of stakeholder engagement is also a critical yet complex part of this process. Leading organizations in this field use advanced, multidimensional measurement frameworks that go beyond traditional quantitative measures. These frameworks encompass advanced qualitative measures, including the depth of stakeholder engagement, the quality of relationships established, the degree of social trust created, the collective capacities developed, and the indirect and long-term impacts of engagement (Grøn, 2025; Subrahmanyam, 2025; Waseel et al., 2025). For example, some leading organizations are using innovative digital storytelling and human-centered techniques to document the human impact of their participatory programs (Waseel et al., 2025). These organizations recognize that the real impact of stakeholder engagements often lies in the personal stories and qualitative changes in the lives of individuals and communities that traditional quantitative metrics cannot measure.

In today's world, where trust in traditional institutions is declining and sustainability challenges are increasingly complex and interconnected, participatory leadership is no longer an option but a critical necessity for organizational survival and growth. Organizations that engage effectively and authentically with a wide range of stakeholders are not only able to design and implement more sustainable and practical solutions but also have much greater flexibility and resilience in the face of crises and external shocks. These organizations have understood that real and sustainable solutions to today's complex challenges depend on the collective cooperation and participation of all social actors. Effective leaders are those who can facilitate, coordinate, and guide this cooperation. Ultimately, participatory leadership demonstrates that the path to a sustainable and equitable future lies in the active, informed, and meaningful participation of all stakeholders. True success is achieved when all members of society participate in and benefit from its benefits. This type of leadership not only guides organizations on the path to sustainability but also helps create systemic and sustainable changes at the community level.

2.5 Digital and Technological Enablers for Sustainable Leadership

In this era of the Fourth Industrial Revolution, digital technologies have become indispensable tools for leaders committed to sustainability. These developments have profoundly transformed traditional management and leadership practices, creating a new paradigm in which data-centricity and advanced technologies complement human judgment and values in strategic decision-making. Leading organizations worldwide have recognized that intelligently integrating these technologies into management processes not only enhances operational efficiency but also enables the simultaneous achievement of economic, social, and environmental goals. The profound transformation in strategic decision-making enabled by artificial intelligence and big data is one such case. Artificial intelligence and big data analytics have become the backbone of informed decision-making in sustainable organizations today. These technologies allow leaders to move beyond intuitive and empirical decision-making to deep, evidence-based analysis (Tasnim, 2024). Advanced AI systems can identify complex patterns in vast amounts of data on environmental performance, social impacts, and economic indicators that are often hidden from human analysts.

In the energy industry, leading companies such as NextEra Energy are utilizing machine learning algorithms to predict energy demand more accurately. By analyzing historical consumption data, weather patterns, and social events, these systems can provide accurate predictions that lead to the optimization of energy production and distribution. The result of this transformation has been a significant reduction in energy waste while simultaneously reducing operating costs. At a broader level, AI-based simulation platforms, such as those developed by companies like Palantir, are enabling leaders to model various sustainability scenarios and assess the long-term impacts of their strategic decisions with unprecedented precision. These technologies are particularly valuable in natural resource management, where predictive systems can analyze consumption patterns and provide optimal solutions to conserve resources. However, numerous ethical and technical challenges arise along the way. The issue of algorithmic bias is one of the most significant, as it can lead to unfair decision-making. Leading leaders are developing comprehensive oversight frameworks that ensure transparency, accountability, and fairness in AI systems (Tasnim, 2024). For example, Microsoft has launched its "Responsible AI" initiative, which includes rigorous ethical principles and independent oversight mechanisms for AI systems.

Blockchain technology, as one of the most transformative digital innovations in the service of sustainability, has also enabled the creation of fully transparent and traceable supply chains. The immutable and decentralized nature of this technology has revolutionized the monitoring of products throughout their entire production and distribution cycle (Bae et al., 2024). In the fashion industry, leading brands such as Stella McCartney are utilizing blockchain platforms to trace the origin of their raw materials fully. These systems enable consumers to track the entire journey of a product—from the cotton field to the store—by scanning a QR code, ensuring that ethical and environmental standards are met at every stage (Qu & Chen, 2021). This level of transparency has not only increased consumer trust but also strengthened the incentive for producers to adhere to sustainable standards.

In the food and agriculture industry, companies such as IBM Food Trust are using blockchain to create transparent food supply chains. These systems can identify the source of contamination or quality issues within seconds, whereas traditional methods may take days or weeks (Theodorakopoulos et al., 2024). This capability not only increases food safety but also significantly reduces food waste. However, the successful implementation of blockchain in the supply chain requires extensive collaboration between all actors in the value chain. Technical challenges, including scalability, energy consumption, and integration with existing systems, are also significant obstacles (Kayikci et al., 2024).

Leading leaders are developing industry standards and innovative collaboration models that allow the full potential of blockchain to be exploited (Ulrich et al., 2024). In smart leadership, the integration of technology and human wisdom is a principle. Visionary leadership in the digital age means strategically and integratively leveraging different technologies to advance sustainability goals. This approach extends beyond the instrumental use of individual technologies, necessitating a comprehensive understanding of how these technologies interact and coordinate to produce transformative impacts (Tasnim, 2024). The Internet of Things is a key technology in this area. Today's vast networks of smart sensors enable leaders to monitor the environmental performance of their businesses in real-time (Clarysse et al., 2022). The primary challenge in this area is to strike a balance between leveraging technology and attending to the human dimensions of sustainability. Smart leaders understand that technology should never replace human connection and direct understanding of social issues. For example, in rural development projects, the use of digital technologies must be accompanied by genuine participation of villagers and attention to local needs.

Another requirement is the transformational integration of digital technologies. Effective integration of these advanced technologies requires fundamental changes in organizational structures and processes (Licite-Kurbe & Chandramohan, 2020). Leading digital sustainability leaders are fundamentally redefining their operating models and management practices. This transformation includes creating interdisciplinary teams of sustainability experts, data scientists, and software engineers. For example, Unilever has established digital sustainability command centers where diverse teams continuously monitor and analyze data related to sustainability performance worldwide. Developing integrated data infrastructures is also essential for this transformation. Leading organizations are creating digital twins that provide accurate virtual models of their physical operations. These models enable the simulation and testing of sustainability solutions prior to their physical implementation (Huang et al., 2022).

Soon, we will see the emergence of a new generation of digital technologies that have the potential to transform sustainable leadership further. Quantum computing has the potential to revolutionize the modeling of complex environmental systems (Sotelo, 2021). The metaverse could offer new ways to engage stakeholders and educate them on sustainability (Alomar & Alatawi, 2025). Next-generation artificial intelligence could provide a deeper understanding of the complex relationships between economic activities and natural systems (Tasnim, 2024). However, smart leaders understand that technology alone is not the end-all be-all. The key challenge facing today's leaders is how to use these technologies in a way that, while increasing economic efficiency, also contributes to a more just society and a healthier environment. Meeting this challenge requires a unique combination of strategic vision, technical skills, and a deep commitment to sustainable values. Leaders who can effectively manage this combination will not only lead their businesses on a sustainable path but will also become role models for the transformation of industries and societies. The future belongs to businesses that can creatively combine digital technologies with human values and social responsibility.

2.6 Measuring Leadership Impact in Sustainable Entrepreneurship

As sustainability has become a fundamental pillar of business, measuring the actual impact of leadership in sustainable entrepreneurship has become a major challenge for researchers and practitioners in this field. Leading organizations have recognized that measuring the effectiveness of sustainable leadership necessitates a multidimensional approach that extends beyond traditional financial metrics and carefully evaluates social and environmental impacts. This transformation in the way performance is measured necessitates the development of new indicators, assessment methods, and a shift in our perspective on the concept of business success. Key performance indicators in the field of sustainable leadership should reflect leaders' commitment to creating a balance between economic, social, and environmental goals (Singh, 2022). Unlike traditional indicators that focus mainly on financial profitability, sustainable leadership indicators should provide a more comprehensive picture of a business's performance. One of the most important indicators is the "employee retention rate", which measures the quality of the workplace and the leader's attention to the well-being of the workforce. Leading businesses in this field have an average employee retention rate of 85–95%, which is much higher than the industry average (Kickul et al., 2018). Another important indicator is the "carbon intensity per unit of production," which measures the amount of greenhouse gas emissions relative to the volume of production (Yiadom et al., 2025). Sustainable leaders continuously improve this indicator by investing in clean technologies and optimizing processes. The percentage of recycled or renewable raw materials in production is another important indicator of the leadership's commitment to the circular economy (Ansari, 2025). In the social dimension, indicators such as "investment rate in local community development" and "number of hours of employee training in sustainability" can provide a clear picture of the impact of sustainable leadership (Guan et al., 2023).

The Triple Bottom Line framework, introduced by John Elkington in the 1990s, has become one of the most fundamental tools for evaluating leadership decisions in sustainable businesses today. This framework emphasizes the three dimensions of people (social), planet (environmental), and profit (economic), helping leaders make decisions that strike an appropriate balance among these three dimensions (Gu et al., 2022; Hasan et al., 2025). On the economic dimension, sustainable leaders consider, in addition to traditional financial indicators such as return on investment and revenue growth, new metrics, including "avoided costs of sustainable actions" (Gu et al., 2021). For example, investing in renewable energy systems may be costly in the short term, but in the long term, it leads to significant savings by reducing energy costs and avoiding environmental penalties. On the social dimension, forward-thinking leaders use tools such as "social impact assessments," which measure the impact of an organization's decisions on various stakeholders (Abhi et al., 2015). On the environmental dimension, in addition to traditional indicators such as carbon footprint and water consumption, sustainable leaders also pay attention to more advanced metrics, including the "value of ecosystem services maintained." This metric calculates the economic value of services provided by healthy ecosystems (such as water purification or pollination) that are maintained as a result of the organization's decisions (Manyise & Dentoni, 2021).

Practical examples of successful sustainable leadership include Interface. Ray Anderson, the legendary CEO of Interface, is a prime example of sustainable leadership in action. When Anderson committed to becoming a "zero-impact company" in 1994, many considered this goal unrealistic. However, by adopting a systematic and innovative approach, Interface was able to reduce its greenhouse gas emissions by 96% by 2019 while simultaneously increasing its profitability. The key to Anderson's success was combining the TBL framework with transformational thinking. He not only incorporated sustainability performance indicators into the executive compensation system but also changed the organizational culture so that all employees felt a sense of ownership over sustainability goals (Li and Wu, 2020). Nestlé's sustainable supply chain management is another example. Paul Bolke, former CEO of Nestlé, demonstrated how sustainable leadership can transform the entire value chain. Through the Nestlé for Sustainable Agriculture program, the company has trained more than 700,000 farmers worldwide and implemented sustainable farming practices on more than 8 million hectares of farmland. This has not only improved farmers' livelihoods but also increased the productivity and quality of raw materials. Bölke used a combination of quantitative (e.g., reduction in water use in agriculture) and qualitative (e.g., improvement in farmers' quality of life) indicators to measure the program's impact (Eugénio et al., 2022).

Tesla is also a company that has revolutionized the energy industry. Although Elon Musk's approach has sometimes been controversial, his unparalleled influence in accelerating the global transition to clean energy is undeniable. By combining bold vision, technological innovation, and a smart business model, Musk has demonstrated how sustainable leadership can have a profoundly positive impact on the environment while creating value for shareholders. Tesla not only transparently reports its sustainability performance indicators but also quantifies its broader impact on reducing greenhouse gas emissions at the industry level (Aybaly et al., 2017). Despite significant progress, measuring the impact of sustainable leadership faces several challenges.

One of the most important of these is the problem of "attribution"—determining what portion of positive outcomes can be directly attributed to leadership actions. For example, if a company is experiencing financial growth while also improving its sustainability indicators, it is challenging to determine the exact contribution of leadership to this success. Another challenge is the potential conflicts between different dimensions of TBL.

In some cases, decisions that are environmentally desirable may lead to reduced profitability in the short term. Effective leaders must be able to manage these conflicts and find creative solutions that benefit all three dimensions in the long term. Another challenge is the problem of time lag. Many of the impacts of sustainable decisions become apparent years later, while performance measurement systems typically focus on short-term outcomes. Leading leaders need to address this problem by developing measurement systems that consider both short-term outcomes and long-term leading indicators.

Shortly, we can expect to see significant changes in the way we measure the impact of sustainable leadership. The emergence of technologies such as blockchain and artificial intelligence will enable sustainability impacts to be tracked and measured with unprecedented precision. Integrated reporting systems that combine financial, social, and environmental data in real-time will help leaders make more informed decisions. We will also see the development of global standards for measuring the impact of sustainable leadership. Initiatives such as the Sustainability Reporting Standards and the Integrated Reporting Framework are evolving to better meet the needs of measuring sustainable leadership performance. Finally, forward-thinking leaders should look to develop sustainability scorecards to visually and in real-time display business performance across all dimensions of TBL. These tools help managers and stakeholders gain a more comprehensive understanding of the impact of sustainable leadership and make more informed decisions. Measuring the impact of leadership in sustainable entrepreneurship is still in its early stages of development; however, recent developments indicate that the field is maturing rapidly. Leaders who can effectively use these new tools and methods will not only lead their businesses toward a sustainable future but will also become role models for the entire industry. The future belongs to leaders who can seamlessly combine economic value creation with social and environmental responsibility, accurately measuring and managing their actual impact.

2.7 Future Trends: Leadership in the Next Era of Sustainable Entrepreneurship

On the eve of profound global transformations, sustainable entrepreneurship is entering a new phase in which business leaders will face unprecedented challenges and opportunities. These transformations necessitate a fundamental reevaluation of the concept of leadership and the development of new skills and approaches that can guide organizations on a path to genuine sustainability. Among these, three key developments will shape the future of leadership in sustainable entrepreneurship: the emergence of purpose-driven and impact-driven entrepreneurs, the increasing responsibility of leaders for addressing climate change, and the evolution of leadership paradigms in response to the changing sustainability landscape.

2.7.1 The Rise of Purpose-Driven and Impact Entrepreneurs

A new generation of entrepreneurs is emerging who are building their businesses not simply on financial profitability but on a deep mission and creating positive social and environmental impact. These purpose-driven entrepreneurs are changing the traditional definition of business success and introducing new metrics for evaluating organizational performance. Recent research shows that more than 60% of new-generation entrepreneurs cite social purpose as their primary motivation for starting a business, compared to less than 30% for the previous generation (Ahmad, 2025). This shift represents a fundamental shift in the mindset of entrepreneurs, with profound implications for the leadership and management styles of organizations. Purpose-driven entrepreneurs seek to create shared value—a concept introduced by Michael Porter that means creating economic value in a way that also creates value for society (Sargani et al., 2020). This approach requires a specific leadership style that combines an ambitious vision, the ability to inspire employees and stakeholders, and the skill to translate social goals into sustainable business models (Suriyankietkaew, 2019). For example, Toms, with its "One-to-One" model, demonstrated how to place social goals at the core of business strategy while achieving significant financial growth. Leaders of these organizations must strike a delicate balance between financial and social goals while navigating their organizations in a highly competitive environment. The main challenge facing these leaders is to maintain a commitment to social goals in the face of market pressures and stakeholder expectations. Experience has shown that organizations deeply and authentically committed to their goals perform better in the long run (Singh & Awasthy, 2025); however, this path is always fraught with obstacles and complex tests. Successful leaders in this area are those who can convince various stakeholders that social impact and financial profitability are not only incompatible but can reinforce each other.

2.7.2 Climate Change and Leadership Responsibility

Climate change has become one of the most defining factors shaping the business environment of the twenty-first century. Business leaders can no longer treat this challenge as a marginal issue or a secondary concern. Scientific reports warn that the world has only a few years to limit global warming to 1.5 °C—the threshold that would prevent the catastrophic consequences of climate change (Purwana et al., 2025; Sharma et al., 2024). In this context, business leaders have an unprecedented moral and professional responsibility to reduce greenhouse gas emissions and adapt to the consequences of climate change. Leading leaders are taking transformative approaches to this challenge. Some companies, such as Unilever and Nestlé, have committed to achieving net-zero carbon emissions by 2030—a goal that goes far beyond the commitments of many governments. This level of ambition requires a fundamental rethinking of all aspects of business operations, from supply chains to product design and distribution. Leading this transformation requires a unique combination of boldness, creativity, and the ability to align diverse stakeholders around a shared vision. One of the key innovations in this area is the emergence of the concept of climate leadership, which is based on three core pillars: rapidly reducing greenhouse gas emissions in one's operations, influencing the entire business ecosystem to accelerate the transition to a low-carbon economy, and actively participating in shaping forward-looking public policies on climate change (Serpa & Sá, 2025). Climate leaders have shown how to create powerful collective movements to address climate change by combining scientific knowledge, diplomatic skills, and transformational leadership (Toukabri & Kalai, 2024).

2.7.3 Future Leadership Paradigms in a Rapidly Evolving Sustainability Landscape

The sustainability landscape is evolving at an unprecedented pace, and these developments necessitate a corresponding evolution of leadership paradigms. Leadership in the new era of sustainable entrepreneurship must be able to navigate the increasing complexity, deep uncertainty, and interdependence of various economic, social, and environmental systems. These conditions have necessitated the emergence of new styles of leadership, including "systemic leadership," "adaptive leadership," and "collective leadership." Systemic leadership is based on a deep understanding of the complex interrelationships between the various components of economic, social, and environmental systems. Rather than focusing on partial, short-term solutions, systemic leaders seek interventions that can create transformational changes in the entire system (Reuveny & Thompson, 1999). For example, Paul Pullman, former CEO of Unilever, recognized that the plastic waste problem could not be solved by the actions of one company alone and launched the Alliance to End Plastic Waste initiative, which brought together more than 80 companies from across the plastics value chain. Adaptive leadership refers to the ability to lead organizations in times of uncertainty and rapid change. In an environment where laws, technologies, and stakeholder expectations are rapidly evolving, leaders must be able to continuously learn, adapt, and prepare their organization to face unexpected circumstances. This type of leadership requires intellectual humility, constant curiosity, and a willingness to experiment with new solutions (Seibel et al., 2023). Collective leadership is also emerging as a new paradigm that emphasizes the power of collective action to solve complex sustainability challenges. In this model, leaders seek collective wisdom and collaboration across organizations, sectors, and geographic boundaries rather than relying on individual perspectives and decisions (Hunter et al., 2012). This approach is particularly critical when facing global challenges such as climate change or social inequalities.

The future of leadership in sustainable entrepreneurship is moving towards more integrated, collaborative, and accountable models. Tomorrow's leaders must be able to operate across multiple dimensions simultaneously: guiding their organizations toward sustainable profitability, creating positive social and environmental impact, and contributing to the solution of global challenges. This level of complexity requires the development of new competencies, including systems literacy, enhanced emotional intelligence, the ability to work in multicultural and multi-stakeholder environments, and the skill to facilitate cross-organizational collaboration. Leadership training and development institutions are also adapting their programs to these new needs. Leading business schools are introducing programs that place sustainability concepts at the core of management and leadership education (Purwana et al., 2025). For example, the Minder Business School in Sweden has designed its entire curriculum around sustainability. These developments represent a fundamental shift in how future leaders are developed. At the same time, expectations of leaders are rising. Stakeholders—from employees and customers to investors and local communities—are increasingly demanding transparency, accountability, and a genuine commitment to sustainable values. Leaders who fail to align with these expectations will quickly lose credibility and influence. In contrast, leaders who can authentically and effectively meet these expectations will strengthen their organizations' market positions and become key players in shaping a more sustainable future.

Leadership in the new era of sustainable entrepreneurship is no longer an option but an inescapable necessity for survival and growth in a world facing unprecedented challenges. Organizations that can cultivate leaders with the vision, courage, and skills to lead this transformation will not only strengthen their competitive position but also make a valuable contribution to building a more just and sustainable future for all. The future belongs to leaders who can seamlessly combine economic value creation with social and environmental responsibility, transforming their organizations into engines of positive change in society.

3 Conclusions

This study demonstrates the significant impact of local leadership on farmers' willingness to participate in sustainable value chain businesses in rural agricultural areas. While individual attitudes, subjective norms, and perceived behavioral control all significantly shape farmers' intentions, local leadership acts as a crucial intermediary that strengthens these effects. Effective local leaders foster trust, social cohesion, and a supportive environment by providing education, coordination, and resource facilitation. Their presence helps reduce psychological and practical barriers, enabling farmers to participate confidently in value chain activities that benefit both their livelihoods and the sustainability of natural resources.

Given these findings, it is recommended that rural development programs and policymakers place strong emphasis on empowering local leaders. Capacity-building initiatives should equip local influencers with skills in sustainable practices, community engagement, and resource management to enhance their ability to guide collective efforts. Establishing participatory governance structures, such as local councils that involve diverse stakeholders, can institutionalize leadership roles and enhance coordination and transparency. Development projects should actively involve local leaders from the planning stage to increase community acceptance and leverage their social capital to motivate farmer participation. Furthermore, policies should support local leaders in connecting farmers to external resources, training, and market opportunities, thereby enhancing perceived behavioral control.

To further strengthen sustainable value chain development, communication strategies led by trusted local leaders should promote positive social norms around participation, helping to overcome resistance within communities. Furthermore, ongoing research is necessary to gain a deeper understanding of the dynamics of leadership styles and their long-term effects on rural economic and environmental sustainability. By integrating these recommendations, rural development efforts can harness the full potential of local leadership to drive sustainable entrepreneurship, ultimately improving livelihoods, fostering social cohesion, and advancing environmental stewardship in rural areas.

4 Case Study

In recent years, the development of small businesses in rural and agricultural areas has been a key strategy for improving the livelihoods of villagers, promoting environmental sustainability, and ensuring food security. The effectiveness of developing such businesses has also been confirmed in various studies. One of the new approaches in this field is the use of sustainable value chains, in which farmers participate not only in production but also in processing, packaging, marketing, and selling products. Success in this path requires the willingness of farmers to actively participate, as well as the presence of effective leadership at the local level, to create coordination, provide education, and facilitate marketing. In this study, we employed the Theory of Planned Behavior (TPB) to examine the willingness of farmers to participate in developing value chain businesses and to analyze the role of local leadership as an intervening variable.

This study is applied in terms of its purpose and employs a descriptive-survey approach in its data collection method. The statistical population comprises all farmers in Fars Province who engage in small-scale agricultural activities. Using the stratified random sampling method, 350 farmers were selected as a sample. The primary instrument was a researcher-designed questionnaire based on the components of the Theory of Planned Behavior (Ajzen, 1991) and the concept of "local leadership." The questionnaire consisted of five main sections: Attitude, Subjective Norms, Perceived Behavioral Control, Behavioral Intention, and Local Leadership. All items were designed on a five-point Likert scale, ranging from "strongly disagree" (1) to "strongly agree" (5) (Table 1).

The results of the present study, conducted within the framework of the Theory of Planned Behavior, showed that all four main variables of this theory, including attitude, subjective norms, perceived behavioral control, and especially local leadership as an extending variable, have a positive and significant effect on farmers' intention to participate in value chain business development. These findings suggest that, beyond individual beliefs and perceptions, the presence of effective leadership at the local level can provide the necessary social and institutional context for farmers' active decision-making. By building trust, strengthening social capital, and creating group cohesion, local leadership can help reduce mental and practical barriers to participation in emerging economic activities.

Table 1 The items used to measure the constructs

No.	Variable	Item
1	Attitudes towards developing value chain businesses	Participating in the agricultural value chain will increase my income
2		This type of business will contribute to the sustainability of natural resources
3		I feel that this type of activity will add more value to the products
4		Developing these businesses will ultimately benefit the local community
1	Subjective norms	My family supports my participation in this type of business
2		My colleagues in the village encourage me to participate in these activities
3		Local authorities have a positive attitude towards these activities
4		Farmers are expected to participate in these types of programs
1	Perceived behavioral control	I possess the necessary skills to participate in this type of business
2		I have access to the necessary resources and information
3		This type of business is easy for me to run
4		I can handle potential obstacles that may arise in this business
1	Intention to perform the behavior (participation in the sustainable value chain)	I plan to enter the agricultural value chain in the following year
2		I am willing to process and market my products
3		I prefer to work in marketing or packaging cooperatives
4		I am highly motivated to learn new ways to sell my products
1	Local leadership	In my area, local leaders encourage people to participate in sustainable businesses
2		Local leaders play an effective role in educating farmers
3		Having an active leader creates better coordination in the value chain
4		Local leaders have provided effective marketing networks for agricultural products
5		Local leaders help farmers with resource management and planning

The focus on local leadership in this study suggests that the influence and guidance of these leaders can play a facilitating role in bridging the gap between psychological and behavioral factors. In areas where local leadership enjoys high social acceptability, farmers are more confident in participating in participatory and risky economic activities such as value chain development. As adequate social references, these leaders can establish positive norms and strengthen supportive attitudes in the community. They also provide a suitable political and economic context for supporting agribusinesses by linking local resources with external institutions, which in turn leads to an increase in farmers' sense of control over their behavior.

Accordingly, it is suggested that rural development and agricultural policymakers prioritize the empowerment of local leadership in their programs. This can be achieved through training local leaders, creating participatory council structures, and providing institutional support for leaders trusted by the community. Also, designing development programs with the active participation of local leaders increases the acceptability and effectiveness of these plans. On the other hand, strengthening the mediating role of regional leadership between farmers' individual beliefs and their participatory behavior can be considered a key strategy for the success of value chain policies in rural areas.

Questions

1. How do different leadership styles at the local level impact farmers' motivation and sustained participation in sustainable value chain initiatives?
2. What are the effects of digital tools and communication platforms on the capacity of local leaders to coordinate and educate farmers in rural value chains?
3. How does the strength of social capital within rural communities influence the effectiveness of local leadership in promoting sustainable agricultural practices?
4. In what ways can policy frameworks be designed to support and institutionalize the role of local leadership in rural development programs?
5. How do cultural and regional differences affect the relationship between local leadership and farmers' willingness to engage in value chain businesses?

References

Abdelwahed, N. A. A., Channa, A., Al Doghan, M. A., & Soomro, B. A. (2025). Building a resilient economy: green initiatives and green entrepreneurship for women's empowerment. *International Journal of Development Issues*.

Abhi, S., Venugopal, V., & Shastri, S. (2015). Social entrepreneurship—Building sustainability through business models and measurement of social impact. In *Entrepreneurial ecosystem: Perspectives from emerging economies* (pp. 295–323).

Adelowo, C. M., & Henrico, A. (2025). Sustainability and determinants of entrepreneurship propensity among young graduates in Nigeria. *SAGE Open, 15*(1).

Ahmad, K. (2025). Sustainability through purposeful leadership: The maybank case study. In *Islamic finance and sustainable development: Balancing spirituality, values and profit* (pp. 192–201).

Ajzen, I. (1991) The theory of planned behavior. *Organizational Behavior and Human Decision Processes, 50*(2), 179–211. https://doi.org/10.1016/0749-5978(91)90020-T

Alomar, J. A., & Alatawi, F. M. (2025). Evaluating the challenges of metaverse-enabled digital entrepreneurship: evidence from Saudi Arabia as an emerging economy. *Journal of Entrepreneurship in Emerging Economies*.

Alzamel, S. (2024). Building a resilient digital entrepreneurship landscape: The importance of ecosystems, decent work, and socioeconomic dynamics. *Sustainability, 16*(17).

Anand, A., Argade, P., Barkemeyer, R., & Salignac, F. (2021). Trends and patterns in sustainable entrepreneurship research: A bibliometric review and research agenda. *Journal of Business Venturing, 36*(3), Article 106092.

Ansari, M. S. A. (2025). Innovation, and green entrepreneurship are catalysts for sustainability in Oman. *Cleaner Waste Systems, 11*.

Aybaly, R., Guerquin-Kern, L., Manière, I. C., Madacova, D., & Holt, J. V. (2017). *Sustainability practices in the luxury industry: How can one be sustainable in an over-consumptive environment?: Sustainability in the automotive world: The case of Tesla.* Paper presented at the Procedia Computer Science.

Bae, S. H., Saberi, S., Kouhizadeh, M., & Sarkis, J. (2024). Evaluating blockchain technology adoption in multi-tier supply chains from an institutional entrepreneurship theory perspective. *Transportation Research Part C: Emerging Technologies, 160*.

Bejarano-Auqui, J. F., Peña, J. C. R., Subia, L. E. C., & Lozano, D. L. A. (2025). Business management praxis, innovation, and entrepreneurship as a predictor of labor productivity in MSEs. *Journal of Educational and Social Research, 15*(2), 247–260.

Belgibayeva, A., Saher, L., & Vakulenko, I. (2023). *Managing green innovations in European Union: Should we expect updates in the concepts of leadership, entrepreneurship, and sustainable development?* Paper presented at the Springer Proceedings in Business and Economics.

Cheng, Z., Jin, X., & Kwak, W. J. (2025). Using the new positive aspect of digital leadership to improve organizational sustainability: Testing moderated mediation model. *Acta Psychologica, 255*.

Clarysse, B., He, V. F., & Tucci, C. L. (2022). How the Internet of Things reshapes the organization of innovation and entrepreneurship. *Technovation, 118*.

Cleveland, S., & Cleveland, M. (2025). Equitable leadership and collaborative strategies: A framework for organizational success. In *Strategic approaches to mindful leadership and change management* (pp. 383–400).

Cruz-Sandoval, M., Vázquez-Parra, J. C., & Carlos-Arroyo, M. (2023). Complex thinking and social entrepreneurship. An approach from the methodology of compositional data analysis. *Heliyon, 9*(2), e13415.

Dote-Pardo, J., Ortiz-Cea, V., Peña-Acuña, V., Severino-González, P., Contreras-Henríquez, J. M., & Ramírez-Molina, R. I. (2025). Innovative entrepreneurship and sustainability: A bibliometric analysis in emerging countries. *Sustainability (Switzerland), 17*(2).

Dzhengiz, T., Henry, L. A., & Malik, K. (2023). The role of partnership portfolios for sustainability in addressing the stability-change paradox: Dong/Orsted's transition from fossil fuels to renewables. *Business & Society, 63*(7), 1518–1557.

Ekanem, I. (2024). *Principles of innovation, entrepreneurship and sustainability: An evidence-based approach*.

Esteves, J., de Haro Rodríguez, G., Ballestar, M. T., & Sainz, J. (2024). Gender and generational cohort impact on entrepreneurs' emotional intelligence and transformational leadership. *International Entrepreneurship and Management Journal, 20*(2), 1295–1322.

Eugénio, T., Rodrigues, S. C., & Gonçalves, M. J. (2022). Sustainability report evolution: The Nestlé case study applicability. In *Research anthology on measuring and achieving sustainable development goals* (Vol. 1, pp. 227–249).

Eynolghozat, M., Rezvani, M., & Ziyae, B. (2025). Corporate entrepreneurship in entrepreneurial city: A new solution for urban sustainability. *Regional Science Policy and Practice, 17*(4).

Fedajev, A., Kojić, M., Mitić, P., & Radulescu, M. (2025). Drivers of entrepreneurship in Europe: the role of digitalization, innovation, capital investments, unemployment, and sustainability. *European Journal of Innovation Management*.

Gao, Y., & Tsai, K. C. (2024). The effect of entrepreneurial leadership on college students' creativity: the mediating role of intrinsic motivation and emotional intelligence. *Environment and Social Psychology, 9*(9).

Gracia-Zomeño, A., García-Toledano, E., García-Perales, R., & Palomares-Ruiz, A. (2025). Leadership and entrepreneurship in education and their impact on the management of sustainable educational environments. *Sustainability (Switzerland), 17*(7).

Le Grange, A., Maas, G., & Johnston, A. (2024). Cases on Transformational Entrepreneurship. In *Chapter 19: Transformational entrepreneurship impact: a complex perspective on measurable contextual theories of change* (pp. 271–286). Edward Elgar Publishing.

Grøn, A. B. (2025). Bridge over troubled waters? Experimental evidence into the influence of leadership on employees' collaborative engagement. *Public Administration Review, 85*(2), 308–325.

Gu, W., Wang, J., Hua, X., & Liu, Z. (2021). Entrepreneurship and high-quality economic development: based on the triple bottom line of sustainable development. *International Entrepreneurship and Management Journal, 17*(1).

Gu, W., Pan, H., Hu, Z., & Liu, Z. (2022). The triple bottom line of sustainable entrepreneurship and economic policy uncertainty: An empirical evidence from 22 countries. *International Journal of Environmental Research and Public Health, 19*(13).

Guan, X., Yao, C., & Zhang, W. (2023). Sustainability of entrepreneurship: An empirical study on the impact path of corporate social responsibility based on internal control. *Sustainability (Switzerland), 15*(16).

Hasan, M., Supatminingsih, T., Tahir, T., Guampe, F. A., Huruta, A. D., & Lu, C. Y. (2025). Sustainable agricultural knowledge-based entrepreneurship literacy in agricultural SMEs: Triple bottom line investigation. *Journal of Open Innovation: Technology, Market, and Complexity, 11*(1).

Hou, B., Yu, Y., Fang, K., Wang, S., Jin, P., & Hong, J. (2025). Social entrepreneurship and employee-related corporate social responsibility: evidence from Chinese privately owned firms. *Chinese Management Studies*.

Hu, S., Liu, S., Li, X., Zhao, J., Chen, J., Chen, W., & Hu, J. (2025). Organizational evidence-based practice culture, implementation leadership, and nurses: A bidirectional mediation model. *International Nursing Review, 72*(2), Article e13054.

Huang, S. H., Day, J. D., Shu, M. H., Huang, H. C., & Huang, J. C. (2022). Construction of virtual marketing interactive platform for digital twin innovation and entrepreneurship based on blockchain. *Scientific Programming, 2022*.

Hunter, S. T., Cushenbery, L., Fairchild, J., & Boatman, J. (2012). Partnerships in leading for innovation: A dyadic model of collective leadership. *Industrial and Organizational Psychology, 5*(4), 424–428.

Iddrisu, I. (2025). Understanding the nexus between organizational culture and trust: The mediating roles of communication, leadership, and employee relationships. *Sustainable Futures, 9*, Article 100555.

Iqbal, Q., Piwowar-Sulej, K., Agarwal, R., & Yaqub, M. (2025). Sustainability-oriented leadership and business strategy: Examining the roles of procedural environmental justice and job embeddedness. *Business Strategy and the Environment, 34*(3), 2917–2933.

Irviani, R. (2025). Fostering sustainability through leadership and employee personality traits. *Sustainable Futures, 9*.

Jamil, M., Stephens, S., & Md Fadzil, A. F. (2025). Sustainability in family business settings: A strategic entrepreneurship perspective. *Journal of Family Business Management, 15*(1), 29–47.

Javed, S., & Al-Mulali, U. (2025). Navigating the innovation–sustainability paradox: The strategic role of leadership and industry 4.0 in oman's manufacturing sector. *Sustainability (Switzerland), 17*(6).

Jayaweera, M., Perera, B. A. K. S., & Liyanawatta, T. N. (2025). Use of emotional intelligence to enhance the leadership skills of project managers in construction: A qualitative delphi study in Greece. *Journal of Legal Affairs and Dispute Resolution in Engineering and Construction, 17*(3).

Karlibaeva, R. H., Lipinsky, D. A., Volokhina, V. A., Gureeva, E. A., & Makarov, I. N. (2024). Sustainable development of entrepreneurship through operational risk management: The role of corporate social responsibility. *Risks, 12*(8).

Kayikci, Y., Gozacan-Chase, N., & Rejeb, A. (2024). Blockchain entrepreneurship roles for circular supply chain transition. *Business Strategy and the Environment, 33*(2), 197–222.

Khan, M. K., Hussain, P., Hussain, M. J., & Michael, K. (2025). Innovation at the nexus of technology and sustainability: CEO leadership and digital transformation in China. *Sustainable Futures, 9*.

Kickul, J., Gundry, L., Mitra, P., & Berçot, L. (2018). Designing with purpose: Advocating innovation, impact, sustainability, and scale in social entrepreneurship education. *Entrepreneurship Education and Pedagogy, 1*(2), 205–221.

Kumar, B., Pimentel, J., Cano-Londoño, N. A., Ruiz-Mercado, G. J., Deak, C. T., & Cabezas, H. (2025). Designing cost-effective supply chains for plastics at the end-of-life. *Journal of Cleaner Production, 501*.

Li, S. H., & Wu, F. S. (2020). *Organizational function boundaries of enterprise innovation in the internet technology environment: A case study on the interface between research and development departments and information departments in the information software service industry.* Paper presented at the Towards the Digital World and Industry X.0—Proceedings of the 29th International Conference of the International Association for Management of Technology, IAMOT 2020.

Li, Z., Chen, H., Ma, Q., & Li, H. (2021). CEO Empowering Leadership and Corporate Entrepreneurship: The Roles of TMT Information Elaboration and Environmental Dynamism. *Frontiers in Psychology, 12*.

Liao, Y. (2022). Sustainable leadership: A literature review and prospects for future research. *Frontiers in Psychology, 13*.

Licite-Kurbe, L., & Chandramohan, A. (2020). Characteristics and challenges of the internet of things in entrepreneurship. *Rural Sustainability Research, 43*(388), 27–34.

Liu, X., Yu, Y., Zhao, X., & Zhang, N. (2022). Top management team boundary-spanning leadership: Measurement development and its impact on innovative behavior. *Frontiers in Psychology, 13*.

Liu, G., Yin, Q., & Zhang, L. (2022). Relations between entrepreneur's social identity and strategic entrepreneurship: Sustainable leadership as mediator. *Frontiers in Psychology, 13*.

Luhas, J., & Mikkilä, M. (2024). Environmental, social, and economic challenges to forest-based micro-entrepreneurship: A comparative case study in Finland. *Canadian Journal of Forest Research, 54*(11), 1198–1212.

Mamani, W. C., Manrique, G. M. L., Madrid, S. D. P. C., Herrera, E. E., Acosta, D. B., Rivas-Diaz, R. R., et al. (2022). The role of entrepreneurship and green innovation intention on sustainable development: Moderating impact of inclusive leadership. *AgBioforum, 24*(1), 134–143.

Manyise, T., & Dentoni, D. (2021). Value chain partnerships and farmer entrepreneurship as balancing ecosystem services: Implications for agri-food systems resilience. *Ecosystem Services, 49*.

Mawardi, M. K., Noor, A. Y. M., Edityastono, L., 'Atikah Zulkiffli, S. N., & Mashahadi, F. (2025). Green entrepreneurship in the era of sustainability: the relationship between intent, institutional support, and student behavior. *Cogent Business and Management, 12*(1).

Mortan, R. A., Ripoll, P., Carvalho, C., & Bernal, M. C. (2014). Effects of emotional intelligence on entrepreneurial intention and self-efficacy. *Revista De Psicologia Del Trabajo y De Las Organizaciones, 30*(3), 97–104.

Muafi, M., Hidayat, A., & Jufri, A. (2025). The role of moderating digital leadership and digital entrepreneurship in enhancing the resilience and independence of MSMEs towards business sustainability. *Review of Integrative Business and Economics Research, 14*(2), 812–827.

Naguib, R., & Barbar, J. (2025). Factors shaping sustainability through female entrepreneurship in the GCC: A systematic review with multi-level and institutional perspective. *Sustainability (Switzerland), 17*(5).

Nakhaei, A., Hadavi, F., Talebpour, M., Farahani, A., & Abbasnejad, E. (2013). The relationship between emotional intelligence and its dimensions with organizational entrepreneurship. *Advances in Environmental Biology, 7*(8), 1356–1365.

Nhu Laursen, L., & Houman Andersen, P. (2023). Resource and supplier interaction in network innovation governance: The case of innovating at Unilever. *Journal of Business Research, 156*.

Oduro Owusu, S., & Gil, M. T. (2025). The role of emotional and cognitive engagement in refugee entrepreneurship and business sustainability: Evidence from Ghanaian SMEs. *Cogent Business and Management, 12*(1).

Pandey, S. C., & Pattnaik, P. N. (2020). Teaching sustainable entrepreneurship and leadership: Case of a taught course—Innovation and sustainability specialization. In *Building an entrepreneurial and sustainable society* (pp. 114–125).

Pett, T., Haddad, G., & Nagpal, G. (2025). Does empowering leadership affect SMEs' sustainability performance through knowledge transfer? *European Management Journal, 43*(2), 211–221.

Purwana, D., Effendi, M. S., Adha, M. A., Musyaffi, A. M., & Suparno, S. (2025). Entrepreneurship education as a catalyst for sustainability: Linking innovation, intention, and business models. *International Review of Management and Marketing, 15*(3), 119–129.

Qu, C., & Chen, L. (2021). *Integrated marketing communications strategy based on data analysis for Stella McCartney.* Paper presented at the Advances in Intelligent Systems and Computing.

Rausser, G., & Strielkowski, W. (2023). *An introduction: Pandemic-inspired policies for leadership, entrepreneurship, and sustainable development.* Paper presented at the Springer Proceedings in Business and Economics.

Reuveny, R., & Thompson, W. R. (1999). Economic innovation, systemic leadership, and military preparations for war: The U.S. case. *Journal of Conflict Resolution, 43*(5), 570–595.

Ribeiro, G. K. A., & Leitão, J. (2024). Sustainable leadership and sustainable entrepreneurship: a systematic literature review. *Discover Sustainability, 5*(1).

Richardson, K., & Burrell, D. N. (2023). Exploring the leadership intersection of social entrepreneurship, sustainability, and environmental public health. In *Transformational leadership styles for global leaders: Management and communication strategies* (pp. 350–366).

Rojas Casarrubias, C., Aparicio López, J. L., Rodríguez Alviso, C., Castro Bello, M., & Villerías Salinas, S. (2025). Community environmental leadership and sustainability: Building knowledge from the local level. *Sustainability (Switzerland), 17*(8).

Sargani, G. R., Zhou, D., Raza, M. H., & Wei, Y. (2020). Sustainable entrepreneurship in the agriculture sector: The nexus of the triple bottom line measurement approach. *Sustainability (Switzerland), 12*(8).

Seibel, M., Kaufman, E. K., Cletzer, D. A., & Elliott-Engel, J. (2023). Advancing adaptive leadership through adaption-innovation theory: Enhancements to the holding environment. *Journal of Leadership Studies, 17*(1), 23–29.

Serpa, S., & Sá, M. J. (2025). Embedding sustainability in organisations through climate, culture, and leadership. *Sustainability (Switzerland), 17*(3).

Sharma, P. K., & Kumra, R. (2025). Emotional intelligence and self-efficacy as mediators in the relationship between transformational leadership and proactive customer service performance. *International Journal of Quality and Service Sciences, 17*(1), 25–47.

Sharma, R., Kumar, S., & Sharma, N. (2024). Assessing the impact of intellectual capital and environment-friendly innovation on sustainability: Evidence of marginalized entrepreneurship in an emerging economy. *Indian Journal of Environmental Protection, 44*(11), 1029–1039.

Silvestri, A., & Veltri, S. (2020). Exploring the relationships between corporate social responsibility, leadership, and sustainable entrepreneurship theories: A conceptual framework. *Corporate Social Responsibility and Environmental Management, 27*(2), 585–594.

Singh, B. (2022). What can young social entrepreneurs tell us about environmental leadership? Perspectives from Brazil and India. In *Rethinking leadership for a green world* (pp. 69–91).

Singh, D., & Awasthy, R. (2025). Spirituality in social entrepreneurship: An ethical approach to sustainability. *Social Enterprise Journal, 21*(3), 492–518.

Sotelo, R. (2021). Quantum computing entrepreneurship and IEEE TEMS. *IEEE Engineering Management Review, 49*(3), 26–29.

Subrahmanyam, S. (2025). Fostering a collaborative culture: The role of leadership. In *Knowledge sharing and fostering collaborative business culture* (pp. 377–396).

Suriyankietkaew, S. (2019). *Sustainable leadership and entrepreneurship for corporate sustainability in small enterprises: An empirical analysis.* Paper presented at the World Review of Entrepreneurship, Management and Sustainable Development.

Suriyankietkaew, S., Krittayaruangroj, K., & Iamsawan, N. (2022). Sustainable leadership practices and competencies of SMEs for sustainability and resilience: A community-based social enterprise study. *Sustainability, 14*(10), 5762.

Tajpour, M., Dekamini, F., SafarMohammadluo, S., Movahed, M. N., & Madadpour, F. (2025). The impact of innovation management on the sustainability of small and medium enterprises with the role of entrepreneurship mediation. *International Entrepreneurship and Management Journal, 21*(1).

Tasnim, M. (2024). Leadership competencies for the age of artificial intelligence. In *Utilizing AI and smart technology to improve sustainability in entrepreneurship* (pp. 20–39).

Teniola, O. T., Muharam, F. M. B., & Ramily, M. K. (2025). Entrepreneurship and sustainability goals: Measuring strategies for offering support services to entrepreneurs in Nigeria. *International Journal of Public Sector Performance Management, 15*(2), 194–207.

Theodorakopoulos, L., Theodoropoulou, A., & Halkiopoulos, C. (2024). Enhancing decentralized decision-making with big data and blockchain technology: A comprehensive review. *Applied Sciences (Switzerland), 14*(16).

Toukabri, M., & Kalai, L. (2024). How does sustainability leadership improve climate change reporting? The choices associated with a sustainable board—A management perspective. *Mitigation and Adaptation Strategies for Global Change, 29*(7).

Ulrich, K., Guaita Martínez, J. M., Carracedo, P., & Soriano, D. R. (2024). Blockchain technology-based crypto assets: New insights into the evolution of the understanding of digital entrepreneurship. *Management Decision, 62*(9), 2836–2854.

Ünal, A. B. (2023). Value-driven leadership in sustainable entrepreneurship. In *De Gruyter handbook of sustainable entrepreneurship research* (pp. 107–122).

van der Duin, P. A., Ortt, J. R., & Aarts, W. T. M. (2014). Contextual innovation management using a stage-gate platform: The case of Philips shaving and beauty. *Journal of Product Innovation Management, 31*(3), 489–500.

Wang, L. (2024). Entrepreneurship vs. mentorship: an analysis of leadership modes on sustainable development with moderation of innovation management. *International Journal of Information Systems and Change Management, 14*(3), 233–249.

Waseel, A. H., Zhang, J., Shehzad, M. U., Saddiqa, A., Liu, J., & Hussain, S. (2025). Does empowering leadership help firms to establish collaborative culture and organizational commitment to stimulate frugal innovation? *Kybernetes, 54*(2), 1289–1318.

Yiadom, E. B., Indome, P. N., Mawutor, J. K. M., & Domfe, G. (2025). Balancing finance and sustainability: The impact of financial access on carbon emissions through innovation and entrepreneurship in a global study. *International Review of Economics and Finance, 98*.

Zadorozhny, B. S., Petrides, K. V., Cheng, Y., Cuppello, S., & van der Linden, D. (2025). Predicting leadership status through trait emotional intelligence and cognitive ability. *Behavioral Sciences, 15*(3).

Zamfir, A.-M., Mocanu, C., & Grigorescu, A. (2018). Resilient entrepreneurship among European higher education graduates. *Sustainability, 10*(8).

Crisis Management and Sustainable Entrepreneurship

Ehsan Khosravi, Nader Naderi, Faranak Karamian, and Seyed Mostafa Razavi

Abstract

This chapter delves into the intricate relationship between crisis management and sustainable entrepreneurship. It begins with a comprehensive analysis of the concepts of crisis and crisis management, outlining the various stages inherent to the crisis lifecycle. Following this, the chapter assesses the vulnerability of businesses operating under crisis conditions, emphasizing the unique challenges they face. Subsequently, it outlines several models of business resilience in these contexts, highlighting the critical variables that impact resilience and adaptability during crises. The discussion culminates in a thorough exploration of the interplay between effective crisis management strategies and the principles of sustainable entrepreneurship. The findings suggest that principled crisis management can establish a foundation for the long-term development of sustainable entrepreneurship initiatives. Conversely, the implementation of sustainable entrepreneurial practices can mitigate the adverse effects of crises, particularly those linked to environmental disruptions, thereby enhancing overall resilience and sustainability in business operations.

Keywords

Crisis management · Business resilience · Business vulnerability · Sustainable entrepreneurship

Learning Objective

- After reading this chapter, readers should be able to describe both crisis and crisis management and explain their different stages.
- After reading this chapter, readers should be able to articulate the vulnerability of enterprises during crises.
- After reading this chapter, readers should be able to describe various models of business resilience in crisis situations.
- After reading this chapter, readers should be able to explain the relationship between crisis management and sustainable entrepreneurship.

E. Khosravi (✉) · N. Naderi · F. Karamian
Razi University, Kermanshah, Iran
e-mail: khosravi.ehsan@razi.ac.ir

N. Naderi
e-mail: n.naderi@razi.ac.ir

F. Karamian
e-mail: karamiyan.faranak@razi.ac.ir

S. M. Razavi
Faculty of Management, University of Tehran, Tehran, Iran
e-mail: mrazavi@ut.ac.ir

M. Akbari et al. (eds.), *Sustainable Entrepreneurship*, University of Tehran Science and Humanities Series, https://doi.org/10.1007/978-3-032-04909-4_21

1 Introduction

The significant events of the last decades in the world from a political, economic, social, geographical, health, and humanitarian perspective show that global crises, with their multifaceted nature, significantly affect economies, businesses, communities, and mental health. These crises may progress to the point of threatening the internal and external interests of societies or the continued survival of a large organization. The interconnectedness of these crises emphasizes the importance of comprehensive approaches to crisis management, integrating economic, social, and environmental considerations to strengthen resilience in the face of ongoing global challenges. Today's turbulent and changing world has become a stage for the struggle for success and advancement of entrepreneurs. Entrepreneurs who are able to operate successfully in changing environmental conditions, who raise their level of knowledge, awareness, and awareness of environmental changes and can manage various crises in a principled manner, recognize and exploit emerging opportunities before other competitors, and in this way not only adapt to environmental conditions but also gain a competitive advantage and overtake their competitors. If crises are managed properly, it has been proven throughout history that not only are the consequences of these crises not a threat to businesses, but they can also be an opportunity to improve and even enhance them compared to pre-crisis conditions through proper planning and management (Goel, 2024).

Although opportunity recognition is a fundamental characteristic of entrepreneurial actions, it is often associated with risk. In this regard, entrepreneurial actions exploit new opportunities that are the source of environmental changes, including crises. If environmental and social goals are prioritized over economic profitability, they form sustainable entrepreneurial businesses. Sustainable development requires sustainable innovation, and entrepreneurs in this field can integrate environmental and social goals with different products or processes in the market in order to attract customers. Considering the factors related to entrepreneurship development and paying attention to the sustainability issue, entrepreneurs should lead the market toward sustainable entrepreneurial actions. Therefore, to progress in this area, sustainable entrepreneurship was introduced (Schaltegger & Wagner, 2011). Paying attention to sustainable entrepreneurship in economic development and integrating it into marketing practices is considered a new approach to creating a comparative advantage in sustainable economic development, which allows businesses to identify new customers and address their diverse and changing demands and needs (Di Vaio et al., 2022). Sustainable entrepreneurship strikes a balance between economic health, social equity, and environmental protection by exploiting entrepreneurial opportunities. The goal of sustainable entrepreneurship is to create value in innovative ways that reduce economic, social, and environmental harms and utilize available resources to meet the needs of current and future society (Batabyal & Yoo, 2017).

Sustainable entrepreneurship has emerged as a critical approach to addressing environmental and social crises while fostering economic growth. This type of entrepreneurship emphasizes the creation of business models that not only seek profit but also prioritize sustainability and social responsibility. Research shows that sustainable entrepreneurs increasingly integrate environmental awareness and social values into their business strategies, which can lead to innovative solutions aligned with the Sustainable Development Goals (SDGs) while capitalizing on emerging opportunities in crisis situations (Manoharan et al., 2024).

In the context of crisis management and sustainable entrepreneurship, it is essential to focus on fostering resilience and adaptive innovation. This encourages businesses to use sustainable technologies and collaborative approaches to develop solutions that address global crises such as climate change and resource scarcity (Khan et al., 2024). Integrating adaptive technologies with environmental crises into sustainable entrepreneurship not only increases operational efficiency but also promotes social equity and environmental responsibility (Odeyemi et al.,

2023). The increasing adoption of sustainable adaptive technologies by businesses will facilitate the realization of a more resilient and sustainable economy, as they can pursue the exploitation of emerging environmental opportunities with an emphasis on sustainability issues and move towards a sustainable future while addressing current and future challenges and crises.

Various studies indicate that the effectiveness of businesses with sustainable entrepreneurial approaches in crises and their improvement and promotion compared to before the crisis have been ignored in most cases and have caused business owners to suffer adverse economic consequences (Lee et al., 2024). Therefore, familiarity with crisis management in businesses with sustainable entrepreneurial approaches is particularly important. In conclusion, crisis management provides the basis for exploiting emerging sustainable entrepreneurial opportunities, and in turn, exploiting sustainable entrepreneurial opportunities provides the basis for managing crises and dealing with them, and significantly reducing their negative consequences, especially in the field of crises such as environmental crises, drought, and climate change. In other words, crisis management and sustainable entrepreneurship have a two-way relationship that must be considered. Accordingly, this chapter presents the theoretical foundations related to crisis management and sustainable entrepreneurship.

2 Background

The complex crises faced by various sectors present significant challenges for businesses worldwide. This issue is particularly acute in developing countries, where companies are disproportionately affected by the severe consequences of these crises and have limited resources for mitigation and recovery. On the other hand, some businesses may unintentionally create conditions that either trigger or worsen crises, often due to a lack of focus on sustainable practices.

Crisis management has been extensively studied and applied across many fields. In the context of entrepreneurial ventures, incorporating strong crisis management frameworks represents a crucial step in enhancing organizational resilience. This approach draws on a wealth of interdisciplinary insights and best practices, ultimately guiding businesses toward sustainable development outcomes. Given the extensive use of crisis management systems in various fields, their role in the context of businesses, especially in businesses with a sustainable entrepreneurial nature, is prominent (Emami et al., 2024a). However, for effective crisis management, it is essential to understand the concepts of crisis and crisis management, their relationship to business resilience and vulnerability, and ultimately, to sustainable entrepreneurship. Accordingly, in the remainder of this chapter, the concept of crisis, types of crises, and crisis management will be examined first. In the second part, the relationship between crisis, resilience, and vulnerability of sustainable entrepreneurial businesses will be discussed by presenting resilience models and defining vulnerability in businesses. In the third part, future research directions in the field of crisis management and sustainable entrepreneurship will be discussed. In the fourth part, a general conclusion will be drawn from the present chapter, and a related case study will be discussed. Finally, this chapter ends with frequently asked questions and definitions of key terms and structures.

3 Crisis Management and Sustainable Entrepreneurship

3.1 The Concept of Crisis

To better understand the nature of a crisis, the term crisis must first be defined. There are various definitions of the term crisis. However, the word crisis comes from the Greek word "Krisis," which means decision or distinction. A crisis can be described as an unwanted, extraordinary, and often unexpected process. A "crisis" is defined as a period of increased difficulty, danger, or uncertainty, often precipitated by unforeseen events that disrupt standard operating frameworks (Barasa et al., 2018).

Henderson (2009) claims that a crisis is a triggering event that causes or has the potential to cause significant change (Henderson, 2009). He believes that each crisis is unique and unexpected and that urgency and danger are specific to them (Tapola & Bagans, 2011). Beirman (2011) defined a crisis as an event or set of conditions that can harm the entire system (Beirman, 2011). Williams et al. (2017) defined a crisis as a process that can culminate in a single event and disrupt normal functioning (Williams et al., 2017). In general, it can be said that crises are events that disrupt the normal functioning of various systems and cause individuals and groups to deviate from their planning systems. In this chapter, crisis refers to a set of conditions and processes that have exposed many sustainable entrepreneurial actions of businesses to serious threats, disrupted their normal functioning, or caused them significant harm and losses.

3.2 Crisis Classification

In general, crises develop over time and can be divided into three stages: potential, latent, and acute. In the first stage, a potential crisis is just an imaginary structure that is uncertain and has no external existence (Tapola & Bagans, 2011). A latent crisis is a state where a crisis has occurred but has not yet progressed much, and its impact can be neutralized through proper management and planning. An acute crisis is one that has devastating consequences not only for the present but also for the future (Tapola & Bagans, 2011).

In terms of duration, crises are also divided into three categories. The first category is called immediate crises, which include crises with very little or no warning. Therefore, organizations are not able to research a problem or prepare a plan before the crisis. The second category is emergent crises, which are slower-moving and may be stopped or limited by organizational actions. The last category is persistent crises, which include crises that may last for weeks, months, or even years (Kukanja et al., 2020).

In terms of nature, crises can be divided into four categories: (1) crises that occur due to technical or economic failure within the organization (such as the Chernobyl nuclear disaster in 1986); (2) crises that occur due to technical or economic failure outside the organization (such as natural disasters); (3) crises that are the result of human error and occur within an organization (such as organizational misconduct and internal sabotage); (4) crises that are the result of human error and occur outside the organization (such as external sabotage and terrorist acts) (Kukanja et al., 2020). Accordingly, it can be said that the cause of a crisis can be an internal or external force.

3.3 Crisis Management

A "crisis" is an event that, in any form, can cause a shock to sustainable entrepreneurial actions and, as a result, a sudden occurrence of an undesirable situation" (Laws & Prideaux, 2006), while crisis management is "strategies, processes, and actions that are planned and implemented to prevent and cope with a crisis" (World Tourism Organization, 2011). Crisis management in organizations, especially within sustainable entrepreneurial businesses, is an emerging field of research (Krishnan et al., 2022). It is defined as the ability to respond effectively to unexpected disruptions and events, ultimately maximizing the speed of organizational recovery. This capability is referred to as dynamic resilience (Annarelli & Nonino, 2016).

Crisis management serves as a strategic response to adverse events, focusing on restoring system stability. The crisis management framework typically encompasses four key stages: (1) prevention and mitigation, which involve proactive measures to reduce vulnerabilities; (2) preparedness, entailing the development of response plans and training; (3) coping and response, where immediate actions are taken to address the unfolding crisis; and (4) recovery, aimed at returning the system to its pre-crisis state or enhancing its resilience for future incidents (Williams et al., 2017). In this chapter, crisis management refers to the process of managing, controlling, and reducing the negative effects and consequences of crises in sustainable entrepreneurial actions in pre-, during, and post-crisis conditions. It is an effort to promote and improve these actions through the categories of organizing, planning, attracting and employing, directing, and efficiently controlling resources (financial, non-financial, and human).

In other words, in this chapter, crisis management is considered an opportunity to improve sustainable entrepreneurial practices. The basis for developing sustainable entrepreneurial actions is provided through crisis management.

In this regard, in the first step, by improving prevention/mitigation measures, the emergence of crises, especially future environmental crises, is prevented.

In the second step, planning based on sustainable entrepreneurial actions provides the basis for preparedness for facing possible crises. In the third step, when crises occur, they are responded to based on sustainable entrepreneurial and innovative actions. Ultimately, under the new normalized conditions, with planning and organization, the recovery attempt will be sustainable through entrepreneurial and innovative actions.

4 Crisis, Resilience, and Vulnerability of Entrepreneurial Businesses

4.1 Business Vulnerability in Crisis Situations

The concept of vulnerability has different meanings and uses in different approaches; therefore, there is no consensus on its meaning, and different definitions have been given depending on the field of research. This concept has been applied separately to the social subsystem, the ecological subsystem (natural or biophysical), or the socio-ecological subsystem (SES) in a combined manner. A Social-Ecological System (SES) encompasses the complex interplay between social (human) and ecological (biophysical) components and their respective subsystems. The term "vulnerability," originating from the Latin "vulnerare," which means "to wound," refers to the potential for adverse effects stemming from exposure to disturbances or stresses (Kasperson et al., 2022). In this context, vulnerability is quantitatively defined as the susceptibility of a system, subsystem, or specific component to sustain damage when subjected to hazards, disturbances, or stressors (Schweikert & Deinert, 2021). In another definition, vulnerability is the inherent characteristics of a social system before an event or occurrence that has the potential to be damaged (Talubo et al., 2022). In essence, vulnerability arises from a combination of exposure, sensitivity, and the ability to adapt. In this regard, various studies show that sustainable entrepreneurial businesses may be more vulnerable to disruptions caused by crises, especially environmental crises, due to resource constraints, dependence on environmental factors, and inadequate support, especially in developing countries (Herbane, 2010).

During a crisis, governments often implement a range of measures to support businesses in overcoming liquidity challenges and ensuring their survival in the short term; however, the long-term survival of businesses in all areas will depend on how they react and cope with the crisis and adapt to new economic and social conditions, and ultimately their resilience. Therefore, analyzing resilience models and focusing on different actions at different stages of crisis management can help businesses not only ensure their continued survival but also create a base for their improvement and promotion. The following analysis of the most important resilience models of businesses with the approach of sustainable entrepreneurs has been carried out.

4.2 Business Resilience Models in Crisis Situations

Resilience can be understood in various ways, including the capacity to absorb disruptions and reorganize while still ensuring system performance (Khosravi et al., 2021; Racca et al., 2016; Reitsema et al., 2017), the ability to bounce back quickly from significant disturbances (John et al., 2016), and the capability to withstand shocks and recuperate from their impacts (Angeon &

Bates, 2015). The concept of business resilience is relatively recent compared to its application in other areas (Mohammadifar et al., 2022; Quenum et al., 2021). Entrepreneurial ventures across all sectors face ongoing risks and crises (Mirakzadeh et al., 2021). These challenging conditions compel entrepreneurs to think about how to manage operational risks and enhance the resilience of their business processes (Zighan et al., 2022). There are various definitions of business resilience. In one definition, business resilience means the ability and capacity of a business to face and deal with crises and challenges and the ability to return to normal conditions (Bhamra et al., 2011). Also, according to Quenum et al. (2021), business resilience broadly refers to the capacity of a business to perform well regardless of the economic or organizational environment (Quenum et al., 2021). This resilience can occur through implementing risk management strategies or by adopting innovative management practices that prepare organizations to deal with inevitable environmental challenges. From another perspective, business resilience is the ability of a business to cope with all disruptions (Apasrawirote & Yawised, 2024). Several models of business resilience in crisis are analyzed below.

4.2.1 Business Resilience Model from the Perspective of Lengnick-Hall and Beck (2009)

Organizational resilience, as noted by Lengnick-Hall and Beck (2009), significantly boosts an organization's capability to thrive when facing unique challenges and to undertake transformative actions during crises. In their model, strategic agility emerges as a multifaceted and varied concept that enhances an organization's potential for growth, as well as the speed at which resilience and systemic dynamism can be implemented. This characteristic is fostered through the sharing of organizational resources, skills, and competencies (Lengnick-Hall & Beck, 2009).

One important aspect included in this model is "resilience capacity." This dimension enables organizations to adapt through various forms of strategic agility and evolve in response to changing conditions. Ultimately, both strategic agility and resilience capacity enable companies to anticipate change, recover from crises and shocks, and acquire valuable skills through their experiences (Lengnick-Hall & Beck, 2009).

4.2.2 Business Resilience Model from the Perspective of Gibson and Tarrant (2010)

Gibson and Tarrant (2010) have explored various models of resilience in businesses through their research. They argue that in recent years, there has been a notable rise in attention to the concept of resilience across various fields. Consequently, a range of definitions, processes, management systems, and measurement tools have emerged in numerous studies addressing this important topic. Gibson and Tarrant (2010) believe that resilience means improving adaptability in the face of uncertainties in critical situations and adversities in the internal and external environment. These researchers have presented models of organizational resilience that indicate that effective resilience occurs when appropriate strategies are adopted in accordance with critical situations that enhance hard and soft organizational capabilities. It has been emphasized that no universal framework, management system, or solution exists for developing a resilient organization. In the following analysis, we will examine the models introduced by Gibson and Tarrant (2010) across different contexts to assess their applicability and effectiveness.

The 'Integrated Functions Model' of Resilience

Initially, concepts of organizational resilience—especially in the UK and USA—focused on rebranding various strategies of business continuity management (BCM) as resilience. This often led to the creation of what were known as 'resilience processes' or 'resilience systems.' Recently, we've seen the emergence of resilience management system cycles, which seem to claim they can achieve what IS09001 has accomplished in quality assurance. However, Gibson and Tarrant warn that these rigid approaches not only fail to capture the essence of resilience but could actually weaken it, particularly when confronting 'black swan' events—those unforeseen occurrences with significant consequences. Therefore, using only one approach cannot ensure organizational resilience in different situations, and it is necessary to build integrated models and solutions around a strong risk management program. In these models, risk management serves as a critical nexus that integrates various organizational capabilities, including contingency planning, business continuity, security protocols, and crisis management strategies (Gibson & Tarrant, 2010). It establishes a common comprehension of how uncertainties stemming from a turbulent environment can impact an organization's strategic objectives and offers a structured approach for leveraging specialized competencies to mitigate those uncertainties effectively.

Attributional Resilience Model

This framework seeks to elucidate the concept of resilience through the lens of the attributes exhibited by exceptionally resilient organizations. It illustrates how specific organizational characteristics enable entities to navigate uncertainty and address challenges effectively. Consequently, these models offer valuable insights into the strategic modifications that organizations can undertake to enhance their resilience capabilities (Gibson & Tarrant, 2010). The "attributional model" identifies key factors contributing to organizational resilience, primarily focusing on "organizational values" and "leadership":

- Organizational Values: These encompass the establishment of commitment, trust, and robust internal alignment, fostering a shared purpose among stakeholders.
- Leadership: Effective leadership involves articulating a clear strategic direction informed by a thorough understanding of risk. Leaders must empower team members to translate the strategic vision into action while simultaneously cultivating trust within the organization.

The interplay of these traits cultivates an organizational culture attuned to both internal dynamics and external environmental shifts. This heightened sensitivity—characterized by an understanding of historical context, diligent monitoring of current conditions, and proactive anticipation of future developments—enables the identification of crucial indicators amid significant change. Such acuity facilitates closer integration across various organizational components and enhances interdependencies within operational chains. Ultimately, this collaborative synergy optimizes the organization's capacity to achieve common objectives (Gibson & Tarrant, 2010).

Composite Resilience Model

The problem with attributional models of resilience is that they neglect the "harder" elements. These elements, it must be said, contribute to the resilience of the system. The composite resilience model provides a different perspective by examining the interplay between both soft and hard elements, such as processes, infrastructure, technology, resources, information, and knowledge. Central to this model is the crucial role of strategy and policy in fostering operational dynamism, enabling organizations to thrive in both typical and challenging circumstances. A notable distinction of the composite model lies in the significance of emergent leadership, which is essential for guiding the adaptation of various organizational elements to effectively respond to a shifting and unpredictable environment.

Herringbone Model of Resilience

The herringbone model synthesizes elements from three prior frameworks, addressing their limitations. It posits that an organization possesses substantial capabilities and engages in diverse activities that enhance its resilience. Furthermore, the model identifies various operational characteristics that significantly impact the efficacy of these capabilities and activities, thereby fostering organizational resilience. While most capabilities, activities, and characteristics are critical to functioning in a normal environment, resilience creates conditions where an organization can adapt to an unusual environment. Certain characteristics contribute to organizational resilience by enhancing performance across all facets of an organization in atypical environments. Key among these traits are insight, tolerance for ambiguity, creativity, agility, stress management, and learning capability (Gibson & Tarrant, 2010).

The Resilience Triangle Model

In summary, previous models suggest that resilience results from a complex interaction of organizational elements or capabilities, and that resilience is facilitated when an organization adapts to a changing environment. The challenge now is how to represent this complexity in a simple model structure. The inspiration for this model is the age-old relationship of the fire triangle (heat + fuel + oxygen = fire). If any side of this triangle is removed, the fire goes out. Hence, in the resilience triangle, if any side of the triangle is removed, resilience is "turned off". More precisely, the model is trying to show that all three types of capabilities, namely (1) process capabilities, (2) infrastructure and resource capabilities, and (3) leadership, people, and knowledge capabilities, are essential for organizational resilience. Therefore, any loss of effectiveness of these capabilities (collectively or individually) potentially reduces resilience (Gibson & Tarrant, 2010).

Resilience Strategies Model

None of the previous models describe resilience itself, but each describes some organizational aspects that can contribute to resilience. However, the question remains about how these models can be implemented in an organization. For each of the models, there are four broad strategic approaches that can be taken to begin building resilience: resilience, reliability, redundancy, and flexibility (Gibson & Tarrant, 2010).

4.2.3 Business Resilience Model from the Perspective of Burnard and Bhamra (2011)

This model focuses on the organization's resilient response and the importance of adapting to the conditions created after the occurrence of crises. According to this model, in conditions of risk and uncertainty, promoting the organization's resilient characteristics increases the organization's success in the face of risk. Therefore, resilient businesses are less damaged in crises (Burnard & Bhamra, 2011).

4.2.4 Business Resilience Model from the Perspective of Bhamra et al. (2011)

This framework explores organizational resilience within the context of complex global dynamics, focusing on the interplay between adaptive capacity, resilience, and vulnerability in business entities. Bhamra et al. characterize vulnerability as the capacity required to sustain the integrity of a system, while resilience is understood as the process of recovery and reconstruction following disruptions and crises. In this model, resilience is delineated as a subsystem of vulnerability that influences business performance under critical conditions by mitigating vulnerability and enhancing adaptive capacity (Bhamra et al., 2011).

4.2.5 Business Resilience Model from the Perspective of Moran and Tame (2012)

Moran and Tame (2012) have presented a model of resilience in which leadership and resilience are introduced as tools for promoting business sustainability. In this model, resilience is introduced as the ability to adapt to challenges, learn from, and experience crises for future business success. In this model, achieving resilience is a complex process that includes increasing awareness, taking ownership, participating in education, and taking action (Moran & Tame, 2012).

4.2.6 Business Resilience Model from the Perspective of Pal et al. (2014)

In this model, by generalizing organizational resilience models, an attempt has been made to analyze and present a resilience model for SMEs in financial crises. Pal et al. (2014) state the resilience model of businesses as influenced by three components: learning and culture, dynamic competitiveness, and assets and resources under the empowerment approach (Pal et al., 2014).

4.2.7 Business Resilience Model from the Perspective of Hashim et al. (2015)

In this model, the elements of resilience in small businesses are analyzed. In the resilience model of Hashim et al (2015), brand is considered an important factor in the success of small businesses. The factors that effectively create brand identity are divided into two categories: internal factors (corporate identity and organizational identity) and external factors (corporate image and reputation). According to this model, branding increases resilience in businesses and ultimately improves business performance (Hashim et al., 2015).

5 Crisis Management and Sustainable Entrepreneurship

The concept of sustainable entrepreneurship integrates principles of sustainable development with the dynamics of entrepreneurial activity. It encompasses a spectrum of related fields, including environmental entrepreneurship, green entrepreneurship, and social entrepreneurship (Gibbs, 2006). Due to the overlapping nature of these categories, defining sustainable entrepreneurship can be challenging. A significant portion of the literature and research tends to focus primarily on environmental entrepreneurship when analyzing the broader concept of sustainable entrepreneurship (Gibbs, 2006). Hockerts and Wüstenhagen (2010) conceptualize sustainable entrepreneurship as the identification and capitalization of economic opportunities arising from market inefficiencies, alongside the initiation of transformations that advance environmental and social sustainability. Consequently, the core inquiry of sustainable entrepreneurship revolves around the capacity of entrepreneurs to navigate and integrate economic viability with environmental stewardship, social equity, and cultural preservation. This necessitates that innovative and creative entrepreneurial activities are characterized by risk-taking and profitability while simultaneously addressing sustainability challenges. Ultimately, the goal of sustainable entrepreneurship is to harness innovative opportunities that yield economic returns, foster social justice, enhance environmental quality, and uphold cultural integrity.

Crisis management within the realm of sustainable entrepreneurship is critical not only for the immediate survival of enterprises but also for facilitating their growth and evolution. Environmental fluctuations frequently give rise to entrepreneurial opportunities, as crises precipitate shifts in the external landscape that can foster sustainable entrepreneurial initiatives. Effectively managing crises through principled frameworks can lay the groundwork for the long-term development of sustainable entrepreneurial pursuits.

Moreover, the cultivation of sustainable entrepreneurship over time serves to mitigate the adverse effects and repercussions of crises, particularly those related to environmental degradation. Therefore, it can be concluded that a robust and significant correlation exists between sustainable entrepreneurship and crisis management. Sustainable entrepreneurship promotes economic growth while simultaneously decreasing environmental vulnerability. In turn, such economic development and diminished ecological impacts create a foundation for enhancing both economic and environmental resilience, which are essential for effective crisis management.

6 Future Research Directions

This chapter offers a broad examination of the interplay between crisis management and sustainable entrepreneurship. While it establishes a foundational understanding, future research should adopt a more focused approach by investigating sustainable entrepreneurial ventures that exhibit heightened vulnerability to crises, particularly those stemming from environmental disruptions. To enhance the practical applicability of this chapter, a management tool should be developed, building upon the theoretical models presented herein and incorporating targeted strategies to bolster the resilience of sustainable entrepreneurial business models. In practical terms, the proposed business model resilience and vulnerability framework should be operationalized to systematically evaluate and monitor the resilience and vulnerability levels of sustainable entrepreneurial business models across diverse organizational contexts. Furthermore, the management toolkit must equip managers and decision-makers with actionable insights to identify structural strengths and weaknesses within sustainable entrepreneurship, particularly in dynamic and crisis-prone environments. By doing so, it will enable organizations to enhance adaptive capacities, mitigate vulnerabilities, and strengthen long-term resilience. Additionally, the adaptive capabilities of sustainable enterprises during crises warrant further scholarly attention. Future studies should also explore the bidirectional dynamics between crisis management and sustainable entrepreneurship through empirical field research, thereby enriching theoretical and practical understanding of their interdependent effects.

7 Conclusion

One of the fundamental concepts and approaches in crisis management is the concept of business resilience. Business resilience in the context of crisis management causes strategies to move from purely reactive approaches to more comprehensive and predictive frameworks. In other words, by infusing an adaptive and forward-looking dimension into the crisis management model, resilience causes actions to go beyond mere reaction and short-term recovery. As a result, it improves adaptation to crisis conditions and even creates a competitive advantage. Ultimately, it enables businesses to continue to survive and even improve during and after crises (Emami et al., 2024b). A review of different models of business resilience shows that resilience means a business's resistance to disruptions and crises, returning to its pre-crisis status, and even improving it. This concept is related to startup and establishment activities, risk and crisis management, business planning, strategic management, reliability, adaptability, vulnerability, etc. Therefore, in the context of crisis management, paying attention to resilience in the first step can help sustainable entrepreneurial businesses to continue their activities in any situation and identify and exploit emerging opportunities for sustainable entrepreneurial actions in the environmental changes resulting from any crisis. Exploiting emerging sustainable entrepreneurial opportunities also provides the opportunity to reduce the negative consequences of crises, especially in the field of environmental crises.

On the one hand, crisis management in sustainable entrepreneurship not only ensures the survival of businesses but also causes their growth and development because crises cause environmental changes. Environmental changes are the source of the emergence of sustainable entrepreneurial opportunities. By creating environmental changes around businesses, crises give rise to opportunities to develop sustainable entrepreneurial actions. Therefore, if crises are managed in a principled manner, they can provide the basis for the development of sustainable entrepreneurial actions in the long term. On the other hand, the development of sustainable entrepreneurial actions in the long term provides the basis for reducing the consequences and negative effects of crises, especially environmental crises. Therefore, what should be emphasized in the long term is the development of sustainable entrepreneurial actions for crisis management and, in turn, the development of crisis management models for the development of sustainable entrepreneurial actions.

8 Case Study

Sustainable entrepreneurship functions within a dynamic landscape characterized by continual change and multiple crises, necessitating a strategic approach to identify and harness emerging environmental opportunities. Effective crisis management is pivotal in fostering business growth while simultaneously mitigating threats. Crises, when leveraged effectively, can catalyze the expansion and development of sustainable entrepreneurship, whereas inadequate management may lead to their downfall. Empirical evidence from various global contexts underscores the importance of exploring the interplay between Crisis Management and Sustainable Entrepreneurship, particularly in enhancing the resilience of entrepreneurial ventures. Businesses adept at optimizing their adaptive capacity in response to crises are positioned for continued survival, growth, and development, while those that fail to do so risk extinction. A pertinent case study is observed in the tourism sector in pristine regions of Iran during the COVID-19 pandemic. Initially, tourism operations faced complete shutdowns due to government-imposed restrictions and the absence of vaccination or effective treatments. However, tourism businesses that adopted a sustainable entrepreneurship framework quickly reoriented their strategies to adapt to the challenging circumstances. By leveraging digital platforms and implementing rigorous health protocols—including the use of masks and sanitizers—these enterprises not only navigated the crisis but also enhanced their market presence, laying the groundwork for future growth and resilience. In other words, sustainable entrepreneurial ventures within Iran's tourism sector successfully ensured their survival, growth, and development during the COVID-19 crisis through adept crisis management and the strategic utilization of their adaptive capacities. Therefore, it is imperative for managers of sustainable entrepreneurial firms to prioritize crisis management to secure the ongoing survival, growth, and strategic advancement of their enterprises by exploiting emergent environmental opportunities and enhancing adaptive capabilities, thus gaining a competitive edge.

Case Questions

- What is crisis management like in sustainable entrepreneurial businesses?
- What is the application of resilience models in sustainable entrepreneurial businesses?
- What does adaptive capacity mean in the context of sustainable entrepreneurial businesses?
- What does vulnerability mean in sustainable entrepreneurial businesses?

9 Key Terms and Definitions—Definitions for the Key Constructs

- Crisis: A process that can culminate in an event and disrupt the normal functioning of a system (Williams et al., 2017).
- Crisis management: Crisis management is considered a response to unexpected disturbances and events to bring the disrupted system to a state of equilibrium (Williams et al., 2017). It can be said that the crisis management process consists of at least four stages: (1) prevention; (2) preparedness; (3) coping; (4) recovery (Williams et al., 2017).
- Business vulnerability: Vulnerability is the inherent characteristics of a system before an incident or event occurs that has the potential to be damaged (Talubo et al., 2022).
- Business resilience: Business resilience refers to the ability and capacity of a business to face and deal with crises and challenges and the ability to return to normal business conditions (Bhamra et al., 2011).
- Sustainable entrepreneurship: Sustainable entrepreneurship is defined as discovering and exploiting economic opportunities through market imbalances and initiating transformations in environmental and social sustainability (Hockerts & Wüstenhagen, 2010).

References

Angeon, V., & Bates, S. (2015). Reviewing composite vulnerability and resilience indexes: A sustainable approach and application. *World Development, 72*, 140–162. https://doi.org/10.1016/J.WORLDDEV.2015.02.011

Annarelli, A., & Nonino, F. (2016). Strategic and operational management of organizational resilience: Current state of research and future directions. *Omega, 62*, 1–18. https://doi.org/10.1016/J.OMEGA.2015.08.004

Apasrawirote, D., & Yawised, K. (2024). The emerging of business resilience plans (BRPs) in dealing with business turbulence. *Management Research Review, 47*, 141–161. https://doi.org/10.1108/MRR-04-2022-0273/FULL/XML

Barasa, E., Mbau, R., & Gilson, L. (2018). What is resilience and how can it be nurtured? A systematic review of empirical literature on organizational resilience. *International Journal of Health Policy and Management, 7*, 491–503. https://doi.org/10.15171/IJHPM.2018.06

Batabyal, A. A., & Yoo, S. J. (2017). On research and development in a model of Schumpeterian economic growth in a creative region. *Technological Forecasting and Social Change, 115*, 69–74. https://doi.org/10.1016/J.TECHFORE.2016.09.020

Beirman, D. (2011). The integration of emergency management and tourism. *Australian Journal of Emergency Management, 26*, 32–36.

Bhamra, R., Dani, S., & Burnard, K. (2011). Resilience: The concept, a literature review and future directions. *International Journal of Production Research, 49*, 5375–5393. https://doi.org/10.1080/00207543.2011.563826

Burnard, K., & Bhamra, R. (2011). Organisational resilience: Development of a conceptual framework for organisational responses. *International Journal of Production Research, 49*, 5581–5599. https://doi.org/10.1080/00207543.2011.563827

Di Vaio, A., Hassan, R., Chhabra, M., Arrigo, E., & Palladino, R. (2022). Sustainable entrepreneurship impact and entrepreneurial venture life cycle: A systematic literature review. *Journal of Cleaner Production, 378*, Article 134469. https://doi.org/10.1016/J.JCLEPRO.2022.134469

Emami, A., Welsh, D. H. B., Hakimian, R., Correspondence, D. H. B., & Welsh, B. (2024a). The impact of entrepreneurial hubris and narcissism on personal social responsibility. *Strategic Change, 33*, 257–274. https://doi.org/10.1002/JSC.2579

Emami, S. G., Lorenzoni, V., & Turchetti, G. (2024). Towards resilient healthcare systems: A framework for crisis management. *International Journal of Environmental Research and Public Health, 21*, 286. https://doi.org/10.3390/IJERPH21030286

Gibbs, D. (2006). Sustainability entrepreneurs, ecopreneurs and the development of a sustainable economy. *Greener Management International*, 63–78. https://doi.org/10.9774/GLEAF.3062.2006.AU.00007

Gibson, C., & Tarrant, M. (2010). A "Conceptual Models" approach to organisational resilience. *Aust. J. Emerg. Manag., 25*, 8–14.

Goel, P., 2024. Crisis Management Strategies: Preparing for and Responding to Disruptions. J. Adv. Manag. Stud. 1, 25–29. https://doi.org/10.36676/JAMS.V1.I1.06

Hashim, S. F., Tajuddin, R. M., & Zainol, A. S. (2015). Creating resilient SME through brand identity (Case study of Bumiputera fashion entrepreneurs in Malaysia). *Int. Acad. Res. J. Soc. Sci., 1*, 264–282.

Henderson, J. C. (2009). Tourism and health crises. *Tourism and Health Crises, 104*. https://doi.org/10.1016/B978-0-7506-7834-6.50008-9

Herbane, B. (2010). Small business research: Time for a crisis-based view. *International Small Business Journal, 28*, 43–64. https://doi.org/10.1177/0266242609350804

Hockerts, K., & Wüstenhagen, R. (2010). Greening Goliaths versus emerging Davids—Theorizing about the role of incumbents and new entrants in sustainable entrepreneurship. *Journal of Business Venturing, 25*, 481–492. https://doi.org/10.1016/J.JBUSVENT.2009.07.005

John, A., Yang, Z., Riahi, R., & Wang, J. (2016). A risk assessment approach to improve the resilience of a seaport system using Bayesian networks. *Ocean Engineering, 111*, 136–147. https://doi.org/10.1016/J.OCEANENG.2015.10.048

Kasperson, J. X., Kasperson, R. E., Turner, B. L., Hsieh, W., & Schiller, A. (2022). Vulnerability to global environmental change. *Social contours of risk. Volume II: Risk analysis, corporations, and globalization* (Vol. 2, pp. 245–285). https://doi.org/10.4324/9781849772556-18/VULNERABILITY-GLOBAL-ENVIRONMENTAL-CHANGE-JEANNE-KASPERSON-ROGER-KASPERSON-TURNER-WEN-HSIEH-ANDREW-SCHILLER

Khan, T., Samiullah, M., Rouf, I. I., Sultana, S., Rahman, S., Rahman, M. B., & Khanum, R. (2024). The nexus of water scarcity and climate change: Understanding interconnected challenges and formulating resilient strategies. *International Journal of Environmental Sciences, 7*, 57–68.

Khosravi, E., Naderi, N., Rezaei, B., & Azadi, H. (2021). Presenting a safe tourism model for the continuation of tourism businesses in the Covid-19 crisis. *Journal of Geography and Environmental Hazards, 10*, 147–165. https://doi.org/10.22067/GEOEH.2021.70107.1052

Krishnan, C. S. N., Ganesh, L. S., & Rajendran, C. (2022). Entrepreneurial Interventions for crisis management: Lessons from the Covid-19 pandemic's impact on entrepreneurial ventures. *International Journal of Disaster Risk Reduction, 72*, Article 102830. https://doi.org/10.1016/J.IJDRR.2022.102830

Kukanja, M., Planinc, T., & Sikošek, M. (2020). Crisis management practices in tourism SMEs during the Covid-19 pandemic. *Organizacija, 53*, 346–361. https://doi.org/10.2478/ORGA-2020-0023

Laws, E., & Prideaux, B. (2006). Crisis management: A suggested typology. *Journal of Travel & Tourism Marketing, 19*, 1–8. https://doi.org/10.1300/J073V19N02_01

Lee, Y., Kim, J., Mah, S., & Karr, A. (2024). Entrepreneurship in times of crisis: A comprehensive review with future directions. *Entrepreneurship Research Journal, 14*, 905–950. https://doi.org/10.1515/ERJ-2022-0366/ASSET/GRAPHIC/J_ERJ-2022-0366_FIG_003.JPG

Lengnick-Hall, C. A., & Beck, T. E. (2009). Resilience capacity and strategic agility: Prerequisites for thriving in a dynamic environment. Work. Pap. 0059, Coll. Business, Univ. Texas San Antonio

Manoharan, A. A., Rangaswamy, G., Muthhukkannu, G., Sabapathi, S., Anandharam, K., & Aazib Afraz, T. N. (2024). Green entrepreneurship for sustainability business: A bibliometric analysis. *Journal of Lifestyle and SDGs Review, 5*, e03204. https://doi.org/10.47172/2965-730X.SDGsReview.v5.n02.pe03204

Mirakzadeh, A. A., Karamian, F., Khosravi, E., & Parvin, F. (2021). Analysis of preventive behaviors of rural tourism hosts in the face of COVID-19 pandemic: Application of health belief model. *Frontiers in Public Health, 9*, 2169. https://doi.org/10.3389/FPUBH.2021.793173

Mohammadifar, Y., Naderi, N., Khosravi, E., & Karamian, F. (2022). Developing a paradigm model for resilience of rural entrepreneurial businesses in dealing with the COVID-19 crisis; application of grounded theory in western of Iran. *Frontiers in Public Health, 10*, Article 833909. https://doi.org/10.3389/FPUBH.2022.833909/BIBTEX

Moran, B., & Tame, P. (2012). Organizational resilience: Uniting leadership and enhancing sustainability. *Sustainability, 5*, 233–237. https://doi.org/10.1089/SUS.2012.9945/ASSET/SUS.2012.9945.FP.PNG_V03

Odeyemi, O., Usman, F. O., Mhlongo, N. Z., Elufioye, O. A., & Ike, C. U. (2023). Sustainable entrepreneurship: A review of green business practices and environmental impact. *World Journal of Advanced Research and Reviews, 21*, 346–358. https://doi.org/10.30574/wjarr.2024.21.2.0461

Pal, R., Torstensson, H., & Mattila, H. (2014). Antecedents of organizational resilience in economic crises—An empirical study of Swedish textile and clothing SMEs. *International Journal of Production Economics, 147*, 410–428. https://doi.org/10.1016/J.IJPE.2013.02.031

Quenum, A., Thorisson, H., Wu, D., & Lambert, J. H. (2021). Resilience of business strategy to emergent and future conditions. *Journal of Risk Research, 24*, 870–888. https://doi.org/10.1080/13669877.2018.1485172

Racca, P., Casarin, R., Squazzoni, F., & Dondio, P. (2016). Resilience of an online financial community to market uncertainty shocks during the recent financial crisis. *Journal of Computer Science, 16*, 190–199. https://doi.org/10.1016/J.JOCS.2016.07.007

Reitsema, L. J., Kozłowski, T., Crews, D. E., Katzenberg, M. A., & Chudziak, W. (2017). Resilience and local dietary adaptation in rural Poland, 1000–1400 CE. *Journal of Anthropological Archaeology, 45*, 38–52. https://doi.org/10.1016/J.JAA.2016.11.001

Schaltegger, S., & Wagner, M. (2011). Sustainable entrepreneurship and sustainability innovation: Categories and interactions. *Business Strategy and the Environment, 20*, 222–237. https://doi.org/10.1002/BSE.682

Schweikert, A. E., & Deinert, M. R. (2021). Vulnerability and resilience of power systems infrastructure to natural hazards and climate change. *Wiley Interdisciplinary Reviews: Climate Change, 12*, Article e724. https://doi.org/10.1002/WCC.724

Talubo, J. P., Morse, S., & Saroj, D. (2022). Whose resilience matters? A socio-ecological systems approach to defining and assessing disaster resilience for small islands. *Environmental Challenges, 7*, Article 100511. https://doi.org/10.1016/J.ENVC.2022.100511

Tapola, H., & Bagans, L. (2011). Crises in the tourism industry and their effects on young travellers. *Tourism Management*, 1–601.

Williams, T. A., Gruber, D. A., Sutcliffe, K. M., Shepherd, D. A., & Zhao, E. Y. (2017). Organizational response to adversity: Fusing crisis management and resilience research streams. *Academy of Management Annals, 11*, 733–769. https://doi.org/10.5465/ANNALS.2015.0134

World Tourism Organization. (2011). Toolbox for crisis communications in tourism: Checklist and best practices. UNWTO. https://doi.org/10.18111/9789284413652

Zighan, S., Abualqumboz, M., Dwaikat, N., & Alkalha, Z. (2022). The role of entrepreneurial orientation in developing SMEs resilience capabilities throughout COVID-19. *The International Journal of Entrepreneurship and Innovation, 23*, 227–239. https://doi.org/10.1177/14657503211046849

Innovative and Interdisciplinary Approaches

Cognitive Neuroscience of Sustainable Entrepreneurship: Cognition and Emotion in Action

Farhad Farkhondeh Tale Navi and Saied Sabaghypour

Abstract

Sustainable entrepreneurship is increasingly recognized as a key driver of economic, social, and environmental progress. This chapter explores the cognitive and emotional mechanisms that influence sustainable entrepreneurial decision-making. Here, we highlight how cognitive neuroscience might provide insights into these processes. By examining the interplay between cognition and emotion, we represent how sustainable entrepreneurs recognize opportunities, assess risks, and make strategic choices that balance profitability with long-term sustainability. The chapter integrates findings from cognitive neuroscience and reviews some of the most important neuroscientific methodologies. Beyond theoretical insights, this chapter highlights practical applications for sustainable entrepreneurship. By identifying neural markers and cognitive biases that influence sustainability-oriented entrepreneurship, neuroscience can inform training programs, enhance ethical leadership, and support long-term business resilience. Finally, we suggest future research directions to better understand the brain processes underlying sustainable entrepreneurship and how they can contribute to creating more responsible and innovative businesses.

Keywords

Sustainable entrepreneurship · Cognitive neuroscience · Decision-making · Emotion and cognition · Entrepreneurial behavior · EEG · fMRI

1 Introduction

Over the past sixty years, researchers have attempted to understand the factors that contribute to sustainable entrepreneurship. Sustainable entrepreneurship reveals fundamental aspects of entrepreneurial cognition and offers insights that extend to broader human decision-making processes (Castellano et al., 2017; Le Loarne Lemaire et al., 2022). Indeed, it has recently been proposed that cognitive and emotional aspects are highly intertwined in entrepreneurial decision-making processes (Baron & Ward, 2004; Hayton & Cholakova, 2012; Jennings et al., 2015). In this respect, cognitive neuroscience has served as a source of inspiration for sustainable entrepreneurship. This chapter's key contribution emphasizes cognitive and emotional engagement as essential mechanisms that empower sustain-

F. Farkhondeh Tale Navi (✉) · S. Sabaghypour
Department of Cognitive Neuroscience, Faculty of Education and Psychology, University of Tabriz, Tabriz, Iran
e-mail: Farkhondeh_F@tabrizu.ac.ir

S. Sabaghypour
Department of Psychology, The University of Western Ontario, London, ON, Canada

M. Akbari et al. (eds.), *Sustainable Entrepreneurship*, University of Tehran Science and Humanities Series, https://doi.org/10.1007/978-3-032-04909-4_22

able entrepreneurs. Here, by integrating insights from cognitive neuroscience, we also explore how memory, decision-making, creativity, and emotional aspects contribute to sustainable business practices and long-term success. By the end of this chapter, readers will be able to understand the role of cognitive neuroscience in sustainable entrepreneurship, recognize how neuroscientific tools can inform entrepreneurial decision-making, and appreciate the interplay between cognition, emotion, and business sustainability. Additionally, we will provide practical applications for neuroscience research to enhance sustainable entrepreneurial practices and explore future directions for advancing the field.

Cognitive neuroscience is a multidisciplinary field that integrates experimental psychology with neuroscience, computer science, philosophy, linguistics, and other disciplines to provide a comprehensive understanding of the relationship between the mind and the brain (Miller & Gazzaniga, 1984; Von Eckardt, 1995). The primary goal of this research field has been to integrate perspectives on cognition and brain function, shedding light on key psychological topics. However, it may also provide valuable insights for other disciplines that draw upon this field. The significant advancements in cognitive neuroscience focus on several areas, including structural anatomy (the 'where'), functional investigations that examine the timing and sequence of cognitive processes (the 'when'), and their possible applications in the real world. Cognitive neuroscience aligns with the naturalistic school of thought (Biswas, 2024), as evident in the methods employed in the field. While such techniques are widely used in fields like economics, marketing, and consumer research, they remain relatively uncommon in entrepreneurship studies. This presents numerous opportunities for advancing the field from a cognitive neuroscience perspective.

On the other hand, despite global economic progress, the world faces growing economic, social, and environmental challenges. The depletion of natural resources and the harmful effects of environmental degradation—such as freshwater shortages, biodiversity loss, and drought—have emerged as pressing concerns (Greco & De Jong, 2017). Furthermore, climate change poses a significant threat to the survival and stability of modern societies. Persistent gender inequalities and disparities in access to opportunities and resources further complicate social and economic challenges (Ploum et al., 2018). These challenges, varying in intensity across different contexts, have heightened the urgency for sustainable entrepreneurship as a means to address economic, social, and environmental concerns. Sustainable entrepreneurship is considered a business strategy that integrates sustainable practices to enhance efficiency and competitiveness while balancing environmental, economic, and social impacts (Di Vaio et al., 2022; Schaltegger & Wagner, 2011). It entails identifying, developing, and considering opportunities that contribute to social and environmental well-being and long-term sustainability within communities (Muñoz & Cohen, 2018). In this perspective, entrepreneurship is analyzed through the lens of entrepreneurs' attitudes, behaviors, and personal characteristics (Cavallo et al., 2019). To achieve this, researchers study entrepreneurship using techniques and methods from cognitive and behavioral sciences. From a cognitive neuroscience perspective, entrepreneurship is examined to understand the cognitive mechanisms that enable entrepreneurs to make appropriate decisions more quickly and efficiently. In the following section, we will further investigate the role of cognitive neuroscience in understanding sustainable entrepreneurship. We will show how cognitive processes shape entrepreneurial decision-making. Then we will highlight key neuroscientific insights that inform sustainable practices and then discuss the relevance of cognitive and emotional mechanisms involved in sustainable entrepreneurship.

2 The Role of Cognitive Neuroscience in Understanding Sustainable Entrepreneurship

Cognitive processes play an essential role in sustainable entrepreneurship (Muñoz, 2018).

These processes influence how entrepreneurs identify opportunities (Chen et al., 2020; Karlesky, 2015), assess risks (Keh et al., 2002; Palich & Bagby, 1995), and make strategic decisions that balance economic, social, and environmental concerns (Baron, 2000; Shepherd et al., 2015; West, 2007). Sustainable entrepreneurs often operate in highly complex and uncertain environments that require them to integrate vast amounts of information, adapt to changing circumstances, and maintain resilience in the face of challenges (Doh et al., 2019; Lv et al., 2018; Ploum et al., 2018). Understanding these cognitive mechanisms from a neuroscientific perspective can provide additional insights into how sustainable entrepreneurs navigate their decision-making processes and drive long-term impact.

The potential for entrepreneurship research to benefit from cognitive neuroscience offers opportunities for enhancing conceptual understanding, constructing hypotheses, and advancing theory development in entrepreneurship (Grégoire et al., 2011). By integrating insights from neuroscience, researchers can refine existing models of entrepreneurial decision-making and cognition, thereby gaining a deeper understanding of the cognitive and emotional mechanisms that drive entrepreneurial behavior. Researchers in this field aim to understand the neurological foundations of entrepreneurial decision-making and identify the factors that influence how entrepreneurs recognize opportunities (Baskaran et al., 2021; Shepherd et al., 2015). Additionally, scientists study the cognitive processes in the entrepreneur's brain that help them navigate uncertainty and respond to opportunities based on neural activity (Clements et al., 2021; McMullen et al., 2014; Shaver et al., 2017). Overall, the final goal in this new field of study is to develop neurocognitive models that explain how entrepreneurship works from a brain-based perspective. Despite the limited number of relevant studies, researchers are increasingly advocating for the integration of neuroscience into the study of cognitive processes within the emerging field of neuroentrepreneurship. They emphasize the importance of examining the cause-and-effect relationships between brain function and various aspects of entrepreneurship, including decision-making abilities, entrepreneurial intentions, competencies, and the capacity to adapt to changing environments.

From a sustainability perspective, the Earth's ecosystem and the human brain share many similarities. Both are highly complex systems made up of trillions of interconnected parts that have evolved over time. They function at different levels, from the overall system (biosphere/central nervous system) down to smaller units (humans/neurons). Both also experience inefficiencies and breakdowns, partly because their subsystems operate independently and sometimes in conflicting ways. Additionally, both require large amounts of energy to function. The human brain, for example, makes up only about 2% of body weight but consumes roughly 20% of the body's energy in the form of glucose (Cunnane, 2005). Given these parallels, it is no surprise that studying the brain has become one of the most promising avenues in entrepreneurship research.

While several traits have been used to explain sustainable decision-making in entrepreneurship, there is no explicit agreement on the importance of cognitive factors in shaping these decisions. In addition, it remains uncertain whether the mental frameworks guiding socially and environmentally responsible choices among new business founders are similar or distinct (Elkington & Hartigan, 2008). Similar to knowledge and mental representations, factors such as attitudes, motivations, and other mental states serve as cognitive resources that individuals rely on to navigate their environment and make decisions about present and future actions (Grégoire et al., 2015). These mental constructs arise from human nature, subjective interpretations, and personal perceptions of the surrounding environment, as well as the accumulation of an individual's unique experiences. In this context, entrepreneurial cognition refers to the knowledge frameworks individuals use to assess, judge, and make decisions related to evaluating opportunities, creating ventures, and driving business growth (Shepherd et al., 2015; West, 2007).

Increasing evidence from cognitive science indicates that individuals vary significantly in their working memory (Baddeley, 1992)—the cognitive system that allows stored knowledge and experiences (essentially, our consciousness) to interact with new information from the external environment. Additionally, research suggests that the more efficiently this system functions—enabling individuals to focus their attention on relevant and task-related information—the better they perform on various complex cognitive tasks, including logical reasoning (Engle, 2010; Oberauer et al., 2000). It is likely that entrepreneurs, particularly those who achieve success, possess a higher memory capacity that enables them to process information efficiently, focus on relevant details, and navigate complex decision-making scenarios more effectively. Furthermore, maintaining a positive attitude and motivation toward sustainable change requires perseverance and determination to bring about meaningful progress. To translate attitude into action, individuals must have confidence in their ability to achieve their objectives and successfully reach their self-defined goals. In this respect, Affective processing—the way emotions are experienced, regulated, and utilized—directly impacts an entrepreneur's ability to persist in the face of uncertainty and adversity (Brundin & Gustafsson, 2013; Hayton & Cholakova, 2012; Shepherd & Williams, 2020).

On the other hand, creativity and innovation are recognized as important drivers of success and are suggested to play vital roles in increasing organizational growth and competitiveness across various industries (Egbu, 2004). Creativity is fundamental for recognizing opportunities and generating value, which positively impacts overall entrepreneurial behavior (Baron, 2007; Baron & Tang, 2011). Therefore, it is reasonable to assume that all creative entrepreneurs naturally incorporate sustainable entrepreneurship principles. Previous reports showed that the levels of entrepreneurial creativity significantly influence intentions to start sustainable businesses (Zemlyak et al., 2022). However, creativity can have a dark side, potentially enabling entrepreneurs to justify environmentally harmful opportunities by disengaging from nature (Qin et al., 2022).

The methodological implications of applying neuroscientific techniques present both possibilities and limitations. Tools such as brain imaging and neurophysiological measurements provide innovative ways to investigate the neural basis of the most critical entrepreneurial cognitive processes, offering objective data on how entrepreneurs process information and regulate emotions. In the following section, we will explore the neuroscientific tools and methodologies that can enhance entrepreneurship research. We will examine key techniques, including functional magnetic resonance imaging (fMRI) and electroencephalography (EEG). We will also discuss their relevance in studying cognitive and emotional processes in entrepreneurial decision-making. We also consider how these methodologies can provide new insights into sustainable entrepreneurship by investigating the relevant neural mechanisms.

3 Neuroscientific Tools and Methodologies for Entrepreneurship Research

With advancements in brain activity measurement technologies, cognitive neuroscience has become increasingly important in recent years. Its main focus is to examine how thought processes influence the execution of tasks and the pursuit of several initiatives. One widely used technique is electroencephalography (EEG), a neuroimaging method that has gained popularity due to its accessibility and minimal testing requirements (Gevins, 1998; Michel & Murray, 2012). EEG is a non-invasive procedure that records the brain's bioelectric activity using electrodes positioned on the scalp. These electrodes detect electrical signals generated by neural activity. These electrodes enable the monitoring of changes in brain activity over time, allowing for the analysis of neural responses to external stimuli.

When evaluating EEG in terms of its usability, it is important to note that it offers high temporal resolution. In the advancement of cognitive neuroscience techniques, one of the most significant developments has been the advancement of functional magnetic resonance imaging (fMRI). This method provides highly detailed images of brain structures and has better spatial resolution than EEG (Smith, 2004). Unlike EEG, which captures the output values transmitted between different brain regions, fMRI focuses on measuring input values as well as the processing of information and stimuli within the brain (Logothetis, 2008). As a result, the images generated through fMRI allow researchers to identify which specific neural inputs and impulses contribute to particular human cognitions and actions. One of the most significant advantages of fMRI is its high spatial accuracy, which enables the precise localization of brain activity and the collection of extensive data for in-depth analysis. This wealth of information allows for a more comprehensive understanding of neural processes and their role in cognition, decision-making, and behavior. Compared to EEG, fMRI provides a more complete and detailed picture of brain functionality, making it very useful in neuroscience research.

However, despite its strengths, fMRI also has notable limitations. One major drawback is its poor temporal resolution, meaning it cannot capture rapid changes in brain activity with the same precision as EEG (Logothetis, 2008; Smith, 2004). Since many cognitive processes, such as decision-making and sensory perception, unfold within milliseconds, fMRI may not always be sufficient for studying their real-time dynamics. Additionally, fMRI scans require participants to remain completely inside a large, enclosed machine. Despite these challenges, the integration of fMRI with other techniques, such as EEG, has opened up new research opportunities, combining high spatial accuracy with improved temporal resolution. This multidisciplinary approach continues to enhance our understanding of the brain's role in entrepreneurship, cognition, and decision-making.

Another widely used technique in cognitive neuroscience is the Galvanic Skin Response (GSR). This method measures electrical changes on the skin's surface in response to various stimuli and provides valuable information about physiological and emotional processes (Ayata et al., 2017; Brocanelli, 2018). Since the skin's electrical resistance is influenced by the level of perspiration, which is controlled by the autonomic nervous system, GSR serves as a useful indicator of an individual's emotional and psychological state. The measurement of the skin's galvanic response is conducted by assessing the electrical resistance of the skin, which fluctuates in relation to its moisture level. This, in turn, is directly linked to the body's reaction to external stimuli, such as stress, excitement, or fear (Satti et al., 2021; Vijaya & Shivakumar, 2013). The process involves passing a low-voltage electrical current between two electrodes attached to the subject's skin, typically on the fingers or palms, where sweat gland activity is most pronounced (Goodman, 1985; Perry, 2007). Despite its usefulness, GSR has some limitations. Since sweat gland activity is influenced by multiple factors, such as temperature, hydration levels, and movement, isolating the effect of specific stimuli can be challenging (Zhan et al., 2023). Additionally, GSR cannot distinguish between different emotions—increased skin conductance may indicate excitement, stress, or fear, requiring additional context or complementary methods (such as facial electromyography or heart rate monitoring) to interpret results accurately.

Regarding the above techniques, it is important to note that researchers are increasingly advocating for the application of cognitive neuroscience methods to explore the neural mechanisms underlying entrepreneurial decision-making (Butler et al., 2016; Robertson et al., 2017). By examining brain and physiological activity, these methods provide valuable insights into how entrepreneurs process information, assess risks, regulate emotions, and respond to challenges in dynamic business environments. At the same time, scholars emphasize that modern entrepreneurship research must integrate neuroscientific perspectives to better understand the fundamental components that shape entrepreneurial behavior. Among these components, the functioning of the entrepreneur's brain and body plays essential roles in determining their ability to identify opportunities, navigate uncertainty, and drive innovation. Decision-making in entrepreneurship is not purely rational; it is influenced by cognitive biases, emotional resilience, and neurophysiological responses that impact how entrepreneurs react to high-pressure situations and evolving market conditions (Hasson Marques et al., 2024; Lichtenstein, 2014). Furthermore, as entrepreneurship continues to evolve in an increasingly complex and competitive landscape, applying advanced neuroscientific techniques, such as EEG, fMRI, and GSR, can help decode the neurological basis of entrepreneurial cognition. Understanding how neural circuits related to memory, creativity, and problem-solving interact allows for the development of more effective training programs, policies, and support systems for entrepreneurs.

Given that research on these issues is still in its early stages, publications on the subject remain relatively scarce. However, despite the limited literature, growing interest in this emerging field highlights its potential to enhance our understanding of how cognitive and emotional processes influence entrepreneurial behavior. With this in mind, the next section will review the available research on key aspects of cognitive neuroscience in sustainable entrepreneurship. First, we will discuss the importance of decision-making for a sustainable future. Following this, we will analyze the role of creativity in sustainable entrepreneurship and the development of sustainable business models. Finally, we will examine emotions in action through affective processing for entrepreneurial resilience. Through this section's review, we aim to provide insights into the neural and psychological foundations that shape entrepreneurial behavior, offering a bridge between neuroscience and sustainable business practices.

4 Thinking and Feeling in Entrepreneurship

The earliest significant contribution to integrating neuroscience techniques into management studies was carried out by Taggart et al. (1985). Their research aimed to explore the connection between decision-making styles and cerebral dominance using EEG. Their findings suggested that traditional psychological measurements capture only a small fraction of actual cerebral processing, indicating an increased need for using advanced methods. As a result, their study paved the way for greater adoption of neuroscientific methods in entrepreneurship research, particularly in decision-making analysis. From a methodological standpoint, Jack et al. (2019) predicted that neuroimaging techniques, particularly fMRI and EEG, would be the most suitable for advancing the field (Jack et al., 2019). As we have seen in the previous section, these technologies offer the potential to enhance our understanding of how cognitive and emotional processes are linked to sustainable entrepreneurship.

Neuroimaging research has identified several brain regions involved in decision-making, including the Anterior Cingulate Cortex (ACC) and the dorsolateral Prefrontal Cortex (dlPFC) (Fleck et al., 2006; McGovern & Sheth, 2017; Rosenbloom et al., 2012). The dorsal ACC is functionally connected to the Prefrontal Cortex (PFC), Parietal Cortex (PC), motor system, and Orbitofrontal Cortex (OFC) (Fleck et al., 2006). It plays a crucial role in processing both top-down and bottom-up stimuli, regulating neural activity by directing control to different brain regions. The dlPFC, a subdivision of the PFC, is particularly associated with higher-order cognitive functions (Jung et al., 2022). It is involved in shifting attention, maintaining and manipulating information in working memory, applying abstract rules, and inhibiting inappropriate responses (Grossberg, 2018; Wang et al., 2018). This region is crucial for executive control, which enables individuals to regulate their behavior and make complex decisions.

On the other hand, the emotional system of the brain mainly consists of the insula, amygdala, and ventromedial Prefrontal Cortex (vmPFC) (Andrewes & Jenkins, 2019; Motzkin et al., 2015; Nejati et al., 2021). The insula is a structure within the cerebral cortex located in the fissure that separates the temporal, parietal, and frontal lobes. It is involved in integrating affective information. The amygdala, an almond-shaped cluster of neurons, is situated on the medial temporal lobe and is central to emotional processing (Markram & Sandi, 2007). It plays a key role in detecting and responding to emotionally significant stimuli. This area of the brain primarily processes threat perception, fear responses, and the formation of emotional memories (LeDoux, 2003). The vmPFC, located in the medial portion of the Prefrontal Cortex, is responsible for a range of social, cognitive, and emotional functions (Bzdok et al., 2013; Hiser & Koenigs, 2018). It is particularly important for regulating negative emotions, as well as encoding reward and value-based decision-making.

Despite clear distinctions between brain areas, neuroscience research suggests that emotion and cognition are inseparable (Gray et al., 2002; Levine, 2022). It has been argued that emotional arousal is the first response to external stimuli, with the amygdala triggering physiological changes. Through the vmPFC, these affective states influence cognition via somatic markers (e.g., heart rate, blood pressure, gut motility). Based on this somatic theory, immediate decision-making is influenced by subcortical mechanisms such as the amygdala, while evaluating long-term consequences relies on the PFC (Bechara & Damasio, 2005; Bechara et al., 2000; Sütterlin et al., 2015). When weighing a decision, conflicting positive and negative somatic signals arise, but ultimately, a dominant 'go' or 'stop' signal determines the choice. Indeed, based on recent advancements in the field, complex cognitive-emotional behaviors emerge from dynamic interactions among brain networks.

Research indicates that entrepreneurs exhibit unique EEG patterns during business decisions, showing interconnected emotional and cognitive processing, though sustainable attitudes were not explicitly addressed in previous works. For example, it has been reported that entrepreneur participants show distinct EEG patterns during specific tasks (Didenko et al., 2021). Recent studies have reported activation in frontal brain regions during tasks involving opportunity recognition, risk assessment, and creative thinking (Aydın et al., n.d.). Furthermore, regarding the relationship between cognition and emotion, it has been reported that entrepreneurship training is associated with changes in EEG markers related to emotional regulation (Egana-delSol et al., 2023). It is not surprising that sustainable thinking brains vary among individuals, given that ensuring sustainability presents complex and unique challenges. These challenges compel the current generation to decide whether to forgo their benefits to support future generations or to prioritize their own gains at the potential cost of future well-being. The interplay between cognition and emotion is thus vital in shaping sustainable entrepreneurship. Entrepreneurs' cognitive biases, such as optimism and overconfidence, significantly influence their emotional responses, which in turn affect their entrepreneurial intentions. Therefore, understanding how the brain processes thinking, emotions, and decision-making is important for sustainable entrepreneurship. As mentioned above, entrepreneurs focused on sustainability must make complex choices, balancing profit, environmental impact, and social responsibility. Insights from neuroscience help explain how emotions, risk perception, and long-term thinking influence these decisions.

5 Implications for Practice in Sustainable Entrepreneurship

Many researchers analyzing why entrepreneurs engage in sustainability practices often adopt organizational or industry-level perspectives (Criado-Gomis et al., 2017; Kraus et al., 2018). In this respect, there is a limited understanding of sustainability-focused practices in the context of neuroscientific research. While various characteristics have been explored to explain sustainable decision-making in entrepreneurship, there remains little agreement on the prominence of emotional and cognitive factors in shaping these decisions. Furthermore, it remains uncertain whether the cognitive and emotional aspects of decisions differ among new business founders or share commonalities. By establishing an empirical and neuroscientific approach to sustainable decision-making in entrepreneurship, scholars can shed additional light on the factors that underpin and directly influence sustainable behavior.

As reviewed above, existing research on human cognition emphasizes the emotional aspects that influence cognition and suggests that cognitive processes may not always be entirely rational (Levine, 2022; Phelps, 2006). When entrepreneurs encounter information overload, uncertainty, novelty, emotional influences, and time constraints, they become particularly vulnerable to various cognitive biases (Baron, 1998). The neuroscience tools we reviewed herein might be considered for exploring the fundamental 'why' questions underlying entrepreneurial decision-making. These questions extend beyond the decision to become an entrepreneur and address the shaping of behaviors. It helps explain what drives them to take risks, recognize opportunities, and navigate uncertainty (Brundin & Gustafsson, 2013; Kochenderfer, 2015; Noebels et al., 2021). Practically, neuroscience research can offer insights into how entrepreneurs process information about sustainability, make ethical choices, and develop long-term strategies that balance profit with social and environmental responsibility.

An integrative approach that examines entrepreneurship from a neural perspective can significantly enhance our understanding of entrepreneurial sustainable intentions and actions (He et al., 2024; N. F. Krueger & Day, 2010). This approach has practical benefits for entrepreneurship education, policymaking, and business support systems. By identifying neural patterns associated with decision-making, training programs can be designed to enhance cognitive flexibility, resilience, and ethical decision-making in entrepreneurs (Hannah et al., 2013; Pless et al., 2017; Yu et al., 2023). Additionally, policymakers and investors can utilize neuroscience-informed insights to create environments that foster sustainable entrepreneurship, considering emotional resilience and mitigating cognitive biases that may hinder responsible business practices (N. Krueger et al., 2011; Shepherd et al., 2015). Neuroscientific research shows that cognitive processes, such as opportunity recognition and risk assessment, are heavily influenced by emotions like fear, excitement, and optimism (Brosch et al., 2013; Welpe et al., 2012) in the context of sustainability. Emotions such as empathy, environmental concern, and social responsibility can shape an entrepreneur's motivation and commitment to long-term, responsible business practices (Fors & Lennerfors, 2019; Jayaratne et al., 2019). Understanding this link can help develop strategies to reinforce positive emotional responses that support sustainable decision-making while mitigating biases that may lead to short-term, profit-driven choices. This perspective can help explain why some entrepreneurs are more inclined toward social and environmental responsibility than others and how cognitive training or behavioral interventions can support more sustainable entrepreneurial ecosystems.

6 Case Study

EcoMind is a startup founded by two neuroscientists to develop eco-friendly consumer products using biodegradable materials. Insights from cognitive neuroscience informed their business strategy. Before launching their product, they conducted EEG-based research on prototype packaging designs to assess subconscious responses and emotional processing. This neuroscience-informed design process helped optimize marketing materials and product presentation. Using EEG signals and GSR measurements, EcoMind identified that minimalist designs with green tones and visual symbols associated with 'nature' consistently triggered stronger positive affective responses. This data not only influenced branding decisions but also guided the team's investment strategy.

7 Questions

1. What is the role of the dorsolateral prefrontal cortex (dlPFC) in entrepreneurial decision-making?
2. Define sustainable entrepreneurship and its three core dimensions.
3. Which neuroscientific methods are most commonly used to study entrepreneurial cognition?

4. How could a startup use EEG or GSR data to refine its pitch for investors?
5. In what ways might cognitive biases such as overconfidence influence sustainable decision-making?
6. To what extent is it ethical for entrepreneurs to use neural data to shape consumer behavior?
7. How might policymakers use findings from neuroentrepreneurship research to design better support systems for sustainable ventures?
8. Considering dual-process theory, how should entrepreneurs balance automatic vs. deliberate thinking when making high-stakes decisions?

8 Future Directions

Building upon the growing body of research on the cognitive and emotional mechanisms underlying sustainable entrepreneurship, several key directions might be considered for future exploration. As cognitive neuroscience continues to shed light on how entrepreneurs process information, regulate emotions, and make strategic decisions, there remains a vast and largely untapped potential for integrating these insights into sustainable business practices.

While traditional entrepreneurial research has relied on psychological and behavioral studies, incorporating neuroscientific methodologies such as EEG, fMRI, and GSR can provide a more objective understanding of the neural basis of sustainable decision-making. Future works might focus on several aspects in the field, including (1) identifying specific neural markers associated with sustainability-oriented entrepreneurial intentions, (2) examining the temporal dynamics of cognitive and emotional processing in sustainable business decision-making, and (3) Investigating the impact of entrepreneurial training programs on brain activity related to sustainability attitudes and behaviors. As mentioned, entrepreneurial decisions are shaped by a complex interplay between rational cognition and emotional engagement. Given that sustainability-driven entrepreneurship often requires long-term commitment, further research might also focus on the role of cognitive biases linked to emotions in shaping sustainable entrepreneurial intentions.

Furthermore, Insights from cognitive neuroscience can inform the design of entrepreneurial education and public policies. The cognitive and emotional drivers of sustainable entrepreneurship may vary across cultures, industries, and personality traits; hence, cultural variations in sustainability attitudes and their neural correlations might also be of interest to future researchers. Overall, as this field continues to evolve, integrating cognitive neuroscience into sustainability research presents significant opportunities for understanding, predicting, and enhancing sustainable practices. By examining the interaction between cognition and emotion in entrepreneurial decision-making, future research can provide new strategies to strengthen ethical leadership, innovation, and resilience in sustainable entrepreneurship.

References

Andrewes, D. G., & Jenkins, L. M. (2019). The role of the amygdala and the ventromedial prefrontal cortex in emotional regulation: Implications for post-traumatic stress disorder. *Neuropsychology Review, 29*, 220–243.

Ayata, D., Yaslan, Y., & Kamaşak, M. (2017). Emotion recognition via galvanic skin response: Comparison of machine learning algorithms and feature extraction methods. *IU-Journal of Electrical & Electronics Engineering, 17*(1), 3147–3156.

Aydın, H. U. D., İrmiş, A., & Demirkundak, B. (n.d.). Comparing brain activity of entrepreneurs and non-entrepreneurs during creative thinking and opportunity recognition. *Journal of Mehmet Akif Ersoy University Economics and Administrative Sciences Faculty, 10*(1), 799–819.

Baddeley, A. (1992). Working memory. *Science, 255*(5044), 556–559. https://doi.org/10.1126/science.1736359

Baron, R. A. (1998). Cognitive mechanisms in entrepreneurship: Why and when entrepreneurs think differently than other people. *Journal of Business Venturing, 13*(4), 275–294.

Baron, R. A. (2000). Psychological perspectives on entrepreneurship: Cognitive and social factors in entrepreneurs' success. *Current Directions in Psychological Science, 9*(1), 15–18. https://doi.org/10.1111/1467-8721.00050

Baron, R. A. (2007). Behavioral and cognitive factors in entrepreneurship: Entrepreneurs as the active element in new venture creation. *Strategic Entrepreneur-*

ship Journal, 1(1–2), 167–182. https://doi.org/10.1002/sej.12
Baron, R. A., & Tang, J. (2011). The role of entrepreneurs in firm-level innovation: Joint effects of positive affect, creativity, and environmental dynamism. *Journal of Business Venturing, 26*(1), 49–60.
Baron, R. A., & Ward, T. B. (2004). Expanding entrepreneurial cognition's toolbox: Potential contributions from the field of cognitive science. *Entrepreneurship Theory and Practice, 28*(6), 553–573. https://doi.org/10.1111/j.1540-6520.2004.00064.x
Baskaran, S., Mahadi, N., & Abd Rasid, S. Z. (2021). Multiple intelligence and entrepreneurial opportunity recognition—A failsafe approach of neuromarketing. *Journal of Research in Marketing and Entrepreneurship, 23*(2), 318–338.
Bechara, A., & Damasio, A. R. (2005). The somatic marker hypothesis: A neural theory of economic decision. *Games and Economic Behavior, 52*(2), 336–372.
Bechara, A., Damasio, H., & Damasio, A. R. (2000). Emotion, decision making and the orbitofrontal cortex. *Cerebral Cortex, 10*(3), 295–307.
Biswas, H. (2024). Teaching philosophy, educational psychology, and cognitive neuroscience. In *Architecture and technological advancements of education 4.0* (pp. 49–74). IGI Global. https://www.igi-global.com/chapter/teaching-philosophy-educational-psychology-and-cognitive-neuroscience/334392
Brocanelli, A. (2018). *Galvanic skin response measurement data processing for user-related information extraction.* https://tesi.univpm.it/handle/20.500.12075/7228
Brosch, T., Scherer, K., Grandjean, D., & Sander, D. (2013). The impact of emotion on perception, attention, memory, and decision-making. *Swiss Medical Weekly, 143*(1920), w13786–w13786.
Brundin, E., & Gustafsson, V. (2013). Entrepreneurs' decision making under different levels of uncertainty: The role of emotions. *International Journal of Entrepreneurial Behavior & Research, 19*(6), 568–591.
Butler, M. J. R., O'Broin, H. L. R., Lee, N., & Senior, C. (2016). How organizational cognitive neuroscience can deepen understanding of managerial decision-making: A review of the recent literature and future directions. *International Journal of Management Reviews, 18*(4), 542–559. https://doi.org/10.1111/ijmr.12071
Bzdok, D., Langner, R., Schilbach, L., Engemann, D. A., Laird, A. R., Fox, P. T., & Eickhoff, S. B. (2013). Segregation of the human medial prefrontal cortex in social cognition. *Frontiers in Human Neuroscience, 7*, 232.
Castellano, S., Khelladi, I., & Menvielle, L. (2017). Unveiling the sustainable facet of the conventional entrepreneur—A cognitive approach. *International Journal of Entrepreneurship and Small Business, 31*(3), 434. https://doi.org/10.1504/IJESB.2017.084848
Cavallo, A., Ghezzi, A., & Balocco, R. (2019). Entrepreneurial ecosystem research: Present debates and future directions. *International Entrepreneurship and Management Journal, 15*(4), 1291–1321. https://doi.org/10.1007/s11365-018-0526-3
Chen, P.-C., Chan, W.-C., Hung, S.-W., & Lin, D.-Z. (2020). How entrepreneurs recognise entrepreneurial opportunity and its gaps: A cognitive theory perspective. *Technology Analysis & Strategic Management, 32*(2), 223–238. https://doi.org/10.1080/09537325.2019.1648790
Clements, E., Stephan, U., Miozzo, M., Hellyer, P., Devonshire, J., & Allesch-Tayalor, S. (2021). Cognitive control in entrepreneurship: The neuroscience of producing novel action despite uncertainty. *PsyArXiv [Preprint]. Doi, 10.* https://scholar.archive.org/work/mafktxbjsfc4zmfkph67pwg654/access/wayback/, https://files.osf.io/v1/resources/ayve7/providers/osf storage/6140c151585ec20153f2c84c?action=download&direct&version=2
Criado-Gomis, A., Cervera-Taulet, A., & Iniesta-Bonillo, M.-A. (2017). Sustainable entrepreneurial orientation: A business strategic approach for sustainable development. *Sustainability, 9*(9), 1667.
Cunnane, S. C. (2005). *Survival of the fattest: The key to human brain evolution.* https://books.google.com/books?hl=en&lr=&id=_MFgDQAAQBAJ&oi=fnd&pg=PR7&dq=The+human+brain,+for+example,+makes+up+only+about+2%25+of+body+weight&ots=wo1GC9cbFp&sig=ZLELCetJvGIdD_lU_Oj5UItJC88
Di Vaio, A., Hassan, R., Chhabra, M., Arrigo, E., & Palladino, R. (2022). Sustainable entrepreneurship impact and entrepreneurial venture life cycle: A systematic literature review. *Journal of Cleaner Production, 378*, Article 134469.
Didenko, A., Golovin, T., Malovitsa, Y., & Morozova, A. (2021). *Divide between entrepreneurs and non-entrepreneurs: Neuroscientifical perspective of the attitude to risk* (pp. 369–378). https://doi.org/10.15405/epsbs.2021.06.04.42
Doh, J. P., Tashman, P., & Benischke, M. H. (2019). Adapting to grand environmental challenges through collective entrepreneurship. *Academy of Management Perspectives, 33*(4), 450–468. https://doi.org/10.5465/amp.2017.0056
Egana-delSol, P., Sun, X., & Sajda, P. (2023). Neurophysiological markers of emotion regulation predict efficacy of entrepreneurship education. *Scientific Reports, 13*(1), 7206.
Egbu, C. O. (2004). Managing knowledge and intellectual capital for improved organizational innovations in the construction industry: An examination of critical success factors. *Engineering, Construction and Architectural Management, 11*(5), 301–315.
Elkington, J., & Hartigan, P. (2008). *The power of unreasonable people: How social entrepreneurs create markets that change the world.* Harvard Business Press. https://books.google.com/books?hl=en&lr=&id=2kWbh9BEicMC&oi=fnd&pg=PT9&dq=it+remains+uncertain+whether+the+mental+frameworks+guiding+socially+and+environmentally+responsible+choices+among+new+business+founders+are+similar+or+distinct+&ots=1lFa2_FIhe&sig=tO13ArKpThwJuezoGhLLGWqANbs

Engle, R. W. (2010). Role of working-memory capacity in cognitive control. *Current Anthropology, 51*(S1), S17–S26. https://doi.org/10.1086/650572

Fleck, M. S., Daselaar, S. M., Dobbins, I. G., & Cabeza, R. (2006). Role of prefrontal and anterior cingulate regions in decision-making processes shared by memory and nonmemory tasks. *Cerebral Cortex, 16*(11), 1623–1630.

Fors, P., & Lennerfors, T. T. (2019). The individual-care nexus: A theory of entrepreneurial care for sustainable entrepreneurship. *Sustainability, 11*(18), 4904.

Gevins, A. (1998). The future of electroencephalography in assessing neurocognitive functioning. *Electroencephalography and Clinical Neurophysiology, 106*(2), 165–172.

Goodman, R. I. (1985). Skin conductance measurement in communication research. *Educational Technology, 25*(4), 23–27.

Gray, J. R., Braver, T. S., & Raichle, M. E. (2002). Integration of emotion and cognition in the lateral prefrontal cortex. *Proceedings of the National Academy of Sciences, 99*(6), 4115–4120. https://doi.org/10.1073/pnas.062381899

Greco, A., & De Jong, G. (2017). *Sustainable entrepreneurship: Definitions, themes and research gaps.* https://research.rug.nl/en/publications/sustainable-entrepreneurship-definitions-themes-and-research-gaps

Grégoire, D. A., Corbett, A. C., & McMullen, J. S. (2011). The cognitive perspective in entrepreneurship: An agenda for future research. *Journal of Management Studies, 48*(6), 1443–1477. https://doi.org/10.1111/j.1467-6486.2010.00922.x

Grégoire, D. A., Cornelissen, J., Dimov, D., & Van Burg, E. (2015). The mind in the middle: Taking stock of affect and cognition research in entrepreneurship. *International Journal of Management Reviews, 17*(2), 125–142. https://doi.org/10.1111/ijmr.12060

Grossberg, S. (2018). Desirability, availability, credit assignment, category learning, and attention: Cognitive-emotional and working memory dynamics of orbitofrontal, ventrolateral, and dorsolateral prefrontal cortices. *Brain and Neuroscience Advances, 2*, 2398212818772179. https://doi.org/10.1177/2398212818772179

Hannah, S. T., Balthazard, P. A., Waldman, D. A., Jennings, P. L., & Thatcher, R. W. (2013). The psychological and neurological bases of leader self-complexity and effects on adaptive decision-making. *Journal of Applied Psychology, 98*(3), 393.

Hasson Marques, R., Violant-Holz, V., & Damião Da Silva, E. (2024). Emotions and decision-making in boardrooms—A systematic review from behavioral strategy perspective. *Frontiers in Psychology, 15*, 1473175.

Hayton, J. C., & Cholakova, M. (2012). The role of affect in the creation and intentional pursuit of entrepreneurial ideas. *Entrepreneurship Theory and Practice, 36*(1), 41–67. https://doi.org/10.1111/j.1540-6520.2011.00458.x

He, J., Wang, Z., Hu, H., & Fan, Z. (2024). Exploring the formation of sustainable entrepreneurial intentions among Chinese university students: A dual path moderated mediation model. *Sustainability, 16*(18), 8069.

Hiser, J., & Koenigs, M. (2018). The multifaceted role of the ventromedial prefrontal cortex in emotion, decision making, social cognition, and psychopathology. *Biological Psychiatry, 83*(8), 638–647.

Jack, A. I., Rochford, K. C., Friedman, J. P., Passarelli, A. M., & Boyatzis, R. E. (2019). Pitfalls in organizational neuroscience: A critical review and suggestions for future research. *Organizational Research Methods, 22*(1), 421–458. https://doi.org/10.1177/1094428117708857

Jayaratne, M., Sullivan Mort, G., & D'Souza, C. (2019). Sustainability entrepreneurship: From consumer concern towards entrepreneurial commitment. *Sustainability, 11*(24), 7076.

Jennings, J. E., Edwards, T., Jennings, P. D., & Delbridge, R. (2015). Emotional arousal and entrepreneurial outcomes: Combining qualitative methods to elaborate theory. *Journal of Business Venturing, 30*(1), 113–130.

Jung, J., Ralph, M. A. L., & Jackson, R. L. (2022). Subregions of DLPFC display graded yet distinct structural and functional connectivity. *Journal of Neuroscience, 42*(15), 3241–3252.

Karlesky, M. J. (2015). *Identifying entrepreneurial opportunities: Cognition and categorization in nascent entrepreneurs* [PhD Thesis]. https://deepblue.lib.umich.edu/handle/2027.42/116674

Keh, H. T., Der Foo, M., & Lim, B. C. (2002). Opportunity evaluation under risky conditions: The cognitive processes of entrepreneurs. *Entrepreneurship Theory and Practice, 27*(2), 125–148. https://doi.org/10.1111/1540-8520.00003

Kochenderfer, M. J. (2015). *Decision making under uncertainty: Theory and application.* MIT Press. https://books.google.com/books?hl=en&lr=&id=hUBWCgAAQBAJ&oi=fnd&pg=PR7&dq=Decision+making+under+uncertainty&ots=546RbkOSR2&sig=lu6NX1jnnvrtGGx_4LPXSJLFMgo

Kraus, S., Burtscher, J., Vallaster, C., & Angerer, M. (2018). Sustainable entrepreneurship orientation: Reflection on status-quo research on factors facilitating responsible managerial practices. In *Sustainable entrepreneurship* (pp. 75–98). Routledge. https://doi.org/10.4324/9781315611495-5/sustainable-entrepreneurship-orientation-sacha-kraus-janina-burtscher-christine-vallaster-martin-angerer

Krueger, N. F., & Day, M. (2010). Looking forward, looking backward: From entrepreneurial cognition to neuroentrepreneurship. In Z. J. Acs & D. B. Audretsch (Eds.), *Handbook of entrepreneurship research* (pp. 321–357). Springer. https://doi.org/10.1007/978-1-4419-1191-9_13

Krueger, N., Hansen, D. J., Michl, T., & Welsh, D. H. (2011). Thinking 'sustainably': The role of intentions, cognitions, and emotions in understanding new domains of entrepreneurship. In *Social and sustainable entrepreneurship* (Vol. 13, pp. 275–309). Emerald Group Publishing Limited. https://doi.org/10.1108/S1074-7540(2011)0000013014/full/html?queryID=1/5408120

Le Loarne Lemaire, S., Razgallah, M., Maalaoui, A., & Kraus, S. (2022). Becoming a green entrepreneur: An advanced entrepreneurial cognition model based on a practiced-based approach. *International Entrepreneurship and Management Journal, 18*(2), 801–828. https://doi.org/10.1007/s11365-021-00791-1

LeDoux, J. (2003). The emotional brain, fear, and the amygdala. *Cellular and Molecular Neurobiology, 23*, 727–738.

Levine, D. S. (2022). Neuroscience of emotion, cognition, and decision making: A review. *Medical Research Archives, 10*(7). https://esmed.org/MRA/mra/article/view/2869

Lichtenstein, B. B. (2014). *Generative emergence: A new discipline of organizational, entrepreneurial and social innovation.* Oxford University Press. https://books.google.com/books?hl=en&lr=&id=IbsBBAAAQBAJ&oi=fnd&pg=PP1&dq=Decision-making+in+entrepreneurship+is+not+purely+rational%3B+it+is+influenced+by+cognitive+biases,+emotional+resilience,+and+neurophysiological+responses+that+impact+how+entrepreneurs+react+to+high-pressure+situations+and+evolving+market+conditions&ots=5K7vf7k6Nv&sig=-STUZVtgk1X1LLwl_BUXkpV7M_g

Logothetis, N. K. (2008). What we can do and what we cannot do with fMRI. *Nature, 453*(7197), 869–878.

Lv, W.-D., Tian, D., Wei, Y., & Xi, R.-X. (2018). Innovation resilience: A new approach for managing uncertainties concerned with sustainable innovation. *Sustainability, 10*(10), 3641.

Markram, K., & Sandi, C. (2007). *The role of the amygdala in emotional memories: A multidisciplinary approach.* EPFL. https://infoscience.epfl.ch/record/88100

McGovern, R. A., & Sheth, S. A. (2017). Role of the dorsal anterior cingulate cortex in obsessive-compulsive disorder: Converging evidence from cognitive neuroscience and psychiatric neurosurgery. *Journal of Neurosurgery, 126*(1), 132–147.

McMullen, J. S., Wood, M. S., & Palich, L. E. (2014). Entrepreneurial cognition and social cognitive neuroscience. In *Handbook of entrepreneurial cognition* (pp. 316–363). Edward Elgar Publishing. https://www.elgaronline.com/downloadpdf/edcollchap/edcoll/9781781006580/9781781006580.00019.pdf

Michel, C. M., & Murray, M. M. (2012). Towards the utilization of EEG as a brain imaging tool. *NeuroImage, 61*(2), 371–385.

Miller, G. A., & Gazzaniga, M. S. (1984). The cognitive sciences. In M. S. Gazzaniga (Ed.), *Handbook of cognitive neuroscience* (pp. 3–11). Springer. https://doi.org/10.1007/978-1-4899-2177-2_1

Motzkin, J. C., Philippi, C. L., Wolf, R. C., Baskaya, M. K., & Koenigs, M. (2015). Ventromedial prefrontal cortex is critical for the regulation of amygdala activity in humans. *Biological Psychiatry, 77*(3), 276–284.

Muñoz, P. (2018). A cognitive map of sustainable decision-making in entrepreneurship: A configurational approach. *International Journal of Entrepreneurial Behavior & Research, 24*(3), 787–813. https://doi.org/10.1108/IJEBR-03-2017-0110

Muñoz, P., & Cohen, B. (2018). Sustainable entrepreneurship research: Taking stock and looking ahead. *Business Strategy and the Environment, 27*(3), 300–322. https://doi.org/10.1002/bse.2000

Nejati, V., Majdi, R., Salehinejad, M. A., & Nitsche, M. A. (2021). The role of dorsolateral and ventromedial prefrontal cortex in the processing of emotional dimensions. *Scientific Reports, 11*(1), 1971.

Noebels, M., Quirós-Tortós, J., & Panteli, M. (2021). Decision-making under uncertainty on preventive actions boosting power grid resilience. *IEEE Systems Journal, 16*(2), 2614–2625.

Oberauer, K., Süß, H.-M., Schulze, R., Wilhelm, O., & Wittmann, W. W. (2000). Working memory capacity—Facets of a cognitive ability construct. *Personality and Individual Differences, 29*(6), 1017–1045.

Palich, L. E., & Bagby, D. R. (1995). Using cognitive theory to explain entrepreneurial risk-taking: Challenging conventional wisdom. *Journal of Business Venturing, 10*(6), 425–438.

Perry, J. C. (2007). *The psychophysiology of risk processing and decision making at a regional stock exchange.* https://dspace.mit.edu/handle/1721.1/37599

Phelps, E. A. (2006). Emotion and cognition: Insights from studies of the human amygdala. *Annual Review of Psychology, 57*(1), 27–53. https://doi.org/10.1146/annurev.psych.56.091103.070234

Pless, N., Sabatella, F., & Maak, T. (2017). Mindfulness, reperceiving, and ethical decision making: A neurological perspective. In S. Savur & S. Sandhu (Eds.), *Research in ethical issues in organizations* (Vol. 17, pp. 1–20). Emerald Publishing Limited. https://doi.org/10.1108/S1529-209620170000017001

Ploum, L., Blok, V., Lans, T., & Omta, O. (2018). Toward a validated competence framework for sustainable entrepreneurship. *Organization & Environment, 31*(2), 113–132. https://doi.org/10.1177/1086026617697039

Qin, X., Shepherd, D. A., Lin, D., Xie, S., Liang, X., & Lin, S. (2022). The dark side of entrepreneurs' creativity: Investigating how and when entrepreneurs' creativity increases the favorability of potential opportunities that harm nature. *Entrepreneurship Theory and Practice, 46*(4), 857–883. https://doi.org/10.1177/1042258720915582

Robertson, D. C., Voegtlin, C., & Maak, T. (2017). Business ethics: The promise of neuroscience. *Journal of Business Ethics, 144*, 679–697.

Rosenbloom, M. H., Schmahmann, J. D., & Price, B. H. (2012). The functional neuroanatomy of decision-

making. *The Journal of Neuropsychiatry and Clinical Neurosciences, 24*(3), 266–277. https://doi.org/10.1176/appi.neuropsych.11060139
Satti, F. A., Hussain, M., Hussain, J., Kim, T.-S., Lee, S., & Chung, T. (2021). User stress modeling through galvanic skin response. In *2021 15th International Conference on Ubiquitous Information Management and Communication (IMCOM)* (pp. 1–6). https://ieeexplore.ieee.org/abstract/document/9377380/?casa_token=c0titwfdYeMAAAAA:fZgA4Fe9Zfpf0syNNCIgO17flvtTXIqgoIN_RJqUoNaPE-wgjxQTyzyqCZnCcpeCx8mkeI1E9w
Schaltegger, S., & Wagner, M. (2011). Sustainable entrepreneurship and sustainability innovation: Categories and interactions. *Business Strategy and the Environment, 20*(4), 222–237. https://doi.org/10.1002/bse.682
Shaver, K. G., Schjoedt, L., Passarelli, A., & Reeck, C. (2017). The cognitive neuroscience of entrepreneurial risk: Conceptual and methodological challenges. In *Handbook of research methodologies and design in neuroentrepreneurship* (pp. 207–240). Edward Elgar Publishing. https://www.elgaronline.com/abstract/edcoll/9781785365034/9781785365034.00018.xml
Shepherd, D. A., & Williams, T. (2020). Entrepreneurship responding to adversity: Equilibrating adverse events and disequilibrating persistent adversity. *Organization Theory, 1*(4), 2631787720967678. https://doi.org/10.1177/2631787720967678
Shepherd, D. A., Williams, T. A., & Patzelt, H. (2015). Thinking about entrepreneurial decision making: Review and research agenda. *Journal of Management, 41*(1), 11–46. https://doi.org/10.1177/0149206314541153
Smith, S. M. (2004). Overview of fMRI analysis. *The British Journal of Radiology, 77*(Suppl. 2), S167–S175.
Sütterlin, S., Olsen, V., & Lugo, R. (2015). The somatic marker theory in the context of addiction: Contributions to understanding development and maintenance. *Psychology Research and Behavior Management, 8*, 187. https://doi.org/10.2147/PRBM.S68695
Taggart, W., Robey, D., & Kroeck, K. G. (1985). Managerial decision styles and cerebral dominance: An empirical study. *Journal of Management Studies (Wiley-Blackwell), 22*(2).
Vijaya, P. A., & Shivakumar, G. (2013). Galvanic skin response: A physiological sensor system for affective computing. *International Journal of Machine Learning and Computing, 3*(1), 31.
Von Eckardt, B. (1995). *What is cognitive science?* MIT Press. https://books.google.com/books?hl=en&lr=&id=E8l48-DZln0C&oi=fnd&pg=PA1&dq=what+is+cognitive+science&ots=XmQmrsat9h&sig=aGkxJV2_PvUKqCShOK2uQ7gUZtg
Wang, J., Tian, J., Hao, R., Tian, L., & Liu, Q. (2018). Transcranial direct current stimulation over the right DLPFC selectively modulates subprocesses in working memory. *PeerJ, 6*, Article e4906.
Welpe, I. M., Spörrle, M., Grichnik, D., Michl, T., & Audretsch, D. B. (2012). Emotions and opportunities: The interplay of opportunity evaluation, fear, joy, and anger as antecedent of entrepreneurial exploitation. *Entrepreneurship Theory and Practice, 36*(1), 69–96. https://doi.org/10.1111/j.1540-6520.2011.00481.x
West, G. P. (2007). Collective cognition: when entrepreneurial teams, not individuals, make decisions. *Entrepreneurship Theory and Practice, 31*(1), 77–102. https://doi.org/10.1111/j.1540-6520.2007.00164.x
Yu, X., Liu, T., He, L., & Li, Y. (2023). Micro-foundations of strategic decision-making in family business organisations: A cognitive neuroscience perspective. *Long Range Planning, 56*(5), Article 102198.
Zemlyak, S., Naumenkov, A., & Khromenkova, G. (2022). Measuring the entrepreneurial mindset: The motivations behind the behavioral intentions of starting a sustainable business. *Sustainability, 14*(23), 15997.
Zhan, X., Jiang, B., & Guo, F. (2023). Emotional responses to spatial design in a regeneration project: A study using virtual reality and galvanic skin response methodology. *Art and Design Review, 12*(1), 105–119.

Further Reading

For readers interested in exploring the intersection of cognitive neuroscience and sustainable entrepreneurship in greater depth, the following resources are suggested:

Damasio, A. (1994). *Descartes' error: Emotion, reason, and the human brain*. HarperCollins.

This book argues that emotions play a crucial role in rational decision-making and challenges traditional views of economic rationality.

Grégoire, D. A., Corbett, A. C., & McMullen, J. S. (2011). The cognitive perspective in entrepreneurship: An agenda for future research. *Journal of Management Studies, 48*(6), 1443–1477.

A review of cognitive research in entrepreneurship, this article discusses key cognitive processes such as decision-making, risk assessment, and information processing.

Shepherd, D. A., & Patzelt, H. (2011). The new field of sustainable entrepreneurship: Studying entrepreneurial action linking 'what is to be sustained' with 'what is to be developed.' *Entrepreneurship Theory and Practice, 35*(1), 137–163.

This article provides an in-depth discussion of sustainable entrepreneurship and its role in balancing economic, social, and environmental goals. It aligns with this chapter by examining how entrepreneurial cognition contributes to sustainability-driven business strategies.

Sustainable Entrepreneurial Ecosystems: Building a Green Future for Business Innovation

Niloofar Salajegheh and Armin Khaleghi Forghani

Abstract

Sustainable entrepreneurial ecosystems (SEEs) have become increasingly vital in addressing global sustainability challenges while fostering economic innovation. This chapter provides a conceptual exploration of SEEs, which integrate environmental stewardship, social equity, and economic viability into entrepreneurial systems. Drawing on theoretical frameworks including the Triple Bottom Line approach, stakeholder theory, and social capital theory, we explain how sustainable entrepreneurial ecosystems (SEEs) differ from traditional entrepreneurial ecosystems in their emphasis on long-term resilience and sustainability-oriented innovation. The discussion highlights key components of SEEs—such as institutional support mechanisms, financing structures, and collaborative networks—while identifying persistent challenges in their development. By synthesizing current research, this descriptive chapter offers a foundation for understanding how SEEs can align business innovation with sustainable development goals, providing insights for scholars and practitioners in the field.

Keywords

Entrepreneurial ecosystem · Sustainable entrepreneurial ecosystem · Tripple bottom line

Learning Objectives

By the end of this chapter, readers will be able to:

- Define SEEs and distinguish them from traditional entrepreneurial ecosystems using the Triple Bottom Line framework.
- Explain the roles of key SEE components, including institutional policies, sustainable finance, human capital, and collaborative networks.
- Analyze how theoretical perspectives, such as social capital theory, stakeholder theory, and process views, apply to SEE development.
- Identify the significant challenges in transitioning to SEEs, including governance fragmentation, cultural barriers, and misaligned incentives.

1 Introduction

Entrepreneurial ecosystems have long been recognized as critical drivers of innovation, economic growth, and regional development (Chaudhary et al., 2023). These complex networks, comprising

N. Salajegheh (✉)
College of Farabi, University of Tehran, Tehran, Iran
e-mail: n.salajegheh@ut.ac.ir

A. Khaleghi Forghani
Faculty of Management, Kharazmi University, Tehran, Iran
e-mail: armin.khaleghi@khu.ac.ir

M. Akbari et al. (eds.), *Sustainable Entrepreneurship*, University of Tehran Science and Humanities Series, https://doi.org/10.1007/978-3-032-04909-4_23

entrepreneurs, investors, institutions, and support organizations, create the conditions for new ventures to emerge and thrive (Acs et al., 2014). However, in an era marked by climate change, resource depletion, and growing social inequalities, traditional models of entrepreneurial ecosystems are increasingly challenged to integrate sustainability principles (Huang et al., 2023; Ninh & Hue, 2025; Shahid et al., 2023). Sustainable entrepreneurial ecosystems (SEEs) have emerged as systems that deliberately integrate environmental stewardship, social equity, and long-term economic viability into their core functioning. The transition from the conventional paradigm to SEEs reflects not only an ethical imperative but also a strategic necessity, as businesses and policymakers recognize that future economic success depends on addressing pressing global challenges (Audretsch et al., 2021).

The concept of sustainability in entrepreneurship extends beyond corporate social responsibility or environmental compliance (Schaltegger et al., 2016). It demands a fundamental rethinking of how entrepreneurial activities are conceived, funded, and scaled. Rising demand for sustainable business practices—from consumers, investors, and regulators—has created both pressures and opportunities for entrepreneurs (Bocken et al., 2014). This shift is evident in the growth of impact investing, environmental, social, and governance (ESG) criteria, as well as sustainable finance (Dorfleitner et al., 2021). Concurrently, technological advancements in clean energy, circular economy models, and digital sustainability solutions are enabling new possibilities for sustainable ventures (Geissdoerfer et al., 2017). Together, these developments suggest that SEEs may represent the next evolutionary stage of innovation systems in the twenty-first century (Theodoraki et al., 2022).

Despite their promise, significant barriers hinder the development of effective SEEs. Entrepreneurs often face higher initial costs when adopting sustainable technologies or business models (Hockerts & Wüstenhagen, 2010). Many regions lack the specialized infrastructure, financing mechanisms, or policy frameworks needed to support sustainability-focused ventures (O'Shea et al., 2021). Additionally, measuring and comparing sustainability impacts across ventures and sectors remains a challenge (Demirel & Parris, 2015; Huang et al., 2023). Nevertheless, the potential benefits of SEEs—from climate change mitigation to inclusive economic development—make them a critical area of study and policy focus (Theodoraki, 2024).

This chapter provides a comprehensive analysis of SEEs, beginning with their conceptual foundations and distinguishing characteristics. It explores the drivers behind their emergence and the key components that constitute these ecosystems, including policy frameworks, financing mechanisms, education systems, and innovation networks. Particular attention is given to the challenges entrepreneurs face in navigating SEEs and the opportunities arising from Sustainability-oriented business models. Finally, the chapter offers insights into how stakeholders (policymakers, investors, educators, and entrepreneurs) can collaborate to build robust SEEs.

By synthesizing current research, this chapter bridges theoretical understanding and practical guidance for developing entrepreneurial ecosystems that advance sustainability goals. While the transition to SEEs presents complex challenges, it also offers unprecedented opportunities to align economic growth with environmental and social progress. In doing so, this analysis contributes to broader discussions on harnessing entrepreneurship as a force for positive change in a Resource-constrained world.

2 From Entrepreneurial Ecosystems to Sustainability

The concept of entrepreneurial ecosystems has become a key reference in strategic management for understanding the contextual factors that enable and constrain entrepreneurial activity across different geographic regions (Stam & Van de Ven, 2021). Rooted etymologically in the Greek words "oikos" (home) and "systema" (complex system), the ecosystem metaphor fundamentally conveys both a sense of hospitable environment and intricate interdependence (Cavallo et al., 2019). This conceptualization builds on Tansley's (1935) original biological formulation of ecosystems as complex systems, where living organisms (biotic elements) interact with their physical environment (abiotic elements). It is adapted to the entrepreneurial context as a dynamic network of interdependent actors and institutional factors.

In strategic terms, these ecosystems represent "*the alignment structure of multilateral partners whose interaction enables focal value propositions to materialize*" (Adner, 2017), emphasizing how coordinated relationships among diverse stakeholders create conditions conducive to venture creation and growth. This systemic perspective represents a significant evolution from earlier individualistic approaches in entrepreneurship research, which focused primarily on the personal characteristics of entrepreneurs.

The ecosystem lens recognizes that successful entrepreneurship requires more than talented individuals; it depends on complex, home-like environments where multiple stakeholders, including investors, policymakers, educational institutions, and support services, interact synergistically (Spigel, 2017). The biological origins of the metaphor remain profoundly relevant, as entrepreneurial ecosystems demonstrate properties similar to those of complex adaptive systems, exhibiting emergent behaviors, non-linear relationships, and co-evolutionary dynamics among their constituent elements (Cavallo et al., 2019). This conceptual shift reflects growing recognition that entrepreneurial outcomes emerge not from isolated actions but from the quality of interactions within these homes for entrepreneurship and their ability to nurture and sustain entrepreneurial activity through collaborative relationships and shared resources.

At its core, an entrepreneurial ecosystem includes all the elements necessary to initiate and sustain entrepreneurial activity within a specific geographic context (Torres & Godinho, 2022). These orchestrated communities bring together diverse stakeholders and external resources that collectively enable the identification, pursuit, and scaling up of new ventures (Autio & Thomas, 2014). The ecosystem includes both formal institutions (such as legal frameworks and financial systems) and informal factors (such as cultural attitudes toward entrepreneurship) that shape the environment for new venture creation (Leendertse et al., 2022).

Importantly, these systems may emerge through different evolutionary pathways; some develop organically through entrepreneur-led initiatives, while others are consciously shaped by government intervention (Fuerlinger et al., 2015; Stam, 2015) or supported by large corporations and academic institutions (Cunningham et al., 2024). The ecosystem approach highlights how these various components interact to foster entrepreneurship as an emergent characteristic of the system as a whole (Stam, 2015).

Unlike traditional industrial clusters, which tend to be sector-specific, entrepreneurial ecosystems are typically industry-agnostic and operate at broader regional or national scales (Cho et al., 2022). These ecosystems often evolve from pre-existing systems such as industry clusters through processes of technological convergence and institutional transformation (Autio & Thomas, 2014). Universities, in particular, play a multifaceted role as ecosystem anchors, generating knowledge, training talent, facilitating technology transfer, and often directly supporting venture creation (Johnson et al., 2019). The dynamic interplay between these various actors and institutions (whether corporate, academic, or governmental) creates the conditions for entrepreneurial activity to flourish across multiple industries simultaneously (Ryan et al., 2021).

The theoretical foundations of entrepreneurial ecosystems draw from multiple disciplines, including economic geography, institutional theory, and network analysis (Cantner et al., 2021). Early work in this area emphasized the importance of "entrepreneurial infrastructure"- the combination of physical, social, and economic resources that facilitate the formation of new businesses. More recent conceptualizations view ecosystems as dynamic, evolving systems where the quality of connections between actors is as important as the actors themselves (Johnson et al., 2019). Mason and Brown (2014) identify several distinguishing features of vibrant ecosystems, including place-based assets, entrepreneurial recycling of resources, supportive cultures, accessible financing, and specialized service providers. Table 1 summarizes key definitions from leading scholars, highlighting the ecosystem's multidisciplinary foundations in economic geography and institutional theory.

Traditional entrepreneurial ecosystems have been predominantly assessed through narrow economic metrics—such as startup rates, employment generation, and GDP contribution—reflecting their foundations in classical economic development theory (Leendertse et al., 2022). While valuable for measuring growth, these conventional indicators fail to account for critical environmental externalities and social inequities, revealing fundamental limitations in addressing contemporary global challenges (Cohen, 2006). This analytical gap has necessitated a paradigm shift toward sustainability-oriented frameworks that reconcile economic development with ecological boundaries and social welfare (Hörisch et al., 2017).

The concept of sustainability, initially rooted in environmental science, emerged to describe the capacity of ecological systems to maintain equilibrium and regenerate after disturbances (Baldin, 2023). Early applications focused on the self-renewal capacities of biological systems, but the concept has evolved into a multidimensional paradigm encompassing economic viability, social justice, and environmental stewardship—the three pillars operationalized through Elkington's (1997) Triple Bottom Line (TBL) framework. This expanded conceptualization acknowledges that long-term prosperity necessitates harmonizing financial objectives with planetary boundaries and intergenerational equity (Baldin, 2023), thereby demanding entrepreneurial ecosystems that simultaneously optimize economic, social, and environmental value creation.

The imperative for integrating sustainability into entrepreneurial ecosystems stems from several interrelated global trends. Climate change represents perhaps the most urgent challenge, with the IPCC[1] warning that current economic systems are incompatible with maintaining a habitable planet (Schaltegger et al., 2016). Simultaneously, the widening of social inequalities within and between nations has highlighted the limitations of growth-oriented economic models, which fail to distribute benefits equitably (Bocken et al., 2014). These systemic challenges have been exacerbated by shifting societal expectations, as consumers, employees, and investors increasingly expect businesses to address environmental and social issues alongside financial performance (Dorfleitner et al., 2021).

Regulatory pressures are also driving the transition toward sustainable entrepreneurial ecosystems. International agreements, such as the Paris Climate Accord and the UN Sustainable Development Goals, have established frameworks that national governments are translating into local policies (Audretsch et al., 2021). These policies range from carbon pricing mechanisms to sustainability reporting requirements, all of which reshape the operating environment for entrepreneurs. At the same time, financial markets are rapidly incorporating ESG[2] factors into investment decisions, creating new opportunities and constraints for startups (Chaudhary et al., 2023).

The concept of sustainable entrepreneurial ecosystems represents an evolution from traditional models, as it explicitly incorporates environmental and social dimensions alongside economic considerations (Theodoraki et al., 2022). Cohen (2006) defined sustainable entrepreneurial ecosystems as "*an interconnected group of actors in a local geographic community committed to sustainable development through the support and facilitation of new sustainable ventures.*" Bischoff and Volkmann (as cited in Ninh & Hue, 2025) define SEEs as interconnected networks of stakeholders that provide support centered on sustainability, enabling entrepreneurs to carry out initiatives that tackle economic, environmental, and social aspects simultaneously.

[1] Intergovernmental Panel on Climate Change.

[2] Environmental, Social and Governance.

Table 1 Review of entrepreneurial ecosystem definitions

Author (year)	Definition
Isenberg (2011)	The entrepreneurship ecosystem is made up of numerous specific components, which can be broadly categorized into six key areas: a supportive cultural environment, effective policies and leadership, access to appropriate finance, quality human capital, venture-friendly markets for products, and a range of institutional and infrastructural supports
Mason and Brown (2014)	A network of interdependent entrepreneurial actors (including existing and would-be entrepreneurs), entrepreneurial organizations (e.g., firms, investors, and supportive entities), institutions (universities, public sector agencies, and financial organizations), and entrepreneurial processes. This last element encompasses key entrepreneurial activities, such as business creation rates, the prevalence of growing firms, breakthrough innovation, the number of serial entrepreneurs, exit strategies, and the overall drive toward entrepreneurial success. These components interact through formal and informal mechanisms to shape, regulate, and enhance the productivity of the local entrepreneurial landscape
Acs et al. (2014)	A continuously evolving process, shaped by institutional frameworks, where people's entrepreneurial attitudes, abilities, and aspirations interact to steer resources toward the formation and growth of new ventures
Mack and Mayer (2016)	Entrepreneurial ecosystems are networks of interdependent elements that support the establishment of new businesses and drive entrepreneurial endeavors within a region
Stam (2015)	An entrepreneurial ecosystem is a set of interconnected players and factors that interact to support and drive productive entrepreneurship in a region
Spigel (2017)	A mix of social, political, economic, and cultural factors in a region that foster the development and growth of innovative startups and support nascent entrepreneurs and other actors to take the risks of launching, funding, and otherwise assisting high-risk ventures
Entrepreneurial Ecosystem Playbook 3.0 (2019)	At its core, an entrepreneurial ecosystem thrives on its human capital and the culture of trust and cooperation that enables effective interaction and collaboration. Such an ecosystem facilitates and accelerates the flow of talent, information, and resources, empowering entrepreneurs to access the necessary support throughout their development journey efficiently. This synergistic environment creates collective value that exceeds what individual components could achieve independently
Ozkazanc-Pan and Muntean (2021)	Entrepreneurial ecosystems can be broadly defined as communities of entrepreneurs who are in exchange relationships with one another and various entrepreneur support organizations, within the context of institutions and institutionalized sociocultural norms

This tripartite approach aligns with Elkington's (1997) triple bottom line framework but applies it at the systemic level rather than to individual firms (Table 2).

Several key differences distinguish SEEs from their traditional counterparts. First, SEEs prioritize long-term viability over short-term growth, recognizing that environmental degradation and social instability ultimately undermine economic prosperity (Tolstykh et al., 2021). Second, they emphasize collaborative innovation across sectors and disciplines to address complex sustainability challenges that cannot be solved by individual actors working in isolation (DiVito & Ingen-Housz, 2021). Third, SEEs incorporate new performance metrics that capture environmental and social outcomes alongside financial returns (Geissdoerfer et al., 2017). Ultimately, they necessitate governance structures that can balance competing interests and mediate trade-offs between various sustainability objectives (Ninh & Hue, 2025; Table 3).

The evolution from traditional to SEEs encounters multiple systemic barriers that hinder progress. Established entrepreneurial systems frequently exhibit entrenched structures and institutional path dependencies that favor conventional business models, creating resistance to sustainable alternatives (O'Shea et al., 2021). Financial systems remain a particularly persis-

Table 2 Review of sustainable entrepreneurial ecosystem definitions

Author (year)	Definition
Cohen (2006)	Sustainable entrepreneurial ecosystems consist of interconnected stakeholders Within a region that collaborates to promote sustainable development by fostering and enabling the growth of new, environmentally and socially responsible businesses
Bischoff and Volkmann (2018)	A collaborative system of stakeholders offers sustainability-oriented support for entrepreneurs, encouraging business activities that balance economic, environmental, and social sustainability while driving the transition toward a greener regional economy
O'Shea et al. (2021)	The SEE concept can be understood as a localized development strategy designed to foster sustainable startups that generate social, ecological, and economic benefits, thereby contributing to a community
Ratajczak-Mrozek et al. (2025)	An interrelated and collaborative system of actors (such as organizations, institutions, and individuals) operating within macro-level institutional, informational, and socioeconomic systems. They offer sustainability-centered support to entrepreneurs while promoting initiatives that holistically balance economic, environmental, and social objectives—the three pillars of the triple bottom line (TBL)

Table 3 Comparison of traditional versus sustainable entrepreneurial ecosystems

Characteristic	Traditional entrepreneurial ecosystems	Sustainable entrepreneurial ecosystems
Primary objective	Economic growth and competitiveness	Economic, social, and environmental balance
Time horizon	Short-to-medium term profitability	Long-term sustainability and resilience
Success metrics	Startup rates, job creation, and GDP contribution	ESG performance, social impact, carbon footprint
Innovation focus	Technological and business model innovation	Sustainable and circular innovation
Resource management	Efficiency-driven	Regenerative and circular principles

tent challenge, as prevailing investment mechanisms prioritize short-term profitability over long-term environmental and social value, leaving sustainability-driven ventures struggling to secure capital (Demirel & Parris, 2015).

Additionally, the absence of universally accepted metrics for measuring ecological and societal impact complicates investment decisions and performance benchmarking (Volkmann et al., 2021). Beyond structural and financial obstacles, the transition demands a fundamental cultural shift, one that redefines entrepreneurial success beyond economic growth to incorporate sustainability as a core value (Hockerts & Wüstenhagen, 2010).

Nevertheless, pioneering examples demonstrate that SEEs are not only possible but are already emerging in various regions. The Nordic countries, for instance, have successfully integrated circular economy principles into their entrepreneurial landscapes, fostering innovation while minimizing resource waste (Audretsch et al., 2021). Urban centers such as Amsterdam and Copenhagen have enacted forward-thinking policies that incentivize green entrepreneurship and align business growth with climate resilience objectives.

Academic institutions are also emerging as critical catalysts for SEE development, spearheading initiatives like MIT's Climate Grand Challenges, which bridge research, education, and sustainable venture creation (Kivimaa et al., 2019). These cases underscore that, despite inherent complexities, the transformation toward sustainability is feasible when stakeholders (governments, investors, universities, and entrepreneurs) collaborate intentionally.

As these examples demonstrate, SEEs represent a distinct paradigm that reimagines entrepreneurial ecosystems through an environmental, social, and economic lens. The following section will examine the foundations of sustainable entrepreneurial ecosystems and their core elements, offering a more transparent framework for understanding how these systems operate and how they differ from traditional models.

3 Theoretical Perspectives of Sustainable Entrepreneurial Ecosystems

Sustainable entrepreneurial ecosystems represent a complex interplay of economic, social, and environmental dimensions, necessitating a multidisciplinary theoretical framework to grasp their dynamics fully. This section examines four key theoretical perspectives that collectively illuminate how SEEs function, evolve, and maintain their impact. Each perspective offers unique insights into the structural, relational, and dynamic aspects of SEEs, providing a foundation for understanding their role in fostering sustainable ventures.

3.1 Triple Bottom Line Framework

SEEs represent an emerging theoretical domain that integrates sustainability principles with the dynamics of entrepreneurial ecosystems. At the core of this discourse is the Triple Bottom Line (TBL) framework, which asserts that SEEs must harmonize economic, social, and environmental dimensions to ensure long-term viability (Cohen, 2006; Ratajczak-Mrozek et al., 2025). Cohen's (2006) foundational definition characterizes SEEs as "*interconnected groups of actors in a local geographic community committed to sustainable development through the support of new sustainable ventures*," emphasizing the systemic collaboration among entrepreneurs, institutions, and policymakers to embed sustainability into entrepreneurial activities.

The economic dimension of the TBL framework examines how entrepreneurial activities influence broader economic systems, ensuring that growth aligns with both present and future financial needs (Fischer et al., 2021). Key indicators of economic sustainability include job creation, income distribution, sectoral revenue generation, and the resilience of businesses to global challenges (Franco & Rodrigues, 2021). A thriving SEE fosters economic adaptability, enabling ventures to scale sustainably while contributing to regional prosperity. For instance, SEEs that prioritize circular economy models often achieve long-term economic stability by minimizing waste and optimizing resource efficiency.

The social dimension underscores the importance of equitable and inclusive business practices that benefit employees, supply chain partners, customers, and local communities (Spence et al., 2011). Social sustainability relies on fostering trust, promoting fair labor practices, and encouraging community engagement, which collectively enhance societal well-being and foster stakeholder loyalty. SEEs that actively address social inequities—such as by supporting underrepresented entrepreneurs or investing in local education—create a more inclusive environment where sustainable ventures can flourish. This dimension highlights the interdependence between entrepreneurial success and societal health, reinforcing the notion that businesses must "give back" to sustain long-term legitimacy and impact.

The environmental dimension addresses the ecological consequences of entrepreneurial activities, emphasizing the need to mitigate environmental degradation and climate change (Dickel, 2017). Sustainable entrepreneurial ecosystems prioritize responsible resource use, reduction of carbon footprint, and adherence to environmental regulations. For example, ventures within SEEs may adopt renewable energy solutions, implement waste-reduction strategies, or develop eco-innovative products that align with planetary boundaries. By integrating environmental stewardship into business models, SEEs ensure that economic growth does not compromise the ability of future generations to meet their needs.

The TBL framework thus provides a holistic lens for analyzing SEEs, demonstrating that economic prosperity, social equity, and environmental health are not mutually exclusive but rather interdependent pillars of sustainability. When these dimensions are balanced,

SEEs can achieve systemic resilience, fostering entrepreneurial activity that benefits both society and the planet (Ratajczak-Mrozek et al., 2025).

3.2 Social Capital Approach

The social capital theory provides a robust lens for analyzing SEEs, emphasizing the role of networks, trust, and shared norms in fostering sustainable ventures. This perspective posits that SEEs thrive when actors (entrepreneurs, incubators, universities, and policymakers) leverage three dimensions of social capital: structural, cognitive, and relational. The structural dimension focuses on the configuration and stability of ties within the ecosystem. For instance, university business incubators (UBIs) serve as hubs, bridging the gaps between researchers, investors, and entrepreneurs, thereby enhancing access to resources and opportunities. In SEEs, dense networks with both strong and weak ties facilitate knowledge spillovers and collaborative innovation, as evidenced by UBIs in France, which connected incubators to regional research centers and funding entities (Theodoraki et al., 2018).

The cognitive dimension underscores shared goals, language, and narratives that align ecosystem actors toward sustainability. UBIs cultivate a common culture by promoting values such as economic development and environmental stewardship, which foster a collective identity. For example, incubator staff and incubates in Montpellier's SEE adhered to shared methodologies and ethical norms, reinforcing a cohesive vision. This alignment mitigates conflicts and enhances legitimacy, critical for scaling sustainable ventures (Theodoraki et al., 2018).

The relational dimension highlights trust, reciprocity, and complementarity among actors. Trust reduces opportunistic behavior, enabling long-term collaborations, such as co-support agreements between UBIs and regional incubators. However, tensions arise when actors prioritize economic gains over social or environmental goals, necessitating governance mechanisms to balance competing interests. Complementarity ensures that diverse actors—like technology parks and social incubators—synergize their strengths, as seen in agri-food SEEs where UBIs partnered with research institutes to commercialize sustainable technologies (Theodoraki et al., 2018).

Social capital theory challenges static views of SEEs by framing them as dynamic, co-evolving networks. The interplay of structural, cognitive, and relational dimensions determines the sustainability of ecosystems. Policymakers can leverage this framework to design interventions that strengthen ties, align goals, and build trust, thereby enhancing SEE resilience. Future research should investigate how digital platforms improve social capital in SEEs and how cultural contexts impact their various dimensions (Theodoraki et al., 2018).

3.3 Stakeholder Theory Approach

Stakeholder theory offers a valuable lens for understanding SEEs by highlighting the interconnected roles of diverse stakeholders in promoting sustainable ventures. Stakeholders are any groups or individuals who can affect or are affected by an organization's objectives, making their inclusion critical for organizational success. In the context of SEEs, stakeholders such as employees, customers, financial institutions, governments, and NGOs collectively shape an environment conducive to sustainable entrepreneurship (Bischoff & Volkmann, 2018). These stakeholders provide tailored support—ranging from financial resources to policy incentives—that aligns with the triple bottom line of economic, social, and environmental sustainability (Elkington, 1997). For instance, higher educational institutions contribute by integrating sustainability into curricula, while governments offer regulatory frameworks that incentivize green business practices (Cohen, 2006).

Stakeholder collaboration is pivotal in SEEs, as it ensures that support mechanisms are coherent and well-coordinated. Freeman et al. (2018) argue that systemic stakeholder engagement fosters the creation of shared value, which is essential for the long-term viability of sustainable ventures. This aligns with ecosystem theory, which posits that interactions among stakeholders enhance innovation and resource allocation (Autio & Thomas, 2014). For example, multi-stakeholder partnerships between NGOs and businesses can drive sustainability initiatives, while local communities influence cultural acceptance of sustainable practices (Bischoff & Volkmann, 2018). However, SEEs face challenges, such as conflicting stakeholder interests and the need for context-specific adaptations. Despite these challenges, stakeholder theory underscores the potential of SEEs to transform regional economies by embedding sustainability at their core.

3.4 Process View

Systems theory further enriches SEE frameworks by conceptualizing ecosystems as dynamic, adaptive systems where components co-evolve. Stam (2015) and Spigel (2017) highlight the interdependence of SEE elements, such as institutions, resources, and cultural norms, arguing that their interactions determine ecosystem performance. For example, the configurational approach (Spigel, 2017) identifies cultural, social, and material attributes as pillars of SEEs, where cultural values (e.g., sustainability-oriented mindsets) shape social interactions (e.g., stakeholder collaboration) and material outcomes (e.g., green innovations). This systemic view challenges static models, advocating for adaptive governance to address emergent sustainability challenges (Audretsch et al., 2021).

The process view of SEEs shifts the focus from static structural elements to the dynamic, coevolutionary nature of entrepreneurial opportunities and actor interactions within these ecosystems (O'Shea et al., 2021; Spigel & Harrison, 2018). Rather than assuming that opportunities are preexisting or independently created, this perspective frames them as emergent phenomena shaped by collective sensemaking, iterative design, and institutional collaboration. For instance, O'Shea et al. (2021) illustrate this through their longitudinal study of the Sustainable Cellulose Ecosystem Project (SCEP) in Helsinki, where entrepreneurial opportunities evolved through continuous stakeholder engagement, pilot testing, and adaptive governance. Initially, actors engaged in informal networking to generate "buzz" around sustainable nanocellulose applications, later formalizing partnerships and embedding sustainability into regional policies—a process reflecting Spigel and Harrison's (2018) phases of buzz creation, confidence-building, and institutionalization.

Central to this perspective is the conceptualization of SEEs as "design artifacts"—dynamic systems intentionally structured through the interventions of actors (O'Shea et al., 2021). A design science approach reveals how sustainability is enacted through iterative cycles of co-intuiting (identifying shared challenges), co-interpreting (refining solutions via feedback), and co-integrating (institutionalizing practices such as circular economy standards). This process mirrors Spigel and Harrison's (2018) emphasis on resource flow analysis, where knowledge from failed ventures is recycled to fuel new opportunities. For example, the SCEP initially explored nanocellulose for packaging but pivoted to biomedical applications after collaborative experimentation and market validation, demonstrating the adaptive nature of SEE development.

The process view also highlights the phased evolution of SEEs, rejecting deterministic models in favor of an agency-driven framework. Early stages involve "buzz" creation, where actors align around a sustainability mission, fostering emotional commitment and informal networks (Cohen, 2006). As confidence grows through prototypes and pilot projects, ecosystems attract investment and policy support, eventually reaching institutionalization, where sustainable practices become embedded in regional value chains. Challenges arise in balancing economic scalability with environmental and social objectives, as seen in the SCEP's tensions between production efficiency and eco-friendly methods (O'Shea et al., 2021). Key enablers include a shared sustainability culture, flexible governance structures, and resource mobility, such as knowledge spillovers from universities to startups (Volkmann et al., 2021). This view captures SEEs as living systems, emphasizing actor agency and longitudinal dynamics over static typologies.

Table 4 Main theoretical perspectives on SEE

Theory	View on SEEs
Triple bottom line framework	Asserts that SEEs must harmonize economic, social, and environmental dimensions to ensure long-term viability
Social capital approach	Emphasizes the role of networks, trust, and shared norms in SEEs
Stakeholder approach	Frames sees ecosystems as places where diverse actors—governments, NGOs, and entrepreneurs—co-create value
Process view	Conceptualize SEEs as adaptive, co-evolving systems. This dynamic view challenges static models, advocating for flexible governance and a shared sustainability culture

Ultimately, the process view redefines SEEs as arenas where sustainability is co-constructed through iterative, collective action; a dynamic that demands continued scholarly and practical attention (Table 4).

Together, these theories provide a comprehensive framework for analyzing SEEs, revealing their structural foundations, relational dynamics, and evolutionary pathways. The TBL framework establishes the foundational pillars of sustainability, while social capital theory elucidates how networks and shared norms facilitate collaborative action among ecosystem actors. Stakeholder theory extends this understanding by demonstrating how diverse actors co-create value through negotiated engagement, and process views capture the dynamic, iterative nature of SEE development. By integrating these lenses, we gain a nuanced understanding of how SEEs emerge, function, and adapt to balance economic, social, and environmental objectives in diverse regional contexts.

4 Key Elements of Sustainable Entrepreneurial Ecosystems

The transition from conventional to sustainable entrepreneurial ecosystems (SEEs) represents a fundamental reconfiguration of how we conceptualize value creation in business innovation. Where traditional ecosystems prioritize economic metrics like startup density and job creation, SEEs introduce a transformative paradigm that harmonizes entrepreneurial activity with environmental stewardship and social equity (Audretsch et al., 2019). This shift manifests through seven interconnected elements that collectively differentiate SEEs from their traditional counterparts. Key aspects of SEEs include supportive institutional frameworks, access to sustainable finance, human capital development, sustainable innovation and knowledge spillovers, infrastructure and digitalization, market demand for sustainable products, and collaborative networks (Roundy et al., 2018). These elements interact synergistically to create an environment that fosters sustainable ventures.

A critical element of SEEs is a supportive institutional framework, comprising policies, regulations, and governance structures that incentivize sustainable business practices (Acs et al., 2014). Governments and regulatory bodies play a pivotal role in shaping SEEs by implementing policies such as tax incentives for green businesses, subsidies for renewable energy projects, and stringent environmental regulations. Institutional support also encompasses legal frameworks that protect intellectual property rights, reduce bureaucratic barriers, and promote transparency, all of which are crucial for fostering trust and long-term entrepreneurial engagement (Autio & Thomas, 2014). Additionally, public–private partnerships can amplify institutional effectiveness by aligning stakeholder interests with sustainability objectives.

Another fundamental element is access to sustainable finance, which ensures that entrepreneurs have the necessary capital to develop and scale sustainable ventures (Demirel & Parris, 2015). Traditional funding mechanisms often fail to account for the long-term value of sustainability-oriented businesses, making alternative financing models crucial (Hörisch et al., 2017). Impact investing, green bonds, and venture capital focused on environmental, social, and governance (ESG) criteria are increasingly important in SEEs (Bocken et al., 2014). Crowdfunding platforms and microfinance institutions also provide accessible funding for early-stage sustainable startups, democratizing access to capital. Financial institutions that integrate sustainability metrics into their risk assessments further reinforce the stability of SEEs by directing capital toward ventures with positive societal and environmental impacts.

Human capital development is another cornerstone of SEEs, as a skilled and sustainability-conscious workforce drives innovation and enterprise growth (Feldman, 2014). Educational institutions, vocational training programs, and entrepreneurship incubators must emphasize sustainability literacy, equipping future entrepreneurs with the knowledge and skills needed to address complex ecological and social challenges (Lans et al., 2014). Universities and research centers contribute to SEEs by fostering interdisciplinary collaboration and producing cutting-edge research on sustainable technologies and business models (Etzkowitz & Leydesdorff, 2000). Lifelong learning initiatives and corporate training programs further ensure that existing professionals remain adaptable to the evolving trends in sustainability.

Sustainable innovation and knowledge spillovers are vital for maintaining the competitiveness of SEEs. Innovation ecosystems that prioritize sustainability generate new technologies, processes, and business models that reduce environmental footprints while enhancing profitability. Knowledge spillovers occur when research findings, best practices, and technological advancements diffuse across firms, industries, and regions, amplifying their impact (Audretsch et al., 2021). Clusters of sustainable enterprises, often concentrated in innovation hubs or green industrial parks, facilitate these spillovers by fostering proximity-based collaboration. Open innovation platforms and digital knowledge-sharing tools further accelerate the dissemination of sustainable solutions (Chesbrough, 2003).

Infrastructure and digitalization serve as enablers of SEEs by providing the physical and technological foundations for sustainable entrepreneurship (Malecki, 2018). Green infrastructure, including energy-efficient buildings, renewable energy grids, and sustainable transportation networks, reduces operational costs for businesses while minimizing environmental harm. Digital technologies, including big data analytics, artificial intelligence, and blockchain, enhance resource efficiency, supply chain transparency, and circular economy practices (Brenner, 2018). Digital platforms also connect entrepreneurs with mentors, investors, and customers, lowering barriers to market entry and fostering inclusivity (Nambisan, 2017).

Market demand for sustainable products is a driving force behind SEEs, as consumer preferences increasingly favor environmentally and socially responsible offerings (Schaltegger & Wagner, 2011). Businesses that integrate sustainability into their value propositions gain competitive advantages through brand differentiation and customer loyalty (Kramer & Porter, 2011). Public awareness campaigns and eco-labeling initiatives further stimulate demand by educating consumers about the benefits of sustainable consumption. Corporate procurement policies that prioritize sustainable suppliers also create stable demand-side incentives for entrepreneurs.

Finally, collaborative networks underpin SEEs by fostering trust, reciprocity, and collective action among stakeholders (Spigel, 2017). Entrepreneurial ecosystems thrive when diverse actors, including startups, corporations, nonprofits, academia, and government agencies, work together toward shared sustainability goals (Roundy et al., 2018). Industry associations, sustainability-focused accelerators, and cross-sector partnerships facilitate knowledge exchange, resource pooling, and joint problem-solving. Social capital, built through repeated interactions and shared norms, enhances the resilience of SEEs by mitigating risks and enabling adaptive responses to disruptions.

5 Challenges of Sustainable Entrepreneurial Ecosystems

The transition from traditional entrepreneurial ecosystems to SEEs presents multifaceted challenges, stemming from the need to balance economic, social, and environmental dimensions while navigating systemic complexities (Ratajczak-Mrozek et al., 2025). SEEs require intentional alignment of diverse actors, including entrepreneurs, institutions, and policymakers, to foster ventures that address sustainability goals (Cohen, 2006; Volkmann et al., 2021). However, achieving this alignment is fraught with obstacles, including governance dilemmas, cultural barriers, and tensions between short-term profitability and long-term sustainability.

One of the foremost challenges lies in balancing the TBL dimensions within SEEs. Unlike conventional EEs, SEEs must integrate economic viability with ecological stewardship and social equity, often leading to conflicting priorities (Volkmann et al., 2021). For instance, entrepreneurs may prioritize economic performance to secure funding, inadvertently marginalizing environmental or social goals (Høgevold et al., 2015). This tension is exacerbated by the lack of consensus on how to operationalize sustainability across all ecosystem actors. While some firms may excel in one TBL dimension, others might lag, raising questions about whether an ecosystem can be deemed "sustainable" if only a subset of actors adheres to TBL principles (Ratajczak-Mrozek et al., 2025). The absence of standardized metrics to measure ecosystem-wide sustainability further complicates this challenge, necessitating frameworks that evaluate both individual and collective TBL performance (Theodoraki et al., 2022).

Governance and performance distribution pose another critical hurdle. SEEs comprise heterogeneous actors with divergent goals, capabilities, and bargaining power, making coordinated action difficult (Stam & Van de Ven, 2021). For example, large corporations may impose sustainability standards on suppliers without sharing the associated costs, disproportionately burdening smaller actors. This imbalance undermines the equitable distribution of costs and benefits, a cornerstone of SEEs. Additionally, the lack of precise governance mechanisms, whether formal (e.g., regulations) or informal (e.g., shared norms), hampers the ecosystem's ability to resolve conflicts and enforce sustainability commitments. Flexible governance models that accommodate actor specializations while ensuring TBL compliance are essential but remain understudied (Ratajczak-Mrozek et al., 2025).

Cultural and communication barriers further impede SEE development. Cultivating a shared culture that values sustainability and entrepreneurship is pivotal but challenging, as attitudes toward sustainability vary regionally (Volkmann et al., 2021). In specific contexts, traditional employment mindsets prevail, discouraging risk-taking and sustainability-oriented innovation (Vicentin et al., 2024). Effective communication is equally critical, as misalignment between internal practices and external messaging can lead to accusations of greenwashing (Bothello et al., 2023). For instance, overemphasizing sustainability achievements without substantive action risks eroding stakeholder trust. Conversely, inadequate communication may obscure genuine efforts, limiting ecosystem legitimacy and resource mobilization (Cloitre et al., 2024). Addressing these challenges requires targeted education and networking initiatives to align the motivations of actors with ecosystem goals (Volkmann et al., 2021).

At the entrepreneurial level, capability gaps hinder the progression of SEE. Sustainable entrepreneurs must navigate uncertain markets, reconcile TBL trade-offs, and advocate for systemic change, tasks requiring specialized skills (DiVito & Ingen-Housz, 2021). However, many lack access to training or mentorship in sustainability-driven decision-making (Pankov et al., 2021). Empowering these actors requires ecosystem-wide support, including incubator programs and funding tailored to green ventures (Theodoraki et al., 2018). Yet, financing remains a bottleneck, as traditional venture capital often prioritizes short-term returns over long-term TBL outcomes (Bergset & Fichter, 2015). Public–private partnerships and impact investing could bridge this gap, but require further exploration.

Finally, the dynamic interplay between individual and ecosystem-level changes complicates SEE transitions. Bottom-up initiatives (e.g., entrepreneur-led sustainability ventures) and top-down interventions (e.g., policy incentives) must co-evolve, creating feedback loops that reinforce sustainability (Ratajczak-Mrozek et al., 2025). For example, while entrepreneurs drive innovation, policymakers must provide enabling conditions, such as tax incentives for circular economy practices. Misalignment between these levels can stall progress, as seen in regions where regulatory frameworks lag behind entrepreneurial sustainability efforts (Pankov et al., 2021).

The development of SEEs confronts systemic barriers that reveal fundamental tensions in contemporary economic systems. Persistent challenges emerge from competing TBL priorities, fragmented governance structures, deeply embedded cultural norms, skill gaps among sustainability entrepreneurs, and disconnects between policy frameworks and grassroots initiatives (Theodoraki et al., 2022). Overcoming these obstacles requires coordinated interventions at multiple levels, from establishing rigorous impact measurement standards and inclusive decision-making processes for transforming educational paradigms and aligning incentive structures across institutional boundaries. These interdependent solutions underscore the complexity of cultivating entrepreneurial environments where economic, ecological, and social value creation become mutually reinforcing rather than competing objectives.

6 Conclusion

Sustainable entrepreneurial ecosystems represent a transformative paradigm that redefines entrepreneurship through the interdependence of economic, social, and environmental outcomes (Cohen, 2006; Theodoraki et al., 2022). This chapter has demonstrated that SEEs are not merely extensions of traditional ecosystems, but rather complex, adaptive systems that require intentional design and multi-stakeholder collaboration (Spigel & Stam, 2018). Theoretical frameworks such as the TBL, social capital theory, and stakeholder theory provide a foundation for understanding SEEs, emphasizing aligned goals, trust-based networks, and inclusive governance (Freeman et al., 2018). Key elements, from policy incentives to innovation clusters, must co-evolve to create enabling conditions for sustainable ventures (Audretsch et al., 2019).

However, the transition to SEEs faces significant challenges, including conflicting stakeholder priorities, cultural resistance, and gaps in entrepreneurial capabilities (Volkmann et al., 2021). Addressing these barriers requires holistic strategies, such as standardized sustainability metrics, equitable resource distribution, and capacity-building programs (Ratajczak-Mrozek et al., 2025). Policymakers play a pivotal role in fostering SEEs through regulatory frameworks that incentivize sustainable practices, while educational institutions must integrate sustainability literacy into curricula to prepare future entrepreneurs (Lans et al., 2014).

The transformative potential of SEEs lies in their ability to reconcile economic growth with planetary boundaries and social equity (Schaltegger et al., 2016). Future research should explore dynamic interactions among SEE elements across diverse contexts, empirically validating best practices for ecosystem resilience (Theodoraki, 2024). As global challenges intensify, SEEs offer a viable pathway to align entrepreneurial activity with the urgent imperatives of sustainability, ensuring prosperity for current and future generations.

7 Questions

1. How do sustainable entrepreneurial ecosystems (SEEs) differ from traditional entrepreneurial ecosystems in terms of goals, structures, and outcomes?
2. How does the Triple Bottom Line framework serve as a conceptual foundation for the development and evaluation of SEEs?

3. How do stakeholder theory, social capital theory, and process views contribute to our understanding of SEEs?
4. What roles do policy frameworks, sustainable finance mechanisms, and educational institutions play in enabling SEEs?
5. What are the principal challenges involved in designing and implementing sustainable entrepreneurial ecosystems, particularly in contexts with limited institutional and financial support?

Further Reading

Theodoraki, C. (2024). Building entrepreneurial ecosystems sustainably. *Foundations and Trends in Entrepreneurship, 20*(4), 384–480.

Cunningham, J. A., Menter, M., O'Kane, C., & Romano, M. (Eds.). (2024). *Research Handbook on Entrepreneurial Ecosystems.* Edward Elgar Publishing.

Chaudhary, S., Kaur, P., Alofaysan, H., Halberstadt, J., & Dhir, A. (2023). Connecting the dots? Entrepreneurial ecosystems and sustainable entrepreneurship as pathways to sustainability. *Business Strategy and the Environment, 32*(8), 5935–5951.

References

Acs, Z. J., Autio, E., & Szerb, L. (2014). National systems of entrepreneurship: Measurement issues and policy implications. *Research Policy, 43*(3), 476–494.

Adner, R. (2017). Ecosystem as structure: An actionable construct for strategy. *Journal of Management, 43*(1), 39–58.

Audretsch, D. B., Belitski, M., & Desai, S. (2019). National business regulations and city entrepreneurship in Europe: A multilevel nested analysis. *Entrepreneurship Theory and Practice, 43*(6), 1148–1165.

Audretsch, D. B., Belitski, M., & Cherkas, N. (2021). Entrepreneurial ecosystems in cities: The role of institutions. *PLoS ONE, 16*(3), Article e0247609.

Autio, E., & Thomas, L. (2014). Innovation ecosystems: Implications for innovation management. In *The Oxford handbook of innovation management* (pp. 204–228). Oxford University Press.

Baldin, M. (2023). A sustainable view on entrepreneurial ecosystems: The study of the Lagos context.

Bergset, L., & Fichter, K. (2015). Green startups—A new typology for sustainable entrepreneurship and innovation research. *Journal of Innovation Management, 3*(3), 118–144.

Bischoff, K., & Volkmann, C. K. (2018). Stakeholder support for sustainable entrepreneurship—A framework of sustainable entrepreneurial ecosystems. *International Journal of Entrepreneurial Venturing, 10*(2), 172–201.

Bocken, N. M., Short, S. W., Rana, P., & Evans, S. (2014). A literature and practice review to develop sustainable business model archetypes. *Journal of Cleaner Production, 65*, 42–56.

Bothello, J., Ioannou, I., Porumb, V. A., & Zengin-Karaibrahimoglu, Y. (2023). CSR decoupling within business groups and the risk of perceived greenwashing. *Strategic Management Journal, 44*(13), 3217–3251.

Brenner, B. (2018). Transformative sustainable business models in the light of the digital imperative—A global business economics perspective. *Sustainability, 10*(12), 4428.

Cantner, U., Cunningham, J. A., Lehmann, E. E., & Menter, M. (2021). Entrepreneurial ecosystems: A dynamic lifecycle model. *Small Business Economics, 57*, 407–423.

Cavallo, A., Ghezzi, A., & Balocco, R. (2019). Entrepreneurial ecosystem research: Present debates and future directions. *International Entrepreneurship and Management Journal, 15*, 1291–1321.

Chaudhary, S., Kaur, P., Alofaysan, H., Halberstadt, J., & Dhir, A. (2023). Connecting the dots? Entrepreneurial ecosystems and sustainable entrepreneurship as pathways to sustainability. *Business Strategy and the Environment, 32*(8), 5935–5951.

Chesbrough, H. W. (2003). *Open innovation: The new imperative for creating and profiting from technology.* Harvard Business Press.

Cho, D. S., Ryan, P., & Buciuni, G. (2022). Evolutionary entrepreneurial ecosystems: A research pathway. *Small Business Economics, 58*(4), 1865–1883.

Cloitre, A., Theodoraki, C., & dos Santos Paulino, V. (2024). Entrepreneurial support organizations in sustainable knowledge-driven ecosystems. *The Journal of Technology Transfer*, 1–25.

Cohen, B. (2006). Sustainable valley entrepreneurial ecosystems. *Business Strategy and the Environment, 15*(1), 1–14.

Cunningham, J. A., Menter, M., O'Kane, C., & Romano, M. (Eds.). (2024). *Research handbook on entrepreneurial ecosystems.* Edward Elgar Publishing.

Demirel, P., & Parris, S. (2015). Access to finance for innovators in the UK's environmental sector. *Technology Analysis & Strategic Management, 27*(7), 782–808.

Dickel, P. (2017). The impact of protectability and proactiveness on the environmental performance of new ventures. *Corporate Governance: THe International Journal of Business in Society, 17*(1), 117–133.

DiVito, L., & Ingen-Housz, Z. (2021). From individual sustainability orientations to collective sustainability innovation and sustainable entrepreneurial ecosystems. *Small Business Economics, 56*(3), 1057–1072.

Dorfleitner, G., Utz, S., & Zhang, R. (2021). The pricing of green bonds: External reviews and the shades of green. *Review of Managerial Science*, 1–38.

Elkington, J. (1997). The triple bottom line. *Environmental Management: Readings and Cases, 2*, 49–66.

Etzkowitz, H., & Leydesdorff, L. (2000). The dynamics of innovation: From National Systems and "Mode 2" to a Triple Helix of university–industry–government relations. *Research Policy, 29*(2), 109–123.

Feldman, M. P. (2014). The character of innovative places: Entrepreneurial strategy, economic development, and prosperity. *Small Business Economics, 43*, 9–20.

Fischer, B., Guerrero, M., Guimón, J., & Schaeffer, P. R. (2021). Knowledge transfer for frugal innovation: Where do entrepreneurial universities stand? *Journal of Knowledge Management, 25*(2), 360–379.

Freeman, R. E., Harrison, J. S., & Zyglidopoulos, S. (2018). *Stakeholder theory: Concepts and strategies*. Cambridge University Press.

Franco, M., & Rodrigues, M. (2021). Sustainable practices in SMEs: Reducing the ecological footprint. *Journal of Business Strategy, 42*(2), 137–142.

Fuerlinger, G., Fandl, U., & Funke, T. (2015). The role of the state in the entrepreneurship ecosystem: Insights from Germany. *Triple Helix, 2*(1), 1–26.

Geissdoerfer, M., Savaget, P., Bocken, N. M., & Hultink, E. J. (2017). The Circular Economy—A new sustainability paradigm? *Journal of Cleaner Production, 143*, 757–768.

Hockerts, K., & Wüstenhagen, R. (2010). Greening Goliaths versus emerging Davids—Theorizing about the role of incumbents and new entrants in sustainable entrepreneurship. *Journal of Business Venturing, 25*(5), 481–492.

Høgevold, N. M., Svensson, G., Klopper, H. B., Wagner, B., Valera, J. C. S., Padin, C., Ferro, C., Petzer, D. (2015). A triple bottom line construct and reasons for implementing sustainable business practices in companies and their business networks. *Corporate Governance, 15*(4), 427–443.

Hörisch, J., Kollat, J., & Brieger, S. A. (2017). What influences environmental entrepreneurship? A multilevel analysis of the determinants of entrepreneurs' environmental orientation. *Small Business Economics, 48*, 47–69.

Huang, Y., Li, P., Bu, Y., & Zhao, G. (2023). What entrepreneurial ecosystem elements promote sustainable entrepreneurship? *Journal of Cleaner Production, 422*, Article 138459.

Isenberg, D. (2011, May 12). *The entrepreneurship ecosystem strategy as a new paradigm for economic policy: Principles for cultivating entrepreneurship* (pp. 1–13). Institute of International and European Affairs, Dublin, Ireland.

Johnson, D., Bock, A. J., & George, G. (2019). Entrepreneurial dynamism and the built environment in the evolution of university entrepreneurial ecosystems. *Industrial and Corporate Change, 28*(4), 941–959.

Kivimaa, P., Boon, W., Hyysalo, S., & Klerkx, L. (2019). Towards a typology of intermediaries in sustainability transitions: A systematic review and a research agenda. *Research Policy, 48*(4), 1062–1075.

Kramer, M. R., & Porter, M. (2011). *Creating shared value* (Vol. 17). FSG.

Lans, T., Blok, V., & Wesselink, R. (2014). Learning apart and together: towards an integrated competence framework for sustainable entrepreneurship in higher education. *Journal of Cleaner Production, 62*, 37–47.

Leendertse, J., Schrijvers, M., & Stam, E. (2022). Measure twice, cut once: Entrepreneurial ecosystem metrics. *Research Policy, 51*(9), Article 104336.

Mack, E., & Mayer, H. (2016). The evolutionary dynamics of entrepreneurial ecosystems. *Urban Studies, 53*(10), 2118–2133.

Malecki, E. J. (2018). Entrepreneurship and entrepreneurial ecosystems. *Geography Compass, 12*(3), Article e12359.

Mason, C., & Brown, R. (2014). Entrepreneurial ecosystems and growth oriented entrepreneurship. *Final Report to OECD, Paris, 30*(1), 77–102.

Nambisan, S. (2017). Digital entrepreneurship: Toward a digital technology perspective of entrepreneurship. *Entrepreneurship Theory and Practice, 41*(6), 1029–1055.

Ninh, T. T. T., & Hue, T. T. (2025). Sustainable entrepreneurial ecosystems: What are the main schools of thought and topical trends? *Discover Sustainability, 6*(1), 115.

O'Shea, G., Farny, S., & Hakala, H. (2021). The buzz before business: A design science study of a sustainable entrepreneurial ecosystem. *Small Business Economics, 56*, 1097–1120.

Ozkazanc-Pan, B., & Muntean, S. C. (2021). *Entrepreneurial ecosystems: A gender perspective*. Cambridge University Press.

Pankov, S., Schneckenberg, D., & Velamuri, V. K. (2021). Advocating sustainability in entrepreneurial ecosystems: Micro-level practices of sharing ventures. *Technological Forecasting and Social Change, 166*, Article 120654.

Ratajczak-Mrozek, M., Hauke-Lopes, A., Sajdak, M., & Wieczerzycki, M. (2025). Guiding entrepreneurial ecosystems towards sustainability: Identifying key changes and future research directions. *Journal of Organizational Change Management, 38*(2), 321–339.

Roundy, P. T., Bradshaw, M., & Brockman, B. K. (2018). The emergence of entrepreneurial ecosystems: A complex adaptive systems approach. *Journal of Business Research, 86*, 1–10.

Ryan, P., Giblin, M., Buciuni, G., & Kogler, D. F. (2021). The role of MNEs in the genesis and growth of a resilient entrepreneurial ecosystem. In *The dynamics of entrepreneurial ecosystems* (pp. 36–53). Routledge.

Schaltegger, S., & Wagner, M. (2011). Sustainable entrepreneurship and sustainability innovation: Categories and interactions. *Business Strategy and the Environment, 20*(4), 222–237.

Schaltegger, S., Hansen, E. G., & Lüdeke-Freund, F. (2016). Business models for sustainability: Origins, present research, and future avenues. *Organization & Environment, 29*(1), 3–10.

Shahid, M., Bakhat, H. F., Shah, G. M., & Murtaza, B. (2023). Recent trends in environmental sustainability. *Environmental Science and Pollution Research, 30*(44), 99198–99201.

Spence, M., Ben Boubaker Gherib, J., & Ondoua Biwolé, V. (2011). Sustainable entrepreneurship: Is entrepreneurial will enough? A north–south comparison. *Journal of Business Ethics, 99*, 335–367.

Spigel, B. (2017). The relational organization of entrepreneurial ecosystems. *Entrepreneurship Theory and Practice, 41*(1), 49–72.

Spigel, B., & Harrison, R. (2018). Toward a process theory of entrepreneurial ecosystems. *Strategic Entrepreneurship Journal, 12*(1), 151–168.

Spigel, B., & Stam, E. (2018). Entrepreneurial ecosystems. In *SAGE handbook of entrepreneurship and small business* (p. 21). Sage.

Stam, E. (2015). Entrepreneurial ecosystems and regional policy: A sympathetic critique. *European Planning Studies, 23*(9), 1759–1769.

Stam, E., & Van de Ven, A. (2021). Entrepreneurial ecosystem elements. *Small Business Economics, 56*(2), 809–832.

Tansley, A. G. (1935). The use and abuse of vegetational concepts and terms. *Ecology, 16*(3), 284–307.

Theodoraki, C., Messeghem, K., & Rice, M. P. (2018). A social capital approach to the development of sustainable entrepreneurial ecosystems: An explorative study. *Small Business Economics, 51*, 153–170.

Theodoraki, C., Dana, L. P., & Caputo, A. (2022). Building sustainable entrepreneurial ecosystems: A holistic approach. *Journal of Business Research, 140*, 346–360.

Theodoraki, C. (2024). Building entrepreneurial ecosystems sustainably. *Foundations and Trends® in Entrepreneurship, 20*(4), 384–480.

Tolstykh, T., Gamidullaeva, L., & Shmeleva, N. (2021). Universities as knowledge integrators and cross-industry ecosystems: Self-organizational perspective. *SAGE Open, 11*(1), 2158244020988704.

Torres, P., & Godinho, P. (2022). Levels of necessity of entrepreneurial ecosystems elements. *Small Business Economics, 59*(1), 29–45.

Vicentin, D. C., de Moraes, G. H. S. M., Fischer, B. B., Campello, B. S. C., do Prado, N. B., & Anholon, R. (2024). The interdependence between the Entrepreneurial Ecosystem and Entrepreneurial Culture—An analysis with sustainable and traditional entrepreneurs. *Journal of Cleaner Production, 466*, 142821.

Volkmann, C., Fichter, K., Klofsten, M., & Audretsch, D. B. (2021). Sustainable entrepreneurial ecosystems: An emerging field of research. *Small Business Economics, 56*(3), 1047–1055.

Behavioral Dimensions of Sustainable Entrepreneurship

Ehsan Chitsaz and Salar Tebyani

Abstract

Sustainable entrepreneurship merges the pursuit of business opportunities with a commitment to environmental stewardship and social well-being. This chapter explores the **behavioral dimensions** underpinning sustainable entrepreneurship—the motivations, decisions, and actions of entrepreneurs who aim to achieve the "triple bottom line" of **people, planet, and profit**. We begin by defining key concepts, including sustainable development and sustainable entrepreneurship, and review foundational theories such as the triple bottom line framework and stakeholder theory. We then examine the behaviors and mindset of sustainable entrepreneurs, drawing on evidence from research and real-world examples. This includes an analysis of how personal values, cognitive factors, and global trends (e.g., the rise of AI and climate innovation) influence entrepreneurial intentions and ventures that prioritize sustainability. Further sections discuss debates and challenges, from balancing profit with purpose to addressing greenwashing and scaling impacts. Practical strategies and tools for sustainable entrepreneurship are highlighted, and a global perspective is taken with examples from different contexts. A case study of Patagonia, the outdoor apparel company, illustrates the chapter's themes in practice. The chapter concludes with key takeaways and reflection questions, reinforcing the learning objectives and encouraging deeper consideration of the role of behavior in driving sustainable entrepreneurial success.

Keywords

Sustainable entrepreneurship · Entrepreneurial behaviour · Triple bottom line · Stakeholder theory · Environmental values · Sustainable business models

Learning Objectives

Behavioral Dimensions of Sustainable Entrepreneurship

By the end of this chapter, readers will be able to:

1. **Define sustainable entrepreneurship** and distinguish it from traditional and social entrepreneurship based on behavioral, economic, and ethical dimensions.
2. **Analyze the psychological and motivational drivers** (e.g., values, identity, intentions) that influence entrepreneurs who pursue sustainability-oriented ventures.

E. Chitsaz (✉) · S. Tebyani
Faculty of Entrepreneurship, University of Tehran, Tehran, Iran
e-mail: Chitsaz@ut.ac.ir

S. Tebyani
e-mail: salartebyani@ut.ac.ir

M. Akbari et al. (eds.), *Sustainable Entrepreneurship*, University of Tehran Science and Humanities Series, https://doi.org/10.1007/978-3-032-04909-4_24

3. **Apply foundational theories**—such as the triple bottom line, stakeholder theory, and market failure logic—to understand decision-making processes in sustainable entrepreneurship.
4. **Evaluate practical strategies and tools** (e.g., sustainable business model design, impact metrics, partnerships) that entrepreneurs use to balance profit with social and environmental goals.
5. **Critically reflect on real-world examples**, including the Patagonia case study, to identify key behaviors, trade-offs, and leadership approaches that drive sustainability in entrepreneurial settings.

1 Introduction

Around the world, entrepreneurs are increasingly motivated by social and environmental concerns when launching new ventures (GEM Consortium, 2025). From climate change to inequality, global challenges have underscored the need for businesses that do more than generate profit—they must also create a **positive impact** on society and the planet. This is the domain of **sustainable entrepreneurship**, where business innovation is coupled with sustainability objectives from the outset. Sustainable entrepreneurs recognize that the health of economies, communities, and ecosystems is intensely interdependent. They seek opportunities to build companies that address environmental and social problems while remaining financially viable (Dean & McMullen, 2007). In doing so, they differ from traditional entrepreneurs (who typically focus on the financial "bottom line" alone) and even from social entrepreneurs (who focus primarily on social mission, sometimes without an environmental emphasis).

Understanding the **behavioral dimensions** of sustainable entrepreneurship is crucial. After all, it is human action—the choices of founders and the culture they instill in their ventures—that ultimately determines how sustainability is integrated into a business. Key questions arise: *What drives entrepreneurs to pursue sustainability-oriented businesses? How do their values and mindsets shape their ventures? What practical behaviors and strategies help them succeed in meeting triple bottom-line goals?* This chapter addresses these questions by examining the psychological and behavioral underpinnings of sustainable entrepreneurship alongside theoretical frameworks and examples. We take a global perspective, recognizing that sustainable entrepreneurship is unfolding in diverse contexts—from tech startups in Silicon Valley deploying AI for climate solutions to grassroots innovators in emerging economies solving local environmental challenges.

The sections that follow will introduce foundational concepts and theories (Sect. 4.1), delve into an in-depth analysis of entrepreneurial behavior and decision-making for sustainability (Sect. 4.2), explore further debates and comparisons (Sect. 4.3), and highlight practical applications and tools (Sect. 4.4). A concluding section will summarize the insights and suggest areas for reflection. Finally, a real-world **case study**—the story of Patagonia—will illustrate how an entrepreneurial venture can embed sustainability at its core, providing material for discussion and learning. Through this chapter, you will gain a comprehensive understanding of how sustainable entrepreneurship is not just a business approach but also a behavioral phenomenon, driven by the passion, vision, and ethics of individuals who aim to make a difference through enterprise.

2 Introduction to Key Concepts, Definitions, and Foundational Theories

2.1 Sustainable Development and Triple Bottom Line

To ground our discussion, we begin with the concept of sustainable development, which the Brundtland Commission famously defined as "development that meets the needs of the present without compromising the ability of future generations to meet their own needs" (World Commission on Environment & Development, 1987). This definition emphasizes a long-term perspective and the need to balance present progress with future well-being. In business contexts, this idea is often operationalized through the **triple bottom line (TBL)** framework. The triple bottom line expands the traditional focus on financial performance to include the social and environmental dimensions of success (Miller, 2020). In other words, a sustainable enterprise measures its achievements not only by profits but also by its impact on people (society) and the planet (environment). John Elkington, who popularized the term in the 1990s, argued that companies should be evaluated on all three "P's" —people, planet, profit—rather than just one (Miller, 2020; Wikipedia, n.d.). This framework encourages entrepreneurs to create business models that generate economic value **while** contributing to social equity and environmental health. It also aligns with **stakeholder theory**, which holds that businesses have a responsibility to a broad set of stakeholders (employees, customers, communities, the environment, etc.), not just shareholders (Freeman, 1984). Together, sustainable development, the triple bottom line, and stakeholder theory form a foundation that guides sustainable entrepreneurs toward creating integrated value.

2.2 What Is Sustainable Entrepreneurship?

Bringing these ideas into entrepreneurship, scholars have proposed formal definitions of sustainable entrepreneurship. A widely cited definition by Dean and McMullen (2007) frames sustainable entrepreneurship as *"the process of discovering, evaluating, and exploiting economic opportunities that are present in market failures which detract from sustainability"* (Dean & McMullen, 2007). In this view, many environmental and social issues (such as pollution or poverty) can be seen as **market failures**—unmet needs or externalities that traditional markets have failed to address. Sustainable entrepreneurs turn these problems into opportunities by innovating new products, services, or processes that resolve market failures, thereby generating profits and improving environmental or social outcomes. For example, an entrepreneur might see the abundance of plastic waste not just as a pollution problem but as an opportunity to create a recycling business that turns waste into valuable products. By addressing a market failure (plastic pollution) through entrepreneurial action, they aim for both economic gain and a sustainable **impact** (Dean & McMullen, 2007).

Another comprehensive definition offered by Shepherd and Patzelt (2011) emphasizes the intent and broad value created by sustainable entrepreneurs. They define sustainable entrepreneurship as **focused on "the preservation of nature, life support, and community in the pursuit of perceived opportunities to bring into existence future products, processes, and services for gain, where the gain is broadly construed to include economic and non-economic gains to individuals, the economy, and society"** (Shepherd & Patzelt, 2011). This definition highlights that sustainable entrepreneurs explicitly aim to protect and improve vital environmental and social systems ("nature, life support, community") as they build their businesses. Importantly, the "gain" they seek is **multi-faceted**—it's not just about personal wealth or investor returns, but also benefits to society at large and the environment. In essence, sustainable entrepreneurship exists at the intersection of traditional entrepreneurship and sustainable development goals, inheriting the opportunity-seeking, innovation, and risk-taking of entrepreneurship while infusing it with a sustainability mission.

2.3 Relation to Social and Environmental (Eco) Entrepreneurship

It is helpful to distinguish sustainable entrepreneurship from closely related concepts. Social entrepreneurship typically refers to ventures that are driven by a primary mission of addressing a social problem (such as poverty, education, or health disparities). Social entrepreneurs often reinvest profits to further their mission, prioritizing social value over financial returns. *Environmental entrepreneurship* (or "ecopreneurship") refers to ventures focused on environmental issues (green products, conservation, renewable energy, etc.), sometimes also called "green entrepreneurship." Sustainable entrepreneurship encompasses elements of both—it typically targets environmental and/or social objectives simultaneously—and emphasizes achieving these in tandem with economic success. Whereas a pure social enterprise might, for instance, operate as a nonprofit or accept lower returns in exchange for greater social impact, a sustainable enterprise generally strives to be financially self-sustaining or profitable, believing that economic success can go hand in hand with mission success. As one analysis notes, in social entrepreneurship, creating social value is paramount, and economic value serves as a means to sustain operations.

In contrast, sustainable entrepreneurship attempts to equally integrate social, environmental, and economic value creation (Dean & McMullen, 2007). In practice, the boundaries blur—many ventures could be seen as both social and sustainable entrepreneurship. The key nuance is that **sustainable entrepreneurship explicitly recognizes the interdependence of the three pillars (social, environmental, and economic)** and attempts to optimize for all rather than focusing on one while treating the others as mere constraints.

2.4 Foundational Theories Influencing Sustainable Entrepreneurship

Several theoretical perspectives provide insight into the behavioral dimensions of sustainable entrepreneurship:

- *Entrepreneurial Opportunity and Innovation*: Building on classic entrepreneurship theory (Schumpeter's innovation and Shane & Venkataraman's opportunity recognition), sustainable entrepreneurship is viewed as a **driver of innovation for sustainable development** (Hart & Milstein, 1999). Hart and Milstein (1999) argued that sustainability challenges can be "one of the biggest business opportunities in the history of commerce," suggesting that creative entrepreneurs will find innovative solutions that transform industries towards sustainability. **Recent evidence also shows that the technology level fundamentally shapes how entrepreneurs configure and leverage different resource types to achieve performance, particularly in high-tech and service ventures** (Chitsaz et al., 2020a, 2020b). This perspective motivates entrepreneurs to proactively seek out opportunities in areas like renewable energy, circular economy, or inclusive finance as both high-impact and high-potential ventures.
- *Market Failures and Externalities*: As mentioned, Dean and McMullen's opportunity-based view is rooted in economics, specifically, the idea that issues such as pollution, resource waste, or social injustice often stem from externalities or imperfect markets. Their theory posits that **entrepreneurs can profit by correcting these inefficiencies** (Dean & McMullen, 2007). For example, if incumbents overlook an environmental cost (such as CO_2 emissions) because it's not priced, a startup that finds a way to reduce or monetize those emissions (through carbon capture, for instance) can create value that incumbents have left on the table. This aligns doing good (reducing harm) with doing well (capitalizing on a market gap).
- *Institutional and Stakeholder Theory*: The context in which sustainable entrepreneurship occurs is shaped by institutions (laws, regulations, and cultural norms) and stakeholder expectations. Institutional theory suggests that the rules and standards of society influence entrepreneurs—for instance, stronger environmental regulations or social norms regarding sustainability can spur the development of more sustainable startups. At the same time, pioneering entrepreneurs can change institutions by introducing new norms or demonstrating new viable business models. **Stakeholder theory** emphasizes that sustain-

able entrepreneurs consider a broad range of stakeholders in their decision-making, often engaging with governments, NGOs, and communities as partners rather than viewing them solely as constraints or sources of pressure. This leads to behaviors such as greater transparency, community consultation, and partnership building, which we will discuss later.

- *Psychology and Values*: Sustainable entrepreneurship is inherently driven by values. A psychological perspective (also highlighted by Shepherd & Patzelt, 2011) examines the personal **values, attitudes, and motivations** of entrepreneurs who pursue sustainability (Shepherd & Patzelt, 2011). Why would an entrepreneur accept the possibly more challenging path of balancing multiple bottom lines rather than the simpler mandate of profit maximization? Answers often lie in individual ethics (e.g., a deep ecological awareness and empathy for a community) and identity (seeing oneself as a steward or change-maker). Foundational psychological theories like the **Theory of Planned Behavior** have been applied to study entrepreneurial intentions toward sustainability—positing that an individual's attitude toward sustainability (how important or rewarding they think it is), subjective norms (perceived social pressure or support), and perceived behavioral control (confidence in their ability to succeed) together shape their intention to start a sustainable venture (Peng et al., 2021). We will see evidence of this in Sect. 4.2, but suffice it to say that a strong pro-sustainability attitude and supportive social context are key antecedents to the behavior of launching a sustainable business.

In summary, sustainable entrepreneurship is grounded in the notion that business can be a powerful catalyst **for sustainable development.** It stands on theoretical pillars that stress holistic value creation (triple bottom line), turning problems into opportunities (entrepreneurial innovation addressing market failures), and the importance of human values and broader responsibility (stakeholder inclusion and ethical drivers). With these concepts and theories established, we can now delve into how they manifest in the behaviors and decision-making processes of entrepreneurs who embark on this path.

3 In-Depth Analysis: Behavioral Drivers, Evidence, and Examples

What motivates an entrepreneur to build a business around sustainability, and how do they behave differently from a conventional entrepreneur? Research in recent years has shed light on the **motivational and cognitive factors** that characterize sustainable entrepreneurs. In this section, we examine these factors—from personal values and intrinsic motivations to risk perceptions and opportunity recognition—and provide evidence and examples illustrating how they influence entrepreneurial action.

3.1 Intrinsic Motivations and Values

Studies consistently find that strong intrinsic motivations and a sense of mission often drive sustainable entrepreneurs. Many are **mission-driven** or purpose-driven individuals, meaning that beyond any financial incentives, they feel a passionate commitment to contributing to a cause (such as combating climate change, reducing waste, or empowering marginalized communities) (Yasir et al., 2021). Personal values play a central role: entrepreneurs with pronounced **pro-environmental values or social compassion** are more likely to find sustainable ventures. A 2021 meta-analysis of 40 empirical samples (with over 190,000 observations) *confirmed that environmental values are indeed positively related to sustainable entrepreneurial intention* (Peng et al., 2021). In other words, individuals who care deeply about environmental issues are significantly more inclined to intend to start a business that addresses those issues. These values often stem from personal experiences or education—

for example, a founder who grew up in an area affected by pollution might develop a lifelong commitment to environmental restoration, which later inspires their business idea.

Beyond values, sustainable entrepreneurs often demonstrate a strong sense of responsibility and ethics. They may feel personally accountable to future generations or to their community to conduct business in a manner that causes no harm (or actively benefits others). This sense of stewardship can manifest as a willingness to sacrifice or go against the grain for the greater good. For instance, many sustainable entrepreneurs impose voluntary constraints on themselves, such as using only fair-trade or eco-certified materials, refusing to engage in corrupt practices, or limiting their company's growth to ensure quality and impact, even if such choices might limit short-term profits. These behaviors are driven by internal moral standards as much as by external rules. However, studies suggest that relying too heavily on extrinsic rewards can lead to a decline in intrinsic motivation over time (Etemadi et al., 2024).

3.2 Attitudes and Entrepreneurial Intentions

In the realm of psychology, the **attitude** an entrepreneur holds toward sustainable business strongly influences their decision to pursue (or not pursue) a sustainable venture. According to the Theory of Planned Behavior (TPB), a positive attitude toward an activity increases the likelihood of engaging in it. Empirical evidence supports this in the context of sustainability: *positive views of being a sustainable entrepreneur have been shown to have a significant effect on the intention to become one* (Yasir et al., 2021). In practical terms, if an entrepreneur believes that running a sustainable business is desirable, feasible, and rewarding, they are much more likely to commit to that path. Success stories and role models often reinforce such attitudes. Research indicates that team-based motivation, combined with a design thinking mindset, significantly enhances innovative behavior among organizational members (Shiralian et al., 2024). Hearing of companies that achieved commercial success and social impact can strengthen one's belief that "doing well by doing good" is achievable.

Social psychologists also highlight the roles of **subjective norms** (perceived societal expectations) and **perceived behavioral control** (self-efficacy) in sustainable entrepreneurial intentions (Peng et al., 2021). For example, if an aspiring entrepreneur is surrounded by a community that values sustainability (peers, family, or mentors who encourage sustainable ideas), they may feel normative pressure and support to start a sustainable business. Likewise, if they have higher confidence in their skills and the viability of sustainable business models (perhaps due to relevant expertise or witnessing others' success), their perceived ability to make it work is higher, which boosts their intention. Interestingly, the meta-analysis mentioned earlier found that factors such as prior entrepreneurial experience can moderate these relationships—for less-experienced entrepreneurs, personal environmental values had a more pronounced effect on intention (Peng et al., 2021). This suggests that newcomers driven by passion can be very bold in starting sustainable ventures, whereas experienced entrepreneurs may temper passion with pragmatism, potentially seeking more evidence or supportive conditions before leaping in.

3.3 Opportunity Recognition and Cognition

Entrepreneurial behavior begins with recognizing an opportunity. Sustainable entrepreneurs tend to have a particular lens for seeing opportunities in places others might overlook or dismiss. Emerging research suggests that higher levels of mindfulness may support this process by enhancing the feasibility of idea generation, albeit at the cost of some novelty (Lotfi et al., 2024). They often engage in what some researchers call **"sustainability-oriented opportunity recognition"**—scanning for ways to solve environmental and social problems profitably. Cognitive studies (e.g., Muñoz & Dimov, 2015; as referenced in later work) have identified that sustainable entrepreneurs may employ different mental models or decision frameworks. For instance, they might simultaneously evaluate potential opportunities on multiple criteria (impact and profit), which requires more complex thinking and creativity to satisfy both. A 2017 study using cognitive mapping identified various "mental models" of sustainable decision-making among entrepreneurs, such as "*purpose-driven*" profiles, where entrepreneurs are very clear on a sustainability mission and determined to achieve it, versus "value-based, vacillating" profiles where entrepreneurs value sustainability but struggle with trade-offs (Hart & Milstein, 1999). Despite different styles, a common thread was that **no single factor alone** (be it values, knowledge, or opportunity attractiveness) was sufficient—it was the configuration of factors that led to the decision to pursue a sustainable opportunity (Hart & Milstein, 1999). This means sustainable entrepreneurs typically integrate multiple considerations (economic viability, ethical alignment, personal fulfillment, stakeholder support) when deciding on a venture idea.

One concrete behavioral manifestation is that sustainable entrepreneurs often engage in **extensive stakeholder** consultation and research early in the process. For example, an entrepreneur developing a clean cooking stove for rural areas might spend a significant amount of time with the target community, NGOs, and scientists to ensure the solution is culturally acceptable, environmentally sound, and economically affordable. This stakeholder-engaged opportunity development is a behavior less emphasized in classic entrepreneurship (which might focus more narrowly on market and customer rather than community and environment as stakeholders).

3.4 Risk Perception and Decision-Making

Every entrepreneur faces risk, but sustainable entrepreneurs may perceive and weigh risks differently. On the one hand, they face additional uncertainties—Will customers pay for a green product? Can we secure funding if our returns to investors might be slower? Are there regulatory risks or benefits?—which could make sustainable ventures seem riskier from a traditional standpoint. On the other hand, their personal commitment can make them more **risk-tolerant** in pursuit of their mission. Many are willing to accept a lower short-term financial reward in exchange for achieving a social or environmental goal, which might be viewed as a personal risk (opportunity cost) they are taking. Conversely, some research using **regulatory focus theory** suggests that sustainable entrepreneurs might adopt a prevention focus in certain decisions—being careful to avoid strategies that could harm stakeholders or the environment (a self-imposed constraint that could be seen as risk-averse behavior in terms of impact)—even while they are promotion-focused about seizing opportunities to do good. In sum, they juggle a dual risk calculus: *financial risk* (which they attempt to mitigate through innovation, lean startup practices, and so on) and mission risk (the risk of failing to deliver impact, which they mitigate by deeply integrating their values into operations). Their decision-making is often characterized by **longer-term thinking**—for example, investing in sustainable supply chains or energy efficiency now for future resilience rather than opting for the quick, cheaper fix. This long-term orientation is a behavioral hallmark aligned with sustainable development principles.

Examples of Behavioral Drivers in Action

To illustrate these points, consider a few real-world examples:

- *Motivation by Mission*: **EcoTech** (hypothetical name) is a startup founded by two engineers who were alarmed by electronic waste. They left high-paying jobs to develop a business around recycling e-waste into reusable materials. Their friends thought it was risky to leave corporate careers, but the founders' **intrinsic motivation** was clear: "We could not watch the e-waste problem grow and do nothing." They set modest salary expectations for themselves and reinvested profits in research and development (R&D) for cleaner recycling techniques.

Here, personal environmental values and a strong sense of purpose drove behaviors such as career risk-taking and reinvesting profits for impact.

- *AI for Sustainability*: In the realm of high-tech, many entrepreneurs are using **AI (Artificial Intelligence) as a tool to amplify sustainable solutions.** For instance, the founders of **Ellipsis Earth** developed AI-driven software to identify and map litter from drone images, helping cities pinpoint where trash accumulates and prevent it from reaching oceans. Their startup is tackling plastic pollution with cutting-edge technology. The behavior of these entrepreneurs demonstrates opportunity recognition in action—they recognized a global concern about ocean plastic and recent advances in AI vision and connected the dots to create a solution. They also embody a **global mindset**: the problem is worldwide, so they collaborate with NGOs and governments to deploy their tech. As the World Economic Forum noted, such innovators are "using AI to tackle global challenges, including the conservation and restoration of nature and protecting the ocean" (Torkington, 2023). The decision to focus on a planetary problem rather than a narrower commercial application of AI reflects their values and long-term vision.
- *Balancing Act (Profit and Principles)*: **Allbirds**, a globally recognized sustainable footwear startup, exemplifies sustainable entrepreneurial behavior in the consumer goods sector. The founders, Tim Brown and Joey Zwillinger, were motivated by the idea of creating a shoe that is not only comfortable and stylish but also environmentally friendly. They committed to using natural materials (like merino wool, eucalyptus fiber, and sugarcane) instead of conventional synthetic materials (Joyce & Paquin, 2016). This commitment came with higher material costs and supply chain challenges—a risk to profitability—but they believed that consumers would value sustainability and that volume and innovation would drive costs down over time. Their **marketing behavior** was also atypical: rather than glossing over their environmental impact, Allbirds has been transparent about their carbon footprint and even labels products with their CO_2 emissions. They engage customers in the mission by explaining how using wool reduces reliance on plastics, among other benefits. This transparency and **educational marketing** show how their values (authenticity, environmentalism) translate into behavior that builds trust and a community of loyal, like-minded customers. Indeed, Allbirds helped "change consumer behavior" by proving there is demand for sustainable fashion (Joyce & Paquin, 2016)—many of their customers report being more conscious of materials and sustainability in other products after being exposed to Allbirds' philosophy. Thus, entrepreneurs not only act according to their own principles; they also actively try to influence the behavior of others to promote a more sustainable lifestyle.
- *Persistence and Resilience*: Sustainable entrepreneurs often face skepticism or additional hurdles (e.g., convincing investors of a double-bottom-line return, navigating uncharted regulatory territory for a new sustainable product). A notable behavioral trait is **resilience and perseverance**. For example, entrepreneurs in the clean energy sector in developing countries, such as those introducing solar home systems to off-grid communities, frequently encounter setbacks like policy changes, technical failures, or community distrust. Many persist through years of pilot projects and refinements, driven by a belief in their impact mission. Their passion translates into patience and learning-oriented behavior. They gather feedback, iterate on their business models (perhaps using tools like the Triple Layered Business Model Canvas to adjust the economic, social, and environmental aspects of their model concurrently), and often adopt a **collaborative approach** (partnering with local cooperatives or microfinance institutions to reach users). This collaborative, systems-oriented behavior is a strategic choice that acknowledges the complexity of sustainable development challenges.

In summary, the behavioral profile of sustainable entrepreneurs tends to feature **high intrinsic motivation, strong prosocial and pro-environmental values, a broadened sense of responsibility, a long-term and stakeholder-inclusive outlook, and a willingness to innovate under constraints.** Empirical evidence supports that these internal drivers (values, attitudes) significantly shape their entrepreneurial actions (Peng et al., 2021). At the same time, external factors (such as community support, role models, and cultural context) and enabling technologies (like AI and clean tech) can significantly influence their trajectory. Understanding these behaviors is key for anyone looking to foster or become a sustainable entrepreneur—it's about mindset as much as strategy. Next, we will explore how these entrepreneurs and their ventures compare to others, what challenges they face, and how differing viewpoints debate the merits and difficulties of sustainable entrepreneurship.

4 Further Exploration: Comparisons, Debates, and Opposing Viewpoints

Having examined the core behaviors and drivers, we now broaden our perspective to compare sustainable entrepreneurship with other models and to examine the debates or critiques that have emerged. Sustainable entrepreneurship, as inspiring as it is, does not escape challenging questions: *Do sustainable ventures truly perform well financially, or do they inevitably sacrifice profit? Is sustainability just a buzzword some entrepreneurs use (i.e., "greenwash") to appeal to the market? Can small startups make a dent in global problems, or do we rely on big companies and policies?* In this section, we address such issues, providing a balanced view.

4.1 Sustainable Versus Traditional Entrepreneurship

A key point of comparison is how decision-making and success metrics differ between sustainable entrepreneurs and traditional entrepreneurs. A traditional entrepreneur might ask: *"What's the market size? How can we maximize profit and growth?"* A sustainable entrepreneur will ask those questions, **plus others**: *"What environmental footprint will this have? How will this benefit or impact the community? Is this solution ethical and future-proof?"* This broader questioning can be viewed in two ways. Proponents argue that it leads to better long-term decisions—by considering social and environmental factors, sustainable entrepreneurs anticipate risks and opportunities that others may overlook (for example, foreseeing that consumers will eventually demand eco-friendly alternatives, or avoiding future regulatory liabilities by being cleaner from the outset). Indeed, many innovative companies have demonstrated that adopting a sustainability lens can drive creativity and efficiency, rather than hinder them (Miller, 2020). For instance, designing a product to use less material and energy can cut costs (profit win) while also reducing environmental impact (planet win). As Harvard Business School research notes, "many firms have reaped financial benefits by committing to sustainable business practices" (Miller, 2020), countering the notion that there is an inherent trade-off between sustainability and profitability.

However, skeptics argue **that trade-offs do exist**. They point out that emphasizing sustainability can divert focus and resources. An opposing viewpoint might cite cases where a company invested heavily in green initiatives and failed commercially, suggesting that the entrepreneur lost sight of market realities. For example, during the early 2010s "cleantech" investment boom, several startups focused on renewable energy or biofuels went bankrupt—critics used this as evidence to claim that perhaps the founders were too idealistic about how quickly they could scale or how willing customers were to adopt new, pricier sustainable technologies. The counter-argument from the sustainable entrepreneurship side is that many of those failures were part of any innovation cycle (and lessons learned have led to today's successes) and that some ventures were ahead of their time rather than fundamentally flawed. Indeed, by the 2020s, a strong climate tech sector is emerging—global climate-tech venture funding reached $70 billion in 2022, nearly double the amount from **the prior year** (HolonIQ, 2023), indicating that markets are catching up with sustainability-driven opportunities.

4.2 Greening Goliaths Versus Emerging Davids

A comparison often discussed in the sustainability literature is between large incumbent firms that improve their sustainability (sometimes dubbed "greening Goliaths") and new entrepreneurial entrants with sustainability at their core ("emerging Davids") (Hockerts, 2006; Hockerts & Wüstenhagen, 2010). Both play a role in the transition to a sustainable economy (Freeman, 1984). Large companies have the resources and scale to make a significant impact if they change (think of a multinational retailer reducing waste across its supply chain), but legacy systems and profit pressures can hinder them. Entrepreneurs, on the other hand, are more agile and can introduce **disruptive innovations**—for example, a startup selling plant-based proteins can challenge the meat industry to reduce its environmental footprint. The interplay between these Davids and Goliaths can be complementary: sometimes, startups partner with or get acquired by big firms to spread sustainable innovations, and sometimes they pressure incumbents by capturing conscious consumers. Researchers Hockerts and Wüstenhagen (2010) theorized that the compounded impact of both is what drives industry-wide sustainable change (Hockerts & Wüstenhagen, 2010).

From a behavioral perspective, entrepreneurs (Davids) often have more radical visions and a **willingness to challenge the status quo** since they have less to lose and everything to gain. Incumbents may adopt a more cautious and incremental sustainability strategy to protect their brand and meet stakeholder expectations. Both approaches have merits, but they can be in tension—for example, an oil company might tout small investments in renewable energy (greening Goliath) while entrepreneurs in clean energy advocate for a rapid shift away from fossil fuels. Critics may accuse incumbents of "greenwashing" —making superficial sustainability claims or minor improvements to deflect more profound change—whereas entrepreneurs might be criticized for being naïve or unproven in their ability to scale up. Recent studies have also warned of the risks of opportunism and ethical ambiguity in emerging domains such as Web3 entrepreneurship, where the decentralization narrative can obscure accountability (Moezzi et al., 2025). The truth often lies in between, with genuine efforts and challenges on both sides.

4.3 Greenwashing and Authenticity

One debate that sustainable entrepreneurs face is ensuring their venture's claims and image are authentic. As sustainability has become a trend, there is concern that some businesses might label themselves "sustainable" without substantive action, essentially **marketing-driven greenwashing**. This puts pressure on truly sustainable entrepreneurs to **demonstrate credibility**. They often go to great lengths to obtain certifications (such as B Corp certification, fair trade labels, and organic certifications) or to publish impact reports. Such behavior is partly defensive (to differentiate from pretenders) and partly mission-driven (transparency and accountability are core values). From a critical perspective, one might ask: *How do we truly know that this startup is making a difference rather than just riding a trend?* Sustainable entrepreneurs respond by embracing third-party audits, engaging with independent evaluators, and basing their practices on scientific principles. For instance, a company claiming a low-carbon product will often conduct a formal life-cycle assessment and share the data. This level of disclosure represents a behavioral shift in entrepreneurship—it means being open not only about successes but also about areas for improvement. Entrepreneurs who truly embrace this tend to build trust with customers and partners, but they also expose themselves to scrutiny. The debate continues in the marketplace: savvy consumers are becoming alert to greenwashing and demand proof, which overall raises the bar and, in the long run, benefits authentic sustainable enterprises.

4.4 Profitability and Growth Versus Impact

A classical debate is whether pursuing sustainability dilutes a venture's focus on profitability and growth. Some investors historically believed that a dual mission could "cloud" an entrepreneur's judgment—for example, would they make a tough decision to pivot the business if the current model isn't yielding profit, even if they are emotionally attached to its social impact? Conversely, many impact investors today argue that considering environmental and social factors reduces certain risks and opens new markets, thus aligning with long-term profit. We are seeing more empirical evidence that **sustainable ventures can be competitive.** Many sustainable startups attract premium customers or enjoy greater customer loyalty. They may also benefit from talent attraction—employees, especially younger generations, often prefer to work for mission-driven companies, providing a labor advantage. These can lead to improved business performance. Additionally, as global policy and consumer trends shift (e.g., carbon pricing, bans on single-use plastics, ESG-oriented procurement), sustainable businesses might have lower regulatory and market risk compared to incumbents clinging to unsustainable practices.

However, it's fair to acknowledge the challenges: sustainable entrepreneurs often face funding gaps, while impact investment is growing, it remains a subset of total venture capital. They may need to work harder to prove their business model to financiers, striking a balance between impact metrics and traditional financial metrics. This has spurred an entire movement in developing impact measurement and management practices, enabling entrepreneurs to present a holistic performance report. A debate in this realm centers on how to measure success: should a sustainable startup report a blended value metric (a composite of financial and impact outcomes)? Traditionalists might resist, preferring pure financial indicators. The field is evolving with concepts like **IRIS metrics** and social return on investment (SROI) to provide structure, but a universal standard has yet to be established.

4.5 Global Perspective and Justice

Another perspective in the global context is the question of who qualifies as a sustainable entrepreneur and who benefits. Some argue that most high-profile sustainable startups come from developed countries or well-resourced founders, and their solutions might not always align with the needs of marginalized populations. For example, a sleek new electric vehicle startup is excellent for cutting carbon emissions among affluent consumers, but does it help a poor farmer adapt to climate change? Conversely, thousands of local entrepreneurs in developing nations are tackling issues such as clean cooking, water purification, and sustainable agriculture—but they often lack access to capital and markets to scale their impact. This raises debates about **inclusive entrepreneurship.** Global development organizations are increasingly supporting entrepreneurship as a means to achieve the UN Sustainable Development Goals (SDGs), with a focus on capacity building and funding for local solutions. A point often made is that sustainability has a justice component: sustainable entrepreneurship should also aim to empower those most affected by environmental and social challenges. This broadens the behavioral expectation—entrepreneurs are encouraged to incorporate equity into their mission (e.g., ensuring their business model benefits low-income communities or involves them as stakeholders). Some ventures adopt models, such as social enterprises or cooperatives, specifically to address this issue. The debate here isn't so much opposed viewpoints as it is an ongoing dialogue about making sustainable entrepreneurship a truly global and inclusive movement rather than a niche or elite endeavor.

4.6 Change Versus Case for Systemic Individual Initiative

Skeptics sometimes argue that *no number of sustainable startups can fix problems like climate change without systemic policy change.* They advocate that while these businesses are nice to have, decisive government action (carbon taxes, regulations, international agreements) is what's truly needed, and sometimes businesses even lobby against such policies. In response, sustainable entrepreneurs often become **advocates for policy change**, using their success stories as evidence that stricter environmental regulations or social standards are feasible and even beneficial. They may join coalitions pushing for higher sustainability standards in their industry. Indeed, many sustainable business leaders support policies like renewable energy incentives, extended producer responsibility laws, or social equity programs that level the playing field. This is a behavior somewhat unique—entrepreneurs are historically known for opposing regulation to maintain flexibility, but sustainable entrepreneurs frequently call for smarter regulation to institutionalize the good practices they voluntarily adopt. A balanced view recognizes that **systemic change and entrepreneurship are complementary**: entrepreneurs can innovate and demonstrate models on a small scale, and policy can amplify and universalize those that work. For example, countless small organic agriculture entrepreneurs helped prove that pesticide-free farming can be viable; policy changes and big players transitioning are now scaling that up.

In weighing these debates, it's clear that sustainable entrepreneurship is not a silver bullet, but it is a vital piece of the puzzle. **The behavioral evidence** suggests that with the right mindset and strategies, sustainable entrepreneurs can achieve competitive advantage and drive change. They often operate with a "business as a force for good" philosophy that is contagious, influencing consumers, competitors, and policy. As more success stories emerge, the old notion that sustainability and profitability are at odds is fading. Still, healthy skepticism ensures that sustainable businesses stay true to their claims and continuously improve. In the next section, we move from theory and debate to practice: what tools and strategies can sustainable entrepreneurs use, and how do they implement their ideals on the ground?

5 Practical Applications: Tools, Strategies, and Real-World Implementation

Translating the ideals of sustainable entrepreneurship into concrete business practice requires strategic tools and innovative approaches. In this section, we highlight practical frameworks and strategies that sustainable entrepreneurs use to build and grow their ventures, along with real-world illustrations of these practices. The focus is on actionable insights—**how to do** sustainable entrepreneurship effectively.

5.1 Sustainable Business Model Design

One of the first practical challenges is designing a business model that delivers triple-bottom-line value. Traditional business modeling tools have been adapted for this purpose. For example, the **Business Model Canvas**, a popular one-page blueprint for startups, has been extended into the **Triple Layered Business Model Canvas (TLBMC)** (Joyce & Paquin, 2016). The TLBMC adds two layers to the standard economic layer: an environmental layer (mapping the ecological costs and benefits of each part of the model, such as raw materials, energy use, and waste outputs) and a social layer (mapping stakeholder impacts social value propositions, employee well-being, etc.) (WDO, 2023). By visualizing all three layers, entrepreneurs can ensure alignment—for instance, if the economic layer relies on selling a product, the environmental layer examines the product's lifecycle (from sourcing to disposal), and the social layer examines how the product affects communities or how the enterprise treats its workers. Using such a tool encourages holistic thinking and helps identify potential conflicts or synergies early. Blockchain-based models, as seen in emerging energy startups, further enhance this systemic alignment (Abedpour et al., 2024). For example, a venture might realize that increasing product longevity (an environmental goal) could reduce repeat sales in the short term (an economic hit), so they pivot to a service or subscription model that benefits from durable products—turning a potential conflict into an innovative solution (as seen in companies that lease apparel or electronics and maintain them, rather than rely on quick turnover).

5.2 Lean Startup with Impact Metrics

Sustainable entrepreneurs often borrow methods from the lean startup methodology, such as build-measure-learn cycles and MVPs (minimum viable products), but integrate **impact metrics** alongside business metrics. When prototyping, they test not only customer acceptance and revenue streams but also monitor key sustainability indicators. For instance, a startup making biodegradable packaging will conduct small-scale pilots to see if the packaging meets customer needs and also decomposes as intended (maybe measuring decomposition rate or reduction in plastic use). Early incorporation of sustainability metrics helps avoid later-stage pivots that could be costly (for example, discovering late that your "green" product isn't that green would be problematic). Many entrepreneurs set up a dual set of KPIs: one for business (e.g., customer acquisition cost, profit margin) and one for impact (e.g., CO_2 saved per product, number of underserved beneficiaries reached). **In resource-constrained environments, such as those under economic sanctions, studies show that individual-level knowledge and networks become disproportionately important for venture performance, especially when financial capital is limited** (Chitsaz et al., 2017). This disciplined approach ensures that as they iterate the product or service, they don't improve one dimension at the expense of others.

5.3 Funding and Impact Investment

On the practical side of financing, sustainable entrepreneurs are increasingly tapping into the growing pool of **impact investors** and sustainable financing options. Impact investors are venture capitalists or funds that explicitly seek social and environmental returns in addition to financial returns. Entrepreneurs craft their pitch to show a compelling impact thesis (the specific change their business will create) along with the business case. Tools such as social impact bonds and **green crowdfunding platforms** have also emerged. For example, a renewable energy entrepreneur might utilize a crowdfunding platform that connects them with citizens who wish to invest in solar projects, thereby effectively raising capital from the community that will benefit from the project. Another strategy is pursuing grants or competitions focused on sustainability or the SDGs—many governments, foundations, and institutions run incubators or provide seed funding for sustainability-oriented startups. While these sources might not fully fund a company's growth, they can be critical in the early stages and often come with mentorship and networks tailored to impact-focused businesses.

5.4 Partnership and Ecosystem Strategy

Sustainable challenges are often too significant or systemic for one startup to tackle alone, so successful sustainable entrepreneurs skillfully build **partnerships.** Collaboration is a practical strategy: for instance, a startup that develops an app to reduce food waste by connecting restaurants with surplus food to charities or consumers (like the app "Too Good To Go") must network with restaurants, supermarkets, local nonprofits, and municipalities. By forming an ecosystem of partners, the startup greatly extends its reach and credibility. We frequently see sustainable ventures forming public–private partnerships, such as working with city governments on pilot programs (e.g., a clean water tech startup partnering with a city to install and test their systems in a neighborhood) or joining forces with NGOs who have grass-roots trust. These partnerships are often win–win: the entrepreneur gains access and legitimacy, and the partner gains an innovative solution to further their mission. Practically, entrepreneurs often dedicate significant time to stakeholder management—conducting workshops, co-design sessions, or MOUs (memoranda of understanding) with partners. While time-consuming, this **inclusive approach** can lead to more robust and accepted solutions. A real-world illustration is the expansion of solar lantern companies in Africa that partner with microfinance NGOs to reach remote villages; by themselves, the companies might struggle to distribute and finance the lanterns, but with partners, they leverage local knowledge and microloan systems to succeed.

5.5 Technology and Innovation as Enablers

Modern sustainable entrepreneurship often leverages cutting-edge technology to achieve impact at scale. We've mentioned AI earlier—entrepreneurs are also utilizing IoT (Internet of Things), blockchain, biotechnology, and other technological domains to promote sustainability. For instance, IoT sensors are used by agri-tech startups to optimize irrigation (saving water) and improve crop yields for farmers. Blockchain is being piloted to create transparent supply chains (e.g., proving a product's fair trade or sustainable origin by tracing it through a tamper-proof ledger) and to build decentralized, sustainable energy solutions (Abedpour et al., 2024). These technologies are tools; the practical skill is in **aligning the right technology to the sustainability problem**. A noteworthy trend is the emergence of digital platforms that enable the sharing economy—entrepreneurs have created platforms for sharing homes (Airbnb), rides (Lyft/Uber), tools, and more, which some argue increases resource efficiency (although debates on the actual sustainability impact exist). Still, the mindset of "access over ownership" that many such startups promote resonates with sustainability by potentially reducing overproduction. Entrepreneurs in this space focus on building user trust and critical mass—the behavior change of consumers (to share rather than own) is as important as the tech itself. Many sustainable entrepreneurs thus see themselves not just as tech providers but as behavior change agents, designing their products and services to nudge users toward sustainable habits. For example, an app might incorporate gamification to encourage users to compete on who saves the most energy or send friendly alerts about the positive impact of their actions, reinforcing the desired behavior.

5.6 Consumer Engagement and Education

Engaging customers in the mission is a practical strategy that not only builds brand loyalty but also amplifies its impact. Sustainable brands often utilize storytelling in their marketing—they share the story of their product's origins (such as the organic farm where cotton was grown or the artisan cooperative that crafted the bags) to create a connection and educate customers on why their purchase matters. Patagonia's famous "Don't Buy This Jacket" campaign is a prime example of unconventional marketing to promote sustainability values (Allchin, 2013). In that 2011 campaign, Patagonia ran ads encouraging consumers not to buy their products unnecessarily, advising them to purchase only what they need and to consider the environmental impact of consumerism (Allchin, 2013). This was part of their **Common Threads Initiative**, urging people to reduce, repair, reuse, and recycle. The bold move built immense brand respect and loyalty, demonstrating that authenticity in values can translate into customer goodwill (and ironically, it reinforced Patagonia's reputation, arguably contributing to sales growth in the long run). The lesson here for entrepreneurs is that **educating the market** may initially seem to discourage consumption, but in reality, it can attract a tribe of dedicated customers who become brand advocates. Many startups hold workshops, publish impact reports for their customers, or create online communities centered around their cause. This deepens the relationship beyond a transactional one. It's a shift from viewing a customer as just a buyer to viewing them as a partner in a movement for change.

5.7 Management and Culture

Internally, sustainable entrepreneurs cultivate company cultures that align with their mission and values. Recent research suggests that coaching-based interventions—especially those grounded in cognitive-behavioral frameworks—can strengthen entrepreneurial self-efficacy and help individuals overcome limiting beliefs that prevent social and business engagement (Ghiafeh Davoodi et al., 2025). This often includes practices such as democratic decision-making, employee volunteer programs, or environmentally friendly workplace operations. This often includes practices such as democratic decision-making, employee volunteer programs, or environmentally friendly workplace operations. For instance, a small eco-startup might implement zero-waste offices, utilize renewable energy, and offer employees paid days to participate in community service. These practices keep the team aligned with the purpose and can also be a selling point in recruitment (attracting talent that shares the values), remarkably when motivation is grounded in shared values rather than external incentives (Etemadi et al., 2024). Culturally, many sustainable startups encourage innovation from all staff members, since sustainability challenges can be complex, frontline employees often have valuable ideas for improvement. **However, as recent research shows, excessive team cohesion—especially under competitive pressure—can lead to an escalation of commitment and poor decision-making** (Khoshsoroor et al., 2019). A continuous improvement ethos (borrowed from quality management and applied to sustainability) can be ingrained; for example, teams routinely brainstorm ways to reduce their footprint further or increase their social impact, creating a cycle of innovation.

5.8 Real-World Illustration—Circular Economy Startup

Consider a startup in the **circular economy** space: it designs furniture made entirely from reclaimed wood and recycled materials and offers a take-back program that allows customers to return products at the end of their lives for the company to refurbish or recycle again. Practically, this venture had to solve several challenges: securing a steady supply of reclaimed materials, designing modular furniture that is easy to repair, and convincing customers of the quality and style of recycled products. The entrepreneurs formed partnerships with demolition companies and city waste departments to source reclaimed wood. They invested in design innovation to make furniture that can be disassembled (facilitating repair/upgrades). They also established an online marketplace to resell refurbished items, generating an additional revenue stream and reducing waste.

To communicate this, they brand themselves around the idea of "furniture that grows forests"—because every item sold helps save trees by using reclaimed wood. They share metrics: "X number of trees saved and Y tons of waste diverted this year," making customers feel part of an environmental impact. Financially, they charge a premium but highlight longevity (buy one table that lasts a lifetime rather than a cheap one you toss in 5 years). This example demonstrates multiple practical strategies, including circular design, partnerships for material sourcing, transparent impact communication, and a service model (take-back and resale) integrated into a product business. It exemplifies how sustainable entrepreneurs often need to **redesign the whole system of a product**, not just the product itself.

5.9 Leveraging Global Initiatives

Globally, frameworks such as the United Nations Sustainable Development Goals (SDGs) are used by many entrepreneurs as a guiding checklist. For example, an agritech startup might align with SDG 2 (Zero Hunger) and SDG 13 (Climate Action) by developing climate-resilient crops. By explicitly aligning with such goals, entrepreneurs can attract support from development agencies or corporate CSR programs, as well as measure their contributions to global targets. Some participate in accelerator programs focused on specific Sustainable Development Goals (SDGs) or in challenges like the **Solar Decathlon** or XPRIZE competitions for sustainability. These programs provide not only funding but also mentorship and recognition, which can serve as springboards for growth.

In implementing all these strategies, a common thread is that sustainable entrepreneurs must be **innovators and integrators**. They integrate environmental and social considerations into every aspect of business strategy—from product design, supply chain, production, and marketing to end-of-life. It's a more complex puzzle to solve, but the rewards are multifold: strong brand loyalty, alignment with emerging market trends, reduced regulatory risk, and, above all, the satisfaction of contributing to solutions for pressing global issues.

With practical approaches in hand, we will now conclude the chapter, summarizing key learnings and looking ahead to how sustainable entrepreneurship might evolve. Following the conclusion, you can apply what you've learned by examining the case study of Patagonia and tackling the discussion and review questions.

6 Conclusion

In this chapter, we have undertaken a comprehensive exploration of the **behavioral dimensions of sustainable entrepreneurship**. We began by defining sustainable entrepreneurship as an approach to business that integrates economic, environmental, and social value creation, aiming for a triple bottom line of profit, people, and planet (Miller, 2020). Foundational concepts, such as sustainable development (World Commission on Environment & Development, 1987) and stakeholder theory, provided context for why this form of entrepreneurship has emerged and why it is globally relevant. Sustainable entrepreneurs, at their core, are individuals who view the world's social and environmental challenges not only as problems but also as opportunities for innovation and positive change (Dean & McMullen, 2007).

One of the key takeaways is that **behavior and mindset are crucial differentiators** of sustainable entrepreneurs. They are typically driven by strong intrinsic motivations, ethical values, and a vision that extends beyond the immediate financial payoff (Peng et al., 2021). We saw evidence that personal attitudes toward sustainability and perceived support from others significantly influence one's intention to start a sustainable venture (Yasir et al., 2021). In practice, these entrepreneurs demonstrate behaviors such as long-term thinking, stakeholder engagement, and resilience in pursuing their mission. They redefine success to include social and environmental impact and often engage in transparent and accountable practices to ensure their businesses live up to their claims.

We also discussed the dynamic interplay between sustainable entrepreneurs and the broader ecosystem. Comparisons between sustainable and traditional entrepreneurship highlighted both the advantages (such as innovation and brand loyalty through purpose) and challenges (including managing dual objectives and avoiding greenwashing) of the sustainable approach. Debates were examined, but importantly, we see that many presumed trade-offs between sustainability and profitability can be reconciled with creative business models and growing market support for sustainable products (Miller, 2020). The rise of impact investment and global sustainability awareness indicates that the environment for sustainable entrepreneurship is becoming more enabling each year.

On a practical level, the chapter provided strategies and tools, including the use of frameworks such as the Triple Layered Business Model Canvas for holistic planning, leveraging technology (AI, IoT, etc.) for scaling impact, and implementing partnerships and community-building as part of business growth. Real-world examples, capped by the Patagonia case study, illustrate that these concepts are not just theoretical—they are actively shaping businesses that succeed in competitive markets while driving positive change. Patagonia's story, in particular, showed how a deep commitment to environmental and social principles can permeate every aspect of a company's behavior, creating a brand that is both highly respected and impactful (Allchin, 2013; McCormick, 2022).

In reinforcing **our learning objectives**, you should now be able to clearly define sustainable entrepreneurship, distinguishing it from other entrepreneurial approaches and relating it to key concepts such as the triple bottom line and sustainable development. You can identify and discuss the **behavioral drivers** (values, attitudes, motivations) that propel sustainable entrepreneurs, backed by research findings (e.g., the role of environmental values in entrepreneurial intentions; Peng et al., 2021). You are aware of the **challenges and debates**, such as the need to maintain authenticity and the question of balancing profit with purpose, and can articulate both sides of these issues. Moreover, you have learned about **practical methods** for creating and running a sustainable venture, from business model innovation to stakeholder engagement and marketing ethics.

Finally, this chapter invites reflection on the future: As global challenges intensify and technologies advance, sustainable entrepreneurship is likely to become even more critical. Entrepreneurs will play a crucial role in achieving climate targets, promoting social justice, and building economic resilience. The behavioral traits discussed—innovation, empathy, ethical leadership, and collaboration—will be in high demand. We encourage you to reflect on how you might apply these insights. Consider the discussion questions and the case study that follows; they are designed to deepen your understanding and enhance your ability to use the chapter's concepts. Whether you aim to be a sustainable entrepreneur yourself or support such initiatives as an investor, policymaker, or consumer, understanding the human and behavioral aspects of sustainable entrepreneurship will empower you to contribute effectively to this global movement for a more sustainable and equitable future.

7 Case Study: Patagonia—Balancing Profit and Purpose at an Outdoor Apparel Company

Background and Context: Patagonia, Inc. is a California-based outdoor clothing and gear company founded in 1973 by **Yvon Chouinard**, a passionate rock climber and environmentalist. What began as a small business making metal climbing pitons eventually grew into a globally recognized brand for outdoor apparel. From the very start, Chouinard infused Patagonia with an unconventional philosophy: *build the best product, cause no unnecessary harm, and use business to inspire and implement solutions to the environmental crisis*. This mission statement (which Patagonia proudly publicizes) set the tone for the company's behavioral ethos (Joyce & Paquin, 2016). Unlike many entrepreneurs of his era, Chouinard did not seek aggressive growth at all costs—in fact, he was skeptical of traditional corporate culture and was reluctant even to call himself a businessman. Yet, over the decades, Patagonia has become highly successful, demonstrating that a company can be profitable while staying true to its **deep environmental values.**

8 Sustainability Initiatives and Entrepreneurial Behavior

Patagonia's history is filled with examples of pioneering sustainable entrepreneurship practices:

- **Product Innovation and Materials**: In the 1980s, Patagonia recognized that conventional cotton farming practices involved the use of harmful pesticides. To reduce this environmental harm, Chouinard made a bold decision: Patagonia would switch to 100% organic cotton for all its cotton clothing lines, even though organic cotton was more expensive and the supply was limited. This transition, completed in the 1990s, was a risky endeavor—it threatened short-term profits and required the establishment of new supply chains. However, it showcased the company's willingness to take financial risks in pursuit of **environmental principles.** Similarly, Patagonia invested in recycling—it was one of the first apparel companies to make fleece jackets from recycled plastic bottles. By focusing on durable, high-quality products and utilizing materials such as recycled polyester and organic wool, Patagonia has aligned its product development with sustainability. As their marketing often notes, "Quality is sustainability"—a jacket that lasts decades is far better for the planet (and customers' wallets) than one that wears out in a year (Allchin, 2013; Joyce & Paquin, 2016).

- **"Don't Buy This Jacket" Campaign**: On Black Friday 2011, Patagonia ran a now-famous full-page ad in The New York Times with the headline **"Don't Buy This Jacket."** The advertisement featured one of Patagonia's best-selling jackets. It implored customers to think twice before purchasing new gear, explaining the environmental costs of making the jacket (water, energy, and carbon emissions) and urging them to buy only if necessary (Allchin, 2013; Medium, 2020). This seemingly paradoxical marketing move was part of Patagonia's **Common Threads Initiative**, a partnership with customers aimed at reducing consumption by repairing and reusing products. Many thought Patagonia was crazy to tell customers not to buy; however, the campaign struck a chord with them. It exemplified authentic communication of values—Patagonia was willing to potentially dampen sales on the biggest shopping day to deliver an environmental message. The result? Patagonia's brand reputation soared for its honesty and integrity. Sales eventually grew, but more importantly, a community of loyal customers grew even stronger, trusting that Patagonia truly prioritized the planet over profit. This case demonstrates how **driving consumer** behavior change (toward conscious consumption) became a part of Patagonia's entrepreneurial strategy (Allchin, 2013).
- **Corporate Activism and Community Engagement**: Patagonia has never been a neutral actor in environmental issues. Chouinard and his team have a history of activism, from campaigning for the protection of wilderness areas to addressing climate change. The company donates 1% of its sales (not just profit, but sales) to environmental nonprofits every year, earning recognition as a leader in corporate philanthropy (they co-founded the "1% for the Planet" alliance). Patagonia employees are encouraged to be activists—the company will bail employees out of jail if they are arrested protesting environmental injustices. This level of encouragement shows how **deeply the values are embedded in the culture**. The company also instituted "Let My People Go Surfing" flextime, acknowledging the importance of work-life balance and allowing employees to go surfing when the waves are good, reflecting Chouinard's belief in the quality of life and connection to nature. These practices attract employees who are passionate and engaged, creating a virtuous cycle: employees join Patagonia because of its mission, and in turn, they strengthen that mission by contributing ideas and energy.
- **Supply Chain Responsibility**: In line with stakeholder principles, Patagonia has worked to ensure fair labor practices throughout its supply chain, participating in initiatives to audit factories for safety and fair wages. They also invest in sustainable technology—for example, improving dyeing processes to reduce water use. While not perfect (like any global company, Patagonia has confronted issues), it tends to respond to problems with transparency. In one case, when the company discovered labor abuses in a supplier's mills, it publicly acknowledged the issue and took corrective steps. This accountability fosters trust with consumers and sets a standard in the industry.
- **Recent Bold Move—Giving the Company Away**: In September 2022, Yvon Chouinard

made global headlines by announcing that he and his family were transferring 100% ownership of Patagonia to two new entities designed to ensure the company's profits are used to combat climate change and protect nature. Essentially, "Earth is now our only shareholder," Patagonia declared (McCormick, 2022). The Chouinard family donated the bulk of the company (valued at roughly $3 billion) to a newly established nonprofit called the Holdfast Collective, which will receive about 98% of Patagonia's future profits to fund environmental work. The remaining shares were placed in a trust to ensure the company stays true to its values and never drifts toward unsustainable practices or traditional public ownership. This unprecedented decision highlights the end of sustainable entrepreneurship behavior—prioritizing purpose so strongly that even ownership and profit distribution are engineered entirely for impact. Chouinard, at 83, said he "never wanted to be a businessman," and with this move, he ensured Patagonia would continue its mission beyond his lifetime (McCormick, 2022). The business continues to operate as a for-profit entity (to remain healthy and competitive), but its ultimate purpose is firmly rooted in being a nonprofit: supporting the planet.

Outcomes and Reflections

Patagonia is today a $ 1 billion+ per year revenue company (approximately), with strong profitability and a beloved brand. It has proved that an unwavering commitment to sustainability can be compatible with business success—indeed, it can be a differentiator that secures customer loyalty and attracts top-tier talent. Patagonia's story provides several key insights:

- **The personal values of the founder set the tone**: Yvon Chouinard's love for nature and reluctance to harm it became Patagonia's North Star. His willingness to break the mold (e.g., telling customers to buy less or giving away the company's wealth) exemplifies how authentic leadership behavior can drive a company to blaze new trails in sustainability.
- **Consistency and transparency build trust**: Patagonia did not just have one green product line or one campaign; it consistently integrated sustainability into everything from materials to marketing to finance. When mistakes occurred, they owned up to them. This consistency over decades earned them credibility that marketing alone cannot buy.
- **Innovation is key to solving trade-offs**: Patagonia encountered many instances where doing the right thing was hard or costly (like switching to organic cotton). They often had to innovate—find new suppliers, develop new materials, or persuade customers to accept changes. Their success came from being persistent and creative in resolving the tension between sustainability and business demands.
- **Influence beyond the company**: Perhaps Patagonia's most significant impact lies in its influence on other businesses and consumers. It helped popularize ideas such as corporate transparency, repairing goods, and high-quality, sustainable products. Many entrepreneurs cite Patagonia as an inspiration, and large companies have been pressured to follow its lead (for example, other apparel brands have increased their use of recycled materials). Patagonia shows that **a mission-driven company can "punch above its weight"** in changing industry norms and public expectations.

9 Discussion Questions (Patagonia Case)

1. *Mission versus. Marketing*: Patagonia's "Don't Buy This Jacket" ad was essentially encouraging customers to buy less. How did this seemingly anti-commercial move ultimately benefit the company? What does this tell you about the importance of authenticity in sustainable entrepreneurship?
2. *Risk and Reward*: Identify a decision Patagonia made that posed a significant business risk in favor of sustainability (for example,

the switch to organic cotton or the donation of company ownership). What were the short-term and long-term impacts of that decision? Would you have made the same choice as an entrepreneur, and why or why not?

3. *Scaling Impact*: Patagonia is a single company, albeit an influential one. In what ways has Patagonia tried to scale its impact beyond its operations (consider initiatives like 1% for the planet or encouraging other businesses and policies)? Discuss how an entrepreneurial company can act as a catalyst for broader change in its industry or community.
4. *Challenges*: Despite Patagonia's success, the company still faces challenges (e.g., producing synthetic fleece creates microplastic pollution, and selling more products can conflict with reducing consumption). What strategies could Patagonia (or a similar company) employ to navigate the inherent tensions between operating a profitable retail business and promoting environmental conservation?

Patagonia's journey illustrates many principles of sustainable entrepreneurship in action—from the powerful role of values-driven leadership to the innovative tactics that can align mission with margin. By examining such cases, aspiring sustainable entrepreneurs can glean lessons on how to lead with purpose and integrity and how to turn bold ideas into practical outcomes.

References

Abedpour, A., Chitsaz, E., & Rostami Asrabadi, S. (2024). Designing a business model for energy management and optimization in Iran using blockchain technology. *Education and Management of Entrepreneurship, 2*(4), 59–74. https://doi.org/10.22126/eme.2023.2505

Allchin, J. (2013, January 23). Case study: Patagonia's 'Don't buy this jacket' campaign. *Marketing Week.* https://www.marketingweek.com

Chitsaz, E., Khoshsoroor, S., Salami, G. E., & Song, M. (2020). Technology role in bundling and configuration resources in new ventures. *International Journal of Innovation Management, 24*(2). https://doi.org/10.1142/S1363919620500237

Chitsaz, E., Khoshsoroor, S., Salami, G. E., & Song, M. (2020b). Technology role in bundling and configuration resources in new ventures. *Journal of Organisational Studies and Innovation, 7*(2), 16–31.

Chitsaz, E., Liang, D., & Khoshsoroor, S. (2017). The impact of resource configuration on Iranian technology venture performance. *Technological Forecasting and Social Change, 122*, 186–195. https://doi.org/10.1016/j.techfore.2016.03.009

Dean, T. J., & McMullen, J. S. (2007). Toward a theory of sustainable entrepreneurship: Reducing environmental degradation through entrepreneurial action. *Journal of Business Venturing, 22*(1), 50–76. https://effectuation.org

Etemadi, M., Chitsaz, E., & Ghodratizadeh, F. (2024). The paradox of rewards: Reconsidering employee satisfaction and performance. *Journal of Sustainable Human Resource Management, 6*(10), 71–87. https://doi.org/10.22080/SHRM.2024.4599

Freeman, R. E. (1984). *Strategic management: A stakeholder approach.* Pitman.

GEM Consortium. (2025). *Global entrepreneurship monitor 2025 global report*. Global Entrepreneurship Research Association. https://www.gemconsortium.org

Ghiafeh Davoodi, Z., Chitsaz, E., & Seyfikar Qomi, M. (2025). Overcoming invisible barriers: How coaching brings new mothers back to the social and entrepreneurial stage. *Women in Development and Politics, 56*(1), 1–20.

Hart, S. L., & Milstein, M. B. (1999). Global sustainability and the creative destruction of industries. *Sloan Management Review, 41*(1), 23–33. https://www.researchgate.net

Hockerts, K. (2006). Entrepreneurial opportunity in sustainable development: A resource-based perspective. In *Frontiers in entrepreneurship research* (Vol. 26, Issue 8, Article 3). Babson College.

Hockerts, K., & Wüstenhagen, R. (2010). Greening Goliaths versus emerging Davids—Theorizing about the role of incumbents and new entrants in sustainable entrepreneurship. *Journal of Business Venturing, 25*(5), 481–492. https://research.cbs.dk

HolonIQ. (2023, January 3). Defying gravity, 2022 climate tech VC funding totals $70.1B, up 89% on 2021. https://www.holoniq.com

Joyce, A., & Paquin, R. L. (2016). The triple layered business model canvas: A tool to design more sustainable business models. *Journal of Cleaner Production, 135*, 1474–1486. https://bookdown.org

Khoshsoroor, S., Liang, D., Ebrahimbay Salami, G., & Chitsaz, E. (2019). Team cohesion: A double-edged sword in the relationship between rival estimation and escalation of commitment. *Social Behavior and Personality, 47*(9), Article e8122. https://doi.org/10.2224/sbp.8122

Lotfi, M. A., Chitsaz, E., Rasoulian, P., & Talebi, K. (2024). Investigating the effect of mindfulness and alexithymia on the generation of new business ideas. *Psychological Researches in Management, 10*(3), 59–83. https://doi.org/10.22034/jom.2024.2036049.1241

McCormick, E. (2022, September 15). Patagonia's billionaire owner gives away company to fight climate crisis. *The Guardian.* https://www.theguardian.com

Medium. (2020). Patagonia's "Don't Buy This Jacket" Campaign: The paradox of anti-consumption branding. https://medium.com

Miller, K. (2020, September 28). The triple bottom line: What it is & why it's important. *Harvard Business School Online.* https://online.hbs.edu

Moezzi, H., Chitsaz, E., & Majidi, M. (2025). The dark side of Web3 entrepreneurship: An exploratory Delphi study on factors and implications. *Journal of Digital Innovation and Ethics, 3*(1), 22–40.

Muñoz, P., & Dimov, D. (2015). The call of the whole in understanding the development of sustainable ventures. *Journal of Business Venturing, 30*(4), 632–654. https://doi.org/10.1016/j.jbusvent.2014.07.012

Peng, H., Li, B., Zhou, C., & Sadowski, B. M. (2021). How does the appeal of environmental values influence sustainable entrepreneurial intention? *International Journal of Environmental Research and Public Health, 18*(3), 1240. https://pmc.ncbi.nlm.nih.gov

Shepherd, D. A., & Patzelt, H. (2011). The new field of sustainable entrepreneurship: Studying entrepreneurial action linking "what is to be sustained" with "what is to be developed." *Entrepreneurship Theory and Practice, 35*(1), 137–163.

Shiralian, M., Ghodratizadeh, F., Talebi, K., & Chitsaz, E. (2024). Team-based motivation and design thinking in sustainable innovation behavior. *Journal of Innovation and Entrepreneurship Studies, 12*(3), 45–62.

Torkington, S. (2023, September 14). Trees, oceans and mental health: 3 ways entrepreneurs are using AI to solve global challenges. *World Economic Forum.* https://www.weforum.org

WDO. (2023). Triple layered business model canvas by World Design Organization. https://bility.wdo.org

Wikipedia contributors. (n.d.). Triple bottom line. *Wikipedia.* https://en.wikipedia.org/wiki/Triple_bottom_line

World Commission on Environment and Development. (1987). *Our common future.* Oxford University Press. https://www.britannica.com

Yasir, N., Mahmood, N., Mehmood, H. S., Rashid, O., & Liren, A. (2021). The integrated role of personal values and theory of planned behavior to form a sustainable entrepreneurial intention. *Sustainability, 13*(16), 9249. https://www.mdpi.com

Future Research Directions

A Holistic Review of the Sustainable Entrepreneurship

Maryam Khodayari, Morteza Akbari, and Pantea Foroudi

Abstract

Recently, there has been growing scholarly interest in sustainable entrepreneurship (SE), particularly in the context of advancing the United Nations' Sustainable Development Goals (SDGs) set for 2030. This study aims to perform a bibliometric analysis of the academic literature on SE. Drawing on data extracted from the WoS database, the research provides a detailed overview of the intellectual landscape within the field. The bibliometric methodology involves analyzing publications from 1900 to 2025, utilizing tools such as VOSviewer, Bibexcel, SPSS, and Datawrapper. By mapping key elements—such as influential articles, prominent authors, leading journals, and major research themes—this study contributes to the existing body of knowledge on SE. It identifies thirteen research clusters using Multidimensional Scaling (MDS) and four through Hierarchical Cluster Analysis (HCA), offering insights into potential directions for future research. The most significant themes in this field include sustainability-oriented innovation, dynamic capabilities, sustainability, and the redefinition of traditional entrepreneurial paradigms. Lastly, the research presents theoretical contributions, acknowledges the current study's limitations, and recommends directions for further exploration.

Keywords

Sustainable entrepreneurship · Bibliometrics · Multidimensional scaling · Hierarchical cluster analysis

Learning Objective

By the end of this chapter, readers will be able to:

- To explore the intellectual foundations of the sustainable SE literature
- To identify the core intellectual traditions that have significantly shaped the sustainable entrepreneurship field

1 Introduction

As discussions around sustainability become increasingly important in society—highlighted by agreements like the Paris Agreement and the United Nations Sustainable Development Goals (SDGs)—this topic is also gaining traction in

M. Khodayari · M. Akbari (✉)
Faculty of Entrepreneurship, University of Tehran, Tehran, Iran
e-mail: mortezaakbari@ut.ac.ir

M. Khodayari
e-mail: Maryamkhodayari@ut.ac.ir

P. Foroudi
Brunel Business School, London, UK
e-mail: pantea.foroudi@brunel.ac.uk

M. Akbari et al. (eds.), *Sustainable Entrepreneurship*, University of Tehran Science and Humanities Series, https://doi.org/10.1007/978-3-032-04909-4_25

academic research. Sustainable entrepreneurship (SE) is vital in promoting a more sustainable environment.

Schaltegger and Wagner define SE as realizing sustainability innovations for the mass market that benefit society (Schaltegger & Wagner, 2011). This leads to an exploration of "sustainable innovations." Boons notes that while there is no definition, sustainable innovation, often linked to eco-innovation, is described as "innovation that improves the performance of sustainability" (Boons & Lüdeke-Freund, 2013), addressing ecological, economic, and social issues. Varadarajan identifies three types of sustainable innovations: decreasing resource use, eliminating resource use, and replacing resources (Varadarajan, 2017). Additionally, Albort-Morant et al. (2016) define it as the integration of conservation and development to ensure the well-being of all people.

SE focuses on preserving nature, supporting life, and fostering community while seeking opportunities to create beneficial products, processes, and services, both economically and otherwise (Muñoz & Cohen, 2018). Cohen and Winn (2007) describe SE as identifying opportunities for goods and services while examining their broader implications. Belz and Binder (2017, p. 2) add that SE involves recognizing and seizing opportunities to create socially, economically, and ecologically beneficial goods and services. A core principle of SE is that entrepreneurs should not harm the social and ecological environments in which they operate (Muñoz & Cohen, 2018). SE often starts by identifying social or ecological challenges and devising innovative solutions (Eller et al., 2020). Sustainability start-ups distinguish themselves from traditional ones through their value-based approach and commitment to driving social and environmental change (Bocken, 2015).

We conducted a bibliometric analysis to assess the state of scientific and academic research in SE. Previous bibliometric studies in this domain include work by Moya-Clemente et al. (2021) and Benavides-Sánchez et al. (2022), which examined research trends up to 2025. This paper aims to provide an updated overview of the literature published in SE through 2025. Given the increasing urgency of this topic in political and societal contexts, over 400 new research papers have been published between 2021 and 2025. Our research aims to provide a more comprehensive and up-to-date representation of this field. Additionally, some bibliometric analyses have targeted specific aspects of SE; for instance, Anand et al. (2021) explored trends and patterns in SE, while Andrade-Valbuena et al. (2022) presented an overview of SE in the tourism and hospitality sectors.

Despite the emergence of recent studies (e.g., Benavides-Sánchez et al., 2022; Moya-Clemente et al., 2021), the present research suggest an updated and comprehensive analysis with two main objectives: (1) to explore the intellectual foundations of the sustainable SE literature through co-citation analysis (CCA), thereby identifying the core intellectual traditions that have significantly shaped the field; and (2) to propose potential avenues for future research. To fulfill these aims, a multi-method bibliometric approach was used. Initially, the current landscape of SE publications was evaluated by analyzing the distribution of documents across countries, identifying the most cited articles, and examining the leading journals through citation analysis (CA). Subsequently, CCA was conducted to map a wide array of thematic areas, utilizing multidimensional scaling (MDS) and hierarchical cluster analysis (HCA) to uncover underlying research structures.

This paper reviews 3331 articles published in March 2025. To overcome the contradiction and fill research gaps, the present study seeks to explore the structure and nature of SE, articulate SE research trends, and examine their implications for academics and practitioners. The findings show that key issues in this area include Sustainability-oriented innovation (SOI), The dynamic capabilities, Sustainability, The process of SE, Sustainable business models, Innovation challenges for sustainable development, SDGs, Entrepreneurial Opportunities arising from Market Failures, The green prison, The strategic link between sustainability and innovation, SE as a systematic response to ecological and social challenges, Institutional transformation and the redefinition of incentive structures, Redefine the concept of entrepreneurial opportunities.

Research on SE examines how enterprises strive to harmonize economic profitability with environmental stewardship and social responsibility. This field provides valuable insights for academics, industry professionals, and policymakers seeking to promote sustainable development through innovative business practices. It enhances theoretical frameworks, fosters interdisciplinary collaboration, and provides practical tools for implementing sustainable practices, addressing challenges like upfront costs, and tackling global issues such as climate change and inequality. SE research also informs structural reforms and public policies, contributing to a sustainable economy while redefining business success as a balance of economic, social, and environmental impact.

2 Method

2.1 Database

Bibliometric analysis employs different techniques to systematically examine the structural characteristics of knowledge domains (Mora et al., 2017). As defined by Groos and Pritchard (1969), it involves "the application of mathematics and statistical methods to books and other media of communication." This method encompasses a range of analytical approaches, including citation analysis (CA), keyword co-occurrence, co-citation analysis (CCA), and co-authorship analysis (Dias, 2019). By adopting a quantitative framework, bibliometric analysis enables the description, evaluation, and interpretation of existing scholarly literature in a comprehensive and objective manner.

The statistics collected in the present study were collected from the Clarivate Analytics–WoS database. Despite many internationally branded databases, such as Scopus and Google Scholar, the Web of Science database (WoS) has the highest research standards (Merigó et al., 2015). In many articles that use the bibliographic method to analyze their specific topics, the WoS database is more popular (Akbari et al., 2020b; Hsiao & Yang, 2011; Rodríguez-Bolívar et al., 2018; Sánchez-Pérez et al., 2020; Wamba & Queiroz, 2020; Xie et al., 2020; Zhu Liu et al., 2020). Covering a variety of bibliometric analyses, the WoS was the main foundation of many studies that provide wide attention in other sciences (Akbari et al., 2020a; Mulet-Forteza et al., 2019).

2.2 Procedure and Data

We began the bibliographic evaluation of SE literature by searching for articles using the keywords "sustain* entrepreneur*", "sustain* venture*", "sustain* start-up", and "sustain* innovat*". The Web of Science (WoS) was selected as the source for article extraction due to its reliability in processing citation data (Khodayari et al., 2025). On March 10, 2025, 4521 records were found. After refining the search, the number of records was reduced to 3441. A further filter for English-language articles resulted in 3331 extracted articles.

2.3 The Analytical Methods and Software

The present RESEARCH utilizes four analytical methods to explore the SE literature. CA serves to assess the research activity within the field, identify influential journals, evaluate the contributions of established scholars, and highlight emerging trends and potential directions for future inquiry. Additionally, CCA provides a comprehensive examination of the intellectual foundations of SE. By analyzing the frequency with which pairs of documents are cited together, this method enables the identification of conceptual linkages and the delineation of key knowledge domains within the field. Two other co-citation techniques, Hierarchical Cluster Analysis (HCA) and Multidimensional Scaling (MDS), are also utilized. These methods address validity concerns while offering insights into the co-citation data. Table 1 discusses the relationships among the various analyses conducted more specifically.

Table 1 Summary of bibliometric analysis

Methods	Description	Pros	Cons	Examples
CA	Evaluate the impact of documents, authors, or Journals by citation rates	Rapidly identify key publications in the field	Because newer publications have had less time to be cited, the number of citations as a measure of influence is biased toward older publications	(Akbari et al., 2020a; Di Guardo & Harrigan, 2012; Dzikowski, 2018; Ertz & Leblanc-Proulx, 2018; Fetscherin & Heinrich, 2015; Rey-Martí et al., 2016)
CCA	Documents, authors, and journals are linked based on their joint appearances in reference lists	This is the most reliable and widely used method of bibliometrics. It effectively connects documents, authors, and journals with source citations Subsequently, citations serve as a measure of impact, providing a means to filter the most significant works	Co-citation occurs in cited papers; therefore, it is not suitable for mapping study fronts Citations take time to accumulate; thus, new documents cannot be linked directly, but only through knowledge base groups Many citations are necessary to map papers; therefore, it is impossible to map papers that are not frequently cited	(Gurzki & Woisetschläger, 2017; Kumar et al., 2019; Lara-rodríguez et al., 2019; van Oorschot et al., 2018)
MDS	Create a map based on the closeness matrices to analyze the primary structure	Identifying the groups being studied in the analysis	There are no definitive rules for interpreting results; it remains limited to a small data set	(Di Guardo & Harrigan, 2012; Foroudi et al., 2020; Marcussen, 2014; Pilkington & Meredith, 2009; Shiau & Dwivedi, 2013)
HCA	To identify the sub-groups in the dataset	It provides the generalization that underlies the result It is an agglomerative method	No definitive rule for interpreting results	(Choi & Seo, 2021; Foroudi et al., 2020; Hepsen & Vatansever, 2012; Zha et al., 2021)

Citation analysis (CA)—After selecting the appropriate keywords, the subsequent steps in bibliographic research involve clearly defining the scope of the study to ensure a comprehensive and focused investigation. In the current study, we used a CA as a bibliographic method. CA helps to illustrate the intellectual landscape of a particular field and enables researchers to identify the foundational documents by highlighting the most frequently cited works (Pasadeos et al., 1998).

Co-citation analysis (CCA)—This study employed CCA to identify subdomains within the field of sustainable entrepreneurship (SE). Relevant documents were sourced from the Web of Science (WoS) database and processed using Bibexcel software to conduct the bibliographic analysis, thereby establishing a robust foundation for CCA. Utilizing Bibexcel, the 30 most frequently cited documents were selected, and a co-citation matrix was constructed. This matrix served as the foundation for performing multidimensional scaling (MDS), a technique that enables researchers to accurately map the intellectual structure of a research field. As highlighted by Chabowski (2017), the co-citation matrix reflects the interrelationships among scholarly publications within a given domain. MDS was applied in this study as a comprehensive method for evaluating citation data and visually presenting the underlying patterns in the bibliometric landscape (Chabowski, 2017).

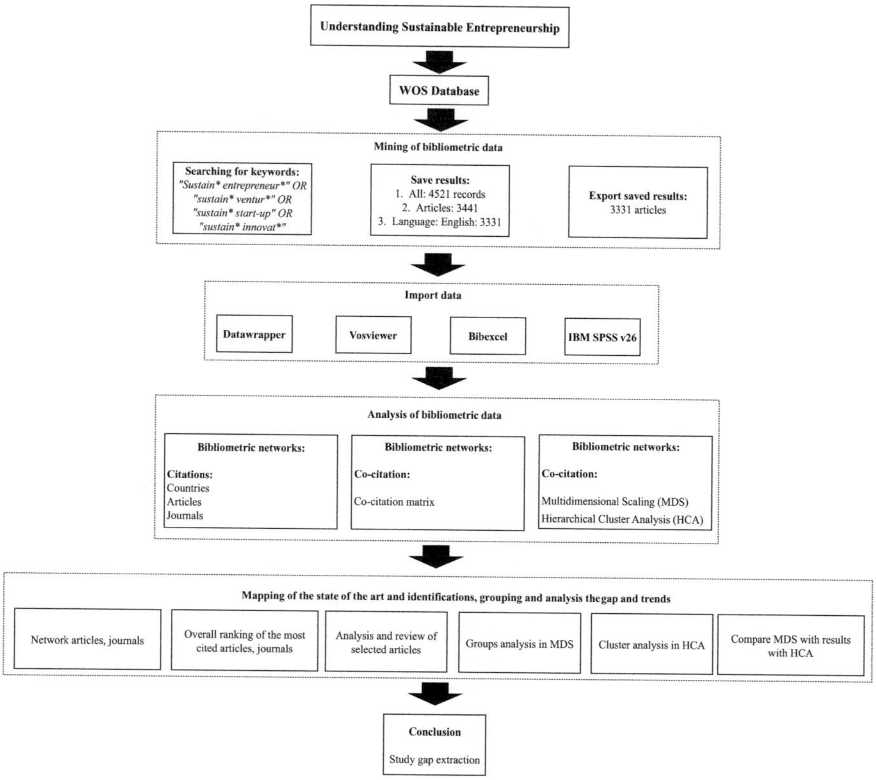

Fig. 1 Methodological approach (Author's presentation)

MDS—MDS is a widely used quantitative analysis technique that examines the interconnections between study areas and the intensity of their relationships (Adams et al., 2020). In MDS analysis, co-citation values indicate proximity among frequently cited publications (Foroudi et al., 2020). This analysis effectively highlights the similarities and distinct issues in various papers. The study was conducted using *IBM SPSS for Windows, version 26.*

HCA—HCA was employed to enhance the accuracy of this study. It is a quantitative technique that delineates the subgroups and academic pathways relevant to the study, identifying similarities among entities (Foroudi et al., 2020). HCA generates a dendrogram that illustrates the variables within each cluster and how these clusters are interconnected (Janssens, 2007). A widely used approach for cluster determination in HCA is Ward's method, which offers researchers transparent and interpretable outcomes (Yari et al., 2020). This analysis used IBM SPSS for Windows v26, focusing on documents from 1900 to 2025. The methodology is depicted in Fig. 1, which outlines all analytical steps and the software utilized at each stage.

Table 2 presents the countries ranked by citation count within the SE, from highest to lowest. The top three countries in terms of citations are the USA, with 311 articles and 9721 citations; Germany, with 233 documents and 9705 citations; and the United Kingdom, with 248 documents and 13,019 citations. Meanwhile, Fig. 2, derived from Datawrapper, illustrates a region figure based on citation numbers, ranked from maximum to minimum. Countries shaded in green have the fewest citations, those in red have the most, and those in gray lack data.

Recognition of the most highly cited publications on the SE is a significant metric for comprehending the writer's ranking and sets a

Table 2 Citation-based ranking of countries in sharing economy studies

Country	Documents	Citations	Country	Documents	Citations
England	13,019	248	Vietnam	276	18
USA	9721	311	Russia	250	24
Germany	9705	233	Turkey	239	19
Netherlands	9459	192	Czech Republic	238	22
Denmark	6859	71	Colombia	237	24
Italy	5527	195	Ireland	226	12
People R china	4936	345	Wales	221	13
Spain	3466	170	Tunisia	207	12
Canada	3452	87	Indonesia	204	25
France	3054	107	Slovenia	200	11
Sweden	3013	77	Ecuador	198	11
Australia	2826	114	Nigeria	180	9
Switzerland	2567	60	Peru	178	12
India	1763	117	Estonia	170	14
Finland	1659	52	Cyprus	169	14
Norway	1579	49	Mexico	166	16
Portugal	1490	60	United Arab Emirates	147	23
Brazil	1445	71	Malta	143	5
Austria	1442	47	Bangladesh	115	11
Pakistan	1148	58	Slovakia	114	9
Malaysia	1065	49	Egypt	105	12
South Korea	862	43	Israel	102	9
Thailand	774	31	Hungary	86	8
New Zealand	764	28	Qatar	82	11
Taiwan	758	45	Azerbaijan	70	5
Belgium	725	38	Kenya	64	5
Greece	645	30	Ukraine	61	15
Chile	569	16	Uganda	59	7
Poland	562	48	Lebanon	54	10
Romania	503	47	Kazakhstan	53	5
South Africa	492	44	Turkiye	51	19
Lithuania	462	19	Jordan	37	13
Japan	428	10	Latvia	27	5
Saudi Arabia	421	43	Palestine	21	5
Scotland	415	29	Argentina	20	9
Ghana	351	11	Serbia	18	5
Singapore	347	17	North Macedonia	16	6
Iran	299	27	Croatia	13	6

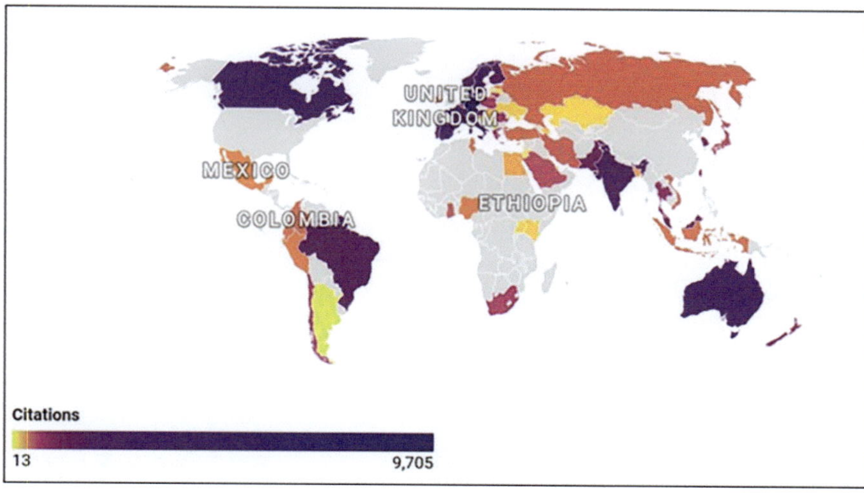

Fig. 2 Map of countries extracted from Datawrapper

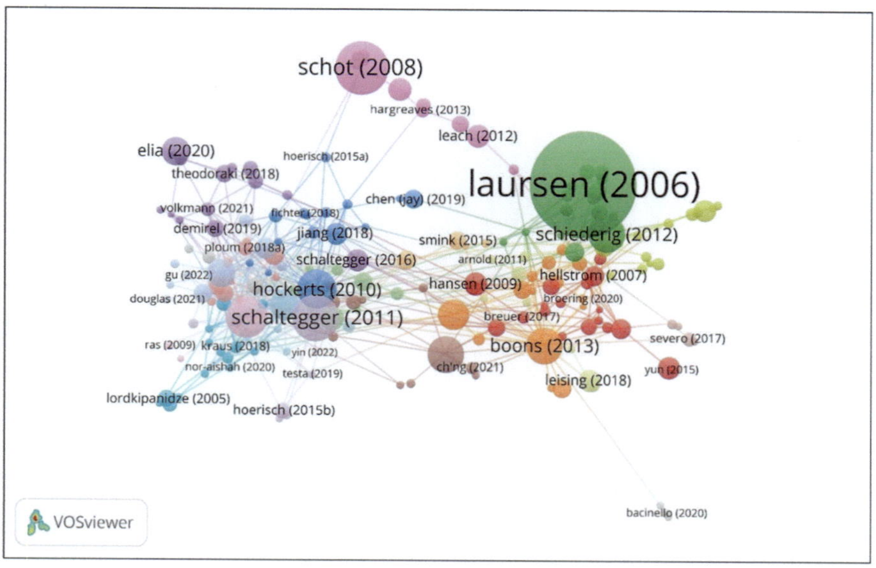

Fig. 3 The most-cited documents

basis for additional investigations. Observing the most highly cited publications on the SE (see Fig. 3), the study by Laursen and Salter (2006) has received the most attention. Boons (2013) wrote the second-most highly cited paper. Schot and Geels (2008) was the third most popular author. Table 3 represents the overview of the most highly cited papers on SE.

Through content analysis, Table 4 and Fig. 4 present the most cited journals on SE, as follows: Journal of Cleaner Production, Sustainability, and Business Strategy and the Environment. From 1900 to 2025, the Journal of Cleaner Production published 150 articles on SE, leading with 9474 citations; Sustainability includes 358 articles (with 5246 citations); and Business Strategy and the Environment has 68 articles (with 5038 citations). Table 4 highlights the leading publications in sustainable entrepreneurship based on citation frequency.

Table 3 The most cited papers in the SE

	Papers	Source title	Drives	Method	Citations
1	Open for innovation: the role of openness in explaining innovation performance among UK manufacturing firms	Strategic Management Journal	This paper explores how the extent of a firm's external search strategies—both in breadth and depth—affects its innovation outcomes and how this connection differs across various innovation types	The study uses data from the UK Innovation Survey based on Eurostat	3909
2	Business models for sustainable innovation: state-of-the-art and steps towards a research agenda	Journal of Cleaner Production	This paper contributes to sustainable innovation research by adopting a business model perspective and proposing directions for future inquiry	Systematic literature review of business model research articles	1318
3	Strategic niche management and sustainable innovation journeys: theory, findings, research agenda, and policy	Technology Analysis and Strategic Management	This article reviews the past decade of empirical results and theoretical advancements in strategic niche management (SNM) research. The SNM framework posits that sustainable innovation pathways can be supported by establishing technological niches—protected environments that foster experimentation with the co-evolution of technology, user behaviors, and regulatory frameworks	Systematic literature review	1221
4	SE and sustainability innovation: categories and interactions	Business Strategy and the Environment	The paper proposes a framework to position SE about sustainability innovation	The document lacks a formal "Methodology" section; instead, it presents a qualitative strategy for creating a framework that classifies companies according to their sustainability focus and market influence	1032

(continued)

Table 3 (continued)

	Papers	Source title	Drives	Method	Citations
5	Toward a theory of SE: reducing environmental degradation through entrepreneurial action	Journal of Business Venturing	The paper introduces the concept of "SE," suggesting that entrepreneurs can mitigate environmental damage by capitalizing on opportunities created by market failures related to the environment. It contends that this method is more effective than depending only on government regulations	The paper lacks detailed methodology or data collection methods. Instead, it is a conceptual work that integrates theories from environmental economics and entrepreneurship literature to create a framework	867
6	Market imperfections, opportunity, and SE	Journal of Business Venturing	The paper contends that market imperfections contribute to environmental degradation, yet they also create opportunities for SE to generate entrepreneurial returns while enhancing environmental and social conditions	The paper does not outline a specific methodology, as it is a conceptual work that develops a theoretical argument instead of presenting an empirical study	834
7	Greening Goliaths versus emerging Davids—theorizing about the role of incumbents and new entrants in SE	Journal of Business Venturing	The paper presents a conceptual framework examining the roles of small firms ("Emerging Davids") and large firms ("Greening Goliaths") in driving industries towards sustainable development, highlighting how their interactions contribute to this transformation	Not mentioned (the paper does not describe any specific methodology)	760
8	Green innovation in technology and innovation management—an exploratory literature review	R&D Management	This paper summarizes current literature on green innovation, defines the term "green innovation" and its associated concepts, and highlights the leading scholars, institutions, and key publications in this area	Literature review using the Google Scholar database	689

(continued)

Table 3 (continued)

	Papers	Source title	Drives	Method	Citations
9	Sustainable innovation, business models and economic performance: an overview	Journal of Cleaner Production	The paper offers an overview of how the concept of the business model can serve as a framework for studying and promoting sustainable innovation	The paper does not describe a specific methodology since it is an introductory and conceptual piece rather than a report on an empirical study	658
10	Business cases for sustainability: the role of business model innovation for corporate sustainability	International journal of Innovation and Sustainable Development	The paper offers a comprehensive definition and exploration of a "business case for sustainability," referring to instances where companies can attain economic success through voluntary environmental and social management activities rather than merely alongside them	The paper does not specify any particular methodology used in the study	654

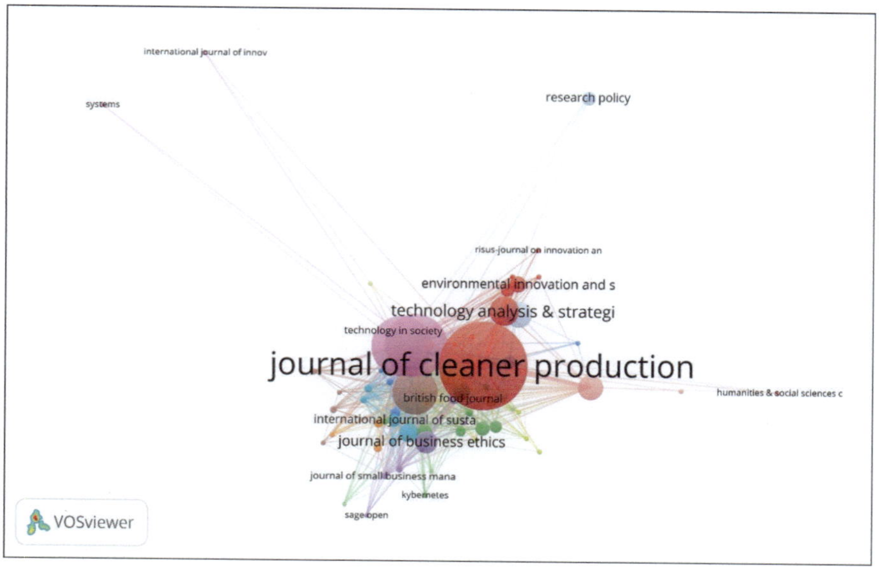

Fig. 4 Mapping of the most cited journals

Table 4 Most cited journals

Source	Total studies (TS)	Total citations (TC)
Journal of Cleaner Production	150	9474
Sustainability	358	5246
Business Strategy and The Environment	68	5038
Journal of Business Venturing	12	3144
Technology Analysis & Strategic Management	14	1542
Technological Forecasting and Social Change	25	1262
Small Business Economics	20	1202
Journal of Business Research	26	1164
Journal of Business Ethics	12	953
International Journal of Entrepreneurial Behavior & Research	25	760
Organization & Environment	10	652
International Journal of Innovation Management	10	619
Corporate Social Responsibility and Environmental Management	26	580
Environmental Innovation and Societal Transitions	12	560
Sustainable Development	22	510

3 Results

3.1 MDS

In the present study, we employed MDS to visualize the similarities, proximities, and relationships between publications in a multidimensional space (Zupic & Čater, 2015). The main dimensions of the fields of the study, identified by different researchers through similarities, dissimilarities, or distances of subjects, are revealed in a multidimensional space (White & McCain, 1998).

After collecting papers during the years 1900 to 2025 and extracting the co-citation matrix, MDS was performed to identify subfields of a specific study area (Cobo et al., 2011). In the MDS analysis, the total number of citations has been used as an indicator of proximity among highly cited publications (Ramos-Rodríguez & Ruíz-Navarro, 2004). MDS was executed by applying IBM SPSS v26 to determine if the data had a good model fit.

Ramo-Rodríguez and Ruíz-Navarro (2004) indicated that assessing the amount of stress is essential for selecting the right model. A stress value of 0 is relevant, while a stress range between 0.10 and 0.20 is considered suitable. The MDS employed a standard distance of 0.25. MDS aims to illustrate the intellectual structure of SE, thereby revealing the intricate network of research groups united by shared research interests and academic traditions (Foroudi et al., 2020). The initial analyses and configurations impact the stress level, indicating that an increase in analyzed cases raises the stress value. Essentially, as the number of mapped items increases, the goodness of fit deteriorates. This was a key factor in determining the number of documents to map (Foroudi et al., 2020). Given that a standard distance of 0.25 was used in the MDS analysis, Fig. 5 illustrates how points within this distance were grouped. Thirteen distinct groups were identified, reflecting that these groups are not interconnected and remain separate (see Table 5).

Group 1 features two publications titled Sustainability-Oriented Innovation (SOI), both exploring the SOI concept. Adams et al. (2016) define SOI as the purposeful adjustments made to an organization's values, philosophy, products, processes, or actions to generate social and environmental benefits alongside economic profits. Meanwhile, Boons et al. (2013) highlight the significance of sustainable innovations in sustainable development. They specifically address the critical role of sustainable business models in promoting sustainable innovation and establishing a connection between individual firms and broader systems. This idea is also implied in the first article, where SOI encompasses organizational changes and processes, potentially including business models. Both studies underscore the necessity of integrating environmental, social, and economic factors into innovation and development strategies. Adams et al. (2016) introduce the concept of the "Triple Bottom Line" (TBL), advocating for businesses to embrace a responsible approach that equally values environmental, social, and economic aspects in their decision-making. Additionally, Boons et al. (2013) assert that sustainable innovation must enhance sustainability performance based on environmental, economic, and social standards.

Sustainable innovations demand deep transformations rather than just incremental changes. Achieving sustainable development necessitates a rethinking of larger aspects of production and consumption systems, as incremental innovations fall short. Overall, the articles in this collection highlight the significance of a sustainability-focused approach to innovation, the role of sustainable business models, and the incorporation of environmental, social, and economic factors, stressing that radical and systemic changes are essential for attaining sustainable development.

Group 2 includes two publications that emphasize dynamic capabilities. The dynamic capabilities framework analyzes firms' resources and methods for wealth creation and competitive advantage amidst rapid technological changes. Central to these papers is dynamic capabilities, which refers to firms' ability to integrate, create, and reconfigure internal and external competencies in response to evolving environments (Barney, 1991). Dynamic capabilities encompass organizational processes, positions (assets), and evolutionary paths of the firm. Organizational processes encompass the firm's procedures, behavioral patterns, and operational methods, playing a crucial role in resource coordination, learning, and reconfiguration. Positions pertain to firm-specific assets, including technology, intellectual property, complementary assets, customer bases, and relationships with suppliers and partners. Such assets can provide a competitive advantage, particularly when they are difficult to replicate or imitate. Paths signify the strategic options available to the firm and the implications of incremental returns and path dependence. Histor-

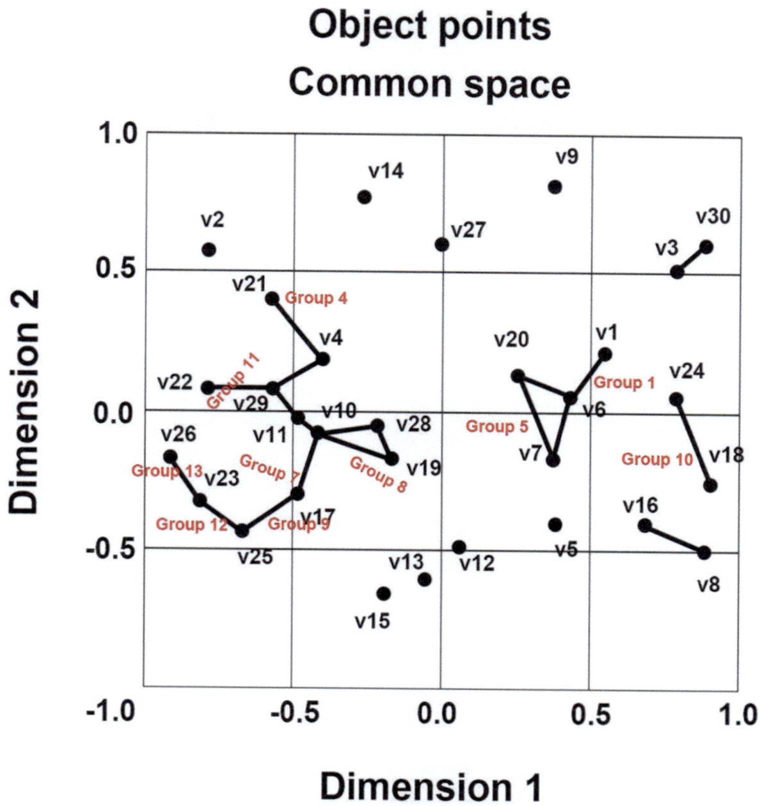

Fig. 5 Multidimensional scaling. Stress value = 0.04472. Standardized distance = 0.25. V1 = (Adams et al., 2016), V2 = (Ajzen, 1991), V3 = (Barney, 1991), V4 = (Belz & Binder, 2017), V5 = (Bocken et al., 2014), V6 = (Boons et al., 2013), V7 = (Boons & Lüdeke-Freund, 2013), V8 = (Carrillo-Hermosilla et al., 2010) V9 = (Cillo et al., 2019), V10 = (Cohen & Winn, 2007), V11 = (Dean & McMullen, 2007), V12 = (Eisenhardt, 1989), V13 = (Eisenhardt & Graebner, 2007), V14 = (Fornell & Larcker, 1981), V15 = (Gioia et al., 2013), V16 = (Hall & Mitsloan, 2003), V17 = (Hall et al., 2010), V18 = (Hansen et al., 2009), V19 = (Hockerts & Wüstenhagen, 2010), V20 = (Klewitz & Hansen, 2014), V21 = (Kuckertz, 2010), V22 = (Muñoz & Dimov, 2015), V23 = (Muñoz & Cohen, 2018), V24 = (Nidumolu et al., 2009), V25 = (Pacheco et al., 2010), V26 = (Patzelt & Shepherd, 2011), V27 = (Podsakoff et al., 2003), V28 = (Schaltegger & Wagner, 2011), V29 = (Shepherd & Patzelt, 2011), V30 = (Teece et al., 1997)

ical decisions and investments may limit future options and impact dynamic capabilities. Distinctive competencies are capabilities that are hard to imitate and provide a competitive edge for the firm (Teece et al., 1997). These competencies typically emerge from a unique blend of organizational processes, situations, and evolutionary paths.

These articles highlight the resource-based view as a theoretical framework that underscores the role of a firm's unique assets and capabilities in generating competitive advantage. However, they contend that this perspective is insufficient on its own to explain how competitive advantage is cultivated and sustained in dynamic contexts, necessitating a focus on dynamic capabilities. Consequently, the articles collectively propose the dynamic capabilities framework as a thorough approach for understanding the creation, development, and utilization of organizational competencies and capabilities to secure sustainable competitive advantage amid change.

Group 3 comprises four articles that outline a shared conceptual framework illustrating the connections among economic, social, and environmental sustainability goals. First, leveraging market failures as entrepreneurial opportunities is a central theoretical foundation. Based on economic theories, Dean and McMullen (2007)

Table 5 Group-specific publications and topics identified through the distance matrix

Groups	Publications	
Group 1	*V1, V6*	*Sustainability-oriented innovation (SOI)*
Group 2	*V3, V30*	*The dynamic capabilities*
Group 3	*V4, V10, V11, V29*	*Sustainability*
Group 4	*V4, V21*	*The process of SE*
Group 5	*V6, V7, V20*	*Sustainable business models*
Group 6	*V8, V16*	*Innovation challenges for sustainable development*
Group 7	*V11, V17*	*SDGs*
Group 8	*V11, V19, V28*	*Entrepreneurial Opportunities Arising from Market Failures*
Group 9	*V17, V25*	*The green prison*
Group 10	*V18, V24*	*The strategic link between sustainability and innovation*
Group 11	*V22, V29*	*SE as a systematic response to ecological and social challenges*
Group 12	*V23, V25*	*Institutional transformation and the redefinition of incentive structures*
Group 13	*V23, V26*	*Redefine the concept of entrepreneurial opportunities*

assert that market failures, such as public goods, external costs, and inadequate information, are not merely barriers to sustainability; instead, they present chances for entrepreneurs to achieve economic profit while aiding in the reduction of environmental harm by creating new markets (like renewable energy) or establishing property rights for natural resources. This perspective aligns with the ideas proposed by Shepherd and Patzelt (2011), who characterizes sustainability as "the conservation of nature and life support systems alongside economic and social development."

Second, the integration of the three sustainability dimensions—economic, social, and environmental—serves as a theoretical foundation for daily applications across all articles. Shepherd and Patzelt (2011), inspired by the World Commission on the Environment report (WCED, 1987), highlights the need for sustainability efforts to address both' what needs to be preserved" (nature, essential resources, and society) and' what needs to be developed" (economic opportunities and enhanced quality of life). This perspective is also echoed in the work of Dean and McMullen (2007), who underscore the critical role of entrepreneurs in mitigating the external costs of environmental degradation, such as water pollution, while also generating economic value. Cohen and Winn (2007), drawing on Williamson's institutional framework and sustainable development concepts, reiterates the importance of aligning economic institutions with environmental objectives. Another shared theme is the significance of innovation and the redefinition of business models. Sustainable entrepreneurs reshape processes, such as material recycling, or develop green technologies like energy-efficient buildings, removing institutional barriers while merging profitability with a lower ecological footprint (Dean & McMullen, 2007). This concept is exemplified in innovative irrigation systems used in sustainable agriculture or green banking initiatives (Gramin), which enhance the livelihoods of local communities while safeguarding natural resources.

Collectively, these four articles merge economic theories (like market failure and external costs), ecological ideas (such as biodiversity conservation), and institutional frameworks to present SE as a foundation for a "balanced development engine." This approach fosters the creation of more resilient and equitable economic systems by turning environmental threats into economic opportunities while simultaneously redefining the social responsibilities of enterprises.

Group 4 includes two articles that examine the process of SE and highlight five main conceptual axes:

1. Triple Bottom Line: Both articles assert that SE necessitates the integration of economic, social, and environmental objectives. Belz and Binder (2017) illustrate this with a six-step model that integrates these objectives. In contrast, Kuckertz and Wagner (2010) argue that focusing on sustainability helps uncover opportunities within imperfect markets.
2. Market imperfections as opportunity sources: Both papers draw inspiration from Cohen and Winn's (2007) framework. Belz and Binder (2017) demonstrate how market inefficiencies, such as resource mispricing, can be turned into entrepreneurial opportunities. Kuckertz and Wagner (2010) further confirm a positive correlation between a sustainability orientation and entrepreneurial intention in flawed markets by analyzing survey data from German students.
3. Experience and education: Belz and Binder (2017) highlight personal experiences with poverty as a significant motivator, while Kuckertz and Wagner (2010) indicate that traditional business experience might lessen the influence of sustainability on entrepreneurial intention, advocating for the integration of sustainability education into university programs.
4. Distinction from social/green entrepreneurship: Both articles present SE as a distinct phenomenon, defining it beyond the narrow views of social or green entrepreneurship, and differentiate it from traditional entrepreneurship by referencing "three-dimensional management."
5. Institutional support: The articles explore the role of governmental support in fostering sustainable markets and emphasize the importance of university education in nurturing entrepreneurs aware of sustainability issues.

Both articles, using distinct methodologies of qualitative case studies and quantitative survey data analysis, confirm three ordinary propositions:

- Sustainability necessitates a balance of economic, social, and environmental factors,
- Market failures present both challenges and opportunities,
- The advancement of SE calls for a reevaluation of policies and education.

These studies enhance each other, illustrate the SE process, and examine individual influences on entrepreneurial intentions. Ultimately, the conclusion is that SE is not just a moral obligation but also a practical approach to addressing sustainable development challenges.

Group 5 includes three publications highlighting the crucial role of sustainable business models in integrating the three sustainability dimensions: economic, social, and environmental. In these collaborative works (Boons et al., 2013; Boons & Lüdeke-Freund, 2013; Klewitz & Hansen, 2014), sustainable innovation is a structured response to economic, environmental, and social challenges. This method emphasizes that sustainable innovation extends beyond minor adjustments to products or processes, prompting significant and systematic transformations in business structures and models. Here, business models serve as an all-encompassing framework that connects individual company innovations with larger production-consumption systems by weaving together value propositions, organizing value chains, and developing revenue models (Boons et al., 2013; Boons & Lüdeke-Freund, 2013).

Conversely, Klewitz and Hansen (2014) explore sustainability-oriented innovations (SOIs) within small and medium-sized enterprises (SMEs) and highlight that these innovations manifest across three dimensions: product, process, and organizational. The article demonstrates that a company's success hinges not only on green technology or product development but also on enhancing organizational strategies and engagement with internal and external stakeholders. Furthermore, these studies underscore the necessity of a holistic, interactive approach, which clarifies the role of business models in merging economic, social, and environmental factors. Nevertheless, further research is essential to understand these components' interactions better and address existing literature gaps.

Group 6 has published two articles addressing innovation challenges for sustainable development. A detailed review indicates that the primary focus is to analyze the challenges and opportunities linked to embedding sustainable innovations within business structures, aiming for a fundamental change in economic, environmental, and social practices (Carrillo-Hermosilla et al., 2010). The authors argue that contemporary businesses can no longer remain passive observers of environmental changes; instead, they must prioritize innovation strategically and systematically. They stress the necessity of re-evaluating traditional business models, which should evolve from linear, one-dimensional approaches to multidimensional systems that concurrently account for economic, social, and environmental factors (Carrillo-Hermosilla et al., 2010; Hall & Mitsloan, 2003).

Additionally, the articles explore various aspects of sustainable innovation, ranging from incremental improvements in products and processes to transformative shifts in organizational structures and value chain networks. The authors contend that achieving sustainability goals hinges not on superficial reforms but on deep-seated changes in management tactics, organizational policies, and the relationships among key stakeholders, which include suppliers, customers, government entities, and research institutions. In summary, the articles highlight that sustainable innovations offer avenues for adding value and can serve as a crucial competitive advantage in rapidly evolving global markets (Carrillo-Hermosilla et al., 2010; Hall & Mitsloan, 2003).

In summary, the articles in this collection urge a reevaluation of business management and structural methods to align innovation strategies with sustainability requirements. This realignment has the potential to instigate profound changes in global markets and reshape economic activities during the sustainable transformation era.

Group 7 consists of two articles highlighting entrepreneurship's crucial role in addressing environmental challenges and promoting sustainable development. Both papers share a foundational perspective, viewing market failures as significant entrepreneurial opportunities. Dean and McMullen (2007) assert that market failures, particularly in environmental contexts, represent challenges and a chance for entrepreneurs to create innovative solutions that generate profits while mitigating environmental harm (Dean & McMullen, 2007). Similarly, Hall et al. (2010) adopt this viewpoint, presenting entrepreneurship as a means to leverage opportunities that arise from market failures within the sustainability sector (Hall et al., 2010).

A common conceptual foundation is the role of entrepreneurs in driving transformation toward sustainability. Both articles emphasize entrepreneurship as an economic activity and a catalyst for systemic changes aimed at sustainable development. Dean and McMullen (2007) suggest that by addressing market barriers and establishing new markets for environmental resources, entrepreneurs can foster the economic institutions essential for well-functioning markets, thereby facilitating a shift toward a more sustainable economy (Dean & McMullen, 2007). Similarly, Hall et al. (2010) highlight that entrepreneurship is increasingly crucial for developing sustainable products and processes to address various social and environmental challenges (Hall et al., 2010).

Ultimately, both articles highlight the significance of grasping the motivations behind sustainable entrepreneurs and the environments that encourage SE. Dean and McMullen's (2007) study leans toward examining theoretical mechanisms and opportunities stemming from market failures. In contrast, Hall et al.'s (2010) research tackles crucial inquiries regarding the factors that propel entrepreneurs toward sustainability and how public policies can expedite this transition (Hall et al., 2010). Both studies, whether directly or indirectly, stress the necessity for additional research on SE and its contribution to realizing the SDGs.

Group 8 encompasses three papers focused on identifying and leveraging entrepreneurial opportunities that arise from market failures. Dean and McMullen (2007) assert that environmental degradation often stems from market inefficiencies, which open avenues for entrepreneurs to develop innovative solutions that yield profits while enhancing environmental quality (Dean & McMullen, 2007). SE can serve as a tool to address market failures, facilitating better resource allocation and reducing adverse environmental effects (Hockerts & Wüstenhagen, 2010; Schaltegger & Wagner, 2011). All three papers highlight, whether directly or indirectly, the pivotal role of entrepreneurs in advancing sustainability. Entrepreneurs function not just as economic agents but as instigators of systemic change, capable of navigating barriers to sustainable development through innovation and creating new markets (Dean & McMullen, 2007). Additionally, a conceptual shift broadens the theoretical framework of entrepreneurship. The articles likely expand beyond the traditional perspective centered on information asymmetry, stressing that entrepreneurial opportunities involve uncovering new information and recognizing and addressing various market failures (Dean & McMullen, 2007, p. 53). Thus, SE is presented as a strategy that aims for economic viability, enhances market efficiency, and fosters environmental and social sustainability.

Group 9 consists of two papers highlighting the green prison idea. Both pieces identify SE as a potential catalyst for economic and social change to mitigate the negative environmental and social effects of businesses (Hall et al., 2010; Pacheco et al., 2010). However, they also recognize the challenges and obstacles related to SE. Hall et al. (2010) discuss the "cure-all hypothesis," which views entrepreneurship as the answer to all sustainability challenges while pointing out the ambiguities and structural constraints that sustainable entrepreneurs encounter. Pacheco et al. (2010) elaborate on these challenges in the "green prison," where insufficient economic incentives hinder entrepreneurs from pursuing sustainable practices.

In their 2010 study, Pacheco et al. propose that entrepreneurs can break free from the "green prison" by establishing new institutions—such as norms, property rights, and legislation—and reshaping competitive frameworks to make sustainability profitable. This perspective resonates in Hall et al.'s (2010) analysis, which underscores the importance of investigating public policies in fostering SE. Both studies, whether directly or indirectly, emphasize the necessity for radical changes in production and consumption behaviors and economic frameworks to attain sustainability (Hall et al., 2010; Pacheco et al., 2010). Ultimately, both pieces highlight the need for continued research on SE and recognize the existing knowledge gaps concerning SE's role and mechanisms in meeting the SDGs.

Group 10 comprises two papers highlighting the strategic connection between sustainability and innovation. Firstly, both papers depict sustainability as a catalyst for innovation. They argue that it should not be seen merely as an added expense but as a valuable asset that can redefine competitiveness and lead to the "next industrial revolution." This perspective is supported by Hansen et al. (2009), who emphasize the urgent need for a "10- to 50-fold increase in resource efficiency to achieve sustainability," alongside the demand for radical solutions. Furthermore, both papers assert that integrating economic, social, and environmental aspects (such as the Triple Bottom Line framework) is not only a moral responsibility but also an excellent opportunity for gaining a competitive edge (Hansen et al., 2009; Nidumolu et al., 2009). The second central point concerns the significance of structured frameworks in implementing sustainability. Nidumolu et al. (2009) introduce a "five-step journey" that starts with perceiving regulatory requirements as opportunities and culminates in business model innovation. This concept is further elaborated by Hansen et al. (2009) through the "Sustainability Innovation Cube," which offers a three-dimensional framework for evaluating the sustainability effects of product innovations (Hansen et al., 2009). Both methodologies stress the necessity of analytical tools to manage directional risks and align innovation with sustainability objectives. Additionally, both papers highlight the dynamic relationship between regulatory pressures and market opportunities. Companies that take the lead in meeting stringent standards gain a "first mover" advantage, elaborated under the terms "regulatory push" and "vision pull," which refer to the influence of new regulations and innovative concepts on shaping market possibilities, respectively (Hansen et al., 2009). Both sources concur that shifting regulatory landscapes and evolving customer preferences crucially redefine the notion of economic value. They argue that achieving sustainability requires transcending incremental enhancements and embracing systems approaches, incorporating technological innovations alongside institutional and behavioral transformations.

Group 11 contains articles on creating a conceptual framework for SE as a systematic response to ecological and social issues. Despite variations in methodology and specific focus areas, both articles align on three central themes. They regard SE as a transformative process necessitating reevaluating the relationship between economic value and environmental-social requirements. In their analysis of 45 sustainable ventures, Muñoz and Dimov (2015) highlight two distinct pathways: the "conformist" path, which operates within supportive sustainability contexts, and the "insurgent" path, which challenges existing unfavorable frameworks. This dynamic perspective is reiterated by Shepherd and Patzelt (2011), who describe SE as an endeavor that integrates "conservation of nature, life-supporting systems, and society" with "socio-economic development."

Both studies highlight the importance of advancing past one-dimensional methods in sustainability analysis. Muñoz and Dimov (2015) demonstrate through fsQCA that intricate causal combinations of concepts, actions, and exchange relationships play a crucial role in the emergence of sustainable ventures. Shepherd and Patzelt (2011) further support this systems perspective, introducing a "meta-theoretical" framework that integrates economics, institutions, and psychology.

Third, both recognize that the tension between market dynamics and sustainability needs drives SE. Insurgent ventures aim to alter institutional dynamics by "confronting existing structures." In contrast, Shepherd and Patzelt (2011) highlight how entrepreneurs serve as a link between "what must be preserved" (the environment) and "what must be developed" (the economy). Both viewpoints concur that redefining value and establishing new institutional mechanisms are essential for success.

Group 12 emphasizes the importance of institutional transformation and redefining incentive structures as essential for generating sustainable value. Pacheco et al. (2010) introduce the term "green prison," describing a scenario where market incentives lead entrepreneurs to engage in unsustainable practices. They argue that overcoming this issue necessitates the establishment of new institutions, including norms, property rights, and laws (Pacheco et al., 2010). This perspective is supported by Muñoz and Cohen (2018), who provide a thorough review of SE literature. They discuss ideas like "institutional innovation" and the need to find a "balance between efficiency and equity," highlighting that a paradigm shift in economic systems is crucial. They view SE as a transformative process that redefines the interplay between economic value and environmental-social needs. Both studies underscore the pivotal role entrepreneurs play in reshaping market rules. Pacheco et al. (2010) illustrate how entrepreneurs integrate economic incentives with sustainability objectives by reforming institutions, including establishing organic standards and renewable energy laws. This idea is echoed in Muñoz and Cohen (2018), where they highlight research such as "SE as a subset of public entrepreneurship" and "the interaction between technology and institutions." Both studies concur that the viability of sustainable business models relies on the simultaneous generation of economic and institutional value, aligning individual profits with broader environmental and social advantages. By merging game theory and institutional economics, Pacheco et al. (2010) propose a framework that elucidates the relationship between personal incentives and overarching sustainability needs. Moreover, Muñoz and Cohen (2018) underscore that addressing structural challenges to sustainability demands a fusion of economic, social, and behavioral perspectives, attainable only through interdisciplinary research and innovative entrepreneurial efforts.

Group 13 features two articles that focus on reinterpreting entrepreneurial opportunities through the lens of sustainability requirements. Both articles contend that SE necessitates a departure from traditional profit-centered models, advocating for incorporating economic, social, and environmental aspects into the value creation process. Muñoz and Cohen (2018) conduct a systematic literature review and characterize SE as an activity that merges "profitable opportunities with participation in sustainable development," highlighting the importance of balancing nature, society, and economic activities. Meanwhile, Patzelt and Shepherd (2011) propose a model for identifying opportunities for sustainable development, indicating that entrepreneurs' awareness of natural and social environments, comprehension of environmental risks, and an inclination towards altruism are essential for recognizing sustainable opportunities (Patzelt & Shepherd, 2011).

The second conceptual foundation critiques simplistic views on entrepreneurship. Muñoz and Cohen (2018) highlight the shortcomings of current research, arguing that social entrepreneurship should not merely fall under the categories of social or environmental entrepreneurship; it necessitates its own independent and interdisciplinary theoretical frameworks. Patzelt and Shepherd (2011) illustrate that acknowledging these opportunities involves focusing on the "triple win" (economic, social, environmental) and blending a specialized understanding of natural and social environments with altruistic goals. Furthermore, both articles stress the transformative power of entrepreneurs in fostering systemic change. Muñoz and Cohen (2018) introduce "sustainability-as-flourishing" as a catalyst for radical changes in socio-economic systems, rather than just incremental ones. Patzelt and Shepherd (2011) expand on this concept with the "third/first person opportunity belief" model, demonstrating that sustainable entrepreneurs can redefine market norms and establish new institutions by merging environmental expertise with metamaterial motivations. Both sources concur that successfully implementing these transformations requires a re-evaluation of research paradigms and crafting theoretical frameworks that accurately represent the complexities of sustainability at both micro and macro levels.

3.2 HCA

HCA is generally suitable for datasets where the number of clusters is uncertain. The principle behind HCA is straightforward: it organizes clusters into sub-clusters or combines them into larger clusters, resulting in a hierarchical arrangement of items within groups (Vijith & Dodge-Wan, 2020). HCA can be utilized when researchers decide how to categorize items into clusters (Foroudi et al., 2020).

This study adopted HCA since it works on any item in any cluster. It replicates mixing the nearest couple of clusters and matching their resemblance till the whole data is credited to a single cluster. By grouping homological items to classify similar characteristics, cluster analysis has broad usage (Vijith & Dodge-Wan, 2020). This was done using IBM SPSS v26. To evaluate similarity in attendance or nonattendance, the Squared Euclidean distance method was used along with *Ward's method* as a connectivity-based clustering method.

Figure 6 shows the outcomes of HCA, which includes four clusters and shows the most similar characteristics.

Cluster 1 includes eight papers centered on "sustainability-driven innovation," presented to tackle environmental, social, and economic issues. Each paper explores various dimensions of sustainable innovation, such as organizational processes, business models, and systemic shifts in industries and markets. Below is a summary of the standard conceptual foundations found in the papers:

The primary focus of both groups of papers centers on the essential role of innovation in promoting sustainable development. According to the authors of the first paper (Nidumolu et al., 2009) in the Harvard Business Review, sustainability has evolved from being just a social or

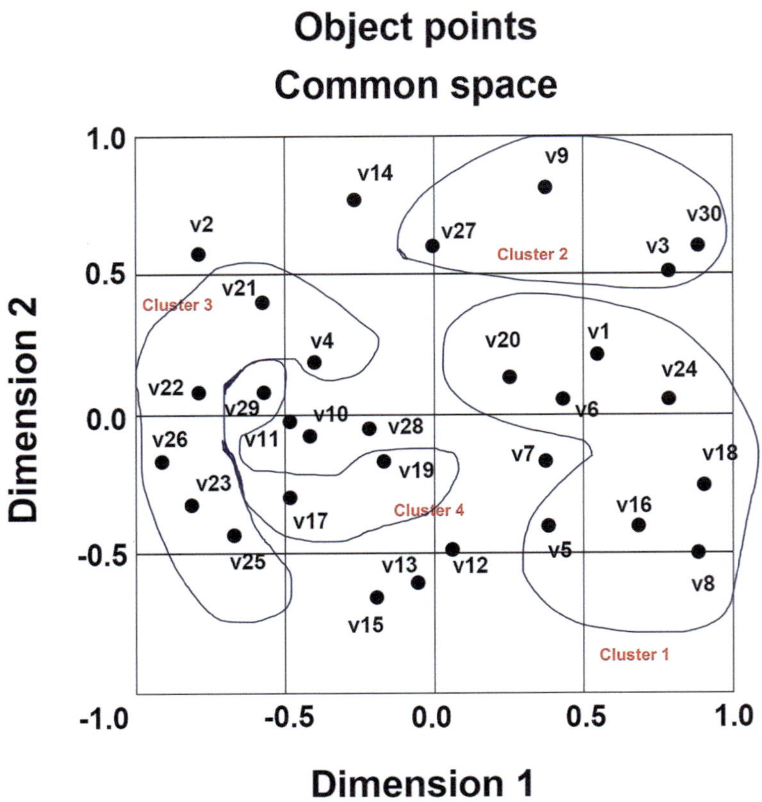

Fig. 6 Hierarchical clustering analysis

environmental obligation to becoming a catalyst for business innovation. Research indicates that when companies implement sustainable practices, they not only cut costs but also open up new avenues for revenue generation and improve their product offerings. Furthermore, this approach to sustainability enables organizations to strengthen their technological and competitive stance in the marketplace.

Conversely, the article by Boons et al. (2013) and Klewitz and Hansen (2014) in the Journal of Cleaner Production highlights the significance of developing sustainable business models. These models are frequently illustrated as a cohesive framework enabling businesses to reap sustainability benefits both within the organization and across the broader socio-economic landscape. This viewpoint also suggests that sustainable innovation must be integrated into strategies that encompass both radical and incremental transformations in organizational processes and practices.

A recurring theme in both article sets is the shift in perspective from exclusively environmental innovations to three dimensions: economic, social, and environmental. The analyzed reports indicate that companies should focus on formulating long-term strategies to create and spread sustainable technologies and products to generate significant changes in the industry, particularly within small and medium-sized enterprises (SMEs) (Klewitz & Hansen, 2014).

Both studies highlight the crucial role of interactions between sustainable innovation and external stakeholders, including customers, policymakers, and public institutions. These interactions enhance companies' marketing strength while promoting wider acceptance of sustainable innovations. For instance, Klewitz and Hansen (2014) emphasize the significance of SMEs' involvement in sustainable supply chains and their collaboration with academic research.

The conclusion emphasizes that sustainable innovation is a complex and evolving process necessitating essential shifts in managerial perspectives, organizational structure, and proactive involvement with crucial external stakeholders. As a result, companies must function not just as resource users but also as catalysts for sustainable transformation within broader economic systems.

Cluster 2 features two papers that explore Dynamic Capabilities and their relevance to strategic management and innovation. Dynamic capabilities signify an organization's ability to recognize external shifts, capitalize on opportunities, and adjust resources and processes to attain a competitive edge in unpredictable and intricate environments. These capabilities are essential for companies aiming to thrive amid rapid environmental changes. In the article "Dynamic Capabilities and Strategic Management" by Teece et al. (1997), this concept is presented as a framework for examining the "sources of wealth creation" and for "sustaining competitive advantage." The framework highlights three critical processes: (1) sensing opportunities and threats, (2) seizing opportunities, and (3) transforming organizational resources and directions. These processes empower organizations to function effectively during technological transitions or market changes.

Klewitz and Hansen's (2014) research on sustainability innovations in SMEs highlights how small businesses can align their strategies with sustainability objectives by utilizing dynamic capabilities. The study illustrates that despite resource constraints, SMEs can improve these capabilities by collaborating with external partners like customers and academic institutions.

Innovation serves as a key conceptual foundation in many papers. Teece et al. (1997) identify technological and organizational innovations as the primary drivers behind dynamic capabilities. These innovations stem from organizational learning processes, allowing firms to recognize future risks and adapt their strategic directions accordingly. Additionally, Klewitz and Hansen (2014) highlight the significance of learning and engaging with external stakeholders in fostering sustainable innovations within SMEs.

All four articles discuss "path dependency" and how an organization's history shapes dynamic capabilities. Teece et al. (1997) contend that accumulated organizational resources—like technological expertise and managerial skills—are challenging to replicate and transfer. These traits are vital for establishing and sustaining a competitive advantage.

The articles also address the intersection of sustainability and dynamic capabilities. Klewitz and Hansen (2014) explore sustainable innovations within SMEs and highlight that organizations can develop innovations that align with environmental and social objectives while simultaneously achieving a competitive edge. Additionally, the article by Cillo et al. (2019) presents dynamic capabilities as foundational for developing sustainable strategies. This paper underscores the necessity of crafting strategies that balance sustaining economic value and addressing environmental and social issues.

Interacting with the external environment is a crucial aspect highlighted in the collaborative papers. Cillo et al. (2019) demonstrate that fostering relational capital and collaborating with stakeholders, such as customers and policymakers, enables organizations to focus their dynamic capabilities on sustainable innovation effectively. Similarly, Klewitz and Hansen (2014) argue that SMEs can leverage market complexities to their benefit by engaging actively in data chains and innovation networks.

These four papers highlight the importance of dynamic capabilities in strategic management and innovation, particularly within unstable environments. Organizations can leverage these capabilities to establish and sustain competitive advantages, foster sustainable innovations, and address environmental and societal challenges. Additionally, factors like path dependence, interactions with external elements, and organizational learning play crucial roles in developing these capabilities.

Cluster 3 includes articles highlighting the integration of the three sustainability goals—economic, social, and environmental—as the core of sustainable innovation and entrepreneurship. These articles illustrate that turning sustainability into a competitive edge necessitates a strategy that transcends merely merging these goals; it also requires rethinking conventional ideas in management and entrepreneurship.

For instance, Belz and Binder (2017) emphasize, in their qualitative study, the incremental process of merging three goals at various phases of SE, which includes identifying social and environmental issues and establishing sustainable markets. This process-focused view corresponds with the conclusions of Adams et al. (2016), who highlight the need to transform organizational philosophy and foster radical innovations in products, processes, and operational practices to attain sustainability in their systematic review. From a conceptual perspective, all articles emphasize the intricate interactions between internal and external factors that influence sustainability.

Cillo et al. (2019) classify existing literature into three categories: internal management, external relations, and performance evaluation. They indicate that while research primarily centers on managerial issues, engaging with external stakeholders—such as customers, suppliers, and policy-making entities—has grown increasingly vital. Furthermore, establishing sustainability performance metrics, particularly in social and environmental domains, is becoming more significant. This observation is consistent with Kuckertz and Wagner (2010), who highlight a positive correlation between sustainability orientation and entrepreneurial intention, especially among those lacking business experience. Nonetheless, the study cautions that extensive experience in conventional business settings can diminish sustainability orientation, as immediate economic concerns often overshadow long-term social and environmental aspirations.

Methodologically, the articles stress the importance of establishing rigorous frameworks to evaluate sustainability impacts. Fornell and Larcker (1981) introduced sophisticated statistical techniques for assessing structural equation models, laying the groundwork for quantifying latent variables associated with sustainability, such as environmental attitudes and social capital. This methodology has been further developed in later studies, including Cillo et al. (2019), which have concurrently produced composite indicators to assess economic, social, and environmental performance. Additionally, Adams et al. (2016) underscore the "lack of operational measures" within sustainability research, emphasizing the necessity to develop tools that accurately capture the complex interactions among the three sustainability dimensions.

Education plays a crucial role in transforming the entrepreneurial ecosystem. Kuckertz and Wagner (2010) suggest that entrepreneurship education programs should emphasize incorporating sustainability concepts into graduate and executive courses. This strategy aims to lessen the adverse effects of conventional business practices. This concept resonates with the research by Belz and Binder (2017), which shows that sustainable entrepreneurs often gain a richer comprehension of sustainability issues through engaging learning with diverse stakeholders, including NGOs and local communities. Thus, sustainability should be viewed not as an "inherent trait" but as a "dynamic process" that necessitates continuous adjustments among the three objectives in response to environmental changes.

Cluster 4 comprises three papers that highlight the necessity of rethinking traditional entrepreneurial models to facilitate systemic change in sustainability. These studies illustrate that progressing towards sustainable development demands a shift from incremental improvements to embracing bold, innovative strategies. For instance, Hall et al. (2010) investigate how entrepreneurship can instigate significant environmental changes. They contend that sustainable entrepreneurs should pinpoint opportunities within existing markets and develop new markets centered around sustainable products and services. This perspective is consistent with Shepherd and Patzelt's (2011) characterization of SE, which emphasizes the connection between what needs conserving (nature and society) and what requires growth (economic opportunities). In this role, entrepreneurs connect ecological systems and economic frameworks.

In the organizational dimension, all three studies explore the dynamic relationship between new and existing actors in the SE ecosystem.

Hockerts and Wüstenhagen (2010) introduce the terms "emerging Davids" for small firms driving sustainable innovations and "green giants" for large companies gradually adopting sustainability. They underscore these groups' complementary roles during various stages of industrial transformation. During the early phases, small firms spearhead disruptive technologies like renewable energy. However, significant participation from large companies with established infrastructures is crucial for these innovations to scale at an industrial level. This aligns with the concerns identified in Hall et al.'s (2010) paper, which discusses institutional barriers like burdensome regulations and insufficient financial incentives that hinder collaboration between these groups. On the other hand, Shepherd and Patzelt (2011) stress that achieving success in SE requires creating "combined value" that addresses economic, social, and environmental aspects. The emerging Davids pursue this strategy through their agile business models, while green giants employ corporate social responsibility practices.

The papers emphasize the importance of incorporating multidisciplinary theoretical frameworks to understand the complexities of SE. By merging concepts from entrepreneurial psychology, such as opportunity recognition, with those from environmental economics, including internalizing external costs, Shepherd and Patzelt (2011) present a framework that helps entrepreneurs balance immediate interests with long-term sustainability goals. This view aligns with Hockerts and Wüstenhagen's (2010) analysis of "institutional innovation," which aims to transform the industry landscape. They contend that sustainable entrepreneurs go beyond merely launching new technologies; they redefine industry standards, consumption behaviors, and legal structures. Moreover, Hall et al. (2010) address "institutional entrepreneurship," underscoring the need for further research on the relationship between entrepreneurs, policymakers, and civil society to cultivate supportive ecosystems.

3.3 Multi-method Analogy

In light of the combined strengths of MDS and HCA, the clusters and corresponding publications reflect diverse forms of understanding within the field. The collections formed through MDS generally concentrate on the attributes of SE, exploring diverse themes related to SE. In contrast, HCA delivers a thorough, thoughtful approach by establishing four clusters, which aim to elucidate the different orientations in which SE can thrive. The groups identified from the MDS align with the HCA. The main MDS group links to cluster 1, where the main topic of the articles in both sections focuses on SOI. The second MDS group is linked to cluster 2, as the central theme of the articles in both sections addresses dynamic capabilities. The third, fourth, tenth, and eleventh MDS groups align with Cluster 3, which revolves around sustainability as the primary focus of the articles in both sections. The twelfth and thirteenth MDS groups are associated with cluster 4 because their central theme is redefining traditional entrepreneurial paradigms.

4 Discussion and Future Direction

This section utilizes the author's keywords to create a strategic graph to better understand the future directions for SE research. The graph plots influence on the y-axis and centrality on the x-axis, following the simple centering algorithm in R packages (Cobo et al., 2011). In this context, centrality indicates the importance of various subjects, while influence reflects the development and relevance of specific issues. Figure 7 presents a strategic diagram divided into four quadrants, based on Cahlik's (2000) typological framework. Notable topics for future research are summarized in Table 6 and illustrated in Fig. 7.

Keywords located in the upper right quadrant represent the primary topics characterized by high

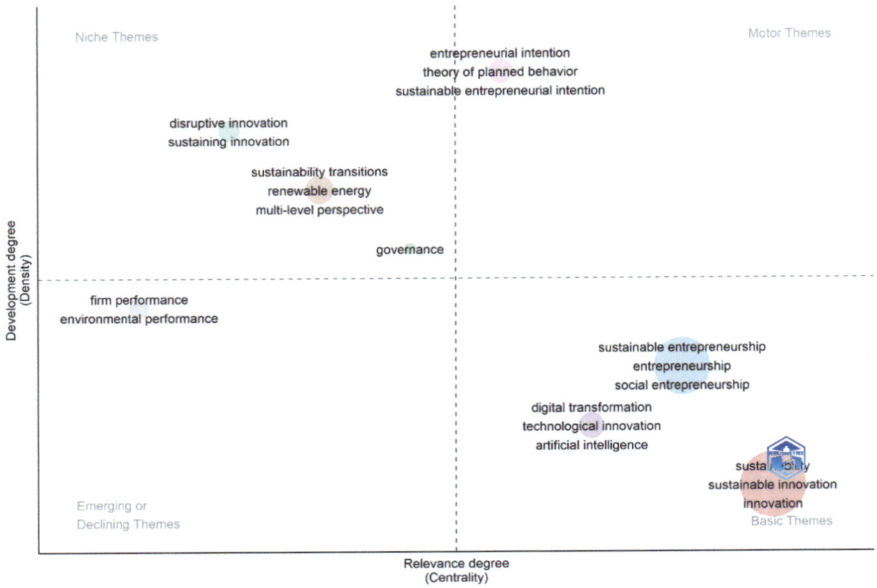

Fig. 7 Strategic diagram of SE research

centrality and significant influence. The keywords in this quadrant—i.e., "entrepreneurial intention," "theory of planned behavior," and "sustainable entrepreneurial intention"—are well-developed and serve as fundamental insights into SE.

Keywords in the lower right quadrant represent topics with high centrality but comparatively low impact. The keywords of this quarter—that is, "SE", "entrepreneurship", "social entrepreneurship", "digital transformation", "technological innovation", "artificial intelligence", "sustainability", "sustainable innovation", and "innovation"—are essential in the field of SE. However, they have not been developed to the required extent.

Keywords in the upper left quadrant correspond to specialized topics characterized by low centrality yet high impact. The keywords in this quadrant—namely, "disruptive innovation," "sustaining innovation," "sustainability transition," "renewable energy," "multi-level perspective," and "governance"—represent well-developed topics that have a substantial impact on the field. However, they remain subtopics rather than core topics. Future research in these areas will likely be fruitful due to their potential impact on the field.

Keywords in the lower left quadrant are identified as emerging topics, characterized by both low centrality and low influence. The keywords of this quadrant—"firm performance" and "environmental performance"—show potential areas of research that would benefit from new and innovative ideas.

The "Future Research Directions" table provides a structured overview of key themes relevant to corporate responsibility, knowledge management, SDGs, higher education, and the impact of digital platforms. Each theme is accompanied by targeted research questions that explore essential aspects of sustainability, entrepreneurial innovation, and knowledge management within organizations.

1. Corporate Responsibility: This theme examines how organizations can meet their ethical duties to society and the environment. Research questions could evaluate the success of corporate social responsibility initiatives, their influence on brand reputation, and effective strategies for incorporating ethical considerations into business practices.
2. Knowledge Management: This theme centers on the ways organizations generate, exchange,

Table 6 Future research directions

Themes	Research questions	References
Corporate responsibility	How can businesses incorporate circular economy principles into their supply chain management to improve sustainability?	Gigauri (2025), Perțicas et al. (2025) and Rosário and Figueiredo (2024)
	What are some practical methods for measuring social and environmental impact, and how can they enhance the evaluation of sustainability reporting frameworks?	
	How can innovation in technology and business models, mainly through startups and entrepreneurial ventures, drive sustainable development?	
	What are some effective strategies for engaging stakeholders and fostering collaboration between businesses and non-profit organizations?	
Knowledge management	How does sustainable innovation contribute to effective knowledge management within organizations?	Al Koliby et al. (2024), Alfiero et al. (2025), and Alkathiri et al. (2024)
	How do leadership practices influence knowledge management strategies in SE?	
	What are the stages of the entrepreneurial opportunity process that can help us understand how digital entrepreneurs develop sustainable business models?	
	How does knowledge management contribute to SE in technology-based businesses?	
	How do the stages of knowledge management influence SE?	
SDGs	How do social enterprises contribute to each of the 17 SDGs across various regions and sectors?	Al-Abbadi et al. (2025), Benavides-Sánchez et al. (2025), and Wisedsind et al. (2025)
	How can SE contribute to achieving the less-researched SDGs?	
	How do the entrepreneurial framework circumstances in various countries impact their progress towards achieving the SDGs?	
	How can SE effectively contribute to achieving the SDGs in practice?	
	How can an economic model effectively illustrate the "give-receive-return" value cycle in SE, and what is its relationship to innovation and the SDGs?	
	How does public policy influence the development of SE and contribute to the attainment of the SDGs?	

(continued)

Table 6 (continued)

Themes	Research questions	References
Higher education	What courses and degree programs focus on SE?	Alimehmeti et al. (2025), Filho et al. (2015), and Vargas-Merino et al. (2025)
	How are cross-disciplinary programs encouraged to integrate business education with environmental science, engineering, and social sciences?	
	What grants and funding opportunities are available for research in SE?	
	How are business incubators and accelerators established or supported for sustainable enterprises?	
	In what ways do collaborations with companies and nonprofits that are leaders in sustainability provide students with real-world projects and experiences?	
	How can local communities engage more effectively to address sustainability challenges and develop sustainable business solutions?	
Digital platforms	What vital practices are needed to establish sustainable business models that leverage digital technologies while addressing potential negative impacts, such as increased energy consumption or social inequality?	Fülöp and Cifuentes-Faura (2025), Gregori et al. (2024), and Nur and Muafi (2025)
	How do emerging technologies, such as generative AI and blockchain, shape SE and business models?	
	What advancements in renewable energy, including solar and wind power, energy storage solutions, and smart grid integration, are currently emerging?	
	How can circular economy practices, including reducing waste, enhancing material recycling efficiency, and developing circular business models, be optimized?	
	In what ways do digital technologies like blockchain, IoT, and data analytics promote sustainability through improved transparency, traceability, and resource efficiency?	
	What is the social impact of SE, and how can inclusive practices benefit local communities?	
	How can emerging technologies like AI and blockchain be utilized to promote social and environmental sustainability, and what are the ethical implications of their use?	
	What are the implications of strengthening digital networking capabilities to enhance dynamic capabilities in SE across various industry contexts?	
	How can integrating Artificial Intelligence and Big Data enhance SE's effectiveness while addressing potential challenges arising from their interaction?	

and leverage knowledge, prompting inquiries into how efficient knowledge management can support sustainability initiatives. Scholars could explore how organizational culture influences knowledge-sharing practices or investigate the channels through which knowledge management fosters innovation in sustainable solutions.

3. SDGs: This theme highlights the importance of aligning organizational goals with the SDGs to promote sustainable development. Research inquiries might investigate how companies assess their contributions to the SDGs, pinpoint obstacles to effective integration, and examine partnerships that utilize shared resources to reach these worldwide objectives.
4. Higher Education: Acknowledging the influence of educational institutions on future leadership, this theme explores how higher education can integrate sustainability into its curricula and programs. Potential research questions may examine the effectiveness of experiential learning in sustainability education or how interdisciplinary studies contribute to a comprehensive understanding of sustainability challenges.
5. Digital Platforms: This developing theme examines the role of digital technologies and platforms in advancing SE and responsible business practices. Research questions may investigate the impact of digital tools on encouraging collaborative consumption, improving transparency in supply chains, or bolstering local economies via digital marketplace solutions.

This table provides a thorough framework for future research to enhance our understanding of corporate responsibility, incorporate sustainable practices into knowledge management, effectively contribute to the SDGs, improve sustainability education, and utilize digital platforms to foster SE. Each area presents numerous opportunities for exploration, highlighting different avenues for further academic investigation and real-world applications.

5 Implications

Research on SE significantly impacts scholars, practitioners, and society by illuminating the complex interplay between sustainability, innovation, and social responsibility. For researchers, these studies define a unique academic frontier that not only examines how businesses can thrive economically but also how they can align their operations with environmental sustainability and social equity. This field invites academics to explore the integration of environmental and social considerations into entrepreneurial frameworks, prompting a critical balance between profitability and sustainable practices.

The insights derived from SE research enhance existing theoretical models related to sustainable value creation and offer practical guidelines for entrepreneurs seeking to innovate responsibly. By emphasizing the importance of ethical considerations in business practices, these studies contribute to the development of new entrepreneurial strategies that prioritize long-term ecological and social benefits. Moreover, SE fosters an interdisciplinary approach that enriches the dialogue between various fields, including economics, environmental science, sociology, management, and policy studies. This collaborative exploration not only helps to reframe traditional notions of business success but also provides novel perspectives on the operations of sustainable enterprises.

In addition to enriching academic discourse, this body of research identifies critical knowledge gaps and points to emerging avenues for further investigation. By highlighting these gaps, researchers encourage the development of effective strategies that inform policy-making, shape educational curricula, and promote responsible innovation practices within both public and private sectors. The broader implications of SE research are profound, as they contribute to the refining of larger socio-economic systems. This work ultimately plays a crucial role in fostering societal well-being and enhancing environmental health, equipping future entrepreneurs with the tools to create businesses that not only succeed in the marketplace but also contribute positively to the planet and society as a whole.

For practitioners, the implications of SE are also invaluable and far-reaching. Research in this area has introduced a rich body of studies that demonstrate how businesses have been able to combine profitability with a commitment to environmental protection and promoting social responsibility. These studies demonstrate that adopting sustainability-based approaches not only facilitates long-term growth but also provides distinct competitive advantages in an increasingly environmentally conscious marketplace. By examining these issues, practitioners gain access to practical frameworks that allow them to implement sustainable practices in their businesses, identify untapped opportunities in emerging markets, and manage the complexities of aligning business objectives with the diverse expectations of stakeholders such as customers, employees, and investors.

Furthermore, the insights derived from these studies empower entrepreneurs to proactively address common challenges to sustainability, such as the high upfront costs associated with green technologies and practices and the often limited awareness among customers and business partners regarding the importance of sustainability. The findings underscore the crucial need for long-term thinking and adaptable strategies, encouraging practitioners to embed sustainability at the core of their business philosophies.

In addition to identifying market trends and evolving consumer expectations, these studies serve as a comprehensive roadmap for entrepreneurs. They provide guidance on how to innovate solutions that tackle pressing global issues such as climate change, resource depletion, and social inequality. By equipping practitioners with the knowledge and tools to implement sustainable practices, this body of research not only fosters individual business success but also contributes to a broader movement towards a more sustainable economy.

SE research encompasses a vast array of implications that reach far beyond the confines of academia and practical business applications, significantly influencing societal norms and informing public policy. This field underscores the critical need for structural reforms within the economy and the business landscape to promote sustainability-driven initiatives effectively. By shifting the focus toward long-term perspectives and adaptive strategies, SE research plays a vital role in shaping collective attitudes toward sustainable development.

Moreover, it acts as a catalyst for governments and institutions to refine and enhance policies that support green innovation. These policies are essential for establishing clear guidelines for sustainable investment, which in turn encourages businesses to adopt practices that align closely with the broader global objectives, such as the United Nations' SDGs. This alignment ensures that entrepreneurial efforts not only seek profit but also contribute positively to societal and environmental well-being.

Ultimately, the studies conducted within this field serve as a critical bridge between theoretical insights and practical implementation. They offer a comprehensive framework for linking entrepreneurial ventures with solutions to pressing global challenges, including poverty alleviation, income inequality, and environmental degradation. By integrating sustainability into the core of business models, SE emerges as a vital framework for fostering a transition toward a more equitable and resilient future. This approach not only positions entrepreneurs as key players in addressing global issues but also redefines the role of business in society, paving the way for a holistic understanding of success that encompasses both economic viability and social responsibility.

6 Conclusion

This research has conducted a bibliometric study to understand the SE domain better. WoS was selected as the database in the first step for its credibility and broad academic acceptance. By searching for the keywords "Sustain* entrepreneur*" OR "sustain* ventur*" OR "sustain* start-up" OR "sustain* innovat*," 4521 records were found on March 10th, 2025. The filtering process included two steps: (1) Document Types: Articles (3441) and (2) Languages: English (3331). The CA was conducted to illustrate the intellectual framework of sustainable entrepreneurship (SE) and to identify key knowledge areas within the scientific domain. To further examine the structure of the field, a CCA was employed to investigate its underlying subdomains. The 30 most frequently cited documents were selected for analysis, and a co-citation matrix was constructed using Multidimensional Scaling (MDS) with the aid of Bibexcel and IBM SPSS v26 software. In the MDS analysis, complete citation data functioned as a proximity index, indicating the degree of relatedness among the most cited publications.

A CA was performed to present the intellectual construction of SE and identify knowledge within the science domain. Additionally, a CCA was utilized to explore subdivisions of the research domain. The 30 most cited documents were selected, and a co-citation matrix (Multidimensional Scaling) was created using Bibexcel software and IBM SPSS v26. In the MDS analysis, the full citations served as a measure of closeness among highly cited publications.

HCA was applied using Ward's method in IBM SPSS v26 to enhance the study's accuracy, and the most cited journals were mapped with VOSviewer software. Through CCA, 13 distinct groups identified by MDS and four clusters derived from HCA were established.

To enhance the analytical precision of the study, HCA was conducted using Ward's method in SPSS v26, while the most frequently cited journals were visualized with VOSviewer software. As a result of the CCA, thirteen distinct research groups were identified through MDS, and four thematic clusters were derived from the HCA. The groups formed by MDS were labeled as: (1) Sustainability-oriented Innovation (SOI), (2) The Dynamic Capabilities, (3) Sustainability, (4) The Process of SE, (5) Sustainable Business Models, (6) Innovation Challenges for Sustainable Development, (7) SDGs, (8) Entrepreneurial Opportunities Arising from Market Failures, (9) The Green Prison, (10) The Strategic Link Between Sustainability and Innovation, (11) SE as a Systematic Response to Ecological and Social Challenges, (12) Institutional Transformation and the Redefinition of Incentive Structures, and (13) Redefining the Concept of Entrepreneurial Opportunities. The clusters labeled in the HCA were: (1) Sustainability-based Innovation, (2) Dynamic Capabilities, (3) Sustainability, and (4) Redefining Traditional Entrepreneurial Paradigms.

7 Limitations

Like many studies, the current research has its limitations. Connecting relevant keywords can produce a wider array of documents. As a result, the findings may vary since MDS results heavily depend on the most frequently cited documents in a sample; any alteration in the sample can influence the outcomes. Furthermore, this study relied solely on one database (WoS), indicating that future research should consider investigating other databases, such as Scopus, and comparing their findings. Although MDS and HCA are primarily utilized for bibliometric assessments of science, integrating another bibliometric analysis type could uncover additional aspects of social networks. Hence, we propose using alternative approaches like EFA to explore future research possibilities. Additionally, various visualization tools, such as Pajek, can be used to understand the scientific research landscape comprehensively.

References

Adams, H., Blumstein, M., & Kassab, L. (2020). Multidimensional scaling on metric measure spaces. *Rocky Mountain Journal of Mathematics, 50*(2), 397–413. https://doi.org/10.1216/rmj.2020.50.397

Adams, R., Jeanrenaud, S., Bessant, J., Denyer, D., & Overy, P. (2016). Sustainability-oriented innovation: A systematic review. *International Journal of Management Reviews, 18*(2), 180–205. https://doi.org/10.1111/ijmr.12068

Ajzen, I. (1991). The theory of planned behavior. *Organizational Behavior and Human Decision Processes, 50*(2), 179–211. https://doi.org/10.1016/0749-5978(91)90020-T

Akbari, M., Khodayari, M., Danesh, M., Davari, A., & Padash, H. M. (2020a). A bibliometric study of sustainable technology research. *Cogent Business and*

Management, 7(1), 1–25. https://doi.org/10.1080/23311975.2020.1751906

Akbari, M., Khodayari, M., Khaleghi, A., Danesh, M., & Padash, H. (2020b). Technological innovation research in the last six decades: a bibliometric analysis. *European Journal of Innovation Management* (ahead-of-print). https://doi.org/10.1108/EJIM-05-2020-0166

Albort-Morant, G., Leal-Millán, A., & Cepeda-Carrión, G. (2016). The antecedents of green innovation performance: A model of learning and capabilities. *Journal of Business Research, 69*(11), 4912–4917.

Al Koliby, I. S., Mehat, N. A. B., Al-Swidi, A. K., & Al-Hakimi, M. A. (2024). Is knowledge management a missing link? Linking entrepreneurial competencies and sustainable performance of manufacturing SMEs. *The Bottom Line, 37*(1), 71–97.

Al-Abbadi, L. H., Alshawabkeh, R., Alkhazali, Z., Al-Aqrabawi, R., & Rumman, A. A. (2025). Business intelligence and strategic entrepreneurship for sustainable development goals (SDGS) through: (Green technology innovation and green knowledge management). *Economics-Innovative and Economics Research Journal, 13*(1).

Alfiero, S., Battisti, E., Nirino, N., & Papa, A. (2025). Exploring sustainability in start-ups: The impact of knowledge management on achieving sustainable goals. *Journal of Knowledge Management*.

Alimehmeti, G., Ndoka, E., & Paletta, A. (2025). Cultivating green pioneers: Examining the antecedents of sustainable entrepreneurial intent in higher education. *International Journal of Sustainability in Higher Education*.

Alkathiri, N. A., Said, F. Ben, Meyer, N., & Soliman, M. (2024). Knowledge management and sustainable entrepreneurship: A bibliometric overview and research agenda. *Journal of Innovation and Entrepreneurship, 13*(1). https://doi.org/10.1186/s13731-024-00387-3

Anand, A., Argade, P., Barkemeyer, R., & Salignac, F. (2021). Trends and patterns in sustainable entrepreneurship research: A bibliometric review and research agenda. *Journal of Business Venturing, 36*(3), 106092.

Andrade-Valbuena, N., Baier-Fuentes, H., & Gaviria-Marin, M. (2022). An overview of sustainable entrepreneurship in tourism, destination, and hospitality research based on the Web of Science. *Sustainability, 14*(22), 14944.

Barney, J. (1991). Barney-Firm-Resources-and-Competitive-Advantage-1991. *Journal of Management, 17*, 99–120. https://doi.org/10.1177/014920639101700108

Belz, F. M., & Binder, J. K. (2017). Sustainable entrepreneurship: A convergent process model. *Business Strategy and the Environment, 26*(1), 1–17. https://doi.org/10.1002/bse.1887

Benavides-Sánchez, E. P., Moya-Clemente, I., & Ribes-Giner, G. (2022). Sustainable entrepreneurship and sustainable development goals: A bibliometric analysis. *Tec Empresarial, 16*(1), 101–122.

Benavides-Sánchez, E. P., Moya-Clemente, I., & Ribes-Giner, G. (2025). Bibliometric analysis and systematic literature review of the relationship between sustainable development goals and sustainable entrepreneurship over time. *Discover Sustainability, 6*(1). https://doi.org/10.1007/s43621-024-00572-0

Bocken, N. M. P., Short, S. W., Rana, P., & Evans, S. (2014). A literature and practice review to develop sustainable business model archetypes. *Journal of Cleaner Production, 65*, 42–56. https://doi.org/10.1016/j.jclepro.2013.11.039

Boons, F., & Lüdeke-Freund, F. (2013). Business models for sustainable innovation: State-of-the-art and steps towards a research agenda. *Journal of Cleaner Production, 45*, 9–19. https://doi.org/10.1016/j.jclepro.2012.07.007

Boons, F., Montalvo, C., Quist, J., & Wagner, M. (2013). Sustainable innovation, business models and economic performance: An overview. *Journal of Cleaner Production, 45*, 1–8. https://doi.org/10.1016/j.jclepro.2012.08.013

Cahlik, T. (2000). Comparison of the maps of science. *Scientometrics, 49*(3), 373–387.

Carrillo-Hermosilla, J., Del Río, P., & Könnölä, T. (2010). Diversity of eco-innovations: Reflections from selected case studies. *Journal of Cleaner Production, 18*(10–11), 1073–1083. https://doi.org/10.1016/j.jclepro.2010.02.014

Chabowski, B. R. (2017). A review of global competitiveness research: Past advances and future directions. *Journal of International Marketing, 25*(4), 1–24. https://doi.org/10.1509/jim.16.0053

Choi, S., & Seo, J. (2021). Trends in healthcare research on visual impairment and blindness: Use of bibliometrics and hierarchical cluster analysis. *Ophthalmic Epidemiology, 28*(4), 277–284. https://doi.org/10.1080/09286586.2020.1863993

Cillo, V., Petruzzelli, A. M., Ardito, L., & Del Giudice, M. (2019). Understanding sustainable innovation: A systematic literature review. *Corporate Social Responsibility and Environmental Management, 26*(5), 1012–1025. https://doi.org/10.1002/csr.1783

Cobo, M. J., López-Herrera, A. G., Herrera-Viedma, E., & Herrera, F. (2011). Science mapping software tools: Review, analysis, and cooperative study among tools. *Journal of the American Society for Information Science and Technology, 62*(7), 1382–1402. https://doi.org/10.1002/asi.21525

Cohen, B., & Winn, M. I. (2007). Market imperfections, opportunity and sustainable entrepreneurship. *Journal of Business Venturing, 22*(1), 29–49. https://doi.org/10.1016/j.jbusvent.2004.12.001

Dean, T. J., & McMullen, J. S. (2007). Toward a theory of sustainable entrepreneurship: Reducing environmental degradation through entrepreneurial action. *Journal of Business Venturing, 22*(1), 50–76. https://doi.org/10.1016/j.jbusvent.2005.09.003

Di Guardo, M. C., & Harrigan, K. R. (2012). Mapping research on strategic alliances and innovation: A co-citation analysis. *The Journal of Technology Transfer,*

37(6), 789–811. https://doi.org/10.1007/s10961-011-9239-2

Dias, G. P. (2019). Fifteen years of e-government research in Ibero-America: A bibliometric analysis. *Government Information Quarterly, 36*(3), 400–411. https://doi.org/10.1016/j.giq.2019.05.008

Dzikowski, P. (2018). A bibliometric analysis of born global firms. *Journal of Business Research, 85*, 281–294. https://doi.org/10.1016/j.jbusres.2017.12.054

Eisenhardt, K. M. (1989). Building theories from case study research. *The Academy of Management Review, 14*(4). https://www.jstor.org/stable/258557

Eisenhardt, K. M., & Graebner, M. E. (2007). Theory building from cases: Opportunities and challenges. *The Academy of Management Journal, 50*(1). https://www.jstor.org/stable/20159839

Eller, F. J., Gielnik, M. M., Wimmer, H., Thölke, C., Holzapfel, S., Tegtmeier, S., & Halberstadt, J. (2020). Identifying business opportunities for sustainable development: Longitudinal and experimental evidence contributing to the field of sustainable entrepreneurship. *Business Strategy and the Environment, 29*(3), 1387–1403.

Ertz, M., & Leblanc-Proulx, S. (2018). Sustainability in the collaborative economy: A bibliometric analysis reveals emerging interest. *Journal of Cleaner Production, 196*, 1073–1085. https://doi.org/10.1016/j.jclepro.2018.06.095

Fetscherin, M., & Heinrich, D. (2015). Consumer brand relationships research: A bibliometric citation meta-analysis. *Journal of Business Research, 68*(2), 380–390. https://doi.org/10.1016/j.jbusres.2014.06.010

Filho, W. L., Shiel, C., & do Paço, A. (2015). Integrative approaches to environmental sustainability at universities: An overview of challenges and priorities. *Journal of Integrative Environmental Sciences, 12*(1), 1–14.

Fornell, C., & Larcker, D. F. (1981). Evaluating structural equation models with unobservable variables and measurement error. *Journal of Marketing Research, 18*(1).

Foroudi, P., Akarsu, T. N., Marvi, R., & Balakrishnan, J. (2020). Intellectual evolution of social innovation: A bibliometric analysis and avenues for future research trends. *Industrial Marketing Management, 93*, 446–465. https://doi.org/10.1016/j.indmarman.2020.03.026

Fülöp, M. T., & Cifuentes-Faura, J. (2025). Digital approaches in sustainable entrepreneurship training: Effects on student well-being and satisfaction. *Education and Information Technologies*, 1–32.

Gigauri, I. (2025). The power of sustainable entrepreneurship to foster women empowerment and family business. In *Sustainable entrepreneurship and family business for women's empowerment* (pp. 1–24). IGI Global Scientific Publishing.

Gioia, D. A., Corley, K. G., & Hamilton, A. L. (2013). Seeking qualitative rigor in inductive research: Notes on the Gioia methodology. *Organizational Research Methods, 16*(1), 15–31. https://doi.org/10.1177/1094428112452151

Gregori, P., Holzmann, P., & Audretsch, D. B. (2024). Sustainable entrepreneurship on digital platforms and the enactment of digital connectivity through business models. *Business Strategy and the Environment, 33*(2), 1173–1190. https://doi.org/10.1002/bse.3551

Groos, V., & Pritchard, A. (1969). Documentation notes. *Journal of Documentation, 25*(4), 344–349. https://doi.org/10.1108/eb026482

Gurzki, H., & Woisetschläger, D. M. (2017). Mapping the luxury research landscape: A bibliometric citation analysis. *Journal of Business Research, 77*, 147–166. https://doi.org/10.1016/j.jbusres.2016.11.009

Hall, J. K., Daneke, G. A., & Lenox, M. J. (2010). Sustainable development and entrepreneurship: Past contributions and future directions. *Journal of Business Venturing, 25*(5), 439–448. https://doi.org/10.1016/j.jbusvent.2010.01.002

Hall, J., & Mitsloan, H. V. (2003). The challenges of innovating for sustainable development. *FALL 2003, 45*(1) REPRINT NUMBER 45113. www.frankenfoods.com

Hansen, E. G., Grosse-Dunker, F., & Reichwald, R. (2009). Sustainability Innovation Cube—A framework to evaluate sustainability of product innovations. *International Journal of Innovation Management*.

Hepsen, A., & Vatansever, M. (2012). Using hierarchical clustering algorithms for Turkish residential market. *International Journal of Economics and Finance, 4*(1), 138–150. https://doi.org/10.5539/ijef.v4n1p138

Hockerts, K., & Wüstenhagen, R. (2010). *Greening goliaths versus emerging davids theorizing about the role of incumbents and new entrants in sustainable entrepreneurship.*

Hsiao, C. H., & Yang, C. (2011). The intellectual development of the technology acceptance model: A co-citation analysis. *International Journal of Information Management, 31*(2), 128–136. https://doi.org/10.1016/j.ijinfomgt.2010.07.003

Janssens, F. (2007). *Clustering of scientific fields by integrating text mining and bibliometrics.*

Khodayari, M., Akbari, M., & Foroudi, P. (2025). The sharing economy: A systematic literature review and research agenda. *International Journal of Consumer Studies, 49*(1), Article e70010.

Klewitz, J., & Hansen, E. G. (2014). Sustainability-oriented innovation of SMEs: A systematic review. *Journal of Cleaner Production, 65*, 57–75. https://doi.org/10.1016/j.jclepro.2013.07.017

Kuckertz, A., & Wagner, M. (2010). *The influence of sustainability orientation on entrepreneurial intentions—Investigating the role of business experience.* http://ssrn.com/abstract=1884076

Kumar, P., Sharma, A., & Salo, J. (2019). A bibliometric analysis of extended key account management literature. *Industrial Marketing Management, 82*, 276–292. https://doi.org/10.1016/j.indmarman.2019.01.006

Lara-rodríguez, J. S., Rojas-contreras, C., Jair, E., Oliva, D., Van Oorschot, J. A. W. H., Hofman, E., Halman, J. I. M., Bartolini, M., Bottani, E., Eric, H., Merediz-solà, I., Bariviera, A. F., Ertz, M., He, M., Zhang, Y., Gong, L., & Zhou, Y. (2019). Discovering emerging research topics for brand personality: A bibliometric analysis. *Technological Forecasting & Social Change, 50*, 294–305. https://doi.org/10.1016/j.jclepro.2019.04.055

Laursen, K., & Salter, A. (2006). Open for innovation: The role of openness in explaining innovation performance among UK manufacturing firms. *Strategic Management Journal, 27*(2), 131–150.

Marcussen, C. (2014). Multidimensional scaling in tourism literature. *Tourism Management Perspectives, 12*, 31–40. https://doi.org/10.1016/j.tmp.2014.07.003

Merigó, J. M., Mas-Tur, A., Roig-Tierno, N., & Ribeiro-Soriano, D. (2015). A bibliometric overview of the Journal of Business Research between 1973 and 2014. *Journal of Business Research, 68*(12), 2645–2653. https://doi.org/10.1016/j.jbusres.2015.04.006

Mora, L., Bolici, R., & Deakin, M. (2017). The first two decades of smart-city research: A bibliometric analysis. *Journal of Urban Technology, 24*(1), 3–27. https://doi.org/10.1080/10630732.2017.1285123

Moya-Clemente, I., Ribes-Giner, G., & Chaves-Vargas, J. C. (2021). Sustainable entrepreneurship: An approach from bibliometric analysis. *Journal of Business Economics and Management, 22*(2), 297–319.

Mulet-Forteza, C., Genovart-Balaguer, J., Mauleon-Mendez, E., & Merigó, J. M. (2019). A bibliometric research in the tourism, leisure and hospitality fields. *Journal of Business Research, 101*, 819–827. https://doi.org/10.1016/j.jbusres.2018.12.002

Muñoz, P., & Cohen, B. (2018). Sustainable entrepreneurship research: Taking stock and looking ahead. *Business Strategy and the Environment, 27*(3), 300–322. https://doi.org/10.1002/bse.2000

Muñoz, P., & Dimov, D. (2015). The call of the whole in understanding the development of sustainable ventures. *Journal of Business Venturing, 30*(4), 632–654. https://doi.org/10.1016/j.jbusvent.2014.07.012

Nidumolu, R., Prahalad, C. K., & Rangaswami, M. R. (2009). Why sustainability is now the key driver of innovation. *Harvard Business Review.*

Nur, S. A., & Muafi, M. (2025). Link between dynamic capability and sustainable entrepreneurship: The mediation role of digital networking. *Jurnal Siasat Bisnis*, 1–14. https://doi.org/10.20885/jsb.vol29.iss1.art1

Pacheco, D. F., Dean, T. J., & Payne, D. S. (2010). Escaping the green prison: Entrepreneurship and the creation of opportunities for sustainable development. *Journal of Business Venturing, 25*(5), 464–480. https://doi.org/10.1016/j.jbusvent.2009.07.006

Pasadeos, Y., Phelps, J., & Kim, B.-H. (1998). Disciplinary impact of advertising scholars: Temporal comparisons of influential authors, works and research networks. *Journal of Advertising, 27*(4), 53–70. https://doi.org/10.1080/00913367.1998.10673569

Patzelt, H., & Shepherd, D. A. (2011). Recognizing opportunities for sustainable development. *Entrepreneurship: Theory and Practice, 35*(4), 631–652. https://doi.org/10.1111/j.1540-6520.2010.00386.x

Perțicas, D. C., Florea, A. G., Borodin, V., Hajjaj, M., & Sabau-Popa, C. D. (2025). The study on corporate sustainability entrepreneurship in Romania: Analysis on dependencies of economical state of the corporation on their green politics through eyes of their management. *Sustainability, 17*(5), 2231.

Pilkington, A., & Meredith, J. (2009). The evolution of the intellectual structure of operations management—1980–2006: A citation/co-citation analysis. *Journal of Operations Management, 27*(3), 185–202. https://doi.org/10.1016/j.jom.2008.08.001

Podsakoff, P. M., MacKenzie, S. B., Lee, J. Y., & Podsakoff, N. P. (2003). Common method biases in behavioral research: A critical review of the literature and recommended remedies. *Journal of Applied Psychology, 88*(5), 879–903. https://doi.org/10.1037/0021-9010.88.5.879

Ramos-Rodríguez, A., & Ruíz-Navarro, J. (2004). Changes in the intellectual structure of strategic management research: A bibliometric study of the Strategic Management Journal, 1980–2000. *Strategic Management Journal, 25*(10), 981–1004. https://doi.org/10.1002/smj.397

Rey-Martí, A., Ribeiro-Soriano, D., & Palacios-Marqués, D. (2016). A bibliometric analysis of social entrepreneurship. *Journal of Business Research, 69*(5), 1651–1655. https://doi.org/10.1016/j.jbusres.2015.10.033

Rodríguez-Bolívar, M. P., Alcaide-Muñoz, L., & Cobo, M. J. (2018). Analyzing the scientific evolution and impact of e-Participation research in JCR journals using science mapping. *International Journal of Information Management, 40*, 111–119. https://doi.org/10.1016/j.ijinfomgt.2017.12.011

Rosário, A. T., & Figueiredo, J. (2024). Sustainable entrepreneurship and corporate social responsibility: Analysing the state of research. *Sustainable Environment, 10*(1). https://doi.org/10.1080/27658511.2024.2324572

Sánchez-Pérez, M., Rueda-López, N., Marín-Carrillo, M. B., & Terán-Yépez, E. (2020). Theoretical dilemmas, conceptual review and perspectives disclosure of the sharing economy: A qualitative analysis. *Review of Managerial Science*, 1–35. https://doi.org/10.1007/s11846-020-00418-9

Schaltegger, S., & Wagner, M. (2011). Sustainable entrepreneurship and sustainability innovation: Categories and interactions. *Business Strategy and the Environment, 20*(4), 222–237. https://doi.org/10.1002/bse.682

Schot, J., & Geels, F. W. (2008). Strategic niche management and sustainable innovation journeys: Theory, findings, research agenda, and policy. *Technology Analysis & Strategic Management, 20*(5), 537–554. https://doi.org/10.1080/09537320802292651

Shepherd, D. A., & Patzelt, H. (2011). The new field of sustainable entrepreneurship: Studying entrepreneurial action linking "what is to be sustained" with "what is to be developed." *Entrepreneurship: Theory and Practice, 35*(1), 137–163. https://doi.org/10.1111/j.1540-6520.2010.00426.x

Shiau, W.-L., & Dwivedi, Y. K. (2013). Citation and co-citation analysis to identify core and emerging knowledge in electronic commerce research. *Scientometrics, 94*(3), 1317–1337. https://doi.org/10.1007/s11192-012-0807-5

Teece, D. J., Pisano, G., & Shuen, A. (1997). Dynamic capabilities and strategic management. *Strategic Management Journal, 18*(7).

van Oorschot, J. A. W. H., Hofman, E., & Halman, J. I. M. (2018). A bibliometric review of the innovation adoption literature. *Technological Forecasting and Social Change, 134*, 1–21. https://doi.org/10.1016/j.techfore.2018.04.032

Varadarajan, R. (2017). Innovating for sustainability: A framework for sustainable innovations and a model of sustainable innovations orientation. *Journal of the Academy of Marketing Science, 45*(1), 14–36.

Vargas-Merino, J. A., Rios-Lama, C. A., Bello-Mamani, D. X., & Panez-Bendezú, M. H. (2025). Sustainable entrepreneurship in HEIs. A systematic review from the perspective of higher education in business. *The International Journal of Management Education, 23*(2), 101137.

Vijith, H., & Dodge-Wan, D. (2020). Spatial and temporal characteristics of rainfall over a forested river basin in NW Borneo. *Meteorology and Atmospheric Physics, 132*(5), 683–702. https://doi.org/10.1007/s00703-019-00714-4

Wamba, S. F., & Queiroz, M. M. (2020). Blockchain in the operations and supply chain management: Benefits, challenges and future research opportunities. *International Journal of Information Management*, 102064. https://doi.org/10.1016/j.ijinfomgt.2019.102064

White, H. D., & McCain, K. W. (1998). Visualizing a discipline: An author co-citation analysis of information science, 1972–1995. *Journal of the American Society for Information Science, 49*(4), 327–355. https://doi.org/10.1002/(SICI)1097-4571(19980401)49:4%3c327::AID-ASI4%3e3.0.CO;2-W

Wisedsind, N., Ruanggoon, J., Opasvitayarux, P., & Naknok, S. (2025). Assessing risks in Otop production: Integrating sufficiency economy philosophy and Sustainable Development Goals (SDGs) for sustainable business practices in Samut Prakan Province. *Journal of Lifestyle and SDGs Review, 5*(2), e04522–e04522.

Xie, H., Zhang, Y., & Duan, K. (2020). Evolutionary overview of urban expansion based on bibliometric analysis in Web of Science from 1990 to 2019. *Habitat International, 95*, Article 102100. https://doi.org/10.1016/j.habitatint.2019.102100

Yari, N., Lankut, E., Alon, I., & Richter, N. F. (2020). Cultural intelligence, global mindset, and cross-cultural competencies: A systematic review using bibliometric methods. *European Journal of International Management, 14*(2), 210–250. https://doi.org/10.1504/EJIM.2020.105567

Zha, D., Foroudi, P., Jin, Z., & Melewar, T. C. (2021). Making sense of sensory brand experience: Constructing an integrative framework for future research. *International Journal of Management Reviews*, 1–38. https://doi.org/10.1111/ijmr.12270

Zhu Liu, M., Su, Y., Meng, X., Han, S., & Duan, Z. R. (2020). A bibliometric analysis of publication of funded studies in nursing research from Web of Science, 2008–2018. *Journal of Advanced Nursing, 77*(1), 176–188.

Zupic, I., & Čater, T. (2015). Bibliometric methods in management and organization. *Organizational Research Methods, 18*(3), 429–472. https://doi.org/10.1177/1094428114562629

FSC
www.fsc.org
MIX
Papier aus verantwortungsvollen Quellen
Paper from responsible sources
FSC® C105338

If you have any concerns about our products,
you can contact us on
ProductSafety@springernature.com

In case Publisher is established outside the EU,
the EU authorized representative is:
Springer Nature Customer Service Center GmbH
Europaplatz 3, 69115 Heidelberg, Germany

Printed by Libri Plureos GmbH
in Hamburg, Germany